DR. CARLA C. JOHNSON

INTERNATIONAL HANDBOOK OF SCIENCE EDUCATION

# International Handbook of Science Education

Part One

*Edited by*

Barry J. Fraser
*Curtin University of Technology, Perth, Australia*

and

Kenneth G. Tobin
*University of Pennsylvania, Philadelphia, U.S.A.*

This special paperback edition is in celebration of the
*Distinguished Contributions to Science Education Award*, 2003
made to Barry J. Fraser by the National Association for Research
in Science Teaching (NARST).

KLUWER ACADEMIC PUBLISHERS
DORDRECHT / BOSTON / LONDON

A C.I.P. Catalogue record for this book is available from the Library of Congress.

Set ISBN 1-4020-1551-8

Published by Kluwer Academic Publishers,
P.O. Box 17, 3300 AA Dordrecht, The Netherlands.

Sold and distributed in North, Central and South America
by Kluwer Academic Publishers,
101 Philip Drive, Norwell, MA 02061, U.S.A.

In all other countries, sold and distributed
by Kluwer Academic Publishers,
P.O. Box 322, 3300 AH Dordrecht, The Netherlands.

*Printed on acid-free paper*

All Rights Reserved
© 2003 Kluwer Academic Publishers
No part of this work may be reproduced, stored in a retrieval system, or transmitted
in any form or by any means, electronic, mechanical, photocopying, microfilming, recording or
otherwise, without written permission from the Publisher, with the exception
of any material supplied specifically for the purpose of being entered
and executed on a computer system, for exclusive use by the purchaser of the work.

Printed in the Netherlands.

# Table of Contents

## PART ONE

**Preface**     xiii

**Section 1: Learning**
*Reinders Duit & David F. Treagust*

REINDERS DUIT & DAVID F. TREAGUST
1.1 Learning in Science – From Behaviourism Towards Social Constructivism and Beyond     3

CLIVE SUTTON
1.2 New Perspectives on Language in Science     27

WILLIAM W. COBERN & GLEN S. AIKENHEAD
1.3 Cultural Aspects of Learning Science     39

JOHN K. GILBERT & CAROLYN J. BOULTER
1.4 Learning Science Through Models and Modelling     53

PHILIP H. SCOTT & ROSALIND H. DRIVER
1.5 Learning About Science Teaching: Perspectives From an Action Research Project     67

KATHLEEN E. METZ
1.6 Scientific Inquiry Within Reach of Young Children     81

CLARK A. CHINN & WILLIAM F. BREWER
1.7 Theories of Knowledge Acquisition     97

JACQUES DÉSAUTELS & MARIE LAROCHELLE
1.8 The Epistemology of Students: The 'Thingified' Nature of Scientific Knowledge     115

**Section 2: Teaching**
*Kenneth Tobin*

KENNETH TOBIN
2.1 Issues and Trends in the Teaching of Science     129

JOHN R. BAIRD
2.2 A View of Quality in Teaching     153

WOLFF-MICHAEL ROTH
**2.3** Teaching and Learning as Everyday Activity  169

WYNNE HARLEN
**2.4** Teaching For Understanding in Pre-Secondary Science  183

PETER W. HEWSON, MICHAEL E. BEETH & N. RICHARD THORLEY
**2.5** Teaching for Conceptual Change  199

PAUL HOBDEN
**2.6** The Role of Routine Problem Tasks in Science Teaching  219

DOROTHY GABEL
**2.7** The Complexity of Chemistry and Implications for Teaching  233

VINCENT N. LUNETTA
**2.8** The School Science Laboratory: Historical Perspectives and Contexts for Contemporary Teaching  249

## Section 3: Educational Technology
*Marcia C. Linn*

MARCIA C. LINN
**3.1** The Impact of Technology on Science Instruction: Historical Trends and Current Opportunities  265

BARBARA Y. WHITE
**3.2** Computer Microworlds and Scientific Inquiry: An Alternative Approach to Science Education  295

DANIEL C. EDELSON
**3.3** Realising Authentic Science Learning through the Adaptation of Scientific Practice  317

NANCY BUTLER SONGER
**3.4** Can Technology Bring Students Closer to Science?  333

ROBERT D. SHERWOOD, ANTHONY J. PETROSINO, XIAODONG LIN
& COGNITION AND TECHNOLOGY GROUP AT VANDERBILT
**3.5** Problem-Based Macro Contexts in Science Instruction: Design Issues and Applications  349

MICHELE WISNUDEL SPITULNIK, STEVE STRATFORD, JOSEPH KRAJCIK
& ELLIOT SOLOWAY
**3.6** Using Technology to Support Students' Artefact Construction in Science  363

*Table of Contents* vii

HORST P. SCHECKER
3.7 Integration of Experimenting and Modelling by Advanced Educational Technology: Examples from Nuclear Physics . . . . . . 383

ANGELA E. McFARLANE & YAEL FRIEDLER
3.8 Where You Want IT, When You Want IT: The Role of Portable Computers in Science Education . . . . . . 399

## Section 4: Curriculum
*Jan van den Akker*

JAN VAN DEN AKKER
4.1 The Science Curriculum: Between Ideals and Outcomes . . . . . . 421

REUVEN LAZAROWITZ & RACHEL HERTZ-LAZAROWITZ
4.2 Cooperative Learning in the Science Curriculum . . . . . . 449

JOHN WALLACE & WILLIAM LOUDEN
4.3 Curriculum Change in Science: Riding the Waves of Reform . . . . . . 471

RODGER W. BYBEE & NAVA BEN-ZVI
4.4 Science Curriculum: Transforming Goals to Practices . . . . . . 487

DONNA F. BERLIN & ARTHUR L. WHITE
4.5 Integrated Science and Mathematics Education: Evolution and Implications of a Theoretical Model . . . . . . 499

MICHAEL R. ABRAHAM
4.6 The Learning Cycle Approach as a Strategy for Instruction in Science . . . . . . 513

## Section 5: Learning Environments
*Barry J. Fraser*

BARRY J. FRASER
5.1 Science Learning Environments: Assessment, Effects and Determinants . . . . . . 527

THEO WUBBELS & MIEKE BREKELMANS
5.2 The Teacher Factor in the Social Climate of the Classroom . . . . . . 565

CAMPBELL J. McROBBIE, DARRELL L. FISHER & ANGELA F. L. WONG
5.3 Personal and Class Forms of Classroom Environment Instruments . . . . . . 581

HANNA J. ARZI
5.4 Enhancing Science Education Through Laboratory Environments: More Than Walls, Benches and Widgets   595

BONNIE SHAPIRO
5.5 Reading the Furniture: The Semiotic Interpretation of Science Learning Environments   609

KENNETH TOBIN & BARRY J. FRASER
5.6 Qualitative and Quantitative Landscapes of Classroom Learning Environments   623

# PART TWO

## Section 6: Teacher Education
*Hugh Munby & Tom Russell*

HUGH MUNBY & TOM RUSSELL
6.1 Epistemology and Context in Research on Learning to Teach Science   643

RONALD W. MARX, JOHN G. FREEMAN, JOSEPH S. KRAJCIK & PHYLLIS C. BLUMENFELD
6.2 Professional Development of Science Teachers   667

BEVERLEY BELL
6.3 Teacher Development in Science Education   681

JEFF NORTHFIELD
6.4 Teacher Educators and the Practice of Science Teacher Education   695

KATHRYN F. COCHRAN & LORETTA L. JONES
6.5 The Subject Matter Knowledge of Preservice Science Teachers   707

THOMAS M. DANA & DEBORAH J. TIPPINS
6.6 Portfolios, Reflection and Educating Prospective Teachers of Science   719

CHAO-TI HSIUNG & HSIAO-LIN TUAN
6.7 Science Teacher Education in Selected Countries in Asia   733

ONNO DE JONG, FRED KORTHAGEN & THEO WUBBELS
6.8 Research on Science Teacher Education in Europe: Teacher Thinking and Conceptual Change   745

## Section 7   Assessment and Evaluation
*Pinchas Tamir*

PINCHAS TAMIR
**7.1** Assessment and Evaluation in Science Education: Opportunities to Learn and Outcomes . . . . . . . . . . . . . . . . . . . . . . . . 761

DREW H. GITOMER & RICHARD A. DUSCHL
**7.2** Emerging Issues and Practices in Science Assessment . . . . . . . 791

PAUL BLACK
**7.3** Assessment by Teachers and the Improvement of Students' Learning . . . . . . . . . . . . . . . . . . . . . . . . . . . . . . . . . . . . . . . . 811

MICHAL S. LOMASK, JOAN BOYKOFF BARON & JEFFREY GREIG
**7.4** Large-Scale Science Performance Assessment in Connecticut: Challenges and Resolutions . . . . . . . . . . . . . . . . . . . . . . . . . . 823

GAALEN L. ERICKSON & KAREN MEYER
**7.5** Performance Assessment Tasks in Science: What Are They Measuring? . . . . . . . . . . . . . . . . . . . . . . . . . . . . . . . . . . . . . . 845

## Section 8:   Equity
*Dale R. Baker*

DALE R. BAKER
**8.1** Equity Issues in Science Education . . . . . . . . . . . . . . . . . . . . 869

LESLEY H. PARKER & LÉONIE J. RENNIE
**8.2** Equitable Assessment Strategies . . . . . . . . . . . . . . . . . . . . . . 897

JAN HARDING
**8.3** Grass Roots Equity Initiatives . . . . . . . . . . . . . . . . . . . . . . . . 911

ROBERTTA H. BARBA & KAREN E. REYNOLDS
**8.4** Towards an Equitable Learning Environment in Science for Hispanic Students . . . . . . . . . . . . . . . . . . . . . . . . . . . . . . . . . 925

ALEJANDRO GALLARD, ELIZABETH VIGGIANO, STEPHEN GRAHAM, GAIL STEWART & MICHAEL VIGLIANO
**8.5** The Learning of Voluntary and Involuntary Minorities in Science Classrooms . . . . . . . . . . . . . . . . . . . . . . . . . . . . . . . . . . 941

ROSE N. AGHOLOR & PETER OKEBUKOLA
8.6 The Junior Engineers, Technicians and Scientists (JETS) Program in Nigeria .......... 955

SHARON E. NICHOLS, PENNY J. GILMER, ANTHONY D. THOMPSON & NANCY DAVIS
8.7 Women in Science: Expanding the Vision .......... 967

## Section 9: History and Philosophy of Science
*Michael R. Matthews*

MICHAEL R. MATTHEWS
9.1 The Nature of Science and Science Teaching .......... 981

ROBERT N. CARSON
9.2 Science and the Ideals of Liberal Education .......... 1001

FABIO BEVILACQUA & ENRICO GIANNETTO
9.3 The History of Physics and European Physics Education .......... 1015

ARTHUR STINNER & HARVEY WILLIAMS
9.4 History and Philosophy of Science in the Science Curriculum .......... 1027

RICHARD A. DUSCHL & RICHARD J. HAMILTON
9.5 Conceptual Change in Science and in the Learning of Science .......... 1047

NANCY W. BRICKHOUSE
9.6 Feminism(s) and Science Education .......... 1067

DOUGLAS ALLCHIN
9.7 Values in Science and in Science Education .......... 1083

VICENTE MELLADO
9.8 Preservice Teachers' Classroom Practice and Their Conceptions of the Nature of Science .......... 1093

PETER C. TAYLOR
9.9 Constructivism: Value Added .......... 1111

**Section 10: Research Methods**
*John P. Keeves*

JOHN P. KEEVES
10.1 Methods and Processes in Research in Science Education     1127

FREDERICK ERICKSON
10.2 Qualitative Research Methods for Science Education     1155

JAY L. LEMKE
10.3 Analysing Verbal Data: Principles, Methods and Problems     1175

JOE L. KINCHELOE
10.4 Critical Research in Science Education     1191

RICHARD T. WHITE
10.5 Decisions and Problems in Research on Metacognition     1207

K. C. CHEUNG & JOHN P. KEEVES
10.6 Modelling Processes and Structure in Science Education     1215

JOHN P. KEEVES & SIVAKUMAR ALAGUMALAI
10.7 Advances in Measurement in Science Education     1229

**Index of Names**     1245

**Index of Subjects**     1263

# Preface

The field of science education has been developing for over half a century and has flourished especially during the previous few decades. It is timely and fitting now that the *International Handbook of Science Education* should be assembled to synthesise and reconceptualise past research and theorising in science education, provide practical implications for improving science education, and suggest desirable ways to advance the field in the future.

This *Handbook* provides a detailed and up-to-date overview of advanced international scholarship in science education. This two-volume, 72-chapter, 1,200+-page work is the largest and most comprehensive resource ever produced in science education for use by researchers, teacher educators, policy-makers, advisers, teachers and graduate students.

In structuring the *Handbook*, we divided the field of science education into the following ten significant areas:

- Learning
- Teaching
- Educational Technology
- Curriculum
- Learning Environments
- Teacher Education
- Assessment and Evaluation
- Equity
- History and Philosophy of Science
- Research Methods

To each section, we appointed a 'section coordinator', who is a leading international scholar in that particular area and who assisted us in identifying authors and topics for a section and in evaluating drafts of chapters and suggesting improvements.

Each of the *Handbook*'s ten major sections contains a longer lead chapter (approximately 12,000 words) that provides an overview and synthesis of that area, together with 5–8 shorter chapters (around 6,000 words) that provide a narrower focus on research and current thinking on selected key issues in that area. In order to enhance the overall quality of the *Handbook*, most chapters were evaluated by a 'chapter consultant' (in addition to the editors and section coordinators) who also made helpful suggestions for improvement.

In designating the *Handbook* as 'international', we wanted to have a book that would have significance to readers from many countries. Consequently, each chapter author was asked to include research from a variety of countries, and also broad geographic coverage was considered when selecting authors. Similarly, chapter consultants were chosen to achieve broad international coverage.

Altogether 173 people (122 authors and 51 chapter consultants) from 25 countries were involved in producing the *Handbook*. While the six countries most frequently represented are the USA (81 people), Australia (23 people), Canada (14 people), the UK (12 people), The Netherlands (6 people) and Israel (6 people), the *Handbook* also involves three people from each of Germany, Italy, New Zealand, Taiwan and Nigeria and two people from each of South Africa and Spain. Other countries represented by one chapter author or chapter consultant are Singapore, Korea, Macau, the Philippines, Greece, France, Colombia, Belize, the West Indies, Costa Rica, Norway and Denmark.

Although the *Handbook*'s authors and consultants span many parts of the world, their distribution among countries also reflects the fact that most published research in science education is written in English and involves researchers from Anglo-Saxon countries (especially the USA).

Because of the current rapid changes within the field of science education, the half-life of a handbook such as the present one will be limited. Therefore, before long, it will be desirable to consider a second edition of the *International Handbook of Science Education* which will allow major updating of selected chapters and the commissioning of completely new chapters in most cases.

A work this large leaves us indebted and grateful for the help and support of many people over quite a few years. Peter de Liefde, Kluwer's Humanities and Social Sciences Publishing Editor at the time, first discussed this project with us during the annual meeting of the American Educational Research Association in Atlanta in 1993. Irene van den Reydt, Kluwer's Assistant to the Humanities and Social Sciences Publishing Editor, has remained helpful and pleasant throughout our frequent e-mail conversations during what sometimes seemed like an endless task.

We especially thank our 122 chapter authors and 51 chapter consultants for being part of this enormous publishing enterprise. Authors willingly modified their chapters based on our suggestions, evolved their chapters through various versions, and were patient with us when we were unable to keep all the balls in the air at once.

At Florida State University, Ken Tobin's editorial role was greatly assisted by Aldrin Sweeney (now at the University of Central Florida) and Rowhea Elmesky, a doctoral student in science education. Additional invaluable help with the indexing was provided by Barbara Tobin, Gena Merliss, Susan Wellhofer and Brian Gauvin from the University of Pennsylvania.

We both are especially indebted to Trudy Tanner at Curtin University of Technology for her enormous contribution to the *Handbook* in many ways over the years. She organised the initial invitation letters to authors, assisted Barry Fraser with the editing, management and wordprocessing associated with several versions of chapters in five *Handbook* sections, and much more. In particular, she checked the penultimate version of each of the 72 chapters, including the thousands of references, converted chapters to Kluwer style, and corresponded with authors about inconsistencies and omissions in references. Without her dedication, thoroughness and sheer hard work, the *Handbook* would not be of such high editorial quality and would not have been finished at the time when it was.

BARRY J. FRASER
*Curtin University of Technology, Perth, Australia*

KENNETH G. TOBIN
*University of Pennsylvania, Philadelphia, USA*

Section 1

Learning

REINDERS DUIT & DAVID F. TREAGUST

# 1.1 Learning in Science – From Behaviourism Towards Social Constructivism and Beyond

REINDERS DUIT
*University of Kiel, Germany*

DAVID F. TREAGUST
*Curtin University of Technology, Perth, Australia*

Although constructivism has the ascendancy among learning theories in the 1990s, this has not always been the situation. In the first half of this century, behaviourism was the dominant learning theory in education, at least in the USA (Schunk 1991). Published research in the USA prior to the late 1950s had a predominantly behaviourist tone, although cognitively-based research did occur without becoming mainstream (see the review by Oakes 1947). How these changes from behaviourist to cognitive theories of learning influenced the science education community can be discerned from observations of the research literature on learning in science education during this period.

In this chapter, we present a brief outline of the developments towards a view of learning that includes issues of mainstream constructivism of the late 1980s and the early 1990s, and issues of social constructivism that have gained increasing attention in science education. With regards to the different views of learning, we believe that rival positions emphasise different aspects of the learning process. Further research should not focus on the differences but present an inclusive view of learning and conceptualise the different positions as complementary features that allow researchers to address the complex process of learning more adequately than from a single position.

Initially, this chapter provides an overview of the various developments in views of learning in science education from behaviourism to constructivism, and describes frameworks for categorising current research on science learning. Secondly, we examine the role of Piagetian ideas of learning in science education, which leads to the third section which addresses conceptual change approaches from the perspectives of learning pathways, conceptual change theory and resistances to change. The fourth section of the chapter focuses on social-constructivist aspects of learning. The final section provides an overview of this chapter and a brief description of the other seven chapters in this section of the *Handbook*.

---

Chapter Consultant: Stella Vosniadou (University of Athens, Greece)

## CONCEPTIONS OF LEARNING

*From Behaviourist to Constructivist Views of Learning in Science Education*

Scientists in the USA in the late 1950s grew increasingly concerned about the poor quality of science education in secondary schools and this concern led directly to the now famous curriculum development projects of the 1960s. This activity and the deliberations of concerned scientists and educators led to the book, *The Process of Education* (Bruner 1960), which 'served as both a reservoir and watershed' (Shulman & Tamir 1973, p. 1098) in changing and shaping the immediate future of science education. The four themes of Bruner's book each focused attention on learning and the learner. The first theme was concerned with the role of the structure of the subject matter in learning, emphasising that learning and teaching of structure is more productive than mastery of facts and techniques. The second theme was readiness for learning, especially how learning of new ideas involves revisiting them in the curriculum so that the learner can use them effectively in progressively more complex forms. The third theme involved intuition and analytical thinking that led to the notions of discovery and inquiry which were so influential for a long time. The fourth theme related to the desire to learn and how it might be stimulated.

These themes greatly influenced activities in science education as was evident in Shulman and Tamir's (1973) review in the *Second Handbook of Research on Teaching*, which identified the central themes in science education during the period 1963–1973 as being the structure of the subject matter of science education and the impact of major curriculum developments. In White and Tisher's (1986) chapter in the *Third Handbook of Research on Teaching*, these same two themes – the learner's acquisition of knowledge and the implementation of curricula – were used as organising themes in their review of research on natural sciences education.

The influence of different learning theories is evident in changes of focus in research in science education. In the decades before Shulman and Tamir's (1973) review, researchers were interested primarily in discovering whether or not changes in a teaching procedure or in a curriculum led to changes in students' performances. Attention to why or how these changes came about was of little interest and was less common. In his seminal paper comparing a quantitative study of student learning (and other output measures) among 72 Harvard Project Physics classes with a qualitative study of science classes in nine schools, Welch (1983) identified how the goal of the research and nature of the research questions changed the essence of the whole research enterprise. These changes towards qualitative studies, similar to those reported by White and Tisher (1986), involved researchers looking for reasons for any effects in learning and examining the details of learning outcomes.

By the late 1960s, the influence of behaviourist theories of learning in science education was waning and Piaget's ideas of intellectual development came into prominence. Even so, the focus of Piaget's research on the development of cognitive structures or cognitive operations by the individual was incorporated initially into research that was still influenced by behaviourism. This research examined Piaget's constructs of concrete and formal thinking and attempted to create conditions and design convenient measuring systems so that students could move from concrete to formal thinking in optimal ways. The major challenge to the focus of Piaget's research into learning in the 1970s came from Novak (1978) and his interpretation of the work of Ausubel (1968). Novak challenged whether children develop general cognitive structures or cognitive operations to make sense out of experience and instead asked whether they acquire a hierarchically-organised framework of specific concepts to allow them to make sense of the experience. Essentially, Novak argued that Ausubel's theory of meaningful reception learning, being dependent on the framework of specific concepts and integration between these concepts, provided a better analysis and explanation of the data from studies than did Piagetian stages.

Although research on students' learning in science from a cognitive perspective was evident in the first half of the 20th century, this interest in students' learning in science became a central aspect of research around the world only in the middle of the 1970s. There appear to be two major reasons for this research development (White 1987). First, the curricula designed in the 1960s and early 1970s had been far less successful in terms of improvements in the standards of science education, particularly in learning outcomes, than was expected from the effort invested in them. Second, various disciplines relevant to science education, such as philosophy of science, cognitive psychology and pedagogy, encompassed the notions of 'constructivism'. Initially, research in the middle of the 1970s focused on investigating students' learning of science phenomena, principles and concepts such as heat, energy, photosynthesis or genetics. The large number of empirical studies provide ample evidence that students' learning in many fields in the science curriculum is substantially different from the scientific concepts held by scientists. Most of these conceptions are held strongly and hence are resistant to change. As a result, research shows that students learn science concepts and principles only to a limited degree, sometimes persisting almost totally with their preinstructional conceptions, sometimes trying to hold onto two inconsistent approaches – one intuitive and one formal – and sometimes possessing genuine alternative conceptions which are unrecognised and undervalued in their potential implications. In research since the middle of the 1970s, science educators treated students' conceptions in isolation, topic by topic. When this led to limited success in modifying students' beliefs, researchers extended the scope of their investigations (Duit 1994).

Learning science is related to students' and teachers' conceptions of science content, the nature of science conceptions, the aims of science instruction, the purpose of particular teaching events, and the nature of the learning process. For example, many students hold limited empiricist views of the nature of science (cf Désautels & Larochelle's chapter in this *Handbook*). Further, many students' views of learning and the learning process are limited in that they conceptualise learning as the transfer of prefabricated knowledge that then is stored in memory. Accordingly, science is primarily learned as an accumulation of facts. (See Sutton's chapter in this *Handbook* for a discussion of how scientific writing reinforces this way of learning.) This passive view of learning influences the students' conceptions of what counts as work in school. Classroom discussions of alternative viewpoints and negotiated consensus are not considered a part of the 'work' of the classroom, and simply are viewed as wasted time that hinders efficient progress (Baird & Mitchell 1986). The social aspects of understanding and learning are increasingly important (Solomon 1987; Taylor 1993) because knowledge construction requires an active process of interpretation within a social and cultural setting by a learner (see Roth 1995 and Metz's chapter in this *Handbook*). In this respect also, models and modelling play an important role in contributing to learning in classrooms and in other contexts (see Gilbert & Boulter's chapter in this *Handbook*).

## *Frameworks for Categorising Research on Science Learning*

The different positions or orientations of learning taken by theorists have important implications for instruction. However, a major problem is the need to place the different positions within a framework so that commonalities and differences can be identified. Eylon and Linn (1988) made such an analysis by choosing four research perspectives referred to as *concept learning, developmental, differential* and *problem-solving perspectives*. Concept learning studies are concerned with the qualitative differences among the conceptions that students use to explain scientific phenomena and examine students' topic-related understanding of scientific concepts. Reviews of such studies include Driver, Squires, Rushworth and Wood-Robinson (1994) and Pfundt and Duit (1994). The developmental perspective offers a more global view of the learner than the concept-learning perspective and examines how individual conceptions change over time, often from a Piagetian or neo-Piagetian perspective to a Vygotskian perspective. In the developmental perspective, a major research focus is on what develops (see Metz's chapter in this *Handbook*).

The differential perspective examines individual differences in abilities and aptitudes and the interaction of these differences with instruction. Of specific interest are scientific proficiency, intellectual skills, psychological aptitudes relevant to scientific proficiency, and distributions of these skills across demographic groups. Studies of this type are no longer so common because the complexity of the interactions vary with other factors such as science knowledge, learning context and social context. The problem-solving perspective comprises studies of the processes or procedures that individuals employ to answer scientific questions. Of particular interest in this perspective is the research on characteristics of novices and expert problem solvers (Reif & Allen 1992) which shows that teaching general problem-solving skills is difficult because science topic-specific concepts influence reasoning and interact with general ability.

Farnham-Diggory (1994) acknowledged the problem of categorising in education, including categorising approaches to learning. She limited learning theories to three mutually-exclusive models which she calls *behaviour, development* or *apprenticeship* models. For her, the essential criterion for distinguishing the behaviour model of learning is a comparison of expert and novice differences on the same scale(s), with any difference observed being transformed by incrementation. Novices are systematically able to accrue her science knowledge types – declarative, procedural, conceptual, analogical and logical – until they reach expert levels. This category denotes learning as training in a particular behaviour. The existing cognitive structure changes only in that something new is added. This categorisation of learning looks essentially like that sought and examined from the large curriculum projects of the 1960s. In the development model, novices and experts are distinguished on the bases of their personal theories and explanations of events and experiences. Personal theories and the concepts and principles to be learned usually are embedded in different qualitative frameworks. Teachers challenge students' personal theories by questioning, contradicting and challenging that theory, in a process which she calls perturbation, so that the student is encouraged to revise it. The result is a qualitative shift in thinking and a reconstruction of her five types of knowledge. This categorisation of learning looks essentially like that introduced from the work of Piaget and the recent constructivist positions and is so broad that it encompasses all four of the learning perspectives described by Eylon and Linn (1988). In the apprenticeship model, the novice learner gets to be an expert through the mechanism of acculturation into the world of the expert. Often this learning of new knowledge is tacit and a novice's learning is facilitated by becoming a member of the culture of the expert.

The categories of Eylon and Linn (1988) and by Farnham-Diggory (1994) both provide valuable frameworks for identifying and describing main themes of science education research on learning and instruction over the past decades. They also appear to be suited to identifying trends of future development in this domain. Farnham-Diggory's three 'models' seem to be distinct at first sight but, in every real teaching and learning situation, there are facets of all three models included (Farnham-Diggory 1994, p. 467). Even constructivist approaches, which fall into the 'development' category, usually include 'training' of certain kinds when, for instance, terms or skills have to be learned. Cognitive apprenticeship approaches unavoidably incorporate issues that fall into the development model as soon as it comes to teaching and learning of certain science concepts and principles. We conclude that progress in teaching and learning science is not achieved when the three models of Farnham-Diggory are viewed as 'rival' approaches. We rather think that it is necessary to find out in which way the *three* models can be harnessed in intelligent ways to address the different facets of learning that science education includes.

## An Inclusive View of Science Learning

Currently, there is an encouraging tendency towards an 'inclusive' view of science learning which brings together approaches of different theoretical orientations. In short, learning is viewed as conceptual development in much the same way as introduced by the seminal work of Piaget (1954). His idea of equilibration of assimilation and accommodation appears to be accepted still as a valuable perspective (Lawson 1994). Piaget also can be viewed as one of the 'fathers' of the variants of constructivism that dominated science education through the 1980s and the first years of the 1990s alike (von Glasersfeld 1995). At the heart of this constructivist view (Steffe & Gale 1995; Tobin 1993; Treagust, Duit & Fraser 1996) is the idea that the conceptions held by each individual guide understanding. A further key aspect of this view is that knowledge about the world outside is viewed as human construction. A reality outside the individual is not denied; rather, it is claimed only that all we know about reality is our tentative construction. Accordingly, learning is not viewed as transfer of knowledge but the learner actively constructing, or even creating, his or her knowledge on the basis of the knowledge already held. In addition, there are social aspects of the construction process; although individuals have to construct their own meaning of a new idea, the process of constructing meaning always is embedded in a particular social setting of which the individual is part. However, in mainstream constructivism in science education throughout the 1980s, there was undoubtedly a tendency to neglect social aspects of the construction process and emphasise the individual's construction instead. Social constructivist perspectives of various kinds that have gained growing attention in science education over the past years stress the significance of social aspects of knowledge construction (Hennessy 1993; Roth 1995; Metz's chapter in this *Handbook*). Undoubtedly these social perspectives have influenced the development and enrichment of the original constructivist view towards an inclusive view that incorporates both social and individual aspects alike.

The mainstream constructivist view of the 1980s and early 1990s, with its focus on qualitative understanding distinguished on the basis of personal theories and explanations and changes in learning that occur by perturbations created by teaching, falls into Farnham-Diggory's (1994) development model. On the other hand, social-constructivist perspectives, with their attention to the influence on the social milieu of knowledge construction, usually fall into her apprenticeship model. Another significant difference between the two constructivist positions concerns their view of knowledge in relation to the influence of the individual or social group on learning. In accordance with the leading cognitive science views of knowledge acquisition, mainstream constructivism has held that mental representations of certain structures or features of the world outside are stored in the human brain. Learning is seen as construction of mental models. Knowledge then is something an individual possesses. Social constructivist perspectives (e.g., Gergen 1995) do not deny that there is something stored in the human brain, but they claim that knowledge has significant 'social' aspects: knowledge can be distributed among the members of a certain community or shared by this community. Knowledge, then, is something that is 'between' the individual and the social.

## THE ROLE OF PIAGETIAN IDEAS OF LEARNING IN SCIENCE EDUCATION

It is hardly possible to overemphasise the impact of Piaget's thinking, including his idea of stages of cognitive development, on our contemporary views of learning despite the many critiques of his approach. In order to do justice to Piaget's way of thinking about learning, it is necessary to take into consideration that his main concern was not psychological but epistemological (Bliss 1995). As Lawson (1994) argued, Piaget wanted to develop epistemology from a mere philosophical enterprise to an empirical domain. Piaget, therefore, has to be viewed as an empirical epistemologist who was interested in the development of knowledge in humans (compare Metz's chapter in this *Handbook*). His epistemological commitments were strongly influenced by Immanuel Kant, and can be called constructivist in the contemporary sense (Lawson 1994; von Glasersfeld 1995). Piaget's original training in biology influenced his views about knowledge construction in that he drew on analogies to adaptation of living beings to their environment. This orientation becomes most obvious in his distinction of assimilation and accommodation and the idea of equilibration which is the kernel of Piagetian thinking.

Assimilation is the process of the individual's adaptation to new sense impressions, with the inputs basically fitting the already-existing cognitive structure. On the other hand, accommodation indicates that, in the adaptation process, restructuring of the already-existing structure is necessary when the inputs do not fit existing cognitive structure. Assimilation and accommodation are always intimately interrelated; there is no assimilation without accommodation and vice versa. If the inputs do not fit, there is a disturbance of the mental balance or, in other words, a cognitive conflict. The balance can be restored by a process which Piaget calls equilibration, that is, by an interplay of assimilation and accommodation. It is easy to find this key Piagetian view in most contemporary constructivist approaches as discussed in the following section of the present chapter. This view also is the kernel of the influential instructional strategy of the 'learning cycle' (see Karplus 1977 and Abraham's chapter in this *Handbook*) which is based on Piagetian epistemology and which has been proven fruitful and successful (Lawson, Abraham & Renner 1989). If the learning cycle strategy is carefully analysed and compared to constructivist approaches of the 1980s that also deliberately employ cognitive conflict (Driver 1989), there are only marginal differences in instruction.

Piaget's stage theory, which has often been discussed and questioned, holds that there is a development of general thinking skills. There are four kinds of logical operations that children and adolescents exhibit in sequence: *sensorimotor* (about the first 18 months of life); *preoperational* (until about seven years); *concrete operational* (after about seven years); and *formal operational* (between 11 and 15 years). There is no doubt that the idea of general, logical thinking skills and their development in certain stages can be valuable in describing cognitive development, but there is a number of difficulties with Piagetian stage theory (Bliss 1995).

First, and foremost, the idea that logical thinking operations are independent from contexts has been seriously challenged. Research clearly showed that there is a strong domain-specific effect. In other words, the student's choice of logical operation depends on the particular science content and the problem's context. If an individual uses formal operational thinking in one domain, it is not certain that the same person would use that kind of thinking in other domains also (Seiler 1973). These findings, that have been supported by numerous studies, call into question not that general logical operations are of significance in learning science but that they are not universally transferred to other tasks once applied in certain tasks. Nevertheless, studies usually show significant positive correlations between Piagetian stages and science achievement (Lawson & Thompson 1988; Shayer & Wylam 1981). Shayer and Adey (1992) also demonstrated that deliberate training in Piagetian logical operations had some general impact in that accelerated learning occurred in content areas not included in the original training.

Lawson (1994, p. 163) provides a comprehensive critique of Piagetian stage theory. He claims that Piaget's belief that thinking patterns are isomorphic with rules of formal propositional logic is the most problematic position in his theory. He proposes to distinguish the terms *intuitive* and *reflective*, with 'reflective' replacing 'formal reasoning' in the Piagetian sense. The reflective adult is able to consider alternative theories and ask which is the most appropriate, whereas the intuitive thinker does not consider the relative merits of alternative theories.

In conclusion, Piagetian ideas still could provide powerful tools for thinking about learning (Bliss 1995). Piaget's view of knowledge acquisition as outlined by the equilibration process still appears to be widely accepted as a useful perspective. Even stage theory could provide valuable orientation if interpreted in a non-orthodox manner.

## CONCEPTUAL CHANGE APPROACHES

'Conceptual change' has become a term that denotes key aspects of the mainstream constructivist approaches of the 1980s and early 1990s. Conceptual change approaches have their roots both in science education research (Duit in press; Posner, Strike, Hewson & Gertzog 1982) and in developmental psychology (Carey 1985; Vosniadou 1994). In the first case, conceptual change theory implies that students' conceptions need to be exchanged for the new science conceptions. This was at the heart of the Posner *et al.* (1982) framework. On the other hand, developmental research on conceptual change is usually descriptive and can only lead to recommendations for what to change or how to bring about conceptual change. Certainly the idea that context is an important variable in this process that needs to be taken into consideration is an important one, and not inconsistent with conceptual change research. As a consequence of this development and prominence, there are many slightly-different or even substantially-different meanings given to this term. Nevertheless, there seems to be some agreement among the key representatives of conceptual change approaches. The term conceptual change denotes that learning of science concepts and principles usually involves major restructuring of students' already-existing preinstructional conceptions. In other words, students' preinstructional conceptions and science conceptions are usually embedded in different qualitative frameworks.

Undoubtedly the term conceptual change is not well chosen as it invites a number of misinterpretations, among which is the idea that students' preconceptions have to be exchanged for the new science conceptions. In the late 1970s and the early 1980s especially, a predominant focus was that students' conceptions (often called 'misconceptions') have to be extinguished and replaced by the correct science view. Research has shown that this is not possible. Indeed, there appears to be no study which found that a particular student's conception could be completely extinguished and then replaced by the science view. Most studies show that the 'old' ideas stay 'alive' in particular contexts. Usually the best that could be achieved was a 'peripheral conceptual change' (Chinn & Brewer 1993) in that parts of the initial idea merge with parts of the new idea to form some sort of hybrid idea (Jung 1993; see also Chinn & Brewer's chapter in this *Handbook* for a discussion of intermediate stages in knowledge change). Further, extinction of old ideas is not only impossible but also undesirable. Many students' everyday conceptions – for instance, conceptions of the process of seeing, the propagation of light or heat phenomena – have proven fruitful and valuable in most everyday situations. The vast majority of adults (even scientists) successfully draw on such conceptions in everyday situations.

Conceptual change approaches therefore hold that the aim of science instruction is not to replace everyday views but to make students aware that, in certain contexts, science conceptions are much more fruitful than their own conceptions. Ideas of 'situated cognition' (Brown, Collins & Duguid 1989) have substantially supported this view of context dependency of conceptions (Hennessy 1993) and it is claimed that every cognition and every learning event is situated (see below). The situated cognition perspective provides a valuable framework for describing and understanding research findings which show that change does not come easily and is limited to particular contexts (Tytler 1994). In relation to context dependency, Hewson and Hewson (1992) view conceptual change as change of status given to the old and the new conceptions: old students' conceptions lose status at the same time that the new science conceptions gain status.

*Learning Pathways – Conceptual Change Versus Conceptual Growth*

The key assumption of conceptual change approaches is that learning has to start from certain already-existing conceptions and that learning pathways (Scott

1992) have to be designed so that they lead from these preconceptions towards the science conceptions to be learned. Learning pathways can be described as being continuous or discontinuous. Continuous pathways of teaching/learning try to avoid the fundamental restructuring that is necessary in the case of the discontinuous pathways of teaching/learning. One kind of instruction using a continuous pathway starts from aspects of students' preinstructional conceptions or frameworks that are at least in part compatible with the science view to be achieved. From there, a basically continuous passage of learning is possible. A second continuous learning pathway is that of 'reinterpretation' (Jung 1986); the strategy is different in that the starting point is a set of students' conceptions that appear to be in contrast to science conceptions. Key facets of the students' conceptions, then, are reinterpreted in such a way that they are basically in accordance with the science conceptions.

In contrast to the above teaching/learning approaches, discontinuous pathways deliberately draw on the conflict between students' conceptions and science conceptions. Cognitive conflict, therefore, is a significant tool in these pathways (Scott, Asoko & Driver 1992). There are three primary kinds of cognitive conflict: students are asked for predictions and then are challenged by the conflicting results of an experiment; there is a conflict between students' and the teacher's ideas; and there is a conflict between the ideas of different students. The theoretical orientation of cognitive conflict usually is Piaget's idea of restoring mental equilibrium by intimate interplay of assimilation and accommodation (Lawson 1994; Rowell & Dawson 1985). Reference also is given to Festinger's (1962) theory of cognitive dissonance (Driver & Erickson 1983). The crucial issue in cognitive conflict strategies is that students need to 'see' the conflict. What appears to be clearly discrepant from the perspective of a teacher can be viewed as only marginally different or might not be considered discrepant at all from the perspectives of the students.

When the issue of conceptual change versus conceptual growth is debated, two features are not given sufficient attention. First, even when students' everyday conceptions and science conceptions are in stark contrast, it is not absolutely necessary to start from these conceptions. It is not even necessary to bring these everyday conceptions explicitly into play in instruction. There are possibilities of finding 'intelligent' teaching/learning pathways that initially bypass, so to speak, students' conceptions of the phenomena affiliated with the science concepts and principles to be learned. In such cases, learning pathways start with general thinking schemata (or with conceptions that are not in contrast with the science view) and they lead to the science conceptions, perhaps via analogies (see Chinn & Brewer's chapter in this *Handbook* for a discussion of how old knowledge is used to construct new knowledge). Second, conceptual growth and conceptual change should be viewed as complementary terms. In every learning pathway from students' conceptions towards science conceptions, there are facets that can be indicated by the two poles of that complementarity (see the previous remarks on the intimate interplay of assimilation and accommodation in Piagetian theory). Studies of learning processes have clearly shown that real learning pathways are very complex and cannot adequately be described by just conceptual growth or conceptual change (Duit, Goldberg & Niedderer 1992; Niedderer 1996). They are quite different for different students of the same groups. Usually, there are 'backwards and forwards' movements, 'dead-end streets', parallel developments and the like. Tytler (1994, p. 311) therefore considers that terms like conceptual change (he uses the term 'theory exchange') can only be useful in describing the thinking of student cohorts. These terms might offer insight into difficulties in attaining new concepts, but they do not offer much explanatory insight into the process of individual construction of understanding.

It is somewhat difficult to come to a clear-cut conclusion regarding the success of conceptual change approaches. A key difficulty is that these approaches often include fundamental restructuring of more traditionally-oriented science instruction. Conceptual change strategies, in other words, often are only one facet within approaches that aim at making science instruction understandable and fruitful for the students in a very comprehensive way. Therefore, it is difficult or even impossible to compare the new approaches with others. All that is possible, then, is to investigate whether conceptual change approaches achieve the aims which they intend. In reviewing advantages and problems of the mainstream constructivism of the 1980s and early 1990s in science education, Solomon (1994) is rather sceptical. However, Wandersee, Mintzes and Novak (1994) come to a much more optimistic conclusion from an analysis of 103 conceptual change studies. Although they found a number of methodological limitations in several studies, they concluded 'even with the aforementioned caveats in mind, we remain impressed by the relative success some researchers have achieved today' (Wandersee et al. 1994, p. 192). Guzetti, Snyder, Glass and Gamas (1993) carried out a meta-analysis of 70 of studies of intervention strategies in science education and in science-related reading education. They included only studies that incorporated quantitative measures comparing treatment and control groups. For this reason, key constructivist conceptual change approaches, such as the Children's Learning in Science (CLIS) project in Leeds (Driver 1989; Scott & Driver's chapter in this *Handbook*) were not included. Guzetti et al. (1993, p. 149) concluded: 'Based on the accumulated evidence from two disciplines [reading and science education], we have found that instructional interventions designed to offend the intuitive conception were effective in promoting conceptual change. The format of the strategy (e.g., refutational text, bridging analogies, augmented activation activities) seems irrelevant, providing the nature of the strategy includes cognitive conflict.'

*The Conceptual Change Theory*

The most influential theory of conceptual change was developed by a group of science educators and philosophers of science at Cornell University (Hewson 1981; Posner et al. 1982; Strike & Posner 1985). The theory has become 'very popular and useful' (Pintrich, Marx & Boyle 1993, p. 169) in science education as well as in a number of other fields, and has been extensively applied and subsequently changed (see the review by Hewson & Thorley 1989). According to the theory, there are four conditions that foster conceptual change. There must be dissatisfaction with current conceptions and any new conception must be intelligible, initially plausible and fruitful.

The theory provides answers to the question: 'How do learners make a transition from one conception, to a successor conception?' The transition is conceptualised in Piagetian terms as equilibration of assimilation and accommodation. The theory establishes analogies between conceptual development in science and in individual learners. The four conditions for conceptual change are derived from the work of philosophers and historians of science, especially Kuhn (1970), Lakatos (1970) and Toulmin (1972). The metaphor of the student (or the child) as scientist, which also is a leading metaphor in several other constructivist approaches, therefore plays a significant role in conceptual change theory [see Driver's (1983) 'pupil-as-scientist' perspective which is reminiscent of Kelly's (1955) idea of 'man-the-scientist']. Another key characteristic of the theory is the term 'conceptual ecology' which denotes that the already-existing cognitive structure of the learner is a system of closely interrelated items which, in several respects, is reminiscent of interactions in eco-systems.

In critically analysing their 1982 theory, Strike and Posner (1992) suggest that the initial theory put too much emphasis on the rational and neglected affective and social issues of conceptual change. They also claimed that students' conceptual ecology should be viewed much more in terms of a dynamic system than as in the initial theory. There, the interaction of prior conceptions and the new conceptions was not sufficiently acknowledged.

Pintrich *et al.* (1993), in addressing deficiencies of the initial theory of conceptual change by Posner *et al.* (1982), use the thermal metaphor of 'cold' to denote their reservation about overly rational approaches and 'cognition only' models of students' learning. The theory of conceptual change, according to the authors, is too much oriented to rational aspects in two ways (compare the critique by Strike & Posner 1992). First, it is based on a philosophy of science perspective that places major emphasis on rationality, or the significance of logical arguments in the process of conceptual development. Compared with the approaches by Kuhn, Lakatos and Toulmin, more recent developments in the philosophy of science (e.g., social constructivist approaches like the one by Knorr-Cetina 1981) have pointed out that manifold 'non-rational' issues play a role also. Second, the rational is also overemphasised in the process of conceptual change in individuals from their initial preinstructional conceptions to the science concepts. The key metaphor of the initial theory of conceptual change, the student as scientist, is undergoing rigorous discussion (Caravita & Halldén 1994). It is questionable that this metaphor in fact provides valuable analogies for understanding the process of conceptual change. The learning communities in science classrooms and the scientific community are very different in that they operate on the grounds of fundamentally different aims and within fundamentally different institutional conditions. For instance, schools are much more driven by the need to maintain bureaucratic and institutional norms rather than scholarly norms. O'Loughlin (1992) constructed a similar critique against constructivist approaches in stating that the culture in science classrooms, with its power structures and discourses, is not adequately taken into account.

In summarising this line of critique, conceptual change has to be viewed as a process of bewildering complexity that is dependent on many closely interrelated variables. Conceptual change, the process of conceptual development from students' prior ideas towards science concepts, has to be embedded in 'conceptual change supporting conditions', including the motivation, interests and beliefs of learners and teachers as well as classroom climate and power structures.

*Resistances to Change*

Learning of key science concepts and principles is difficult because there is resistance to conceptual change due to everyday experiences, possible biological predispositions, and the complexity of the learning task. Learning science is especially difficult in fields in which students' preinstructional conceptions are deeply rooted in daily life experiences. Conceptions that are based on empirical evidence through sense experiences (like the process of seeing, thermal phenomena, and conceptions of forces and motions) fall into this category as do everyday ways of speaking about natural and technical phenomena. A further resistance to change is a biological predisposition to interpret empirical evidence in ways related to how the human mind has evolved. Vosniadou and Brewer (1992) introduced the term of 'entrenched' belief in the sense that beliefs are presuppositions organised in complex interrelated structures. They explicitly go beyond science content and include conceptions like 'ontological beliefs' (i.e., beliefs about fundamental categories and properties of the world) and 'epistemological commitments' (i.e., beliefs about what scientific knowledge is and what counts as good scientific theory).

The initial theory of conceptual change by Posner *et al.* (1982) provides explanations of why conceptual change often is so difficult. If a conception is deeply rooted (entrenched) and has proven successful in most previous daily life situations, there is no dissatisfaction with this conception. Further, if there is no conception available that is intelligible and plausible from a student's perspective, a change is most unlikely. Students are frequently unable to understand the new theory, because their old conceptions provide the interpretation schemata, the goggles so to speak, for looking at the new science conceptions. Hence, the new conceptions do not become intelligible and plausible to students, who are unable to understand the new view because they do not posses sufficient 'background knowledge' (Chinn & Brewer 1993; Schumacher, Tice, Wen Loi, Stein, Joyner & Jolton 1993; Strike & Posner 1985). Without a certain amount of background knowledge, the arguments in favour of the new conceptions might not be understood. There is a certain dilemma which has similarities to Bereiter's (1985) learning paradox that a new conception becomes understandable only if there is already some knowledge about that conception available. The condition of 'intelligible' does not 'guarantee' conceptual change. There are several cases in the literature of students understanding a new theory but not believing it (e.g., Jung 1993).

Schumacher et al. (1993) discuss motivational factors that impede conceptual changes and claim that, 'if a misconception is held in an area where students have little interest, they will be unlikely to invest the cognitive resources' (p. 4). In other words, dissatisfaction substantially depends also on affective features. These authors review research on resistance to change in several domains such as studies of human judgement and decision making, psychotherapy and attitude change. They conclude that resistance to change as found in science misconceptions appears to be a very common human trait. There usually are important benefits to having stable conceptions, beliefs or attitudes. These conceptions, beliefs and attitudes have been formed by the individual in processes of adaptation to life-world experiences and usually provide valuable frames for behaviour. To give them up usually is a loss of stability for the students.

Chi, Slotta and de Leeuw (1994) developed a theory of conceptual change for learning science concepts. It assumes that conceptual change occurs when a concept has to be reassigned to an ontologically distinct category. They distinguish the three 'trees' of categories of 'matter' (or things), 'processes' and 'mental states', and they provide examples across the sciences for which learning of key concepts includes changes of ontological categories. They hold that the most difficult concepts to be learned require a change of ontological categories. There is no doubt that, when students have severe difficulties in accepting science concepts, a change of an ontological category often is necessary. The physics concept of force, for instance, falls into the category of relations between objects (namely, interactions) and not into the category of properties of things as in daily life. Here force is usually seen as something that strong humans and animals possess. Chi et al. (1994) therefore point to important barriers to learning science concepts but their present theory is limited (Duit 1995). First, their choice of categories is somewhat arbitrary. The ontological change in the case of the concept of force needs a more elaborated set of categories than the three categories used, namely, the change from a property of objects to relations between objects. Second, learning of key science concepts often is not adequately described by such changes from one category to another. In the case of heat concepts, the undifferentiated heat concept of daily life has to be differentiated and unfolded into the concepts of temperature, heat energy, internal energy and entropy. The naive everyday concept of heat includes facets of all of the aspects indicated by the physics heat concepts mentioned (Kesidou, Duit & Glynn 1995). Third, Chi et al. argue at the level which Pintrich et al. (1993) call 'cold' conceptual change, and they do not consider affective issues. Fourth, the theory presents only a syntactic, not a semantic, explanation of conceptual change (Vosniadou 1994). Therefore, the perspective of entrenched beliefs in the above meaning appears to be a more inclusive position because it includes not only ontological changes of the type that Chi *et al.* discuss, but also changes of a broader nature.

There are many studies of students' science conceptions which indicate that counter-evidence does not necessarily change students' points of view, as is explicitly or implicitly assumed in many teaching and learning approaches that emphasise cognitive conflict. A paradigmatic example stems from a study by Tiberghien (1980). A 12-year-old girl is asked to find out if an ice block wrapped in aluminium foil will melt faster than an ice block wrapped in wool. The girl believes that the iceblock wrapped in wool will melt first, because wool is warm and therefore will give heat to the ice. When the ice block wrapped in aluminium foil melts first, the girl's initial conception is not shaken and she invents a number of protective arguments in favour of her idea.

Chinn and Brewer (1993) provide a review of the role of anomalous data in knowledge acquisition, especially in science. They describe seven ways in which students deal with discrepant evidence by ignoring anomalous data; by rejecting anomalous data; by excluding anomalous data; by holding anomalous data in abeyance; by reinterpreting anomalous data; by peripheral theory change; and by theory change. They also discuss conditions under which anomalous data can occur and identify how people respond to anomalous data. Among these are characteristics of the prior knowledge like its 'entrenchment', ontological beliefs, epistemological commitments and, as previously mentioned, background knowledge (see also Chinn & Brewer's chapter in this *Handbook*).

## INDIVIDUAL AND SOCIAL CONSTRUCTION – SOCIAL-CONSTRUCTIVIST ISSUES OF LEARNING

As outlined above, there is a growing line of critique against mainstream 'conceptual change' approaches in science education that address the tendency to overemphasise the individual's learning and neglect social issues in knowledge-construction processes, and to view knowledge primarily as something stored in the individual mental system, as mental models of the world outside. Marton (1986) developed a phenomenological counterposition to mainstream constructivist perspectives which he calls a 'phenomenographic approach' (Lybeck, Marton, Stroemdahl & Tullberg 1988; Rennstroem 1987). He distinguishes between a 'mental model based perspective' and an 'experientially based perspective' of conceptions. The first perspective views conceptions as mental representations (i.e., as tangible constructs in the learner's head), whereas the latter perspective depicts conceptions as being characterisations of categories of descriptions reflecting person-world relationships. From the perspective of mainstream constructivism, conceptual change takes place within a person's head. From the phenomenographic perspective, conceptual change is achieved by changing one's relationship with the world. In discussing 'challenges to conceptual change' from the phenomenographic perspective, Linder (1993) emphasises that students need to develop meaningful relationships with the new conceptions in particular contexts. He concludes that less emphasis should be put on:

> ... efforts to change segments of students' existing repertoires of conceptualizations and more efforts on enhancing students' capabilities to distinguish between conceptualizations in a manner appropriate to some specific context – in other words, being able to appreciate the functional appropriateness of one, or more, of their conceptions in a particular context, making science education into a functional base from which to view the world. (Linder 1993, p. 298)

Similar views have been developed within social-constructivist approaches which draw, not only on phenomenological ideas, but also on the work of Vygotsky (1986) and the Soviet school of activity theory based on Vygotsky's work (Wertsch & Toma 1995; Metz's chapter in this *Handbook*). Other sources are social-constructivist studies of the genesis of knowledge in scientific communities (Knorr-Cetina 1981) and empirical studies of everyday mathematics and science (Lave & Wenger 1991). In social-constructivist approaches that have been employed in science education, the idea of situated cognition usually plays a key role (Hennessy 1993; Roth 1995). Brown, Collins and Duguid (1989) describe the basic ideas of situated cognition as follows:

> The activity in which knowledge is developed and deployed, it is now argued, is not separable from or ancillary to learning and cognition. Nor is it neutral. Rather, it is an integral part of what is learned. Situations might be said to co-produce knowledge through activity. Learning and cognition, it is now possible to argue, are fundamentally situated. (p. 32)

From the perspective of situated cognition, learning means change from one sociocultural context, usually the everyday context, to a new, science context or, in other words, changes from the practice of one culture to another (for crossing between cultures, see Cobern & Aikenhead's chapter in this *Handbook*). As language is a key aspect of culture in the sense used here, science learning also is viewed as change of languages or change of language games. Learning science means to learn 'talking science' (Lemke 1990). 'Authentic' learning situations according to Roth (1995) are dominated by open-inquiry activities and play a key role in change in classrooms. 'Cognitive apprenticeship' is often seen as the best method for introducing the learner into the new culture as the expert guides the apprentice (the novice). By developing participation in activities within the community in question, step-by-step the apprentice becomes a member of that community. The metaphor of apprenticeship provides a different flavour to the continuous and discontinuous learning pathways in conceptual change approaches. The process of acculturation, the slowly growing understanding, in numerous hermeneutical circles can be viewed as a promising counter-position to conceptual change strategies which sometimes appear to be instructional engineering. Briefly summarised, situated cognition, authentic learning situations and cognitive apprenticeship are closely interrelated.

Social-constructivist ideas have gained growing attention in science education over the past years. On the one hand, there are attempts to employ these ideas to address limitations and one-sidenesses of mainstream constructivist approaches and to further develop them (Anderson, Belt, Gamalski & Greminger 1987; Scott 1995; Ueno & Arimoto 1993). On the other hand, there is a number of studies that explore the potential of the social-constructivist perspective to investigate and support meaning construction in learning communities. Roth's (1994) work on open experimenting in science instruction and his studies on collaborative design (McGinn, Roth, Boutonné & Woszczyna 1995) could be taken as examples (see Tytler 1994 for a discussion on the social-constructivist perspective as a theoretical framework for interpreting students' learning processes in science). Clearly, the focus of empirical research is studying collaborative meaning construction and learning (i.e., guided inquiry in small groups). In this field, the potential of the social-constructivist views of knowledge becomes obvious as being 'distributed' and 'shared' rather than being the property of individuals.

To view learning science as change from one culture to another, or as initiation into a new culture, opens avenues that appear to go beyond making science learning merely more effective. For example, Cobern and Aikenhead (in this *Handbook*) provide a broad view of what they call 'cultural aspects of learning science'. Their approach is based on the social-constructivist perspective but is given particular characteristics by two aspects. The first is the idea of 'anthropological' learning (Aikenhead 1996) which involves viewing science learners as anthropologists who enjoy, and are capable of, constructing meaning out of the foreign culture of science. The second aspect is a view of 'crossing over' between an everyday life culture to the science culture in societies or parts of societies for which the everyday culture is far from being science and technology-oriented (e.g., Aborigines and Native Indians). Science learning in science-oriented 'Western' societies can be viewed as analogous to that kind of 'crossing over' in non-Western cultures where deep-rooted difficulties of learning science go far beyond the issues of knowledge construction.

## CONCLUSIONS

Domain-specific preinstructional knowledge has proven to be *the* key factor determining learning and problem solving in research in all science domains. Ausubel's (1968, p. vi) famous dictum, 'The most important single factor influencing learning is what the learner already knows . . .', has been corroborated many times since it was written. Although this dictum concerns learning science in a particular way based on what the learners already know, science instruction frequently is not designed for the science perspectives to be learned effectively. This is true for science content (i.e., for science concepts and principles) and for science epistemologies and ontologies (i.e., for views on the nature of science). Learning science is only successful if learning pathways are designed to lead from certain facets of preinstructional knowledge towards the science perspective.

The conceptual change approaches to learning which are critically reviewed in this chapter must be considered whenever science instruction is designed. Conceptual change strategies of some kind have to be embedded in what has been called 'conceptual change supporting conditions' which incorporate issues such as students' interests, motivation and self-concepts, or the classroom climate and power structures in school. Social constructivist views of learning appear to contribute primarily to the provision of such conceptual change supporting conditions. It is an important challenge for science education research on learning and instruction in the future to investigate relations between issues of conceptual change (of designing adequate learning pathways) and the many supporting conditions. Little research knowledge is available on this so far. Only during such a process of further development and refinement of the inclusive view, will it be possible to see whether the implementation of this view actually leads to more effective learning than using previous, more limited views. The inclusive, social constructivist view of learning appears to be suited also to guiding our thinking about science education beyond the aim of making science teaching and learning more effective. It could lead to an inclusive view of science education in a much broader sense in that it also incorporates considerations of future aims of science instructions.

The views of learning science discussed in the present chapter all are on a 'macro' or phenomenological level. The rapid advances in the neuro-sciences regarding the structure and function of the human brain so far have not contributed substantially to views of effective learning. In underpinning key facets of his revised Piagetian view of learning with recent findings on the architecture of the human brain, Lawson's (1994) attempt to provide a 'micro' view of learning by referring to neurological issues is an exciting first step that appears to provide promising insights for the future. However, at the moment, it seems that the explanatory and predictive power of the 'neuro' view is still too limited to contribute to the development of powerful teaching and learning strategies.

The other seven chapters of this section of the *Handbook* on Learning provide a comprehensive picture of the essential areas that we believe are important in understanding learning in science classrooms: language and science; cultural aspects of learning science; learning science through models and modelling; learning about science teaching; what young children can be expected to know about science; theories of knowledge acquisition for identifying gaps in our knowledge as well as guiding teaching; and students' representation of the nature of scientific knowledge.

In the chapter entitled 'New Perspectives on Language in Science', Sutton explores the relationship between language and learning in science by arguing that the impersonal nature of today's scientific writing does not help students connect their scientific understanding with their own human concerns and those of other people. As well as providing illustrations of historical changes in science writing, Sutton makes recommendations for introducing a personal voice back into science and what is meant by communication.

In their chapter, entitled 'Cultural Aspects of Learning Science', Cobern and Aikenhead provide a cultural perspective on science education, illustrate it with examples from secondary science on how students' culture can affect their learning of science, and identify related issues for teaching and research. The authors show how, for many students in school, learning science is not a straightforward process but often involves a variety of cultural border crossings.

Gilbert and Boulter examine 'Learning Science Through Models and Modelling' and show how models can and do contribute to learning in classrooms and other contexts. A major aspect of this chapter is a description of the use of models and narratives in the classroom and in other contexts involving computers, educational television and museums.

In their chapter, entitled 'Learning About Science Teaching: Perspectives From an Action Research Project', Scott and Driver provide an account of an action research project whose aim was to draw upon a constructivist view of learning in developing approaches and materials for teaching particular concepts in high school. The chapter focuses on curriculum design and pedagogy, as well as the nature of the teachers' involvement in the project.

In her chapter entitled 'Scientific Inquiry Within Reach of Young Children', Metz examines emergent literature about the process and products of children's scientific inquiry, children's domain-specific knowledge and children's collaborative cognition. This literature shows that children between six and 13 years of age can engage in independent empirical investigations. While the literature has gaps for older children and adults and includes some inadequate experimental designs, it supports both fruitful theory construction and improvement of the inquiry process itself.

The chapter by Chinn and Brewer entitled 'Theories of Knowledge Acquisition' analyses the problem of knowledge acquisition and presents eight core questions that are the basis of a framework for assisting researchers, theorists and teachers to identify the gaps in current knowledge. The chapter shows that there is a wide range of theoretically-important questions that have not been adequately investigated and that teachers could develop more effective instruction if better answers existed.

The epistemology of school science is the issue addressed by Désautels and Larochelle in their chapter entitled 'The Epistemology of Students: The "Thingified" Nature of Scientific Knowledge'. The authors show that an understanding of the nature of models, laws and theories appears to change little as a result of schooling and that empirical perceptions of scientific phenomena dominate over theoretical and personal perspectives. A major feature of students' epistemologies of science is the tendency for students to give science a material entity.

## REFERENCES

Aikenhead, G.S.: 1996, 'Science Education: Border Crossing into the Subculture of Science', *Studies in Science Education* 26, 1–52.

Anderson, C.W., Belt, B.L., Gamalski, J.M. & Greminger, J.E.: 1987, 'A Social Constructivist Analysis of Classroom Science Teaching', in J. Novak (ed.), *Proceedings of the Second International Seminar 'Misconceptions and Educational Strategies in Science and Mathematics'* (Volume II), Cornell University, Ithaca, NY, 11–24.

Ausubel, D.P.: 1968, *Educational Psychology: A Cognitive View*, Holt, Rinehart and Winston, New York.

Baird, J.R. & Mitchell, I.M. (eds.): 1986, *Improving the Quality of Teaching and Learning: An Australian Case Study – The PEEL Project*, Monash University, Melbourne, Australia.

Bereiter, C.: 1985, 'Toward a Solution of the Learning Paradox', *Review of Educational Research* 55, 201–226.

Bliss, J.: 1995, 'Piaget and After: The Case of Learning Science', *Studies in Science Education* 25, 139–172.

Brown, J.S., Collins, A. & Duguid, P.: 1989, 'Situated Cognition and the Culture of Learning', *Educational Researcher* 18(1), 32–42.

Bruner, J.S.: 1960, *The Process of Education*, Vintage, New York.

Caravita, S. & Halldén, O.: 1994, 'Re-framing the Problem of Conceptual Change', *Learning and Instruction* 4, 89–111.

Carey, S.: 1985, *Conceptual Change in Childhood*, MIT Press, Cambridge, MA.

Chi, M.T.H., Slotta, J.D. & de Leeuw, N.: 1994, 'From Things to Processes: A Theory of Conceptual Change for Learning Science Concepts', *Learning and Instruction* 4, 27–43.

Chinn, C.A. & Brewer, W.F.: 1993, 'The Role of Anomalous Data in Knowledge Acquisition: A Theoretical Framework and Implications for Science Education', *Review of Educational Research* 63, 1–49.

Driver, R.: 1983, *The Pupil as Scientist?*, Open University Press, Milton Keynes, UK.

Driver, R.: 1989, 'Changing Conceptions', in P. Adey (ed.), *Adolescent Development and School Science*, Falmer Press, London, 79–99.

Driver, R. & Erickson, G.: 1983, 'Theories-in-Action: Some Theoretical and Empirical Issues in the Study of Students' Conceptual Frameworks in Science', *Studies in Science Education* 10, 37–60.

Driver, R., Squires, A., Rushworth, P. & Wood-Robinson, V.: 1994, *Making Sense of Secondary Science*, Routledge, London.

Duit, R.: 1994, 'Research on Students' Conceptions – Developments and Trends', in H. Pfundt & R. Duit (eds.), *Bibliography: Students' Alternative Frameworks and Science Education* (fourth edition), Institute for Science Education at the University of Kiel, Kiel, Germany, xxii–xlii.

Duit, R.: 1995, 'Constraints on Knowledge Acquisition and Conceptual Change – The Case of Physics', Paper presented at the Symposium 'Constraints on Knowledge Construction and Conceptual Change: A Look Across Content Domains', 6th European Conference for Research on Learning and Instruction, Nijmegen, The Netherlands.

Duit, R.: in press, 'Conceptual Change in Science Education', in M. Carretero, W. Schnotz & S. Vosniadou (eds.), *Conceptual Change*, Lawrence Erlbaum, London.

Duit, R., Goldberg, F. & Niedderer, H. (eds.): 1992, *Research in Physics Learning: Theoretical Issues and Empirical Studies*, Institute for Science Education at the University of Kiel, Kiel, Germany.

Eylon, B.-S. & Linn, M.C.: 1988, 'Learning and Instruction: An Examination of Four Research Perspectives in Science Education', *Review of Educational Research* 58, 251–301.

Farnham-Diggory, S.: 1994, Paradigms of Knowledge and Instruction. *Review of Educational Research* 64, 463–477.

Festinger, L.: 1962, *A Theory of Cognitive Dissonance*, Stanford University Press, Stanford, CA.

Gergen, K.J.: 1995, 'Social Construction and the Educational Process', in L. Steffe & J. Gale (eds.), *Constructivism in Education*, Lawrence Erlbaum, Hillsdale, NJ, 41–56.

Guzetti, B.J., Snyder, T.E., Glass, G.V. & Gamas, W.S.: 1993, 'Promoting Conceptual Change in Science: A Comparative Meta-Analysis of Instructional Interventions from Reading Education and Science Education', *Reading Research Quarterly* 28, 116–159.

Hennessy, S.: 1993, 'Situated Cognition and Cognitive Apprenticeship: Implications for Classroom Learning', *Studies in Science Education* 22, 1–41.

Hewson, P.W.: 1981, 'A Conceptual Change Approach to Learning Science', *European Journal of Science Education* 4, 383–396.

Hewson, P.W. & Hewson, M.G.: 1992, 'The Status of Students' Conceptions', in R. Duit, F. Goldberg & H. Niedderer (eds.), *Research in Physics Learning: Theoretical Issues and Empirical Studies*, Institute for Science Education at the University of Kiel, Kiel, Germany, 59–73.

Hewson, P.W. & Thorley, N.R.: 1989, 'The Conditions of Conceptual Change in the Classroom', *International Journal of Science Education* 11, 541–553.

Jung, W.: 1986, 'Alltagsvorstellungen und das Lernen von Physik und Chemie' [Everyday Conceptions and Learning Physics and Chemistry]. *Naturwissenschaften im Unterricht – Physik/Chemie* 34(April), 2–6.

Jung, W.: 1993, 'Hilft die Entwicklungspsychologie dem Naturwissenschaftsdidaktiker' [Is Developmental Psychology of Any Help for a Physics Educator?], in R. Duit & W. Gräber (eds.),

*Kognitive Entwicklung und Lernen der Naturwissenschaften*, Institute for Science Education at the University of Kiel, Kiel, Germany, 86–108.
Karplus, R.: 1977, 'Science Teaching and the Development of Reasoning', *Journal of Research in Science Teaching* 14, 33–46.
Kelly, G.A.: 1955, *The Psychology of Personal Constructs*, Norton, New York.
Kesidou, S., Duit, R. & Glynn, S.: 1995, 'Conceptual Development in Physics: Students' Understanding of Heat', in S. Glynn & R. Duit (eds.), *Learning Science in the Schools: Research Reforming Practice*, Lawrence Erlbaum, Mahwah, NJ, 179–198.
Knorr-Cetina, K.: 1981, *The Manufactor of Knowledge: An Essay on the Constructivist and Contextual Nature of Science*, Pergamon, New York.
Kuhn, T.: 1970, *The Structure of Scientific Revolutions*, University of Chicago Press, Chicago, IL.
Lakatos, I.: 1970, 'Falsification and the Methodology of Scientific Research', in I. Lakatos & A. Musgrave (eds.), *Criticism and the Growth of Knowledge*, Cambridge University Press, Cambridge, UK, 91–196.
Lave, J. & Wenger, E.: 1991, *Situated Learning: Legitimate Peripheral Participation*, Cambridge University Press, Cambridge, MA.
Lawson, A.E.: 1994, 'Research on the Acquisition of Science Knowledge: Epistemological Foundations of Cognition', in D. Gabel (ed.), *Handbook of Research on Science Teaching and Learning*, Macmillan, New York, 131–176.
Lawson, A. E., Abraham, M. & Renner, J.: 1989, *A Theory of Instruction: Using the Learning Cycle to Teach Science Concepts and Thinking Skills* (NARST Monograph Number One), National Association for Research in Science Teaching, University of Cincinnati, Cincinnati, OH.
Lawson, A.E. & Thompson, L.D.: 1988, 'Formal Reasoning Ability and Misconceptions Concerning Genetics and Natural Selection', *Journal of Research in Science Teaching* 27, 589–606.
Lemke, J.: 1990, *Talking Science*, Ablex, Norwood, NJ.
Linder, C. J.: 1993, 'A Challenge to Conceptual Change', *Science Education* 77, 293–300.
Lybeck, L., Marton, F., Stroemdahl, H. & Tullberg, A.: 1988, 'The Phenomenography of the Mole Concept: An Example of How Students' Understanding Can Contribute to the Advancement of Science and Science Education', in P. Ramsden (ed.), *Improving Learning: New Perspectives*, Kogan Page, London, 81–108.
Marton, F.: 1986, 'Phenomenography – A Research Approach to Investigate Different Understandings of Reality', *Journal of Thought* 21(3), 28–49.
McGinn, M.K., Roth, W.-M., Boutonné, S & Woszczyna, C.: 1995, 'The Transformation of Individual and Collective Knowledge in Elementary Science Classrooms that are Organized as Knowledge-Building Communities', *Research in Science Education* 25, 163–189.
Niedderer, H.: 1996, 'Überblick über Lernstudien in Physik' [Review of Learning Process Studies in Physics], in R. Duit & Ch. von Rhöneck (eds.), *Lernen in den Naturwissenschaften*, Institute for Science Education at the University of Kiel, Kiel, Germany, 119–144.
Novak, J.D.: 1978, 'An Alternative to Piagetian Psychology for Science and Mathematics Education', *Studies in Science Education* 5, 1–30.
Oakes, M.E.: 1947, *Children's Explanations of Natural Phenomena*, Columbia University, Teachers College, New York.
O'Loughlin, M.: 1992, 'Rethinking Science Education: Beyond Piagetian Constructivism Toward a Sociocultural Model of Teaching and Learning', *Journal of Research in Science Teaching* 29, 791–820.
Pfundt, H. & Duit, R.: 1994, *Bibliography: Students' Alternative Frameworks and Science Education* (fourth edition), Institute for Science Education at the University of Kiel, Kiel, Germany.
Piaget, J.: 1954, *The Construction of Reality in the Child*, Basic Books, New York.
Pintrich, P.R., Marx, R.W. & Boyle, R.A.: 1993, 'Beyond Cold Conceptual Change: The Role of Motivational Beliefs and Classroom Contextual Factors in the Process of Conceptual Change', *Review of Educational Research* 6, 167–199.
Posner, G.J., Strike, K.A., Hewson, P.W. & Gertzog, W.A.: 1982, 'Accommodation of a Scientific Conception: Toward a Theory of Conceptual Change', *Science Education* 66, 211–227.
Reif, F. & Allen, S.: 1992, 'Cognition for Interpreting Scientific Concepts: A Study of Acceleration', *Cognition and Instruction* 9, 1–44.
Rennstroem, L.: 1987, 'Pupils' Conceptions of Matter: A Phenomenographic Approach', in J. Novak (ed.), *Proceedings of the Second International Seminar 'Misconceptions and Educational Strategies in Science and* Mathematics' (Volume III), Cornell University, Ithaca, NY, 400–414.

Roth, W.-M.: 1994, 'Experimenting in a Constructivist High School Physics Laboratory', *Journal of Research in Science Teaching* 31, 197-223.
Roth, W.-M.: 1995, *Authentic School Science: Knowing and Learning in Open-Inquiry Laboratories*, Kluwer, Dordrecht, The Netherlands.
Rowell, J.A. & Dawson, C.J.: 1985, 'Equilibrium, Conflict and Instruction: A New Class-Oriented Perspective', *European Journal of Science Education* 5, 203-215.
Schumacher, G.M., Tice, S., Wen Loi, P., Stein, S., Joyner, C. & Jolton, J.: 1993, 'Difficult to Change Knowledge: Explanations and Interventions', Paper presented at the Third International Seminar on Misconceptions and Educational Strategies in Science and Mathematics, Cornell University, Ithaca, NY.
Schunk, D.H.: 1991, *Learning Theories: An Educational Perspective*, Macmillan, New York.
Scott, P.: 1992, 'Conceptual Pathways in Learning Science: A Case Study of the Development of One Student's Ideas Relating to the Structure of Matter', in R. Duit, F. Goldberg & H. Niedderer (eds.), *Research in Physics Learning: Theoretical Issues and Empirical Studies*, Institute for Science Education at the University of Kiel, Kiel, Germany, 203-224.
Scott, P.: 1995, 'Social Interactions and Personal Meaning Making in Secondary Science Classrooms', Paper presented at the First European Conference on Research in Science Education, University of Leeds, Leeds, UK.
Scott, P., Asoko, H. & Driver, R.: 1992, 'Teaching for Conceptual Change: A Review of Strategies', in R. Duit, F. Goldberg & H. Niedderer (eds.), *Research in Physics Learning: Theoretical Issues and Empirical Studies*, Institute for Science Education at the University of Kiel, Kiel, Germany, 310-329.
Seiler, T.B.: 1973, 'Die Bereichsspezifizität formaler Denkstrukturen – Konsequenzen für den pädagogischen Prozeβ' [On Domain-Specific Issues of Formal Thinking – Consequences for the Educational Process], in K. Frey & M. Lang (eds.), *Kognitionspsychologie und naturwissenschaftlicher Unterricht*, Huber, Bern, Switzerland, 249-283.
Shayer, M. & Adey, P.: 1992, 'Accelerating the Development of Formal Thinking in Middle and High School Students 3: Testing the Permanency of Effects', *Journal of Research in Science Teaching* 29, 1101-1115.
Shayer, M. & Wylam, H.: 1981, 'The Development of the Concept of Heat and Temperature in 10-13 year olds', *Journal of Research in Science Teaching* 18, 419-434.
Shulman, L. & Tamir, P.: 1973, 'Research on Teaching in the Natural Sciences', in R.M.W. Travers (ed.), *Second Handbook of Research on Teaching*, Rand McNally, Chicago, IL, 1098-1148.
Solomon, J.: 1987, 'Social Influences on the Construction of Pupils' Understanding of Science', *Studies in Science Education* 14, 63-82.
Solomon, J.: 1994, 'The Rise and Fall of Constructivism', *Studies in Science Education* 23, 1-19.
Steffe, L. & Gale, J. (eds.): 1995, *Constructivism in Education*, Lawrence Erlbaum, Hillsdale, NJ.
Strike, K.A. & Posner, G.J.: 1985, 'A Conceptual Change View of Learning and Understanding', in L. West & L. Pines (eds.), *Cognitive Structure and Conceptual Change*, Academic Press, Orlando, FL, 211-231.
Strike, K. & Posner, G.: 1992, 'A Revisionist Theory of Conceptual Change', in R. Duschl & R. Hamilton (eds.), *Philosophy of Science, Cognitive Psychology, and Educational Theory and Practice*, State University of New York, Albany, NY, 147-176.
Taylor, P.C.: 1993, 'Collaborating to Reconstruct Teaching: The Influence of Researcher Beliefs', in K. Tobin (ed.), *The Practice of Constructivism in Science Education*, American Association for the Advancement of Science, Washington, DC, 267-298.
Tiberghien, A.: 1980, 'Modes and Conditions of Learning – An Example: The Learning of Some Aspects of the Concept of Heat', in F. Archenhold, R. Driver, A. Orton & C. Wood-Robinson (eds.), *Cognitive Development: Research in Science and Mathematics*, University of Leeds, Leeds, UK, 288-309.
Tobin, K. (ed.): 1993, *The Practice of Constructivism in Science Education*, American Association for the Advancement of Science, Washington, DC.
Toulmin, S.: 1972, *Human Understanding: An Inquiry into the Aims of Science*, Princeton University Press, Princeton, NJ.
Treagust, D., Duit, R. & Fraser, B. (eds.): 1996, *Improving Teaching and Learning in Science and Mathematics*, Teacher College Press, New York.
Tytler, R.W.: 1994, *Children's Explanations in Science – A Study of Conceptual Change*, PhD thesis, Monash University, Faculty of Education, Melbourne, Australia.

Ueno, N. & Arimoto, N.: 1993, 'Learning Physics by Expanding the Metacontext of Phenomena', *The Quarterly Newsletter of the Laboratory of Comparative Human Cognition* 15(2), 53–63.
von Glasersfeld, E.: 1995, 'Sensory Experiences, Abstraction, and Teaching', in L. Steffe & J. Gale (eds.), *Constructivism in Education*, Lawrence Erlbaum, Hillsdale, NJ, 369–383.
Vosniadou, S.: 1994, 'Capturing and Modeling the Process of Conceptual Change', *Learning and Instruction* 4, 45–69.
Vosniadou, S. & Brewer, W.F.: 1992, 'Mental Models of the Earth: A Study of Conceptual Change in Childhood', *Cognitive Psychology* 24, 535–585.
Vygotsky, L.: 1986, *Thought and Language*, MIT Press, Cambridge, MA.
Wandersee, J.H., Mintzes, J.J. & Novak, J.D.: 1994, 'Research on Alternative Conceptions in Science', in D. Gabel (ed.), *Handbook of Research on Science Teaching and Learning*, Macmillan, New York, 177–210.
Welch, W.W.: 1983, 'Experimental Inquiry and Naturalistic Inquiry: An Evaluation', *Journal of Research in Science Teaching* 20, 95–103.
Wertsch, J. & Toma, C.: 1995, 'Discourse and Social Dimensions of Knowledge and Classroom Teaching', in L. Steffe & J. Gale (eds.), *Constructivism in Education*, Lawrence Erlbaum, Hillsdale, NJ, 159–174.
White, R.T.: 1987, 'The Future of Research on Cognitive Structure and Conceptual Change', Paper presented at the annual meeting of the American Educational Research Association, Washington, DC.
White, R.T. & Tisher, R.: 1986, 'Research on Natural Sciences', in M. C. Wittrock (ed.), *Handbook of Research on Teaching* (third edition), Macmillan, New York, 874–905.

# 1.2 New Perspectives on Language in Science

CLIVE SUTTON
*University of Leicester, UK*

PERSONAL LETTERS AND THE START OF SCIENTIFIC PERSUASION

When Charles Darwin or Michael Faraday wanted to share their ideas with others, they often reached for pen and paper to compose a letter to a trusted friend or colleague. Many of their letters have survived and we can learn a lot from them that might make us better managers of learning and communication in the classroom.

The earliest of Faraday's letters which we still have were written in the summer of 1812 when he was not quite 21 years old (Williams 1971). They show him writing with great enthusiasm to his friend, Benjamin Abbott, about a wide range of topics – the value of letters, the cost of sheet zinc, how to use it to make a 'Voltaic pile', and sundry thoughts and adventures which he had experienced while running through London in heavy rain. They provide a vivid picture of a young man actively sorting out ideas and thoroughly animated about Humphry Davy's views on the new green gas, 'chlorine'. That's what Davy was calling it, even though others had insisted on naming it 'oxy-muriatic acid'.

Faraday used his letters to rehearse the arguments. Was the green gas truly a simple elementary substance as Davy maintained? If so, then the more well-known 'steamy' gas from salt which people called 'smoking spirit of salt' or 'muriatic acid gas' might be renamed 'hydrogen chloride' and recognised as a compound of two things only. The trouble was that to think and talk about it in that way would mean abandoning Lavoisier's idea that all acids contain 'oxy-gen', the 'acid-begetter'. Abbott had expressed objections to the 'chlorine as a simple element' scheme of thought, and Faraday was keen to persuade him to think again. Here are some extracts from his second attempt on 1 September 1812 (see Williams 1971, p. 21):

---

Chapter Consultant: Daniel Gil-Pérez (University of Valencia, Spain)

*Dear Obliging Abbott,*

*... In my last [letter] if I remember rightly I gave as proofs that Chlorine gas contained no Oxygen experiments in the following import*

- *That the Carbonaceous part of a taper would not burn in it.*
- *That ignited Carbon ... would not burn in it.*
- *That when Chlorine and Hydrogen were united no results containing Oxygen appeared. ...*

*... [M]y arguments in favour of the simplicity of Chlorine will be drawn from the nature of the results when it ... is made the supporter of combustion and when it is combined with other bodies -- I have already observed one of these combinations ... -- Hydrogen when united to chlorine forms a pure unmixed uniform binary compound Muriatic acid and no water is produced ...*

*Look into your Lavoisier into your Nicholson into your Fourcroy and what other chemical books you have at the article Oxy-Muriatic acid ... they will tell you that its Oxygen is held by an affinity so weak that the combustibles burn in it very easily and compounds of the oxygen ... are obtained as well ...*

*You will think me bold dear A if I deny all these authorities but Davy has done so and I will do it too ...*

*In my next I will continue the subject but positively will first hear from you so that I may know my opponent & his objections. ...*

*Adieu, dear Ab ...*

Several features of this letter are important for what I want to say about sharing new ideas in science. First, there is a strong personal voice – we can hear Faraday 'speaking'. Second, he wants to share his viewpoint and to persuade his reader. Third, he writes with conviction because he has a clear image in his mind's eye.

Personal involvement and use of a new point of view linked with a new way of talking are key features of the initial stages of scientists' communication. Indeed that is what the word itself means in this context – *commun*-ication is an attempt to create a *commun*-ity of thought, a shared understanding. Teachers engage in a similar activity when they say 'Try looking at it like this' and the learners are invited to see something with new eyes. However, viewpoint sharing is not just a matter of passing over information. It involves winning agreement with a certain perspective, and winning attention to the points which matter from that perspective. When it is successful, the participants start to see the relevance of the evidence presented, and they come to possess both the new way of seeing and the new way of talking.

Nowadays letters take their place alongside other informal kinds of communication in scientific research, including both electronic mail and face-to-face discussion at scientific meetings. Above these, we have another layer of more formal written communication, whose form and function has received a lot of attention recently from historians, as part of a general trend in the history of science to recognise that communicative activities are central to the scientific endeavour. Experiment is a part of science, but so is writing and talk. It is through publication and discussion in learned societies that some of the new insights and claims of individual researchers get transformed into what becomes accepted scientific knowledge. So, writing for a journal or attending a congress is just as much a part of science as the more practical activities of handling apparatus and conducting experiments. A current problem with school science is that a heavy emphasis on experiments is in danger of giving students an unbalanced picture of the range of what a scientist does.

Even in writing, awareness of the range is incomplete. Although the formal article for a journal has often been held to embody the ideals of how writing in science 'should' be done, it is only one of the genres that scientists learn to use and not the only way in which they write. For a fuller understanding, both teachers and learners need to see those journal articles in relation to other kinds of writing, ranging from letters at one end to textbooks at the other.

## HOW THE HUMAN VOICE OF THE SCIENTIST FADES

Despite the impersonal image of scientific writing, there is always a detectable personal voice when a scientist writes about something for the first time. When William Harvey put forward the idea of a circulatory movement of blood, he wrote: 'I began to think whether there might not be a motion, as it were, in a circle' (Harvey 1628, ch. 8). Nearer to our own time, James Watson and Francis Crick began their most famous paper with the words: 'We wish to propose a structure for the salt of deoxyribose nucleic acid (DNA). The structure has novel features which are of considerable biological interest' (Watson & Crick 1953, p. 737).

The sense of personal identification with a new viewpoint is captured in the expressions 'I began to think' and 'we wish to propose'. Even in more run-of-the-mill reports of experiments, fellow scientists know that the authors are making a *claim* which they hope will be taken into the body of accepted fact, and that such a contribution is deeply connected with the thoughts, hopes and fears of particular human beings. What historians and literary scholars have been exploring recently is how the personal connection is gradually diminished as the new science becomes established science, and the claim becomes no longer just 'So-and-so's idea' or interpretation, but something worthy to be called 'So-and-so's *discovery*'. We can trace the fading of personal attribution in a sequence of different kinds of writing – journals first, then research reviews and then textbooks. The journal account might be cited in a review alongside other researchers' claims, or mentioned in a *Handbook of Recent Research in XXology* which every researcher in that field will read. If it continues to be sustained, it then gets into the textbooks, and in this handing-on process phrases such as 'it is thought that' or 'So-and-so has suggested that' are gradually reduced or omitted. Ideas and claims which could be identified with individuals are made into common property. They are converted into agreed public knowledge which merits the status of 'fact', 'fact for the time being' or at least 'best available theory, which to all intents and purposes we can assume to be correct'. (For a fuller account of this process, see Sutton 1996.)

The texbook account as the end-product of this process of assimilation is important because it expresses the consensus about what is important in a particular branch of science. It also gives a powerful sense of 'what we have found out about how the world works'. Nevertheless, there are losses from an educational point of view because the definiteness of the language and its detachment from human beings can give a very misleading impression of how the knowledge was established.

## WHAT ARE STUDENTS LEARNING – A DISTORTED CONCEPTION OF SCIENCE?

### The Influence of Textbook Knowledge

It might seem strange that the textbook account, which is the product of successful science, can be a source of misunderstanding, but the problem is that learners encounter this product without experiencing any of the uncertainty and controversy that was involved in establishing it. 'Chlorine *is* an element', says the textbook. 'Air *is* a mixture of nitrogen and oxygen.' Just like that. These useful summaries of what we know today are not wrong, but what they fail to explain is that most of the words in those sentences were human inventions, hotly debated before they became an accepted part of current science. The definiteness of the textbook account sweeps away the memories of doubt or difficulty over what might be taken as true, and makes 'the facts' appear to be completely outside human agency. It is as if people had no part in shaping the facts or arguing what exactly could be believed. Facts were not argued into existence; language was not involved in creating them; the scientists' role was just to find them, ready-made.

To put the problem another way, constant exposure to long-accepted accounts of scientific knowledge can give too simple an idea of what a fact is. Learners pick up a Baconian view of the scientist as a 'fact gatherer' (Driver, Leach, Scott & Millar 1994), a person who goes out and makes discoveries by 'seeing what happens', rather than by any process of imaginative effort and painstaking construction. 'Atoms are made of protons, neutrons and electrons', we say, but without elaboration such an expression encourages an uncritical reification of these human constructs and implies that they were simply 'found' rather than being suggested, invented or built up as part of helpful systems of explanation. A statement like 'every atom has a nucleus' can be deeply misleading if the learner has never understood one of the fuller formulations such as 'Ernest Rutherford suggested that every atom has a nucleus'.

The problem for a teacher is to achieve an appropriate blend of definiteness and tentativeness. If we restore the human authorship and re-admit uncertainty and the possibility of argument, we can help students to gain a better idea of science-in-the-making. On the other hand, it is also a source of pride that scientists *have* managed to detach knowledge from individuals. They *have* got beyond 'So and so's opinion' and obtained agreement about facts which are understood as common to all – 'universal', 'public' or 'objective' knowledge as it is called, even by those who insist that it is consensual and more usefully thought of as 'intersubjective'.

*The Influence of Custom and Form*

Slavish imitation of a detached style of writing is another way in which learners pick up a distorted view of science, and it is here that historians can help teachers by showing the origins of scientific styles, and how they have changed since they were first invented.

'Detached' writing was first developed within the courtesies and customs of the early scientific societies, as they considered how to keep argument and discussion within manageable bounds. Shapin and Schaffer (1985) argue that, after the civil war in England, Robert Boyle and his cofounders of the Royal Society created what was then a new way of presenting and sharing ideas, and they were able to gain agreement about 'matters of fact' which would stand in sharp contrast to the political and religious 'enthusiasms' which had been socially corrosive in previous decades.

The need to stop philosophical disputes getting out of control was not a purely academic problem at that time, but one of war and peace, imprisonment or mutual toleration. Leading thinkers of the day discussed whether toleration of different views was compatible with the maintenance of civic order. In such a climate, the new approach to *'natural* philosophy' was a contribution to an urgent social problem. Its proponents developed what Shapin and Schaffer call a new 'literary technology' – a way of writing which separates the opinions of the writer from the 'matters of fact' being reported. It was accompanied by a 'material technology' (making use of instruments), so that investigators could withdraw their personal voices from a part of the report by saying that 'it is not we who say this but our equipment'. In addition, there was a 'social technology' of conventions and manners within the Society (e.g., in the courteous 'receiving' of a report of experiments). Through these ways of working, it was understood that 'data' could be something separate from 'speculations', and a challenge to the reliability of members' equipment could be separated from a challenge to their honour. This new approach was successful in recruiting people who were eager to collaborate in the search for 'natural knowledge' without fragmenting prematurely into factions.

One of Boyle's first techniques was to list the distinguished witnesses who had been present at an experiment, so that the reader more readily would agree that the phenomena must be as he said they were. Later, he developed a way of reporting the event in such detail that the reader became in effect a 'virtual witness' and the written account became acceptable as its own authority. From that small beginning, we can trace the idea of a separation of 'methods' from 'results' and 'discussion'. Journals did not immediately adopt a fixed format enshrining such divisions, but respect for the different components nevertheless became an important part of the ethic of all the societies. Whatever part of a report might be questioned or attacked, there is usually another part which can be respectfully accepted, or which can lead someone else into further experiments. The societies thus achieved a method of placing ideas before other people in a way which assists semi-collaborative inquiry. Bazerman (1988) shows that, in the first 150 years of the Royal Society, articles in its *Philosophical Transactions* kept changing in form, but belief in the value of separating 'findings' from other parts of one's writing grew in strength. By the 20th century, some journal editors were insisting on a standard format, and unfortunately this now features in the public imagination of 'how science should be done'. In the biological sciences, it includes such headings as Introduction, Previous Work, Methods, Results and Conclusions. Peter Medawar (1974, p. 14) called it 'a totally mistaken conception, even a travesty, of the nature of scientific thought', because it glosses over 'why we were doing this' and 'what our initial hunches were' and it pretends that thought (i.e., the discussion) is done mainly after the factual results have been collected. This criticism should be taken seriously when educating citizens at large about science, although the pretence has been successful in the research community.

Gradually, scientific writers found that various linguistic devices also could be helpful in creating distance between investigators and their 'findings'. One such device is to use the passive voice, as in 'Measurements were taken' rather than 'I measured' and 'Experiments were conducted' rather than 'My colleague and I carried out the experiments'. Another was to create new abstract nouns and noun phrases such as in 'The ray suffered linear refraction' rather than 'The light bent along a new line'. Halliday and Martin (1993) argue that the constant development of nouns and elaborated noun phrases is crucial to science because it separates the investigator from Nature. Generations of scientists have taken verbs like 'flowing' and chosen the noun form as their object of study (e.g., 'the rate of flow' or 'the current'). In the 20th century, this nominalisation of language has intensified in all academic discourse. Modern citizens require some competence in reading and understanding that kind of expression, but their science lessons should not be dominated by it.

The greatest misunderstanding about writing occurred when science was professionalised in the 19th century, and recruits to science were trained to write using conventions that in part were derived from the above devices and traditions. Schools inherited rituals for writing in 'objective' ways and holding back acknowledgement of one's own thought and involvement. In chemistry, for example, qualitative analysis was to be written up under the headings Test, Observation and

Inference. 'What I thought beforehand' or 'why I wanted to try this test' were absent. If such systems were ever justified in schools, it was in the limited context of training technicians for the routines of laboratory life; but they still influence ideas about what is permissible, even though guidance to students is now much less rigid.

Custom and form in scientific writing should be understood, rather than followed in a ritual manner. What's involved in writing a definition? Why are there so many '-tion' words in science? What's the best way to present a formal report? Students who address those questions with their teachers will be able to see the distinctively scientific genres as forms of expression which have useful functions in a certain context, rather than as something fixed and arbitrary (which you must use all the time, otherwise it's not science).

## WHAT ELSE ARE STUDENTS LEARNING – A DISTORTED VIEW OF LANGUAGE?

Important as it is that learners should know how science works, it is even more important that they should sense the multiple purposes of language and be able to use it well in their own learning. A highly 'factual' account, distanced from human beings, is not a good model for them because it offers *giving and receiving of information* as the main thing that we do with language, as in the middle column of Table 1. With a more balanced view of human language, it should be clear that new learning involves employing language firstly as an *interpretive* tool, a means of making sense of what we see and what we think other people are saying (the right-hand column of the Table 1). Learners need to hear language used in that way – as expression of thought rather than as statement of disembodied fact.

Another idea about language which is sometimes held by students and teachers is that language in science *describes* what we see in experiments. This is misleading because what anyone is able to select with their eyes depends on the mental organisation which they already have. Strictly speaking, science is not about describing, but about *re-describing*. At points of change in scientific theory, the breakthrough usually involves a new way of talking. Someone starts to talk about enzymes as fitting their substrates 'like a lock and key'. Leaves are spoken of as 'chemical factories'. Water in a high lake is referred to as a 'store' of 'potential' energy, and mountains take part in a 'cycle' of erosion and re-building. In other words, scientists try out a new talk pattern in which they select a new metaphor in an attempt to figure out what is happening. If the re-description is successful, they are able to elaborate it into a testable model. For new learners, the teacher's job is to help them to get on the inside of the models and ways of talking which organise the different branches of science – current-talk for circuits, field-talk for magnets, and molecular bombardment-talk for understanding air pressure. Each of these began as a successful re-description of something which previously had defeated human comprehension. In each case, *the language is the theory* (see the right hand column of Table 1).

Table 1: Differences between language as a system for transmitting information and as an interpretative system

| Characteristic | Role of language | |
| --- | --- | --- |
| | A system for transmitting information | An interpretive system for making sense of experience |
| 1. What the speaker or writer appears to be doing | Describing, telling, reporting | Persuading, suggesting, exploring, figuring |
| 2. What hearers or readers think that they are doing | Receiving, noting, accumulating | Making sense of the other person's intended meaning |
| 3. How language is thought to work in learning | Clear transmission from teacher to learner is needed; the teacher's speech is important. | The main process is active re-expression of ideas by the learner; the learner's speech is important. |
| 4. How language is thought to work in communication generally | Like Morse Code in a wire. If the message is clearly sent and received, it will be an accurate copy, unchanged. | What the hearer constructs can approximate the speaker's intention, but communication is seldom complete. The important part is how to decode the Morse. |
| 5. What language seems to do vis à vis the world of Nature | Words correspond in a simple way to features of the external world, and generally there is one correct word for one thing. | There is always more than one possible word. A choice of words highlights certain features for further thought. For example, a mule could be called a 'hybrid', a 'herbivore' or just a 'beast of burden' according to context. |
| 6. How language is thought to work in scientific discovery | We find a fact, label it, and report it to others. Words stand for things. | We choose words which influence how we see things. For example, the scientist re-describes a 'Scottish glen' as a 'glaciated valley' and thus attends to previously neglected features. |

Re-description is another word for change of metaphor and the traditional idea that figurative language is the enemy of good science is now in doubt. Many studies, summarised in Sutton (1992), show the centrality of figurative language in scientific theorising and how it provides the starting points for new thought. A phrase such as 'the orbit of the electron', which began as a figure of speech, sometimes gets treated as if it were a label for reality, while for other people it remains much less literal and no more than a working model; students need to get a sense of the interpretive origins of such terms and to understand that *a human choice of language* was involved in shaping them. Much research on metaphors and analogies in the science classroom has been about their use as teaching aids (Treagust, Duit, Joslin & Lindauer 1992), but an equally urgent task is to help the learners to appreciate the metaphors which already lie *within* major scientific theories.

Because the distinction between the figurative and the literal is not sharp, we cannot reliably separate out 'scientific' language for reporting observations uncontaminated by theory. The hope for a clear division was strong in the 17th century and has persisted since (Sutton 1994), but the demarcation is no longer tenable because even the most elementary descriptions of what we see are shaped partly by language. One continuous language works *both* as an instrument of figurative interpretation *and* as a means of attempting to transmit to others what we think. Science needs both aspects, and so does education.

Unfortunately, the dominant traditions of the classroom have given too much prominence to one aspect, and learners quickly pick up what is allowable and 'proper' for science. Lemke (1990) describes what students in some American high schools take as the rules of the game, and traces the alienating effects of 'the one right way to talk science'. He describes it as serious and dignified, verbally explicit, full of detached universalised generalisations, and always about things rather than about people. It also involves continually correcting the everyday word in order to replace it by a technical term (not 'bent' but 'refracted'; not 'bounced' but 'reflected'). Lemke suggests that this 'sets up a pervasive and false opposition between a world of objective, authoritative, impersonal, humorless scientific fact and the ordinary, personal world of human uncertainties, judgments, values and interests' (p. 120).

Such alienation should not be taken lightly, for however much a minority of students are excited by science, there are many more for whom the ideas of science become neither their own ideas nor a part of their everyday tools of thought. Those students come to regard science as just 'information' which belongs to specialists, and they react to it mainly with indifference, fear and disengagement. Can we avoid the over-use of the language forms which create this detachment of 'knowledge' from 'people'? The styles of journals and textbooks are important in science but not appropriate as the dominant experience in school, where the re-working of ideas requires much greater flexibility.

Certain cultural trends at this end of the 20th century have taken the question of language and alienation from science much further. Consider for example the feminist critique of 'man' as the 'interrogator of Nature' or the concerns of environmentalists who point to the past mistakes of an over-confident technocracy. Writers on these topics are angry at those who, in the name of being scientific, appear to claim a privileged access to 'how things really are'. They suggest that, unless the conclusions of science can be presented better as a set of well-grounded but still potentially fallible *interpretations*, science might be rejected even more forcefully (Gough 1993).

## TOWARDS A POLICY ON LANGUAGE FOR LEARNING

The problems outlined in the previous sections cannot be solved by science teachers alone, because they are entangled with general beliefs about the status of science in society and its claims for independence from the wider culture. Those beliefs are changing, however, and teachers can contribute to a better public understanding by developing a school language policy to recover the human voice and personal expression of thought. Some themes to include in such a policy are suggested below.

### *The Learner's Voice*

Science teachers should guide students about the role of language in their own learning. Learners should experience language as a medium for conversation about ideas, not just for receiving 'the truth'. Students should re-work scientific ideas and practise using those ideas in argument and discussion.

### *The Scientist's Voice*

We should present the language of scientists in such a way that students are conscious of human authors. The thoughts behind a particular choice of words should be explored, with the emphasis being on 'what these people thought and why they thought it', not just on 'what we know'.

### *The Teacher's Interpretive Voice*

The teacher should set an example in the use of language for purposes other than handing on information. When we rephrase an idea and express it in a different way, we should draw attention to that process in order to show that language is *not* a fixed set of labels. Everyday expressions and technical terms should be put together, and students should discuss how well each succeeds in capturing a particular idea. Science teachers already do this informally by explaining, negotiating and joking with students to help them to wrestle with new concepts, but at the cost, in Lemke's terms, of breaking the assumed rules of what is 'proper'. Science teaching is humane at the informal level, but much less so in the formal structure of what people believe a science lesson should consist of. A language policy should attempt to make the formal aspects of lessons more consistent with the best practice of teachers in their spontaneous informal approaches.

*Science as Story*

One way to bring out the voice of the learner, the scientist and the teacher as personal expressions of thought is to present the lesson as a critical discussion of a scientific story (e.g., 'in this circuit, most scientists now think that . . .'). To work in this way places the ideas in a form in which they are open to discussion as well as to experimental test. Students are likely to explore the story and to respond with a personal voice (e.g., 'I think that what they mean is . . .'). Learners then have an incentive to work on their own thoughts so that a dialogue can be achieved.

The word 'story' has many advantages in comparison with 'fact' or 'truth'. It involves learners and invites them to think 'Is this reasonable?' It corrects the idea that science deals only in certainties and admits the students into a more genuine discussion of areas of doubt. At another level, it also gives a better idea of science by loosening the rigid division of fact from theory.

Science often is represented as uncontroversial, but this is true only of the most stable scientific stories. If students are to connect their science with the issues of the day and learn the discipline of using evidence in structured argument, some less certain stories also should be included (e.g., the one about whether the atmosphere is warming up or not). Gough (1993) argues that such topics involve a complex interaction of values and knowledge, and also that more overtly fictional narratives might offer a good way to draw students' minds onto the problem. To handle such work without letting it degenerate into fatuous talk, teachers need a new range of class management skills in addition to the conventional repertoire of practical work, demonstrations and question-and-answer sessions. Teachers need the confidence to organise debate and role play and to coach students in how to write clearly for a variety of audiences. Resources for such activities include *Active Teaching and Learning in Science* (Centre for Science Education, Sheffield Polytechnic 1992) and *Science and Technology in Society* (Association for Science Education 1986–1988). Textbooks in the conventional form are gradually being displaced by such materials. In the British course called Salters' Advanced Chemistry, the leading book is entitled *Chemical Storylines*.

## CONCLUSION

The primary aim of this chapter has been to show that language in science evolves from personal reports in which investigators wrestle with uncertainty, and moves towards impersonal objectivised accounts of nature which lack the human voice and are alienating for some learners. From over-exposure to such accounts in science lessons, it is possible to pick up a distorted concept of science and a very limited idea of language as a system for transmitting unproblematic information. Students who are limited by that experience are likely to be less effective as learners and less skilled at connecting scientific understanding with their own human concerns. Much of this is unnecessary because the assumed 'rules' about how to talk and write about science are largely self-imposed as a result of misunderstandings of what science is. We can address the whole cluster of problems by developing a school language policy that emphasises the learner's voice, the scientist's voice, the teacher's personal voice, and science as story.

## REFERENCES

Association for Science Education: 1986–1988, *Science and Technology in Society*, Author, Hatfield, UK.
Bazerman, C.: 1988, *Shaping Written Knowledge – The Genre and Activity of the Experimental Article in Science*, University of Wisconsin Press, Madison, WI.
Centre for Science Education, Sheffield Polytechnic: 1992, *Active Teaching and Learning Approaches in Science*, Collins Educational Books, London.
Driver, R., Leach, J., Scott, P. & Millar, R.: 1994, *Students' Understanding of the Nature of Science, Working Papers 1 to 11*, Centre for Science and Mathematics Education, University of Leeds, Leeds, UK.
Gough, N.: 1993, 'Environmental Education, Narrative Complexity and Postmodern Science/Fiction', *International Journal of Science Education* 15, 607–625.
Halliday, M.A.K. & Martin, J.R.: 1993, *Writing Science: Literacy and Discursive Power*, Falmer Press, London.
Harvey, W.: 1628, *Exercitatio Anatomica de Motu Cordis et Sanguinis in Animalibus*, (English translation in Willis, R.: 1847, *The Works of William Harvey*), reprinted 1965 by Johnson Reprint Corporation, New York.
Lemke, J.L.: 1990, *Talking Science: Language, Learning and Values*, Ablex, Norwood, NY.
Medawar, P.B.: 1974, 'Is the Scientific Paper a Fraud?', in E.W. Jenkins & R.C. Whitfield (eds.), *Readings in Science Education*, McGraw Hill, London, 14–16.
Shapin, S. & Schaffer, S.: 1985, *Leviathan and the Air Pump: Hobbes, Boyle and the Experimental Life*, Princeton University Press, Princeton, NJ.
Sutton, C.R.: 1992, *Words, Science and Learning*, Open University Press, Buckingham, UK.
Sutton, C.R.: 1994, 'Nullius in Verba and Nihil in Verbis: Public Understanding of the Role of Language in Science', *British Journal for the History of Science* 27, 55–64.
Sutton, C.R.: 1996, 'Beliefs About Science and Beliefs About Language', *International Journal of Science Education* 18, 1–18.
Treagust, D.F., Duit, R., Joslin, P. & Lindauer, L.: 1992, 'Science Teachers' Use of Analogies: Observations from Classroom Practice', *International Journal of Science Education* 14, 413–422.
Watson, J.D. & Crick, F.H.C.: 1953, 'A Structure for Deoxyribose Nucleic Acid', *Nature* 171, 737.
Williams, L.P.: 1971, *The Selected Correspondence of Michael Faraday* (Volume 1), Cambridge University Press, Cambridge, UK.

# 1.3 Cultural Aspects of Learning Science

WILLIAM W. COBERN
*Western Michigan University, Kalamazoo, USA*

GLEN S. AIKENHEAD
*University of Saskatchewan, Saskatoon, Canada*

Over the past few decades, perspectives on learning science have evolved (Aikenhead 1996; Cobern 1991, 1996; Solomon 1994). Earlier psychological perspectives on the individual learner, such as Piaget, Ausubel and personal constructivism, have expanded to encompass sociological perspectives that contextualise learning in social settings, including social constructivism, science for specific social purposes, and situated cognition (Goodnow 1990; Hennessy 1993). This chapter addresses the next stage in the evolution of our thinking on learning science – an anthropological perspective that contextualises learning in a *cultural* milieu.

Although psychological and sociological approaches are useful in science education, a more encompassing perspective from cultural anthropology can provide fresh insights into familiar problems associated with students learning science. Despite sociologists' appropriation of ideas from cultural anthropology, the two disciplines of sociology and anthropology differ dramatically, even in their definitions of such fundamental concepts as society, culture and education (Traweek 1992). For example, from a sociologist's point of view, teaching chemistry tends to be seen as socialising students into a community of practitioners (chemists) who express in their social interactions certain 'vestigial values' and puzzle-solving exemplars. On the other hand, an anthropologist tends to view chemistry teaching as enculturation via a rite of passage into behaving according to cultural norms and conventions – especially the way in which the group makes sense of the world – held by a community of chemists with a shared past and future (Hawkins & Pea 1987). One consequence of the disparity between sociology and anthropology is the realisation that the anthropological perspective described in this chapter represents a significant change in our thinking about students learning science. Unfortunately, terms such as 'culture' and 'sociocultural' are found in the science education literature without a definition of what culture means in the context. Because an invocation of terms does not clarify the process of learning science, this chapter introduces appropriate terms from cultural anthropology to conceptualise cultural aspects of learning.

---

Chapter Consultant: Peter Taylor (Curtin University of Technology, Australia)

An anthropological viewpoint for science education was proposed in 1981 by Maddock when he wrote that 'science and science education are cultural enterprises which form a part of the wider cultural matrix of society and that educational considerations concerning science must be made in the light of this wider perspective' (p. 10). The anthropologist, Geertz, metaphorically characterised cultural enterprises when he suggested that people are animals suspended in a web of significance which they themselves have spun. 'I take culture to be those webs, and the analysis of it is not an experimental science in search of law but an interpretative one in search of meaning' (Geertz 1973, p. 5). This chapter represents an interpretative way of thinking about students learning science. Because learning is about making meaning within a cultural milieu, several questions must be asked. Within a cultural milieu of a particular student, what knowledge is important? What knowledge is meaningful? How does scientific knowledge relate to a student's cultural milieu?

Since Maddock (1981) articulated his anthropological perspective for science education, a body of literature on multicultural and crosscultural science education has accumulated. Pomeroy (1994) synthesised this literature into nine research agendas described later in this chapter. All these investigations dealt with students who studied in non-Western countries or in indigenous societies, or with students who comprised minority groups within Western countries (i.e., groups under-represented in the professions of science and technology). In this chapter, we broaden this anthropological perspective for science education by including Western students in industrialised countries in our cultural view of learning science (Cobern, Gibson & Underwood 1995) and recognising that these students cross cultural borders from the worlds of their peers, community and family into the worlds of science and school science (Costa 1995). We assume that typical science classroom events are crosscultural events for many Western and non-Western students.

The purpose of this chapter is to discuss cultural aspects of learning science by sketching a cultural perspective for science education, by illustrating empirically how a student's culture can affect his/her learning science, and by identifying issues for research and teaching. Accordingly, the chapter is organised into several main sections: culture and science education; subcultures and students; views of the nature and learning science; issues for research; and issues for teaching.

## CULTURE AND SCIENCE EDUCATION

In cultural anthropology, teaching science is viewed as cultural transmission (Spindler 1987) and learning science as culture acquisition (Wolcott 1991), where culture means 'an ordered system of meaning and symbols, in terms of which social interaction takes place' (Geertz 1973, p. 5). For example, we talk about a Western culture or an Oriental culture because members of these groups generally share a system of meaning and symbols for the purpose of social interaction. Geertz's definition of culture is used in this chapter.

Other definitions of culture that have guided research in science education suggest the following attributes of culture: communication (psycholinguistic and sociolinguistic), social structures (authority, participant interactions, etc.), skills (psychomotor and cognitive), customs, norms, attitudes, values, beliefs, expectations, cognition, conventional actions, material artefacts, technological know-how and worldview. In past studies, different attributes of culture have been selected to focus on a particular interest in multicultural or crosscultural science education (Baker & Taylor 1995; Barba 1993). For instance, Maddock (1981, p. 20) listed 'beliefs, attitudes, technologies, languages, leadership and authority structures', Ogawa (1986) addressed a culture's view of humans and nature and its way of thinking, and Aikenhead (1996) conceptualised culture according to the norms, values, beliefs, expectations and conventional actions of a group.

Within every culture, there exist subgroups that are commonly identified by race, language, ethnicity, gender, social class, occupation, religion, etc. A person can belong to several subgroups at the same time (e.g., a Native American female middle-class research scientist or a Euro-American male working-class technician). Large numbers and many combinations of subgroups exist due to the associations that naturally form among people in society. In the context of science education, Furnham (1992) identified several powerful subgroups that influence the learning of science, including the family, peers, the school, the mass media and the physical, social and economic environment. Each identifiable subgroup is composed of people who generally embrace a defining system of meaning and symbols, in terms of which social interaction takes place. In short, each subgroup shares a culture, which we designate as 'subculture' to convey its identity with a subgroup. For example, one can talk about the subculture of females, of our peers, of a particular science classroom and of science.

Science itself is a subculture of Western or Euro-American culture (Horton, 1994; Pickering 1992) and so 'Western science' also can be called '*subculture science*'. Scientists share a well-defined system of meaning and symbols in terms of which social interaction takes place. Because science tends to be a Western cultural icon of prestige, power and progress (Adas 1989), Western science often can permeate the culture of those who engage it with ease. Assimilation or acculturation can threaten non-Western cultures, thereby causing Western science to be seen as a dominating power (Battiste & Barman 1995; Simonelli 1994). Similarly, but much more subtly, attempts to assimilate Western students into the subculture of science can create alienation and an anti-science element in Western countries (Appleyard 1992).

Closely aligned with Western science is school science, whose main goal has been cultural transmission of the subculture of science and cultural transmission of the country's dominant culture (Stanley & Brickhouse 1994). Thus, the subculture of school science is comprised of a dynamic integration of at least two major cultural influences (Apple 1979; Fensham 1992). Transmitting a scientific subculture to students can either be supportive or disruptive. If the subculture of science generally harmonises with a student's everyday culture, science instruction supports the student's view of the world, and the result is *enculturation* (Hawkins & Pea 1987). When enculturation occurs, scientific thinking enhances a person's everyday thinking.

If the subculture of science is generally at odds with a student's everyday world, as it is with many students (Costa 1995; Ogawa 1995), science instruction tends to disrupt the student's view of the world by forcing the student to abandon or marginalise his/her indigenous way of knowing and reconstruct in its place a new (scientific) way of knowing. The result is *assimilation* (Jegede 1995) which has highly negative connotations as evidenced by such epitaphs as 'cultural imperialism' (Battiste 1986, p. 23), the 'arrogance of ethnocentricity' (Maddock 1981, p. 13) and 'racist' (Hodson 1993, p. 687). Students struggle to negotiate the cultural borders between their indigenous subcultures and the subculture of science. But, in doing so, students often come to reject important aspects of their own natal culture. For example, in a series of studies between 1972 and 1980, Maddock (1983) found that science education in Papua New Guinea had an alienating effect that separated students from their traditional culture; 'the more formal schooling a person had received, the greater [was] the alienation' (p. 32). When assimilation occurs, scientific thinking dominates a person's everyday thinking. Assimilation has caused oppression throughout the world and has disempowered whole groups of people (Battiste & Barman 1995; Gallard 1993; Urevbu 1987).

Although the cultural function of school science traditionally has been to enculturate or assimilate students into the subculture of science (Aikenhead 1996; American Association for the Advancement of Science 1989), many students persistently and creatively resist assimilation (Driver 1989), some by playing a type of school 'game' that allows them to pass their science course without learning in any meaningful way the content assumed by the teacher and community. The game can have explicit rules which Larson (1995) named as 'Fatima's Rules' after an articulate student in a high school chemistry class. Latour (1987) anticipated the phenomenon when he noted that one of the cultural expectations of school science is that 'most schooling is based on the ability to answer questions unrelated to any context outside of the school room' (p. 197). Fatima's Rules tell us how to do just that without understanding the subject matter meaningfully.

Conventional science education has produced enculturation, assimilation and Fatima's Rules as three avenues for 'learning' science. When cultural analysis of learning science is extended to include *crosscultural* learning, the new avenues of autonomous acculturation and 'anthropological' learning emerge (Aikenhead 1996). Autonomous acculturation is a process of intercultural borrowing or adaptation of attractive content or aspects of another culture and incorporates (assimilates) them into one's indigenous (everyday) culture. Examples include Haden's (1973) use of traditional Ugandan iron-smelting procedures as a basis for secondary school chemistry lessons and George's (1995) case study of a Trinidadian woman who combined aspects of Western medicine with her indigenous folk medicine. A paradigm of educational practice is found in Snively's (1990) case study of a Canadian First Nations boy in grade 6 studying the seashore. The phrase 'autonomous acculturation' attempts to avoid the negative connotations associated with acculturation and assimilation (described above).

Autonomous acculturation is not the only process that nurtures learning. Students do not need to modify features of their indigenous culture to understand the subculture of science (Solomon 1987). In other words, the conceptual modification associated with autonomous acculturation is set aside in favour of conceptual proliferation dictated by specific social or practical contexts. By analogy, cultural anthropologists do not need to accept the cultural ways of their 'subjects' in order to learn and engage in some of those ways (Traweek 1992). A different type of learning called 'anthropological' learning (Aikenhead 1996), puts students in a position not unlike an anthropologist. 'Anthropological' learning is associated with students who enjoy and are capable of constructing meaning out of the 'foreign' subculture of science, but who do not assimilate or acculturate science's cultural baggage. Somehow, these students easily negotiate the transitions between their everyday worlds and the subculture of science.

Cultural transitions are endemic to crosscultural learning. Inspired by Giroux's (1992) *Border Crossings*, Pomeroy (1994, p. 50) suggested that Western teachers and their non-Western students should become 'cultural border crossers'. Aikenhead (1996) applied this idea to Western students studying science, and described students' classroom experiences in terms of students crossing cultural borders from the subcultures of their peers and family into the subcultures of science and school science. Border crossing, therefore, becomes a crucial cultural aspect of learning science.

## SUBCULTURES AND STUDENTS

Only a few researchers have studied the phenomenon of individuals moving back and forth from their indigenous subcultures to the subculture of science. Jegede's (1995) collateral learning theory explains how students might benefit from being guided through a progression of types of collateral learning, a progression that appears to move from 'anthropological' learning to 'autonomous acculturation'. Peat (1994) provides a personal account of his transitions from a theoretical physics worldview into a Native American worldview. The capacity and motivation to participate in diverse subcultures are well known human phenomena.

Capacities and motivations to participate in other subcultures are not shared equally among all humans, as American anthropologists (Phelan, Davidson & Cao 1991) discovered when they investigated students' movement (cultural border crossings) between the worlds of students' families, peers, schools and classrooms. School success largely depends on how well a student learns to negotiate the boundaries separating these cultural worlds. Phelan and colleagues identified four patterns to the cultural border crossings between these multiple worlds: congruent worlds support *smooth* transitions; different worlds require transitions to be *managed*; diverse worlds lead to *hazardous* transitions; and highly discordant worlds cause students to resist transitions which therefore become virtually *impossible*.

Costa (1995) provides a link between Phelan et al.'s (1991) anthropological study of schools and the specific issues faced by science educators. Based on the words and actions of 43 high school science students enrolled in two Californian schools with diverse student populations, Costa concluded:

> Although there was great variety in students' descriptions of their worlds and the world of science, there were also distinctive patterns among the relationships between students' worlds of family and friends and their success in school and in science classrooms. (p. 316)

These patterns in the ease with which students move into the subculture of science were described in terms of familiar student characteristics and were clustered into five categories (summarised here in a context of border crossing). First, 'Potential Scientists' cross borders into school science so *smoothly* and naturally that the borders appear invisible. Second, 'Other Smart Kids' *manage* their border crossing so well that few express any sense of science being a foreign subculture. Third, 'I Don't Know Students' confront *hazardous* border crossings but learn to cope and survive. Fourth, 'Outsiders' tend to be alienated from school itself and so border crossing into school science is *virtually impossible*. Fifth, 'Inside Outsiders' find border crossing into the subculture of school to be *almost impossible* because of overt discrimination at the school level, even though the students possessed an intense curiosity about the natural world.

Above, we have summarised several theoretical frameworks that provide a coherent cultural perspective on learning science. The perspective is illustrated in the next section by research investigating one attribute of culture, namely, one's view of nature.

## VIEWS OF NATURE AND LEARNING SCIENCE

Based on a decade of research, Jegede (1995) described five major cultural inhibitions to learning science in Africa (authoritarianism, goal structure, traditional worldview, societal expectations and sacredness of science). A traditional worldview 'holds the notion that supernatural forces have significant roles to play in daily occurrences' (p. 114). Such a view of nature makes it difficult for African students to cross the cultural border into the mechanistic reductionistic rationalism of Western science to construct meaning of natural phenomena. Native Americans traditionally analyse nature rationally and empirically, but their rationalism and empiricism are guided by spirituality, holism, mystery and survival (Battiste & Barman 1995; Knudtson & Suzuki 1992). This disparity between Native American cultures and Western science creates *hazardous* or *impossible* border crossings for Native American students. A Japanese cultural perspective illustrates potential difficulties for Japanese students who also hold a view of nature at odds with the Western scientific view (Kawasaki 1990; Ogawa 1986). Their sense of harmony with nature contrasts with Western scientific images of power and dominion over nature. The crassness of our scientific conceptualisation of nature can be offensive to Japanese students and can inhibit border crossings.

Only recently, interpretive research has focused on American students' conceptualisations of nature. If one assumes that all ideas including scientific ones are expressed within a cultural milieu, then one must ask questions such as how do the cultural ideas of the science teacher compare with the ideas of students in the classroom. Cobern *et al.* (1995) examined this question by analysing transcripts of interviews with 14-year-old students and their science teachers. The interviews simulated naturalistic conversation on the topic of *nature* by inviting informants to respond to a series of elicitation devices. The rationale here is that it is one thing to be able to give (or not give) correct answers during a science examination, but it is quite another thing to use appropriate scientific knowledge in the absence of any kind of science prompt or cue. For instance, we expect that science holds meaning for a science teacher so strongly that science knowledge comes readily to mind (without prompting) when the teacher talks about nature. Similar expectations are reasonable for Costa's (1995) Potential Scientists. But, what about the students who are not potential scientists? The following subsections introduce a science teacher and four students from the Cobern *et al.* (1995) and Costa (1995) research to illustrate the different patterns for border crossings.

*Views of a Science Teacher*

Mr Hess, a science teacher, without prompting immediately began to discuss nature in terms of scientific processes and concepts:

> Nature is orderly and understandable . . . [and] the planets and the stars are governed by physical forces and any deviations are simply because we have not yet discovered the other part of nature's orderliness . . . As a science teacher I feel that with enough scientific knowledge all things are understandable . . . I think that the more we understand about matter itself, and the more we know about how to make things, the more predictable nature will be. Scientific or reductionistic thinking is very powerful. I feel that, once we know enough about the minutia of the world, breaking it down by using the scientific method (with scientists tearing it apart and analysing the parts of nature and seeing how they interact), we will be able to predict just about anything about nature. (Cobern *et al.* 1995, p. 18)

Mr Hess' conceptualisation of nature is essentially monothematic. His comments about nature are focused and have an explicitly scientific emphasis.

*Views of a Potential Scientist*

Howard, a student, considers himself to be very scientific:

> I think that nature is understandable. We don't understand all that there is to nature at this moment but we will understand more and more as time goes on. Most things about nature are somewhat orderly and/or have a pattern to them. Because of this, the study of science allows us to explain what is going on in nature. The orderliness also lets us predict many things that are going to happen in nature, like the weather, for example . . . I think that most things in nature can be explained by science. Matter, both living and non-living, follows basic laws. (Cobern *et al.* 1995, p. 17)

As with Mr Hess, a significant portion of Howard's thoughts about nature is scientific. Moreover, his other thoughts about curiosity about nature and resources in nature are both connected to his scientific thoughts about nature. As with Mr Hess, Howard's comments are focused, explicit and about science. The point of these examples is that they show a science teacher and student who think similarly. Though this represents only a small window into the lives of these two people, the cultural border crossing for this student in this teacher's science class is likely to be smooth.

*Views of Other Smart Kids*

A second pattern for border crossings is followed by students who manage the border crossings between the cultural worlds of family, peers or community, and school, even though the world of science is largely irrelevant to the students' personal worlds. Ann was such a student in Mr Hess' science class. She is bright and also has been a good science student. In the following excerpt, Ann clearly speaks about nature as something that one can know about through science:

> Nature is knowable . . . We can learn to understand many things about nature through personal experience, school and science. Science itself provides us with technology which in turn increases our scientific knowledge. Technology helps to provide us with many wants which, of course, increases our pleasure. It also uses resources. (Cobern *et al.* 1995, p. 24)

This appreciation of science, however, is not where her narrative begins:

> *To me, nature is beautiful and pure because it is God's creation.* Nature provides both *aesthetic and emotional pleasure* and I need it for self renewal. I like to go where you can't see any influence by man. When I'm out *in nature, I feel calm and peaceful.* It is a *spiritual feeling* and it helps me understand myself . . . This leads me to ask questions that I'd like to find answers to. The *pleasure* I get from nature is enhanced by the mysteries I see in it. (Cobern *et al.* 1995, p. 24; emphasis added)

Ann's conceptualisation of the natural world has significant aesthetic and religious elements. Moreover, when Ann was asked about Mr Hess' science class, she made it quite clear that the class was not about nature. Nature in her view is something friendly of which you can joyously be part. What impressed her about the physical science class was the teacher's warning about the dangerous chemicals that students would be handling during the course. One might be tempted to dismiss this student's aversion to dangerous chemicals as temporary and solely a result of insufficient conceptual understanding. She does not yet understand that there is danger in nature but, with proper understanding and technique, this danger need not be viewed as a threat. From a cultural perspective, however, Ann's aversion is rooted in an aesthetic sense of nature that has more scope and force than the science teacher's assurances and explanations. Thus, for example, conceptual change research might lead to instructional approaches that would help this student do better in her physical science course. But it is difficult to see how the improvement could be anything like what science education aspires to achieve (American Association for the Advancement of Science 1989), given the student's distaste for the context given to the concepts being taught.

Ann's problem is not with science but with the context which her science teacher chose to give science. In contrast to Howard and Mr Hess, Ann's thoughts about nature are not dominated by canonical scientific thinking. Her thoughts form a synoptic view of nature in which several themes (not one) or large concepts have scope and force. Ann has a sense of wonder about nature that is grounded in her fundamental view of nature as beautiful and pure. This sense of wonder leads her to ask questions about nature and thus adds to her understanding of nature, including scientific and technological understanding. During the interview, Ann volunteered accurate information from science and technology as part of her discussion of what one can know about nature. She is interested in scientific concepts, but her foundation, the cultural frame that gives meaning to that interest, is in conflict with the classroom frame provided by the teacher.

*Views of 'I Don't Know' Students*

The third border crossing pattern is followed by students whose cultural worlds show even greater disparity, thus making border crossings hazardous. A student from Costa's (1992, 1995) research provides a good example. Rattuang attends an American high school. He 'spends his free time singing and playing rhythm guitar for a heavy metal band, "Rattfinks", often writing his own music . . . [Rattuang is] a Hawaiian with long, thick black hair, neatly groomed, usually wears black leather pants and jacket, with boots and a dark t-shirt' (Costa 1992, p. 26). When Rattuang graduates, he aspires to be a professional recording artist with his band. Ruttuang made the following comments about science:

> If you want to learn science, you should be in a class. *Society doesn't really need it*. Really, it's up to you. But, it's kind of good, 'cause you learn things. At least you know some knowledge about science. You don't want to be a dummy in school. (Costa 1995, p. 322; emphasis added)

Rattuang is ambiguous about science. He says that science is good to know about but it is something society does not really need. He seemed to be saying that science is fine if you like it, but that he has other interests like music. One would not expect to find Rattuang in science classes beyond the school's minimum requirements. His world is oriented to the culture of his peers.

*Views of Students Who are Outsiders*

The fourth type of cultural border crossing identified by Phelan *et al.* (1991) and Costa (1995) is impenetrable or insurmountable because students are alienated from school and science. Art, a ninth grader from Cobern *et al.*'s study (1995), provides a good example. Art has been in and out of high school largely depending on his own whims. His teachers say that he is a nice person who is opposed to the organised structures of society. He has strong inclinations for the aesthetic and mystical aspects of life and his concept of knowing about nature has little to do with canonical scientific views. He sees some value in science-related knowledge such as the study of origins, but he also links scientific study with pollution and exploitation. Moreover, he strongly links knowledge of nature with his mystical, religious views of nature. Art's comments below can be contrasted with those of Mr Hess:

> I believe that man does not stand separate from nature but is part of it . . . Man has changed the natural world by exploiting its resources and polluting the environment . . . Nature is a source of knowledge . . . At the present time, our knowledge of the natural world is limited. Many things that we perceive to be complex and confusing because we don't understand them are actually quite simple and orderly. The construction of a spider web, for example, is quite a complicated operation to us but, to the spider, building the web is a simple procedure. As we gain in understanding of the diversity and power of nature, we understand the perfect balance of everything in nature. We will also begin to understand our place within nature. It is more important to have a spiritual understanding of nature than just scientific knowledge. That understanding can't be gained from school. You have to spend time in nature and learn to feel it. Then you will understand it. There is a spiritual aspect to nature to many people . . . Animals are very important to me, I can feel things through animals. The American Indian culture has the kind of understanding for nature that encourages preservation rather than destruction. Scientists, also, are people who understand the need to preserve and protect . . . Unfortunately scientists and scientific knowledge are also increasing our tendency to pollute, destroy and clutter up the earth and space. They are trying to destroy it and study it at the same time. (Cobern *et al.* 1995, pp. 19–20)

This paragraph is important because it shows the student's alienation from science. It is all the more interesting because the student is an American who grew up in what is considered a highly Western scientific and technological society. Yet, there is still the alienation at the root of which is a serious cultural mismatch. Art is a true modern day heretic who resolutely refuses to accept the meaningfulness of canonical science.

## CULTURAL RESEARCH IN SCIENCE EDUCATION

The preceding cases exemplify four patterns of cultural border crossings experienced by students studying school science. For most students, the crossings are easiest when the cultures of family, peer or community, school and school science are congruent. The greater the disparities, the greater the border crossing hazards; and the chances for successful student achievement in science are reduced. When struggling students do succeed, the chances of natal cultural alienation are increased (Waldrip & Taylor in press).

Cultural border crossing implies multicultural or crosscultural science education. Based on a review of this literature, Pomeroy (1994) abstracted the nine distinct research agendas of (1) science and engineering career support projects, (2) an indigenous social issues context for science content, (3) culturally sensitive instruction strategies, (4) historical non-Western role models, (5) demystifying stereotype images of science, (6) science communication for language minorities, (7) indigenous content for science to explain, (8) compare and bridge students' worldviews and the worldview of science, and (9) explore the content and epistemology of Western and non-Western knowledge of the physical world. Agendas one to seven can lead to the assimilation of students into Western science, while agendas eight and nine challenge us to conceive of alternatives to assimilation (and to Fatima's rules). Two described earlier in the chapter are 'autonomous acculturation' and 'anthropological' instruction.

Obviously more research is needed to understand the diverse experiences of students managing with, coping with, or being repelled by their border crossings into the subculture of science or school science. Pomeroy's (1994) nine research agendas provide a framework for defining many research programs. She gives high priority to item nine because she believes that crosscultural science education has the greatest potential to engage students in a way that has scope and force for them. For example, how can teachers recognise cultural aspects of their students learning science? How can we train science teachers to be cultural brokers? What happens when student cultures, a teacher's culture and the culture of science meet face to face in the classroom? What are the political ramifications of a crosscultural science curriculum characterised by Pomeroy's ninth research agenda? A plethora of new and significant research ideas emerge, which is a phenomenon associated with any new paradigm of research called *cultural aspects of learning science*.

## IMPLICATIONS FOR SCIENCE TEACHING

Science is a system of meaning and symbols with which social interaction takes place. Because science has great explanatory power for natural phenomena, it is invasive of other systems of meaning. The question to educators involves the extent to which science's system of meaning is compatible with, or attractive to, students' culturally-based systems of meanings. Though there is good evidence that cultural compatibility improves education, the honest and unfortunate response is that all too often the extent of compatibility is limited. How can we deal with students who traditionally have been the victims of scientific assimilation? Students need a contextualised approach to teaching science that draws upon the cultural worlds of students and makes sense in those worlds. We need to develop teaching methods that allow the incorporation of the content or aspects of another culture into a student's everyday culture (autonomous acculturation) and enable students to enjoy and construct meaning out of Western science without the need to assimilate science's cultural baggage (anthropological instruction).

O'Loughlin (1992, p. 791) proposed as an ideal goal that students 'master and critique scientific ways of knowing without, in the process, sacrificing their own personally and culturally constructed ways of knowing'. The capacity and motivation to master and critique scientific ways of knowing seem to depend on the ease with which students cross the cultural borders between their everyday worlds and the world of science. One implication for teaching, therefore, is that instructional methods and materials should: (1) make border crossings explicit for students, (2) facilitate these border crossings, (3) promote discourse so that *students*, not just the teacher, are talking science, (4) substantiate and build on the legitimacy of students' personally and culturally constructed ways of knowing, and (5) teach the knowledge, skills, and values of Western science *in the context* of its societal roles (social, political, economic, etc.). This implication for teaching strengthens Pomeroy's (1994) ninth research agenda involving exploration of the content and epistemology of Western and non-Western knowledge of the physical world.

In brief, cultural aspects of learning science mean that learning results from the organic interaction among the personal orientations of a student, the subcultures of a student's family, peers, community, tribe, school, media, etc., the culture of his or her country or nation, and the subcultures of science and school science. Much more research and development is needed to understand this organic interaction more clearly and to provide appropriate learning experiences for all students.

## REFERENCES

Adas, M.: 1989, *Machines as the Measure of Man: Science, Technology, and Ideologies of Western Dominance*, Cornell University Press, Ithaca, NY.
Aikenhead, G.S.: 1996, 'Science Education: Border Crossing into the Subculture of Science', *Studies in Science Education* 27, 1–52.
American Association for the Advancement of Science: 1989, *Project 2061: Science for All Americans*. Author, Washington, DC.
Apple, M.: 1979, *Ideology and Curriculum*, Routledge & Kegan Paul, London.
Appleyard, B.: 1992, *Understanding the Present – Science and the Soul of Modern Man*, Anchor Books Doubleday, New York.

Baker, D.A. & Taylor, P.C.S.: 1995, 'The Effect of Culture on the Learning of Science in Non-Western Countries: The Results of an Integrated Research Review', *International Journal of Science Education* 17, 695–704.

Barba, R.H.: 1993, 'A Study of Culturally Syntonic Variables in the Bilingual/Bicultural Science Classroom', *Journal of Research in Science Teaching* 30, 1053–1071.

Battiste, M.: 1986, 'Micmac Literacy and Cognitive Assimilation', in J. Barman, Y. Herbert & D. McCaskell (eds.), *Indian Education in Canada, Vol. 1: The Legacy*, University of British Columbia, Vancouver, BC, 23–44.

Battiste, M. & Barman, J. (eds.): 1995, *First Nation Education in Canada: The Circle Unfolds*, University of British Columbia, Vancouver, BC.

Cobern, W.W.: 1991, *World View Theory and Science Education Research* (NARST Monograph No. 3), National Association for Research in Science Teaching, Manhattan, KS.

Cobern, W.W.: 1996, 'Worldview Theory and Conceptual Change in Science Education', *Science Education* 80, 579–610.

Cobern, W.W., Gibson, A.T. & Underwood, S.A.: 1995, 'Everyday Thoughts About Nature: An Interpretive Study of 16 Ninth Graders' Conceptualizations of Nature', Paper presented at the annual meeting of the National Association for Research in Science Teaching, San Francisco, CA.

Costa, V.B.: 1992, 'Student Responses to School Science', Paper presented at the annual meeting of the American Educational Research Association, San Francisco, CA.

Costa, V.B.: 1995, 'When Science is "Another World": Relationships Between Worlds of Family, Friends, School, and Science', *Science Education* 79, 313–333.

Driver, R.: 1989, 'Students' Conceptions and the Learning of Science', *International Journal of Science Education* 11, 481–490.

Fensham, P.J.: 1992, 'Science and Technology', in P.W. Jackson (ed.), *Handbook of Research on Curriculum*, Macmillan, New York, 789–829.

Furnham, A.: 1992, 'Lay Understanding of Science', *Studies in Science Education* 20, 29–64.

Gallard, A.J.: 1993: 'Learning Science in Multicultural Environments', in K.G. Tobin (ed.), *The Practice of Constructivism in Science Education*, Lawrence Erlbaum, Hillsdale, NJ, 171–180.

Geertz, C.: 1973, *The Interpretation of Culture*, Basic Books, New York.

George, J.M.: 1995, 'Health Education Challenges in a Rural Context', *Studies in Science Education* 25, 239–262.

Giroux, H.: 1992, *Border Crossings: Cultural Workers and the Politics of Education*, Routledge, New York.

Goodnow, J.J.: 1990, 'Using Sociology to Extend Psychological Accounts of Cognitive Development', *Human Development* 33, 81–107.

Haden, J.: 1973, 'Iron and Education in Uganda', *Education in Chemistry* 10(2), 49–51.

Hawkins, J. & Pea, R.D.: 1987, 'Tools for Bridging the Cultures of Everyday and Scientific Thinking', *Journal of Research in Science Teaching* 24, 291–307.

Hennessy, S.: 1993, 'Situated Cognition and Cognitive Apprenticeship: Implications for Classroom Learning', *Studies in Science Education* 22, 1–41.

Hodson, D.: 1993, 'In Search of a Rationale for Multicultural Science Education', *Science Education* 77, 685–711.

Horton, R.: 1994, *Patterns of Thought in Africa and the West*, Cambridge University Press, Cambridge, UK.

Jegede, O.J.: 1995, 'Collateral Learning and the Eco-Cultural Paradigm in Science and Mathematics Education in Africa', *Studies in Science Education* 25, 97–137.

Kawasaki, K.: 1990, 'A hidden Conflict Between Western and Traditional Concepts of Nature in Science Education in Japan', *Bulletin of the School of Education Okayama University* 83, 203–214.

Knudtson, P. & Suzuki, D.: 1992, *Wisdom of the Elders*, Stoddart, Toronto, Canada.

Larson, J.O.: 1995, 'Fatima's Rules and Other Elements of an Unintended Chemistry Curriculum', Paper presented at the annual meeting of the American Educational Research Association, San Francisco, CA.

Latour, B.: 1987, *Science in Action*, Harvard University Press, Cambridge, MA.

Maddock, M.N.: 1981, 'Science Education: An Anthropological View Point', *Studies in Science Education* 8, 1–26.

Maddock, M.N.: 1983, 'Research Into Attitudes and the Science Curriculum in Papua New Guinea', *Journal of Science and Mathematics Education in Southeast Asia* 6(1), 23–35.

Ogawa, M.: 1986, 'Toward a New Rationale of Science Education in a Nonwestern Society', *European Journal of Science Education* 8, 113–119.

Ogawa, M.: 1995, 'Science Education in a Multiscience Perspective', *Science Education* 79, 583–593.
O'Loughlin, M.: 1992, 'Rethinking Science Education: Beyond Piagetian Constructivism Toward a Sociocultural Model of Teaching and Learning', *Journal of Research in Science Teaching* 29, 791–820.
Peat, D.: 1994, *Lighting the Seventh Fire*, Carol Publishing Group, New York.
Phelan, P., Davidson, A.L. & Cao, H.T.: 1991, 'Students' Multiple Worlds: Negotiating the Boundaries of Family, Peer, and School Cultures', *Anthropology & Education Quarterly* 22, 224–249.
Pickering, A. (ed.): 1992, *Science as Practice and Culture*, University of Chicago Press, Chicago, IL.
Pomeroy, D.: 1994, 'Science Education and Cultural Diversity: Mapping the Field', *Studies in Science Education* 24, 49–73.
Simonelli, R.: 1994, 'Sustainable Science: A Look at Science Through Historic Eyes and Through the Eyes of Indigenous Peoples', *Bulletin of Science, Technology & Society* 14(1), 1–12.
Snively, G.: 1990, 'Traditional Native Indian Beliefs, Cultural Values, and Science Education', *Canadian Journal of Native Education* 17(1), 44–59.
Solomon, J.: 1987, 'Social Influences on the Construction of Pupil's Understanding of Science', *Studies in Science Education* 14, 63–82.
Solomon, J.: 1994, 'The Rise and Fall of Constructivism', *Studies in Science Education* 23, 1–19.
Spindler, G.: 1987, *Education and Cultural Process: Anthropological Approaches (2nd Ed.)*, Waveland Press, Prospect Height, IL.
Stanley, W.B. & Brickhouse, N.W.: 1994, 'Multiculturalism, Universalism, and Science Education', *Science Education* 78, 387–398.
Traweek, S.: 1992, 'Border Crossings: Narrative Strategies in Science Studies and Among Physicists in Tsukuba Science City, Japan', in A. Pickering (ed.), *Science as Practice and Culture*, University of Chicago Press, Chicago, IL, 429–465.
Urevbu, A.O.: 1987, 'School Science in West Africa: An Assessment of the Pedagogic Impact of Third World Investment', *International Journal of Science Education* 9, 3–12.
Waldrip, B.G. & Taylor, P.C.: in press, 'The Permeability of Students' World Views to Their School Views in a Non-Western Developing Country', *Journal of Research in Science Teaching*.
Wolcott, H.F.: 1991, 'Propriospect and the Acquisition of Culture', *Anthropology & Education Quarterly* 22, 251–273.

# 1.4 Learning Science Through Models and Modelling

JOHN K. GILBERT and CAROLYN J. BOULTER
*The University of Reading, UK*

A model can be defined as a representation of an idea, an object, an event, a process or a system. The role of models and modelling in the learning of science is worthy of a distinctive focus for a number of reasons. Firstly, the terms are used ubiquitously in science education to describe representations ranging from an individual's transient ideas to grand objects found in museums. The range of things represented, the scope and degree of the changes brought about in the formation of representations, and the wide-ranging existential nature of the resulting models ensure that the meanings of the processes involved are clouded with uncertainty. Secondly, by being more perceptually accessible than theories, models play a key role in the conduct of scientific inquiry. They more readily permit the consequences of theories to be deduced and tested by experiment. The widespread desire to provide science education which is more obviously related to the conduct of science suggests that the nature of models and modelling must be directly addressed. Thirdly, the cognitive psychology representation of learning, including that in science education, is predicated on the formation and development of models by an individual within the context of a social group (Harré & Gillett 1994). An understanding of learning in science education thus necessarily involves an understanding of the nature of models and modelling. Fourthly, models play a substantial role in everyday classroom teaching and this role should be further explored.

The main aims of this chapter are to discuss the word 'model' when used in science education and to show how models can and do contribute to learning in classrooms and other contexts. The first section of this chapter examines the relationships between the words 'model', 'theory' and 'concept'. The second section describes the nature of learning in 'situations', 'narratives' and 'models'. Having provided a framework for the remainder of the chapter, the third and fourth sections, respectively, examine the use of models and narratives in the classroom, and in other contexts involving computers, television and museums. The penultimate section identifies three forms of argumentation and illustrates how each governs the roles played by different types of models in science education. The chapter concludes with a summary and suggestions for further research.

---

Chapter Consultant: Shawn Glynn (University of Georgia, USA)

## MODEL, THEORY AND CONCEPT

A *model* of a *target* (that which is to be represented) is produced from a *source* (some other object, event or idea) by the use of metaphor in which the target is seen, if only initially for the sake of argument and for a short time, as being very similar to the source. Taking the interactive view of metaphor (Black 1962), the elements of which the source is composed are projected onto the target. Those which seem to have an evident value in representing the target are altered to fit the special circumstances of the target by the drawing of analogy (Hesse 1966; Thagard 1992).

### Model and Theory

Nagel (1987) has attempted to clarify the distinction between model and theory by stating that there are:

> ... three components in a theory: (1) an abstract calculus that is the logical skeleton of the explanatory system, and that 'implicitly defines' the basic notions of the system; (2) a set of rules that in effect assign an empirical content to the abstract calculus by relating it to the concrete materials of observation and experiment; and (3) an interpretation or model for the abstract calculus, which supplies some flesh for the skeletal structure in terms of more or less familiar ... visualizable materials. (p. 90)

A model thus can be viewed as an intermediary between the abstractions of theory and the concrete actions of experiment, therefore helping to make predictions, guide inquiry, summarise data, justify outcomes and facilitate communication. Nagel (1987, p. 110) suggests that, on some occasions, one theory and an attendant model, together, can serve *as a model* for the development of another theory plus model. For example, Newton's laws and the attendant billiard ball model were the sources from which the kinetic theory and model of gases was derived. Here a model is established at the same time as the theory. On other occasions, just the theory is established by analogy to an existing theory. Nagel (1987, p. 111) quotes the case of Maxwell's recognition of the similarity between the structure of the mathematics of gravitation theory and those of the theory of heat conduction; the accompanying model evolved later. Harré (1978) has discussed the relationship between a theory, the process of production of a model, and the model itself, in respect of a modern version of the Darwin-Wallace Theory of Natural Selection. These relationships deserve greater elaboration.

There have been some recent inquiries into the relationship between models and theories as perceived by students. From an interview-based study of understanding of the word 'theory' among 30 pairs each of 9-, 12- and 16-year-old students in the UK (Scott, Driver, Leach & Millar 1993), it can be inferred that models: are used from an early age; act as a way of understanding theories more clearly; form a bridge between theory and behaviour by enabling predictions to be generated and empirically tested; and can be transferred between contexts. The evidence is that students construct a narrative about the nature of science which is different from that of their teachers; inevitably, the notion of what a model is forms part of this divergence.

*Model and Concept*

Like 'model', the notion of 'concept' is widely used in science education, again without a generally-agreed operational definition of the word. Carroll (1962) suggests that, for any individual, a particular concept is: an abstract generalisation which emerges from experiences with more than one example of an event or object; evaluated in terms of its relative adequacy for an individual's purpose; constantly being revised by that individual; governed in nature by the specific experiences which led to its formulation; and one of many that can exist for a particular object or event in a population of people.

The formation of a concept is difficult to justify by use of the language of representation. The target of the concept, or what is to be represented in the formation of the concept, involves the elements perceived to be common to a range of objects or events. Yet how can these regularities be recognised without the use of the concept itself? How can experience exist without being recognised as such through the use of the concept? Two possibilities exist for the source of the concept (i.e., that from which the representation is derived). It can be another concept, which is seen to act as a metaphor through which experience is recognised and acted upon. Or the source can lie within the many possible elements of a crudely-perceived experience, which are compared with each other until positive elements of analogy of convincing strength are identified. Even within a seemingly common and homogeneous social environment (e.g., a well-disciplined and purposeful school classroom), the concept formed by a given student on a specific taught topic will deviate both from the socially-sanctioned concept that the teacher is trying to teach and from those formed by other students, thus giving rise to so-called 'alternative conceptions' (Gilbert & Watts 1983). Concepts seem to involve the formation of propositions, and models make use of images.

## LEARNING IN SITUATIONS, NARRATIVES AND MODELS

A broad view of constructivism has been influential in perceptions of teaching, learning and research in science education over the past decade or so. In the 'situated cognition' version of constructivism:

> ... learning is a process of enculturation or individual participation in socially organised practices, through which specialized local knowledge, rituals, practice and vocabulary are developed. The foundation of actions in local interactions with the environment is no longer an extraneous

problem but the essential resource which makes knowledge possible and actions meaningful. (Hennessy 1993, p. 2)

Within this approach, learning takes place in particular situations. But what constitutes a 'situation'? Rodrigues and Bell (1995) point out that, in the science education literature, the word 'context' can mean a variety of things, such as 'the classroom', the 'learning environment' or the 'relevance of an activity'. However, what is important in their view is the personal meaning attached to new or established knowledge when used to understand the external, physical situation. A situation, then, is a specific external environment which is turned into a context at a particular time by the mental activity of an individual, often acting within a group.

When a situation is used deliberately for an educational activity, the four factors of where it takes place, the focus of the activity, the educational purposes being addressed, and the people involved remain the same throughout the activity. It allows for an *event* to take place. The language which is used within the confines of an event is the *text* of the event. An analysis for participation in the text of an event can show who said what to whom, how that person responded, and what followed next. A text carries a series of *narratives*, or stories composed by one or more people for others to pay attention to. A text consists of a series of interwoven narratives, such as those produced by the teacher as the demands of the curriculum are being presented to the students, or those being produced by individual students as they ascribe their own meanings to what is going on. The text of an educational event, usually spoken but often with a substantial written component and sometimes an element of physical action, can be analysed to reveal some of the narratives being produced. This set of relationships is summarised in Figure 1.

Models play a major part in the narratives of science education. It is possible to differentiate between *target systems* (which exist in common experience and are to be represented), *mental models* (personal, private, representations of a target), *expressed models* (which are expressed by an individual through action, speech or writing), *consensus models* (expressed models which have been subjected to testing by a social group, such as ones drawn from the science community, and which have been agreed by some as having some merit), and *teaching models* (specially-constructed models used to aid the understanding of a consensus model).

The great value of many models is that they enable ideas, objects, events, processes or systems – which are complex, on a different scale to that which is normally perceived, abstract, or some combination of these – to be rendered either visible or more-readily visualised. This is done by simplifying the target (selecting only some entities and relations between them for representation), perhaps altering the scale and using a medium which is widely accessible. As a given model is initially produced for a specific purpose, the selection (and hence the suppression) of features in the target for representation is highly specific.

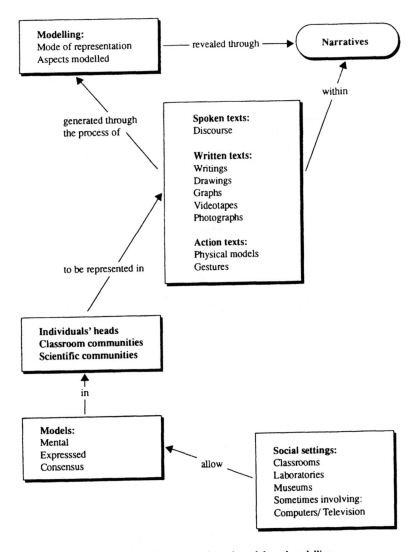

Figure 1: Learning science through models and modelling

In science education, students often encounter one or more distinctive situations, such as the conventional classroom, computers, educational television (usually in the form of videotape), museums (as well as zoos and galleries) or the external natural environment. Each student will construct a narrative for each event in a given situation. If the events which a student encounters are intended by the teacher to have a similar or related subject content, then the narratives constructed in the various situations will interact with each other in the production of an extended narrative of the phenomenon being studied. Models, of whatever type, which are produced, expressed or encountered in one situation, serve as the basis for the construction of narratives in the next and subsequent situations experienced. For example, students studying 'the Apollo landings on the Moon' in the classroom construct a mental model of what took place and of the scientific and technological ideas involved. This serves as a way of interpreting what is shown on a videotape of the landings and is modified by that experience. This modified mental model then could be used in a science museum and be further developed in the light of exhibitions of the objects involved, or replicas of those objects or consensus models of them that are on display there.

A fuller understanding of how models contribute to the integration of learning across situations can arise only from a clearer perception of what happens in each situation.

## MODELS AND NARRATIVES IN THE CLASSROOM

In this section, we examine how the narratives of the conventional classroom relate to models and modelling in three ways, namely, by learning consensus models, learning about models, and learning how to model. Each of these ways of relating narratives to models is described below.

### Learning Consensus Models

Teachers build narratives around the most important of the consensus models produced by scientists in the past not only because of their value *per se*, but also because they can provide an efficient way of learning the multitudinous specific 'facts' of science. The narratives provided across a range of events used to explain a given phenomenon to students often make use of the available consensus models of it in the sequence of their historical invention by scientists. Thus, in chemistry, the structural relationship between maleic acid and fumaric acid is taught by a recapitulation of consensus models in the order in which they were invented (van Hoeve-Brouwer & de Vos 1994). This 'narrative recapitulatory' approach is often crude. Frequently, neither a narrative of the development of a given model nor a narrative of its abandonment in scientific research is discussed. Given their divergent experiential bases, students construct their own, individual and perhaps divergent narratives of the phenomenon being taught. The issue is the 'gap' between the narratives of teacher and taught. Unintended or incomplete narratives of a phenomenon can be constructed by students if they are presented with a consensus model which is inappropriate for the purposes being addressed by the teacher.

Specially-developed teaching models often are used by teachers and textbooks to ease an intellectual path for students towards an understanding of consensus models, or towards the construction of an 'appropriate narrative'. This path consists of: the selection of a source, known to the students, from which to construct a teaching model; and the use of that teaching model as a source from which to develop an acceptable mental model of the consensus model. Although the use of such teaching models varies widely between topics and teachers, it is possible to identify good practice in their construction (Treagust, Duit, Joslin & Lindauer 1992) and use (Dagher 1995). Thus, a good teaching model:

- contains major features which approximately match in number those in the consensus model, with a high degree of similarity existing between the two so that equivalences of meaning can be perceived readily;
- serves as an introduction to a consensus model which students both believe to be important and find difficult to understand;
- is based on a source with which the students are fully acquainted, preferably at an hands-on, experiential level;
- can be used in combination with other such models in respect of a given consensus model.

For students to be able to have a greater chance of building a narrative involving a consensus or teaching model which is of a kind which the teacher intends, they must be able to answer all the following questions:

- From what source was the teaching/consensus model drawn?
- What are the major elements in the source and what are the relationships between them?
- What aspects (structure, behaviour, mechanism) of the source have been used in forming the model?
- How strong is the similarity between the source and the target in respect of these aspects?
- What aspects of the source have not been used in forming the model?
- What aspects of the target are not being represented in the model?
- What are the 'codes of interpretation' of the particular representation being used?

In short, the student must be able to form a view of the scope and limitations of the model as a representation *per se* (Dagher 1994; Duit 1991; Glynn 1991). If the model is presented in text form (Thiele & Treagust 1995), students benefit from specific support in the formation of a mental model. There is evidence that students can generate what are, in effect, their own teaching models (Cosgrove 1995; Wong 1993). They could need to do this unless the teacher recognises that there is a problem of understanding (Abell & Roth 1995) and produces a teaching model. The theme of teaching models, otherwise known as teaching analogies, has formed the focus of a special issue of the *Journal of Research in Science Teaching* (Lawson 1993).

*Learning about Models*

The narrative which students are required to construct involves learning about the codes of interpretation of consensus models, the scope and limitations of models of all types, the role of expressed models in scientific methodology, and how the various models of a given phenomenon were introduced, how they evolved and why they were displaced by others in the practice of science. The evidence available suggests that an understanding of the nature of 'a model' is slow to develop. When Grosslight, Unger, Jay and Smith (1991) interviewed students of 12–13 and 16–17 years of age, as well as 'experts' (university teachers) in the USA, they identified three broad levels of thinking about the formation, nature and use of models. At Level 1, models were thought of as either toys or copies of reality which could be incomplete because the producer of them wished it to be so. At Level 2, models were thought of as consciously produced for a specific purpose, with some aspects of reality being omitted, suppressed or enhanced. The emphasis is still on reality and its modelling rather than on the ideas portrayed. At Level 3, a model is seen as being constructed to service and develop ideas, rather than being a copy of reality. The modeller plays an active role in the modelling process, manipulating and testing it in order to develop ideas. Although there was a drift from Level 1 to Level 2 with increasing age in the sample, no school student demonstrated a Level 3 appreciation of models, which is the exclusive province of the 'experts' (Grosslight et al. 1991). Given this divergence of narrative concerning the notion of model itself, it is hardly surprising that students construct narratives involving specific models which diverge greatly both from each other and from that of the teacher. Similar results were found amongst university students in South Africa (Smit & Finegold 1995). Students' understanding of the meaning the word 'model' can be advanced by specific teaching, in one form or another, about the issue. Where this has been attempted, there is evidence of success (Raghaven & Glaser 1995).

*Learning How to Model*

Within the cognitive view of learning, the individual scientist (and hence the student of science) needs to form mental models of a phenomenon and to share these with others in the form of expressed models. An application of general constructivist approaches to science education (Gilbert 1993) suggests that the capacity of students to form and express mental models should be enhanced if they both are provided with explicit opportunities to become aware of their mental models and are given explicit opportunities to express those models. Where such provision is made, there is evidence that it is successful (Paton 1996). These circumstances offer support for the construction of narratives which link past experience to present learning.

The broad nature of practical work undertaken in the classroom relates most closely to the narratives of 'learning how to model' that are constructed there. Simon and Jones (1992) have produced a typology of practical work which is built around three dimensions involving the extent to which (1) the problem is defined by the teacher or the students, (2) the methods are pre-defined or selected and (3) one or several solutions are possible. An approach which emphasised the confirmation of consensus models would sit at the first stated aspect of these dimensions.

Taken together, the second stated aspect of the three dimensions would support the production and articulation of mental models which involve phenomena which are of intrinsic interest to students, and which make predictions about the models' consequences, the design and conduct of inquiries to test these, and the review of the models in the light of these inquiries. This 'open work', as Simon and Jones (1992) call it, comes much closer to the way in which science seems to be actually conducted (i.e., a realistic 'narrative of science'). It could be possible to achieve some 'open work' within the strait-jacket of the school timetable and the limited resources of even the best-equipped school laboratory (Gott & Duggan 1995). There certainly seems to be considerable scope for the introduction of such holistic inquiries within structured extra-curricular activities (Tytler & Swatton 1992).

## MODELS AND NARRATIVES IN COMPUTER, TELEVISION AND MUSEUM CONTEXTS

*Computers*

Computers can be integrated into the classroom, but are more likely to be used by science students, both individually and in small groups, both in school and in other locations (e.g., at home). Books which combine theoretical insights into, and practical applications of, their use already have appeared (e.g., Mellor, Bliss, Boohan, Ogborn & Tompsett 1994; Scaife & Wellington 1993). With respect to models and modelling, computers can be used in science education in two distinctive modes: exploratory modes (known as simulation modes); and expressive modes (known as modelling modes).

In the exploratory mode, students are presented with a definite model already completely programmed into the software. Their task is to 'run' the simulation by inputting numbers or other information so that the program processes them in accordance with the model to produce an output. Examples include: direct copies of existing laboratory activities (e.g., titrations); simulations of industrial processes (e.g., the production of sulphuric acid); simulations of processes which are too dangerous, too slow or fast, or too large or small to be directly experienced in the school environment; simulations of non-existent entities (e.g., friction-free surfaces); and simulations of the behaviour of specific models (e.g., the wave model of light).

The exploratory mode is at the conservative end of the continuum of activity types involving models and modelling in science education. The purpose is limited to showing how a given consensus model behaves. Students usually do not know what the program does with the input figures, and often they do not know what the model actually is. In these circumstances, students cannot appreciate the scope and limitations of the model, are not led to a clearer appreciation that any model is an idealised representation, and confuse the model with reality. In short, there is a much lower likelihood that students' narratives about and using models are linked to those of their teachers. Moreover, the narratives that they do construct can be dysfunctional in terms of future experience.

The expressive or modelling use of the computer contains content-free software. The student therefore is able to specify what the variables are and, within the constraints of the software, to investigate how these variables can be related. Some applications model the process of change in a system (e.g., how predator-prey number relations change under different environmental circumstances, or how nuclear decay sequences operate in terms of the numbers of atoms involved). Other applications focus on the operations contained in the model (i.e., on what needs to be done to achieve a specific outcome) (Scaife & Wellington 1993, p. 54). This use of the computer places much more emphasis on the production of mental models by students, on their articulation and use to make predictions, and on how they evolve towards consensus models. By requiring the 'envisioning' of systems and their behaviour, it makes much more use of, and encourages the formation of, mental models. The narratives constructed will be both more internally integrated and more closely linked to that of students' past experience and to that of the teacher.

*Television*

Few studies have investigated the contribution made by television to the construction of narratives in science education. Priest (1993) interviewed 47 students aged 14–15 years, individually and in groups, as well as their teachers, in ten schools in the UK. The classes were observed watching one of three widely-used science videotapes, within a regular science class, and students' responses to it obtained immediately afterwards. The videotapes themselves were not designed with an overt commitment to models and modelling, let alone to the construction of narratives *per se,* and Priest's (1993) inquiry did not explicitly address these themes. However, it is possible to make a number of deductions, both by viewing the videotapes and by reinterpreting the data. The videotapes made extensive use of a wide variety of models including a picture of a piece of scientific equipment or industrial plant, a block diagram of its component pieces, presented both statically and dynamically, and a cut-away material model showing the mode of operation. They did not usually include a symbolic (mathematical) model of the processes and outcomes of the use of the equipment or plant, an omission which was regretted by the students. The construction of mental models was supported by the provision of extended sequences of carefully-selected and carefully-designed transactions. Learning from models was supported by the use of a variety of consensus and teaching models, of a range of types, related to a given phenomenon. Moreover, due emphasis was placed on an appreciation of the scope and limitations of a given model by demonstrating its use in a wide variety of personal, social and school curriculum contexts. Learning about models was facilitated to an extent by the provision of a wide range of surrogate experiences. Students were not required, by the commentary, either to make predictions using mental models or to observe phenomena and construct explanations by use of consensus models. However, many students did so on their own initiative after the lesson. Learning how to model was supported by a requirement to form, to become explicitly aware of, and to challenge mental models. It was also encouraged by the general emphasis on explanation, by the progressive building of models and by the requirement to transfer models across a wide range of contexts.

*Museums*

Museums contain consensus and teaching models (with a full range of interactivity), objects (both original and replica) and text. From the situated cognition point of view, the museum exhibit which best supports learning could be one designed around specific topics, particularly those of known significance to visitors. The evidence available (Priest & Gilbert 1994) suggests that students visiting museums learn by interacting with each other, with the teacher and adult helpers, with the display material, and by relating experience to their existing knowledge. Visually stimulating displays, whilst attracting attention, do not hold that attention and motivate learning unless the students could both humanise them in some way, for example, by touching them, and obtain answers to their personal questions by building up a personal or group narrative. Objects and static models can meet these criteria, but they do not always do so. Interactive exhibits, whilst proving interesting, require considerable cognitive input by the user if they are to contribute to learning at the time of a visit. A balance has to be struck between interaction with the exhibits and interaction with other students. Text is only read when absolutely necessary, and then only skimmed. A bold display using colour seems the most successful format in attracting attention and in provoking engagement, but not in sustaining it.

## MODELS, NARRATIVES AND ARGUMENTATION

Producing narratives of a phenomenon in science and in science education, particularly where models are being constructed and tested, involves the use of argument, both intra- and inter-personally, by those involved (Sutton 1992). Boulter and Gilbert (1995) identified three forms of verbal argumentation which, in varying proportions, can be in use in an educational context at any one time.

In 'didactic' argument, the purpose of the teacher is to transmit established knowledge to the students, who are largely passive receptors. Where this form is used, and that is probably for the majority of science class time in many secondary schools, the emphasis with respect to models and modelling will be placed on the teaching of consensus models, either directly or via teaching models. The focus will be on the teacher's narrative, and students will be given little encouragement to be aware of their own personal narratives.

In 'Socratic' argument, the teacher asks specific questions of the students in anticipation of being able to select and use particular answers to establish predetermined truths; this forms the basis of the so-called 'discovery' methods of learning. Where this form is used, students are certainly encouraged to form mental models, but the substance and use of these will largely be governed by the teacher's inputs and evaluations. Students will be 'guided', often at high speed, towards consensus models. Again, the construction of students' narratives will be marginalised.

In 'dialogic' argument, on the other hand, the control of the discussion is shared by the teacher with the students. The former takes action largely to ensure that rules of participation are adhered to and that the focus of attention remains on the issue under discussion. Where this form is used, largely by teachers with a strong commitment to a constructivist view of knowledge and of learning, the formation of mental models, and their articulation with a view to testing them, takes pride of place. In these circumstances, it is more likely that students will express their own models and build their own narratives about what is under consideration in the class.

The balance of use of these three forms of argumentation in a given classroom governs the roles played by mental, expressed, consensus and teaching models in the science education of students. In turn, this defines the appreciation developed by them of the nature, scope and limitations of the various types of model.

## DISCUSSION AND SUGGESTIONS FOR RESEARCH AND PRACTICE

Inevitably, given that it is an eclectic field which draws on the history, philosophy and sociology of science, as well as on cognitive psychology, much more needs to be known about models and modelling in science education. Some inquiries will be theoretical (e.g., further work on the relationship between the meanings of the words 'model', 'theory' and 'concept'). Other inquiries will require investigations in classrooms into the practical consequences of the ideas outlined in this chapter. Some of the classroom inquiries suggested by the discussion in this chapter include the following:

- How are expressed models developed and used in 'open work?'
- What relationships are possible between the patterns of argumentation used in classroom discourse, the narrative structures to which they lead, and the nature of the expressed models that students put forward?
- What implications does the construction of 'extended narratives' across contexts have for the ways in which mental models are developed and changed?
- What implications for students' understanding of the nature of science has the use of an 'historical recapitularity' approach to the teaching of consensus models?
- What impact does the use of particular teaching models, in an attempt to support students' understanding of a given consensus model, have on their

development of an understanding of more complex, often historically later, models of the same phenomenon?
- What can be done to help students acquire an acceptable mental model of 'model'?

Improved educational provision with respect to models and modelling is likely to result if the research questions outlined above are addressed in collaboration with practitioners. An understanding of the various meanings attached to the word 'model' (mental, expressed, consensus, teaching) is likely to enable the contribution of models and modelling to both teaching and learning to be more clearly appreciated. In turn, this will enable model-related work in a wide variety of situations (classroom, museum, computers, television, the natural environment) to be more closely integrated. By use of a wider range of representational forms (gesture, speech, drawing, writing and other symbolic forms such as mathematics), a more diverse range of narrative forms will be encouraged. Measures such as these would improve cognitive aspects of science education.

## REFERENCES

Abell, S. & Roth, M.: 1995, 'Reflections on a Fifth-Grade Science Lesson: Making Sense of Children's Understanding of Scientific Models', *International Journal of Science Education* 17, 59–74.
Black, M.: 1962, *Models and Metaphors*, Cornell University Press, Ithaca, NY.
Boulter, C. & Gilbert, J.: 1995, 'Argument in Science Education', in P. Costello & S. Mitchell (eds.), *Competing and Consensual Voices*, Multilingual Matters, London, 84–94.
Carroll, J.: 1962, 'Words, Meanings, and Concepts', *Harvard Educational Review* 32, 178–202.
Cosgrove, M.: 1995, 'A Study of Science-in-the-Making as Students Generate an Analogy for Electricity', *International Journal of Science Education* 17, 295–310.
Dagher, Z.: 1994, 'Does the Use of Analogies Contribute to Conceptual Change?', *Science Education* 78, 577–600.
Dagher, Z.: 1995, 'Analysis of Analogies Used by Science Teachers', *Journal of Research in Science Teaching* 32, 259–270.
Duit, R.: 1991, 'On the Role of Analogies and Metaphors in Learning Science', *Science Education* 75, 649–672.
Gilbert, J.K. (ed.): 1993, *Models and Modelling in Science Education*, Association for Science Education, Hatfield, UK.
Gilbert, J.K. & Watts, D.M.: 1983, 'Concepts, Misconceptions and Alternative Conceptions: Changing Perspectives in Science Education', *Studies in Science Education* 10, 61–98.
Glynn, S.: 1991, 'Explaining Science Concepts: A Teaching-with-Analogy Model', in S. Glynn, R. Yeany & B. Britton (eds.), *The Psychology of Learning Science*, Lawrence Erlbaum, Hillsdale, NJ, 219–240.
Gott, R. & Duggan, S.: 1995, *Investigative Work in the Science Curriculum*, Open University Press, Buckingham, UK.
Grosslight, L., Unger, C., Jay, E. & Smith, C.: 1991, 'Understanding Models and Their Use in Science: Conceptions of Middle and High School Students and Experts', *Journal of Research in Science Teaching* 29, 799–822.
Harré, R.: 1978, *Principles of Scientific Thinking*, Macmillan, London.
Harré, R. & Gillett, G.: 1994, *The Discursive Mind*, Sage, London.
Hennessy, S.: 1993, 'Situated Cognition and Cognitive Apprenticeship', *Studies in Science Education* 22, 1–41.
Hesse, M. (ed.): 1966, *Models and Analogies in Science*, Sheen and Ward, London.
Lawson, A. (guest ed.): 1993, 'Special Issue: The Role of Analogy in Science and Science Teaching', *Journal of Research in Science Teaching* 30, 1211–1348.
Mellor, H., Bliss, J., Boohan, R., Ogborn, J. & Tompsett, C. (eds.): 1994, *Learning with Artificial Worlds: Computer Based Modelling in the Curriculum*, Falmer Press, London.

Nagel, E.: 1987, *The Structure of Science*, Hackett, Indianapolis, IN.
Paton, R.: 1996, 'On an Apparently Simple Modelling Problem in Biology', *International Journal of Science Education* 18, 55–64.
Priest, M.: 1993, *Learning Science from Television in the Classroom: A Case Study of the Interaction Between the Science Message, the Message of the Medium, the Pupil, and the Teacher*, Unpublished PhD thesis, University of Reading, Reading, UK.
Priest, M. & Gilbert, J.: 1994, 'Learning in Museums: Situated Cognition in Practice', *Journal of Education in Museums* 15, 16–18.
Raghaven, K. & Glaser, R.: 1995, 'Model-Based Analysis and Reasoning in Science: The MARS Curriculum', *Science Education* 79, 37–62.
Rodrigues, S. & Bell, B.: 1995, 'Chemically-Speaking: A Description of Student-Teacher Talk During Chemistry Lessons and Building on Students' Experiences', *International Journal of Science Education* 17, 797–809.
Scaife, J. & Wellington, J.: 1993, *Information Technology in Science and Technology Education*, Open University Press, Buckingham, UK.
Scott, P., Driver, R., Leach, J. & Millar, R.: 1993, *Students' Understanding of the Nature of Science: Working Papers 1–11*, Children's Learning in Science Research Group, University of Leeds, Leeds, UK.
Simon, S. & Jones, A.: 1992, *Open Work in Science*, Centre for Educational Studies, King's College, University of London, London.
Smit, J. & Finegold, M.: 1995, 'Models in Physics: Perceptions Held by Prospective Physical Science Teachers Studying at South African Universities', *International Journal of Science Education* 17, 621–634.
Sutton, C.: 1992, *Words, Science and Learning*, Open University Press, Buckingham, UK.
Thagard, P.: 1992, 'Analogy, Explanation and Education', *Journal of Research in Science Teaching* 29, 537–544.
Thiele, R. & Treagust, D.F.: 1995, 'Analogies in Chemistry Textbooks', *International Journal of Science Education* 17, 783–797.
Treagust, D.F., Duit, R., Joslin, P. & Lindauer, I.: 1992, 'Science Teachers' Use of Analogies: Observations from Classroom Practice', *International Journal of Science Education* 14, 413–422.
Tytler, R. & Swatton, P.: 1992, 'A Critique of Attainment Target 1 Based on Case Studies of Students' Investigations', *School Science Review* 74, 21–38.
van Hoeve-Brouwer, G. & de Vos, W.: 1994, 'Chemical Bonding or Chemical Structure?', in H.-J. Schmidt (ed.), *Problem Solving and Misconceptions in Chemistry and Physics*, International Council of Associations for Science Education, Hong Kong, 238–245.
Wong, E.: 1993, 'Self-Generated Analogies as a Tool for Constructing and Evaluating Explanations of Scientific Phenomena', *Journal of Research in Science Teaching* 30, 367–380.

# 1.5 Learning About Science Teaching: Perspectives From an Action Research Project

PHILIP H. SCOTT
*University of Leeds, UK*

ROSALIND H. DRIVER
*King's College London, UK*

The findings of research carried out internationally over the last 20 years have led to widespread recognition that young people have informal or spontaneous ideas about natural phenomena which are commonly considered in science instruction (Carmichael, Driver, Holding, Phillips, Twigger & Watts 1990; Driver, Guesne & Tiberghien 1985; Pfundt & Duit 1994). Teaching and learning studies have shown that the evolution of students' ways of thinking about phenomena tends to be a slow and piecemeal process. In some cases, students' informal ideas have been shown to persist through extended tuition in science whilst, in others, the ideas presented in instruction are assimilated by students in ways which are in keeping with their current conceptions. Further instruction then can reinforce misunderstandings of this hybrid nature. Such findings draw attention to a significant problem and challenge for science teaching.

In response to this challenge, science educators, over the last decade, have been focusing attention on approaches to teaching science which take account of students' informal conceptions (Black & Lucas 1993; Duckworth, Easley, Hawkins & Henriques 1990; Fensham, Gunstone & White 1994; Glynn, Yeany & Britton 1991; West & Pines 1985). In the early 1980s in the UK, research findings from studies carried out by the Children's Learning in Science (CLIS) Research Group, on students' conceptions in science, were gaining attention and questions were being asked by science teachers and researchers about possible implications for teaching. It was with this focus that the CLIS Research Group initiated a research project into teaching and learning for conceptual understanding of science in classroom settings. In this chapter, we offer an overview of the research, present the main findings and review these in light of our present thinking about developments in the field.

The research reported here was grounded in the realities of teaching and learning science in high school classrooms in the UK and involved researchers from the

---

Chapter Consultant: Andrée Tiberghien (CNS University of Lyons, France)

CLIS Research Group working alongside groups of local science teachers. The overall strategy was to work with teachers as collaborators in exploring ways of improving students' conceptual understanding of selected topics in science. The project had two strands: (1) the development of curriculum and pedagogy to promote conceptual understanding; and (2) teacher professional development and change. This chapter focuses principally on the first of these, namely, the design, development and evaluation of curriculum sequences for teaching specific scientific concepts and the features which need to be taken into account to ensure that these are responsive to what we know about students' learning. The chapter also draws attention to the changes in teachers' practice which were required by the project and how teachers responded to these.

## A COLLABORATIVE PROJECT

An initial open meeting with local science teachers was held at the University of Leeds early in 1984 to outline and discuss the central aim which was to 'devise, implement and evaluate teaching materials and strategies which attempt to promote conceptual change in selected science topic areas' and to base teaching on a constructivist view of learning (Driver & Oldham 1986). The three topic areas of *plant nutrition*, *particulate theory of matter* and *energy* were selected because (1) research on children's conceptions indicated that they presented particular conceptual problems for learners, (2) they are all concepts which are central to the sciences and (3) they represent important concepts in both the biological and physical sciences. An educational case therefore could be made for investing time and effort in them.

In the overall planning of the project, the issue of what part teachers should play had been the subject of careful thought prior to the first meeting. A number of different ways of working were considered, each reflecting a different relationship between researchers and teachers. There appeared to be two extreme positions which could be taken. The first focuses on the researchers who develop curriculum units, based on specific principles, which they themselves then teach in selected schools. In this model, the teachers do no more than provide access to their classrooms and students. The advantage of such an approach is that the people who are most intimately aware of the rationale behind new curricula and associated pedagogy (the researchers) take the lead in field testing. However, researchers do not know the students and are not aware of the work practices of particular schools. Nevertheless, such an approach has been taken in similar research projects (Adey & Shayer 1994). At the other extreme, researchers work with teachers in a non-interventionist way, observing teaching of specified concepts and identifying, documenting and developing best practices. Such an approach would use an action-research methodology to report back to teachers on what learning is happening in their classrooms and hence give them the opportunity to reflect on and modify their practice. This course of action is essentially conservative in providing a mirror to teachers' existing good practice, and it tends to reveal implicit craft knowledge of teaching.

Neither of these approaches matched the aims of the CLIS project which set out to introduce a new theoretical perspective in devising teaching approaches which explicitly took account of, and promoted change in, students' existing ways of thinking. Although such approaches were radically different to existing practice for most teachers, the aim was to develop materials and expertise that could be used by experienced teachers in the day-to-day context of science teaching in high schools. Therefore, it was decided to adopt an interventionist approach in which researchers worked alongside teachers so that the theoretical perspective might be brought to bear on the design and development of teaching schemes which would be practicable in UK high schools (taking into account factors such as class size, available equipment and lesson time allocation). In this approach, the research team had responsibility for setting the aims of the project, for providing a theoretical steer, for providing guidance in developing curriculum materials and for monitoring and evaluating classroom trials. The teachers were centrally involved in developing the materials which they field tested in their own classrooms.

A key issue for the teachers was that they needed to develop an understanding of the theoretical perspective on teaching and learning which underpinned the curriculum development work. Just as their students would be challenged to develop an alternative way of knowing about natural phenomena, so too the teachers faced the challenge of reconstructing their views of teaching and learning science. Just as the teaching schemes took account of students' initial thinking, the curriculum development process needed to help teachers to reflect on and develop their own views of teaching and learning.

With these points in mind, working groups of about 15 teachers (all of the teachers volunteered to become involved in the project) were set up in each of the three concept areas. The thinking here was that groups of teachers would provide the professional and social support which would be needed in a project of this kind. Furthermore, teachers from particular schools were encouraged to enrol in pairs to allow for support both within and across the schools represented in the working groups. In addition, it was anticipated that this number of teachers would bring a breadth of ideas and skills to the task of curriculum development and also would offer a range of different kinds of schools in which the materials might be tried out.

## THEORETICAL PERSPECTIVES INFORMING THE PROJECT

The project was based on a constructivist view of learning which was discussed extensively by researchers and teachers in early meetings and subsequently was outlined in a position paper (Driver & Oldham 1986). The main features of this theoretical perspective are described below.

## Scientific Knowledge

Science as a form of public knowledge is conjectural and socially constructed, but is also constrained by empirical and rational criteria. As the position paper indicates, 'Science as public knowledge is not so much a discovery as a carefully checked construction. It follows that science in secondary schools involves not just knowledge about events and phenomena in the natural world, but an appreciation of theories as imaginative human constructions' (Driver & Oldham 1986, pp. 109–110). The paper further argues that, although scientific knowledge is a human construction, it aims to represent a shared physical world and thus, contrary to assumptions implicit in some current critiques (Matthews 1992), it assumes a realist not a relativist ontology. An important consequence of this position is that teaching schemes have learning goals which are both conceptual and epistemological. The conceptual goals help students to understand and be able to use scientific frameworks in specific topic domains. The epistemological goals help students to recognise that scientific knowledge is conjectural and to gain an appreciation of the rational criteria (such as consistency, coherence, parsimony) which are drawn on by the scientific community in generating and validating knowledge claims.

## Students' Learning

Individuals construct their personal knowledge through social interaction and experiences with the physical environment. Learning, therefore, is a purposive activity on the part of the learner and requires active engagement. Furthermore, individuals' existing conceptions influence the meanings which they construct in a given situation (whether a lecture, demonstration or practical activity), and what is learned results from an interaction between the learner's present conceptions and the various linguistic and sensory experiences provided. Designing teaching schemes to support science learning is therefore inherently problematic in that it requires some appreciation of the prior knowledge that students are likely to bring with them to the learning situation, whilst recognising that individual learners make sense of learning experiences in personal ways. It is not possible, therefore, to design learning sequences in such a way that learning outcomes can be fully anticipated.

## Curriculum Development

Following this perspective on teaching and learning science, curriculum was interpreted as being a set of learning experiences which enable learners to develop understanding towards a scientific view (Driver & Oldham 1986, p. 112). This was quite different to prevailing (and current) views which frame curricula in terms of what is to be taught (statements of the scientific end-points). Interpreting curriculum as a set of learning experiences also leads to a different view of the process of curriculum development. It is not possible *a priori* to anticipate the ways in which students (with their teachers) will engage with particular learning activities. If curriculum development is to involve development of activities to support learning, that process must include an empirical, reflexive step. In other words, the learning activities are first developed by drawing on best available knowledge about students' thinking in particular topic domains, and then those activities are tried out and refined in the light of those trials. Trialling is an essential part of the development process rather than an optional evaluation step. As we have argued elsewhere, effective curriculum development is essentially a research activity (Driver & Scott 1996), a view which is supported by others (Lijnse 1995). Such curriculum research results both in teaching schemes which are adapted to the needs of learners and documentation of the ways in which young people are likely to respond to the material (hence making it easier for future users of the schemes to anticipate how learners might interact with the materials).

## DESIGN AND IMPLEMENTATION OF THE PROJECT

The project program was planned to extend over two years and was divided into two parts, the *Preparatory* and *Intervention* phases, each lasting one year.

### Preparatory Phase

The preparatory phase of the program focused upon existing approaches to teaching in the three concept areas. Teachers in each working group taught the selected topic in their usual way, monitored the learning of one class of students (aged 13–14 years) using diagnostic tests, and kept a reflective journal of their lessons. In addition, the lessons of two teachers in each group were documented in detail by a researcher who observed and audiotaped all lessons and held discussions with the teacher and students in between the lessons. Case studies of those lessons were prepared as working reference documents for the groups and were published (Bell 1985; Brook & Driver 1986; Wightman 1986).

The purpose of the preparatory phase was to help both researchers and teachers to frame the curriculum problem under scrutiny by reflecting systematically on existing practice. Within group meetings, questions were raised about current approaches to teaching each topic, how students responded to those approaches, the particular problems that students encountered, and how teaching approaches might be revised to address those perceived problems. Thus, steps were taken to problematise the taken-for-granted nature of existing teaching approaches and to question well-rehearsed strategies which had become routinised through their familiarity.

*Design of Teaching Schemes*

One week was set aside towards the end of the first year to review findings from the Preparatory Phase and to draft outline plans for new teaching schemes in each of the concept areas. Teachers were given release from school and worked full-time with researchers during this 'writing week' to generate outline teaching schemes which would be used by participating teachers in the next year's Intervention Phase. As a result of the writing week, the three schemes were prepared in draft form, with each being designed to take about two hours per week for five weeks with classes of 13–14-year-old students. Although the teaching schemes were developed in separate working groups, the groups shared a common approach which involved: (1) identifying the target concept(s) to be taught (the science way of knowing); (2) considering the nature of students' existing thinking in this area; (3) analysing the differences between the science way of knowing and the students' existing thinking; and (4) developing teaching activities to help students to develop the science way of knowing.

Traditionally, the design of curriculum materials has been based on the science to be taught, with planning decisions made in terms of how the science concepts might be presented so that they follow logically. For example, concepts of mass and volume are addressed prior to introducing notions of density. This approach makes good sense as far as it goes, but it does not consider how teaching might be planned to address the problems which students have in developing an understanding of each of the concepts, given their conceptual starting points. There is a need to analyse the differences between the science way of knowing and the students' existing ideas (referred to as the *learning demand* of coming to understand a particular concept or conceptual area) (Leach & Scott 1995). The learning demand is a description of the cognitive changes or developments which the learner must undergo in developing a scientific perspective on a given topic, given the conceptual and epistemological starting point of their existing knowledge structures. For example, the learning demand in understanding about energy involves the student in shifting from seeing energy as a quasi-fuel which gets used up as things happen, to accepting that it is conserved and can be used as an accounting device in describing changes.

Having identified the learning demands, the next step was to develop teaching approaches to address the demands through the appropriate selection and sequencing of learning activities. Initially the ordering of activities was guided by a generalised teaching sequence which included the following phases: orientation; elicitation of students' ideas; restructuring of students' ideas; application of new ideas; and review of learning, including comparison of new ideas with prior ideas. This sequence has been reported and cited widely in the literature as characterising 'constructivist teaching'. However, we regard it as misleading to view the sequence in this way and regard it as misleading to view the sequence as a kind of algorithm or recipe for designing constructivist teaching approaches. The core of a constructivist position is that an individual's prior knowledge influences subsequent learning. The consequence of this for teaching for conceptual understanding is that learning activities, whatever their nature (reading, discussion, practical activity, teacher presentation), need to be planned to take account of students' perspectives and, through feedback and dialogue, to support understanding of the scientific view. In other words, learning activities should address the *learning demands* of particular students in particular topics, thereby acknowledging the *differences* between students' existing knowledge and the scientific perspective. Just as there is a wide range of types of learning demands, so too there needs to be different types of learning activities to address these. In other words, there is no simple 'recipe' for designing teaching schemes.

A central guiding principle for planning each of the schemes was that students should have the opportunity to reflect upon and talk about their own understandings as the teaching progressed. This principle led to the inclusion of strategies such as small-group discussion work, student poster presentations, and use of student learning diaries. Other strategies involved the ways in which teachers interact with students and included teachers using open rather than closed questions, encouraging student contributions in class discussions, and supporting the evaluation of different views in the light of evidence rather than focusing discussion on 'right answers'.

*Intervention Phase*

During the Intervention Phase, the new schemes were trialled by teachers in each of the groups. They monitored student learning using diagnostic tests and kept a reflective journal of their lessons. In addition, the lessons of two teachers in each group were documented in detail by a researcher who observed and audiotaped all lessons.

During these trials, regular meetings allowed teachers and researchers to review each instructional activity in the light of the sense that students had made of it and the problems that they had encountered. These meetings were of fundamental importance to the project in facilitating the crucial review and evaluation step of the curriculum development cycle. What soon became apparent was that similar conceptual problems and issues were being identified across the classrooms from different schools when the same instructional activity was trialled. For example, in the particles scheme, a core lesson required students to work in groups and develop a model to account for the different properties of ice, water and steam. Results from the trial classes showed that it was common for students to generate particle models and that there were also some common features in students' reasoning about these models (such as the problem of what occupies the space between the molecules of a gas). Such results also have been reported in other teaching studies (Nussbaum 1985). In some cases, feedback from the teaching indicated that activities were misleading and that modifications would be helpful.

Case studies of these lessons (Johnston & Driver 1991; Oldham, Driver & Holding 1991) were prepared for the groups as working documents and include encouraging results of diagnostic tests used with trial classes. The data generated in the trials were drawn upon in reviewing, revising and writing up each of the schemes. The published schemes (Children's Learning in Science 1987) not only provide directions for the teacher, but also offer information about how students responded to each of the activities. We believe that inclusion of such information is consistent with viewing each activity as a focus for a 'thinking interaction' between students and teacher. The curriculum materials provide teachers with insights into the likely nature of that interaction.

After completion of the project, a two-year dissemination program, involving both project teachers and researchers, was set up to bring the teaching schemes and the theoretical ideas underlying them to the attention of teachers throughout England and Wales. The program involved both project teachers and researchers and was based on a 'cascade model' which involved representatives from regional centres attending five-day training courses which were developed to prepare the representatives for working with teachers in their own localities.

## PERSPECTIVES ON THE PROJECT

In this section, we reflect on the teachers' involvement in the project and issues relevant to curriculum development.

### *Perspectives on Teacher Professional Development*

The project provided an intense learning activity for participating teachers, all of whom were called upon to review their existing approaches to planning and teaching, to develop new approaches and to put these into action. Planning teaching to take account of students' thinking offered both problems and rewards.

In general, the aspects of the approach about which teachers were positive and which they generalised to other parts of their practice included finding out about students' prior ideas, using small-group discussion activities and using group posters as a way of providing a focus for whole-class discussion and feedback. In addition, the insights offered by research into alternative conceptions provided teachers with a rationale for thinking about teaching and learning and, for many, it was the first time since their initial teacher training that they had looked at practice from a reflective, theoretical stance.

The aspects which teachers considered problematic involved the fundamental issue of developing lessons in a way which was more responsive to students' understandings. Indeed some teachers found the move away from pre-set lesson planning stressful. In particular, teachers lacked confidence and expertise in allowing the direction and development of lessons to be set, at least in part, by the questions and contributions of the students. Some were prepared to move in this direction but found it intellectually much more demanding as it required not only being aware of a range of different ways in which students conceptualise given tasks, but also responding appropriately in each case. Perhaps for the first time, teachers were being called upon not only to consider in detail what they would do in lessons but also to anticipate, as best they could, how their students would respond.

The management of whole-class discussions in ways that respected the contributions of individuals while maintaining a clear direction was seen as particularly demanding. It also was recognised that girls tended to be less comfortable than boys in contributing to this phase of lessons (Carmichael 1990). As well, teachers identified aspects of students' small-group discussion work as being potentially problematic. Some teachers felt that the groups would only make progress if they closely monitored them. In fact, researchers' audio-recordings showed that interventions by the teachers frequently disrupted group discussion and slowed down progress. An associated issue was that, when teachers *did* allow sufficient time for group discussion, this often resulted in lessons coming to an end without the teacher sensing that 'closure' had been achieved with the task in hand. As teachers gained experience with the approach, however, they became more confident about giving students sufficient time to allow development of understanding of major concepts.

Overall, most teachers felt that the development of the schemes had been a worthwhile exercise. During the two-year project, few teachers withdrew and there was considerable evidence that teachers were drawing upon this new way of conceptualising and implementing their teaching across the science curriculum. An account of insights into teachers' professional development through the project is given in Johnston (1991).

*Perspectives on Curriculum Development*

This chapter has presented the general rationale which guided the design of teaching sequences in this project, namely, the identification of *learning demands* and the development of teaching approaches to address those demands. We also have argued that curriculum materials, which are effective in supporting students' science learning, need to be developed through a reflexive research process. Subsequent to this project, members of the CLIS Research Group devised and tried out further teaching schemes based on these principles in a number of science topics (Johnston & Scott 1991; Scott, Asoko, Driver & Emberton 1994). In planning these and other teaching sequences, it is useful to distinguish between *instructional activities/sequences* and *pedagogical strategies*.

Each instructional activity can be thought of as a teaching approach to address a particular learning demand. As the nature of the learning demand differs in different topic areas, the nature of appropriate instructional activities will change. Within an instructional sequence, a number of different instructional activities is used, with each being matched to a particular learning demand. A range of instructional activities was used in the teaching schemes and these are listed in the following paragraph. Each type of instructional activity is referred to in terms of the learning demand addressed.

The first type of instructional activity is designed to broaden the range of application of a conception. Students' existing conceptions can offer a starting point which can be extended in coming to the scientific view. The second type aims to differentiate conceptions. In many areas, students' conceptions can be global and ill-defined, and therefore activities are needed to help them to differentiate ideas such as heat and temperature and weight and mass. The third type involves building experiential bridges to a new conception. In some cases, students do not have the necessary experiential evidence to allow them to make sense of a particular scientific idea. In such a case, additional experience is needed. The fourth instructional activity aims to construct new conceptions. In some cases, because students' prior ideas are incommensurate with scientific conceptions, attempting to shape students' notions into the scientific ideas only leads to problems. In a case of this kind, students' ideas are acknowledged and discussed. We then indicate that scientists have a different view and present an alternative model. Students have the opportunity later to evaluate the scientific model in relation to their prior ideas.

*Perspectives on Science Pedagogy*

Whereas the instructional activities outlined in the previous section were selected to address specific learning demands in planning teaching, pedagogical strategies are more general in nature and can be thought of as the styles of classroom interaction which support learners in the processes of construction, reflection on, and evaluation of ideas and through which the instructional activities are mediated in the classroom. Already attention has been drawn to the range of pedagogical strategies (such as group discussion work, poster presentations, open questioning by the teacher) which were included in the schemes.

As a result of this research program and the continuing research into classroom learning, our perspective on science learning in classroom settings has developed to incorporate a more sociocultural view of the classroom (Driver, Asoko, Leach, Mortimer & Scott 1994). In essence, this involved coming to focus on the pedagogical strategies employed by teachers just as much as the instructional approaches which they mediate.

Language plays a fundamental role in the ways in which teachers support learning in the classroom. In developing ways of analysing classroom teaching and learning of science concepts, we drew upon Vygotsky's (1978) concept of internalisation and carried out detailed case study work to identify and characterise the various moves made by teachers in attempting to develop shared meanings on the interpsychological plane of the classroom. This work characterises the fine detail of the pedagogical strategies, referred to in this chapter, which mediate presentation of instructional activities. Recent work

(Scott 1996) has led us to portray the interlinking sequences of instructional activities and pedagogical strategies as the development, extended over time, of a 'teaching narrative' introducing the science way of knowing.

The process of internalisation does *not* simply involve transfer of concepts, via language, to the individual. The learner reorganises and reconstructs experiences of the social plane. The process of internalisation is not the transfer of an external activity to a pre-existing, internal plane of consciousness, but the process in which this plane is formed. In this respect, Vygotskian theory acknowledges the Piagetian perspective that the child cannot be a passive recipient of knowledge and instruction. The Vygotskian analysis therefore combines both social and personal perspectives, centred on the process of internalisation. Our analysis of language use on the inter-psychological plane is complemented by examination of the personal meanings developed by students as instruction progresses.

## FINAL COMMENTS

This chapter has reviewed a major project of the Children's Learning in Science Research Group and has reflected on its contribution to the theory and practice of teaching and learning science. More recent critiques of the field have tended to focus on two aspects of the work: the theoretical position based on constructivist epistemology; and the practical implications for science pedagogy. The main criticism of the theoretical position has been made by Matthews (1992) who argues that constructivism is essentially empiricism revisited and fails to take account of the symbolic nature of scientific knowledge. As far as our own interpretation of constructivism is concerned, Matthews is erecting a straw man. As earlier sections of this chapter make clear, we acknowledge the differences between the realities of the physical world and the symbolic scientific knowledge used to describe that world. Hence we recognise the need to provide students with insight into science as a 'new way of knowing'. Our interpretation of constructivist science teaching does not involve leaving students to make sense of their physical experiences in their own terms.

Solomon (1992) raises the related pedagogical issue that, although students might discuss their own naive mini-theories for considerable periods of time, there is no evidence that this helps them to change their ideas. In response, we point out first that, in some situations, student discussion about specific physical tasks can enhance scientific understanding (Howe, Tolmie & Rodgers 1990). However, student discussion by itself will not necessarily result in changes in students' understandings. There are occasions when the scientific 'way of knowing' is not represented within the discourse of a group (and when it cannot be induced from observation of the phenomenon itself). Here the teacher plays a central role in introducing the science view. In such cases, it is helpful to provide opportunities for discussion in which students clarify their initial ways of thinking, relate these ways of thinking to the science view and consider the implications of the science view in different contexts. The pedagogical approaches enable students to become involved in discussions in which meanings are outlined, checked, challenged and supported. These are the kinds of discussion that we all value in coming to develop our own understandings of new ideas. Also our understanding of how students learn scientific concepts in classroom settings can be advanced through careful examination of patterns of talk, both in whole-class and small-group situations (Barnes & Todd 1995; Edwards & Mercer 1987; Hicks 1995; Lemke 1990; Moll 1990).

Finally, we acknowledge that a particular view of learning does not lead directly to a particular approach to teaching, a point that has been made elsewhere (Millar 1989). However, knowledge of students' existing understandings and their likely influence on subsequent learning provide an important basis for making decisions in planning and implementing teaching. In particular, the notion of '*learning demand*' outlined earlier is central to our approach to teaching. It has been suggested (Matthews 1994) that such an approach leads to teaching which is not unusual and certainly not unique to constructivism. Experienced teachers have developed approaches to teaching which are sensitive to students' understandings without any explicit recourse to a constructivist perspective. However, explicit recognition of a constructivist perspective and the consequent notion of learning demand leads teachers to a point in their own thinking at which the essentially *dialogic* nature of teaching and learning science becomes inescapable. This is a dialogue, involving different ways of knowing, which acknowledges everyday understandings and scientific views. Once the teacher is aware of the differences between these ways of knowing, as they are represented by teacher and students in the classroom, instructional activities can be planned to address those differences and pedagogical approaches can be chosen to support meaning making. Once the teacher is aware of those differences, science teaching becomes transformed.

## REFERENCES

Adey, P. & Shayer, M.: 1994, *Really Raising Standards: Cognitive Intervention and Academic Achievement*, Routledge, London.

Barnes, D. & Todd, F.: 1995, *Communication and Learning Revisited: Making Meaning Through Talk*, Boynton/Cook, Portsmouth, UK.

Bell, B.F.: 1985, *The Construction of Meaning and Conceptual Change in Classroom Settings: Case Studies in Plant Nutrition*, Children's Learning in Science Project, Centre for Studies in Science and Mathematics Education, University of Leeds, Leeds, UK.

Black, P.J. & Lucas, A.M.: 1993, *Children's Informal Ideas in Science*, Routledge, London.

Brook, A. & Driver, R.: 1986, *The Construction of Meaning and Conceptual Change in Classroom Settings: Case Studies on Energy*, Children's Learning in Science Project, Centre for Studies in Science and Mathematics Education, University of Leeds, Leeds, UK.

Carmichael, P.: 1990, *An Evaluation of the Learning Experiences of Boys and Girls in Secondary School Science Classes Involved in a Curriculum Development Project*, Unpublished PhD thesis, University of Leeds, Leeds, UK.

Carmichael, P., Driver, R., Holding, B., Phillips, I., Twigger, D. & Watts, M.: 1990, *Research on Students' Conceptions in Science: A Bibliography*, Centre for Studies in Science and Mathematics Education, University of Leeds, Leeds, UK.

Children's Learning in Science: 1987, *CLIS in the Classroom: Approaches to Teaching*, Children's Learning in Science Project, Centre for Studies in Science and Mathematics Education, University of Leeds, Leeds, UK.
Driver, R., Asoko, H., Leach, J., Mortimer, E. & Scott, P.: 1994, 'Constructing Scientific Knowledge in the Classroom', *Educational Researcher* 23(7), 5–12.
Driver, R., Guesne, E. & Tiberghien, A.: 1985, *Children's Ideas in Science*, Open University Press, Milton Keynes, UK.
Driver, R. & Oldham, V.: 1986, 'A Constructivist Approach to Curriculum Development in Science', *Studies in Science Education* 13, 105–122.
Driver, R. & Scott, P.: 1996, 'Curriculum Development as Research. A Constructivist Approach to Science Curriculum Development and Teaching', in D.F. Treagust, R. Duit & B.F. Fraser (eds.), *Improving Teaching and Learning in Science and Mathematics*, Teachers College Press, New York, 94–108.
Duckworth, E., Easley, J., Hawkins, D. & Henriques, A.: 1990, *Science Education: A Minds on Approach to the Elementary Years*, Lawrence Erlbaum Associates, Mahwah, NJ.
Edwards, D. & Mercer, N.M.: 1987, *Common Knowledge: The Growth of Understanding in the Classroom*, Methuen, London.
Fensham, P.J., Gunstone, R.F. & White, R.T. (eds.): 1994, *The Content of Science: A Constructivist Approach to its Teaching and Learning*, Falmer Press, London.
Glynn, S., Yeany, R. & Britton, B.: 1991, *The Psychology of Learning Science*, Lawrence Erlbaum, Hillsdale, NJ.
Hicks, D.: 1995, 'Discourse, Learning and Teaching', in M.W. Apple (ed.), *Review of Research in Education*, American Educational Research Association, Washington, DC, 49–95.
Howe, C., Tolmie, A. & Rodgers, C.: 1990, 'Physics in the Primary School: Peer Interaction and the Understanding of Floating and Sinking', *European Journal of Psychology of Education* 4, 459–475.
Johnston, K.: 1991, 'High School Science Teachers' Conceptualisations of Teaching and Learning: Theory and Practice', *European Journal of Teacher Education* 14(1), 65–77.
Johnston, K. & Driver, R.: 1991, *A Constructivist Approach to the Teaching of the Particulate Theory of Matter: A Report on a Scheme in Action*, Children's Learning in Science Project, Centre for Studies in Science and Mathematics Education, University of Leeds, Leeds, UK.
Johnston, K. & Scott, P.: 1991, 'Diagnostic Teaching in the Science Classroom: Teaching and Learning Strategies to Promote Development in Understanding about Conservation of Mass on Dissolving', *Research in Science and Technological Education* 9, 193–213.
Leach, J. & Scott, P.: 1995, 'The Demands of Learning Science Concepts: Issues of Theory and Practice', *School Science Review* 76(277), 47–51.
Lemke, J.L.: 1990, *Talking Science: Language, Learning and Values*, Ablex, Norwood, NJ.
Lijnse, P.: 1995, 'Developmental Research as a Way to an Empirically Based "Didactical Structure" of Science', *Science Education* 79(2), 189–199.
Matthews, M.: 1992, 'Constructivism and Empiricism: An Incomplete Divorce', *Research in Science Education* 22, 299–307.
Matthews, M.: 1994, 'Discontent with Constructivism', *Studies in Science Education* 24, 165–172.
Millar, R.: 1989, 'Constructive Criticisms', *International Journal of Science Education* 11, 587–596.
Moll, L.C. (ed.): 1990, *Vygotsky and Education: Instructional Implications and Applications of Sociohistorical Psychology*, Cambridge University Press, Cambridge, UK.
Nussbaum, J.: 1985, 'The Particulate Nature of Matter in the Gaseous Phase', in R. Driver, E. Guesne & A. Tiberghien (eds.), *Children's Ideas in Science*, Open University Press, Milton Keynes, UK, 124–144.
Oldham, V., Driver, R. & Holding, B.: 1991, *A Constructivist Approach to Teaching Plant Nutrition: A Report on the Scheme in Action*, Children's Learning in Science Project, Centre for Studies in Science and Mathematics Education, University of Leeds, Leeds, UK.
Pfundt, H. & Duit, R.: 1994, *Bibliography: Students' Alternative Frameworks and Science Education* (fourth edition), Institute for Science Education at the University of Kiel, Kiel, Germany.
Scott, P.: 1996, 'Social Interactions and Personal Meaning Making in Secondary Science Classrooms', in G. Welford, J. Osborne & P. Scott (eds.), *Research in Science Education in Europe*, Falmer Press, London, 325–336.
Scott, P., Asoko, H., Driver, R. & Emberton, J.: 1994, 'Working from Children's Ideas: An Analysis of Constructivist Teaching in the Context of a Chemistry Topic', in P.J. Fensham, R.F. Gunstone & R.T. White (eds.), *The Content of Science: A Constructivist Approach to its Teaching and Learning*, Falmer Press, London, 201–220.

Solomon, J.: 1992, 'Of Science Teaching', *Education in Science* 148, 12–13.
Vygotsky, L.S.: 1978, *Mind in Society: The Development of Higher Psychological Processes*, Harvard University Press, Cambridge, MA.
West, L. & Pines, A. (eds.): 1985, *Cognitive Structure and Conceptual Change*, Academic Press, Orlando, FL.
Wightman, T.: 1986, *The Construction of Meaning and Conceptual Change in Classroom Settings: Case Studies in the Particulate Theory of Matter*, Children's Learning in Science Project, Centre for Studies in Science and Mathematics Education, University of Leeds, Leeds, UK.

# 1.6 Scientific Inquiry Within Reach of Young Children

KATHLEEN E. METZ
*University of California, Riverside, USA*

The international education community has long shared a concern that science instruction for young children should be well matched to their young minds. Realisation of this goal raises a number of demanding issues, including an understanding of children's knowledge and ways of knowing, and how these map onto the content and processes of science instruction. From the perspective of current issues in the field of science education, a particularly important aspect of this match is the extent to which children can engage in scientific investigations.

In its search for a theoretical base to constrain the match between science process objectives and children's minds, many corners of the science education community have relied on cognitive developmental theory, particularly Piaget's stages of logical-mathematical structures (e.g., Karplus 1977). Whereas interpretations of Piaget's concrete operational thought have been used in the conceptualisation of science inquiry processes within reach of children in the first six or seven years of school, interpretations of formal operational thought have been used as a window on their limitations. This chapter critiques reliance on Piagetian stages of logical-mathematical structures for this purpose and poses another approach in its stead.

There is a number of problems in relying on Piagetian theory in conceptualising scientific inquiry within reach of young children. First, Piaget focused on development, not learning. Indeed, it was only at the end of his life that his interests moved towards the interaction of development and learning (Piaget 1974). However, educators need a theory that describes how children can think and what children can understand, *given effective instruction*. It is problematic to derive science curricular frameworks or instructional objectives from a theory that describes stages of children's thinking apart from instructional support.

A second problem in relying on Piagetian theory for this purpose is that his research was restricted to the thinking of the child working alone. Although examination of the full scope of Piaget's writings indicates that he was convinced of the powerful effect of the social context on the development of thought (e.g., Piaget 1928), study of collaborative cognition was largely outside of the scope of Piaget's research agenda. However, if we are to take seriously contemporary calls

---

Chapter Consultant: Joan Bliss (University of Sussex, UK)

to narrow the gap between science as practised by scientists and science as practised by children in classrooms (e.g., Burbules & Linn 1991; Rutherford & Ahlgren 1990), we need to integrate collaborative work into children's science instruction. The scientific enterprise as practised by scientists is fundamentally a collaborative activity (Dunbar 1995; Latour & Woolgar 1986). Furthermore, there is strong evidence that strategic use of collaborative cognition can lower significantly the cognitive load experienced by the individual participant (e.g., Brown, Ash, Rutherford, Nakagawa, Gordon & Campione 1993; Resnick, Levine & Teasley 1991) and thus enable the child to engage successfully in more complex intellectual tasks.

Finally, use of Piagetian descriptions of logical-mathematical structures at the concrete level (rather than at the formal operational level) to frame the possibilities of children's inquiry appears problematic, given that researchers, including Piaget himself, have voiced serious doubts that the model actually captures what develops. Piaget (1980) expressed deep concerns about the adequacy of the model on the grounds that it failed to consider how the semantics of the situation influenced the course of scientific inquiry. In this same vein, Piaget's close colleague, Guy Cellérier (1972), criticised the distance that the model presumes between the symbols at which reasoning supposedly takes place and the phenomenology of the situation. British logician, Charles Parsons (1960), identified internal inconsistencies in how the symbolism is used and pointed out that the logic that Piaget attributed to the formal operational thinker only applies to a restricted set of domains meeting specific conditions. Furthermore, analyses of the process of scientific inquiry (e.g., Simon 1992) bring into doubt that any model of logic and proof is sufficient to capture this complex process. For a more extended analysis of the limitations of relying on Piagetian theory for deriving constraints for primary school science education, see Metz (1995).

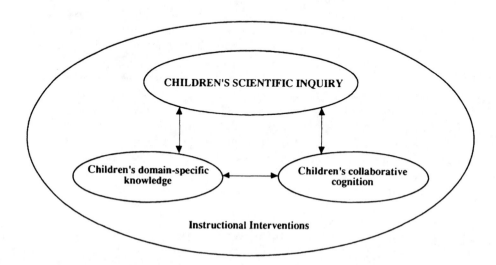

Figure 1: Conceptual frame for analysing scientific inquiry within children's reach

For the purpose of conceptualising powerful science curriculum that is both within reach of young children and that capitalises on their emergent abilities, this chapter posits a model focusing on the analysis of children's scientific inquiry, in conjunction with analysis of the impact of domain-specific knowledge and the effect of different forms of collaborative cognition, contextualised within the sphere of instructional interventions (see Figure 1). As the information-processing research tradition provides a rich literature for each of these issues and, to a lesser extent, their interactions, the chapter frames the model in terms of this research base.

Given space constraints, the analysis of the model and related literature is necessarily delimited. This examination is restricted to the learning and teaching of primary school children, thus omitting analysis of even seminal work with older students (cf Pea 1993; Roth & Bowen 1994). Furthermore, as the chapter cannot examine every interaction, four major sections focus, respectively, on (1) the process and products of children's scientific inquiry with varying degrees of scaffolding, (2) the effect of domain-specific knowledge on scientific inquiry, (3) the effect of collaboration cognition on scientific inquiry and (4) the connection between domain-specific knowledge and collaborative cognition.

## THE PROCESS AND PRODUCTS OF THE CHILD'S SCIENTIFIC INQUIRY

The information-processing research literature has identified many weak aspects of the child's independent science inquiry process. For example, children frequently design experiments that cannot support definitive conclusions (Dunbar & Klahr 1989; Schauble & Glaser 1990). The evidence that students consider sufficient to support a hypothesis is frequently inadequate (Dunbar & Klahr 1989; Kuhn, Amsel & O'Loughlin 1988; Metz 1985; Schauble & Glaser 1990), and they tend not to be troubled by disconfirming evidence (Dunbar & Klahr 1989; Kuhn, Amsel & O'Loughlin 1988). Children's goal structure could be different in that they could be working towards purposes more engineering-like than scientific (Schauble, Klopfer & Raghavan 1991; Tschirgi 1980). Not surprisingly, most theories developed by the young and scientifically naive manifest fundamental deficiencies which, for example, reflect simplifications (Karmiloff-Smith & Inhelder 1974), cosmologies diverging from older students and adults (Vosniadou & Brewer 1992) or different 'ontological cores' (Carey 1985b). Nevertheless, from a number of perspectives, the scientific inquiry within reach of young children looks much brighter.

*Positive Aspects of Children's Unassisted Inquiry*

This literature also has identified strengths of children's unassisted scientific inquiry. Studies have documented children's understanding of the scientific inquiry process, even if they frequently resort to a more engineering stance. In this vein, Dunbar and Klahr (1989, p. 140) conclude that children understand 'what the scientific reasoning process should look like'; they realise that they have to elicit and observe behaviour and work towards a 'summary statement that captures the behavior in a universal and general fashion' (p. 141). Similarly, Sodian, Zaitchek and Carey (1991) conclude that, by seven years of age, children understand the goal of hypothesis testing.

Researchers have identified various other parameters of strength in children's scientific inquiry. Dunbar and Klahr (1989) report that a large proportion of the experiments run by the children were similar to the experiments run by the adults. Klahr, Fay and Dunbar (1993) conclude that most of their sixth graders and some of their third graders differentiated theory from evidence, contradicting previous work that claimed that children of this age did not make this distinction (Kuhn, Amsel & O'Loughlin 1988).

Researchers have also documented marked similarities between the theorising and theories of scientists and children. Samarapungavan (1992) concludes that, by seven years of age, children grasp a number of scientists' metaconceptual criteria for theory selection, including range, empirical consistency and logical consistency. Karmiloff-Smith (1988) contends that, by six years of age, children attempt to construct a unified theory to account for all events. Brewer and Samarapungavan (1991) conclude that children's theories resemble scientific theories in the sense that they: transcend what is concrete and directly perceptible; are internally consistent from the child's perspective; frequently involve an attempt to integrate different sources of information (e.g., what they have heard from adults and their own observations); and have explanatory power.

We also can identify numerous studies that document fundamental improvements in children's scientific inquiry and domain-specific knowledge, with scaffolding limited to the experimenter's strategic selection of materials and framing of a functional goal. Under these conditions, Karmiloff-Smith and Inhelder (1974) report that many of their subjects constructed the principle of the law of moments and Metz documents fundamental improvements in preschool children's understanding of mechanical equilibrium (Metz 1993) and fifth graders' understanding of gears (Metz 1985). Similarly, Schauble (1990) reports that her ten-year-old and 11-year-old subjects were able to discover almost all the factors involved in two physics microworlds through their self-regulated experimentation.

*Improvements Through Instructional Interventions*

In analysing the weaknesses of children's investigations, we need to recognise that they are operating at a distinct disadvantage, due to the fact that their scientific cognition takes place apart from the knowledge, culture and accumulated methodological traditions of science (Brewer & Samarapungavan 1991). Other works identify directions in which instructional support potentially might redress these handicaps. Current ideas for instructional intervention manifest a broad spectrum of emphases, from the purpose of investigations, to the scaffolding of theory formation and support of data analysis. The sections below briefly describe a few of these ideas, to illustrate how instructional interventions can raise our sense of scientific inquiry within children's reach.

In a three-week instructional intervention, Carey, Evans, Honda, Jay and Unger (1989) were able to develop 12-year-old students' understanding of the nature of science and, more specifically, the aim and process of scientific inquiry. In the beginning, Carey et al.'s students' experiments were more engineering-like than scientific. The authors report that, at the end of the intervention, 'over half of the students saw experiments as tests of ideas, and some could articulate how unexpected results lead to revisions of ideas' (Carey et al. 1989, p. 524).

White's (1993) *ThinkerTools* project is designed to foster children's understanding of Newtonian mechanics, as it scaffolds their inquiry process and generation of explanations. White's classroom-based instructional intervention includes children's explorations with computer microworlds and the real world, where they formulate their ideas into laws and conceptual models. The teachers scaffold the children's critical examination of the laws that they construct, using the scientific aesthetic of generality and parsimony. White's project has had remarkable results, with 11-year-old and 12-year-old students outperforming high school physics students in transfer tasks.

Hancock, Kaput and Goldsmith (1992) scaffold children's investigations through a computer tool for data representation and analysis. The software *Tabletop* supports several processes found to pose challenges to the young scientist: designing a structure for organising the data; keeping track of their accumulating data sets; and subsequent analysis of trends. These researchers studied classrooms of children aged eight to 13 years, using the tool for their investigations across the school year. Hancock et al. report that the children could successfully tackle investigations involving ill-structured problems and messy data sets.

*Educational Implications*

Children at the first level of schooling are capable of independent empirical investigations that, while falling short of older children and adults in adequacy of experimental design and inference, support both fruitful theory construction and the improvement of the inquiry process itself. Even before instruction, many aspects of children's theorising resemble the theorising of scientists. Furthermore, current instructional interventions document the possibility of strengthening children's scientific inquiry through the support of computer aids or instruction in different aspects of the scientific process.

THE DOMAIN-SPECIFIC KNOWLEDGE FACTOR

This section examines the impact of domain-specific knowledge on the adequacy of information-processing in general and scientific inquiry in particular. It analyses this key relation from several perspectives, including a re-examination of relevant inferences within the school-age cognition literature, a comparative analysis of preschool and school-age cognition literature, and the adult expert/novice physics cognition literature.

*The Connection Between Domain-Specific Knowledge and Scientific Inquiry*

On the basis of her analysis of the literature, Carey concludes that the most important factor differentiating the thinking of children from adults is adequacy of domain-specific knowledge. 'Children are novices in almost every domain in which adults are experts' (Carey 1985a, p. 514). Carey also documents this essential confounding in Inhelder and Piaget's (1958) work, *The Growth of Logical Thinking*:

> These experiments confound knowledge of particular scientific concepts with scientific reasoning more generally. It is well-documented that before the ages of 10 or 11 or so the child has not fully differentiated weight, size, and density and does not have a clear conceptualization of what individuates different kinds of metals (density being an important factor). If these concepts are not completely clear in the child's mind, due to incomplete scientific knowledge, then the child will of course be unable to separate them from each other in hypothesis testing and evaluation. (p. 498)

Brown (1990) argues that Piaget's emphasis on structure and de-emphasis on content have strongly influenced the school-age cognition literature. She contends that this influence has resulted in this literature's confounding of children's cognitive process deficiencies with weak domain-specific knowledge. In turn, this has resulted in a tendency to underestimate children's capabilities.

Ironically, the research literature examining preschoolers' scientific investigations frequently conveys an image of greater competency than the equivalent young school-age literature. Indeed, an underlying agenda of much of the preschool research has been to rebut the theoretical position of a fundamental shift in adequacy of cognition from preschool to school age children (Bullock 1985). A key tactic in achieving this end has been a keen attention to children's familiarity with domains in the design of experimental procedures, in order to avoid confounding inadequacy of domain-specific knowledge with poor information-processing capacities. A pay-off has been the documentation of cognitive capacities traditionally attributed to much older children.

Thus, Goswami and Brown (1989) document three-year-olds' reasoning on the basis of higher-order relations, a competency commonly associated with adolescents and formal operational reasoning. Similarly, whereas a common interpretation of Piagetian theory is that school-age children at the concrete operational stage reason on the basis of the perceptual and concrete, researchers (Brown 1990; Gelman & Markman 1986) have found that preschoolers can reason instead on the basis of 'deep structural principles' when they have the requisite domain-specific knowledge.

The key importance of domain-specific knowledge on adequacy of inquiry also emerges in the expert/novice literature in physics cognition (e.g., Chi, Feltovich & Glaser 1981; Larkin 1983; Larkin, McDermott, Simon & Simon 1980). These studies point to a kind of fundamental bootstrapping between domain-specific knowledge and inquiry processes, according to Glaser:

> Our interpretation is that problem-solving difficulty of novices can be attributed largely to the inadequacies of their knowledge bases and not to limitations in their processing capacities such as the inability to use problem-solving heuristics. (Glaser 1984, p. 99)

*Educational Implications*

The deep connection between levels of domain-specific knowledge and information-processing has important implications for science education. The science education community has long vacillated between an emphasis on subject matter and an emphasis on process. John Dewey (1910) identified this kind of pendulum swing in the 19th century, arguing himself for an emphasis on science process. Dewey (1910) contended: 'In the order of both time and importance, science as method precedes science as subject matter' (p. 188). Millar and Driver (1987) have identified this same kind of vacillation from the 19th century to contemporary science education. According to Linn (1982), interpretations of Piagetian theory have influenced science curriculum designers in the direction of emphasis on structure and de-emphasis of domain-specific knowledge.

This research about the domain-specific knowledge factor indicates that school science instruction needs to pay deeper attention to the *interaction* of evolving domain-specific knowledge and evolving scientific methods. A common practice in the teaching of written composition is to structure various prewriting activities in order to 'activate' children's knowledge of the domain, based on the assumption that accessible knowledge is critical to the process of written composition (Bereiter & Scardamelia 1982). Although the research base linking the quality of domain-specific knowledge and the quality of process appears stronger in science cognition, this connection is generally poorly reflected in science curricula.

This finding of the interaction between domain-specific knowledge and adequacy of scientific inquiry supports current recommendations for deeper study of fewer domains. Most children's science curriculum is not designed to support in-depth knowledge of any sphere. Rather, as Brown, Campione, Metz and Ash (in press) assert for the majority of children's science texts:

> There is a striking lack of cumulative reference (volcanoes following magnets, following a unit on whales, etc.). This lack of coherent themes

or underlying principles all but precludes systematic knowledge building on example, analogy, principle, or theme or theory. (p.16)

This research suggests that strategic selection of a small number of domains for children's science, together with the scaffolding of scientific inquiry in interaction with the scaffolding of relatively deep understanding within these spheres, more effectively can support children's construction and refinement of scientific knowledge and ways of knowing, as it gives them an understanding of the nature of scientific knowledge and the scientific enterprise.

## THE COLLABORATIVE COGNITION FACTOR

Many educators have recommended that phases of the instructional process be structured in terms of small groups of students working together. Assumptions concerning why small groups should be used and the ways in which they should be organised have reflected fundamental changes. This section examines changes in these assumptions. Building on this analysis, how selected instructional interventions extend children's scientific inquiry through collaborative cognition is considered.

### Instructional Uses of Collaboration

In the USA, much of the grouping for instructional processes in the 1970s and 1980s was inspired by the work of Johnson and Johnson. In a meta-analysis of 122 studies investigating the effects of cooperative learning, Johnson, Maruyama, Johnson, Nelson and Skon (1981) assert:

> Cooperation is superior to individualistic efforts in promoting achievement and productivity. It may be hypothesized that the superiority of cooperation is enhanced when the task is other than a rote decoding or correcting task, when peer tutoring is encouraged, and when the task does not require a division of labor. Interestingly cooperation was relatively better in studies that lasted shorter periods of time. (p. 57)

Johnson and Johnson were not concerned with changing the curriculum *per se*, but with supporting children's motivation and mastery. From their perspective, the tasks that work best for student collaboration are short ones without any kind of parsing of component aspects to different individuals.

In her recent review of this literature, Cohen (1994) argues that we need to examine the cooperative learning literature in terms of the factors that support productive discourse within cooperative groups and, more specifically, the kinds of discourse that foster different types of learning. Cohen differentiates group work on routine, well-structured tasks from group work on ill-structured tasks beyond the reach of the individual working alone:

> Whereas limited exchange of information and explanation are adequate for routine learning in collaborative seatwork, more open exchange and elaborated discussion are necessary for conceptual learning with group tasks and ill-structured problems. (Cohen 1994, p. 1)

Although investigations even at the secondary level have frequently been presented to students in well-structured form (Tobin, Tippins & Gallard 1994), scientific inquiry is inherently ill-structured beyond artificial cookbook problems. Consequently, Cohen's analysis suggests that we should not assign roles to group members, but rather encourage a socially more fluid construction of ideas.

The writings of Vygotsky and his contemporary interpreters constitute a rich theoretical perspective on the instructional value of students working together that further supports the desirability of using collaboration in children's science instruction. Vygotsky argues that social interaction is fundamental to learning and the realisation of optimal performance. He contends that, for purposes of instruction and assessment, we need to consider not only children's current capacities, but also their 'zone of proximal development' (i.e., how far their capacities can extend), given the support of the teacher *and their peers*. This socially manifested competence paves the way for its eventual internalisation in the individual participants. According to Vygotsky (1978):

> Every function in the child's cultural development appears twice: first, on the social level, and later on the individual level; first *between* people (*interpsychological*), and then *inside* the child (*intrapsychological*) . . . All higher functions originate as actual relations between human beings. (p. 57)

> Learning awakens a variety of environmental processes that are able to operate only when the child is interacting with people in his environment and in cooperation with peers. Once these processes are internalized, they become part of the child's independent developmental achievement. (p. 90)

The theories of Vygotsky, in conjunction with contemporary theories of information-processing and learning as apprenticeship, have formed the basis of current theories of 'socially-distributed expertise', conceptualised in terms of how competence can be situated in and manifested by a group working closely together (Resnick, Levine & Teasley 1991). Underlying this theoretical orientation is the premise that the knowledge needed for a complex task can reside in a collaborating group, as opposed to the individual. Thus knowledge is in part shared, in part 'distributed'.

This process simultaneously supports making explicit and reflecting on component processes of the intellectually complex enterprise. Novices can take on less-demanding aspects of the task, while still participating in the activity as a whole. Thus, decomposition of the task is avoided. Furthermore, parsing of task aspects across the collaborating group can be flexible, allowing for increased responsibility to novices as they are ready. We can observe instances of socially-distributed expertise in many contexts, including teams navigating a ship (Hutchins 1991), children playing baseball (Heath 1991) and some forms of collaborative groups in the classroom (Brown & Palinscar 1989).

*Extending Children's Scientific Inquiry Through Collaborative Cognition*

There are now a number of classroom-based experiments that aim to apply these ideas to children's science instruction. For example, Rosebery, Warren and Conant (1992) have developed an approach to science instruction that emphasises collaboration. These researchers frame their approach in terms of social-distributed expertise theory, in conjunction with their vision of the work of scientists:

> Cognitively, students share the responsibility for thinking and doing, distributing their intellectual activity so that the burden of managing the whole process does not fall on any one individual. . . . Collaborative inquiry creates powerful contexts for constructing scientific meanings . . . In challenging one another's thoughts and beliefs, students must be explicit about their meanings; they must negotiate conflicts in belief or evidence; and they must share and synthesize their knowledge in order to achieve a common goal. (Rosebery *et al.* 1992, pp. 2–3)

In accordance with the researchers' goal that children actually 'do science', these students engaged in the full scope of scientific inquiry, with a remarkable degree of regulation on the part of the children themselves. They generated their own questions, planned their research, collected and interpreted data, and developed and refined their theories. The researchers reported growth in both scientific inquiry and scientific knowledge.

The Brown and Campione project, *Fostering a Community of Learners* (FCL), is also grounded in collaborative cognition. Their project uses the principle of socially-distributed expertise to enable the simplification of scientific inquiry as experienced by the individual, while maintaining the integrity of the enterprise as a whole (Brown, Ash, Rutherford, Nakagawa, Gordon & Campione 1993).

FCL emphasises three different 'participatory structures' that organise social-distributed expertise. It employs *reciprocal teaching* for the purpose of scaffolding expert text comprehension and analysis, in the service of conducting text-based research. In this technique, the component cognitive processes of expert reading (e.g., clarifying and summarising) are initially distributed across the small group, monitored and aided as necessary by the teacher. Group members trade aspects for which they take responsibility, in preparation for each eventually taking on all aspects in their independent strategic reading. In *jigsawing*, different groups take responsibility for researching different aspects of a complex problem or domain (such as adaptation or interdependence) and subsequently teaching their part to the rest of the class in the context of shuffled groups. *Majoring* connotes children's specialisation in some cognate sphere or technology related to the science curriculum. Children are encouraged to major in a particular area of their choosing and subsequently function as resident experts for the 'community of learners'.

From his observations of the project in action, Jerome Bruner (1996) offers an analysis of how collaboration in various forms of social interaction supports the development of scientific inquiry in FCL classrooms:

> Ann Brown has joined agency and collaboration together in her design of classroom culture. Kids not only generate their own hypotheses, but negotiate them with others – including their own teachers. But they also take the role of teacher – offering their expertise to those with less. This is how it works *structurally*. You argue with your benchmates about the best way to get oil off a polluted sea bird caught in the *Exxon Valdez* oil spill, or how it could have happened in the first place, and you learn about explanatory and interpretative accountability in the process. (Bruner 1996, pp. 93–94)

Ongoing assessments of the project indicate that this approach enables remarkably rich scientific reasoning on the part of young children. Above and beyond their success in attaining significant gains in domain-specific knowledge, Brown and Campione (1994) found evidence of increasing sophistication in critical thinking within the domain, causal explanation and complexity of argumentation.

*Educational Implications*

Analysis of possible uses of social collaboration appears crucial to our conceptualisation of scientific inquiry within reach of children. Current research, in both cooperative groups and socially-distributed expertise, supports the premise that capitalisation on collaborative cognition can extend the possibilities of children's scientific inquiry. The high cognitive demand of scientific inquiry can be adjusted within reach of different learners by having them take on different aspects of the inquiry process while, at the same time, remaining a participant in the enterprise as a whole. This approach allows us to avoid the decomposition of the inquiry process and its step-by-step introduction to the young child. It appears inadequate to assess children's capacities apart from the context of collaborative cognition.

## CONNECTION BETWEEN COLLABORATIVE COGNITION AND DOMAIN-SPECIFIC KNOWLEDGE

The model posited in this chapter presumes an interaction between collaborative cognition and domain-specific knowledge. This section briefly considers the dynamic between these two aspects, from the perspective of scientists' practice and instructional interventions.

Anthropologists, Latour and Woolgar (1986), went into the research laboratory to study 'tribes of scientists' and their 'production of science' (p. 17). Their rich account tells a story of the fundamental social basis of the construction of scientific knowledge:

> Arguments between scientists transform some statements into figments of one's subjective imagination, and others into facts of nature . . . By observing artefact construction, we showed that reality was the *consequence* of the settlement of a dispute rather than its *cause*. (p. 236)

Latour and Woolgar's work emphasises the function and dynamics of scientists' social discourse in the enterprise of constructing new knowledge.

From a cognitive science perspective, Kevin Dunbar's (1995) research documents how social interactions of scientists support their scientific inquiry. Dunbar conducted his research in four leading molecular biology laboratories, in order to maximise the probability that he would be able to witness the construction of significant theoretical advancements. In all four laboratories, the social context proved pivotal to seeing problems in a new way, transcending impasses and achieving conceptual change. Significant advancements were frequently made when a researcher presented a problem and subgroups addressed aspects of the problem, which were then integrated into the larger solution by the presenter and problem-originator. Whereas individual scientists from a given research group frequently would interpret inconsistent findings in terms of error, the groups operating as a whole were much more likely to analyse closely the inconsistency and posit alternative hypotheses. In fact, this dynamic constituted a common route towards conceptual change.

This dynamic is approximated in the participatory structures of socially-distributed expertise delineated by Brown & Campione. In the research team meetings observed by Dunbar and in the small-group work at the core of Brown and Campione's interventions, problems and processes are made explicit, subject to reflection by the proposer and others. The scientists' division of problems into subproblems attacked by different groups, which subsequently reassemble to address the problem as a whole, is echoed in the jigsawing instructional technique. In their use of 'majoring', Brown and Campione exploit the idea of specialisation inherent in any scientific research team. Brown, Campione, Metz and Ash (in press) emphasise that 'all [of these structures] rely on the fact that the participants are trying to understand *deep disciplinary content*' (p. 19).

This work suggests a rich interaction between collaboration, cognition and domain-specific knowledge that we are just beginning to understand and just beginning to tap into for science instruction. We can conceptualise the educational implications of this dynamic in terms of drawing the knowledge base for relatively informed scientific inquiry from a range of students and/or parsing the complex inquiry task across these individuals, as well as the importance of building a classroom-based collaborative community that reflects the values and practices of scientific interpretation and argumentation. The potential benefits are intriguing, including the support of individual's interests, making explicit the emergent knowledge-base and reasoning processes, the opportunity to build scientific discourse, and the possibility of enriching and extending both scientific knowledge and process.

## CONCLUSIONS

The chapter analysed scientific inquiry within reach of young children by examination of the developmental scientific inquiry literature, studies of the impact of domain-specific knowledge on cognition, and the collaborative cognition literature. This analysis indicates that children can design and implement experiments and, in so doing, develop more adequate knowledge and improve their inquiry processes.

The literature indicates that domain-specific knowledge has a large impact on the adequacy of children's scientific inquiry. This finding has both methodological and pedagogical implications. We cannot infer that students are incapable of some form of inquiry without making sure that they have adequate understanding of the domain in which we investigate this capacity. It appears that science instruction that scaffolds children's deep exploration of relatively few domains will support more effectively the development of their scientific inquiry, as it simultaneously opens students' eyes to the nature of scientific knowledge.

In analysis of scientific inquiry within reach of children, we need to consider the possibilities of collaborative cognition. Above and beyond capitalisation on children's cooperative and competitive goal structures, collaborative work can raise the level of the cognitive task that the children can undertake, as it begins to mirror the essential working together and sharing of ideas at the heart of the scientific community.

Although the research literature documents the interaction of children's information-processing with their domain-specific knowledge and collaborative cognition, we need more research investigating the dynamics of this interaction in the context of children's science instruction. The field of children's scientific cognition and instruction needs to investigate how curriculum can best take advantage of these complex interactions, including how to capitalise most effectively on the social context to support and extend children's scientific inquiry, and how to bootstrap children's emergent domain-specific knowledge and scientific inquiry strategies.

## REFERENCES

Bereiter, C. & Scardamelia, M.: 1982, 'From Conversation to Composition: The Role of Instruction in a Developmental Process', in R. Glaser (ed.), *Advances in Instructional Psychology* (Volume 2), Lawrence Erlbaum, Hillsdale, NJ, 1–64.

Brewer, W. & Samarapungavan, A.: 1991, 'Children's Theories Versus Scientific Theories: Differences in Reasoning or Differences in Knowledge?', in R.R. Hoffman & D.S. Palermo (eds.), *Cognition and the Symbolic Processes: Applied and Ecological Perspectives*, Lawrence Erlbaum, Hillsdale, NJ, 209–232.

Brown, A.L.: 1990, 'Domain-Specific Principles Affect Learning and Transfer in Children', *Cognitive Science* 14, 107–133.

Brown, A.L., Ash, D., Rutherford, M., Nakagawa, K., Gordon, A. & Campione, J.C.: 1993, 'Distributed Expertise in the Classroom', in G. Salomon (ed.), *Distributed Cognitions*, Cambridge University Press, New York, 188–228.

Brown, A.L. & Campione, J.C.: 1994, 'Guided Discovery in a Community of Learners', in K. McGilly (ed.), *Classroom Lessons: Integrating Cognitive Theory and Classroom Practice*, MIT Press/Bradford Press, Cambridge, MA, 229–270.

Brown, A.L., Campione, J., Metz, K.E. & Ash, D.: in press, 'The Development of Science Learning Abilities in Children', in A. Burgen & K. Härnqvist (eds.), *Growing up with Science: Developing Early Understanding of Science*, Academia Europaea, Göteborg, Sweden.

Brown, A.L. & Palinscar, A.S.: 1989, 'Guided, Cooperative Learning and Individual Knowledge Acquisition', in L.B. Resnick (ed.), *Knowing, Learning, and Instruction: Essays in Honor of Robert Glaser*, Lawrence Erlbaum, Hillsdale, NJ, 393–451.

Bruner, J.: 1996, *The Culture of Education*, Harvard University Press, Cambridge, MA.

Bullock, M.: 1985, 'Causal Reasoning and Developmental Changes over the Preschool Years', *Human Development* 28, 169–191.

Burbules, N.C. & Linn, M.C.: 1991, 'Science Education and Philosophy of Science: Congruence or Contradiction?', *International Journal of Science Education* 13, 227–241.

Carey, S.: 1985a, 'Are Children Fundamentally Different Kinds of Thinkers Than Adults?', in S. Chipman, J. Segal & R. Glaser (eds.), *Thinking and Learning Skills: Volume 2: Research and Open Questions*, Lawrence Erlbaum, Hillsdale, NJ, 485–517.

Carey, S.: 1985b, *Conceptual Change in Childhood*, MIT Press, Cambridge, MA.

Carey, S., Evans, R., Honda, M., Jay, E. & Unger, C.: 1989, 'An Experiment Is When You Try It and See If It Works: A Study of Grade 7 Students' Understanding of the Construction of Scientific Knowledge', *International Journal of Science Education* 11, 514–529.

Cellérier, G.: 1972, 'Information Processing Tendencies in Recent Experiments in Cognitive Learning – Theoretical Implications', in S. Farham-Diggory (ed.), *Information Processing in Children*, Academic Press, New York, 115–123.

Chi, M.T.H., Feltovich, P.J. & Glaser, R.: 1981, 'Categorization and Representation of Physics Problems by Experts and Novices', *Cognitive Science* 5, 121–152.

Cohen, E.G.: 1994, 'Restructuring the Classroom: Conditions for Productive Small Groups', *Review of Educational Research* 64, 1–35.

Dewey, J.: 1910, 'Science as Subject-Matter and as Method', *Science* 31, 121–127.

Dunbar, K.: 1995, 'How Scientists Really Reason: Scientific Reasoning in Real-World Laboratories', in R.J. Sternberg & J. Davidson (eds.), *Mechanisms of Insight*, MIT Press, Cambridge, MA, 365–395.

Dunbar, K. & Klahr, D.: 1989, 'Developmental Differences in Scientific Discovery Strategies', in D. Klahr & K. Kotovsky (eds.), *Complex Information Processing: The Impact of Herbert A. Simon*, Lawrence Erlbaum, Hillsdale, NJ, 109–144.

Gelman, S.A. & Markman, E.M.: 1986, 'Categories and Induction in Young Children', *Cognition* 23, 183–209.

Glaser, R.: 1984, 'Education and Thinking: The Role of Knowledge', *American Psychologist* 39, 93–104.

Goswami, U. & Brown, A.L.: 1989, 'Melting Chocolate and Melting Snowmen: Analogical Reasoning and Causal Relations', *Cognition* 35, 69–95.

Hancock, C., Kaput, J.J. & Goldsmith, L.T.: 1992, 'Authentic Inquiry with Data: Critical Barriers to Classroom Implementation', *Educational Psychologist* 27, 337–364.

Heath, S.B.: 1991, 'It's About Winning!: The Language of Knowledge in Baseball', in L. Resnick, J.M. Levine & S.D. Teasley (eds.), *Perspectives on Socially-Shared Cognition*, American Psychological Association, Washington, DC, 101–126.

Hutchins, E.: 1991, 'The Social Organization of Distributed Cognition', in L. Resnick, J.M. Levine & S.D. Teasley (eds.), *Perspectives on Socially-Shared Cognition*, American Psychological Association, Washington, DC, 283–307.

Inhelder, B. & Piaget, J.: 1958, *The Growth of Logical Thinking from Childhood to Adolescence* (translated from French by E.A. Lunzer & D. Papert), Basic Books, New York.

Johnson, D., Maruyama, G., Johnson, R., Nelson, D. & Skon, L.: 1981, 'Effects of Cooperative, Competitive, and Individualistic Goal Structures on Achievement: A Meta-Analysis', *Psychological Bulletin* 89, 47–62.

Karmiloff-Smith, A.: 1988, 'The Child is a Theoretician, not an Inductivist', *Mind and Language* 3, 183-195.
Karmiloff-Smith, A. & Inhelder, B.: 1974, 'If You Want to Get Ahead, Get a Theory', *Cognition* 3, 195-212.
Karplus, R.: 1977, 'Science Teaching and the Development of Reasoning', *Journal of Research in Science Teaching* 14, 169-175.
Klahr, D., Fay, A.L. & Dunbar, K.: 1993, 'Heuristics for Scientific Experimentation: A Developmental Study', *Cognitive Psychology* 25, 111-146.
Kuhn, D., Amsel, E. & O'Loughlin, M.: 1988, *The Development of Scientific Thinking Skills*, Academic Press, New York.
Larkin, J.: 1983, 'The Role of Problem Representation in Physics', in D. Gentner & A. Gentner (eds.), *Mental Models*, Lawrence Erlbaum, Hillsdale, NJ, 53-73.
Larkin, J., McDermott, L., Simon, D.P. & Simon, H.A.: 1980, 'Models of Competence in Solving Physics Problems', *Cognitive Science* 4, 317-345.
Latour, B. & Woolgar, S.: 1986, *Laboratory Life: The Social Construction of Scientific Facts*, Princeton University Press, Princeton, NJ.
Linn, M.: 1982, 'Theoretical and Practical Significance of Formal Reasoning', *Journal of Research in Science Education* 19, 743-760.
Metz, K.E.: 1985, 'The Development of Children's Problem Solving in a Gears Task: A Problem Space Perspective', *Cognitive Science* 9, 431-472.
Metz, K.E.: 1993, 'Preschoolers' Developing Knowledge of the Pan Balance: From New Representation to Transformed Problem Solving', *Cognition and Instruction* 11, 31-93.
Metz, K.E.: 1995, 'Re-assessment of Developmental Assumptions in Children's Science Instruction', *Review of Educational Research* 65, 93-127.
Millar, R. & Driver, R.: 1987, 'Beyond Processes', *Studies in Science Education* 14, 33-62.
Parsons, C.: 1960, 'Inhelder and Piaget's The Growth of Logical Thinking II: A Logician's Viewpoint', *British Journal of Psychology* 51, 75-84.
Pea, R.: 1993, 'Learning Scientific Concepts Through Material and Social Activities: Conversational Analysis Meets Conceptual Change', *Educational Psychologist* 28, 265-277.
Piaget, J.: 1928, 'Logique Genetique et Sociologie', *Revue Philosophique de la France et l'Etranger* 3/4, 161-205.
Piaget, J.: 1974, Foreword to Inhelder, B., Sinclair, H. & Bovet, M., *Learning and the Development of Cognition*, Routledge & Kegan Paul, London.
Piaget, J.: 1980: 'The Constructivist Approach', *Cahiers de la Fondation Archives Jean Piaget* 1, 3-7.
Resnick, L., Levine, J.M. & Teasley, S.D. (eds.): 1991, *Perspectives on Socially-Shared Cognition*, American Psychological Association, Washington, DC.
Rosebery, A.S., Warren, B. & Conant, F.R.: 1992, *Appropriating Scientific Discourse: Findings from Language Minority Classrooms* (TERC Working Paper 1-92), TERC (Technical Education Research Center), Cambridge, MA.
Roth, W.-M. & Bowen, G.M.: 1994. 'Mathematization of Experience in a Grade 8 Open-Inquiry Environment: An Introduction to the Representational Practices of Science', *Journal of Research in Science Teaching* 31, 293-318.
Rutherford, F.J. & Ahlgren, A. (eds.): 1990, *Science for All Americans*, Oxford University Press, New York.
Samarapungavan, A.: 1992, 'Children's Judgements in Theory Choice Tasks: Scientific Rationality in Childhood', *Cognition* 45, 1-32.
Schauble, L.: 1990, 'Belief Revision in Children: The Role of Prior Knowledge and Strategies for Generating Knowledge', *Journal of Experimental Psychology* 49, 31-57.
Schauble, L. & Glaser, R.: 1990, 'Scientific Thinking in Children and Adults', *Human Development* 21, 9-27.
Schauble, L., Klopfer, L.E. & Raghavan, K.: 1991, 'Students' Transition from an Engineering Model to a Science Model of Experimentation', *Journal of Research in Science Teaching* 28, 859-882.
Simon, H.A.: 1992, 'Scientific Discovery as Problem Solving', *International Studies in the Philosophy of Science* 6, 3-14.
Sodian, B., Zaitchik, D. & Carey, S.: 1991, 'Young Children's Differentiation of Hypothetical Beliefs from Evidence', *Child Development* 62, 753-766.
Tobin, K., Tippins, D.J. & Gallard, A.J.: 1994, 'Research on Instructional Strategies for Teaching Science', in D.L. Gabel (ed.), *Handbook of Research on Science Teaching and Learning*, Macmillan, New York, 43-93.

Tschirgi, J.E.: 1980, 'Sensible Reasoning: A Hypothesis about Hypothesis', *Child Development* 51, 1-10.
Vosniadou, S. & Brewer, W. F.: 1992, 'Mental Models of the Earth: A Study of Conceptual Change in Childhood', *Cognitive Psychology* 24, 535-585.
Vygotsky, L.S.: 1978, *Mind in Society: The Development of Higher Psychological Processes* (translated and edited by M. Cole, V. John-Steiner, S. Scriber & E. Souberman), Harvard University Press, Cambridge, MA.
White, B.Y.: 1993, 'ThinkerTools: Causal Models, Conceptual Change, and Science Education', *Cognition and Instruction* 10, 1-100.

# 1.7 Theories of Knowledge Acquisition

CLARK A. CHINN
*Rutgers University, New Brunswick, USA*

WILLIAM F. BREWER
*University of Illinois at Urbana-Champaign, USA*

This chapter presents a framework for understanding and evaluating theories of knowledge acquisition. The framework consists of eight questions that any theory of knowledge acquisition should address. The framework can assist researchers and teachers by: pointing out where many current theories of knowledge acquisition (including our own) are incomplete; providing a conceptual structure for comparing different theories and clarifying where theorists agree and where they disagree; directing work which contrasts knowledge acquisition in different scientific domains; and providing a guide for developing instruction. The framework applies to social constructivist as well as cognitive theories of knowledge acquisition; the same eight questions can be asked both of theory change within whole communities of scientists or learners and of knowledge change in individuals. The framework does not, however, address emotional issues in knowledge change.

The chapter is organised according to the eight questions related to knowledge acquisition:

- What is the nature of the knowledge change?
- Are there intermediate stages in knowledge change?
- What initiates knowledge change?
- What factors influence knowledge change?
- What is the fate of the old knowledge and the new information after knowledge change occurs?
- What is the relationship between belief and knowledge?
- What factors influence belief change?
- What changes in meta-awareness occur during knowledge change?

## WHAT IS THE NATURE OF THE KNOWLEDGE CHANGE?

The first question in the framework addresses the differences between snapshots of what people know at two different times. Answers to this question either focus

---

Chapter Consultant: Silvia Caravita (National Research Council, Rome, Italy)

on large-scale global changes or local changes, typically occurring over a short period of time. *Global changes* are discussed in the next section, and this is followed by consideration of *local changes*.

*Global Changes*

Researchers concerned with global knowledge changes have been concerned with how whole systems of new knowledge are related to systems of old knowledge. The learner typically is viewed as ending up with structured knowledge of some kind, such as knowledge about dynamics or photosynthesis. Proposed ways in which this new knowledge is related to the old knowledge are considered below.

No Knowledge to Structured Knowledge

Lawson (1988) argues that, when students learn about cells, they have no prior knowledge about cells. Therefore, the learning process involves a shift from no knowledge at all to structured knowledge. The extreme version of this position is the classical British empiricist view that the mind starts as a *tabula rasa*.

Fragmented Knowledge to Structured Knowledge

Knowledge change can also be characterised as a shift from fragmented knowledge to more structured knowledge. According to diSessa (1993), naive learners in the domain of physics start out with a multitude of intuitions about the physical world called 'p-prims'. Common p-prims include the idea that greater force produces greater effects and the notion of equilibrium. Expert knowledge does not override these intuitions. Instead, as students learn, their fragmented knowledge is gradually refined into a structured whole.

Simple Core Knowledge to Elaborated Knowledge Built Around This Core

Spelke (1991) has argued that, in the absence of special instruction in physics, knowledge about the physical world is simply a matter of elaborating but not changing core conceptions. For instance, infants already have the core conceptions that objects are solid and that objects have continuous existence even when they cannot be seen. These conceptions remain at the core of later physics knowledge, which elaborates and adds to the core conceptions without changing them.

Structured Knowledge to Conceptually-Consistent Structured Knowledge

When students who believe that dinosaurs were warm-blooded are presented with contrary evidence, they sometimes change their minds and say that dinosaurs were cold-blooded (Brewer & Chinn 1994). This change does not seem to involve any changes in explanatory concepts (i.e., learners don't change their ideas about what it means to be cold-blooded or warm-blooded). There is a change of theory that does not involve any changes to conceptions.

Structured Knowledge to Conceptually Incommensurate Structured Knowledge

T. Kuhn (1962) argues that major theory changes in science, such as the Copernican and chemical revolutions, involve a shift between incommensurate theories (i.e., theories that employ fundamentally different, incomparable conceptions). Most social constructivists who investigate theory change in scientific communities have adopted this position either explicitly or implicitly (e.g., Knorr-Cetina 1981).

Kuhn's notion also has been applied to individual cognition. Carey (1991) argues that, when children learn about weight and mass, there is a fundamental change in conceptions such that the conceptual system completely changes. For young children, some forms of matter can be weightless, whereas for adults, matter is coextensive with weight. Carey argued that the concepts of weight for children and adults are embedded in qualitatively different theoretical systems.

Researchers frequently disagree about how to characterise particular instances of knowledge change. For instance, Vosniadou and Brewer (1987) argue that children's evolving understanding of the earth's shape (from flat to spherical) involves a change between conceptually inconsistent systems of knowledge. Larreamendy-Joerns and Chi (1994), by contrast, believe that this shift does not involve a fundamental change in conceptions.

The issue of the nature of knowledge change is crucial for teachers. A teacher who believes that learning about chemical reactions is a matter of elaborating core conceptions probably would prepare lessons differently from one who believes that learning about reactions involves a shift from one system of thought to an incommensurable one.

*Local Changes*

Researchers have characterised local knowledge changes into several types (Klahr 1984). *Generalisation* consists of inducing an abstract principle from instances or applying a principle to a greater range of instances. *Specialisation* refers to creating two principles or conceptions where only one existed before. Students who know only that stars are like the sun specialise their knowledge when they learn that there are several different types of stars, including white dwarfs and red giants. Other local changes include *addition*, in which new knowledge is simply added to old knowledge, and *deletion*, in which old knowledge is deleted or suppressed. In Klahr and Siegler's (1978) model of how children learn about the balance scale, the difference between old and new knowledge appears to be that children delete an old rule and add a new rule. Exchanging one piece of knowledge for another (Hewson 1981) also could be an example of simultaneous addition and deletion.

## ARE THERE INTERMEDIATE STAGES IN KNOWLEDGE CHANGE?

Given that global changes in knowledge occur, a second question about the knowledge acquisition process is whether there are any intermediate stages of knowledge. Several positions on this issue are discussed in this section.

### No Intermediate Stages

Gestalt psychologists portray knowledge change as happening all at once. At one moment, the learner holds one knowledge structure and, an instant later, a complete new structure snaps into place. T. Kuhn's (1962) discussions of knowledge change in scientists often are interpreted as being of this form.

### No Stable Knowledge Structures

Another position is that there is an intermediate stage of confusion. In other words, there is a period when students are learning a new theory, but there is no coherence or systematicity in their intermediate knowledge. In some of the work on the development of conservation, researchers identify a transition stage (e.g., Perret-Clermont 1980). It is possible that some students in the transitional phase are in a state of temporary confusion.

### One or Few Intermediate Knowledge Structures

A third position on the issue of intermediate states is that students pass through one or a few well-defined intermediate knowledge structures as they move from an initial knowledge state to an expert theory. The intermediate stages are often incorrect from the scientists' point of view rather than being simpler versions of the scientist's theory. Vosniadou and Brewer (1992) have argued that children learning about the earth's shape begin believing that the earth is flat and end up believing that the earth is round. In between, many students adopt at least one intermediate model, such as the belief that the earth is round like a coin or the belief that there are two earths, the flat one on which we live and the round one seen in pictures.

*Many Intermediate Stages*

A final position is that students progress through numerous intermediate knowledge structures as they move from an initial naive theory to a final theory (Chi 1992). In this view, knowledge is altered in many small steps from the initial theory to the final theory. For researchers with this view, there is an important additional question. What is the step-by-step sequence of local knowledge changes that leads to the final knowledge? For instance, are long sequences of local changes, such as generalisation, specialisation, addition and deletion, sufficient to produce global knowledge changes, or are other kinds of local knowledge change needed? Such questions have not yet been addressed.

For teachers, the issue of whether there are intermediate knowledge changes can have important implications. For instance, a teacher who believes that intermediate knowledge states are unnecessary or undesirable will treat students' misunderstandings differently from a teacher who believes that intermediate knowledge states are a natural aspect of knowledge acquisition.

## WHAT INITIATES KNOWLEDGE CHANGE?

Empirically, researchers answer the third question about what initiates knowledge change by trying to determine what events reliably precede instances of knowledge change. Researchers infer that knowledge change is triggered by events involving *new data, new conceptions, reflection* and *social pressures*. These four classes of events provide the organisation of discussion in this section.

*New Data*

The term *data* is used in a broad sense, which includes results of experiments, observations of the world and instances or examples. In procedural learning, data correspond to problems that the learner is trying to solve. A number of researchers have emphasised the role of data in knowledge change. Data can either be presented to the learner by another person (such as a teacher or a scientist who holds a rival theory) or be discovered by the unaided learner. The relationship between the new data and current knowledge is crucial. New incoming data can have one of five relationships to old knowledge, as discussed below.

Inconsistent (Anomalous) Data

Inconsistent data are data that contradict one's current beliefs. The procedural counterpart is a problem for which current knowledge yields an incorrect solution. Many science educators and psychologists have emphasised the role of inconsistent data in knowledge change (Chinn & Brewer 1993). Posner, Strike, Hewson and Gertzog (1982) argued that inconsistent data lead students to be dissatisfied with their current theories, which paves the way for theory change. Sometimes researchers have proposed that a single piece of data leads to knowledge change, as when a single instance of a black swan prompts revision of the belief that all swans are white. More commonly, researchers have urged the use of multiple instances of anomalous data to promote changes in theory (Gorsky & Finegold 1994).

Some social constructivists, such as Collins (1985), deny that anomalous data can trigger theory change. Because anomalous data always can be explained away by an opponent, Collins argues that data cannot be the real cause of theory changes.

Unexplainable Data

Unexplainable data are those that cannot be explained using current knowledge. Although unexplainable data do not conflict with current knowledge, current knowledge is insufficient to generate a complete explanation. For instance, many people cannot produce any explanation – right or wrong – to explain the phases of the moon. In procedural learning, the counterpart is a problem that simply cannot be solved, either correctly or incorrectly. Such problems are said to produce an impasse. VanLehn, Jones and Chi (1992) argue that impasses are a driving force in knowledge acquisition.

Awkward Data

Awkward data are data that can be explained using current knowledge, but only in an inelegant or clumsy manner. The equivalent in procedural learning is a problem that can be solved but only in an inefficient manner. For instance, counting on one's fingers as a strategy for adding numbers is an inefficient solution strategy for problems such as 68 + 3 = ?, even though it can yield the correct answer. Siegler and Jenkins (1989) propose that such inefficiencies initiate the development of new counting strategies.

Consistent Data

Consistent data can be explained using current knowledge. An explanation for consistent data does not pre-exist in current knowledge, but can be constructed using current knowledge. For instance, students learning electronics might be able to work out an explanation for a new type of circuit that they had never encountered before, and then they can add the new explanation to current knowledge. The procedural counterpart is a novel problem that can be solved, but for which a solution procedure did not already exist.

Matching Data

The simplest form of knowledge acquisition occurs when data exactly match prior knowledge. When a person who understands simple circuits sees a simple circuit with a light bulb, the person has encountered a new instance of old knowledge. The new instance can be stored in memory, but the explanatory knowledge does not change. Schema researchers have referred to this process as 'schema instantiation' (Rumelhart & Ortony 1977).

*New Conceptions*

The term 'new conception' refers to any sort of externally provided abstract knowledge, such as a rule or a theory, that counts as an explanation for some data. New conceptions are typically presented by another person, such as a teacher, a peer or a textbook author. New conceptions, which can be accompanied by new data, can be of two broad types. First, new conceptions can contradict current knowledge. Hashweh (1986) argued that conflicts between prior conceptions and scientific conceptions (not just between prior conceptions and new data) can lead students to change their theories. Second, new conceptions can be consistent with current knowledge. According to Lawson (1988), children learn about cells by being presented with new information that does not contradict any preconceptions but, instead, is consistent with prior knowledge.

*Reflection*

In some cases, reflection upon current knowledge can initiate knowledge change. This occurs because humans do not have automatic access to all information stored in long-term memory and do not have automatic processes for resolving all potential inconsistencies. In this way, humans are different from many current artificial intelligence systems. Knowledge change through reflection can be triggered by internal inconsistency among current theoretical conceptions (Levin, Siegler & Druyan 1990), inconsistency between current theories and known data (Dreyfus, Jungwirth & Eliovitch 1990), excessive complexity of current conceptions, or the inability of the current theory to explain known data.

*Social Pressures*

Many sociologists of science have argued that social pressures initiate theory change in science. For instance, Forman (1971) contended that the idea of indeterminacy in quantum mechanics arose not because of any encounter with data or with other scientific theories, but because scientists desired to bring the content of their theories into line with mainstream German thought in other fields, which stressed the role of mystery and nondeterminism.

As teachers try to help students learn widely-accepted scientific explanations, they need to identify reliable ways to initiate the process of knowledge change. Thus, the question of what initiates knowledge change is one of the basic research questions with the most direct impact on instruction.

## WHAT FACTORS INFLUENCE KNOWLEDGE CHANGE?

Once knowledge change is triggered, a fourth question addresses the factors that influence the way in which the changes unfold. Most research on knowledge acquisition has examined three sets of factors, discussed in the subsections below, that facilitate or impede knowledge change: *prior knowledge*; *characteristics of input information* (if any); and *processing strategies*.

### Prior Knowledge

Most researchers agree that prior knowledge has powerful effects on the learning process. Below we discuss just a few important positions concerning aspects of prior knowledge.

Entrenchment of Prior Conceptions

A conception is entrenched to the extent that it has strong evidentiary support, participates in a broad range of explanations and satisfies strong personal or social goals (Chinn & Brewer 1993). According to Chinn and Brewer (1993), entrenchment of prior conceptions makes theory change in response to anomalous data and inconsistent theories much less likely. Krems and Johnson (1995), by contrast, argue that entrenchment does not influence the response to anomalous data. Social constructivists who investigate theory change among scientists have argued that social goals are the only source of entrenchment. For instance, MacKenzie and Barnes (1979) contend that early 20th century English scientists defended the theory that evolution proceeded by small continuous variations, not because of biological evidence, but because this view dovetailed with their social interest in eugenic engineering.

Quality of Background Knowledge

Background knowledge refers to knowledge that is not part of a particular theory, but that can come into play when a person learns a theory. For instance, knowledge about billiard balls is not a part of a learner's prior knowledge about the nature of air, but it can be used to comprehend analogies that compare gas molecules to billiard balls. Chinn and Brewer (1993) propose that background knowledge can influence people's response to anomalous data. The quality of background knowledge also can influence learning from analogies.

*Learners' Naive Philosophy of Science*

Learners' conception of what science is and what scientists do can influence how they learn. For instance, the belief that science is an enterprise of accumulating facts (rather than an enterprise of constructing theories to account for data) can inhibit students' learning of new theories (Songer & Linn 1991).

*Characteristics of Input Information*

Input information can include new data, new rules, new explanatory models or new theories. Many complex issues arise regarding how the characteristics of information interact with prior knowledge to influence learning. Here we mention only a few of these issues.

One important set of issues revolves around how descriptions of theories should be presented to students. Chinn (1995) proposes that highly explicit presentations of theories foster understanding of the new theories. By contrast, McNamara, Kintsch, Songer and Kintsch (1996) contend that high-knowledge students learn more from implicit texts than from explicit texts. Some proponents of discovery learning have argued that students learn best by discovering ideas themselves through exploration without being presented with any exposition (e.g., Bruner 1961).

A second issue arises when learners learn from examples or instances. One common position is that numerous examples are needed to induce regularities. By contrast, Ahn, Brewer and Mooney (1992) contend that high-knowledge learners can learn new ideas from a single example.

A third important issue concerns the effect of analogies in learning. One widely held position is that analogies are an effective means of teaching scientific ideas (Suzuki 1994). However, Donnelly and McDaniel (1993) present evidence that analogies have at best marginal effects, and Spiro, Feltovich, Coulson and Anderson (1989) argue that analogies sometimes promote misconceptions.

*Processing Strategies*

The type of processing strategy employed by the learner can have strong influences on the course of knowledge acquisition. The choice of strategies is determined both by the goals that learners set for themselves when they are learning and their knowledge about strategies that are effective for achieving these goals.

In science and mathematics education, there has been great interest in self-explanation as a means of promoting knowledge change. Chi, Bassok, Lewis, Reimann and Glaser (1989) report that the difference between successful learners and unsuccessful learners of physics problems was that the successful learners worked hard at explaining the example problems to themselves. Ferguson-Hessler and de Jong (1990) and other researchers also have reported that such self-explanations facilitate learning. However, not all agree that self-explanation or reflection is always beneficial. Schooler, Ohlsson and Brooks (1993) report that, in some instances, discovery of new ideas can be undermined by self-explanation.

## WHAT IS THE FATE OF THE OLD KNOWLEDGE AND THE NEW INFORMATION AFTER KNOWLEDGE CHANGE OCCURS?

The fifth question about knowledge acquisition is considered in terms of a scenario in which a person holds a structured or semi-structured body of knowledge A, and then encounters a new knowledge structure B. After any knowledge change occurs, what is the fate of A and B? Several of the possible positions on this issue are as follows:

- *B replaces A, with A being forgotten or ignored.* Hewson (1981) proposes a process called 'conceptual exchange', in which one scientific idea is abandoned and a new one replaces it.
- *A is reinterpreted within the framework of B.* According to Vosniadou and Brewer (1992), children who come to believe that the earth is round must give up a presupposition that the earth is flat, which is based on perceptual evidence. Presumably, they must reinterpret the old concept of the earth's flatness so that it can mean the approximate flatness of the surface of an extremely large curved surface.
- *B is reinterpreted within the framework of A.* This option appears to be closest to Piaget's notion of assimilation. Sivaramakrishnan and Patel (1990) reported that Indian mothers who learned about the modern biomedical conceptions of health and disease reinterpreted them within their initial knowledge framework of Indian traditional medicine. The mothers incorporated some of the causal ideas of modern medicine without changing their ideas about the proper treatments as prescribed by traditional medicine.
- *A is incorporated into B.* The difference between reinterpreting A within B and incorporating A into B is whether A is changed in the process. A is changed during reinterpretation but not during incorporation. Smith, diSessa and Roschelle (1993/1994) propose that old conceptions in physics are incorporated into a more refined, structured theory. The conditions of application of the principles in A change but at least some principles remain the same. In this view, principles of A do not change but are refined into more systematic knowledge structures.
- *A and B are compartmentalised.* Brown and Clement (1992) argue that, when children learn about physics, they distinguish between laws of physics that apply

to outer space and different laws of physics that apply to their surroundings on earth. The common observation that many students do not apply scientific ideas learned in school to their everyday experiences is another example of compartmentalisation.

A teacher's position on the issue of the fate of the old and new knowledge has important implications for instruction. If one believes, for instance, that A is simply replaced by B, then one need not worry about A interfering with later learning. On the other hand, if one believes that A continues to exist and influence the understanding of B, one must design instruction that tries to reduce the strength of A gradually.

## WHAT IS THE RELATIONSHIP BETWEEN BELIEF AND KNOWLEDGE?

The sixth question concerns the relationship between knowledge and belief. Researchers (including ourselves) sometimes have confounded belief and knowledge when investigating knowledge acquisition. In contrast to philosophers who view knowledge as justified true belief, we use the term in a psychological sense. A person can have knowledge about Ptolemaic cosmology without believing it. In this section, several possible relationships between knowledge and belief are considered in terms of the situation of a student who has a belief in Theory A and who is provided with information about Theory B.

### Old Knowledge Always Being Believed

One obvious possibility is that belief never changes. Students continue to believe A and reject B. Champagne, Gunstone and Klopfer (1985) note that some students who are explicitly told that objects of different weights fall at the same rate reject this idea and retain the idea that heavier objects fall faster.

### Belief Change Occurring Immediately as Knowledge Begins to Change

Many theorists have treated belief and knowledge as inseparable; students will not construct new knowledge unless they believe it. For instance, students learning about relativity might add to knowledge only those ideas (or reinterpreted ideas) that they believe. This is a common but untested assumption in much research on conceptual change.

### Knowledge Change Followed by Belief Change

Some researchers suggest that students can construct complete theories that they do not believe and switch belief at a later time. For example, a child can have an excellent understanding of the spherical model of the earth's shape but not believe it. If this child later changes belief, it will be after he or she already has constructed an accurate understanding of the new theory.

*Both Systems of Knowledge Being Believed*

Other researchers have argued that the old knowledge always continues to be believed, but in different contexts (Brown & Clement 1992). This corresponds to compartmentalisation, as described in the previous section.

*Period of Withheld Belief*

It is also possible that students pass through a stage in which they don't know what to believe and believe neither A nor B.

*Belief Changes Without Knowledge*

Finally, it is possible to believe a theory one does not understand. Presumably most nonscientists who believe that quantum mechanics is a valid theory do not understand it. Linn and Songer (1991) report that students learning about heat and temperature often seem to believe the theory which they were learning without understanding it.

Different positions on how belief change and knowledge change are related can lead to fundamentally different approaches to instruction. For instance, a teacher who assumes that belief change occurs simultaneously with knowledge change would focus on dislodging belief in the old theory before teaching the new theory. By contrast, a teacher who assumes that knowledge change precedes belief change might encourage learners to try to understand a new theory even if they do not yet believe it. Instead of trying to refute the old theory, this teacher might assume that beliefs will naturally change later on.

## WHAT FACTORS INFLUENCE BELIEF CHANGE?

In many instances of learning, learners at least briefly consider two or more possible knowledge structures as possible explanations for some data. The knowledge structures could be whole theories or rival hypotheses within a theory. The seventh question in the framework addresses factors that influence the learner's decision about which structures to accept as valid. Several answers that have been given to this question are discussed below.

*Structures Conforming to Characteristics of Good Theories*

The classical position is that individuals believe those knowledge structures conforming to the characteristics of good theories proposed by philosophers of science. For instance, T. Kuhn (1977) states that theories are preferred to the extent that they are accurate, consistent, simple, fruitful and capable of explaining a broad range of data. Samarapungavan (1992) presents evidence showing that children, as well as scientists, prefer theories that meet these criteria. Many sociologists of science, by contrast, deny that such criteria can account for theory choice; because scientists interpret these criteria flexibly to suit different occasions, the criteria cannot constrain theory choice (Mulkay & Gilbert 1991).

*Structures Provided by a High Status Individual*

An individual might adopt a knowledge structure because it is advocated by a high-status person or institution. Scientists appear willing to defer to experimental conclusions of scientists regarded as rigorous, proficient experimenters (Woodward 1989). Science students might believe scientific theories simply because respected adults, such as teachers, parents or textbook authors, say that the theories are correct.

*Structures Leading to Desired Social Goals*

Many sociologists of science have advocated the view that the choice between scientific theories has little or nothing to do with constraints imposed by the natural world (Latour & Woolgar 1986). Rather, people choose theories that help them achieve greater status or some other social goal.

*Useful Structures*

Some researchers take the pragmatist view that scientific theories are not true or false, but rather more or less useful. If a knowledge structure is capable of helping the individual solve valued problems, then the knowledge will be adopted.

The relevance of the issue of belief change to teachers is obvious. If teachers hope to foster belief in scientific theories, teachers must know the basis on which students choose what to believe.

## WHAT CHANGES IN META-AWARENESS OCCUR DURING KNOWLEDGE CHANGE?

The eighth question in the framework is directed at changes in the learner's awareness. Many specific positions can be taken on how meta-awareness changes; three broad possibilities are discussed below.

## Continuing Nonawareness

One position is that learners remain unaware of their knowledge throughout. D. Kuhn (1989) argues that most children and some adults are not able to reflect on their theories or to reason explicitly about the relationships between theory and data. Instead, beliefs about the world remain largely at an implicit level, even as these beliefs change.

## Continuing Awareness

Another position is that learners are aware of their knowledge at the beginning and end of the knowledge change process. This is the picture of the self-reflective scientist who is completely aware of two theories and deliberately chooses to adopt a new theory.

## Change from Nonawareness to Awareness

A third position is that there is a change from nonawareness to awareness. Karmiloff-Smith (1992) advocates this position as a general account of development. As children learn, they become increasingly aware of the principles that govern their knowledge and become increasingly able to reflect on their knowledge.

## SUMMARY AND CONCLUSIONS

This chapter has considered eight questions that should be addressed by theories of knowledge acquisition. Rather than advocating any particular positions, we have pointed out the diversity of positions that a variety of theorists have taken on each issue. The framework has several potential benefits for theorists, researchers and teachers.

First, the framework helps to point out what a complete theory of knowledge acquisition should include. The framework suggests that there are few if any comprehensive theories of knowledge acquisition at present. Rather, most current theories are fragments of theories that address only one, two or three of the issues.

Second, the framework helps orient researchers to unresolved issues. On some issues, there is little systematic research in any scientific domain. On other issues, there is research in only one or a few domains. Because the answers to the eight questions can differ across scientific domains such as mechanics, genetics, cell theory and states of matter, there is a need for empirical research to clarify when the various positions on each issue hold for various domains of scientific knowledge.

Third, finding answers to these questions is not merely an academic exercise, but is also necessary for developing effective instruction in each scientific domain.

Science teachers' beliefs about how knowledge is acquired can have important implications for the way in which science content is presented and how learning is organised. Researchers and teachers who consider the eight questions in this framework will be in a better position to diagnose students' difficulties in learning and design curricula to take into account the manner in which changes in learning occur. For example, the ways in which teachers use anomalous data in the classroom would vary according to the teacher's positions on the eight questions.

REFERENCES

Ahn, W., Brewer, W.F. & Mooney, R.J.: 1992, 'Schema Acquisition From a Single Example', *Journal of Experimental Psychology: Learning, Memory, and Cognition* 18, 391–412.

Brewer, W.F. & Chinn, C.A.: 1994, 'The Theory-Ladenness of Data: An Experimental Demonstration', in A. Ram & K. Eiselt (eds.), *Proceedings of the Sixteenth Annual Conference of the Cognitive Science Society*, Lawrence Erlbaum, Hillsdale, NJ, 61–65.

Brown, D.E. & Clement, J.: 1992, 'Classroom Teaching Experiments in Mechanics', in R. Duit, F. Goldberg & H. Niedderer (eds.), *Research in Physics Learning: Theoretical Issues and Empirical Studies*, Institute for Science Education at the University of Kiel, Kiel, Germany, 380–397.

Bruner, J.S.: 1961, 'The Act of Discovery', *Harvard Educational Review* 31, 21–32.

Carey, S.: 1991, 'Knowledge Acquisition: Enrichment or Conceptual Change?', in S. Carey & R. Gelman (eds.), *The Epigenesis of Mind: Essays on Biology and Cognition*, Lawrence Erlbaum, Hillsdale, NJ, 257–291.

Champagne, A.B., Gunstone, R.F. & Klopfer, L.E.: 1985, 'Instructional Consequences of Students' Knowledge About Physical Phenomena', in L.H.T. West & A.L. Pines (eds.), *Cognitive Structure and Conceptual Change*, Academic Press, Orlando, FL, 61–90.

Chi, M.T.H.: 1992, 'Conceptual Change Within and Across Ontological Categories: Implications for Learning and Discovery in Science', in R. Giere (ed.), *Minnesota Studies in the Philosophy of Science: Vol. XV. Cognitive Models of Science*, University of Minnesota Press, Minneapolis, MN, 129–186.

Chi, M.T.H., Bassok, M., Lewis, M.W., Reimann, P. & Glaser, R.: 1989, 'Self-Explanations: How Students Study and Use Examples in Learning to Solve Problems', *Cognitive Science* 13, 145–182.

Chinn, C.A.: 1995, *Constructing Scientific Explanations from Text: A Theory with Implications for Conceptual Change* (Technical Report No. 626), Center for the Study of Reading, Champaign, IL.

Chinn, C.A. & Brewer, W.F.: 1993, 'The Role of Anomalous Data in Knowledge Acquisition: A Theoretical Framework and Implications for Science Instruction', *Review of Educational Research* 63, 1–49.

Collins, H.M.: 1985, Changing Order: Replication and Induction in Scientific Practice, Sage, London.

diSessa, A.A.: 1993, 'Toward an Epistemology of Physics', *Cognition and Instruction* 10, 105–225.

Donnelly, C.M. & McDaniel, M.A.: 1993, 'Use of Analogy in Learning Scientific Concepts', *Journal of Experimental Psychology: Learning, Memory, and Cognition* 19, 975–987.

Dreyfus, A., Jungwirth, E. & Eliovitch, R.: 1990, 'Applying the "Cognitive Conflict" Strategy for Conceptual Change – Some Implications, Difficulties, and Problems', *Science Education* 74, 555–569.

Ferguson-Hessler, M.G.M. & de Jong, T.: 1990, 'Studying Physics Texts: Differences in Study Processes Between Good and Poor Performers', *Cognition and Instruction* 7, 41–54.

Forman, P.: 1971, 'Weimar Culture, Causality, and Quantum Theory, 1918–1927: Adaptation by German Physicists and Mathematicians to a Hostile Intellectual Environment', in R. McCormmach (ed.), *Historical Studies in the Physical Sciences* (Volume 3), University of Pennsylvania Press, Philadelphia, PA, 1–115.

Gorsky, P. & Finegold, M.: 1994, 'The Role of Anomaly and of Cognitive Dissonance in Restructuring Students' Concepts of Force', *Instructional Science* 22, 75–90.

Hashweh, M.Z.: 1986, 'Toward an Explanation of Conceptual Change', *European Journal of Science Education* 8, 229–249.

Hewson, P.W.: 1981, 'A Conceptual Change Approach to Learning Science', *European Journal of Science Education* 3, 383–396.

Karmiloff-Smith, A.: 1992, *Beyond Modularity*, MIT Press, Cambridge, MA.
Klahr, D.: 1984, 'Transition Processes in Quantitative Development', in R.J. Sternberg (ed.), *Mechanisms of Cognitive Development*, Freeman, New York, 102–139.
Klahr, D. & Siegler, R.S.: 1978, 'The Representation of Children's Knowledge', in H.W. Reese & L.P. Lipsitt (eds.), *Advances in Child Development and Behavior* (Volume 12), Academic Press, New York, 61–116.
Knorr-Cetina, K.D.: 1981, *The Manufacture of Knowledge: An Essay on the Constructivist and Contextual Nature of Science*, Pergamon, Oxford, UK.
Krems, J. & Johnson, T.R.: 1995, 'Integration of Anomalous Data in Multicausal Explanations', in J.D. Moore & J.F. Lehman (eds.), *Proceedings of the Seventeenth Annual Conference of the Cognitive Science Society*, Lawrence Erlbaum, Hillsdale, NJ, 277–282.
Kuhn, D.: 1989, 'Children and Adults as Intuitive Scientists', *Psychological Review* 96, 674–689.
Kuhn, T.S.: 1962, *The Structure of Scientific Revolutions*, University of Chicago Press, Chicago, IL.
Kuhn, T.S.: 1977, *The Essential Tension: Selected Studies in Scientific Tradition and Change*, University of Chicago Press, Chicago, IL.
Larreamendy-Joerns, J. & Chi, M.T.H.: 1994, 'Commentary', *Human Development* 37, 246–256.
Latour, B. & Woolgar, S.: 1986, *Laboratory Life: The Construction of Scientific Facts* (second edition), Princeton University Press, Princeton, NJ.
Lawson, A.E.: 1988, 'The Acquisition of Biological Knowledge During Childhood: Cognitive Conflict or Tabula Rasa?', *Journal of Research in Science Teaching* 25, 185–199.
Levin, I., Siegler, R.S. & Druyan, S.: 1990, 'Misconceptions About Motion: Development and Training Effects', *Child Development* 61, 1544–1557.
Linn, M.C. & Songer, N.B.: 1991, 'Teaching Thermodynamics to Middle School Students: What are Appropriate Cognitive Demands?', *Journal of Research in Science Teaching* 28, 885–918.
MacKenzie, D. & Barnes, B.: 1979, 'Scientific Judgment: The Biometry-Mendelism Controversy', in B. Barnes & S. Shapin (eds.), *Natural Order: Historical Studies of Scientific Culture*, Sage, Beverly Hills, CA, 191–210.
McNamara, D.S., Kintsch, E., Songer, N.B. & Kintsch, W.: 1996, 'Are Good Texts Always Better? Interactions of Text Coherence, Background Knowledge, and Levels of Understanding in Learning from Text', *Cognition and Instruction* 14, 1–43.
Mulkay, M. & Gilbert, G.N.: 1991, 'Theory Change', in M. Mulkay (ed.), *Sociology of Science: A Sociological Pilgrimage*, Indiana University Press, Bloomington, IN, 131–153.
Perret-Clermont, A.N.: 1980, *Social Interaction and Cognitive Development in Children*, Academic Press, London.
Posner, G.J., Strike, K.A., Hewson, P.W. & Gertzog, W.A.: 1982, 'Accommodation of a Scientific Conception: Toward a Theory of Conceptual Change', *Science Education* 66, 211–227.
Rumelhart, D.E. & Ortony, A.: 1977, 'The Representation of Knowledge in Memory', in R.C. Anderson, R.J. Spiro & W.E. Montague (eds.), *Schooling and the Acquisition of Knowledge*, Lawrence Erlbaum, Hillsdale, NJ, 99–135.
Samarapungavan, A.: 1992, 'Children's Judgments in Theory Choice Tasks: Scientific Rationality in Childhood', *Cognition* 45, 1–32.
Schooler, J.W., Ohlsson, S. & Brooks, K.: 1993, 'Thoughts Beyond Words: When Language Overshadows Insight', *Journal of Experimental Psychology: General* 122, 166–183.
Siegler, R.S. & Jenkins, E.: 1989, *How Children Discover New Strategies*, Lawrence Erlbaum, Hillsdale, NJ.
Sivaramakrishnan, M. & Patel, V.L.: 1990, 'Explanations of Nutritional Concepts: Role of Cultural and Biomedical Theories', in *Proceedings of the Twelfth Annual Conference of the Cognitive Science Society*, Lawrence Erlbaum, Hillsdale, NJ, 931–938.
Smith, J.P., III, diSessa, A.A. & Roschelle, J.: 1993/1994, 'Misconceptions Reconceived: A Constructivist Analysis of Knowledge in Transition', *The Journal of the Learning Sciences* 3, 115–163.
Songer, N.B. & Linn, M.C.: 1991, 'How Do Students' Views of Science Influence Knowledge Integration?', *Journal of Research in Science Teaching* 28, 761–784.
Spelke, E.: 1991, 'Physical Knowledge in Infancy: Reflection on Piaget's Theory', in S. Carey & R. Gelman (eds.), *The Epigenesis of Mind: Essays on Biology and Cognition*, Lawrence Erlbaum, Hillsdale, NJ, 133–169.
Spiro, R.J., Feltovich, P.J., Coulson, R.L. & Anderson, D.K.: 1989, 'Multiple Analogies for Complex Concepts: Antidotes for Analogy-Induced Misconception in Advanced Knowledge Acquisition', in S. Vosniadou & A. Ortony (eds.), *Similarity and Analogical Reasoning*, Cambridge University Press, Cambridge, 498–531.

Suzuki, H.: 1994, 'The Centrality of Analogy in Knowledge Acquisition in Instructional Contexts', *Human Development* 37, 207–219.
VanLehn, K., Jones, R.M. & Chi, M.T.H.: 1992, 'A Model of the Self-Explanation Effect', *The Journal of the Learning Sciences* 2, 1–59.
Vosniadou, S. & Brewer, W.F.: 1987, 'Theories of Knowledge Restructuring in Development', *Review of Educational Research* 57, 51–67.
Vosniadou, S. & Brewer, W.F.: 1992, 'Mental Models of the Earth: A Study of Conceptual Change in Childhood', *Cognitive Psychology* 24, 535–585.
Woodward, J.: 1989, 'Data and Phenomena', *Synthese* 79, 393–472.

# 1.8 The Epistemology of Students: The 'Thingified' Nature of Scientific Knowledge

JACQUES DÉSAUTELS and MARIE LAROCHELLE

*Laval University, Québec, Canada*

The imagination which students display as they make sense of the concepts, laws and theories encountered in their science courses knows no limits. For instance, in a study of the concept of particle among junior-college students, Benyamna, Désautels and Larochelle (1993) noted a manifest tendency to endow scientific concepts with an ontology which is most original, to say the least. In images ranging from the golf ball to a constellation of tiny ball bearings, the concept of particle became imbued by a majority of students with what Moscovici and Hewstone (1984) term 'a layer of reality'; from the outset, this notion assumed dimensions and a form which, moreover, it was not alone in acquiring – electrons, atoms, molecules and photons all took on a family resemblance.

At first glance, it would appear that students tend to impart meaning to the 'relational world' of scientific knowledge by transposing the latter into the world of everyday materiality, by 'thingifying' this knowledge in a certain way, as the research program on students conceptions in science education has also illustrated (Pfundt & Duit 1994). But what if students *have good reasons to believe what they believe*, particularly in light of their everyday experiences (Hills 1989)? And again, what if the experience of schooling had something to do with the tendency of students to picture as the *lessons of things* and *tangible effects* what otherwise in scientific knowledge remains tentative?

In this chapter, we wish to provide an angle from which to view such issues which, to paraphrase Hodson (1985), could be termed the epistemology of school science. Using the discourse of high school students and undergraduates in science who will go on to teach science, we attempt to characterise this epistemology in the first section in terms of scientific knowledge, scientific laws and theories, and the production of scientific knowledge. The second section situates this epistemology within the schooling context and considers its consequences, both social and cognitive.

---

Chapter Consultant: Norman Lederman (Oregon State University, USA)

## STUDENTS' EPISTEMOLOGY: RECONNOITRING THE TERRAIN

For decades, researchers have used a range of perspectives in examining students' answers to the question of what science is (Lederman 1992). In this chapter, we focus on studies of student representations and their relationship to the school context. We leave to one side surveys of students' *attitudes* with regard to science which proceed from a differentialist type of preoccupation. Instead, we examine studies which shed light on the epistemological features of students' *conceptions* of science, and which permit us to situate individual conceptions within the system of social relationships encompassing them, particularly in a school setting. In other words, this more constructivist-type corpus of research, which tends to draw on *situated* subjects, appears to hold much promise, especially because it could identify the preoccupations of subjects' discourses (i.e., delimit which 'reality' the subject is reacting to).

This second research approach generally covers qualitative procedures, but recently quantitative tools have been 'co-generated' by researchers and the persons for whom use is intended (e.g., Views on Science, Technology and Society questionnaire developed by Aikenhead & Ryan 1989). Not only does this questionnaire present the viewpoints of students instead of those of 'certified philosophers', it also takes a look at 'science in action' rather than at science which is 'over and done with'. However, this type of instrument also presents difficulties concerning the meaning that respondents ascribe to different points of view. For example, does the person who opts for the statement 'Models change with time and with the state of knowledge, as theories do', adhere to a relativist perspective or to a realist point of view which is asymptotic?

The next section considers what lessons from this research approach can be applied to students' epistemology. Is scientific knowledge invented or discovered? Is it a collective or individual enterprise? What is the social nature of its practices?

### *Scientific Knowledge: Discovery or Invention?*

> Scientific knowledge is there, it's just waiting for people to discover it. All of it's already there. (A student's comment from Edmondson 1989, p. 124)

The heading of this section refers to a debate spanning centuries. As Bynum, Browne and Porter (1983) have noted, between the 16th and 18th centuries, an opposition arose between two methods of classifying living organisms, one of which was said to be natural and the other artificial, depending on whether one considered that the classification reflected real affinities or differences in nature, or constituted a convention. Hence, even at that time, there was an ongoing epistemological debate concerning the status of scientific knowledge. Does this knowledge make it possible to elucidate a hidden order in the universe, or does it proceed from the necessity which humans feel to create an order which makes their universe intelligible?

It would be presumptuous to consider the debate closed and to dismiss either underlying position as having undergone a 'rise and fall' (Larochelle, Désautels & Ruel 1995). However, the sociology and anthropology of science has sketched the outline of a representation of scientific knowledge which differs radically from the representation associated with positivism (Pickering 1992). In this context, representations of science have left off their empirical certainties and moved on to a constructivist position according to which science is inscribed in larger social projects, as either process or product.

What about the point of view of students on these matters? Clearly, in the case of the student whose words prefaced this section, his opinion already has been worked out on the subject: scientific knowledge is waiting to be divulged! But are such words in a class by themselves or are they representative of a point of view? What do his classmates think? How do students picture scientific laws and theories?

*Scientific Laws and Theories*

> A law of nature means for example some phenomenon (event) about which man in principle cannot do anything. (A student's comment from Engeström 1981, p. 50)

The concept of law has proven to be an epistemological Pandora's box. According to some, scientific laws are a counterpart to the laws governing a preorganised deterministic universe. Accordingly, the probabilistic laws of quantum mechanics are a stop-gap measure which is indicative of our ignorance of the phenomena. According to others, the causal character ascribed to scientific laws stems from *a priori* principles of understanding which provide a framework of intelligibility through which to make sense of the contents of sensory experience. According to others, scientific laws are the proven explanations of empirical invariances established by scientists. In short, any theory which is propounded on this subject supposes that one has consciously or unconsciously made a decision concerning metaphysical propositions which in principle are undecidable (von Foerster 1992): Is chance constitutive of the universe or not?

In the study of Engeström (1981), student discourse presents variation in terms of how the representations associated with the concept of law are expressed. Nevertheless, the epistemological underpinnings of these representations differ little from one another and appear to blend with the metaphysical option which holds that scientists are able to decipher 'the real, as though it were veiled' and the implacable laws of which it is constituted. The notion of implacability, moreover, is a recurrent theme in the views of students concerning laws, as illustrated in the following excerpt. 'If it wasn't [that way], things wouldn't have fallen before gravity was discovered and the sun wouldn't have worked because nuclear fusion hadn't been discovered. If this statement was false then humans would be the creator of the world as they assigned laws to it' (quoted in Roth & Roychoudhury 1994, p. 14). In other words, everything seems to proceed as if a law had ontological correlates. It is inscribed in the nature of things; it is fact which demands no more than to be observed; furthermore, it has the force of law – there's nothing for it but to comply, as is evidenced in the following students' comments. '*Laws are definite, can't change*; they guide us in everyday life just like government laws' (quoted in Griffiths & Barman 1995, p. 252). 'I think it's the opposite of theory. *A law is something you can't deny*. A law is a law, and you have laws that are right there. . . . *It's more clear-cut than a theory is*. . . . It's like gravity, I think. You can see it: your pencil drops. So you say, "There's attraction, *it can't be anything else*". . . . A law, the term "law" is that. *You obey a law*' (quoted in Désautels & Larochelle 1989, p. 117).

As the last quotation also illustrates, the status which many students accord to scientific theories tends to buttress, indeed engrain, this naturalistic conception of laws. In effect, whereas laws enjoy a status which shields them from the vicissitudes of human fortunes, theories present an entirely different case altogether. Theories are contingent, and therefore subject to change. In the absence of discourse which would allow us to contextualise this fluid character of theories, one might think that the tentative nature of the knowledge in question was the object of some sort of recognition, as Lederman and O'Malley's (1990) study has suggested. However, it seems that we are dealing with recognition of a kind which occurs 'despite oneself' rather than an acknowledgement of the logical impossibility of attempting any one-to-one correspondence between a representation and the reality supposedly being represented (von Glasersfeld 1995). Thus, in most cases, the incompleteness of theories would be a question of human nature and would stem from the limits of human organs of sensory perception and the technology used. Sometimes it is the speculative character of theories which is called into question, with the result that these become loaded with the notions of volatility of which laws apparently are devoid, as suggested by the following students' comments. 'Theories change all the time, but laws come out the same way all the time and so we know they are right' (quoted in Lederman, 1995, p. 661). 'Theories and truths are two separate things' (quoted in Edmondson 1989, p. 136).

This definition of theories involving everything that they are *not* is expressed with particular eloquence in the following reasoning sequence, and provides evidence of a mode of representation that is widespread among students (Ryan & Aikenhead 1992). For many students, the various components of scientific knowledge could be ordered according to a 'hierarchy of credibility', as Lederman and O'Malley (1990) have termed it, in which theories represent laws that are just then budding. Just as a hypothesis from the student's viewpoint can become a theory, so too a theory can aspire to a more prestigious status if it is able to prove, empirically and visually, what it suggests. When that is the case, it undergoes an unusual transformation. It is no longer a theory but has become a law, that is, 'a much higher order, because you can always count on them' (quoted in Edmondson 1989, p. 136).

This misappreciation of the model-using nature of scientific knowledge is carried over into the discourses of degree-holders in science (Roberts & Chastko 1990). During a long-term study (Désautels, Larochelle & Pépin 1994) in which preservice science teachers reflected on their own epistemology, 88 percent of the participants initially expressed viewpoints concerning the relationship of model to reality which ranged from specular realism ('Models are constructed on the basis of what was seen. What more do you need!') to asymptotic realism ('These models evolve until they produce a resemblance with this reality.'). A similar realist tendency also can be observed in the discourse of scientists in good standing, if the following conclusion by Grosslight, Unger, Jay and Smith is to be believed: 'In summary, experts agree that . . . the validity of a model can be tested by comparing its implications to observations and measurements in the *real world*' (1991, p. 816, italics added).

In addition, more than 50 percent of the participants in our study were in agreement with the hierarchy of credibility previously alluded to – a conclusion which cannot fail to raise doubts about the type of socialisation in scientific practice being favoured at the university level. But what do students have to say about scientific practice? How do they visualise scientific modes of working and decision making? Do they think that 'the properties of the observer [do] not enter into the description of his [or her] observations' (von Foerster 1992, p. 10), as suggested by the following student's comment: '[Science] is based on experiments and fact, and so it's not really opinions at all' (quoted in Moje 1995, p. 362)?

*Production of Scientific Knowledge: A Particularistic or a Social Practice?*

> You do an experiment and you observe . . . [I]f you didn't observe what was happening you wouldn't have anything to start with. Observation comes before theory because to look at theory you need to have some sort of observation. (A student's comment from Griffiths & Barman 1995, p. 250)

The representations of the production of scientific knowledge are worked out in accordance with epistemological positions which underlie the concepts being drawn upon to construct such representations. For example, in a realist and objectivist perspective, observation more or less amounts to the attentive sensory activity of a spectator on the lookout for reality, which itself is conceived of as being preorganised: 'What you see is what you get'.

This conception has been challenged by Hanson (1958) who, to the great dismay of positivists, showed that all observing activity is inevitably dependent upon the theoretical framework for which the observer consciously or unconsciously opted. Moreover, sociologists and anthropologists of science (Pickering 1992) have shown that it is no longer possible to conceive of observation within a personalistic perspective of the production of knowledge. Instead, it is necessary to move beyond a theory of the subject to a theory of the social actor as a member of a community whose conceptual and methodological positions he or she shares at least in part. Furthermore, this actor must negotiate the facts that he or she produces by submitting them to the judgement of peers, a process which can give rise to controversies. Resolution of these disputes is not a function of so-called scientific rationality alone, but is produced in conjunction with a set of various rules, resources and beliefs, as can be observed in the recent controversies surrounding the 'cold fusion'. Within this socio-constructivistic perspective, the concept of observation reveals a dual break with objectivism and with the sovereign subject of idealism or the *information processor* of cognitivism, because observation and its results are inevitably indebted to the whole set of alliances and activities which combine to provide the former with shape and social recognition. But what do the students have to say on the subject?

For many students, observing basically amounts to perceiving an object or a phenomenon that is posited as existing *per se*, although this activity is motivated by the ideas and previous knowledge of the observer: 'It's watching the reactions which may come about, observing even the smallest detail. [It's] knowing how to detect the things that can't simply be watched. It's watching attentively. If a scientist is there, and I'm there, and we're both watching attentively, *I can see the same things he does, but we won't see them the same way*' (Désautels & Larochelle 1989, p. 100). In other words, the observer is only a peripheral element in the observation process. In effect, regardless of who the observer is, it is always the same object which is involved and, if there is a divergence of interpretation, this would be a circumstantial phenomenon influenced by individual differences in terms of the organs of sensory perception or previous knowledge. Very few students view this stock of knowledge as being constituted by the researchers' paradigmatic affinities which contribute to frame a phenomenon, to define the operating conditions under which its observation is carried out, etc. It is worth noting that the students' tendency to locate interpretation at the end of the observation process testifies to the atheoretical aspect which they ascribe to observation and, sometimes, with no small degree of optimism: 'When you look, you see, and *after that* you think' (Désautels & Larochelle 1989, p. 92). On that score, they are joined by a number of preservice science teachers, as is illustrated in the following excerpt: 'In order to be a good, objective observer, *you have to look at things without even thinking*. You register them the way they come, you don't try to see the relationships' (Guilbert & Meloche 1993, p. 18). But what position is to be adopted when a comparison is made between 'competent' scientists who adhere to different theories?

According to research by Aikenhead and Ryan (1989), only 15 percent of students recognised that the results of observations vary according to what scientists think and believe, whereas 20 percent attributed this to method. Given their particular theoretical positions, scientists will not conduct the same experiments and will note different things as a consequence. When we queried prospective science teachers on this theme, they expressed views which scarcely differed from those of high school students. For instance, a striking parallel can be observed between the preferences of the preservice science teachers and those of the some 600 French-speaking seniors in the Québec high school system who replied to this question (Aikenhead & Ryan 1989). In each case, 70 percent opted for the statements which held that observation is not motivated by a theoretical project.

As things stand, for many students, observation consists of a face-to-face encounter between subject and object, in which the outcome is not a foregone conclusion because the personal characteristics of observers can bias the process or the subsequent interpretation. Students' conception of observation seems to be based not only on the empiricist model of cognition, but also on a personalistic model of the observer. But is this so with other aspects of the production of scientific knowledge?

Students are thought to be aware that scientists work in teams and participate in various forums in which they discuss the results of their research. However, according to Aikenhead and Ryan (1989), this reality is far from endangering their personal conceptions. Students respond as though the various contingencies (sociopolitical, economic, etc.) affecting scientific practices are extraneous to these practices. For example, according to nearly 70 percent of the students surveyed, the underlying factors that motivate scientists to adopt one theory in favour of another are: the facts; the logic of the theory being made use of; the way in which it explains all the facts simply; and the number of times it has been tested. Only a minority of the respondents replied that the personal feelings of scientists, their ways of interpreting the theory, and the quest for prestige or financial gain might affect this decision. While several students admitted that the developers of a new theory must persuade their peers of the theory's relevance, about 30 percent of them ascribe the success of persuasion to the presentation of data *which prove that the theory is true*, whereas 45 percent view this process in terms of reaching a consensus which makes it possible either to revise the theory or to make it more precise. In this perspective, is it surprising that more than 80 percent of these students perceive scientific discoveries as a cumulative series of investigations which are logically dependent on one another, even if chance occasionally plays a role therein?

If these viewpoints are taken together with other perspectives on the status of laws and theories, the epistemological profile which results clearly suggests an empiricist and individualistic perspective on scientific practice. To be sure, these students have some awareness of the collective character of the production of scientific knowledge, but they seem to interpret this more readily as deriving from a collection of individuals than as social practice in which a network of actors and alliances is at work. In accordance with the psychological cast of this perspective, they do recognise that personal projects could affect the production of scientific knowledge; however, in keeping with their conceptions of laws and theories, such projects would amount to nothing more than dross which will be progressively eliminated in favour of the phenomenon, or the disclosure of reality, as Driver, Leach, Millar and Scott (1996) also have emphasised.

## SCIENCE CLASSES: RECONNOITRING THE TERRAIN

As we have briefly outlined, student epistemology appears to be based on a realist and empiricist rhetoric in which knowledge is seen as the mere reflection of reality. Next, we attempt to show that it is not enough merely to portray this type of epistemology, because there is a risk that 'the actor's attitude or disposition' will be identified as the sole culprit. We then concentrate on the lessons which are to be had from research in science education in order to show that the patterns of interaction favoured in science teaching have something to do with the representations in question.

## The 'Thingifying' Message in Science Teaching

Student: It's as if there are two different stories. The Canadians say it's from the States, and the States' guys say that you don't know where it's coming from.

Teacher: But how do they get those two different conclusions based on the same observations?

Student: They're two different countries.

Teacher: But they're scientists . . . and scientists should be people who look at the facts rather than look at the . . . other things. They should be basing it on the facts. (Geddis 1991, p. 179)

Generally, the empirico-realistic position seems to be the predominant representation of science in the school system today. Research into the representations of science conveyed by textbooks and school curricula show the fictive character of the representation which they serve to put forward: the facts speak for themselves; knowledge is the reflection of an ontological reality; and scientific knowledge and truth are one and the same thing, and the result of using 'sound method' to guarantee the ideological immunity of knowledge (Millar 1994). Studies of teachers' epistemology also point in the same direction: Gallagher (1991) brought out the decidedly empirico-realistic tendency of the views held by teachers (25 out of 27) on the subject that they were teaching. Among the interpretations which might possibly explain such a situation is one which challenges the narrowly disciplinary way in which teaching and learning of science is conducted. From high school until the end of university, emphasis is placed primarily on the body of knowledge, in the traditional sense of the term, rather than on the conditions of its production. In the widespread absence of any critical epistemological reflection on scientific knowledge and science learning, teachers tend to teach as they were once taught (Lortie 1975), and contribute to the reproduction of a form of schoolroom science ideology, a close relative of the conventional empiricist ideology of science. Research that focuses on the explicit contract of communication which regulates the interactions of those involved in an educational situation is equally insightful (e.g., the hold exercised by the rhetoric of empiricism over science teaching practices; Lemke 1990).

Recent studies by Sutton (see Sutton's chapter in this *Handbook*) and by Fourez (1988) provide convincing illustrations of the hidden epistemological message of teaching practices and textbooks and the way in which these might contribute to the development of students' ideas of science. The issues covered by these authors also are present in the studies by Robinson (1969) who, as early as the 1960s, called into question the way in which the language of teaching works its subterranean epistemological effects in the form of student 'thingification' of scientific concepts. For example, as Robinson has noted, when a student asks a teacher what a gene is, how should the teacher reply? Should teachers' explanation reflect 'the *construct* gene, postulated to account for certain experimental data, or does their language reify the construct and make it into a "thing"? . . . In relating DNA to inheritance, does the verbal discourse include the construction by Watson and Crick of a physical model that accounted for certain experimental data? . . . Or, does the [teacher's] mediation reduce DNA to a factual particle, a new "discovery" of scientists?' (1969, p. 100).

The teaching episodes reported by Lemke (1990) and Geddis (1988) are in keeping with the previous findings, and show clearly that it is not only students who search for the 'substance behind every substantive', as Wittgenstein (1965) once wrote. Thus, upon hearing the official explanation of the formation of the earth's crust provided by their teacher, a number of students expressed their scepticism by underscoring the speculative character of the explanation – in terms, moreover, which recall the comments of students reported above: 'It's just a theory . . . it can't be a fact. There's no proof that the earth was raised up, unless they took measurements' (Lemke 1990, p. 141). After asserting that measurements had been taken, the teacher attempted to convince the recalcitrant students that they were dealing with facts and not speculation, which, by the same stroke, validated the depreciatory status which students assign to theories. As part of that attempt, the teacher appealed to a principle (the geological principle of Uniformitarianism), which he transmuted into an 'existing state of things' whose basis was the same epistemological position as the students' (i.e., scientific facts are essentially a matter of visual evidence). '*So by looking, by looking at geologic formations, we can tell if things were uplifted, uplifted, or things subsided. OK, just by looking at them*' (p. 141, italics added).

Working in a similar epistemological and linguistic mode, a certain Mr Winters will extract himself from the impasse into which his teaching strategy has led him (Geddis 1988). As a self-respecting empiricist, Mr Winters maintains an unflagging belief in the organs of sight, or at least he does so discursively. If his students didn't 'see' the chemical transformation (the change in colour of silver chloride) that they were supposed to see, this was because they were very mediocre observers; he attempts, moreover, to convince them of this with varying degrees of subtlety. Furthermore, because Mr Winters is not keen on the customary precautions, he brought this particular class episode to a close by asserting that a chemical reaction actually had taken place, without making any allowance for the theoretical and deliberative context in which the interaction in question acquires intelligibility and its scientific credentials. Orthodoxy has been rescued: the chemical reaction was an undisputed fact! But what kind of relationship with science are students able to develop in a teaching context of this sort?

*Conceptions of the Nature of Scientific Knowledge from Classrooms*

The answer to this question is far from simple. However, as we have elaborated elsewhere (Désautels, Larochelle, Gagné & Ruel 1993), the interaction which confines students to a status of reciter is what leads them to form such a monolithic notion of what scientific work is, even if they have received no explicit lesson to that effect. It is also plausible that, *via* this pattern of interaction which focuses only on carrying out a teaching program and reaffirming already established knowledge, students will form a depreciatory image of themselves as knowing subjects. This is all the more true in that students will be reminded constantly throughout their schooling that their common-sense knowledge is erroneous and has little in common with scientific activity. We are dealing here with an instance of symbolic violence (Bourdieu 1980), to the extent that students gradually and unconsciously are led to apply the dominant (scientific or pseudoscientific) criteria of evaluation to their own practices of knowledge-constructing, and to think that the production of a symbolic capital is the preserve of a minority of gifted individuals. Appreciable social issues and considerations are at stake here, because the understanding which an actor has of a situation serves to orient his or her behaviour in this situation (Giddens 1984).

From a learning point of view, the stakes are no less significant. The temptation of empirical evidence, and the substantialisation of concepts to which this penchant almost invariably leads, create major problems of understanding in, for example, the concepts of electric charge, relative mass, or genes and evolution. As Lijnse (1995) has pointed out, it is important to look into ways of fostering an approach among students which is not based on the restatement of knowledge, but which instead consists in an epistemological process that enables students to delve deeper into knowledge games and to move from the exploration of one game to another in an informed but liberated manner. For instance, as shown by Vosniadou (1994), the comprehension by children of the spherical model of the earth as a cosmic body is not a matter of abandoning their prior models of flat earth (rectangular or round) which are viable in their everyday experiences. This implies examining the implicit epistemological assumptions which underlie the construction of these models as well as the assumptions of scientific models. In the same vein, to understand the concept of heat, students need to reconsider the ontological status of heat conceived of as a 'thing' in order to think of it as a process (Chi, Slotta & de Leeuw 1994), within a convention-bound language that has been devised by scientists.

## CONCLUSION

In this chapter, we have examined students' epistemology about scientific knowledge and its production. We have sketched a major feature of this epistemology, namely, the tendency to give knowledge an existential status, which we have termed a tendency to 'thingify'. We also have tried to show ways in which this epistemology relates to the prevailing type of *'rapport au savoir'* (relationship to knowledge) often being promoted in the school and which inevitably bears both social and cognitive consequences.

For teachers and researchers, these results suggest that taking into account these features is neither a luxury nor, to paraphrase Cobern (1993), an irrelevant distraction from science. Moreover, as Driver, Asoko, Leach, Mortimer and Scott (1994, p. 11) have advocated, 'to make these epistemological features an explicit focus of discourse and hence to socialize learners in a critical perspective on science as a way of knowing' seems to be not only a prerequisite for enhancing learning among students, but also a necessary ingredient in the development of a different relation to scientific knowledge and to its emissaries.

## NOTE

For institutional support, the authors wish to express their indebtedness to the Faculty of Education (Laval University) and the CIRADE research centre (University of Québec at Montréal). They also would like to thank Françoise Ruel, Professor at the University of Sherbrooke, for the useful comments which she offered on an earlier draft of this chapter, and Donald Kellough, who made this translation with vigilance and diligence.

## REFERENCES

Aikenhead, G.S. & Ryan, A.: 1989, *The Development of a Multiple Choice Instrument for Monitoring Views on Science-Technology-Society Topics* (Research Report), Social Sciences and Humanities Research Council of Canada, Ottawa, Canada.

Benyamna, S., Désautels, J. & Larochelle, M.: 1993, 'Du Concept à la Chose: La Notion de Particule dans les Propos d'Étudiants à l'Égard de Phénomènes Physiques', *Revue Canadienne de l'Éducation/Canadian Journal of Education* 18, 62–78.

Bourdieu, P.: 1980, *Le Sens Pratique*, Minuit, Paris, France.

Bynum, W.F., Browne, E.J. & Porter, R.: 1983, *Dictionary of the History of Science*, Macmillan, London.

Chi, M.T.H., Slotta, J.D. & de Leeuw, N.: 1994, 'From Things to Process: A Theory of Conceptual Change for Learning Science Concepts', *Learning and Instruction* 4, 27–43.

Cobern, W.W.: 1993, 'College Students' Conceptualizations of Nature: An Interpretive World View Analysis', *Journal of Research in Science Teaching* 30, 935–951.

Désautels, J. & Larochelle, M.: 1989, *Qu'est-ce que le Savoir Scientifique? Points de vue d'Adolescents et d'Adolescentes*, Laval University Press, Québec, Canada.

Désautels, J., Larochelle, M., Gagné, B. & Ruel, F.: 1993, 'La Formation à l'Enseignement des Sciences: Le Virage Épistémologique', *Didaskalia* 1, 49–67.

Désautels, J., Larochelle, M. & Pépin, Y.: 1994, *Étude de la Pertinence et de la Viabilité d'Une Stratégie de Formation à l'Enseignement des Sciences* (Research Report), Social Sciences Research Council of Canada, Ottawa, Canada.

Driver, R., Asoko, H., Leach, J., Mortimer, E. & Scott, P.: 1994, 'Constructing Scientific Knowledge in the Classroom', *Educational Researcher* 23(7), 5–12.

Driver, R., Leach, J., Millar, R. & Scott, P.: 1996, *Young People's Images of Science*, Open University Press, Buckingham, UK.

Edmondson, K.: 1989, 'College Students' Conceptions of the Nature of Scientific Knowledge', in D.E. Herget (ed.), *The History and Philosophy of Science in Science Teaching*, Florida State University, Tallahassee, FL, 132–142.

Engeström, Y.: 1981, 'The Laws of Nature and the Origin of Life in Pupils' Consciousness: A Study of Contradictory Modes of Thought', *Scandinavian Journal of Educational Research* 25, 39–61.

Fourez, G.: 1988, 'Ideologies and Science Teaching', *Bulletin of Science, Technology & Society* 8, 269–277.

Gallagher J.: 1991, 'Prospective and Practicing Secondary School Science Teachers' Knowledge and Beliefs about the Philosophy of Science', *Science Education* 75, 121-133.
Geddis, A.N.: 1988, 'Using Concepts from Epistemology and Sociology in Teacher Supervision', *Science Education* 72, 1-18.
Geddis, A.N.: 1991, 'Improving the Quality of Science Classroom Discourse on Controversial Issues', *Science Education* 75, 169-183.
Giddens, A.: 1984, *The Constitution of Society*, Polity Press, Cambridge, UK.
Griffiths, A.K. & Barman, C.R.: 1995, 'High School Students' Views about the Nature of Science: Results from Three Countries', *School Science and Mathematics* 95, 248-255.
Grosslight, L., Unger, C., Jay, E. & Smith, C.L.: 1991, 'Understanding Models and Their Use in Science: Conceptions of Middle and High School Students and Experts', *Journal of Research in Science Teaching* 28, 799-822.
Guilbert, L. & Meloche, D.: 1993, 'L'Idée de Science Chez des Enseignants en Formation: Un Lien Entre l'Histoire des Sciences et l'Hétérogénéité des Visions?', *Didaskalia* 2, 7-30.
Hanson, N.R.: 1958, *Patterns of Discovery*, Cambridge University Press, Cambridge, UK.
Hills, G.: 1989, 'Students' "Untutored" Beliefs About Natural Phenomena: Primitive Science or Commonsense?', *Science Education* 73, 155-186.
Hodson, D.: 1985, 'Philosophy of Science, Science, and Science Education', *Studies in Science Education* 12, 25-57.
Larochelle, M., Désautels, J. & Ruel, F.: 1995, 'Les Sciences à l'École: Portrait d'Une Fiction' [Special Issue: Science et Société au Québec], *Recherches Sociographiques* 36, 527-555.
Lederman, N.: 1992, 'Students' and Teachers' Conceptions of the Nature of Science: A Review of Research', *Journal of Research in Science Teaching* 29, 331-359.
Lederman, N.: 1995, 'The Influence of Teachers' Conceptions of the Nature of Science on Classroom Practice: The Story of Five Teachers', in F. Finley, D. Allchin, D. Rhees & S. Fifield (eds.), *Third International History, Philosophy and Science Teaching Conference* (Proceedings, Volume 1), University of Minnesota, Minneapolis, MN, 656-663.
Lederman, N. & O'Malley, M.: 1990, 'Students' Perceptions of Tentativeness in Science: Development, Use, and Sources of Change', *Science Education* 74, 225-239.
Lemke, J.L.: 1990, *Talking Science: Language, Learning and Values*, Ablex, Norwood, NJ.
Lijnse, P.: 1995, ' "Developmental Research" as a Way to an Empirically Based "Didactical Structure" of Science', *Science Education* 79, 189-199.
Lortie, D.: 1975, *Schoolteacher: A Sociological Study*, University of Chicago Press, Chicago, IL.
Millar, R.: 1994, 'What is Scientific Method and Can it be Taught?', in R. Levinson (ed.), *Teaching Science*, Routledge & Open University, London, 164-177.
Moje, E.B.: 1995, 'Talking about Science: An Interpretation of the Effects of Teacher Talk in a High School Science Classroom', *Journal of Research in Science Teaching* 32, 349-371.
Moscovici, S. & Hewstone, M.: 1984, 'De la Science au Sens Commun', in S. Moscovici (ed.), *Psychologie Sociale*, Presses Universitaires de France, Paris, France, 539-566.
Pfundt, J. & Duit, R.: 1994, *Bibliography: Students' Alternative Frameworks and Science Education* (fourth edition), Institute for Science Education at the University of Kiel, Kiel, Germany.
Pickering, A. (ed.): 1992, *Science as Practice and Culture*, University of Chicago Press, Chicago, IL.
Roberts, D.A. & Chastko, A.M.: 1990, 'Absorption, Refraction, Reflection: An Exploration of Beginning Science Teacher Thinking', *Science Education* 74, 197-224.
Robinson, J.T.: 1969, 'Philosophy of Science: Implications for Teacher Education', *Journal of Research in Science Teaching* 6, 99-104.
Roth, W.-M. & Roychoudhury, A.: 1994, 'Physics Students' Epistemologies and Views about Knowing and Learning', *Journal of Research in Science Teaching* 31, 5-30.
Ryan, A.G. & Aikenhead, G.S.: 1992, 'Students' Preconceptions about the Epistemology of Science', *Science Education* 76, 559-580.
von Foerster, H.: 1992, 'Ethics and Second-Order Cybernetics', *Cybernetics & Human Knowing* 1, 9-19.
von Glasersfeld, E.: 1995, *Radical Constructivism: A Way of Knowing and Learning*, Falmer Press, London.
Vosniadou, S.: 1994, 'Capturing and Modeling the Process of Conceptual Change', *Learning and Instruction* 4, 45-69.
Wittgenstein, L.: 1965, *Le Cahier Bleu et le Cahier Brun*, Gallimard, Paris, France.

# Section 2

# Teaching

**KENNETH TOBIN**

# 2.1 Issues and Trends in the Teaching of Science

### KENNETH TOBIN
*University of Pennsylvania, Philadelphia, USA*

My research in science education in the 1970s assumed that there is a best way to teach science and an optimal way to learn. On that basis, I set out to identify a monolithic 'one size fits all' solution to the problems of teaching and learning science. Looking back at the research questions and one-directional causal models that framed my first decade of research, it might be possible to conclude that the studies were naive and relatively atheoretical. However, that was not the case. The program of research was built on an evolving conceptual framework comprised of social and psychological components linked with recent experiences as a high school teacher of physics, chemistry and general science. The results from each study were used recursively to reflect on limitations of prior research and to consider new studies. Piagetian ways of thinking about teaching and learning gave way to radical behaviourism, and then to neo-Piagetian models. Although I was dissatisfied with the theories used to re-present teaching and learning and the causal reasoning that permeated my analyses and interpretations, neither of these catalysed a decision to abandon my constantly evolving program of research. The impetus for change was an awareness that, no matter what I might disseminate to practitioners, forces operating within the educational system sustained traditional practices.

In what appeared to be a radical departure from my earlier approaches to research, I employed an interpretive research design to explore teaching and learning of science (Tobin & Gallagher 1987). Because of the explicit use of qualitative data, the change seemed revolutionary at the time. When viewed historically, however, the studies reflect a gradual evolution in the theoretical underpinnings of the substantive foci and methods used to conduct research. For example, use of a hermeneutic approach to research encouraged me to include multiple perspectives on what happens in science classrooms and the associated rationales. Even so, it took me some time to realise that my own stories were no more salient than those of other stakeholders, a fact that is most evident in a study of an exemplary science teacher. Tobin, Espinet, Byrd and Adams (1988) acknowledge the voice of the teacher, but do not present his accounts as potentially viable and the perspectives of the authors are presented in a more compelling manner. Changes to the

---

Chapter Consultant: Aldrin Sweeney (University of Central Florida, USA)

theoretical frame of my research continued to the present time. The appeal of radical constructivism led to a strong focus on the ways in which teachers and learners made sense of their roles and goals and implemented curricula to enhance learning. Because intact classrooms were involved in my studies, the theoretical frame was extended to include social and cultural theories as new windows that allowed me to see other issues as having salience to teaching and learning.

One way to think of research is as a bricolage of theories, methods and prior histories that shape the narratives of what is learned from a study. These research narratives are particularised and are neither applicable across contexts nor privileged ways to describe experience. On the contrary, when research is done well, the associated research texts are potentially useful as sources of perturbation for practitioners in a variety of educational niches and, in my view, educational researchers can serve the educational community best by employing a variety of theoretical frames and methods in the conduct of research. In addition, rather than always interpreting data to produce patterns to reflect coherence and parsimony, it is beneficial to retain a diversity of perspectives and complexity in the multi-layered ways of describing teaching and learning.

This chapter is based largely on my own research program in the past two decades that has focused primarily on the teaching of science. Although a variety of perspectives has been employed in that time, the research primarily focuses on two perspectives. The first views teaching from a hermeneutic perspective in terms of the minds of teachers. The second perspective provides a more phenomenological orientation that describes teaching in terms of the *doings* of a classroom community. The perspectives are complementary and part of a bricolage that sets the stage for the seven chapters that follow in this section of the *Handbook*. The remaining eight sections of this chapter address the following: an overview of the constituent chapters of this section of the *Handbook*; metaphors and the teaching of science; teacher knowledge, beliefs and actions; learning science through co-participation; language and the learning of science; social class; and conclusions.

## OVERVIEW OF OTHER CHAPTERS IN THE TEACHING SECTION

The chapters that comprise this section of the *Handbook* exemplify the position that teaching and learning can be viewed productively through different theoretical lenses. Even though the theoretical frames differ, contradictions are rare and, for the most part, recommendations form a coherent set. Baird's chapter discusses students' actions and interactions in terms of metacognitive activities that link thinking to behaviours within a classroom community. Baird regards productive learning environments as those perceived by students to have appropriate challenge in terms of the cognitive demands of tasks and the extent to which students are interested and motivated to participate in them. Such activities provide students with control over their own learning and make them aware of their cognitive and affective actions. If cognitive demand and/or level of motivation in participating in the activities are less than optimal, then grudging compliance, boredom or frustration can characterise a community. Baird's research suggests that cognitive and affective outcomes in science can be improved when students ask and seek answers to evaluative questions (e.g., What am I doing? Why am I doing it? What do I already know? What else needs to be done?). He reports that assisting students to ask questions like these enables them to become aware of their thinking, assume responsibility for learning, and increase the incidence of actions associated with informed and meaningful learning.

Baird's framework for teacher learning and the improvement of teaching incorporates a model of teachers conducting research in their own classrooms and meeting regularly to reflect on what they learn, decide what to do next, provide mutual support for colleagues, and receive the encouragement needed to maintain motivation and confidence to improve the quality of learning environments. Baird advocates the use of questions like those listed for students to promote metacognitive activities focused on teachers' roles to enhance metacognitive activities for students. He adds two questions to be asked by teachers. How does my teaching promote the types of good learning behaviours that I have identified as appropriate for students? How can I act with students to develop an appropriate level of challenge in science classwork?

The chapter by Roth has the two interconnected foci of learning to teach science and learning science, and both developed from a perspective of learning that is grounded in participation in authentic practices. In a transition from legitimate peripheral to core participation, newcomers learn community-specific ways of talking about, viewing, acting on and interacting within a community. Collaboration between members with different levels of expertise and competence produces groups with overlapping distributions of knowledge and expertise. For learning to be authentic, a learner should engage in a community to learn its practices from a guide who can model and, as necessary, set examples, correct by showing how and provide scaffolding, feedback, assurances and reassurances. According to Roth, activities that promote learning allow students to employ the language of science over long periods of time and produce, manipulate and interpret inscription devices such as concept maps, graphs, diagrams and designs for objects to be built. Roth reports that students develop positive attitudes towards science when they have autonomy regarding the curriculum and can make critical choices. Using the yardstick of authentic practice, he recommends that learning environments be science-like and include ill-defined problems, high levels of uncertainty and ambiguity, a distributed nature of knowing, and a reliance of learning on current knowledge.

Learning to teach is analogous to learning other practices. Roth describes tacit knowledge as an important component of being a teacher and notes that all knowledge cannot be judged against criteria that can be stated as propositions, procedures and heuristics. Experienced teachers have an array of knowledge that guides what they do in particular circumstances and their interpretations of the context shapes what they actually do. In a practice setting, the appropriateness of actions and practical choices is achieved by reflections in action and viable teaching moves are constructed in an iterative fashion by fine tuning teaching to fit the context. The theoretical position that participants can learn by practising at the elbows of peers is exemplified in a study involving team teaching. The teachers coordinated their actions in response to non-verbal cues from one another and began to develop similar questioning styles, mannerisms, and attitudes towards students. In some instances, teachers were conscious of what they tried and could make deliberative decisions on what worked and what did not. Because they shared an experience, they could engage in pedagogically relevant reflection on action.

Harlen's chapter addresses the present-day reasons for teaching science in primary schools in terms of three issues. First, children's ideas of the world around them are developing from the moment when they become curious, whether or not they are being taught science. Harlen observes that these ideas are frequently unscientific, can be difficult to change, and can make it difficult for students to learn science at a later time. Second, the interdependence of science content and science processes suggests that there is value in exposing students to scientific ideas through the use of science processes at an early age. Third, students might not continue to study science at higher levels because of negative impressions about science that develop at an early age. Harlen argues that it is better to allow young children to become informed about the nature of science by providing them with first-hand experience of science as an enjoyable, comprehensible and useful activity.

Harlen uses a constructivist perspective to distinguish between rote learning and understanding in terms of children being able to apply what has been learned to make sense of other contexts. Learning is conceptualised as the change of ideas by testing them against new experience. At each point, what is understood fits the experience of the individual. The significance of social interaction is evident in numerous examples which Harlen provides to emphasise that students fully participate in constructing and testing ideas. A rationale for including science in the curriculum of primary students is that the development of process skills is critical to the construction of viable knowledge; however, children do not automatically test their ideas rigorously and have to learn how to use process skills.

In the next chapter, Hewson, Beeth and Thorley advocate a curriculum in which claims for the status of different ideas are made explicit to enable students to explore fully whether warrants are based on internal (i.e., self) or external (e.g., discipline of science, textbooks, teacher, peers, parents) sources of authority. Metacognition enables students not only to think with ideas, but also to think about their ideas, and to allow explanations to become objects of cognition. Students' knowledge is assumed to play a significant role in learning in a context in which what students know is frequently discrepant from canonical science. The authors note that efforts to teach students science are made difficult by the robust nature of students' ideas. Teachers are encouraged to make students' ideas central and create a discourse that features making choices based on evidence. The roles of teachers should enable students to assume control of their own learning, become aware of their epistemological commitments, re-present their conceptions to teachers and peers, and monitor their own interpretations of scientific phenomena and the views expressed by others.

Conceptual changes occur when the status of ideas varies rather than when some ideas become extinct while fruitful ideas are retained. If conceptual changes are to be promoted, it is necessary for interactions in the classroom to allow students to negotiate effectively the meanings of ideas and their status, where status involves the intelligibility, plausibility and fruitfulness of ideas. Because ideas are often interconnected in a conceptual array, a change in the status of one idea might lead to a ripple effect whereby the status of many ideas within the array is changed. Activities intended to raise the status of particular ideas should occur in conjunction with others intended to lower the status of less fruitful ideas. This can be accomplished when teachers' and students' ideas are an explicit part of a discourse that is explicitly metacognitive, the status of ideas is negotiated, and justification of ideas and their status is explicit. Discussion about the status of what is known allows students to make informed choices.

Hobden's chapter highlights the pervasive presence of routine problem tasks in senior high school physical science courses. Routine problem tasks refer to work assigned by a teacher that involves calculation based on the use of one or more formulae and algebraic manipulations of a number of given variables. Although routine problem-solving activities are widespread, Hobden points out that competence in solving routine problems is not necessarily associated with improved understandings of science. Hobden calls for a transformation of the curriculum to emphasise understanding and the qualitative dimensions of quantitative problems. He illustrates how traditional tasks can be adapted to emphasise understanding by taking the time to make sense of the problem situation, use multiple representations (such as qualitative descriptions and diagrams together with mathematical representations when solving problems), and provide opportunities to reflect on problem solutions by providing written solutions that are detailed and focus on understanding the solution strategy. Hobden suggests that teachers can contextualise physics problems by allowing students to introduce contexts from their everyday lives and select examples from the news.

In her chapter, Gabel suggests that learning chemistry is complex because it can be described at a macroscopic level, is composed of particles, and is represented symbolically. Teachers assist students to learn at the atomic or molecular levels by involving students in the planning of laboratory activities, allowing them to acquire manual skills prior to commencing a laboratory investigation, and encouraging them to apply concepts from long-term memory by participating in problem-solving activities. Gabel describes a conceptual change approach in which all students describe their own understanding of a scientific concept and restructure it during interactions between students and between the student and the teacher. Students then can compare new understandings with previous understandings and repeat the process until a fit with canonical science emerges. Gabel presents evidence to show that increased social interaction is associated with improvements in the learning environment, higher achievement, and a reduced incidence of non-scientific understandings. Gabel describes how analogues and conflict training can reduce misconceptions and the use of models and pictorial diagrams can improve understandings of chemistry.

Lunetta's chapter explores the roles of teachers in laboratory activities whose purposes range from verifying principles or relationships to engaging in inductive activities in which students identify relationships from data which they gather. Learning is enhanced when laboratory activities incorporate metacognitive experiences and focus on the manipulation of ideas instead of just materials, and practical activities incorporate discussions of competing explanatory models, analogies, diagrams, graphs and simulations to promote scientific understandings. Lunetta emphasises the importance of social interaction as students engage in pre-laboratory and post-laboratory discussions and participate fully in activities in which teacher negotiation mediates in the development of understandings that fit with those of canonical science.

## METAPHORS AND THE TEACHING OF SCIENCE

The potential significance of teacher metaphors became apparent to me in a study involving Peter, a teacher of grade 10 general science in Australia (Tobin, Kahle & Fraser 1990). Peter had two quite distinctive styles of teaching and, when we spoke to him, he described himself as a 'captain of the ship' and at other times as an 'entertainer'. The metaphors were used as referents for his interactions with students and, as the metaphor changed, so too did many of his actions and the associated actions of students. Peter's decision to switch metaphors was not necessarily conscious, because his experience allowed him to know instinctively when each of the metaphors was appropriate.

I was particularly interested in knowing if teachers could generate a metaphor, apply it to their teaching, and change the enacted curriculum. In a collaborative study involving Sarah, a beginning middle and high school teacher, a research group assisted her to address severe management problems by building a new metaphor that was consistent with her beliefs about the nature of knowledge and how students learn (Tobin & LaMaster 1995). By reflecting on her actions as a teacher, Sarah built a mental model consisting of her beliefs about knowledge and learning, images of her preferred approach to teaching, and a metaphor that would apply to her management of the classroom. After several weeks, she developed a coherent mental model that incorporated a metaphor of 'teacher as social director'. Sarah argued that a social director can only invite guests to a party. If the invitation is declined, then the hostess can make the invitation more attractive. However, if guests choose to come, they should come to have fun (i.e., learn) and ought not to disrupt the fun (i.e., learning) of others. In addition, guests (i.e., students) should be courteous to the hostess (i.e., the teacher) and other guests. Sarah explained the new rule structure to the class and then she used the social director metaphor as a referent for her actions in the classroom. During the enactment of the curriculum, students had to adapt their actions to fit those of the teacher as a new culture evolved.

Because of the severe nature of the initial problems in the classroom, the new culture did not reach equilibrium without further changes being introduced. Sarah decided that her metaphor for assessment, of 'fair judge', needed to be reconceptualised to be consistent with her new social director metaphor for management, beliefs about constructivism, and a growing awareness that students should have greater autonomy in her class. She accepted a metaphor of assessment as 'windows into students' minds', providing opportunities for them to show what they knew. Adoption of a new metaphor necessitated some explanation to students, because it required them to adapt their actions to be consistent with her new methods of assessment. Within a relatively short period of time, a new culture stabilised with students and the teacher interacting in ways that facilitated their attainment of goals. The classroom culture that reflected the new metaphors, 'teacher as social director' and assessment as 'windows into the minds of students', was profoundly different than the previously dysfunctional culture that had supported neither teaching nor the learning of science.

## TEACHER KNOWLEDGE, BELIEFS AND ACTION

A growing number of studies has highlighted links between the beliefs and classroom practices of science teachers (e.g., Ritchie, Tobin & Hook 1997; Tobin, Kahle & Fraser 1990; Tobin, Tippins & Hook 1994). For example, McRobbie and Tobin (1995) described how Jacobs' teaching roles were defined in terms of identifying the most important facts, presenting them, and assisting students to memorise them. He had a similar approach to assisting students to memorise significant procedures (by rote memory if necessary) and he developed an array of memory joggers and 'cognitive hooks' to assist students to recall information and previously memorised procedures. Understanding was regarded as something that happened when facts were interconnected, a physiological process over which the teacher had no control. All that had to happen was for students to memorise a critical mass of facts and the interconnections, and associated understandings were thought to emerge. The teacher's failure to build a mental model for understanding made it difficult for him to alter his practices to assist students to build undertandings of chemistry. He did not have teaching strategies to enhance understanding, nor did he know, beyond exhortations to try harder, what practices to induce in learners.

Tobin and Tippins (1996) reported that Jacobs was unable to explain what he understood by learning. As he endeavoured to explain the relationships between teaching, learning and knowledge, he used gesture as if he were re-presenting images of his thinking. Several weeks later he announced that he was ready to discuss what he understood by learning and sketched a large swimming pool containing a teacher directing water from a hose towards students who used empty pails in attempts to catch drops of water. The teacher then used the sketch to generate a narrative account of teaching and learning and, in the process, the sketch became an object on which he could reflect and generate additional narratives which also became sources for reflection.

Over a period of several weeks, Jacobs was able to return to the sketch and associated narratives and use them generatively to modify his understandings of teaching and learning. In a recursive process, he related the swimming pool model to his reflections on practice and, in so doing, generated propositional knowledge in the form of beliefs about teaching, learning and knowledge transmission. In this process, Jacobs applied his emerging understandings to extensive experience of teaching and learning and, as necessary, adapted his oral accounts until there was coherence between his reflections on teaching and learning, the swimming pool model, and expressed beliefs about teaching, learning and knowledge transmission. In the absence of signs that his teaching could change to enhance learning, Jacobs refined the sketch and oral accounts to achieve a coherence with his perceptions of his practices.

The enacted curriculum also can be constrained by factors associated with the culture in which teaching and learning occur. Tobin and McRobbie (1996a) describe how cultural myths (Britzman 1991) of science teachers constitute a referential system that favours the retention of traditional practices. Examples include those of: time being a scarce commodity; a high priority goal being to cover prescribed content; a need to prepare students for the next educational level and succeed on examinations; and specification of the syllabus being the prerogative of external agencies. Adherence to these cultural myths placed the teacher in a position of relative disempowerment in regard to enacting changes to the curriculum. Not only could he justify why he could not make decisions, he also adhered to a belief that he was a guardian of the discipline and would not make changes that, in his view, would decrease the rigour of a chemistry course.

Tobin, McRobbie and Anderson (1997) reported that a beginning teacher, Morrell, believed that students constructed knowledge and should retain a high level of autonomy with respect to the tasks they undertook, with whom they worked, and the pace at which they worked. His high school physics class was innovative in terms of arranging students and permitting them to learn from one another and other resources such as textbooks. However, when it came to the understanding of physics, students and the teacher were relatively disempowered and accepted knowledge as right or wrong on the basis of the authority of the discipline. When perturbations emanated from a clash between the answers offered by physics and common sense, the teacher and students suppressed common sense in favour of canonical explanations. Neither the teacher nor the students appeared too concerned at not understanding physics. Their goals were directed towards learning correct answers so that they could succeed on examinations. The culture of the classroom, even though it supported autonomy, also valued knowing correct answers and covering content in sufficient time to maximise success on tests and examinations. Even though Morrell espoused a social constructivist view of learning and wanted to teach in ways that were consistent with constructivism, the significance of current understandings was diminished in comparison to the powerful voice of canonical physics.

Longitudinal research in classes taught by a middle and high school science teacher (Ritchie, Tobin & Hook 1997; Tobin, Tippins & Hook 1994) provides evidence of changes in teaching to reflect the contexts in which the curriculum was enacted. At different times during his career, the teacher expressed a strong value for student autonomy, constructivism and inquiry-oriented approaches to learning. At other times, he was more concerned with issues such as covering content to meet the perceived expectations of colleagues, maintaining control of his students, and exposing students to a variety of ideas. Knowing and valuing particular theoretical constructs was no guarantee of their use as a referent for teaching. Even though the teacher demonstrated that he could teach according to a variety of referents, he employed strategies to fit the needs of the moment as he learned and thought about teaching and learning. Tobin, Tippins and Hook (1994) describe how the teacher reverted to traditional modes of teaching when he endeavoured to use constructivism as a referent. Similarly, Ritchie, Tobin and Hook (1997) described how the same teacher framed his teaching in terms of beliefs that students can learn by being exposed to ideas and, to a less explicit extent, that the teacher should be in control of students. Even though the teacher justified his actions in terms of exposure and control, there is no guarantee that these reasons are why he taught as he did. At the time of the study undertaken by Ritchie *et al.*, the teacher was head of department and had a demanding assignment to teach computer technology. Frequently, he appeared unprepared to teach science and used his experience to plan activities while the lesson was in progress. The extent to which a teacher is prepared to teach is a component of context that must be considered in analyses of teaching and learning. The time available for effective planning can be diminished when teachers are asked to teach out of field, teach too many classes in a day, or teach large classes that require extensive reviewing of students' work. In such circumstances, the focus of the teacher can be on survival rather than the learning needs of students.

## LEARNING SCIENCE THROUGH CO-PARTICIPATION

Science is regarded as a discourse that evolves as participants in a community endeavour to make sense of a universe of phenomena. Discourse involves ways of talking, acting, interacting, valuing and believing, as well as the spaces and the artefacts which a community uses to carry out its social practices. Discourse is a 'social activity of making meanings with language and other symbolic systems in some particular kind of situation or setting' (Lemke 1995, p. 8). Lemke noted that:

> Instead of talking about meaning making as something that is done by minds, I prefer to talk about it as a *social practice* in a community. It is a kind of *doing* that is done in ways that are characteristic of a community, and its

occurrence is part of what binds the community together and helps to constitute it as a community. In this sense we can speak of a community, not as a collection of interacting individuals, but as a system of doings, rather than a system of doers. (p. 9)

If students are to learn science as discourse, it seems imperative that they adapt their language resources as they practice science in settings in which others who know science assist them to learn by engaging activities in which co-participation occurs (Roth 1995; Schön 1985; Tobin 1996). As it is used here, 'co-participation' implies the presence of a shared language that can be accessed by all participants to communicate with one another so that meaningful learning occurs. The shared language is negotiated and evolves as participants learn from others as they engage in the activities of a classroom community. In a setting in which co-participation occurs, knowledge claims that make no sense are clarified and discussion occurs until learners are satisfied that they now understand, students have the autonomy to ask when they do not understand, and the focus always is on what students know and how to re-present what they know. Students do not feel that they cannot understand and that their only recourse is to accept what is said as articles of truth based on a faith that others understand the warrants for the viability of the claim.

One way to think of the roles of the students and teacher in an environment in which co-participation is occurring is that students come to know by engaging at the elbows of those who know. Students receive opportunities to practice and observe others practice and, at different times, a person might be a teacher or a learner with respect to others. Co-participation implies a concern for facilitating one another's learning, and peer teaching is a critical element of such an environment. During interactions, respect is shown for the knowledge of others and efforts are made to find out why particular claims are regarded as viable. Within an evolving discourse community, concern is shown for what is known by learners at any given time and how they can re-present what they know. A teacher can structure activities in which students engage so that they can use their existing knowledge to make sense of what is happening and build new understandings on a foundation of current knowledge. The mediating role of the teacher is focused not only on what students know and how they can re-present what they know, but also on the identification of activities that can continue the evolutionary path of the classroom community towards the attainment of a discourse that is increasingly scientific.

## LANGUAGE AND THE LEARNING OF SCIENCE

Effective learning necessitates that students have chances to try out the language of science in its broadest terms. It is important for students to have opportunities to use the language of science in social settings so that they can receive feedback on the extent to which their emerging competence is appropriate. A productive classroom environment enables the teacher and students to assist others to adapt their knowledge and apply it in ways that are consistent with canonical science. When understandings and practices are not quite right, others can identify a need for corrective actions and act accordingly. Thus, it is appropriate for any participant in a community to be a teacher with respect to others and interact to promote learning. An expectation can be set up in a classroom that peers and the teacher will engage in discussions about the viability of the discursive practices, and that all participants in a community will have roles as teachers and learners. From a hermeneutic perspective, it makes sense that teaching roles are first directed to finding out why individuals regard certain knowledge and practices as viable. Thus, a person making an assertion that others feel might be incorrect would first be required to provide evidence to support the viability of the assertion or practice. Following the presentation of evidence, alternative assertions or practices could be considered and the community can act to decide what is to be regarded as viable.

Gee (1990) distinguished between primary and secondary discourses. Whereas a primary discourse reflects the native language and cultural history of a student, secondary discourses are constructed as students enter and learn to participate in a variety of communities which each is characterised by a discourse. The extent to which a student can develop a secondary discourse that is scientific reflects his/her own extant knowledge and whether teachers in the classroom community can facilitate the development of a shared language that continually evolves into a form that is more science-like. Depending on their language and cultural histories, teachers can assume different roles in mediating learning through co-participation. Teachers and students from diverse backgrounds bring with them their own ways of looking at the world that are reflective of their cultural histories. Problems can occur when teachers do not recognise the viability of the habits of mind that students apply to their efforts to learn science, do not view them as cultural capital, and do not respect those ways of thinking.

*Limited English Proficiency and the Learning of Science*

The role of language in the learning of science is critical because of the relatively high proportion of students for whom the language spoken by a majority is not a native language for a minority. For example, in countries in which English is a majority language, students who do not speak English as a native language often experience difficulties in learning science. Lee, Fradd and Sutman (1995) noted that students from non-mainstream backgrounds had more difficulty with science knowledge and vocabulary compared to monolingual English students. Some did not have requisite science knowledge or experiences, and others had the science concepts but lacked the specific vocabulary to communicate meaning. Many words were needed by limited English proficient (LEP) students to express ideas that could be presented by English proficient students as short, concise statements. There is evidence that a form of linguistic imperialism (Phillipson 1992) underlies decisions to use English to enact the curriculum in linguistically diverse classrooms.

Sweeney and Gallard (1996) noted that science teachers in the USA believe that LEP and cultural minority students should learn the English language in order to be successful within the educational system and mainstream society.

Tobin and McRobbie (1996b) reported that Asian-Australians elected to study courses in science and mathematics based on a belief that an understanding of English is not as critical to success in these subjects as it is in other subjects that might be selected. An important part of the habitus of Australian schools is an unspoken referent that Australia is an English-speaking country and therefore, to be successful, one needs to read and write English at an acceptable level. Accordingly, the problems faced by Chinese students with limited proficiency in speaking standard English were seen in terms of a deficiency, and there was not an acknowledgement of the positive aspects of having well developed language skills in Cantonese. Solutions to learning difficulties were cast in terms of overcoming deficiencies in English rather than using all available language tools to make sense of chemistry. Although the teacher was empathetic to the problems of non-native speakers of English, he did little to respond to their difficulties in understanding chemistry or to prepare assessments in a language other than English.

In the process of addressing the obstacles of LEP, Chinese students were helped by a cultural capital associated with commitment to succeed and applied such effort as necessary to meet their goals. Thus, some of the Chinese students were prepared to cooperate and assist one another even in a context in which their competitive tendencies might otherwise lead to individualised efforts. Competence in Cantonese, preparedness to use their native language even in a milieu where its use was not encouraged, high levels of task orientation, and a will to attain goals were salient features of a cultural capital that was available to support learning. Without the teacher knowing it, the students often spoke quietly to one another in Cantonese and sometimes wrote their notes in their native language. When they collaborated in groups, they cooperated and persisted to get the tasks done and to learn. After school and during lunch breaks, their inscriptions, in Cantonese and English, allowed them to focus on topics of importance in their interactions with Cantonese-speaking peers and with books from the library.

Even though more than 20 languages were spoken in the school, we did not see texts or reference books in chemistry, for example, in a language other than English. LEP Chinese students had no access to chemistry books written in their own language and, in a context of teachers being unable to speak their language, the provision of text resources might have been a significant learning aid. In making this suggestion, I recognise that educators might argue that, because students have to know and be assessed in English, their learning also ought to be in English. Imperialistic stipulations such as these have been referents for decisions associated with extreme learning difficulties experienced by LEP students. I believe that it is time to allow all LEP students, in the process of building understandings about science, to utilise fully their native language tools.

Teemant, Bernhardt, Rodriguez-Muñoz and Aiello (1995) drew the attention of teachers to three significant gaps that ought to be considered when teaching science to LEP students: between what they can understand in English and what they can speak in English; between what they can write in English and what they can write in their own native languages; and how they conceptualise in English and how they conceptualise in their own native languages. Teemant *et al.* claim that students cannot wait until they speak English fluently to be deemed 'ready' for science learning. The implication here is that it should be recognised that building proficiency in a second language, such as English, will take many years. Teachers should realise that helping bilingual students does not mean compromising science content. Because of the language limitaticns of LEP students, it might be necessary to make special provisions for learning science, such as placing them with others who speak their native language. If that is not possible, it still might be beneficial to make it possible for students to use their native language in making summaries of what happened, in raising questions, and in providing answers to those questions. For LEP students, the learning of science could be a bilingual experience during which they are encouraged to use all of their language tools to make sense of what they are learning. Teemant *et al.* caution that students' conversational fluency does not guarantee fluency in the type of language needed for making sense of science. The chief implication of this is to allow students to decide on the extent to which they employ their native language in their initial efforts to describe what is happening, to interpret data, and to interconnect their growing understandings of science to their primary discourses.

*Sources of Symbolic Violence*

Bourdieu (1992) uses the term 'symbolic violence' to describe the problems of peripheral participants within a community, and concludes that their cultural capital is not valued and is seemingly 'worthless' in the context of insufficient scaffolding to permit them to participate in a community. In a process of progressive devaluation, symbolic violence occurs when individuals encounter situations in which their cultural capital is not viable. A common example of symbolic violence occurs when students are unable to use their own language resources to make sense of science. This happens most frequently when the focus of a curriculum is on coverage of subject matter, often relying heavily on the chapters of a textbook. In such circumstances, the teacher might focus on content coverage and only attend to the extant knowledge of students when mistakes are made, and then only to correct those mistakes. Scaffolds are not provided to enable students to employ their cultural capital as they learn to participate in authentic ways.

The potential for science to be a source of symbolic violence exists for all students, but is strongest for those who do not belong to the majority culture for whom the curriculum tends to be tailored. Accordingly, the potential for symbolic violence is greatest for students who have a different language from the majority (e.g., speakers of Spanish, Haitian Creole, Ebonics, etc.). Belonging to a different culture is a source of disempowerment for minorities because the habitus associated with life in their home culture is no longer viable for them. One does not easily participate in a new culture and has to learn new routines, and this takes time.

Instead of allowing the inevitable disempowerment of being unable to co-participate and the symbolic violence of realised failure, teachers might be proactive in identifying the discursive resources that minority learners have to shape their goals, actions and interactions in a community.

The tendency of some teachers to regard the habitus of certain students as a deficiency to be overcome before learning of science can occur is understandable. Some students come to class from homes where survival cannot be assumed from one day to the next. Whereas most students can relax at home in an environment that prepares them for life in the middle class, a significant minority experiences violence, hunger, sickness and fear. The language resources and coping mechanisms learned at home and in the communities which students inhabit out of the classroom also prepare them for lives in societies and, because they are survivors, what they know and can do enables them to meet their goals in out-of-school contexts. How can these resources be applied to the goal of co-participating to learn science? From the perspective of the students, what they know and can do are capital to invest in learning. A challenge for teachers is to perceive the capital of those resources because, for most teachers, there is a significant gap between their own lives and those of the students whom they teach. Even in the case of the Hispanic majority in Dade County, Florida, the teachers mainly originate from Cuba or Puerto Rico whereas most Hispanic students originate from Nicaragua or Colombia (Tobin 1997). Although there is a shared language, the cultural experiences of teachers and students are significantly different. When a common native language is not shared, differences in the cultural histories of teachers and students are even greater and increase the difficulty of regarding what students know and can do as capital, rather than a deficit that must be overcome before learning can occur.

*Activities to Mediate Student Learning*

The teacher's role as a mediator can take many forms. For example, as one who understands canonical science, the teacher has an important role in demonstrating the discursive characteristics that students are to learn. That is, the teacher can make visible the knowledge that students are expected to construct, and provide opportunities for students to act similarly. This does not imply that teachers should provide students with answers to problems, but that the conventions that characterise science (which can be quite a mystery to novices) are displayed and are brought to the attention of those who are to learn. Furthermore, teachers should ensure that students do not struggle to the point that frustration builds and begins to impact negatively on their motivation to learn. There is a fine balance between maintaining a climate of productive challenge, and abandoning students to solve problems using personal resources and collaborative efforts with peers. At the first signs of frustration, or even prior to that, an intervention by a teacher can facilitate the learning process. Such interventions can take the form of conversations between a teacher and students during which there can occur a review of what has been undertaken and learned to that point in time. Effective teaching is more than telling students whether they are right or wrong, and usually it involves reviewing procedures and strategies for tackling particular tasks. To mediate the process of learning effectively requires teachers to diagnose what is happening and frame their roles based on their perceptions of the needs and interests of learners.

The role of the teacher in assisting students to develop canonical understandings of science is to mediate between two communities, namely, a classroom which is gradually evolving to become more science-like over time, and the community of science. The teacher, who is a legitimate participant in both communities, is able to infuse new ideas and practices into the classroom at appropriate times, and assist in the dissemination of appropriate ideas and practices as they are introduced by others in the classroom. Roth (in press) has shown that discursive practices introduced by students spread quickly through a classroom, whereas those introduced by a teacher spread less quickly. This might be because those introduced by students are necessary for making progress on the tasks at hand, whereas those introduced by the teacher are not perceived as relevant to the task. Showing students how to frame experience is a critical component of what teachers need to do. Because the patterns of canonical science do not exist in books and demonstrations, it is necessary for teachers to select appropriate times for introducing ideas to frame experience in subsequent activities.

During whole-class activities, the provision of short periods of silence for thinking, and longer breaks for clarifying what has been said and connecting new ideas with current knowledge, can benefit learning. All speakers should speak clearly and in such a way that co-participation is possible. The use of long wait time in whole-class settings makes sense because, on the average, most listeners will benefit from three-five seconds of silence at the end of sentences, paragraphs and between different speakers (Tobin 1987). Brief periods of silence can allow learners to make sense of the discourse of others, relate it to what is known already, and figure out how what has been said relates to their goals. Similarly, at the end of about ten minutes of instruction, it makes sense to break from a whole-class setting to allow for interactions between students in smaller groups. Rowe (1983) described how a 10–2 method, in which every interval of ten minutes of whole-class activity is followed by two minutes of small-group interaction can benefit students. These longer periods of silence allow students to clarify what has happened in the interval immediately prior to the small-group activity, test emerging understandings, and link what is known to new understandings. Interspersing whole-class and small-group activities in this way allows all learners to practise the discourse at a time that is close to their initial encounters.

The issue of equity in terms of whole-class interactions has been raised over the years. For example, in my own research program, we identified 'target' students as those, usually the more able males, who dominate whole-class activities and interactions with the teacher (Tobin & Gallagher 1987). From an equity standpoint, this might be seen as deleterious to females and other minorities who do not get the chance to engage and benefit from engaging at a whole-class level. Possible inequitable learning opportunities should be closely considered by teachers, especially if a high proportion of the class time is spent in whole-class activities. However, if whole-class activities are just a small proportion of total time, then non-target students might not be at a serious disadvantage, particularly if they would prefer not to participate at a whole-class level. Roth (1996) reported that female students in his study preferred not to engage in whole-class settings. If co-participation is occurring and students are not overtly engaged, then there is no problem. However, if a student wants to participate, and feels a need to do so, then failure to encourage his/her participation could result in frustration and feelings of disempowerment. This is a scenario of which teachers need to be aware so that efforts can be made for students to monitor their own co-participation. Thus, an important part of being an effective teacher in whole-class settings might be to teach students how to monitor their own co-participation and signal to the teacher and others when they are not able to engage as they feel they ought. Communicating an inability to co-participate is a responsibility that all members of a learning community might take seriously.

If the route to facilitating learning is via co-participation, then it makes sense to organise students for learning so that they can interact with others and use the shared language that they are constructing. Group learning comprises a number of quite different activities, all of which assume that learning requires discussion among group members before completing a task. Linn and Burbules (1993) distinguished cooperative learning for which a task is divided into parts and group members each complete a part, collaborative learning for which two or more students work together to arrive at an agreed-upon solution to a problem, and tutored learning for which one student teaches another. They indicated that any small-group activity might involve a combination of one or more of the above types.

Noddings (1989) examined some of the theoretical and empirical factors associated with small-group learning and concluded that the issues to be considered prior to implementing group activities included the purposes of the groups, group membership, the roles of teacher and students, the nature of the activities in which students engage, and the manner in which group outcomes are evaluated. According to Linn and Burbules, group learning is usually effective for brainstorming and generating ideas and is usually ineffective for planning activities. However, they cautioned that successful groups are trained to generate ideas and to accept and elaborate the ideas of others without criticising them. Learning in a social setting, including a small group, can occur when students appropriate the ideas of others by building on someone else's idea to create an outcome that they could not have created alone. Linn and Burbules (1993) noted:

> Collaboration succeeds when students are effective at communicating their ideas and able to help other group members see why their idea contributes to the group goal. It also depends on group adherence to a form of discourse that values argument, reliance on evidence, and explanation. (p. 112)

In a study of discourse and science learning in a grade 10 integrated science class, Richmond and Striley (1996) examined the extent to which students understood and made effective use of scientific tools and ideas in their execution and interpretation of experiments and in their discourse with others. Engagement in problem-solving tasks was conceptualised in terms of the students' psychological investments in efforts directed towards learning, understanding or mastering what was to be learned. Over a three-month period, there were increases in the number of students in each group participating in design, implementation and interpretation activities, as well as in the frequency of their participation in discussion and laboratory-related preparations. As students became better at developing and articulating scientific arguments, their level of engagement also increased. One of the most significant changes was that students became more adept at identifying a relevant problem, collecting useful information, stating a testable hypothesis, collecting and summarising data, and discussing the meaning of data.

Differentiated social roles emerged as students worked together in groups. Of considerable importance was the emergence of different types of leaders, who usually were capable students who generated the group's action plan, coordinated assignments, and liaised between students and teachers. Inclusive leaders were able to provide all group members with opportunities to participate and learn. In groups with persuasive leaders, engagement was high for the leader but rarely for others. As is implied by the label, alienating leaders tended to create conflict which reduced the opportunities to participate and learn.

Johnson and Kean (1992) reported that teachers expected students to have stronger group skills than they actually possessed, thus leading teachers to become impatient and frustrated when students acted inappropriately. Battistich, Solomon and Delucchi (1993) asserted that group efforts often can involve negative interactions that reduce interpersonal attraction and actually can impede the learning and achievement of some students. Battistich *et al.* indicated that pressures towards conformity and concurrence-seeking in groups can lead to unproductive collaboration and unreflective decision making that solidify misconceptions or lead to compromises that combine the worst of, rather than the best of, members' ideas. They identified a need to incorporate procedures for improving group interaction and management skills and for creating and maintaining group norms for interpersonal cooperation and concern. They suggested that, throughout lessons, teachers might emphasise the interpersonal values that were relevant to group activity (e.g., fairness, concern for others, responsibility).

Linn and Burbules (1993) commented that students who ask questions that are answered benefit from group learning. They then raised a question about students who are ignored. This question relates directly to research undertaken in mathematics by King (1993) who observed forms of student passivity during small-group cooperative learning, especially among low-achieving students. Low expectations for the performance of low achievers was fuelled by self perceptions and the reinforcing perceptions of their high-achieving peers. In small-group activities, the engagement of low achievers was mediated by the dominant leadership style of some high achievers, interpersonal relations among group members, their relative inability to make a positive impact on the progress of the group, and the reinforcing effects of their negative self-perceptions with regard to personal progress in mathematics. King noted that, because low achievers enjoy learning and working in small groups, educators need to overcome the problems associated with small-group learning. King suggested that: students should rotate roles within groups; tasks should be selected to enable progression from simple to complex and from short duration to relatively long duration; students should be permitted to practise assigned roles; and success experiences for small groups should be promoted. Teachers were encouraged to monitor the progress of low achievers closely and to alter prevailing status differentials, with the goal of establishing more equal rates of participation among low and high achievers.

Corno (1992) observed that one reason why students enjoy group activities is that they often mix learning and relationship-building activities. She advocated that teachers provide support and create activities for students to challenge themselves and others while they pursue goals and grow comfortable with criticism. Corno suggested that teachers might mediate less in the learning of students in small-group activities because 'the value of peers over teachers in delivering external prompts appears to rest on students' common worldviews at various ages, which lead to developmentally appropriate queries and comments' (p. 79). The suggestion highlights the place of student verbal interaction in the negotiation of a shared discourse. Accordingly, teachers need to strike a balance between effective mediation and over-engineering, by gradually relinquishing control of certain processes and objectives. The assumption is that students assume responsibility for the control relinquished by the teacher and employ their greater autonomy to pursue activities to enhance learning and engagement in relevant and meaningful tasks. If the increased autonomy leads to co-participation for students, then small groups can indeed promote meaningful learning.

## SOCIAL CLASS

Social class comprises such factors as occupational status, income level and relationships to the system of ownership of physical and cultural capital, the structure of authority at work and in society, and to the content and process of the work in which an individual engages. Bourdieu (1977) contends that students from middle and upper class families accrue an advantage over lower class students because of relatively greater exposure to cultural institutions such as libraries, museums and property that better prepare them for success in schools. In addition, the primary discourses of middle and upper class families might be better suited to the acquisition of a scientific discourse. There is a dynamic notion of social class in which families are not regarded as fixed in a static class location, but actively construct and negotiate their futures, often under overwhelming constraints. Although one of the key sites for this negotiation is the school, a study reported by Anyon (1981) suggests that the expectations of teachers might play a significant role in cultural reproduction.

Even though all schools were subject to the same State requirements, striking differences, reflecting variations in social class, were observed in the goals and ways in which science curricula were enacted (Anyon 1981). In working class schools, the expectations of teachers for the performance of students were lower than those of teachers in other schools. Although some instances were reported in which teachers from working class schools endeavoured to go beyond simple facts and skills, there was evidence that they believed that students would not benefit from the activities and would use such knowledge infrequently. In working class schools, there was an emphasis on physical control and few efforts were made to motivate students. Anyon described passive resistance (refusal to engage or less than enthusiastic engagement in learning activities) and active resistance of students to the enacted curriculum. She concluded that what counts as knowledge in working class schools is fragmented facts which are isolated from context, connection to each other and wider bodies of meaning. The curriculum consisted mainly of procedures to permit students to carry out tasks that were largely mechanical. In contrast, Anyon described a science curriculum from a middle class school as projecting education as a valuable commodity for getting a job or for gaining admission to university. Knowledge was regarded as more conceptual than in working class schools and was perceived as separate from the efforts of students to learn. In contrast, in an affluent professional school, learning to think was valued and it was felt that students should not just regurgitate facts but should immerse themselves in ideas. There was an awareness that knowledge was constructed by individuals and served social goals. The science curriculum focused on individual expression and development through the acquisition of knowledge that was regarded as conceptual and open to discovery, construction and meaning making. Finally, teachers at an elite school adopted excellence as a theme and taught more subject matter and more difficult concepts than in the other schools. Knowledge was not regarded as a product of an individual's attempts to make sense but was considered as academic, intellectual and rigorous, and a product of disciplined inquiry and rational thought. Children were more intense in competition than in other schools, were under pressure to excel, and were exhorted to produce top-quality performance and seek admission to the best universities.

An important component of the cultural capital that students bring to school reflects the discourses associated with their lives at home. White (1996) suggested that African American students had difficulties in making connections between their primary discourse and the secondary discourse of science. She questioned the value of using a language in the classroom that is often inaccessible to most students. White advocated the use of a negotiated language that enabled students to learn science using written and oral language, incorporate 'personal and colloquial' terms and show scientific knowledge as connected to knowers. To benefit students who do not use standard English in the home, White emphasised the importance of teachers making explicit thematic relationships that might previously have been regarded as implicit in the language. White noted that, for students to become adept in its use, they should be given chances to use the language of science in contexts in which a teacher is available to assist in building bridges between primary and secondary discourses. According to White, through explicit actions of the teacher, the language and relationships encountered in the science curriculum could become 'less mystical' and within the grasp of a majority of students.

White advocated that teachers find ways for the school community to organise better its customs, practices and language in order to reduce the gap between the discourses of the home and school. She commented that, although each child spent more than a half of every day at home, teachers rarely made efforts to learn about the home environment and how it could promote the learning of science. At best, there was a one-way transmission of information of how the home can be adapted to improve the success of a currently unsuccessful child. The home environment was not seen as a source of capital to enhance learning but something that needed to be changed. Accordingly, for most students, a gulf between the discourse of the home and the school was a detriment to their learning.

## CONCLUSIONS

The research on metaphors, beliefs, images, narratives and actions provides insights into how teachers can conceptualise their goals and roles and translate what they know into professional practice. Becoming aware of referents enables them to become cognitive objects that can be used to reflect in and on action and become a basis for changing classroom cultures. The research discussed in this chapter illustrates the promise of creating conceptual objects from conversations about experience. Jacobs gestured with his hand, produced a sketch from which narratives evolved, and opened the possibility of examining metaphors, images and propositional knowledge embedded in those narratives. Through the use of similar techniques, teachers can begin to transform their knowledge from one mode of re-presentation to another and make changes to accord with other beliefs that they might have about teaching and learning. Such an approach makes it possible to link practice to a variety of ways of re-presenting knowledge of teaching and learning conceptually and, in particular, to connect theory to practice. Negotiating a shared language and co-participation in a discourse community is a framework that illuminates critical issues in the teaching and learning of science. Because of the dependence on primary discourse of the cultural history of individuals, the framework brings issues of equity into the foreground and invites investigations of the role of native language, social class and culture in the acquisition of a secondary discourse of science. The significance of the cultural capital of students provides a directive for teachers to bring the knowledge of students into the foreground of curriculum planning and enactment and to consider the cultural histories of learners not as deficits to be overcome, but as a form of capital to invest in the building of scientific discourses. Establishing classroom environments that allow students to employ the habitus of their primary communities appeals as opportunities for *all* students to achieve high levels of performance and escape the spiral of cultural reproduction in which only those from medium and high socioeconomic strata attain functional levels of scientific literacy.

As a mediator of learning, the teacher of science can view his/her role in terms of facilitating the development of a scientific discourse that emanates from the primary discourses of students in a classroom community. The vehicle for attaining this is to assist the students to negotiate a language that enables every student to co-participate in the activities of the community. Because co-participation implies power sharing, teachers need to develop strategies to allow equitable participation for all in the classroom community and provision for each participant to negotiate and employ a shared language that becomes more science-like.

The implications of co-participation depend on the activity types. In whole-class settings, the emphasis should be on students learning from interactions. Use of an extended wait time and the 10–2 method are examples of successes that are possible. The contexts in which whole-class activities are deemed appropriate will determine what precautions should be adopted to ensure that learning opportunities are maximised. The study by Richmond and Striley (1996) suggests that, when the level of motivation to learn is high, students can accomplish world class standards. However, the challenge for teachers is to attain co-participation by linking the cultural capital of students to the instructional activities. At the present time, there are few images of the nature of co-participation in the diverse environments in which most science is taught.

Research on small groups is yet another example of there being no panacea in education. The 'one size fits all' claim of modernity must be contextualised whether the claims relate to the use of an extended wait time or to cooperative learning. It is certainly not the case that learning is optimised in all circumstances through the use of a wait time of between three-five seconds. Similarly, small-group cooperative learning is not optimal for all types of learners and outcomes. The context dependence of approaches to teaching and learning is apparent in the research reviewed here and in a plethora of studies undertaken throughout the world. Rather than advocate small-group activities for all circumstances, it is wiser to examine the fit between students and teachers and to explore carefully what is to be learned in relation to all students' cultural capital on which learning is to be built.

I am suggesting that teachers become aware of theoretical frames that can be used to describe their extant practices and consider the frames and practices in a recursive relationship in which either or both can be adapted. Reflection and reconstruction of conceptual objects is no guarantee that a curriculum can be reformed in the ways envisioned by a teacher. The students are important stakeholders in determining what happens in classrooms and so too are a host of others including parents, colleague teachers, school board members and institutional administrators. If changes are to be successfully introduced into a classroom, it is essential for effective communication to occur with stakeholder groups such as these. There is cause for optimism in that dysfunctional environments, such as those described by Tobin and LaMaster (1995), can be transformed by the efforts of a determined teacher and her students. At the same time, there is reason for caution in that reversion to traditional practices is also well documented (e.g., Ritchie, Tobin & Hook 1997).

The diverse contexts in which science is taught internationally seem to favour a form of teaching and learning that is crafted from transmission models of knowledge transfer. Research on teaching and learning suggests that we know what our science classrooms should be like. The challenge for science teachers everywhere is to research their students in order to identify the capital which they bring to the classroom and to mediate in the development of a language that *everyone* can use to create a trajectory emanating from a host of primary discourses to secondary discourses that are scientific in nature.

## REFERENCES

Anyon, J.: 1981, 'Social Class and School Knowledge', *Curriculum Inquiry* 10, 55–76.
Battistich, V., Solomon, D. & Delucchi, K.: 1993, 'Interaction Processes and Student Outcomes in Cooperative Learning Groups', *The Elementary School Journal* 94, 19–32.
Bourdieu, P.: 1992, *Language and Symbolic Power*, Harvard University Press, Cambridge, MA.
Bourdieu, P.: 1977, 'Cultural Reproduction and Social Reproduction', in J. Karabel & A.H. Halsey (eds), *Power and Ideology in Education*, Oxford University Press, New York, 487–511.
Britzman, D.P.: 1991, *Practice Makes Practice: A Critical Study of Learning to Teach*, State University of New York Press, Albany, NY.
Corno, L.: 1992, 'Encouraging Students to Take Responsibility for Learning and Performance', *The Elementary School Journal* 93, 69–83.
Gee, J.P.: 1990, *Social Linguistics and Literacies: Ideology in Discourses*, Falmer Press, New York.
Johnson, J. & Kean, E.: 1992, 'Improving Science Teaching in Multicultural Settings: A Qualitative Study', *Journal of Science Education and Technology* 1, 275–287.
King, L.H.: 1993, 'High and Low Achievers' Perceptions and Cooperative Learning in Two Small Groups', *The Elementary School Journal* 93, 399–416.
Lee, O., Fradd, S.H. & Sutman, F.X.: 1995, 'Science Knowledge and Cognitive Strategy Use Among Culturally and Linguistically Diverse Students', *Journal of Research in Science Teaching* 32, 797–816.
Lemke, J.L.: 1995, *Textual Politics: Discourse and Social Dynamics*, Taylor & Francis, London.
Linn, M.C. & Burbules, N.C.: 1993, 'Construction of Knowledge and Group Learning', in K. Tobin (ed.), *The Practice of Constructivism in Science Education*, Lawrence Erlbaum, Hillsdale, NJ, 91–119.
McRobbie, C.J. & Tobin, K.: 1995, 'Restraints to Reform: The Congruence of Teacher and Student Actions in a Chemistry Classroom', *Journal of Research in Science Teaching* 32, 373–385.
Noddings, N.: 1989, 'Theoretical and Practical Concerns About Small Groups in Mathematics', *The Elementary School Journal* 89, 607–623.
Phillipson, R.: 1992, *Linguistic Imperialism*, Oxford University Press, Oxford, UK.
Richmond, G. & Striley, J.: 1996, 'Making Meaning in Classroom: Social Processes in Small-Group Discourse and Scientific Knowledge Building', *Journal of Research in Science Teaching* 33, 839–858.
Ritchie, S.M., Tobin, K. & Hook, K.S.: 1997, 'Viability of Mental Models in Learning Chemistry', *Journal of Research in Science Teaching* 34, 223–238.
Roth, W.-M.: 1995, *Authentic School Science: Knowing and Learning in Open-Inquiry Science Laboratories*, Kluwer, Dordrecht, The Netherlands.
Roth, W.-M.: 1996, 'Teacher Questioning in an Open-Inquiry Learning Environment: Interactions of Context, Content, and Student Responses', *Journal of Research in Science Teaching* 33, 709–736.
Roth, W.-M.: in press, *Designing Communities*, Kluwer, Dordrecht, The Netherlands.
Rowe, M.B.: 1983, 'Getting Chemistry Off the Killer Course List', *Journal of Chemical Education* 60, 954–956.
Schön, D.: 1985, *The Design Studio*, RIBA Publications, London.
Sweeney, A. & Gallard, A.J.: 1996, *Language, Limited English Proficiency, and the Learning of Science*, Florida State University, Tallahassee, FL.

Teemant, A., Bernhardt, E.B., Rodriguez-Muñoz, M. & Aiello, M.: 1995, *Bringing Science and Second Language Learning Together: What Every Teacher Needs to Know*, National Center for Science Teaching & Learning, Columbus, OH.

Tobin, K.: 1987, 'The Role of Wait Time in Higher Cognitive Level Learning', *Review of Educational Research* 57, 69–95.

Tobin, K.: 1996, 'Cultural Perspectives on the Teaching and Learning of Science', in M. Ogawa (ed.), *Traditional Culture, Science and Technology and Development – Toward a New Literacy for Science and Technology*, University of Ibaraka, Mito City, Japan, 75–99.

Tobin, K.: 1997, 'The Mediational Role of Culture on the Enacted Science Curriculum', Paper presented at the annual meeting of the American Educational Research Association, Chicago, IL.

Tobin, K., Espinet, M., Byrd, S.E. & Adams, D.: 1988, 'Alternative Perspectives of Effective Science Teaching', *Science Education* 72, 433–451.

Tobin, K. & Gallagher, J.J.: 1987, 'The Role of Target Students in the Science Classroom', *Journal of Research in Science Teaching* 24, 61–75.

Tobin, K., Kahle, J.B. & Fraser, B.J. (eds.): 1990, *Windows into Science Classrooms: Problems Associated with Higher-Level Learning*, Falmer Press, London.

Tobin, K. & LaMaster, S.: 1995, 'Relationships Between Metaphors, Beliefs and Actions in a Context of Science Curriculum Change', *Journal of Research in Science Teaching* 32, 225–242.

Tobin, K. & McRobbie, C.J.: 1996a, 'Cultural Myths as Restraints to the Enacted Science Curriculum', *Science Education* 80, 223–241.

Tobin, K. & McRobbie, C.J.: 1996b, 'Significance of Limited English Proficiency and Cultural Capital to the Performance in Science of Chinese-Australians', *Journal of Research in Science Teaching* 33, 265–282.

Tobin, K., McRobbie, C.J. & Anderson, D.: 1997, 'Dialectical Constraints to the Discursive Practices of a High School Physics Community', *Journal of Research in Science Teaching* 34, 491–507.

Tobin, K. & Tippins, D.: 1996, 'Metaphors as Seeds for Learning and the Improvement of Science Teaching', *Science Education* 80, 711–730.

Tobin, K., Tippins, D.J. & Hook, K.S.: 1994, 'Referents for Changing a Science Curriculum: A Case Study of One Teacher's Change in Beliefs', *Science & Education* 3, 245–264.

White, R.: 1996, *Barriers to Learning Science: Bridging Gaps Between the Languages of the Home and School Science Classrooms*, Florida State University, Tallahassee, FL.

## 2.2 A View of Quality in Teaching

JOHN R. BAIRD
*University of Melbourne, Australia*

Schools commonly espouse the aim to educate students to be aware, responsible and capable of independent action. If this aim is paraphrased using the recent term *metacognition*, such students would be metacognitive regarding their learning in that they *know* about the nature of effective learning, they are *aware of* the purpose and progress of current learning and they effectively *control* this learning by making productive decisions. Often, however, there is a mis-match between what the school *wants* for students and what the school *does* for students, because everyday classroom perspectives and practices do not explicitly foster students' metacognitive development.

In order to improve classroom practices to provide more adequately for students' metacognitive development, the teacher also might need to develop metacognitively, in order to *know* more about the nature of effective teaching, to be more *aware of* the purpose and progress of current teaching, and to *control* this teaching more effectively by making productive decisions. Many teachers wish to teach more effectively, but they do not know what to do or how to start. In this chapter, I present a view of effective learning (based on metacognitive enhancement) that teachers can use to structure their own metacognitive development regarding their teaching. This view has emerged from my involvement in two Australian research projects on classroom teaching and learning, namely, the *Project for Enhancing Effective Learning* (PEEL) which started in 1985 and still continues, and *Teaching and Learning Science in Schools* (TLSS) which ran for the four years during 1987–1990. PEEL focused on teaching and learning across the curriculum, whereas TLSS was limited to science.

I believe that effective teaching and learning require focused, systematic reflection and action on personal practice, and that these processes should be performed collaboratively with others. In order to improve their practice, teachers could need to reconceptualise their own classroom roles and responsibilities. I have organised the sections to follow according to different aspects of such a reconceptualisation. In fact, the sections are organised according to the sequence outlined in Figure 1.

The first section of this chapter argues that much poor learning is due to learning habits that preclude adequate metacognition. Through training in asking

---

Chapter Consultant: Valda Kirkwood (Victoria University of Wellington, New Zealand)

> Poor Learning Habits
>
> *Replace with:*
>
> Good Learning Habits
>
> *Foster by practising:*
>
> Good Learning Behaviours
>
> *Stimulate such behaviours by applying:*
>
> Good Teaching Procedures
>
> (*that should **challenge** students in their learning*)

Figure 1: Sequence for improving learning and teaching

appropriate questions, however, such habits can be replaced by purposeful inquiry based on active reflection. In the second section, such purposeful inquiry is linked to demonstration of particular Good Learning Behaviours (GLBs) that can be fostered by teachers and practised by students. Fostering GLBs in students requires teachers to apply particular types of teaching procedures. The third section introduces a general classroom focus for helping teachers to stimulate desired learning through the concept of *Challenge,* which has both cognitive and affective elements. Each section starts with a summary of the reasoning upon which the section is based, and each includes some implications of this reasoning for classroom action to foster students' metacognitive development.

## FOSTERING PURPOSEFUL INQUIRY FOR ENHANCED METACOGNITION

This section rests on the assumption that poor learning can be due to specific bad learning habits rather than to general intellectual inadequacy. These bad learning habits relate to the learner *not asking* evaluative questions (such as, 'What am I doing?' and 'Why am I doing it?'). The outcome of not asking such questions is inadequate metacognition that leads, in turn, to passivity, dependence and dissatisfaction. Practice in asking evaluative questions and answering them improves both the cognitive and affective outcomes of learning.

In an early research project, I monitored several university science students as they learned. I found that each person's progress and understanding were hindered by specific bad learning habits (see Figure 1) that I called 'Poor Learning Tendencies' (PLTs). These habits included Superficial Attention (scanning material without becoming adequately intellectually involved), Impulsive Attention (attending to certain information and ignoring other information) and Premature Closure (acting as if the task is completed when in fact parts remain undone). I found seven PLTs altogether, with different learners displaying idiosyncratic combinations of two or more of them (Baird & White 1982). Although the habits varied in their nature, each stemmed from a common cause, as it was a different manifestation of the person not adopting an inquiring approach to the work. In each case, the learner did not ask such evaluative questions – such as 'What am I doing?', 'Why am I doing it?', 'What do I already know about this topic?' and 'What else needs to be done?' – and then endeavour to find answers. Premature Closure, for example, results from failing to ask the last of these questions. Commonly, students seem much better at answering questions than they are at asking them (e.g., Arzi & White 1986). In many cases, this occurs because students do not conceive question-asking as part of their classroom role and responsibilities; rather, this is what the teacher does.

Asking questions such as those above forms part of the process of *reflection*. Dewey's concept of reflective thinking (Dewey 1933) closely relates *thinking*, through a state of doubt, hesitation, perplexity, mental difficulty and *acting*, as an act of searching, hunting and inquiring to find material that will resolve the doubt and settle the perplexity (p. 12). His proposed three attitudes to reflective learning (open-mindedness, whole-heartedness and responsibility) and five phases of reflection (suggestions, problem, hypothesis, reasoning and testing) all support this integration of thinking and doing. Even though there have been many reviews of reflection in the context of teaching (e.g., Calderhead & Gates 1993; Sparks-Langer & Colton 1991), Russell (1993) argues that the term could be the most misunderstood word in teacher education (p. 252). Below, I limit my discussion of reflection to the general Deweyan notion of the term.

My research project showed that, in the same way in which these bad learning habits have a common cause, they also have a common outcome. If such evaluative questions are not asked, the student does not achieve an adequate level of metacognition regarding learning. Poor Learning Tendencies and inadequate metacognition are commonplace for students in classrooms. In this situation, students are neither willing nor able to accept responsibility or control over their learning. As mentioned below, metacognition is assuming an increasingly important place in studies of learning and development.

Many writers relate reflection and metacognition because awareness of cognition arises from the essentially human ability to step back and consider one's own cognitive operations as objects of thought, or to reflect on one's own thinking (Brown 1987). In the literature, metacognition is usually considered to comprise the two main components of knowledge about cognition (metacognitive knowledge) and self-regulation of cognition (Brown 1987; Brown, Bransford, Ferrara & Campione 1983; Nelson 1992). Self-regulatory aspects of metacognition are commonly described under the general and somewhat interrelated headings of planning, monitoring and regulating (Corno 1986; Pintrich & Schrauben 1992). Here, I subsume within the self-regulatory component of metacognition the two terms used earlier – *awareness* through questioning and *control* through decision making. By so doing, I emphasise the central role of purposeful, conscious reflective questioning in generating awareness of the nature, purpose and progress of the current learning task. Awareness is a necessary condition for taking control over learning, as it is required in order to make informed, purposeful decisions regarding what to do and how to do it.

Recently, attention has been paid to the relationship of affect and cognition in metacognitive processes and outcomes (e.g., Lester, Garofalo & Kroll 1989; Pintrich & Schrauben 1992). As discussed later, affective elements contribute significantly to effective self-regulation and control of cognition.

In a subsequent study (Baird 1986), I worked for six months with a grade 9–11 science teacher and three of his classes. The teacher and I attempted to replace students' poor learning habits with good ones (Figure 1), through protracted practice in systematic, purposeful inquiry. Students were guided to ask routinely evaluative questions which related to the nature, purpose and progress of classroom activities and personal learning. Near the end of each lesson, students completed a sheet recording the questions which they had asked and their perceptions of their success at finding answers. This training led to marked improvements in students' attitudes and behaviours regarding active, informed learning. Many students commented that, for the first time, they clearly understood the lesson's purpose and progress and therefore could work towards controlling their learning. The extent of improvement in students' metacognition, however, was limited by the teacher. Even though he fully supported the goals of the study, he found considerable difficulty in changing his teaching approach and behaviours to ones that allowed students adequate time and opportunities to become more inquiring. He needed to ask the same types of evaluative questions that the students were asking, such as 'Why am I in this classroom?', 'What am I doing?', and 'Why am I doing it?', but expressly as they related to fostering students' metacognitive development. Thus, he needed to develop metacognitively with regard to his teaching in the same way in which students needed to develop metacognitively regarding their learning. He found this change demanding and he achieved only limited success.

These studies generate my first perspective regarding how teachers should approach improvement in their teaching. This perspective requires the teacher to ask the self-evaluative question: 'How does what I do in class encourage students to acquire habits of asking evaluative questions that, when answered, will foster their awareness of and control over their own learning?' In the next section, I develop a framework for action to answer this question.

## FOCUSING ON PARTICULAR GOOD LEARNING BEHAVIOURS

It is assumed in this section that, in order to foster students' metacognitive development systematically, the teacher should first identify particular Good Learning Behaviours that are appropriate to current student and topic needs. The teacher then can select and implement teaching procedures that explicitly stimulate these Good Learning Behaviours.

As with the previous studies, the aim of the *Project for Enhancing Effective Learning* (PEEL) (Baird & Mitchell 1986; Baird & Northfield 1992) was to have students *purposefully inquire* about lesson nature, purpose and progress, by engaging in routine reflection and action. Through this process, students would become more aware and in control of what they do – that is, they would become more metacognitive. PEEL was designed to provide more extensive support for teacher and student change than had occurred in the previous study. It started in 1985 as a two-year project at one Melbourne secondary school and involved ten teachers in its first year and more thereafter. At the end of these two years, the teachers elected to continue because of the positive outcomes for themselves and their students. The project continues today in this school and many other Victorian government and private schools, having involved hundreds of teachers (from all subject areas) and thousands of students in scores of schools. It also is being adopted currently in many schools in Sweden and Denmark.

In PEEL, the method of *collaborative action research* was very important for providing teachers with guidance and support for their metacognitive development. By this method, teachers joined with each other and, in the first years of the project especially, with university academics in researching classroom teaching and learning. The reflection, action, observation and evaluation cycle of action research provided the teachers with practice in strategies for enhancing their knowledge of, and awareness and control over, their own teaching. Weekly teacher group meetings provided a forum for reflecting on current activities to plan what to do next (cognitive conditions for change), and for mutual support and encouragement to maintain motivation and confidence (affective conditions for change).

Collaborative action research arises from the notion of action research which began in the 1940s and in which problem-based action on social issues can contribute to the emergence of theory of social change. The term has become increasingly popular in education since the late 1960s, largely because of the advocacy of school-centred inquiry and teacher-as-researcher (e.g., Schaefer 1967; Stenhouse 1975) and because of the advantages of collaborative endeavour (e.g., Agyris & Schön 1976). Collaborative action research typically is distinguished by its focus on practical problems, its emphasis on professional development, and its need for a project structure which provides participants with time and support for open communication (Oja & Smulyan 1989, p.12). While significant differences occur in the nature and extent of collaboration, depending on the purpose and context of the research and the types of participants (Erickson 1989), collaborative research allows teachers to contribute to coherent advances in theory and practice (e.g., Tikunoff & Ward 1983).

Early in the PEEL project, teachers developed a framework for structuring the replacement of poor learning habits with good ones. This framework for development of students' question-asking through appropriate teacher action focused on the notion of *Good Learning Behaviours* (GLBs; Figure 1). GLBs are observable learning behaviours that teachers wanted their students to show more often in lessons. Examples of these behaviours were 'actively attends in class', 'attempts tasks correctly and completes them', 'tells teacher what they don't understand' and 'when "stuck", refers to earlier work *before* asking for help'. This seemingly obvious step of listing GLBs was an important step in helping teachers to orchestrate improvements in classroom practice, as it translated a somewhat formless intention of trying to teach better into an operational basis for pedagogical practice. A teacher then could identify from this list perhaps two or three GLBs that required development in a particular class, and then work towards enhancing these behaviours. Teachers then could focus on improvements in learning that were necessary, observable and of a manageable scope.

The next stage of PEEL – to devise and implement teaching procedures for the express purpose of stimulating selected GLBs (see Figure 1) – occupied much of the teachers' time and efforts over the succeeding years. One chapter in Baird and Northfield (1992) records the product of this work by outlining over 75 teaching procedures adapted or created by the PEEL teachers. Selection and application of particular teaching procedures are always purpose-driven. A teacher can choose and apply a procedure for several main reasons: concern about an aspect of students' learning (e.g., that students are not contributing enough of their own ideas); a desire to overcome one or more PLTs that students are commonly exhibiting in class (such as Premature Closure); an attempt to promote a particular aspect of class work (e.g., to solve numerical problems more efficiently). In each case, teaching procedures are suggested to address these aspects. Based on longitudinal data in science classes, Mitchell (1993) reports that application of PEEL teaching procedures clearly changes students' lesson behaviours towards the GLBs considered above. After a period of training, he recorded a frequency of over 50 GLBs per lesson, compared with only two or three GLBs per lesson in science classes that did not have PEEL teachers.

Figure 2 outlines the conceptual basis for the metacognitive development of teachers and students that has emerged from PEEL. The figure promotes an active, generative (Wittrock 1974) view of a learner, based on the premise that students' *perceptions* – their thoughts, beliefs, and feelings about themselves, other persons, and events (Schunk 1992, p. 3) – directly influence learning *behaviours* that, in turn, influence learning *outcomes*. Schunk and Meece (1992) review the diverse research on student perceptions in the classroom (e.g., Zimmerman & Martinez-Pons 1992). Much of this research is structured according to theories related to the interrelated constructs of attributions, social cognition and goals of learning. As described later, I consider students' classroom perceptions in terms of groups of common classroom factors, with a conceptual focus for these groups of factors being the notion of personal Challenge. Figure 2 also highlights the importance of purposeful inquiry for metacognitive development, the benefits of collaboration with others, and the close relationship between cognition and affect in learning perceptions, behaviours and outcomes. Each of these aspects is considered further below.

The findings of PEEL lead to the second perspective regarding a self-evaluative question to be asked of a teacher: 'How does what I do in class explicitly foster the particular Good Learning Behaviours that I have identified as ones that science students at this level should acquire?'

In the next section, I emphasise another aspect of the relationships shown in Figure 2. The focus is on how students perceive their classroom learning, and how these perceptions influence their behaviours and learning outcomes. For this, I consider findings from another long-term collaborative action research project.

## CHALLENGING STUDENTS TO EXHIBIT GOOD LEARNING BEHAVIOURS AND TO ENHANCE METACOGNITION

This section assumes that, if students feel challenged by what they do, they will learn actively. As defined, personal Challenge is a combination of cognitive and affective components. Both components are influenced by major aspects of the classroom teaching/learning situation. If the teacher works with the students to modify these aspects so as to optimise personal Challenge, science classroom activities can be carried out in a spirit of Shared Adventure.

Figure 2 shows an essential relationship pertaining to effective teaching and learning. This relationship is the influence of a person's *perceptions* of the characteristics of the task on the *behaviours* that this person exhibits, and thus the nature and extent of success achieved. *Teaching and Learning Science in Schools* (TLSS) was a four-year project that had as its focus students' perceptions of actual and desired science class work. In this section, I outline briefly the nature of TLSS and focus on a concept that could significantly influence desirable and effective teaching and learning – the concept of personal and productive Challenge.

The focus of TLSS was to understand reasons for the common diminution in many students' perceptions of and attitudes to science as they progress through grades 7–10 (e.g., Cannon & Simpson 1985; Kelly 1986). Over its four years, the project produced a large quantity of qualitative and quantitative data (Baird 1992a, 1992b, 1993, 1994a; Baird, Fensham, Gunstone & White 1989, 1991). The project comprised over 30 discrete but related research studies. It started with nine teachers at two Melbourne secondary schools, and spread to nine schools, involving over 40 teachers and hundreds of students. As with PEEL, the levels were mainly grades 7–10, but the context was limited to science. Research collaboration occurred in groups, by teachers and students meeting with members of the TLSS project team regularly in out-of-class meetings, or by teachers and students working together during lessons. The project demonstrated that teachers and their students together can identify changes that will improve students' enjoyment and

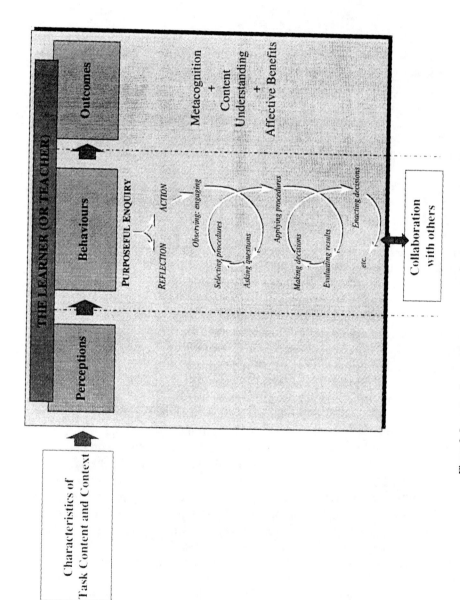

Figure 2: Learning (or teaching) approach, behaviours and outcomes

understanding in lessons, and then work as a team to implement these changes in a constructive and supportive manner (Baird, Fensham, Gunstone & White 1991).

One way of interpreting the large amount of data from the TLSS project was in terms of perceived personal Challenge. Many students become disaffected with school science because it fails to challenge them sufficiently. The notion of Challenge that arose from the TLSS data comprises both a *cognitive/metacognitive* (thinking) *Demand* component, and an *affective* (feeling) *Interest/Motivation* component. Productive Challenge is generated only when the learner perceives *both* Demand and Interest/Motivation to be at desirable levels. Often, however, one or both of these components is not perceived in this way, and the resultant lack of challenge leads to grudging compliance, boredom or frustration. Data indicated that various classroom-based factors influence students' perceived level of Challenge and thus their preparedness to invest the time and energy required to apply GLBs. These factors are included within my model of effective learning shown as Figure 3.

Links between reflection, motivation and metacognition, centring on the notion of Challenge, are central to the model. A convergence of motivational and metacognitive perspectives is advocated by Weinert (1987):

> These two research traditions, metacognition and motivation, have as yet been largely independent, with little common ground . . . [I emphasise] the importance of coordinating studies of the cognitive, metacognitive, and motivational determinants of learning behavior and performance. (pp. 11–13)

Integration of cognition and affect is a feature of various other contemporary classroom learning perspectives and models (e.g., Lester, Garofalo & Kroll 1989; Pintrich & Schrauben 1992). In their extensive review of developmental cognition, Brown, Bransford, Ferrara and Campione (1983) conclude that 'the emotional cannot be divorced from the cognitive, nor the individual from the social' (p. 149). As Challenge is the basis for the integration in Figure 3, I discuss this notion more fully.

The term challenge, when used in the educational literature, usually connotes extension (for the noun) or confrontation (for the verb), usually through higher-order thinking. My notion of Challenge (Figure 3), with its cognitive Demand and affective Interest/Motivation components, is more inclusive than the cognitive use of the term (e.g., Stevenson 1990). Cognitive challenge, however, often is related to affect, especially motivation (e.g., Deci & Ryan 1992). Stevenson (1990) reports that students want to be challenged cognitively by the work that they do, and that they value most the intrinsic rewards associated with feelings of satisfaction and success. Such competence motivation (White 1959) is related to the principle of optimal (cognitive) challenge, in which individuals expend most effort and potentially derive most benefit from tasks that are moderately demanding and that increase personal efficacy (Deci & Ryan 1992; Stipek 1993). However, students' desire for intellectual challenge diminishes across the secondary school years (Igoe & Sullivan 1991).

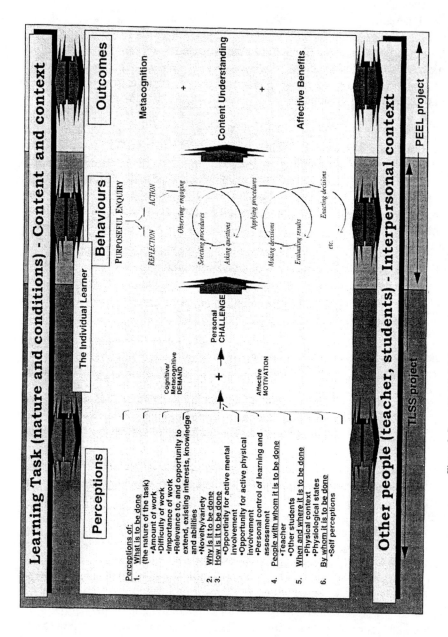

Figure 3: Learning Model that was the conceptual basis for the study

From the extensive literature on motivation, two contrasting orientational traits emerge. These are a *performance* (or *achievement,* but extrinsically-centred) goal orientation and a *mastery* (intrinsic) goal orientation. Challenge can arise with both orientations, but my Interest/Motivation component of Challenge is associated with interest and curiosity, and predisposes towards a motivation to learn by investing effort in the current learning task (Boekaerts 1986). In her survey of research in motivation, information processing and metacognition, Boekaerts (1986) concludes:

> The emergence of a comprehensive theory of motivation to learn (as different from theories of achievement motivation and intrinsic motivation) should be seen as part of the broad trend towards synthesis that is occurring throughout the scientific community. (p. 141)

In Figure 3, the relationship between Demand, Interest/Motivation and personal Challenge forms part of a synthesis of learning approach, progress and achievement. In the figure, students' perceptions of various aspects of classroom and task conditions are important determiners of personal Challenge. These groups comprise perceptions of what is to be done (the nature of the task), why it is to be done, how it is to be done, people with whom it is to be done, when and where it is to be done, and by whom it is to be done (self perceptions). Data from the TLSS project (Baird 1992b, 1993; Baird, Fensham, Gunstone & White 1989) indicate that each of the particular factors grouped within these categories (Figure 3) influenced students' predilection to engage in active, purposeful learning.

Some of these factors are included in other work on motivation in learning. For example, intrinsically motivating tasks should build on students' interests, include novelty and variety, be of moderate difficulty, provide for some degree of self-control, and take place in a positive social context (Corno & Rohrkemper 1985; Stipek 1993). Epstein (cited in Ames 1992) groups different classroom aspects in a manner analogous to that in Figure 3. Her acronym TARGET groups six areas of the classroom that can be manipulated by the teacher to enhance students' intrinsic motivation for learning: Task (design of classroom activities to interest and involve students); Authority (opportunities for students to develop personal independence and control over learning); Recognition (types and use of rewards); Grouping (student-student cooperative arrangements); Evaluation (methods used to assess student learning), Time (workload, pace of instruction, time allotted to complete work).

Another perspective on students' perceptions of school science is included in Claxton's (1991) book *Educating the Enquiring Mind: The Challenge for School Science.* He groups students' perceived failings of science under such headings as its *bittiness* (fragmentary nature), *invisible structure* and *pointlessness* (lack of clear nature and purpose), and consequent *illogicality* and *falsification* (the pursuit of the right answer at the expense of reason). These failings turn students away from science.

Next, I consider how a teacher might use the perspectives from the research discussed above in order to improve the quality of teaching and student learning. My argument is that, by ascertaining students' perceptions regarding *specific aspects of classwork* and then acting to optimise personal Challenge, the class will surmount these failings while applying themselves and enjoying and understanding science more. Stipek (1993) reviews various general ways in which a teacher can generate positive student beliefs regarding their learning and thereby optimise motivation (and thus challenge). I attempt a similar goal, but through action according to the components of my learning model in Figure 3.

Each of the factors in the figure can be manipulated by the teacher. Thus, by focusing explicitly on the influence of each factor on the nature and extent of student Challenge, the teacher has a structure to approach improvement of teaching. Structured diagnosis of student perceptions can be achieved by first having students evaluate recent class work in terms of the factors, followed by class discussion of collated results. Based on these results, the teacher and students together can formulate a plan for action to overcome perceived shortcomings in factors identified as unsatisfactory. In recent research (Baird 1994b), I base such a diagnosis and discussion on a form for completion by students. This form requires students to evaluate the perceived level of each of the major elements of the model in Figure 3. For example, for the factors of Amount and Difficulty of work in the student's science lessons for a particular week, questions include:

- How *much* work was there to do? (Lots/A fair bit/ Not much/Very little)
- How do I *feel* about this level?
- How *difficult* was the work? (Really difficult/Fairly difficult/Not very difficult/ Not difficult at all)
- How do I *feel* about this level?

Data arising from this work indicate that collated student responses to this form provides teacher and students with a useful focus for diagnosis and action to improve classroom practice.

Teaching according to these perspectives is complex and demanding. It requires the teacher to return continually to the three key questions considered earlier. Why am I in this classroom? What am I doing? Why am I doing it? These questions can provide a practice-based, personally relevant focus for pursuing the broader question of 'What are we teaching science for?' Claxton (1991) develops various possible perspectives for answering this question, including those related to knowledge transmission, effective action in society, the improvement of learning and reasoning, and the training of future scientists. Thus, by focused reflection upon particulars of personal practice, a teacher can work towards a meaningful philosophy of professional action.

The conceptualisation shown in Figure 3 forms the basis for a new term to summarise quality in teaching and learning. This term is *Shared Adventure*. Shared Adventure exists where productive *personal* Challenge occurs in a context of supportive *interpersonal* collaboration (as in Figure 3). In the classroom, the teacher and each student are engaged in an endeavour that generates appropriate levels of personal Demand and Interest/Motivation, and that is shared with others.

Based on the TLSS work on fostering personal Challenge for students, a guiding self-evaluation question for teachers could be 'How can I act with students to develop the level of Challenge in science classwork?'

## CONCLUSION

Teacher development is central to school improvement. In order that teachers improve the effectiveness of their teaching, they must acquire competence and confidence to undergo metacognitive development regarding themselves and their practice. Focused professional reflection and dialogue (e.g., Rudduck 1987) as part of teacher collaborative action research can provide opportunities, guidance and support for changes in personal practices and beliefs. In this chapter, I have presented the notions of individual Challenge and Shared Adventure as two foci for such collaborative reflection and action. Collaboration between adults and children in investigating cognitive activities can assist the metacognitive development of both the learner (Scardamalia & Bereiter 1983) and the teacher (Baird 1986; Baird, Fensham, Gunstone & White 1991). Purposeful collaborative inquiry to diagnose and improve individual Challenge and Shared Adventure, indeed, could become the basis for a school environment that, by emphasising thoughtfulness and collegiality (Costa 1992), improves the quality of education for all.

## REFERENCES

Agyris, C. & Schön, D.A.: 1976, *Theory in Practice: Increasing Professional Effectiveness*, Jossey-Bass, San Francisco, CA.

Ames, C.: 1992, 'Achievement Goals and the Classroom Motivational Climate', in D.H. Schunk & J.L. Meece (eds.), *Student Perceptions in the Classroom*, Lawrence Erlbaum, Hillsdale, NJ, 327–348.

Arzi, H.J. & White, R.T.: 1986, 'Questions on Students' Questions', *Research in Science Education* 16, 82–91.

Baird, J.R.: 1986, 'Improving Learning Through Enhanced Metacognition: A Classroom Study', *European Journal of Science Education* 8, 263–282.

Baird, J.R.: 1992a, 'Collaborative Reflection, Systematic Inquiry, Better Teaching', in T. Russell & H. Munby (eds), *Teachers and Teaching: From Classroom to Reflection*, Falmer Press, London, 33–48.

Baird, J.R. (ed.): 1992b, *Shared Adventure: A View of Quality Science Teaching and Learning*, Monash University, Melbourne, Australia.

Baird, J.R. (ed.): 1993, *Exploring Quality in Science Learning*, Monash University, Melbourne, Australia.

Baird, J.R.: 1994a, 'A Framework for Improving Educational Practice: Individual Challenge; Shared Adventure', in J. Edwards (ed.), *Thinking: International and Interdisciplinary Perspectives*, Hawker Brownlow Education, Melbourne, Australia, 137–146.

Baird, J.R.: 1994b, 'Classroom Collaboration to Improve Teaching and Learning', *RefLecT: The Journal of Reflection in Learning and Teaching* 1(1), 20–24.

Baird, J.R., Fensham, P.J., Gunstone, R.F. & White, R.T.: 1989, *Teaching and Learning Science in Schools: A Progress Report*, Monash University, Melbourne, Australia.

Baird, J.R., Fensham, P.J., Gunstone, R.F. & White, R.T.: 1991, 'The Importance of Reflection in Improving Science Teaching and Learning', *Journal of Research in Science Teaching* 28, 163–182.

Baird, J.R. & Mitchell, I.J. (eds.): 1986, *Improving the Quality of Teaching and Learning: An Australian Case Study – The PEEL Project*, Monash University, Melbourne, Australia.

Baird, J.R. & J.R. Northfield (eds): 1992, *Learning from the PEEL Experience*, Monash University, Melbourne, Australia.

Baird, J.R. & White, R.T.: 1982, 'Promoting Self-Control of Learning', *Instructional Science* 11, 227–247.

Boekaerts, M.: 1986, 'Motivation in Theories of Learning', *International Journal of Educational Research* 10, 129–141.

Brown, A.L.: 1987, 'Metacognition, Executive Control, Self-Regulation, and Other More Mysterious Mechanisms', in F. Weinert & R. Kluwe (eds.), *Metacognition, Motivation and Understanding*, Lawrence Erlbaum, Hillsdale, NJ, 65–115.

Brown, A.L., Bransford, J.D., Ferrara, R.A. & Campione, J.C.: 1983, 'Learning, Remembering and Understanding', in J.H. Flavell & E.M. Markman (eds.), *Handbook of Child Psychology Volume 3: Cognitive Development* (fourth edition), Wiley, New York, 77–166.

Calderhead, J. & Gates, P. (eds.): 1993, *Conceptualizing Reflection in Teacher Development*, Falmer Press, London.

Cannon, R.K. & Simpson, R.D.: 1985, 'Relationships Among Attitude, Motivation and Achievement of Ability Grouped, Seventh-Grade, Life Science Students', *Science Education* 69, 121–138.

Claxton, G.: 1991, *Educating the Enquiring Mind: The Challenge for School Science*, Harvester Wheatsheaf, Hertfordshire, UK.

Corno, L.: 1986, 'The Metacognitive Control Components of Self-Regulated Learning', *Contemporary Educational Psychology* 11, 333–346.

Corno, L. & Rohrkemper, M.M.: 1985, 'The Intrinsic Motivation to Learn in the Classroom', in C. Ames & R. Ames (eds.), *Research on Motivation in Education* (Volume 2), Academic Press, New York, 53–90.

Costa, A.L.: 1992, 'An Environment for Thinking', in C. Collins & J.N. Mangieri (eds.), *Teaching Thinking: An Agenda for the 21st Century*, Lawrence Erlbaum, Hillsdale, NJ, 169–182.

Deci, E.L. & Ryan, R.M.: 1992, 'The Initiation and Regulation of Intrinsically Motivated Learning and Achievement', in A.K. Boggiano & T.S. Pittman (eds.), *Achievement and Motivation: A Social-Developmental Perspective*, Cambridge University Press, Cambridge, UK, 9–36.

Dewey, J.: 1933, *How We Think*, Heath, New York.

Erickson, G.: 1989, 'Collaborative Inquiry and the Professional Development of Science Teachers', Paper presented to the CATE pre-conference on Collaborative Approaches to Teacher Education, Quebec City, Canada.

Igoe, A.R. & Sullivan, H.: 1991, 'Gender and Grade-Level Differences in Student Attributes Related to School Learning and Motivation', Paper presented at the annual meeting of the American Educational Research Association, Chicago, IL.

Kelly, A.: 1986, 'The Development of Girls' and Boys' Attitudes to Science: A Longitudinal Study', *European Journal of Science Education* 8, 399–412.

Lester, F.K., Garofalo, J. & Kroll, D.L.: 1989, 'Self-Confidence, Interest, Beliefs and Metacognition: Key Influences on Problem-Solving Behavior', in D.B. McLeod & V.M. Adams (eds.), *Affect and Mathematical Problem Solving. A New Perspective*, Springer-Verlag, New York, 75–88.

Mitchell, I.J.: 1993, *Teaching for Quality Learning*, Unpublished PhD thesis, Monash University, Melbourne, Australia.

Nelson, T.O. (ed.): 1992, *Metacognition: Core Readings*, Allyn and Bacon, Boston, MA.

Oja, S.N. & Smulyan, L.: 1989, *Collaborative Action Research: A Developmental Approach*, Falmer Press, London.

Pintrich, P.R. & Schrauben, B.: 1992, 'Students' Motivational Beliefs and Their Cognitive Engagement in Classroom Academic Tasks', in D.H. Schunk & J.L. Meece (eds.), *Student Perceptions in the Classroom*, Lawrence Erlbaum, Hillsdale, NJ, 149–183.

Rudduck, J.: 1987, 'Partnership Supervision as a Basis for the Professional Development of New and Experienced Teachers', in M.F. Wideen & I. Andrews (eds.), *Staff Development for School Improvement: A Focus on the Teacher*, Falmer Press, London, 129–141.

Russell, T.: 1993, 'Learning to Teach Science: Constructivism, Reflection and Learning from Experience', in K. Tobin (ed.), *The Practice of Constructivism in Science Education*, AAAS Press, Washington, DC, 247–258.

Scardamalia, M. & Bereiter, C.: 1983, 'Child as Coinvestigator: Helping Children Gain Insight Into Their Own Mental Processes', in S.G. Paris, G.M. Olson & H.W. Stevenson (eds.), *Learning and Motivation in the Classroom*, Lawrence Erlbaum, Hillsdale, NJ, 61–82.

Schaefer, R.J.: 1967, *The School as the Center of Inquiry*, Harper and Rowe, New York.

Schunk, D.H.: 1992, 'Theory and Research on Student Perceptions in the Classroom', in D.H. Schunk & J.L. Meece (eds.), *Student Perceptions in the Classroom*, Lawrence Erlbaum, Hillsdale, NJ, 3–23.

Schunk, D.H. & Meece, J.L.: 1992, *Student Perceptions in the Classroom*, Lawrence Erlbaum, Hillsdale, NJ.
Sparks-Langer, G.M. & Colton, A.B.: 1991, 'Synthesis of Research on Teachers' Reflective Thinking', *Educational Leadership* 48(6), 37–44.
Stenhouse, L.: 1975, *An Introduction to Curriculum Research and Development*, Heinemann, London.
Stevenson, R.B.: 1990, 'Engagement and Cognitive Challenge in Thoughtful Social Studies Classes: A Study of Student Perspectives', *Journal of Curriculum Studies* 22, 329–341.
Stipek, D.J.: 1993, *Motivation to Learn* (second edition), Allyn and Bacon, Boston, MA.
Tikunoff, W.J. & Ward, B.A.: 1983, 'Collaborative Research on Teaching', *Elementary School Journal* 83, 453–468.
White, R.: 1959, 'Motivation Reconsidered: The Concept of Competence', *Psychological Review* 66, 297–333.
Wienert, F.E.: 1987, 'Introduction and Overview: Metacognition and Motivation as Determinants of Effective Learning and Understanding', in F. Weinert & R. Kluwe (eds.), *Metacognition, Motivation and Understanding*, Lawrence Erlbaum, Hillsdale, NJ, 1–19.
Wittrock, M.: 1974, 'Learning as a Generative Process', *Educational Psychologist* 11, 87–95.
Zimmerman, B.J. & Martinez-Pons, M.: 1992, 'Perceptions of Efficacy and Strategy Use in the Self-Regulation of Learning', in D.H. Schunk & J.L. Meece (eds.), *Student Perceptions in the Classroom*, Lawrence Erlbaum, Hillsdale, NJ, 185–207.

## 2.3 Teaching and Learning as Everyday Activity

**WOLFF-MICHAEL ROTH**
*University of Victoria, Canada*

Vignette 1: A friend of mine often told me about her dilemma in assessing the performance of preservice and inservice teachers by means of a widely used teacher assessment instrument. Using the instrument, she found that she rated one teacher much higher than another, though a 'gut feeling' told her that the second was a much better teacher and showed a much more convincing performance. To date, she has not been able to resolve this dilemma.

Vignette 2: During a recent visit to my bank, I was served by two customer representatives, one who is a specialist in loans and the other who specialises in retirement plans and investments. Their organisation had paired them up so that they could teach each other; the loans officer learned about retirement plans and investments; and the retirement plans and investment expert learned about loans. They taught each other in the context of the ongoing daily and ordinary activity of banking, at each other's elbows, so to speak. Their teaching was part of the ongoing activity.

The first vignette is representative of the widely differing perspectives on science teaching and teacher education which rifts the field. On the one hand, there are those who believe that teachers' knowledge can be stated in the form of propositions and heuristics and classified (e.g., into subject matter knowledge, pedagogical content knowledge and general pedagogical knowledge). The assumption is that, as soon as teacher knowledge can be stated in the form of propositions, procedures or heuristics, it is explicit and can be assessed more or less reliably. The knowledge and performance of teachers then can be compared against some desirable standard. On the other hand, there are those who believe that some essential aspects of teaching are inherently tacit. This tacit knowledge, by its very nature, cannot be assessed against some external criteria, although expertise and excellence are readily recognised by other competent members of the community.

---

Chapter Consultant: Gilberto Alfaro (National University, Costa Rica)

The second vignette is an example from everyday life in a big corporation where very effective methods of teaching and learning are often practised without ever becoming known to – let alone affecting the life and practice of – those who have made education their career. A significant body of literature suggests that useful teaching occurs in practice, on the job, as newcomers work together with more competent old-timers. Authentic practices are difficult to teach in an explicit way because so much of them are tacit and embodied in the actions of experienced practitioners. Understanding teaching is difficult because it, like the practice to be taught, has aspects which are tacit and thus cannot be brought into the domain of discourse.

In our work on teaching and learning, the notion of *practice* has become central. (I use the plural 'we' to include various teachers with whom I collaborated over the years.) Here, it operates at two levels. First, we designed learning environments that allowed us to teach science as practice. At a second level, we considered science teaching to be a practice in itself. Understanding science teaching is difficult in a double sense because it has to deal with the tacit aspects at both levels. In the first section of this chapter, I outline theoretical perspectives on practice, including the notions of practice, reflective activity of practitioners and the dilemmas of teaching a practice. The second section provides some examples from our own research on the practice of teaching science. Finally, I consider science teacher education.

## THEORETICAL PERSPECTIVES ON PRACTICE

*Practice*

Human practices, whether these relate to laboratory activities in microbiology, construction of Transversely Excited Atmospheric Pressure $CO_2$ lasers in physics, or teaching science, have elements which are inaccessible to description (Collins 1982; Jordan & Lynch 1993; Roth 1996a). That is, it is impossible to render through analysis or teach directly some important elements of these practices, because scientific analysis represents practices recipe-like, as series of discrete steps. However, these 'recipes' have little to do with masterful practices themselves – it is well known that a chef's recipe describes at the same time the production of outstanding cuisine and also humbling home-cooking experiences – and are inherently ambiguous (Suchman 1987).

A second problem of scientific analysis is that it provides a synoptic view of practice by drawing together aspects that are never brought together simultaneously in practice. These aspects, though they might be *logically* incompatible, are compatible *in practice*. This gives rise to apparently irrational, inconsistent, illogical or conflicting actions when they are compared across settings. However, the linearity and temporality of practice and the principle of the economy of logic allow two seemingly contradictory applications of a scheme to be found in one and the same person, because they are never brought face to face in the same practical situation (Bourdieu 1990).

Teaching authentic practices in purely explicit ways is further made difficult because they very often are situation-specific and differ drastically even between quite similar settings. Most dishearteningly for objectivist views on practice, the same technique can give rise to significantly different practices when the same individuals change their work sites. However, good science teachers (*qua* practitioners) know that their practices cannot be the same in two different sections of the same course, or even for different students in the same class. They might have to draw on different practices to teach a concept such as interacting with one student over and about an emerging vector diagram on the blackboard, mounting a demonstration on an air track for another, sending a third group to the library to do some background reading, and so on. They also might generate and try new practices 'on the spot' to facilitate the learning of students when their old practices are insufficient.

*Practice and Reflection*

Competent practitioners are characterised by their dispositions and practical competence to make decisions 'on the spot', 'in the twinkling of an eye' or 'in the heat of the moment', (i.e., under conditions which exclude distance, perspective, detachment and reflection). An architect who aesthetically places a staircase, a dairy worker who figures the correct price of a delivery in an instant, or a science teacher who appropriately contributes to a student-centred conversation all employ 'know-how', or sets of practical principles that orient choices at once minute and decisive. The appropriateness of actions and practical choices in the face of complex settings are achieved by *dispositions* (Lave 1988) or *reflection-**in**-action* (Schön 1987). Competent practitioners apply in their actions these dispositions, which appear to them only in action and in their relationships with lived situations; here, practical hypotheses based on past experience produce an infinite number of practices that are relatively unpredictable but extremely flexible in framing and reframing situations.

The most important aspects of competent practice cannot be brought into the domain of discourse (Dreyfus 1992). There is every reason to think that, as soon as practitioners reflect *on* their practice, adopting a quasi-theoretical posture, they lose any chance of completely expressing their practice because reflection and discourse leave unsaid that 'which goes without saying' (Bourdieu 1990; Schön 1987). When practitioners reflect *on* action, they systematically and deliberately think back on what they have done and on how their *dispositions* have operated to shape the setting. *Reflection-**on**-action* is characterised by a 'stop-and-think' attitude which permits the distancing necessary to 'learn about learning'. When this happens in practice, practice itself is disrupted, because practice excludes all attention to itself and all formal concerns. As professionals and lay who try to improve their game (grip on the golf club, body position with respect to the tennis ball) or their interactions with students in the science classroom know, performance actually can be lower as long as one *consciously* attends to the practice.

Much of the literature on reflective teaching and reflective teacher education has concerned itself with *reflection-on-action* (Munby & Russell 1992). What this literature on teacher change through *reflection-on-action* has backgrounded is the fundamental difference of experiencing a situation while in action versus looking back at it. This overemphasises hindsight over action in real time. Hindsight is an illusion of the nature of life, because it makes life appear as the realisation of an essence that pre-exists and transcends everything (Bourdieu 1990). Thus, a change in practice which we can deduce by retrospectively analysing teaching situations (i.e., by *reflection-on-action*) could prove to be inappropriate. At best, appropriate teaching moves, like sociological objects, are constructed in 'a protracted and exacting task that is accomplished little by little, through a whole series of small rectifications and amendments inspired by what is called *le métier*, the "know-how", that is, by the set of practical principles that orients choices at once minute and decisive' (Bourdieu & Wacquant 1992, p. 228, emphasis in the original).

*Dilemmas of Teaching a Practice*

There are some fundamental dilemmas in the teaching of authentic practices such as doing and talking science or teaching science. First, excellence or practical mastery ceases 'to exist once people start asking whether it can be taught, as soon as they seek to base "correct" practice on rules extracted, for the purpose of transmission, as in all academicisms, from the practices of earlier periods and their products' (Bourdieu 1990, p. 103). Any practice can be described and thus symbolically mastered, but the resulting representation is inadequate in some fundamental ways. It takes more than the 'recipes' that are communicated in lectures, how-to books, reference and instruction manuals, or videotapes to achieve practical mastery of microbiological analysis, tennis, cooking, research design in sociology, using a photocopier (even if it has an 'intelligent' help system) or teaching science (Bourdieu & Wacquant 1992; Jordan & Lynch 1993; Roth 1995, 1996b; Suchman 1987).

A second dilemma lies in the implicit and tacit aspects of masterful practices which give these practices an element of immediacy, surprise and unpredictability. It is impossible to teach these aspects – which practitioners often express as 'feel', 'gut feeling' or 'touch' – especially in modes of transmission which reign in current educational paradigms (from kindergarten to university). Thus, methodologies and methods courses are oxymorons, for they miss the very essence of the practices which they were designed to develop in students. In Bourdieu and Wacquant's (1992) view, 'methodology is like spelling which the French call *la science des ânes* ("the science of the jack-asses"). It is a compendium of errors of which one can say that you must be dumb to commit most of them' (p. 244, italics in the original). A third dilemma arises because the implicit and tacit nature of masterful practices resist measurement. Yet competent individuals recognise masterful practices and can separate them from impostors, just as my friend in Vignette 1 recognised the better performance, although it did not appear on the quantitative scales of the assessment instrument. A fourth dilemma of learning a practice lies in the fact that students must engage in practice *before* they can know what they do and *before* the language of old-timers makes any sense (Schön 1987). It has been suggested that participation in practice provides an avenue out of these dilemmas. By participating in practice, students become new members of communities of practice.

*Teaching and Participation in Practice*

Historians and philosophers of science, and especially scientists themselves, have observed that a good part of scientific practice is acquired by modes of transmission which are thoroughly practical. There is no substitute for learning the fundamental principles of a practice – sociological research, Mayan midwifery, architectural design or science teaching – than by engaging in it alongside a practitioner as guide or coach, who models and scaffolds, provides feedback and corrects *in situ* and with reference to the specific case (Bourdieu & Wacquant 1992; Lave & Wenger 1991; Schön 1988). The implied *pedagogy of silence* leaves little room for explication of both the theoretical knowledge transmitted and the mechanism of the teaching-learning process, especially in those domains, such as teaching, in which the contents of knowledge and the modes of thinking and acting are themselves less explicit and codified. What is to be communicated consists essentially of a *modus operandi*, a mode of production which presupposes ways of perceiving and talking about the world. There is no way to acquire it other than to make people see and experience it in practical operation. 'Do as I do' and 'follow me' are for Bourdieu and Schön the overriding principles of teaching and are central aspects of this activity.

In most communities of practice, teaching and learning are ubiquitous in members' situated activity. Situated activity always involves changes in knowledge and action which are central to what is meant by 'learning' (Lave 1993). Understanding – an open-ended and partial process arising from interaction with and structuring of the world – and participation in ongoing activity replace traditional notions of 'learning'. In this way, much of the non-articulated knowledge of the practitioner can be 'picked up' by the newcomer. Over time, these newcomers become old-timers themselves in that they are recognised as being core members of the community. During this trajectory from legitimate peripheral to core participation, newcomers learn community-specific ways of talking about, viewing, acting on and interacting with the world. In communities of practice, the collaboration of members with different expertise and different competence levels is used to achieve the formation of groups with overlapping distributions of knowledge and expertise. By engaging together in practices (such as in team navigation or collaborative medical diagnosis and as exemplified with the banking officers in Vignette 2), members of these groups receive job training, job performance and safety-enhancing redundancy in a single activity (Cicourel 1990; Hutchins

1995). While such work practices have been shown to be efficient in many domains, they have yet to be utilised in the domain of science teaching (from student and teacher perspectives).

## TEACHING SCIENCE AS PRACTICE AND PRACTICES OF SCIENCE TEACHING

Because the work of education in modern Western-type societies is institutionalised, science teachers find themselves at the intersection of two communities. First, science teachers are members of an autonomous community of practice to which the reproduction of society has been delegated, and which shares similarities beyond subject matter boundaries. Second, they also are representatives of the scientific community which has its own intellectual practices that students should learn (Brown, Collins & Duguid 1989). As such, they are to induct students into the authentic practices, ways of doing, talking and thinking science that scientists' exemplify. Shulman (1987) has tried to capture this double membership, this double role, as practitioner in his distinction between pedagogical knowledge (science teacher *qua* pedagogue) versus content knowledge (science teacher *qua* scientist). Discussing a science teacher's actions in terms of two practices is a matter of heuristics. In the daily lives of science teachers, these two practices are irremediably nested and enfolded within each other. Science teachers always act as whole persons, drawing from repertoires of practices, or generating new ones, to structure their settings in appropriate ways. This heuristic, however, has helped me to think through some difficult issues in my teaching practice.

### *Science Teacher* qua *Pedagogue*

Recent research in the domain of situated cognition has shown that school knowledge is essentially brittle and not useful in everyday out-of-school life; students learn forms of mathematics which have little to do with practices that people develop in their everyday lives, whether as just plain folks, mathematicians or scientists. In the sciences, the same discontinuity exists which is illustrated by an extensive literature on students' understandings (or discourses) which differ in essential ways from those of scientists. Although students learn (evidenced by differences between pretest and posttest, increases on standardised tests over the years, etc.), this knowledge has little to do with everyday scientific practices. Traditional teaching has blamed students who 'failed to apply what they had learned'. However, our schooling systems and all teachers and administrators who subscribe to it need to share this blame. In this function as pedagogue, teaching often has little to do with inducting students into authentic ways of knowing and more with control and the production and reproduction of social inequalities along the lines of gender, race or socioeconomic class; teachers are part of a mechanism to construct failure, limit social mobility or stratify society (Brookhart Costa 1993; Eckert 1990; Lave 1993; Lemke 1990).

On the other hand, we have been able to document the positive attitudes that develop when students structure their science learning, that is, when they take charge of their learning in ways in which people do everyday outside schools (Roth 1994b, 1996a; Roth & Roychoudhury 1994a). As pedagogues, we have helped students to become functioning members of a democratic society, independent, responsible and in charge of their activities, including learning. For example, some of our students decided the topic and nature of their experiments by designing and conducting series of experiments in content areas such as ecology or motion (Roth 1994a; Roth & Bowen 1995). Others designed an electricity unit entirely on their own; one group planned and conducted a lesson for grade 5 students; a second group designed a series of experiments in conductivity, including semiconductivity and superconductivity for which they also organised the purchase and transport of liquid nitrogen; in a third situation, a student decided to do an individual library study on body electricity, for which he collected information from the school, community and university libraries. For the assessment of student learning, 65 percent of the grade derived from self and peer evaluation (with few exceptions, they were lower than my assessments). Students in these classes also contributed to research reports. For example, an article on students' views of concept mapping was edited by 25 percent of the participating students who were representative of the class in terms of gender and achievement (Roth 1994b); another report was coauthored by one of the participating students (Roth & Alexander 1997). Thus, science teachers can contribute, *qua* pedagogues, in students' emancipatory endeavours towards a more democratic society.

Embedded in these examples is another aspect of a science teachers' reflective practice. We were interested in improving teaching and learning in our classrooms. Videotaped classroom events became resources for reflecting *on* our actions and interactions in the classroom (Roth 1993, 1995; Roth & Roychoudhury 1994b) and the kind of learning which happened in the context of various learning environments, such as open-inquiry science laboratories (Roth & Bowen 1993), collaborative concept mapping (Roth & Roychoudhury 1992), modelling with computer microworlds (Roth 1994a) or designing and arguing about simple machines (Roth 1996d). By reflecting *on* classroom events, we adjusted teaching techniques or interactional practices by consciously making changes. We monitored subsequent classroom events to see what (if any) changes there were in our teaching (*in* action). Through successive cycles of engaging *in* teaching and reflecting *on* teaching, we succeeded in making changes to our teaching which led to better learning environments.

*Science Teacher* qua *Scientist*

Science teachers also are representatives of the scientific community, located somewhere along the trajectory from limited participation (e.g., a middle school teacher with little training and experience in the ways of scientists) to core participation (a high school, college or university teacher with an undergraduate or graduate degree in science and/or working experience as a scientist). As such, they have the opportunity to act as practitioners and to induct students into scientific practices. This means, however, that students and teachers have to engage in the same kind of open-ended and ill-defined settings which characterise the everyday practices of scientists.

In the past, we have used the metaphor of the teacher as a reflective practitioner (especially in the guise of the cognitive apprenticeship metaphor) to plan and research learning environments which provided students with opportunities to experience some essential characteristics of scientific practice (Roth 1993, 1994a; Roth & Bowen 1995; Roth & Roychoudhury 1993). In this role, we acted *qua* scientist, helping students in the context of specific and interesting problems to talk and do science in more canonical ways. By necessity, to have any authenticity at all, our learning environments needed to share some fundamental characteristics with the everyday work of scientists. Among these characteristics are the (1) ill-defined nature of problems, (2) irremediable uncertainty and ambiguity of settings, instructions and discourse, (3) distributed nature of knowing and learning and (4) predication of learning on current knowledge. While our students worked in such learning environments, our metaphor of the reflective practitioner (scientist) provided us with patterns for interacting with students. Depending on the situation, we modelled expertise-in-use, coached students or scaffolded their initial attempts in a new skill on a *need-to-know* or *just-in-time* basis, or served as advisors in the students' independent projects. There also was a considerable number of situations in which we, trained physicists or biologists, did not have any indication of possible answers. We took these situations as opportunities to be scientists, working on-line and irreversibly, with all the mumbles and stumbles and unfortunate moves arising from – but also all the affordances provided by – tinkering, exploitation of contingency and professional intuition.

*Science Teaching as Discourse*

In my work as teacher and researcher, I realised that those teaching-learning situations were crucial which allowed, first, students to develop and practice their science talk and, second, teachers to engage in understanding and redirecting (through modelling) students' conversations to allow canonical science discourse to emerge. In our work, we found that those activities worked well which allowed the collective production, manipulation and interpretation of a variety of inscriptions in the form of concept maps, graphs, animated diagrams in computer microworlds, or design artefacts (Roth 1995, 1996a; Roth & Bowen 1995; Roth & Roychoudhury 1992, 1994b). We found that each of these settings encouraged students to engage in science talk which they maintained over long periods of time. This allowed teachers to listen in on and, when necessary, participate in students' conversations. As conversational participants, teachers could identify students' ways of seeing and talking about specific phenomena. They then could contribute to the conversation to facilitate changes in the discourse. These studies also showed that in spite of the teachers' competencies, their contributions were neither necessary nor sufficient to bring about specific forms of discourse.

*Researching the Practice of Science Teaching*

In our recent work, we have begun to bring our theoretical understandings of practice back into practice itself. In the context of a study on sociocultural aspects of learning in elementary science classrooms, graduate students and university-based researchers with academic backgrounds in science and experience in teaching science have begun to engage in joint teaching efforts with classroom teachers. We think that such joint teaching affords learning opportunities that resemble those of the individuals in the banking vignette, or any other team of co-participating practitioners (Roth 1998). Here we see opportunities to study what the collaboration of practitioners – who literally work at each other's elbows – affords to their respective understandings of science teaching. The following are some initial observations in a new and exciting area of theory, research and practice in science teaching and teacher education (Roth 1996b, 1996c).

As we observed Tammy and Gitte, who team-taught a grade 4/5 science unit, we learned important lessons about the tacit and overt aspects of teaching. Some aspects of teaching were never made explicit, although we noted changes in the behaviour of both teachers. Observers noted how, for example, the questions related to content asked by Tammy became more and more like those asked by Gitte; the questions Gitte asked and the techniques which she employed to facilitate collaborative student work increasingly resembled the pedagogical techniques employed by Tammy. Our field notes and annotations of the transcripts reflected this sense in remarks such as, 'Tammy is asking a question, but I hear Gitte' or 'Gitte is doing what Tammy would have done'. More so, Tammy and Gitte resembled each other in a very striking way in their mannerisms, a supportive stance towards some students and more overbearing towards others; an individual, pensive movement of a hand to the chin followed the same motion of the hand by the other; a turn of the head or the whole body of one reflected in the movements of the other. This process was in part unconscious so that Tammy thought she had '*probably* copied all [of Gitte's] questions'. Experiencing productive questioning in action allowed Tammy to 'soak it up' and the type of questioning to 'go in'.

In the team-teaching experience, both soon became aware of differences in their knowing-in-action because they observed the other reacting differently to specific situations. This allowed them to become conscious of specific teaching practices and bring these into the domain of their discourse (reflecting *on* action). They began to 'use the words' of the other, or 'tried on [a technique] for size'. Because it was conscious, they could reject a technique when 'it didn't feel quite like yourself' or adapted it 'to suit me better if it doesn't feel quite right'.

There were three contexts in which Tammy and Gitte became *conscious* of alternative courses of actions. First, while teaching, the actions and reactions of the other were unexpected, leading them to reflect on what was happening. Second, they brought up these issues in conversations after the lesson so that they could talk about their intentions of acting in specific ways. Often, these conversations happened in the classroom over and about the objects or students involved in the situation of interest. Here, the immediacy of the experience, presence of relevant artefacts, and presence of another practitioner who had shared the experience, allowed Tammy to make conscious some aspects of her questioning practices and to engage in a pedagogically useful reflection-*on*-action.

The third important context for becoming conscious of their actions were the videotapes which we made available to Gitte and Tammy. Here they began to see themselves with 'different eyes', as different persons. The process allowed them to attend consciously to their practices. The videotape provided these practitioners with the luxury of time which they did not have during teaching because, as Tammy said, 'I was so much into my agenda in giving my directions that I didn't notice what was going on around the class'. Making practices conscious through reflecting *on* action allowed them to recognise certain behaviours in action and to try to change them because 'you are into them more'. However, because initially of the complexity of the irreversibly unfolding and unstoppable, lived situation, 'you kind of shift away from it until you notice it again'. However, the changes which Tammy and Gitte brought about consciously were by necessity iterative, because of the earlier mentioned problem created by the perspective of hindsight.

## TEACHING SCIENCE TEACHING

Current practices in science teaching and science teacher education and development are in stark contrast with practices in other fields. Most teachers have to work alone, in their own classrooms, separated from the activities of their peers; they rarely see – let alone engage alongside – peers in teaching. Much of their training occurs by taking classes in science content and science teaching methods, rather than in situations of actual, scientific or pedagogical practice. The closest they get is during their student teaching. However, even here, instead of engaging in the practice *with* their master teacher, they either watch or teach the class themselves. However, the outlined theory of practice implies that there is no replacement for learning the fundamental principle of a practice other than engaging in it 'alongside a kind of guide or coach, who provides assurance and reassurance, who sets an example and who corrects by putting forth, *in situation*, precepts applied directly to the *particular case* at hand' (Bourdieu & Wacquant 1992, p. 221). Under these conditions, Bourdieu points out, one can supervise only a small number of newcomers; anyone who pretends to supervise a large number does not do what he/she claims to be doing. These recommendations are applied to practice without great difficulties. In the past, I have had several opportunities to work with preservice and beginning teachers in this way. They began their first steps alongside me, participating in the very practices which they were to learn. Then they took on more and more of the task until they accomplished it on their own. In all cases, the co-participants found the experience unique and irreplaceable because, as they variously indicated, they learned what some theoretical precepts meant in a 'real' classroom. In the same way, my most successful research designs in education courses are with small groups which allow us to engage together in designing studies. Here, students experience experimental design in action which, as their course evaluations show, is one of the most helpful aspects of their graduate program.

Most courses, however, have more than a handful of students enrolled. In this respect, there are alternatives to traditional approaches. One of my colleagues structures his science 'methods' course so that he teaches with the preservice students a summer program for primary students (MacKinnon 1993). Although his set-up is more ideal than traditional courses, the large numbers of preservice students preclude the kind of collaboration which Bourdieu or Schön consider. The theory of practice outlined here implies teacher education programs which have as essential components the collaborative practice of experienced teachers and preservice and beginning teachers; that is, future teachers need to spend an essential component of their training by working *with* competent classroom teachers in team-teaching situations. This collaboration also would eliminate some of the negative aspects of practice and student teaching ('lack of control'), because the classroom teachers with their established routines are still present.

Collaborative practice shows great potential for teacher development. As Tammy pointed out, two months of team-teaching improved her teaching more than she could have done by taking three university courses. Thus, team-teaching could be the ideal vehicle for professional development. One can easily envision inservice efforts in which one competent science teacher team-taught with about three teachers at a time. A simple calculation shows that such arrangements would be less expensive and, if our experience is any indication, much more beneficial for the professional development of science teachers than current modes of delivery based on summer workshops in which teachers get but more science and science methods courses.

We are only beginning to understand the (collaborative) practice of science teaching. Some of the open questions follow. 'How are the collaborations structured by the participating teachers?' 'How do school and classroom settings interface with the teaching experience?' 'To what extent do teachers need additional preparation and release time?' 'What are the relative contributions of imitation and mimesis in teacher change through collaborative teaching?' As soon as science teacher educators begin to arrange for collaborative teaching involving preservice teachers, an entirely new domain of questions will open up. 'What are the roles of identification and empathy toward the practitioner in the student-teacher's development?'

'To what extent do experienced practitioners become conscious of their own practices when they work with beginners?' 'To what extent and at which point do student-teachers develop their own repertoires of teaching practices?'

From a research design perspective, the outlined theory of practice also raises new questions. If teachers like other practitioners know more than they can say, then how is this knowing-in-action to be assessed? How can researchers track tacit knowing-in-action as it is 'picked up' by newcomers? How can researchers access *reflection-in-action* without having to rely on teachers' and students' *reflection-on-action*? Much work still remains to be done in the area of research on science teaching and teacher education to elicit some of the tacit aspects of the practices that distinguish outstanding from mediocre performances. By looking around in other domains, we might find research practices which could help us to deal with questions in our own field; and, by pushing the boundaries of science teacher education and development towards education and training *in* practice, new ways of asking and answering questions will emerge.

## ACKNOWLEDGEMENTS

This work was made possible in part by two grants from the Social Sciences Research Council of Canada, SSHRC # 812-93-0006 and # 410-93-1127. My sincere thanks go to Sylvie Boutonné for the significant help during data collection, transcription of the videotapes and preparation of the manuscript.

## REFERENCES

Bourdieu, P.: 1990, *The Logic of Practice*, Polity Press, Cambridge, UK.
Bourdieu, P. & Wacquant, L.J.D.: 1992, *An Invitation to Reflexive Sociology*, University of Chicago Press, Chicago, IL.
Brookhart Costa, V.: 1993, 'School Science as a Rite of Passage: A New Frame for Familiar Problems', *Journal of Research in Science Teaching* 30, 649-668.
Brown, J.S., Collins, A. & Duguid, P.: 1989, 'Situated Cognition and the Culture of Learning', *Educational Researcher* 18(1), 32-42.
Cicourel, A.V.: 1990, 'The Integration of Distributed Knowledge in Collaborative Medical Diagnosis', in J. Galegher, R.E. Kraut & C. Egido (eds.), *Intellectual Teamwork: Social and Technological Foundations of Cooperative Work*, Lawrence Erlbaum, Hillsdale, NJ, 221-242.
Collins, H.M.: 1982, 'Tacit Knowledge and Scientific Networks', in B. Barnes & D. Edge (eds.), *Science in Context: Readings in the Sociology of Science*, Massachusetts Institute of Technology Press, Cambridge, MA, 44-64.
Dreyfus, H.L.: 1992, *What Computers Still Can't Do: A Critique of Artificial Reason*, Massachusetts Institute of Technology Press, Cambridge, MA.
Eckert, P.: 1990, 'Adolescent Social Categories – Information and Science Learning', in M. Gardner, J.G. Greeno, F. Reif, A.H. Schoenfeld, A. diSessa & E. Stage (eds.), *Toward a Scientific Practice of Science Education*, Lawrence Erlbaum, Hillsdale, NJ, 203-217.
Hutchins, E.: 1995, *Cognition in the Wild*, Massachusetts Institute of Technology Press, Cambridge, MA.
Jordan, K. & Lynch, M.: 1993, 'The Mainstreaming of a Molecular Biological Tool: A Case Study of a New Technique', in G. Button (ed.), *Technology in Working Order: Studies of Work, Interaction and Technology*, Routledge, London, 162-178.

Lave, J.: 1988, *Cognition in Practice: Mind, Mathematics and Culture in Everyday Life*, Cambridge University Press, UK.
Lave, J.: 1993, 'The Practice of Learning', in S. Chaiklin & J. Lave (eds.), *Understanding Practice: Perspectives on Activity and Context*, Cambridge University Press, UK. 3–32.
Lave, J. & Wenger, E.: 1991, *Situated Learning: Legitimate Peripheral Participation*, Cambridge University Press, UK.
Lemke, J.L.: 1990, *Talking Science: Language, Learning and Values*, Ablex Publishing, Norwood, NJ.
MacKinnon, A.: 1993, 'Examining Practice to Address Policy Problems in Teacher Education', *Journal of Educational Policy* 8, 257–270.
Munby, H. & Russell, T.: 1992, 'Frames of Reflection: An Introduction', in T. Russell & H. Munby (eds.), *Teachers and Teaching: From Classroom to Reflection*, Falmer Press, London, 1–8.
Roth, W.-M.: 1993, 'Metaphors and Conversational Analysis as Tools in Reflection on Teaching Practice: Two Perspectives on Teacher-Student Interactions in Open-Inquiry Science', *Science Education* 77, 351–373.
Roth, W.-M.: 1994a, 'Experimenting in a Constructivist High School Physics Laboratory', *Journal of Research in Science Teaching* 31, 197–223.
Roth, W.-M.: 1994b, 'Student Views of Collaborative Concept Mapping: An Emancipatory Research Project', *Science Education* 78, 1–34.
Roth, W.-M.: 1995, 'Affordances of Computers in Teacher-Student Interactions: The Case of Interactive Physics™', *Journal of Research in Science Teaching* 32, 329–347.
Roth, W.-M.: 1996a, 'Art and Artifact of Children's Designing: A Situated Cognition Perspective', *The Journal of the Learning Sciences* 5, 129–166.
Roth, W.-M.: 1996b, 'Learning to Teach Through Participation in Practice', Paper presented at the annual convention of the National Science Teachers Association, St Louis, MO.
Roth, W.-M.: 1996c, 'Teacher Questioning in an Open-Inquiry Learning Environment: Interactions of Context, Content and Student Responses', *Journal of Research in Science Teaching* 33, 709–736.
Roth, W.-M.: 1996d, 'Thinking With Hands, Eyes and Signs: Multimodal Science Talk in a Grade 6/7 Unit on Simple Machines', *Interactive Learning Environments* 4, 170–187.
Roth, W.-M.: 1998, 'Science Teaching as Knowledgeability: A Case Study of Knowing and Learning during Coteaching', *Science Education* 82(2).
Roth, W.-M. & Alexander, T.: 1997, 'The Interaction of Students' Scientific and Religious Discourses: Two Case Studies', *International Journal of Science Education* 19, 125–146.
Roth, W.-M. & Bowen, G.M.: 1993, 'An Investigation of Problem Solving in the Context of a Grade 8 Open-Inquiry Science Program', *The Journal of the Learning Sciences* 3, 165–204.
Roth, W.-M. & Bowen, G.M.: 1995, 'Knowing and Interacting: A Study of Culture, Practices and Resources in a Grade 8 Open-Inquiry Science Classroom Guided by a Cognitive Apprenticeship Metaphor', *Cognition and Instruction* 13, 73–128.
Roth, W.-M. & Roychoudhury, A.: 1992, 'The Social Construction of Scientific Concepts, or The Concept Map as Conscription Device and Tool for Social Thinking in High School Science', *Science Education* 76, 531–557.
Roth, W.-M. & Roychoudhury, A.: 1993, 'The Development of Science Process Skills in Authentic Contexts', *Journal of Research in Science Teaching* 30, 127–152.
Roth, W.-M. & Roychoudhury, A.: 1994a, 'Student Views about Knowing and Learning Physics', *Journal of Research in Science Teaching* 31, 5–30.
Roth, W.-M. & Roychoudhury, A.: 1994b, 'Science Discourse through Collaborative Concept Mapping: New Perspectives for the Teacher', *International Journal of Science Education* 16, 437–455.
Schön, D.A.: 1987, *Educating the Reflective Practitioner*, Jossey-Bass, San Francisco, CA.
Schön, D.A.: 1988, 'Coaching Reflective Teaching', in P.P. Grimmett & G.L. Erickson (eds.), *Reflection in Teacher Education*, Teachers College Press, New York, 19–29.
Shulman, L.S.: 1987, 'Knowledge and Teaching: Foundations of the New Reform', *Harvard Educational Review* 57, 1–22.
Suchman, L.A.: 1987, *Plans and Situated Actions: The Problem of Human-Machine Communication*, Cambridge University Press, UK.

# 2.4 Teaching For Understanding in Pre-Secondary Science

### WYNNE HARLEN
*Scottish Council for Research in Education, Edinburgh, UK*

This chapter is about the teaching of science at the primary level of education, which is taken to mean the years of schooling up to the age of 12 years. In some educational systems, parts of this education could take place in schools described as 'primary' or 'middle', but the age of the students and the organisation of the curriculum are more important as defining features than the label. One central characteristic of primary education is that most, if not all, teaching is in the hands of general teachers rather than subject specialists.

Decisions about what and how to teach at any level derive from views of the nature of learning and of the subject matter (Harlen & Osborne 1985). In addition, the teaching of a subject such as science, which was not traditionally included in the curriculum before the secondary school, has to be justified in some convincing way. Thus, the first section of this chapter briefly indicates how the arguments justifying the inclusion of science in the primary curriculum have changed. The second section concerns the notion of learning with understanding and current views on how understanding is constructed by learners. The third and largest section describes some important aspects of teaching. Finally, the chapter ends with a checklist for teacher self-evaluation which embodies indicators of effective practice and which can be used in various ways to improve the teaching of primary science.

## THE CHANGING RATIONALE FOR PRIMARY SCIENCE

The way in which science is taught at the primary level reflects views of its purpose and contribution to children's education. It is important for teachers to have a clear notion of why they should teach science so that they do so in a way which furthers its value in children's education. But the answer to the question of 'why teach science at the primary level?' has not always had the same answer. Over the past three decades, it is possible to find considerable shifts in the rationale for including science in the primary curriculum, paralleled by changes in the advocacy of different approaches to teaching it.

---

Chapter Consultant: Michael Kamen (Auburn University, USA)

The impetus to introduce science into the primary curriculum followed the concern about the state of school science and technology education in the West in the late 1950s. It was this concern that led to sweeping change in the secondary school science curriculum brought about directly or indirectly by projects such as the Nuffield secondary science in the UK and by the Biological Sciences Curriculum Study (BSCS) and Physical Sciences Study Committee (PSSC) in the USA. The parallel concern to improve what science there was in the primary school attracted less public attention. Any science taught in the primary school in the 1950s was largely nature study, and whole-class demonstration at that. The first sign of official discontent with this situation in the UK was the publication by the then Ministry of Education (1961) of a pamphlet entitled *Science in the Primary School*, calling for a broadening of content and the adoption of an inquiry approach. The Association for Science Education (1963) which played a prominent part in the new developments in the 1960s, showed its interest in primary science by establishing its Primary Schools Sub-Committee and issuing a policy statement. At about the same time, a research project into scientific development in children was funded by the Froebel Foundation, and this was supported by the British Association (Isaacs 1962, 1966) and the setting up of two curriculum development projects in primary science. One of these was sponsored directly by the Ministry of Education and involved producing a book for teachers describing activities relating to four science concept groups (Redman, Brereton & Boyes 1969). The other was the better known Nuffield Junior Science Project, which produced a teacher's guide and other books for teachers (Wastnedge 1967). The ideas of this project were exported quite widely and influenced developments in Canada and Africa particularly. The child-centred, inquiry approach which it embraced was shared by its successor, Science 5–13 (1972–1975), by the African Primary Science Program (1969) and by early programs developed in the USA, such as the Elementary Science Study published by McGraw-Hill in 1966, Science – A Process Approach published by Xerox in 1967 and the Science Curriculum Improvement Study published by Rand McNally in 1967.

These developments had in common a view that children learn through first-hand experience and therefore must have materials from their environment to explore. In the UK, this view was whole-heartedly endorsed by the Plowden Report (Central Advisory Council for Education 1967), with its resounding phrase that 'At the heart of the education process lies the child'. The fervour with which this child-centred philosophy was embraced can be seen, in retrospect, as a reaction to the restrictive regime in primary schools which had held sway from the 1930s to the 1950s when the curriculum was in the grip of the end-of-primary examination (the 11+) which tested mainly English and arithmetic. The purpose of this examination was selection for entry to different types of secondary school. With the introduction of comprehensive secondary schools, this examination died away in all but a few staunchly conservative areas of England and the broadening of the primary curriculum in the 1960s became possible. The introduction of science was at the cutting edge of this broadening, embracing aims of developing children's ability to investigate rather than to acquire knowledge about what they were investigating. The Association for Science Education's policy statement on primary science also supported the view that, at the primary level, 'we are concerned more with the developing of an enquiring mind than with the learning of facts' (Association for Science Education 1963).

The early arguments for including science in the primary school were a combination of benefits to the individual and to society. The former were given more prominence, in such terms as 'exercising the child's mental powers' (Association of Women Science Teachers 1959) whilst the latter reflected the perception in the 1960s of the need for more scientists, for children to be able to function in an increasingly scientific and technological world, and for citizens who would apply scientific thinking in all parts of their lives (Isaacs 1962). These justifications were echoed worldwide as both first and third world countries recognised science as basic to education from the start. A meeting on the incorporation of science and technology in the primary school curriculum, held by UNESCO in 1980, brought together these views (Harlen 1983):

- Science can help children to think in a logical way about everyday events and to solve problems. Such intellectual skills will be valuable to them wherever they live and whatever they do.
- Science, and its applications in technology, can help to improve the quality of people's lives. Science and technology are socially useful activities with which we would expect young children to become familiar.
- As the world is increasingly becoming more scientifically and technologically oriented, it is important that future citizens should be equipped to live in it.
- Science, well taught, can promote children's intellectual development.
- Science can positively assist children in other subject areas, especially language and mathematics.
- Primary school is terminal for many children in many countries and this is the only opportunity they could have to explore their environment logically and systematically.
- Science in the primary school can be real fun. Children everywhere are intrigued by simple problems, either contrived or real ones from the world around them. If science teaching can focus on such problems, exploring ways to capture children's interests, no subject can be more appealing or exciting to young children.

These plausible and persuasive points were borne out by some individual cases for which there were enthusiastic, energetic and well-supported teachers. In general, however, they reflect the desire for reform in the 1960s rather than convincing evidence. Indeed, they were challenged in the recessionary 1970s when it became clear that hard evidence of a positive impact of the developments of the 1960s was almost entirely lacking (Martin 1983). Today, the author still knows no studies which have produced conclusive evidence relating to the claims in the UNESCO list. However, in the 1980s, evidence from research began to lend support to the idea of beginning science in the primary school. The evidence came from three different sources.

The first was the growing body of research, mostly at the secondary level, into children's scientific ideas, which were at first described as 'misconceptions' (e.g., Johnstone, MacDonald & Webb 1977) and later as 'alternative frameworks' (Driver 1985) or 'children's ideas' (Osborne & Freyberg 1985). One important implication of this research was recognition that children's ideas of the world around them are being built up during the primary years, whether or not they are taught science (Gilbert & Watts 1983). It became clear that, without intervention to introduce a scientific approach in their exploration of the world, the ideas that children develop could be non-scientific and could obstruct learning at the secondary level. Research by groups in the USA, Canada, the UK, France, Australia and New Zealand, working mainly in the 1980s with secondary students, reached remarkably similar conclusions about the ideas which students bring with them to their secondary science education. Thus, there was a role for primary science in reducing the gap between the ideas children have and the ones that would enable them to derive greater benefit from their later science education. However, this would require a shift from the 1960s concentration on processes, with its relative neglect of content and development of science concepts. This shift, however, was already happening, and is the second area in which research played a part.

There was both empirical evidence and philosophical argument pointing to the interdependence of scientific processes and concepts. The empirical evidence came from work on assessment in science. In the 1970s, a large-scale program of monitoring achievements in education was mounted in England, Wales and Northern Ireland, run by the Assessment of Performance Unit (APU) from within the Department of Education and Science. The APU was set up to answer questions about what was happening to standards of education as a result of the changes in the expansive 1960s (Johnson 1989). In science, five annual surveys of performance were carried out during 1980 to 1984 among students aged 11, 13 and 15 years (Assessment of Performance Unit 1988). A practical mode of testing was used for three of the six, mainly process-based, categories at the primary level. However, as the attempt to assess scientific processes in any valid way necessarily involved some science content, it became clear that this content influenced the result (even if no knowledge of content was actually required). Thus performance in, say, planning an experiment, would be affected considerably by the subject matter even if all required information was given. The finding that students could use a skill in one situation did not mean that the same students could deploy the same skill in another.

Given the philosophical arguments about the conceptually driven nature of observation, this interaction is no longer surprising. Indeed, it now seems common sense. In making an observation, planning an investigation, interpreting results or giving an account of conclusions, there has to be some content and what that content is, and particularly what is already known about it, makes a considerable difference to how someone interacts with a problem or task. This was in direct conflict with the view that the subject matter is unimportant in the use and development of processes. Moreover, it was realised that the dependence is two-way: not only does the use of the processes depend on the subject matter, but the development of understanding depends on the use of processes (a point to be developed in the next section). In terms of the rationale for including science in the primary curriculum, the evidence of this interdependence supported the *importance of exposing children to scientific ideas using a process approach*. However, the reason for process-based activities was no longer in terms of unsubstantiated claims of 'promoting intellectual development', but based on evidence of how learning takes place.

The third source of evidence was research into the interests of children in science and their attitudes towards it (Ormerod & Duckworth 1975). Kelly's research into factors affecting secondary school students' choice of subjects at the age of 13 years in the 1950s showed that the scientists had long-standing stable attitudes favourable to science which were formed more than two years earlier, whilst non-scientists did not make up their mind until nearer the time of choice (Kelly 1959). This was confirmed by Duckworth (1972), who showed that girls seem to form attitudes in relation to science even earlier than boys. Even though recent curriculum reforms mean that students no longer can choose not to continue science in the later years of their compulsory schooling, this will not necessarily have changed their attitudes towards it. Indeed, the continuing decline in the numbers of those wishing to study science in higher education remains a cause for concern and suggests that the attitudes which are formed so much earlier are not favourable to science. One reason for the early formation of attitudes to science could be that, compared with history or other school subjects, it has a high profile in the media and in daily life. Children, therefore, might think they know what it is, but this image is often of something very mysterious, complex and, for some, sinister. It is important, therefore, for young children to have first-hand experience of science as an enjoyable, comprehensible and useful activity so that their early views are informed. The value for society in this is not primarily in encouraging more students into scientific careers, but in fostering scientific literacy throughout the population.

## LEARNING FOR UNDERSTANDING

To teach towards understanding requires a view of what it means to learn with understanding and how this differs from learning without understanding. Fortunately, this does not mean that it is necessary to be able to define understanding, but rather to identify its major characteristics. White and Gunstone (1992) discuss, in the context of attempting to assess understanding, the difficulties of identifying understanding and conclude that:

> ... understanding of a concept or of a discipline is a continuous function of the person's knowledge, is not a dichotomy and is not linear in extent. To say whether someone understands is a subjective judgement which varies with the judge and with the status of the person who is being judged. Knowledge varies

in its relevance to understanding, but this relevance is also a subjective judgement. (p. 7)

It is important to keep this caution in mind but, at the same time, to recognise that a statement of what characterises understanding must say how it can be recognised or at least how it differs from rote learning. Understanding something depends upon having some information about it, but information alone is not sufficient. For example, information about what certain living things need to keep alive is required for understanding the interdependence of living things, but knowing how particular living things obtain food, air and water does not guarantee understanding of how they depend on other living things. It is when one piece of knowledge is linked to certain others to form a concept embracing them all that the product is usable in a range of contexts. However, in theory, it would be possible to have knowledge about the interdependence of living things that was not derived from linking knowledge of particular living things but was learned by rote. A person with such knowledge could recite a verbal definition but not explain or apply it in particular cases. By this failure to *use* the information, we would conclude that their *understanding* was poor. Rote learned knowledge is not usable in forms or in contexts different from that in which it was obtained. By contrast, understanding implies that the knowledge can be transformed, applied in other contexts and used in various ways, such as in making predictions, attempting explanations or making connections between one thing and another.

From this discussion, it emerges that concepts which are understood are the result of creating links during learning, whilst these links are not made in rote learning. Observations of anyone faced with a new situation which they are trying to understand shows that what they do is to search, often unconsciously, in their minds for existing knowledge that can be used in explaining it. The previous experience of children, of course, is limited and relevant knowledge might not be available to them. Therefore, the knowledge that they link might not be appropriate and the result is that the ideas at which they arrive are the non-scientific ones, or the 'alternative ideas' that research has uncovered. For example, faced with the evidence that varnished cubes of wood stick to each other when wet, several groups of 11-year-olds concluded that the blocks became magnetic when wet (Harlen 1993, p. 20 ). The resemblance of a block sticking to the underside of another, without anything to hold them together, to a magnet picking up another magnet or a piece of iron was clearly very strong. An equally good alternative explanation was not available to the children and so they held on to their view of magnetism, modifying it to accommodate the observation that the blocks only stuck together when wet by concluding that 'they're magnetic when they're wet'. A considerable body of research into the scientific ideas of primary school students has been completed by the Science Processes and Concepts Exploration (1993) project.

Linking to previous knowledge is only the start of the learning process. The existing idea that has been linked might not be useful in helping students to understand the new phenomenon, as we have seen in the case of the children with the wooden blocks. It therefore should be tested to see if it fits further aspects of the new phenomenon. As a result, the initially linked idea might be confirmed, rejected or accepted after modification. If the idea is rejected, then another idea from previous experience has to be sought and the linking and testing process repeated.

What is described here is an essentially constructivist view of learning. Ideas emerging from each learning experience are derived from those held at the start of the learning experience and, in turn, these themselves are used in understanding further phenomena. Learning thus is conceptualised as the *change* in ideas through testing them against new experience. As the process goes on, more ideas are linked together and the emerging ones gradually become applicable to a greater range of phenomena, more powerful in helping students to understand experience, and more abstract. An important consequence of this way of developing ideas is that, at each point, what is understood fits the experience of the individual. As experience changes, so ideas also may have to change but, if this is worked out by the learners through their reasoning, then each new idea or version of an idea becomes 'owned' by the learners and will be used because it makes sense.

It is important to note that, in the process of testing ideas, what emerges will depend upon how the processes of hypothesising, prediction, investigation, interpretation and the drawing of conclusions are carried out. This is the point, referred to earlier, about the dependence of ideas upon skills. When ideas are tested, the outcome in terms of changed or rejected ideas will depend on the *way* in which the testing is carried out. It was assumed in the description of learning above that the testing of ideas was rigorous and systematic, in the way associated with scientific investigation. When this is so, then ideas which do not fit the evidence will be rejected and those which do will be accepted and strengthened. But, if it is not so, then ideas might be retained which should be modified or rejected and there is the danger of perpetuation of non-scientific ideas. Thus, the skills of linking and testing have a crucial part to play in the development of ideas.

Research shows that young children do not investigate with rigour and that the ability to use process skills has to be developed. For instance, they often are extremely selective in their observation, tending to focus on those features to which their preconceptions direct attention; in interpreting evidence they might ignore data which do not fit an assumed relationship; their predictions might be no more than statements of what they already know. So, unless these skills are developed so that their investigations are systematic and scientifically valid, the ideas which emerge from the children's activities and thinking might not be the ones that really fit the evidence from the world around. Thus, the emerging rationale for the attention to processes in primary science is based on their role in the development of understanding rather than as in previous years, on a supposed value in promoting a general intellectual development.

## THE TEACHER'S ROLE

There are several consequences of putting this view of learning into practice. In preparing the details of the classroom activities relating to a topic, within the overall school program, the teacher's plans must include (1) finding out the existing ideas and skills of the children, (2) using appropriate strategies to ensure that students' ideas are being tested (this might involve adapting activities if a published scheme is being used), (3) using appropriate strategies to help students to develop process skills and (4) appraising progress and engaging students in assessing what they have learned. How these plans can be implemented is considered for each of these in turn in this section.

*Finding out the Existing Ideas and Skills of the Children*

Techniques for gathering information about children's current ideas and skills have been identified by research and from observation of effective classroom practice (Harlen 1991, 1992a; Science Processes and Concepts Exploration 1993; Scottish Council for Research in Education 1994; Treagust, Duit & Fraser 1996). They include on-the-spot-observation of children's actions, asking questions designed to probe understanding of ideas and processes, and collecting products such as children's reports, notes, drawings, graphs and other artefacts. Used in combination, these techniques provide evidence both in the form of tangible products of students' work (which can be perused after the event) and as ephemeral evidence (which has to be assimilated and interpreted at the time unless some kind of note or record is made by the teacher). In gathering evidence, however, the teacher's role is not passive. The usefulness of what is found for revealing children's thinking depends on what the children are responding to (i.e., the questions asked and the tasks set).

Questioning by the teacher has a crucial part to play in all aspects of teaching and many of the following points, made in relation to enabling students to express their ideas and reveal their thinking processes, apply equally to helping them to develop their ideas and skills. There are two aspects of questions to consider: the form of the question; and the content (i.e., what is asked).

The form of question effective for gaining access to children's thinking is described as being *open* and *person-centred*. *Open questions*, in contrast to *closed* ones, invite an extended response rather than a short or one-word answer. Questions to which the only answers are 'yes' or 'no' are very unlikely to be useful. A closed question such as 'Will the toy car go further if it starts from higher up the slope?' can be rephrased to be open: 'What difference do you think it will make if the car starts from higher up the slope?' Although this distinction is a well-known one, classroom research shows that open questions form a small proportion of the questions that teachers ask. Galton, Simon and Croll (1980) found that 5 percent of teachers' question were open, whilst 20 percent were closed and a further 30 percent asked only for specific facts.

*Person-centred* questions, which are contrasted with *subject-centred* ones, are phrased to ask overtly for the students' ideas rather than the 'right' answer. A subject-centred question, such as 'Why did your plant grow more quickly in the cupboard than by the window?', suggests that there is a 'right' answer to be supplied. If this is rephrased as a more person-centred question (e.g., 'Why do you think your plant grew more quickly . . . ?'), then it invites the student to give his/her own ideas and shows interest in what these are.

Turning to the content of the question, again the concern is to encourage the children to express and use their ideas. The most useful questions will be those which ask for *explanations* and those which ask for *predictions*. Explanations are often requested through 'why' questions, although a completely open 'why' question is not as useful as one which indicates what is to be explained. For example, when a child has been handling a piece of polystyrene and noticing that it feels warmer than a piece of metal, the question should be 'Why do you think that the polystyrene feels warmer than the metal?' rather than 'Why is this?' Questions which ask for predictions are useful because they require students to use their ideas and, in doing so, sometimes they show more clearly the underlying ideas than do attempts to explain it directly. For example, asking students 'What could we do to dry these wet clothes more quickly, and why do you think that would work?' might reveal more about their ideas of evaporation than asking directly about what they think happens to water when it seems to disappear.

The products of children's work, such as writing, drawing or artefacts, can provide evidence of their thinking and use of skills, but the value for this purpose will depend on the tasks set. For example, if children who have been watching the incubation of eggs in their classroom are asked simply to draw the eggs, the results will be of much less interest in terms of the children's ideas than drawings which result from the teacher asking the children to draw what they think is inside the eggs during incubation.

Children will express their ideas in their writing and through their drawings if they are encouraged to do so by instructions such as 'Make sure that you show (or write about) how you think it works' and 'Use labels and arrows to show what you think each part does'.

On occasion, it might be appropriate with older primary children to use more formal written questions to assess children's thinking. These written questions should have the same form as the oral ones already discussed. The interest is in the open answers which children give, because only in these is their own thinking revealed. When there is a closed part to a question, this is preliminary to an open one asking for an explanation or prediction, as in the following example from the Scottish Council for Research in Education (1994). 'If you want to keep a hot pie warm for as long as possible, which of these would you use to wrap it in: metal foil; bin liner plastic; cardboard; or a padded bag? Explain why you think the one you have chosen would be best.'

*Enabling Students to Test Their Ideas*

The purpose of revealing students' ideas is so that they can be starting points for constructing further learning. This is not necessarily a basis for arguing that each child has to be treated individually and provided with separate activities suited to his/her unique ideas. Not only would this be impossible in practice, but it denies the role of social interaction in learning. Further, as already noted, research shows that across the world there are remarkably similar patterns in the development of children's ideas, and in the non-scientific ideas which form part of this development. Thus, in practice, the same ideas will be held by several children.

Teachers will have in mind the ideas that they wish their children as a class to develop through certain activities. Individual ideas can be accommodated in the way in which the topic is introduced and in the flexibility allowed in the activities. The topic might be introduced through discussion of past experience, or through materials which have been put out for display over a period of time to engage the children's interest and curiosity, or through a visit outside the classroom. A whole-class discussion of these objects or events, initiated by open, person-centred questioning, will reveal various ideas relating to the concept that the teacher wants to help the children to develop. Depending on the range of ideas, the children might work in groups on different ideas which they have suggested. Depending on their age, children might be asked to discuss the ideas further in groups and to decide how to test the ideas through an investigation.

If the teacher is following a published scheme of activities, these could be modified or extended to allow the children's ideas to be tested. For example, a teacher might be intending to develop ideas about air resistance through children making and investigating parachutes made from sheets of plastic and string. They could do this following the instructions in a book or workcard. Having done so, the children might have various ideas about what makes one parachute fall more quickly or more steadily than another – the weight, the area, the presence of a hole in the canopy, the shape of the canopy, etc. They can test these ideas by making parachutes of different sizes, shapes, etc. The first step might be to plan what they will do, giving the teacher opportunity to assess the extent to which they are developing the skills required to devise a 'fair' test and to help them in this development. When practical work is in progress, the teacher again will be able to assess and help the development of practical skills, including those relating to measurement and the recording of results.

After the completion of practical investigations, it is essential for the class to get together to exchange findings (giving an opportunity for children to communicate their results in a formal manner and for an evident purpose) and so that the teacher can ask students to reflect on whether their ideas have changed or have been confirmed. They also can articulate what they have learned about carrying out an investigation. Making these things conscious, open and shared magnifies the learning from any one activity.

Many primary teachers find themselves lacking in confidence in carrying out this part of their role because it involves discussion of scientific ideas which they might feel that they do not fully understand themselves. Lack of background knowledge of teachers for many years has been identified as a factor inhibiting effective science teaching at the primary level and this has been confirmed by recent research into teachers' concepts of science by the Primary School Teachers and

Science Project (1992). This project found that 'many primary teachers lack knowledge of the key concepts of science and that where there is some acquaintance with such concepts, a majority of the teachers hold views of them that are not in accord with those accepted by scientists' (Primary School Teachers and Science Project 1993). Guided by their research, the Primary School Teachers and Science Project team devised inservice study materials to help teachers to improve their understanding of basic science concepts. However, this is an area where more research is needed, for it is clear that knowledge of science, in itself, is not sufficient for teaching it. For many primary teachers, knowledge about how to teach science goes a long way towards compensating for lack of scientific knowledge. The distinctions made by Shulman (1987) among various kinds of pedagogical knowledge, curriculum knowledge and content knowledge could help to focus further research aimed at clarifying this issue and guiding the provision of well-targeted teacher education.

*Strategies for Helping Students to Develop Process Skills*

The most important prerequisite for the development of process skills is the opportunity to use them. Secondly, there must be deliberate steps taken to encourage critical review of how activities have been or are being carried out. A third requirement is access to the techniques needed for advancing skills, such as how to use measuring instruments, to know the conventions of drawing charts and graphs, and to use symbols. When applied to the development of specific skills and their coordination in investigations, these general points lead to a range of strategies from which a teacher can select what is appropriate for a particular situation. Examples only are given in this chapter and fuller accounts can be found in other sources (Harlen 1992b, 1993).

To encourage *the use of the senses* as ways of finding information, children need real objects to handle and phenomena to explore. These can be provided not only for the class activities, but also for observation at other times, in the form of objects on an 'interest' table or displayed on the wall. But some students need more than the *opportunity* to observe things, as is provided by a display or interest table; they need *invitations* to observe. These can be provided by placing cards next to objects on display, which pose questions ('How many different kinds of shells can you find here?') or suggest activities ('Make a higher note and a lower note by changing the amount of water in the bottle.'). Observation is also encouraged by holding discussions of what is displayed, so that ideas are pooled and students who have not noticed something are likely to go back and look more carefully. The display also can be the opportunity to develop skill in using equipment for assisting the senses and for measuring. For example, a card giving a drawing showing its correct use could be placed next to a hand lens along with a bunch of grass (with minute seed heads) with the invitation to see if they are all the same.

Help in *planning* can begin, paradoxically, from reviewing an investigation which has been completed (whether or not the children planned it themselves), helping them to go through what was done, and identifying the structure of the activity through questions such as:

- What were they trying to find out?
- What things did they compare (identifying the independent variable)?
- How did they make sure that it was fair (identifying the variables which should be kept the same)?
- How did they find the result (identifying the dependent variable)?

When planning a new investigation, the lessons learned from reviewing can be recalled, when perhaps variables were not controlled or initial observations were not taken when they should have been. Planning continues throughout an investigation and indeed the initial plan can change as the work progresses and unforeseen practical obstacles emerge. This leads to the review of plans which is such an important part of developing this skill.

Children often implicitly use patterns or hypotheses in *making predictions* but fail to recognise that they in fact do so. When asked to make a prediction of how far a wind-up toy travelled after different numbers of turns, a girl said that she had guessed. Further probing, however, led to her describing that she thought that it would be a bit more than for three and a bit less than for five, suggesting that she was implicitly using a relationship of 'more winds means further'. Becoming aware of the pattern that she was using enabled her to predict other distances with more confidence and indeed satisfaction. Discussion played a central part in bringing about this awareness.

In order to *draw conclusions* and to consider to what extent ideas have been confirmed or challenged, children need to interpret what they find. That is, they must go further than collecting individual observations to see patterns and to relate various pieces of information to each other and to ideas. For example, children measuring the length of the shadow of a stick placed in the ground at different times of the day must go beyond just collecting the measurements if the activity is to have value for developing their ideas. There are more significant outcomes of this activity which depend on using rather than just collecting results. For example, the children could detect the pattern of decreasing and then increasing length of the shadow and consider whether this accords with what they expected; they could consider the possibility of using the pattern to make predictions about the length at times not measured, or the time of day from the measurement of the shadow; or they might reconsider their ideas about how shadows are formed. It is important that the teacher ensures that results *are* used and that children do not rush from one activity to another without talking about and thinking through what their results mean.

*Appraising Progress and Engaging Students in Assessing What They Have Learned*

The techniques for revealing children's ideas and skills outlined previously are equally applicable to assessing progress. If the assessment is to be used for reporting achievement to parents or other teachers, then the techniques need to be used with more rigour and more attention to systematic data gathering and record-keeping than when they are used in the context of finding where students are in order to guide immediate action. It also could require interpretation of the data in terms of criteria of achievement (sometimes expressed as targets, goals or standards) which can be set locally or nationally. This form of teacher-based criterion-referenced assessment requires moderation (quality assurance and control) if the results are used for making comparisons between children or for other 'high stakes' purposes. Research shows that one of the most effective ways of providing this quality assurance is through providing assessed and annotated examples of students' work (such as those published in England and Wales by the School Examination and Assessment Council in 1993 and in New Zealand as part of the achievement initiative) and through 'agreement panels' where teachers meet together to discuss the assessment of examples of their own students' work (James 1994).

There is growing interest in the participation of students in the assessment of their own achievements; indeed this is an important part of helping students to take some responsibility for their learning and to recognise what it is that they are learning. The knowledge of the processes of learning and awareness of one's own learning is the basis of metacognition. White and Mitchell (1994) have put forward persuasive arguments for training in metacognition as a means of giving students control of their learning. Opportunities for this can seem limited in the case of young children, but this is a field in which much interesting work remains to be done. Steps can be taken even with six-year-olds to eight-year-olds to enable them to choose 'best' pieces of work and to discuss what is good about them. Older primary children can become involved in reviewing their work more formally, often as part of a record of achievement (Johnson, Hill & Tunstall 1992).

## TEACHER SELF-EVALUATION

There are many aspects of the provision for primary science and of the teacher's role which have not been mentioned, or mentioned only briefly here. However, all need to be included when teachers review their work. The following checklist is intended for this purpose to help teachers answer the question 'How am I doing?' and to use the results formatively in improving their teaching. The list borrows from several which have been published (Harlen 1992b; Harlen & Elstgeest 1992; Science Processes and Concepts Exploration 1993). The suggested use is by teachers, individually and privately, in looking back over the activities which they have provided and the role that they have taken in the students' activities during a specific period of about two or three weeks. A teacher should ensure that there is evidence for those items for which the answer is 'yes' and should reflect critically and constructively about items for which the answer is 'no'. Questions for teacher self-evaluation follow:

(1) Were the children given opportunity to explore/play with/investigate materials?
(2) Were children encouraged to ask questions?
(3) Did the teacher respond to questions by suggesting what the children might do to find out rather than providing a direct answer?
(4) Did the teacher ask open and person-centred questions which invited the children to talk about their ideas?
(5) Were there suitable sources of information for the children to use to help them to answer their own questions?
(6) Were children given opportunities to talk about their ideas informally in groups?
(7) Did the teacher deliberately keep silent at certain times and listen to what the children were saying?
(8) Were they asked for writing, drawings or other artefacts through which they expressed their ideas?
(9) Were the children challenged to test their ideas?
(10) Were the children helped to develop their investigative skills?
(11) Did the children have opportunity to describe their investigations to others orally, in writing or both?
(12) Did the children hear or read about what other children had done or found?
(13) Did the children know what they were doing and work purposefully and with confidence?
(14) Was the teacher aware of changes in the children's ideas?
(15) Did the teacher interpret the children's work in terms of their ideas and skills, rather than marking it as correct or not?
(16) Did the teacher keep records of what children had done and of their developing ideas and skills?
(17) Were the children involved in discussing and reviewing their work and assessing their own achievements?
(18) Did the teacher avoid bias in activities which could have disadvantaged certain children because of gender, ethnic origin, religion, language or physical disability?

# REFERENCES

African Primary Science Program: 1969, *Activities for Lower Primary: An Introduction*, African Primary Science Project, Newton, MA.

Assessment of Performance Unit: 1988, *A Review of APU Surveys 1980–84*, Her Majesty's Stationery Office, London.

Association for Science Education: 1963, *Policy Statement Prepared by the Primary Schools Science Committee*, Author, Cambridge, UK.

Association of Women Science Teachers: 1959, *Science in the Primary School*, John Murray, London.

Central Advisory Council for Education: 1967, *Children and their Primary Schools* (Plowden Report), Her Majesty's Stationery Office, London.

Driver, R.: 1985, *The Pupil as Scientist?*, Open University Press, Milton Keynes, UK.

Duckworth, D.: 1972, *The Choice of Science Subjects by Grammar School Pupils*, Unpublished PhD thesis, University of Lancaster, Lancaster, UK.
Galton, M., Simon, B. & Croll, P.: 1980, *Inside the Primary School*, Routledge and Kegan Paul, London.
Gilbert, J. & Watts, M.: 1983, 'Concepts, Misconceptions and Alternative Conceptions: Changing Perspectives in Science Education', *Studies in Science Education* 10, 61–98.
Harlen, W. (ed): 1983, *New Trends in Primary School Science Education*, UNESCO, Paris, France.
Harlen, W.: 1991, 'Pupil Assessment in Science at the Primary Level', *Studies in Educational Evaluation* 17, 323–340.
Harlen, W.: 1992a, 'Research and the Development of Science in the Primary School', *International Journal of Science Education* 14, 491–503.
Harlen, W.: 1992b, *The Teaching of Science*, David Fulton, London.
Harlen, W.: 1993, *Teaching and Learning Primary Science* (second edition), Paul Chapman, London.
Harlen, W. & Elstgeest, J.: 1992, *UNESCO Sourcebook for Science in the Primary School*, UNESCO, Paris, France.
Harlen, W. & Osborne, R.J.: 1985, 'A Model for Learning and Teaching Applied to Primary Science', *Journal of Curriculum Studies* 17, 133–146.
Isaacs, N.: 1962, 'The Case for Bringing Science into the Primary School', in W.H. Perkins (ed.), *The Place of Science in Primary Education*, The British Association for the Advancement of Science, London, 4–23.
Isaacs, N.: 1966, *Children Learning Through Scientific Experience*, Froebel Foundation, London.
James, M.: 1994, 'Experience of Quality Assurance as Key Stage 1', in W. Harlen (ed.), *Enhancing Quality in Assessment*, Paul Chapman, London, 116–138.
Johnson, S.: 1989, *National Assessment: The APU Science Approach*, Her Majesty's Stationery Office, London.
Johnson, G., Hill, B. & Tunstall, P.: 1992, *Primary Records of Achievement: A Teacher's Guide to Recording and Reviewing*, Hodder and Stoughton, London.
Johnstone, A.H., MacDonald, J.J. & Webb, G.: 1977, 'Misconceptions in School Thermodynamics', *Physics Education* 12, 248–251.
Kelly, P.J.: 1959, *An Investigation of the Factors which Influence Grammar School Pupils to Prefer Science Subjects*, Unpublished MA thesis, University of London, London.
Martin, M-D.: 1983, 'Recent Trends in the Nature of Curriculum Programmes and Materials', in W. Harlen (ed.), *New Trends in Primary School Science Education*, UNESCO, Paris, France, 55–67.
Ministry of Education: 1961, *Science in the Primary School*, Her Majesty's Stationery Office, London.
Ormerod, M.B & Duckworth, D.: 1975, *Pupils' Attitudes to Science*, National Foundation for Educational Research, Windsor, UK.
Osborne, R. & Freyberg, P.: 1985, *Learning in Science: The Implications of 'Children's Science'*, Heinemann Educational, London.
Primary School Teachers and Science Project: 1992, *Collected Working Papers 1988–92*, Oxford University Department of Educational Studies, Westminster College, Oxford, UK.
Primary School Teachers and Science Project: 1993, *Working Paper No. 18*, Oxford University Department of Educational Studies, Westminster College, Oxford, UK.
Redman, S., Brereton, A. & Boyes P.: 1969, *An Approach to Primary Science*, Macmillan Educational, London.
School Examination and Assessment Council: 1993, *Children's Work Assessed (KS1)* and *Pupils' Work Assessed (KS3)*, Author, London.
Science 5–13: 1972–1975, *Units for Teachers* (various titles), MacDonald Educational, London.
Science Processes and Concepts Exploration: 1993, *Teachers' Handbook*, Collins Educational, London.
Scottish Council for Research in Education: 1994, *Taking a Closer Look at Science: Diagnostic Procedures for Primary Schools*, Author and Scottish Office Education Department, Edinburgh, Scotland, UK.
Shulman, L.: 1987, 'Foundations of the New Reform', *Harvard Educational Review* 5, 1–22.
Treagust, D.T., Duit, R. & Fraser, B.J. (eds.): 1996, *Improving Teaching and Learning in Science and Mathematics*, Teachers College Press, New York.
Wastnedge. E.R (ed.): 1967, *Nuffield Junior Science Project: Teachers' Guide 1*, Collins Educational, London.
White, R.T. & Gunstone, R.: 1992, *Probing Understanding*, Falmer Press, London.
White, R.T. & Mitchell, I.J.: 1994, 'Metacognition and the Quality of Learning', *Studies in Science Education* 23, 21–37.

## 2.5 Teaching for Conceptual Change

PETER W. HEWSON
*University of Wisconsin-Madison, USA*

MICHAEL E. BEETH
*The Ohio State University, Columbus, USA*

N. RICHARD THORLEY
*The Harley School, Rochester, USA*

A major foundation of teaching for conceptual change is the assumption that effective teaching needs to be rooted in an understanding of how students learn. This assumption does not play out in a straightforward manner because the relationship between learning and teaching is not simple, not one-to-one, not unique and certainly not causal (e.g., Driver, Asoko, Leach, Mortimer & Scott 1994; Hewson 1991). A learning model does not prescribe a unique set of teaching sequences and strategies; and a particular teaching strategy does not determine the type of learning that will occur. Instead, in Fenstermacher's (1986) terms, the relationship is ontological: teaching without the intent that learning will occur is a contradiction in terms. Learning models thus can provide a set of general guidelines that can be used in the design of different teaching approaches. Outlining guidelines of this nature is our goal in this chapter.

Much is now known about how students learn. The critical role that students' knowledge plays in learning is widely accepted. So too are the findings that there is considerable diversity in students' knowledge – a diversity frequently at odds with accepted ideas – much of which seems unresponsive to instruction. This points to the need to consider learning, not purely as an accumulation of bits of information, but as an active, interactive, connective process requiring change of different kinds such as addition, linkage, rearrangement and exchange. In this chapter, we call it 'learning as conceptual change'.

To think about what learning as conceptual change means, we use a model of learning as conceptual change initially proposed by Posner, Strike, Hewson and Gertzog (1982) and discussed further elsewhere (Hewson 1981, 1982; Hewson & Thorley 1989; Strike & Posner 1985, 1992; Thorley 1990). The central concepts of the model are status and conceptual ecology. The *status* that an idea has for the

---

Chapter Consultant: Peter Taylor (Curtin University of Technology, Australia)

person holding it is an indication of the degree to which he or she knows and accepts it: status is determined by its intelligibility, plausibility and fruitfulness to that person. The idea of a *conceptual ecology* deals with all the knowledge that a person holds, recognises that it consists of different kinds, focuses attention on the interactions within this knowledge base, and identifies the role that these interactions play in defining niches that support some ideas (raise their status) and discourage others (reduce their status.) Learning something, then, means that the learner has raised its status within the context of his or her conceptual ecology. These two concepts and the explanation that they provide of learning are the organising themes of the chapter.

We use the term 'teaching for conceptual change' to mean teaching that *explicitly* aims to help students experience conceptual change learning, and *meets* guidelines consistent with the conceptual change model. Our use of this term refers to a family of teaching models rather than to one particular way of teaching because there are different ways of meeting these guidelines. Also, we see the term defining a broader, more comprehensive set of teaching activities than do many studies that focus only on changes in students' content answers. In our view, such a focus is incomplete because it does not explicitly take account of significant aspects of conceptual change learning (Hewson & Thorley 1989). The number of studies using this broader set of teaching strategies is reduced correspondingly.

In using the term, 'teaching for conceptual change', we do not imply that conceptual change *learning* only happens in classrooms that *teach* for conceptual change. On the contrary, learning can and does occur in all manner of classrooms. Nevertheless, we believe that the evidence presented below shows that learning will occur *more* frequently for *more* students in classrooms that meet these criteria. Finally, we do not claim that these criteria are either a complete or an exclusive set, nor that we have synthesised the considerable body of research on conceptual change into a comprehensive explanatory model. While we believe that they are necessary to characterise teaching for conceptual change, they are not sufficient. For example, there are features of the teacher's role, students' roles and classroom climate that conceptual change teaching shares with other forms of teaching. For reasons of limited space, these are not considered in detail in this chapter.

In what follows, we first discuss briefly how our view of conceptual change is related to other discussions. We then present a set of guidelines that characterise significant aspects of teaching for conceptual change and illustrate them with examples from the literature. Finally, we discuss some implications of these guidelines for teaching.

## THE CONTEXT OF CONCEPTUAL CHANGE RESEARCH

Approaches to thinking about conceptual change have mirrored aspects of the current debate about constructivism in science and mathematics education (e.g., Cobb 1994; Phillips 1995). Is learning science either an activity performed by individuals who construct their own knowledge, or a process of social enculturation? Is knowledge situated in the mind of the individual or in inter-subjective activity (e.g., Lave & Wenger 1991)? Or is a combination of perspectives necessary? These questions also arise in Scott, Asoko and Driver's (1992) selective review of the conceptual change literature, in which they identify different approaches to, and significant issues concerning, teaching for conceptual change. Common to all is the importance given to students' knowledge prior to instruction. Several studies also give explicit consideration to standards by which conceptions are evaluated. There are different views of the role of conflict and whether it is explicitly needed in instruction. Some authors see conflict as a means of helping students recognise that their conceptions might be problematic, while others regard students' conceptions as an opportunity – the foundation for future learning – and play down the importance of conflict. There also are different explanations of how scientific conceptions originate: whether generated by students themselves in the process of making sense of their experiences, or developed by the social community of science and thus in need of transmission to students.

The focus of attention in personal constructivist perspectives on conceptual change is the individual who experiences the process of conceptual change. An important goal here has been to represent key elements of a person's conceptions and to use these representations to consider the nature of conceptual change. For example, Chi (1992) defines conceptual change as the occurrence of changes either within or between existing knowledge structures, with a radical conceptual change involving a shift between two epistemologically distinct categories (e.g., from thinking of force as an entity to force as an event, a process). This view is mirrored in the history of science by Thagard's (1992) description of conceptual change as including 'branch jumping' (between branches of a hierarchical tree) and 'tree switching' (changing the organising principle of a hierarchical tree.) Dykstra, Boyle and Monarch (1992) take a similar approach in representing the acquisition of Newtonian conceptions of force in terms of shifts of concepts between states, processes, attributes and relations.

Another approach is to think in terms of the whole individual rather than to focus solely on a person's cognition. This arises from critics of the model (e.g., Pintrich, Marx & Boyle 1993; Strike & Posner 1992) who point out that an overly rational view of science learning results in a neglect of important motivational and contextual factors. Pintrich, Marx and Boyle (1993) argue for the need to build connections between cognitive and motivational components of student learning. More specifically, they describe motivational constructs (e.g., goal orientation, values, efficacy beliefs, control beliefs) that can serve to mediate the process of conceptual change. They also discuss the importance of learners' intentions, goals and beliefs in driving and directing their thinking, and they argue for a need to see conceptual change as more than a 'cold, rational' process because satisfaction will be influenced by affective variables and value beliefs. As well, they recognise the influence of the classroom context on motivational and cognitive components and their interaction.

A different emphasis is central to social constructivist and 'cognitive apprenticeship' perspectives on conceptual change, in which the end goal is the competent participation of the learner in a 'community of practice' (Newman, Griffin & Cole 1989). As outlined by Driver, Asoko, Leach, Mortimer & Scott (1994, p. 7), 'a social constructivist perspective recognises that learning involves being introduced to a symbolic world' with the teacher's task being 'to provide appropriate experiential evidence and make the cultural tools and conventions of the science community available to students'. This perspective also has been emphasised by Anderson (1993) who, with his colleagues, has developed teaching strategies and materials with the curricular goal of encouraging conversations between teachers and students about natural phenomena that would be understood and explained by scientists using the concepts and theories that we want our students to understand.

In considering these different approaches, Tobin and Tippins (1993, p. 6) have advocated a 'synthesis position that knowledge is personally constructed but socially mediated'. Cobb (1994) has also argued that it is not necessary to force a choice between individual and social views of constructivism; rather, these views address different, but complementary aspects of learning. In considering different perspectives on conceptual change, we come to a similar conclusion. On one hand, we place emphasis on the individual learner, situated in a social and physical context, and locate knowledge primarily in the mind of the individual. Such a perspective, we believe, confers on the individual both the power and the responsibility to take control of her own learning, become aware of her personal epistemological commitments, represent her conceptions to her peers and teacher clearly, and monitor her own interpretations of scientific phenomena and the expressed views of others. The metacognitive nature of these actions is explicitly recognised in the guidelines presented below. On the other hand, it is clear that such activities are mediated by a social environment in which their importance is recognised, accepted and encouraged. In this regard, we recognise the key role of teachers in creating such an environment, as well as making available to their students scientific standards of evidence, forms of explanation, methods of inquiry and outcomes of investigation. Thus, the view of classroom activity that follows from our perspective is in a great many respects consistent with that advocated by educators whose primary focus is 'the social'.

## GUIDELINES FOR TEACHING FOR CONCEPTUAL CHANGE

The guidelines that follow characterise significant aspects of teaching for conceptual change. They have emerged from the literature as a consequence of a cyclical process of development in which there are repeating, connected phases of grounded, inductive work and principled, deductive work (Martin & Sugarman 1993). In this chapter, we present the guidelines as principles that have emerged from, are supported by and elaborated with illustrations from the literature. In doing so, we are conscious that this conveys an incomplete representation of the whole process. Discussing in detail the cyclical nature of the process in a short chapter, however, is not possible. In presenting and discussing the guidelines, we do not imply that the examples that we use are necessarily complete instantiations of the guidelines; rather, they are meant to illustrate salient aspects.

*Ideas (Students' and Teachers' Ideas as an Explicit Part of Classroom Discourse)*

In teaching for conceptual change, it is necessary that the range of ideas related to the topic held by different people in the class are made explicit. These ideas need to be contributed by both teacher and students. In the process, people become aware of, understand and possibly become committed to ideas that they had not previously encountered or considered seriously. This guideline is not unique; on the contrary, it is central to most recent efforts to reform the teaching of science, mathematics and other subjects.

One part of this guideline is found in typical teaching, because teachers always have made their ideas explicit in teaching the goals of the topic. What is significantly different from common practice is that teachers need to make students' ideas an explicit part of instruction as well. This practice recognises that existing knowledge plays an important role in people's learning. In common practice, students' ideas are not explicitly considered. Teachers give different reasons for not doing so (e.g., it takes more time; or it will confuse other students). When other ideas are not considered, however, students can be aware that their own individual ideas are different from those of the teacher, but can be unaware of the views of other students. A possible consequence is that there is no encouragement for students to take their own ideas seriously, possibly leading to the unintended consequence of students devaluing their own ideas. A further consequence is that the basis on which a given idea is accepted is seldom, if ever, addressed.

A second significant difference from common practice is that students' ideas should be considered in similar ways to the teacher's idea. When this happens, it means that students have the opportunity of choosing between different ideas on the basis, not of who said them, but of how good an explanation each provides. In doing so, they should come to recognise that the source of authority for a given idea should not be the teacher's undoubted position of power, but standards of evidence explicitly stated and discussed. Including such discussions in classroom discourse, as advocated here and in the fourth and last guideline, would represent a significant departure from common practice.

When students' ideas are valued sufficiently that they become part of classroom discourse, a proliferation of ideas will suggest the need to choose between them, leading perhaps to a status reduction for some. There is an apparent paradox here because, while students' ideas are valued, the status of some is reduced. The paradox disappears, however, in the recognition that, in moving from ideas to the basis for choosing between ideas, the nature of the discourse has changed significantly.

The literature contains many different ways of eliciting students' conceptions.

Some are suitable for use in a classroom, while others are better suited in a research setting because they are labour-intensive. Techniques that have been widely used in classrooms include pre-instructional quizzes (Minstrell 1982) and small-group posters (Children's Learning in Science Project 1987). White and Gunstone's (1992) book provides a broad array of techniques that allows teachers to probe different aspects of student understanding.

When classroom discourse includes students discussing their own and others' ideas and explanations, it blurs the distinction between elicitation of students' ideas and other aspects of teaching. An illustration of such an approach is provided by Smith, Blakeslee and Anderson (1993) in a study of teaching strategies associated with conceptual change learning (used by American middle school teachers) which includes a detailed list of questioning strategies that ask students, for example, to construct explanations in their own words, make predictions that might contradict naive conceptions, or respond to one another. Strategies such as these mean that students' ideas are effectively being elicited in the context of the development of the topic rather than as a one-time event at its start.

An illustration of students' ideas being central to classroom discourse comes from Thijs (1992) who evaluated the effect of a constructivist teaching approach in bringing about a conceptual change in students with regard to the concept of *force* in several grade 9 equivalent classes in The Netherlands. A range of different strategies was used to make students' ideas explicit (at one point it was clear that there were two diametrically opposed opinions) and to help students to articulate their own ideas in classroom debates and group discussions. One conclusion was that the course mainly benefited those students who valued the opportunity of discussing their ideas with others.

An approach used by Hennessey (1991b), in her American primary school classrooms, goes one step further. A significant part of her curriculum is centred on students' ideas about the current topic, as distinct from the topic content itself. She requires that her students are able to state their ideas, talk about the attraction and limitations of their ideas, question whether they are consistent, explain them using physical models, accept that they might need to change and apply the ideas of intelligibility, plausibility and fruitfulness to their ideas. In other words, in this classroom, students' ideas about a chosen topic are the central focus of attention.

*Metacognition (Discourse of the Classroom Being Explicitly Metacognitive)*

Metacognition is concerned with knowledge of one's own cognitive processes and products (Flavell 1976); it is the knowledge, awareness and control of one's own learning (Baird 1990). In other words, people are metacognitive when they make their own thoughts 'objects of cognition', a phrase coined by Kuhn, Amsel and O'Loughlin (1988), who claim that people's ability to think not only *with* their ideas but *about* them is the crucial step in the development of scientific thinking. While metacognition's cognitive process aspects have proved to be important in areas such as reading (Palincsar & Ransom 1988) and are important for any forms of learning, knowledge of one's own cognitive products (Flavell 1976) is particularly important in conceptual change learning. Thorley (1990) employed a helpful distinction, using the terms 'metacognitive' and 'metaconceptual' to refer to reflection on, respectively, cognitive processes and 'the content of conceptions themselves'.

Why should metacognition, in general, and metaconception, in particular, be guidelines of teaching for conceptual change? First, as indicated above, they have greatly facilitated instruction in other areas. Second, they are inherent in the process of conceptual change. When students give different explanations of a particular phenomenon or set of phenomena in a classroom, in effect, they are laying out the explanations themselves as objects of cognition. Commenting on, comparing and contrasting these explanations, considering arguments to support or contradict one or other explanation, and choosing one of these possible explanations are all metaconceptual activities (Hewson 1991; Hewson & Thorley 1989). Third, they are the only means by which students provide evidence of the status of ideas learned, thus allowing the teacher to monitor the learning process in a unique and powerful way.

The critical relationship between metacognition and conceptual change also has been recognised by Australian scholars (Gunstone 1992, 1994; White 1993; White & Gunstone 1989) in work that has grown out of extensive studies of metacognition (e.g., Baird 1986, 1990; Baird & Mitchell 1986; Baird & Northfield 1992). As Gunstone (1994) has pointed out, a learner needs to be metacognitive in order to go through the conceptual change process. Gunstone and Northfield (1992) formulated this process as a learner recognising his or her existing ideas, evaluating them and deciding whether to reconstruct them on the basis of dissatisfaction with and/or fruitfulness of them. Gunstone (1994) has also provided an invaluable list of assertions about metacognition that are complementary, not alternative, to one another.

- Metacognition can be learned and taught, and can influence the achievement of normal learning goals.
- Some metacognitive views can be at odds with conceptual change learning goals; there can be tension between metacognitive knowledge, awareness and control. Also, inadequate metacognition can result in poor learning tendencies.
- Metacognitive learners are able to monitor, integrate and extend their own learning, and are likely to use good learning behaviours.

A related question is relevant here. About what should a student be metaconceptual? Scientific conceptions (e.g., of respiration, rusting, erosion) are prototypical objects of cognition; but others suggested as being important include conceptions of the nature of both learning (Gunstone 1994; Scott 1987) and science and its epistemology (Carr, Barker, Bell, Biddulph, Jones, Kirkwood, Pearson & Symington 1994; Scott, Asoko & Driver 1992; Smith 1987). This suggests that some or even all of these should be included explicitly in the curriculum. Perkins (1992), for example, advocates a metacurriculum that includes knowledge about how to get knowledge and understanding, knowledge about thinking and how to think well, and knowledge about the way in which the subject matter works.

Hennessey (1991a) provides an image that epitomises metacognition:

> Tammy sat cross-legged, leaning back against the wall, with palms cupped together as if holding something in her hands, gazing down at them with a meditative expression on her face. Her teacher asked her what she was thinking about: 'Oh, I was just sitting here holding my thoughts in my hands. Not really, I was just pretending as if I really could hold what I was thinking... I thought that, if I could sort of hold my ideas in my hands, I could better look at them to see why my ideas are intelligible, plausible and fruitful to me'. With that, Tammy took her cupped hands, placed them on her head as if returning her thoughts to her mind, stood up and continued the conversation about her thoughts with her teacher. (p. 1)

A simple strategy for encouraging metacognition is to ask students to consider their own recorded responses to some form of pretest. There are different ways of doing this. Trumper (1990, 1991) grouped similar statements that Israeli high school students had made about energy, presented these groups of statements to them, and asked them to point out the common feature in all the statements. The successfully demonstrated intent of this metacognitive comparative strategy was for students to develop a more generalised concept based on their own initial ideas that were, for the most part, poorly formulated and incomplete. In Jiménez-Aleixandre's (1992) study, Spanish high school students first discussed in small groups their own pretest answers on evolutionary theories and also considered counterexamples. Their teacher then used a whole-group discussion strategy to consider inconsistencies between different students' answers and school science. Both of these are metacognitive strategies. The intent was that students would come to see that Darwinian and Lamarckian theories of evolution were incompatible, rather than different forms of the same explanation. Seeing their incompatibility created the need to choose between them.

Another strategy requiring metacognition is to engage students in discussing whether two situations are analogous to one another (Clement 1993). The goal of the strategy is to bridge students' intuitive understandings that are in accord with accepted theories ('anchors') and those that are not ('targets'). In Clement's study, American high school students, who were actively engaged for extended periods in evaluating whether the examples are analogous or not and in finding the best way to view the target situation, made significant gains in understanding.

Another strategy leading to metacognition is direct questioning that involves students in reflecting on their learning experiences. Gustafson (1991) studied a fourth grade classroom in Canada during the teaching of a topic on sound. After each lesson, she asked five chosen students a variety of questions that were designed to explore: what they had learned; what they had understood; what they believed about the lesson; and how their ideas had changed. Asking students, for example, what they believed about the lesson is asking for them to be metacognitive. The questioning strategy also allowed a detailed understanding of the different ways in which students in the study changed their ideas.

While the questions in this study were asked by the researcher rather than the teacher or even the students, questions such as these can be included in regular classroom practice. This is illustrated by Hennessey (1991b) who required that, rather than a teacher or a researcher questioning students, they question themselves and one another. An illustration of this is discussed in the next guideline.

*Status (The Status of Ideas Needing to be Discussed and Negotiated)*

The status of an idea is an indication of the degree to which the person holding it knows it, accepts it and finds it useful (Hewson 1981). There are different aspects of status. A prerequisite is *intelligibility*. Does the person know what it means? Does it make sense? Can the person represent it? If someone finds an idea intelligible, other aspects to consider are its *plausibility* and its *fruitfulness*. An intelligible idea can become plausible if the person believes it to be true, or finds it consistent with and is able to reconcile it with other ideas that he or she has accepted. It can become fruitful if it achieves something of value, or solves otherwise insoluble problems, or suggests new possibilities, directions or ideas. The more that an idea meets these conditions, the higher is its status.

Teaching for conceptual change is likely to facilitate students considering different ideas, leading to the need for them to make informed choices. Possible choices might be a continuing preference for their prior ideas, an acceptance of more than one idea, a combination of their ideas with others, or a preference for a different idea at the expense of their prior ideas. Making any of these choices does not require the extinction of rejected ideas. Rather, in the choice process, students are likely to find that some ideas become more, and others less, acceptable to them. In other words, the status of these ideas changes, with the status of some being raised and of others being lowered.

Because an essential part of the concept of both status and conceptual ecology is the connectedness of knowledge, when the status of one idea changes, there can be increases and/or decreases in the status of related ideas. This occurs when a person chooses between two explanations of the same phenomena that she or he sees as mutually exclusive. Another type of example recognises that each person's conceptual ecology itself contains many ideas of different kinds, and that the status of each one of these ideas could be the focus of that person's attention on different occasions. Thus, when a person changes his or her idea of what constitutes a good explanation, this carries with it the potential for influencing any previously accepted explanation.

Activities aimed at *raising* the status of particular ideas therefore should be a part of teaching for conceptual change. In this respect, there is much in common with normal teaching. These activities might aim to present and develop the ideas, to provide examples of them, to apply them in other circumstances, to give different ways of thinking about them, to link them to other ideas, etc. Activities aimed at *lowering* the status of other ideas also should be a part of teaching for conceptual change. These might aim to explore their unacceptable implications, to consider experiences which they are unable to explain, or to find ways of thinking about them that point to their inadequacies. For any status lowering activity to work for a particular student, she or he needs to see the inadequacy of an idea; a common problem is that teachers often mistakenly assume that the discrepancy is as obvious to their students as it is to them. It is important to note that status raising and status lowering activities can occur simultaneously.

Two comments are relevant here. First, considering the status of one's ideas involves being metacognitive because the ideas being examined become objects of cognition. In other words, this guideline is closely related to the previous one, while highlighting a different aspect of teaching for conceptual change. Second, there are different ways in which the status of a student's ideas can be determined (Hewson & Hewson 1992). While there always needs to be some communication from the student, this could occur in many different settings, whether specifically designed for the purpose (e.g., an interview) or not (e.g., classroom discourse). Another variable is whether or not students use conceptual change language. The status also could be determined by the students themselves or another person (e.g., a teacher). Some of these possibilities are illustrated below.

Some authors, in their studies of the effect of particular instructional approaches on conceptual change, have used discourse about science to determine the status of students' conceptions (i.e., explicit status terminology was not a part of students' conversations). After Australian grade 10 students had completed a topic in optics that used a carefully chosen analogy, Treagust, Harrison and Venville (1993) interviewed them about the role that the analogy had played in their understanding of the topic. The researchers identified factors related to intelligibility, plausibility and fruitfulness in the interview transcripts to determine the status of each student's conception of refraction. Basili and Sanford (1991) recorded the discourse of American community college students within small cooperative groups, analysing and coding it for the use of behaviours that 'promote' and 'impede' desired conceptual change.

One strategy for eliciting the status of students' ideas is direct questioning, as illustrated by Gustafson's (1991) study described above. The distinctions between learning, understanding and believing that students were asked to make provided a way for them to express status differences between their ideas in ways that are not part of normal classroom discourse.

In the above examples, students need not have been aware that the status of their ideas was a matter of concern. An example for which status was a central issue in classroom discourse is provided by Thorley (1990) who analysed a videotaped lesson taught by Minstrell (1982) in an American high school classroom. On this occasion, conceptual change language was not used explicitly. The topic was whether or not a table exerts an upward force on a book it is supporting.

T: How are you conceiving of force that's different from the way some of the other people are conceiving of force?
Jane: I think that pushing the book is force. But, the table's not pushing the book up. It's just there and keeping it from falling down . . .
Beth: I think that Jane's and a lot of people's conception of force and mine too is something active, like pushing or pulling . . . But it's hard to conceive of something just kind of being there as . . . exerting a force. (Thorley 1990, p. 127)

These brief extracts of a much longer classroom interchange provide a vivid illustration of students recognising different ideas (i.e., they were intelligible) and expressing their preference for one (i.e., only some were plausible or fruitful).

Requiring students to use the status language of the conceptual change model explicitly goes a step further in implementing this guideline. Hennessey (1991b) demonstrated that American grade 6 students were able to do this effectively. This is illustrated in a case study of Alma, a student who was able to use the language reliably and with meaning (Hewson & Hennessey 1992). For example, in considering the possibility that there might be no forces acting on a book at rest on a table, Alma said: 'Of course this is not plausible . . . I could not possibly believe in no forces acting on anything because there is always gravity' (p. 184).

In using status language as part of her discourse Alma was able to demonstrate the conceptual changes that she experienced in arriving at an impressive qualitative understanding of the topic while, at the same time, revealing unexpected uncertainties that showed the depth and complexity of her understanding.

Becoming accustomed to using status language provides a way of thinking about ideas that can permeate a discussion in which status considerations are implied, but status language is not used. While studying one of Hennessey's grade 5 classes, Beeth (1993, pp. 126–130) recorded an extended classroom dialogue in which the class was considering Don's explanation of a falling parachute:

Don: I think that there are [two] equal forces . . . because it's going in a pretty straight line and at consistent speed. Regarding those two arrows, this one's gravity and the other one is friction.
Kitt: OK, why do you have equal arrows? I don't think that it would be moving if they had equal arrows.

The implication was that Don should have used a different explanation (unequal forces) for a moving parachute. Kirsty and Ellen then took a different tack:

Kirsty: If that parachute was at rest what would the arrows look like?
Don: Probably nothing [different].
Ellen: Then it would be floating.
Don: [at rest] on the ground?
Ellen: No, in the air.

Underlying these questions was the implication that there could not be the same explanation (equal forces) for different circumstances (a parachute either at rest or steadily falling with a constant speed). Kirsty and Ellen knew what he was saying, but didn't agree with him. Don, however, saw a consistency rather than a difference between the two circumstances and he stated that, 'if a thing is at rest, it's still going in a straight line at a consistent speed'.

It seems clear that this discussion was centrally about the differing status of Don's ideas in his and other students' eyes. A particular idea – Don's explanation – was metaphorically laid out on the table. In a key respect for Don, the two circumstances were the same and therefore could have the same explanation (i.e., it was fruitful for him). It was intelligible to Kitt, Kirsty and Ellen, but their comments and critiques clearly point out that for them that it was neither plausible nor fruitful.

*Justification (The Justification for Ideas and for Status Decisions as an Explicit Component of the Curriculum)*

In conceptual change learning, students need to determine the status that an idea has for them. In order to do this, they have to know what the idea is (i.e., find it intelligible) before deciding whether or not they find it plausible and/or fruitful. In order to provide a justification for such a status decision, students will bring to bear one or more criteria. These criteria, and the knowledge required to apply them, are significant components of each person's conceptual ecology (along with many ideas of different kinds). The status of each one of these ideas could be the focus of that person's attention on different occasions, and thus should be explicitly considered in the curriculum. While this guideline is implicit in the previous one (the necessity of explicit status consideration), it is included to emphasise the essential role that different constituents of a person's conceptual ecology play in conceptual change learning.

The intent of the guideline is that the justification process and relevant components of students' conceptual ecologies should be a part of the curriculum of the classroom, because of the essential role that status determination plays in conceptual change learning. Smith (1987) was an early advocate for such a guideline. In doing so, he recognised a significant parallel with the learning of science content when he argued that 'the development of explanatory *ideals* is itself a change from more naive patterns of explanation' (p. 429) (author's emphasis). Others have also agreed with this recommendation. Scott, Asoko and Driver (1992, p. 324) urged that 'teaching in science . . . needs to introduce explicitly the epistemological assumptions underlying the "language game" of science' and Carr *et al.* (1994) argued that 'open discussion of the "rules of the game" of science would contribute to better learning in the classroom, since learners would be better equipped to change their existing concepts, by knowing more about the nature of science itself' (p. 147).

That students from the earliest grades on can do the tasks required by this guideline is shown in the above example from Beeth (1993). Samarapungavan (1993) has also demonstrated that American children in grades 1, 3 and 5 were able to choose between competing theories; in doing so, in the context of carefully designed tasks, they were able to use significant epistemological criteria such as *consistency* between a theory and empirical evidence, *range* of observations explained by a theory, and the *conceptual coherence* or internal consistency of a theory.

Explanatory ideals are critical components of the conceptual ecology of chemistry (as are chemical knowledge and conservation reasoning), according to Hesse and Anderson (1992) who claimed that, when students haven't mastered these components, chemical equations are much less meaningful to them. In a study of American high school students' conceptions of chemical change, they found that many high school students preferred 'explanations based on superficial analogies with everyday events (e.g., rusting is like decay) over explanations based on chemical theories' (p. 289). To illustrate, Sue was asked to evaluate her own analogy-based explanations by comparing it with one using a balanced chemical equation:

> I bet half . . . the class would take [the analogy] but [in] a group of scientists . . . the formulas would be accepted the most . . . because [the analogy] is for people who don't know how much of what elements are involved. So they are just trying their best to think of how it works . . . the scientists would know how much of what and I think that they could just look at the formula. (p. 289)

An example of a teacher explaining the basis for a physicist's point of view is included in Minstrell's (1982) videotaped lesson referred to above (Thorley 1990). This extract comes later in the same lesson:

> The question is . . . whether we should also say . . . that there's a force exerted by the table. Now, the physicist wanting to come up with a logical sort of definition that makes rational sense as you go across different situations says, 'look, I see something in common in these situations. This book [on the hand] is . . . at rest . . . and this book [on the table] . . . is at rest. So I want to say . . . that here [on the hand] the book is at rest and what keeps it at rest is a down force [*gesture down*] exerted by the earth and an up force [*gesture up*] exerted by the hand, and those two balance each other.' Then the physicist says, 'shoot, I guess that means that I'd better think of force . . . as something that the table can do as well . . ., but I want to think of that . . . as sort of a passive support that the table does . . . rather than something really active muscular-wise. But if I want to be logical about it, then I want the same kind of explanation here [table] as what I have over here [hand].' (Thorley 1990, p. 134)

In other words, the teacher has outlined an epistemological commitment to the value of consistency between instance and explanation and applied it to this example.

An illustration of a different way in which the basis for justification enters classroom discourse comes from Beeth's (1993) study of Hennessey's classroom, discussed above. An essential component of Don's explanation was his use of a consistency argument. Once he had recognised an essential similarity between the falling parachute and the book on the table (neither was speeding up or slowing down), he argued for the same explanation (equal and opposite forces) by saying, in effect, that consistency requires the same explanation for the same effect. In other words, a significant component of Don's conceptual ecology was his epistemological commitment to consistency.

Further similar examples are provided by Hennessey (1991b). In this study, a student rejected the idea that a table exerts an upward force on a book to support it because she could not imagine how 'a dead table knows how much to push up'. This idea has low status for her; she knows what it means, but she doesn't believe it. In other words, a need for her explanations to provide acceptable causal mechanisms is an important component of her conceptual ecology. Another student accepted this idea because he explained the book's state of rest by using balanced forces, an explanation seen in other examples. In other words, his epistemological commitment – similar examples require similar explanations – was instrumental in raising the status of this idea; it was the criterion that he used in making his choice.

Another level of metacognition in meeting this guideline is illustrated in a study of Canadian high school students (Larochelle & Desautels 1992) who constructed explanations of various 'black box' phenomena in which inputs produced known outputs, without a mechanism or process being provided. The students were encouraged not only to create mechanisms that explained how inputs and outputs might be related (i.e., to produce knowledge), but also to reflect on the process by which this happened in what was termed a 'metalogue', or a conversation both about the problem and the structure of the conversation itself.

## CONCLUSION: IMPLICATIONS FOR TEACHING

Teaching for conceptual change builds on the central concepts of the conceptual change model of learning – status and conceptual ecology – within the context of individual and social constructivism. A powerful way of integrating these ideas is through the notion of authority as it relates to the question of who, or what, decides what counts as valid knowledge in the classroom. By *authority*, we mean the basis on which a person decides to accept or reject information in the process of acquiring knowledge. To which sources of authority a person pays attention and the relative importance of those sources in making the decisions involved in acquiring knowledge are critical factors in understanding learning.

There are different sources of authority that are relevant in teaching for conceptual change. While these different forms of authority overlap one another, they are not automatically congruent with one another because they serve different purposes and have arisen under different constraints. On one hand, there is an intrinsic, direct authority that comes when someone makes personal sense of an idea; it is a *sine qua non* of understanding. This type of authority is of primary importance in personal constructivism. We call it *internal authority*.

On the other hand, there are different forms of *external authority*. One form is the *disciplinary authority* that arises from the types of method used for producing knowledge and the sets of values for validating the knowledge produced that have been developed by individuals and the professional communities and organisations within which they work. It is these methods and values that give authority to the knowledge produced by these individual researchers and their disciplines. A critical question to address is how such authoritative knowledge passes between disciplines and students. Another form is *curriculum authority*. In practice, disciplinary authority is mediated to students by other individuals, groups and organisations, such as teachers, curriculum committees, professional societies and publishers, and it is manifested in products such as worksheets, textbooks, videos and other forms of media. Because of both the mediation process and the many other demands on a school curriculum, the notions of the external disciplinary authority of science that are acquired by students can be partial, biased or even absent. In effect, teachers and textbooks become the most prominent sources of external authority for most students. A third form of external authority is *cultural authority*. Underpinning the society that includes students, parents, teachers and schools are sets of cultural norms, ways of knowing and values that have developed and accrued in many different ways and over different time spans. They are less likely than disciplinary and curriculum authority to be codified and consistent, and more likely to be contextualised and heterogeneous. Their value lies in the definition, the focus and the identity that they provide to the communities in the society.

The first two guidelines outlined above – Ideas and Metacognition – predominantly address the internal authority of understanding by providing the opportunity for all to express their ideas and examine them critically. The third guideline – Status – constitutes a bridge between internal and external authority. Here students explore the reasons for the appeal of different ideas. But, considering a range of views, particularly if they are divergent, suggests the need to choose between them (i.e., the need for a different, external, authority that complements the former). The fourth guideline – Justification – predominantly addresses external authority. Implicit within all the activities suggested by these guidelines is their social nature: the obvious way in which to study someone else's ideas is to interact with him or her.

In giving practical effect to the importance of both internal and external sources of authority, a major challenge for teachers to meet is helping students to integrate these several notions of authority with each other. On one hand, teachers need to introduce curriculum and adopt teaching strategies that open up the connections between disciplinary, curriculum and cultural sources of authority. For example, teachers need to ensure that disciplinary authority is visible, undistorted and fully represented within their curriculum authority; and that overlaps with and divergences from cultural authority are identified and explored. Students need to see where these forms of external authority can be compatible and even coherent with each other. On the other hand, teachers need to help students to reconcile their internal authority with the different sources of external authority. This means that teachers need to help students: to make sense of, understand, accept and be able to use both their own and others' ideas; to understand and be able to articulate the reasons underlying their own and others' ideas; to be willing and able to reconsider and, if necessary, change their ideas; and to be able to justify a continuing challenge of others' ideas when these remain in conflict with personal ideas. In other words, in addition to the curriculum's focus on its central ideas, their status needs to be considered. Once again, this is the point of the Status and Justification guidelines.

Traditional schooling seldom explicitly addresses sources of authority, let alone acknowledging that there can be differences between them. In practice, the main source of authority is curriculum authority that is expressed in general goals in the guise of disciplinary authority. Internal authority is assumed to follow automatically from teaching. Cultural authority is unrecognised. This justifiably has led to criticisms of scientific imperialism in which Western science, implicitly assumed to be superior, is imposed on students in non-Western societies without regard to their cultural heritage (e.g., Baker & Taylor 1995; Hewson 1988). The question remains of how to respond to these criticisms. We do not believe that satisfactory answers are either teaching indigenous science by itself, or teaching a menu list in which equal time is given to, for example, evolutionary science and creation science. Rather, we believe that a strong case can be made for Western science, with its rational approach to searching for generalisable and internally consistent concepts, laws, principles and theories that can describe, explain and predict divergent phenomena. Further, we believe that, when properly presented, students will see its value and will want to use it as a powerful way of knowing about the world, not to the exclusion of other forms of knowledge, but in dialogue with them in order to harmonise them (Ogunniyi 1988). Mariana Hewson (1988) has discussed the synergism and hybrid vigour of such a dialogue. However, if the case for Western science (or any other way of knowing, for that matter) carries little conviction for students, then its role in the curriculum needs to be questioned.

By contrast with traditional schooling, teaching for conceptual change recognises the importance of both internal and external sources of authority and seeks to make them an explicit part of the classroom discourse. We acknowledge differences between sources of authority, we plan to examine their relative pros and cons, and we have the goal of reconciling them where possible. In this way, we intend that students will be able to make informed decisions about how they understand natural phenomena, rather than having to accept given information as unquestioned truth.

Implementing these four guidelines requires significant changes in science classrooms. One concerns the teacher's role: teaching for conceptual change (taking responsibility for explicitly encouraging conceptual change learning and implementing the guidelines) would require changes in curriculum, instructional strategies and classroom interactions and in forms of assessment. Another change concerns the students' role: they would carry greater responsibility and would become more active and participatory. A third change concerns the nature of classroom discourse and, by implication, the curriculum: attention would be focused not solely on the factual and conceptual knowledge of science (the products of scientific inquiry), but also on the way in which that knowledge is validated and accepted (the processes and values of scientific inquiry).

A conceptual change teacher thus has different roles to play. One role is setting goals for instruction, creating appropriate contexts for classroom activities, and posing problems that have relevance and meaning to the students. Another role is facilitating the different levels of discourse needed in the classroom for science topics, for the values and methods of science, and for being metacognitive about these issues. A further role is establishing a classroom environment that provides opportunities for students to explore their own and others' ideas individually and collectively without fear of ridicule or sanction. A critical role is monitoring classroom activities and deciding if, when and how to intervene.

Teaching for conceptual change requires a great deal of teachers. With respect to content, they need to know the content of the science curriculum, its associated pedagogical content knowledge, the range of ideas that their students are likely to hold about the content topic, an understanding of conceptual issues significant in the historical development of the topic, and the empirical underpinnings of the content. They also need to be well founded in philosophical issues related to the nature of scientific knowledge (e.g., its methods of inquiry and epistemological foundations). With respect to learning, teachers need to know about the conceptual cnange model of learning and the role and function of components of a learner's conceptual ecology in assessing the status of ideas (e.g., anomalies, analogies, metaphysical beliefs, images of real world objects, exemplars of phenomena, epistemological commitments). With respect to instruction, teachers need to know a variety of pedagogical techniques. In addition to those in the typical repertoire of good teachers, there will be others related to and outlined in the discussion of the guidelines presented above. Implicit in all of these are the teacher's conceptions of the nature of science, learning and teaching that support teaching for conceptual change. These have been outlined elsewhere (Hewson & Hewson 1988).

Dedicated teachers over the years have developed science activities and employed teaching strategies that have challenged many students and facilitated much science learning: without doubt, teaching for conceptual change benefits from this huge base of expertise. Nevertheless, we believe that the combination of the guidelines presented above represents a change in current practice that is not incremental. In comparison with most current teaching, teaching for conceptual change is a radically different enterprise.

## REFERENCES

Anderson, C.W.: 1993, 'Teaching Content in a Multicultural Milieu', Paper presented at the Didactics and/or Curriculum Symposium, Institute for Science Education at the University of Kiel, Kiel, Germany.

Baird, J.R.: 1986, 'Improving Learning Through Enhanced Metacognition: A Classroom Study', *European Journal of Science Education* 8, 263-282.
Baird, J.R.: 1990, 'Metacognition, Purposeful Enquiry and Conceptual Change', in E. Hegarty-Hazel (ed.), *The Student Laboratory and the Science Curriculum*, Routledge, London, 183-200.
Baird, J.R. & Mitchell, I.J. (eds.): 1986, *Improving the Quality of Teaching and Learning: An Australian Case Study - The PEEL Project*, Monash University, Melbourne, Australia.
Baird, J.R. & Northfield, J.R. (eds.): 1992, *Learning from the Peel Experience*, Monash University, Melbourne, Australia.
Baker, D. & Taylor, P.C.S.: 1995, 'The Effect of Culture on the Learning of Science in Non-Western Countries: The Results of an Integrated Research Review', *International Journal of Science Education* 17, 695-704.
Basili, P. & Sanford, J.P.: 1991, 'Conceptual Change Strategies and Cooperative Group Work in Chemistry', *Journal of Research in Science Teaching* 28, 293-304.
Beeth, M.E.: 1993, *Dynamic Aspects of Conceptual Change Instruction*, Unpublished doctoral dissertation, University of Wisconsin-Madison, WI.
Carr, M., Barker, M., Bell, B.F., Biddulph, F., Jones, A., Kirkwood, V., Pearson, J. & Symington, D.: 1994, 'The Constructivist Paradigm and Some Implications for Science Content and Pedagogy', in P. Fensham, R. Gunstone & R. White (eds.), *The Content of Science*, Falmer Press, London, 147-160.
Chi, M.T.H.: 1992, 'Conceptual Change Within and Across Ontological Categories: Examples from Learning and Discovery in Science', in R. Giere (ed.), *Cognitive Models of Science: Minnesota Studies in the Philosophy of Science*, University of Minnesota Press, Minneapolis, MN, 129-186.
Children's Learning in Science Project: 1987, *Approaches to Teaching the Particulate Theory of Matter*, University of Leeds, Leeds, UK.
Clement, J.: 1993, 'Dealing with Students' Preconceptions in Mechanics', in J.D. Novak (ed.), *Proceedings of the Third International Seminar on Misconceptions and Educational Strategies in Science and Mathematics*, Misconceptions Trust, Ithaca, NY. (electronically accessed at website http://www2.ucsc.edu/mlrg/mlrgarticles.html)
Cobb, P.: 1994, 'Where is the Mind? Constructivist and Sociocultural Perspectives on Mathematical Development', *Educational Researcher* 23(7), 13-20.
Driver, R., Asoko, H., Leach, J., Mortimer, E. & Scott, P.: 1994, 'Constructing Scientific Knowledge in the Classroom', *Educational Researcher* 23(7), 5-12.
Dykstra, D.I., Jr., Boyle, C.F. & Monarch, I.A.: 1992, 'Studying Conceptual Change in Learning Physics', *Science Education* 76, 615-652.
Fenstermacher, G.D.: 1986, 'Philosophy of Research on Teaching: Three Aspects', in M.C. Wittrock (ed.), *Handbook of Research on Teaching*, Macmillan, New York, 37-49.
Flavell, J.H.: 1976, 'Metacognitive Aspects of Problem Solving', in L.B. Resnick (ed.), *The Nature of Intelligence*, Lawrence Erlbaum, Hillsdale, NJ, 231-235.
Gunstone, R.F.: 1992, 'Constructivism and Metacognition: Theoretical Issues and Classroom Studies', in R. Duit, F. Goldberg & H. Niedderer (eds.), *Research in Physics Learning: Theoretical Issues and Empirical Studies*, Institute for Science Education at the University of Kiel, Kiel, Germany, 129-140.
Gunstone, R.F.: 1994, 'The Importance of Specific Science Content in the Enhancement of Metacognition', in P. Fensham, R. Gunstone & R. White (eds.), *The Content of Science: A Constructivist Approach to its Teaching and Learning*, Falmer Press, London, 131-146.
Gunstone, R.F. & Northfield, J.: 1992, 'Conceptual Change in Teacher Education: The Centrality of Metacognition', Paper presented at the annual meeting of the American Educational Research Association, San Francisco, CA.
Gustafson, B.J.: 1991, 'Thinking about Sound: Children's Changing Conceptions', *Qualitative Studies in Education* 4, 203-214.
Hennessey, M.G.: 1991a, 'Analysis and Use of the Technical Language of the Conceptual Change Model for Revealing Status: 6th graders' Conceptions of Force and Motion', Paper presented at the annual meeting of the National Association for Research in Science Teaching, Fontane, WI.
Hennessey, M.G.: 1991b, *Analysis of Conceptual Change and Status Change in Sixth Graders' Concepts of Force and Motion*, Unpublished doctoral dissertation, University of Wisconsin-Madison, WI.
Hesse, J.J., III & Anderson, C.W.: 1992, 'Students' Conceptions of Chemical Change', *Journal of Research in Science Teaching* 29, 277-299.
Hewson, M.G.A'B.: 1988, 'The Ecological Context of Knowledge: Implications for Learning Science in Developing Countries', *Journal of Curriculum Studies* 20, 317-326.

Hewson, P.W.: 1981, 'A Conceptual Change Approach to Learning Science', *European Journal of Science Education* 3, 383–396.

Hewson, P.W.: 1982, 'A Case Study of Conceptual Change in Special Relativity: The Influence of Prior Knowledge in Learning', *European Journal of Science Education* 4, 61–78.

Hewson, P.W.: 1991, 'Conceptual Change Instruction', Paper presented at the annual meeting of the National Association for Research in Science Teaching, Fontane, WI.

Hewson, P.W. & Hennessey, M.G.: 1992, 'Making Status Explicit: A Case Study of Conceptual Change', in R. Duit, F. Goldberg & H. Niedderer (eds.), *Research in Physics Learning: Theoretical Issues and Empirical Studies*, Institute for Science Education at the University of Kiel, Kiel, Germany, 176–187.

Hewson, P.W. & Hewson, M.G.A'B.: 1988, 'An Appropriate Conception of Teaching Science: A View from Studies of Science Learning', *Science Education* 72, 597–614.

Hewson, P.W. & Hewson, M.G.A'B.: 1992, 'The Status of Students' Conceptions', in R. Duit, F. Goldberg & H. Niedderer (eds.), *Research in Physics Learning: Theoretical Issues and Empirical Studies*, Institute for Science Education at the University of Kiel, Kiel, Germany, 59–73.

Hewson, P.W. & Thorley, N.R.: 1989, 'The Conditions of Conceptual Change in the Classroom', *International Journal of Science Education* 11, 541–553.

Jiménez-Aleixandre, M.P.: 1992, 'Thinking about Theories or Thinking with Theories?: A Classroom Study with Natural Selection', *International Journal of Science Education* 14, 51–61.

Kuhn, D., Amsel, E. & O'Loughlin, M.J.: 1988, *The Development of Scientific Thinking Skills*, Academic Press, San Diego, CA.

Larochelle, M. & Desautels, J.: 1992, 'The Epistemological Turn in Science Education: The Return of the Actor', in R. Duit, F. Goldberg & H. Niedderer (eds.), *Research in Physics Learning: Theoretical Issues and Empirical Studies*, Institute for Science Education at the University of Kiel, Kiel, Germany, 155–175.

Lave, J. & Wenger, E.: 1991, *Situated Learning: Legitimate Peripheral Participation*, Cambridge University Press, Cambridge, UK.

Martin, J. & Sugarman, J.: 1993, 'Beyond Methodolatry: Two Conceptions of Relations between Theory and Research in Research on Teaching', *Educational Researcher* 22(8), 17–24.

Minstrell, J.: 1982, 'Explaining the "At Rest" Condition of an Object', *Physics Teacher* 20, 10–14.

Newman, D., Griffin, P. & Cole, M.: 1989, *The Construction Zone: Working for Cognitive Change in School*, Cambridge University Press, Cambridge, UK.

Ogunniyi, M.B.: 1988, 'Adapting Western Science to Traditional African Culture', *International Journal of Science Education* 10, 1–9.

Palincsar, A.S. & Ransom, K.: 1988, 'From the Mystery Spot to the Thoughtful Spot: The Instruction of Metacognitive Strategies', *The Reading Teacher* 41, 784–789.

Perkins, D.: 1992, *Smart Schools: From Training Memories to Educating Minds*, The Free Press, New York.

Phillips, D.C.: 1995, 'The Good, the Bad and the Ugly: The Many Faces of Constructivism', *Educational Researcher* 24(7), 5–12.

Pintrich, P.R., Marx, R.W. & Boyle, R.A.: 1993, 'Beyond Cold Conceptual Change: The Role of Motivational Beliefs and Classroom Contextual Factors in the Process of Conceptual Change', *Review of Educational Research* 63, 167–199.

Posner, G.J., Strike, K.A., Hewson, P.W. & Gertzog, W.A.: 1982, 'Accommodation of a Scientific Conception: Toward a Theory of Conceptual Change', *Science Education* 66, 211–227.

Samarapungavan, A.: 1993, 'What Children Know about Metascience', in J.D. Novak (ed.), *Proceedings of the Third International Seminar on Misconceptions and Educational Strategies in Science and Mathematics*, Misconceptions Trust, Ithaca, NY. (electronically accessed at website http://www2.ucsc.edu/mlrg/mlrgarticles.html)

Scott, P.: 1987, 'The Process of Conceptual Change in Science: A Case Study of the Development of a Secondary Pupil's Ideas Relating to Matter', in J. D. Novak (ed.), *Proceedings of the Second International Seminar: Misconceptions and Educational Strategies in Science and Mathematics*, Cornell University, Ithaca, NY, 404–419.

Scott, P.H., Asoko, H.M. & Driver, R.H.: 1992, 'Teaching for Conceptual Change: A Review of Strategies', in R. Duit, F. Goldberg & H. Niedderer (eds.), *Research in Physics Learning: Theoretical Issues and Empirical Studies*, Institute for Science Education at the University of Kiel, Kiel, Germany, 310–329.

Smith, E.L.: 1987, 'What Besides Conceptions Needs to Change in Conceptual Change Learning?', in J.D. Novak (ed.), *Proceedings of the Second International Seminar: Misconceptions and Educational Strategies in Science and Mathematics*, Cornell University, Ithaca, NY, 424–433.

Smith, E.L., Blakeslee, T.D. & Anderson, C.W.: 1993, 'Teaching Strategies Associated with Conceptual Change Learning in Science', *Journal of Research in Science Teaching* 30, 111–126.
Strike, K.A. & Posner, G.J.: 1985, 'A Conceptual Change View of Learning and Understanding', in L.H.T. West & A.L. Pines (eds.), *Cognitive Structure and Conceptual Change*, Academic Press, Orlando, FL, 211–231.
Strike, K.A. & Posner, G.J.: 1992, 'A Revisionist Theory of Conceptual Change', in R.A. Duschl & R.J. Hamilton (eds.), *Philosophy of Science, Cognitive Psychology, and Educational Theory and Practice*, State University of New York Press, Albany, NY, 147–176.
Thagard, P.: 1992, *Conceptual Revolutions*, Princeton University Press, Princeton, NJ.
Thijs, G.D.: 1992, 'Evaluation of an Introductory Course on "Force" Considering Students' Preconceptions', *Science Education* 76, 155–174.
Thorley, N.R.: 1990, *The Role of the Conceptual Change Model in the Interpretation of Classroom Interactions*, Unpublished doctoral dissertation, University of Wisconsin-Madison, WI.
Tobin, K. & Tippins, D.: 1993, 'Constructivism as a Referent for Teaching and Learning', in K. Tobin (ed.), *The Practice of Constructivism in Science Education*, AAAS Press, Washington, DC, 3–21.
Treagust, D.F., Harrison, A. & Venville, G.: 1993, 'Using an Analogical Teaching Approach to Engender Conceptual Change', Paper presented at the annual meeting of the American Education Research Association, Atlanta, GA.
Trumper, R.: 1990, 'Being Constructive: An Alternative Approach to the Teaching of the Energy Concept – Part I', *International Journal of Science Education* 12, 343–354.
Trumper, R.: 1991, 'Being Constructive: An Alternative Approach to the Teaching of the Energy Concept – Part II', *International Journal of Science Education* 13, 1–10.
White, R.T.: 1993, 'Insights on Conceptional Change Derived from Extensive Attempts to Promote Metacognition', Paper presented at the annual meeting of the American Educational Research Association, Atlanta, GA.
White, R.T. & Gunstone, R.F.: 1989, 'Metalearning and Conceptual Change', *International Journal of Science Education* 11, 577–586.
White, R.T. & Gunstone, R.F.: 1992, *Probing Understanding*, Falmer Press, London.

# 2.6 The Role of Routine Problem Tasks in Science Teaching

**PAUL HOBDEN**
*University of Natal, Durban, South Africa*

The solving of problems is central to the teaching and learning of physical science in the secondary school classroom. It is a routine activity occupying a large proportion of curriculum time and plays a central role in students' experience of classroom life. From the first days of science instruction, sets of routine problem tasks assigned by the teacher have been part of classroom life. As a teaching strategy, they largely have been used uncritically (Watts & Gilbert 1989). It would appear that nearly all physical science education, and especially the physics component, seems to be based on the optimistic assumption that success with numerical problems breeds an implicit conceptual understanding of science (Osborne 1990).

This chapter attempts to explore the role of these routine problem tasks in the teaching and learning of secondary school physical science. The chapter is written from the perspective of a developing country typified by strong central control of syllabi coupled with a high stakes examination system. Studies on problem solving in science are not reviewed here, but are reviewed extensively elsewhere (Gabel & Bunce 1994; Lavoie 1995; Maloney 1994). In this chapter, I discuss (1) why routine problems are used in teaching, (2) the typical context in which they are used, (3) whether using routine problem tasks is effective and (4) point to ways of transforming traditional practice.

The label 'problem' can have multiple meanings that make it an ambiguous term. Meaning can range from standard exercises to ill-structured problems (Sternberg & Davidson 1994). For the purposes of this chapter, I assume a very limited view of problem solving, reflecting the current practice in schools in Southern Africa and most other regions with high stakes examinations. The term 'routine problem task' will refer to a piece of work assigned by the teacher that invariably involves the calculation of some quantity through use of a formula and algebraic manipulation of a number of given variables. The following example from an examination paper (Department of Education and Culture 1992) refers to a 'problem task':

The equilibrium constant for the following reaction is 7.5 at 25C:

$$2 NO_2 (g) \rightleftharpoons N_2O_4 (g)$$

---

Chapter Consultant: Tony Lorsbach (Illinois State University, USA)

2 moles of $NO_2$ are placed in a 2 $dm^3$ tube and then sealed. The contents of the tube are then allowed to attain equilibrium at the same constant temperature. Calculate the concentration of the $NO_2$ at equilibrium.

The problem task is well defined, narrow in focus, and could be solved by use of an algorithm. Depending on the students' familiarity with this task, it could be problematic or routine (Bodner 1990).

## WHY ARE ROUTINE PROBLEMS USED IN TEACHING?

*Beliefs Underlying the Use of Routine Problems*

What are the different purposes for which teachers use problems in their practices? What are the underlying beliefs of teachers about the value of engaging students in these problem-solving tasks? Expanding on the three general themes characterising the role of problem solving in the curriculum identified by Stanic and Kilpatrick (1988), the beliefs or assumptions discussed below are suggested as underlying the use of routine problem tasks as a teaching strategy.

*Learning science is about solving routine problems.* Looking back at past classroom activities, teachers see that doing science at school or university always has involved solving routine problems. Therefore, teachers believe that it is their duty to teach students how to solve these problems.

*Expertise at solving routine problem tasks is a necessary preparation for later studies in science.* Students need to know how to solve basic problems as an entry to further study. Many tertiary courses of study in science assume that students can solve these problems and build on that base.

*Problems are a means for developing an understanding of the subject matter.* Solving problems is seen as a vehicle by which a new concept or skill might be learned. Teachers believe that a sequence of routine problem tasks provides a context in which students will come to understand new subject matter.

*Solving routine problems gives an indication of understanding.* Teachers believe that, if you understand the concepts and principles, you can work the problems. Although learning to solve particular types of problems is not a stated goal in the syllabi, success at solving particular types of problems is taken to be the primary indicator of understanding (Dufresne, Leonard & Gerace 1992).

*Practice at solving routine problems prepares students for examinations and other forms of high stakes assessment.* Teachers are under pressure to prioritise preparation for examinations so they provide practice in the problems that they expect to appear in the examinations (Hobden 1995b).

*Solving routine problems is seen as one of a number of general problem-solving skills to be mastered.* Problems must be assigned so that the skills can be acquired through imitation and practice (Schoenfeld 1988).

In most of the themes identified, problem tasks are employed as vehicles in the service of other curricular goals. They are used as a means to one or more ends.

They are not a goal in themselves, or seen as a valuable curriculum end. (Rarely are routine problem tasks seen as useful for dealing with everyday problems or experiences.) This has important implications. If the use of routine problem tasks does not achieve our goals, we can discard them or transform them to meet our goals.

*The Influence of the Existing System*

The existing system's tradition, assessment practices and teacher education also each influence the use of routine problems in teaching.

Tradition

Current practices exist because teachers obtain their beliefs about the role of problems from working within the system and are content with current practice. Examining any physical science text from early in this century shows that sets of routine problem tasks long have occupied a central place in the school science curriculum. What is interesting is that, while the world of science has undergone dramatic changes, the problems in school curricula have not. Problems involving pulleys and sliding blocks, or calculations of percentage composition, are still present. Examples from both chemistry and physics texts in use 50 years ago (e.g., Johns, Ware & Rees 1937) are virtually identical to current problems (e.g., Brink & Jones 1986). The reasons for this lack of change perhaps can be linked to the reigning conception of learning science which involves delineating the desired subject matter content as clearly as possible, carving it into small pieces (diSessa 1988) and providing explicit instructions and practice on each of those pieces so that students master them (Schoenfeld 1992). Sets of routine problem tasks fit conveniently into this conception of curriculum and pedagogy.

Assessment

What roles are assessment practices playing in determining the problems used and the contexts planned by teachers? Looking at the system in Southern Africa, for example, shows that instructional practices are embedded within a tightly bounded system in which a final examination containing routine problem tasks is all important. While the preamble to the syllabi used in schools (Department of Education and Culture 1988) shows that there is plenty of scope for a variety of activities, approaches and interpretations, the system constrains most activities. Because the teachers feel that they are locked into the examination system (Hobden 1994), they sacrifice learning with understanding for the immediate goals of drilling the students in the things for which they will be held accountable (Webb

1988). Similar practices occur wherever centralised examinations are given significant importance (Contreras 1993; Deacon 1989; Helgeson 1993).

Teacher Education

To compound this situation, many teachers are not even qualified to teach science. For example, in South Africa, more than 50 percent of physical science teachers are underqualified (Foundation for Research Development 1993). The option of reducing a problem situation to a mathematical representation that is easily manipulated is attractive to teachers who have difficulty themselves with the content, or who are teaching in a second language and therefore find qualitative explanations of phenomena difficult. That which we *do* teach is limited by that which we *can* teach (Osborne 1990).

## THE CONTEXT WITHIN WHICH ROUTINE PROBLEMS ARE USED

*Description of the Classroom Tasks*

Evidence suggests that science classroom practices are very similar in most secondary schools. Similarly, the routine within which students encounter problem tasks is quite standard (Contreras 1993; Gallagher 1989; Hobden 1994; Tobin & Gallagher 1987). The following could be a typical sequence of events in a senior secondary physical science classroom. The teacher provides a context or scene setting for what is to follow; normally this is a description of a phenomenon or problem situation. The problem task is then introduced. It is stripped of detail, topic specific, and focused to display its essence. The teacher introduces the associated formula. The solution technique is illustrated by the teacher with limited student participation and one or more simple examples are attempted. The students then are assigned a number of similar problems from worksheets or textbooks for practice. Some of these are worked in class and others are assigned for homework. When the tasks have been completed, some or all are gone over in class, or at least the answers are provided. A test or examination gives students opportunities to show that they have mastered particular problem types. At a later date, the solutions to the examination problems are provided or demonstrated by the teacher.

This process involves the development and practice of skills or algorithms for solving particular types of standard problems. The solution strategies are demonstrated and then practised under the assumption that learners absorb them in this way. The implicit purpose of the sequence is for students to be able to solve the problems (Hammer 1989). The students constantly are motivated to participate in this classroom sequence through the teacher referring to tests and examinations (Tobin, Tippins & Gallard 1994). They are led to believe that success at solving problems will indicate conceptual understanding. The basic pedagogical assumption underlying this series of activities is that, by working these sets of problem tasks, students add to their physical science knowledge and understanding in some unspecified way. From the teacher's point of view, this sequence is normal instructional practice and appears effective in that students pass the examinations. Teachers have met their responsibilities to the students, school and community.

*Didactical Contract*

What are the participants' expectations, and why are things as they are in the classroom? The following analysis, borrowing from the work of Balacheff (1988), attempts an understanding of the relationship between the participants and the problem tasks; it is an attempt to situate the problems in the classroom life. We have seen that teachers use problems to serve different functions. For each different function served by a problem task (e.g., for evaluation of learning), a different social interaction is required. The interaction is governed by a didactical contract, or a set of rules that often are unspoken but understood by those in the community of classroom life. The contract determines the place of the problem task in the teaching-learning situation. For example, the task of finding the stopping distance of a moving car becomes a fundamentally different entity depending upon its situation in social relations (Erickson & Schultz 1992; Reusser 1988). The same problem given as an examination question would be transformed from a task of solving a meaningful situation to one of showing the teacher that he/she knows science. This meaning is not fixed by the problem text itself. One of the consequences of the contract is that the student, for most of the time, is making sense of the classroom situation rather than the science. They give solutions acceptable to the context or implicit didactical contract. They are not necessarily constructing knowledge. Also, because students are different and have different experiences of science, what on the surface appears to be the same classroom task is very different experientially for different students.

*Students' Expectations of Problems*

An alternative way of coming to an understanding of the role of routine problem tasks could be to look at them from the students' perspectives. As a consequence of their classroom experiences, students have the expectations discussed below of the problems that they will be required to solve.

*Students expect problems to be familiar.* Students' understandings of what a problem task entails is merely recognition that it is similar (isomorphic) or identical to a previously encountered problem task, and that the solution can be obtained if they apply the appropriate procedure (Schoenfeld 1988). Anything not in a familiar context, or immediately solvable with an algorithm practised in class, is considered a trick question (Hobden 1994).

*Students expect problems to have answers that are nice numbers.* Students often consider whether the numerical values that they are calculating come out to be nice numbers (e.g., 4.0 as opposed to 3.67) as a clue as to whether they are using the correct strategy or not (Reusser 1988).

*Students expect problems to be well defined.* In most routine problems, the task is to determine the specific value for one object or system. The students expect that this will be clearly asked for in the problem. The problem text will contain all the information required to solve the problem and will not contain extraneous information. As a result, they expect that all numbers given somehow should be used to get to the solution.

*Students expect problems to be solvable within a relatively short time.* Students' experiences of watching the teacher solve problems, and what is expected of them in the short time available in tests, tells them that solutions should be found relatively quickly. The consequence of this is that students are unlikely to struggle with a problem task whose solution path is not fairly obvious.

*Students expect problems to be specific to a topic.* Because most problems are stripped of confusing contexts when used as vehicles for instruction, students expect that future problems will be similar. They are taught to solve problems within the context of a topic (e.g., momentum) and to use the context as a cue to which solution technique to use (Reif 1981).

*Students expect problems to have one correct solution method.* While students accept that other solution paths will be given credit, they suspect that there is one path in the marking memorandum that is accepted as the most correct path. Consequently, there is a tendency to search for this path as opposed to solving the problem in ways that make sense to them.

*Students expect problems to have different status.* Students assign the highest status and attention to problems from the most recent past examination because they expect problems of this type to be in their examination.

These expectations emphasise the narrow scope of the problem tasks encountered in courses as well as some of the unintended strategies that students learn from doing such tasks.

## ARE ROUTINE PROBLEM TASKS AN EFFECTIVE INSTRUCTIONAL TOOL?

There are few studies of the effect or use of routine problem tasks as an instructional strategy in the learning of physical science. This perhaps reflects a firm belief among teachers that solving these tasks is the way to learn physical science. However, the evidence supporting this belief as the foundation of classroom practice is wanting. There is growing evidence that traditional problem-solving activities are ineffective for promoting the learning of science. If teachers' primary goal is only for students to solve some classes of problems reliably, then the strategy of encouraging students to follow teacher-provided procedures probably will succeed (Hammer 1989).

What is not clear at first glance is the consequence of this practice. Many students, while able to produce solutions to the routine examples, do not necessarily understand the strategies, algorithms or the science involved. The instructional process followed results in the students memorising solution procedures that are simply recalled by recognition of the example type or other surface features when it appears in an examination (Webb 1988). They will solve certain well-specified problems with apparent understanding, but in reality have acquired only a superficial appearance of competence. This is most obvious when what appears to be a minor deviation from a routine problem confuses a student, who then is unable to recall the standard solving procedure or adapt it (Johsua & Dupin 1991).

*Construction of a Knowledge Base*

Despite the instructional practice of providing sets of similar problem tasks, many students fail to solve these physical science problems correctly. This is not surprising because it is generally recognised that even solving routine problems is a complex and involved process that requires a significant, well-indexed knowledge base (Hauslein & Smith 1995). Unfortunately, the evidence indicates that traditional practices result in many students having unconnected and fragmented understanding of the subject matter despite gaining proficiency at certain kinds of procedures (diSessa 1988; Pickering 1990).

One area that would seem to be central to the argument for using problems is the development of conceptual understanding. Given that students spend a lot of curriculum time engaged in routine problem solving, we can ask questions about its contribution to the development and understanding of science knowledge. There is now convincing evidence (Clement 1981; Heller & Hollabaugh 1992; Sawrey 1990) that traditional school problem-solving activity, while mentally taxing and time consuming, is an inappropriate vehicle for developing a well-organised knowledge base. Sweller (1989) takes an even stronger line and argues that the typical problem tasks assigned in physics courses can be counterproductive for learning physics.

*Development of General Problem-Solving Skills*

Another belief that promotes the use of problems in teaching is that students gain skills to become competent scientific problem solvers (i.e., with non-routine or authentic problems). However, there is little evidence to support this belief, and our own experience tells us that students have great difficulty transferring skills to new contexts (Lemke 1993). Even when teachers use routine problems primarily to encourage general problem solving, a number of instructional practices militate against the skills being developed. First, while knowledge and skills that are usable in many circumstances are required, the problems tend to be context dependent, resulting in skills that are only usable when clearly marked by context (Campione,

Brown & Connell 1988). Second, while the ability to construct solution paths is essential, students are provided with answers and methods leading to a state of learned helplessness when asked to attack new problems (Noddings 1988). Third, while real-world scientific problems do not have clear-cut answers and are not always solvable in a short time through error free and mechanical performance, students' classroom experience is to the contrary (Webb 1988). Fourth, while the ability to identify ambiguity can be important and extremely difficult for real-world scientific problems, students get almost no experience solving ill-defined problems (Reusser 1988).

*Students' Attitude to Science*

One of the more serious consequences of basing instruction largely on sets of routine problems is that students develop distorted perspectives regarding the nature of science that could impede their acquisition and use of other science knowledge. Their belief is that doing physical science consists largely of dealing with routine problem tasks which can be solved, by the application of formulas and algorithms, without significant understanding of the subject matter. A consequence is that students appear to learn and act in a manner which we might think of as unreflective. They do this not because they are disinterested or unintelligent, but because – as demonstrated by what they see and do continually in the classroom – that is what they think the subject entails. Also, for most students, success in science is a means to an end. Solving routine problem tasks is simply one of the many tasks they have to complete to achieve the goal of getting good marks. The facts and algorithms that are learned in the course will soon be forgotten (Hobden 1994). One succeeds in school by performing the tasks. Once school is over, the tasks lose all meaning. Sense making as a goal makes no sense given their educational milieu.

## TRANSFORMING TRADITIONAL PRACTICE

What types of problems are required and what learning contexts should be created if students are to be concerned with sense making and not with mastery of facts and algorithms? The evidence demands a change from traditional routine problem tasks to problem tasks that move beyond algorithms and place more emphasis on understanding.

*Strategies to Transform Traditional Tasks*

More recent research is showing that, if certain strategies are implemented, we can adapt traditional problem tasks to develop conceptual understanding (e.g., Heller & Hollabaugh 1992; van Heuvelen 1992). The following strategies are proposed as useful when commencing the process of transforming what could be considered pedagogically-limiting routine problem tasks into tasks that encourage learning experiences that begin to fit with our goals (Hobden 1995a). Teachers should expand problems into comprehensive problem situations with a carefully structured set of questions to promote understanding of the entire situation with which the original problem was concerned (Schuster 1993). These questions guide students, prompting them to deal with all the relevant aspects of the situation. The focus of the task moves from obtaining a numerical answer to making sense of the problem situation. Teachers must encourage students to use multiple representations such as qualitative descriptions and diagrams together with mathematical representations when solving problems (van Heuvelen 1992). The emphasis must be shifted to understanding the problem situation qualitatively, before calculating numerical answers. To promote understanding of specific problem-solving strategies, students must be given opportunities to reflect on them after they have been modelled by the teacher. This can be done by providing students with detailed written solutions to problems (Sweller 1989). These solutions should include qualitative explanations of the principles underlying the solution plan, reasons for choosing particular mathematical representations, labels for all symbols used, annotated steps in the solution plan, etc. The emphasis should be on understanding the solution strategy as a whole rather than on memorising a number of steps.

*Using Relevant Problem Contexts*

While the above strategies will begin to transform the tasks given to students, we also need to transform the problem contexts themselves. Students must be encouraged to bring their contexts to the classroom. Then, collaboratively with the teacher, they can construct problem situations that touch their lives. Learning in science will then involve the reconstruction of personal experience. Teachers need to use their judgement in selecting problems that are likely to lead to potentially rich learning contexts. There will be tension between using problems of real-world complexity that might be too difficult and complex to meet the instructional needs, and using problems carefully designed by the teacher to meet them. Despite this tension, teachers should not be tempted to continue to adopt the traditional reductionist approach, which strips the problem of everything but the simplest detail.

Another approach could be to use relevant news reports from local newspapers. For example, an article dealing with a person surviving a fall from the seventh floor of a high-rise building but seriously damaging a parked car, could be used as a problem context. However, it should be presented to students as a goal-free situation (Sweller 1988). Students should be given non-specific goals, such as to find out all they can about this situation, rather than a specific task, such as finding the velocity when the person hit the car. This approach can be used to encourage students to bring all their knowledge to bear on the situation, promoting the linking of concepts and the production of an integrated knowledge structure.

*Creating a Supportive Atmosphere*

When students encounter new problems, they find themselves at the limit or, possibly completely out of depth, in terms of their capacity to perform adequately. Students are not comfortable when being scrutinised while attempting to master new knowledge (Erickson & Schultz 1992). When students make mistakes, they must be considered opportunities for reconstruction of their knowledge rather than the consequence of ignorance or uncertainty. Teachers should model and discuss problem solving, showing students that it is not mechanical algorithm implementation, but rather a process similar to that of an art (Schoenfeld 1992). Teachers also should provide opportunities to deal with the conflict that can arise between students' intuitive solutions and more generally acceptable solutions (Hammer 1989). Consequently, teachers must provide a supportive classroom atmosphere in which students can gain confidence, risk showing their incompetence, and know their efforts at solutions are valued.

## CONCLUSION

There does not seem to be a pedagogically defensible role for sets of routine problem tasks in the teaching and learning of science. There is convincing evidence that the use of these tasks as an instructional strategy is not effective in achieving our goals. It is not effective in developing conceptual understanding, general problem-solving skills, or positive attitudes to science. In an attempt to understand this situation, attention was given in this chapter to the context within which routine problems are used. It was seen that the meaning given to the task of problem solving is determined by an interplay between the task, the individual and the social setting. The students' interpretations of these problem tasks result in flawed understandings of their roles as learners. Consequently, they adopt strategies enabling them to attempt to solve problems without understanding through focusing on algorithms and predictable characteristics of the problems.

We need to change students' and teachers' conceptions of problem solving from 'routine task' (i.e., ordinary, uninteresting and predictable) and present it as a creative, interesting and useful activity. This process will require us to commit ourselves to a course of transformation. As a first component in the process of transformation, instructional practices must be grounded in analyses of what it means to do school science rather than acceptance of what has always been traditional practice. Our practices must reflect and promote the ways of thinking and sense making that we consider to be important. The second component requires us to create new learning environments and different experiences if we hope to reconceptualise the role of problem tasks as vehicles for developing understanding and positive attitudes. We know that students develop their understanding and sense of science from their experiences in the classroom. We have seen the results of traditional routine problem tasks. What we need to focus on are problem tasks that touch students' lives and consequently give meaning to their experiences of science.

The third component requires us to challenge the status of quantitative methods of representation and algebraic methods of solution in the learning of physical science. We should transform our problem tasks so that they place more emphasis on the development of qualitative understanding. The fourth component requires us to redefine what achievement means. We must not place such high currency on performance measured through tests of routine problem tasks. Assessment practices must be designed to reward and encourage the kinds of thinking that we consider valuable. If all these components are implemented within a supportive classroom atmosphere, there will be a pedagogically defensible role for the use of problem tasks in the teaching and learning of science.

## REFERENCES

Balacheff, N.: 1988, 'Cognitive Versus Situational Analysis of Problem-Solving Behaviours', in H. Burkhardt, S. Groves, A. Schoenfeld & K. Stacey (eds.), *Problem Solving: A World View* (Proceedings of the Problem Solving Theme Group, 5th International Congress on Mathematical Education), Shell Centre for Mathematical Education, University of Nottingham, Nottingham, UK, 168–173.

Bodner, G.M.: 1990, 'A View from Chemistry'. in M.U. Smith (ed ), *Towards a Unified Theory of Problem Solving*, Lawrence Erlbaum, Hillsdale, NJ, 21–34.

Brink, B.P. & Jones, R.C.: 1986, *Physical Science Standard 10*, Juta, Cape Town, South Africa.

Campione, J.C., Brown, A.L. & Connell, M.L.: 1988, 'Metacognition: On the Importance of Understanding What You Are Doing', in R.I. Charles & E.A. Silver (eds.), *The Teaching and Assessing of Mathematical Problem Solving* (Volume 3), Lawrence Erlbaum, Hillsdale, NJ, 93–114.

Clement, J.: 1981, 'Solving Problems with Formulas: Some Limitations', *Engineering Education* 72, 158–162.

Contreras, A.: 1993, 'The Situated Nature of Middle School Science Teaching: An Interpretive Study in a Ninth Grade Classroom', Paper presented at the annual meeting of the National Association for Research in Science Teaching, Boston, MA.

Deacon, J.: 1989, 'Forces which Shape the Practices of Exemplary High School Physics Teachers', in K. Tobin & B.J. Fraser (eds.), *Exemplary Practice in Science and Mathematics Education*, Curtin University of Technology, Perth, Australia, 59–67.

Department of Education and Culture: 1988, *National Examinations Syllabus for Physical Science: Standard 10 (Higher Grade)* (Code 304/0/1/1/1/88), Author, Pretoria, South Africa.

Department of Education and Culture: 1992, *Senior Certificate Examination, Physical Science, Paper 2, Chemistry*, Author, Durban, South Africa.

diSessa, A.A.: 1988, 'Knowledge in Pieces', in G. Forman & P. Putall (eds.), *Constructivism in the Computer Age*, Lawrence Erlbaum. Hillsdale, NJ, 49–70.

Dufresne, R.J., Leonard, W.J. & Gerace, W.J.: 1992, 'Designing Instructional Materials for Promoting Conceptual Understanding', Paper presented at an informal workshop in Physics Department, University of Natal, Durban, South Africa.

Erickson, F. & Schultz, J.: 1992, 'Students' Experience of the Curriculum', in P.W. Jackson (ed.), *Handbook of Research on Curriculum*, Macmillan, New York, 465–485.

Foundation for Research Development: 1993, 'High School Education in Mathematics and Science', in Foundation for Research Development (ed.), *South African Science and Technology Indicators*, Author, Pretoria, South Africa, 5–16.

Gabel, D.L. & Bunce, D.M.: 1994, 'Research on Problem Solving: Chemistry', in D.L. Gabel (ed.), *Handbook of Research on Science Teaching and Learning*, Macmillan, New York, 301–326.

Gallagher, J.J.: 1989, 'Research on Secondary School Science Teachers Practices, Knowledge and Beliefs: A Basis for Restructuring', in M.L. Matyas, K. Tobin & B.J. Fraser (eds.), *Looking into Windows: Qualitative Research in Science Education*, American Association for the Advancement of Science, Washington, DC, 43–57.

Hammer, D.: 1989, 'Two Approaches to Learning Physics', *The Physics Teacher* 27, 664–670.
Hauslein, P.L. & Smith, M.U.: 1995, 'Knowledge Structures and Successful Problem Solving', in D.R Lavoie (ed.), *Toward a Cognitive-Science Perspective for Scientific Problem Solving* (Monograph of the National Association for Research in Science Teaching, Number Six), Ag Press, Manhattan, KS, 51–79.
Helgeson, S.L.: 1993, *Assessment of Science Teaching and Learning Outcomes*, National Center for Science Teaching and Learning, Ohio State University, Columbus, OH.
Heller, P. & Hollabaugh, M.: 1992, 'Teaching Problem Solving Through Co-operative Grouping. Part 2: Designing Problems and Structuring Groups', *American Journal of Physics* 60, 637–644.
Hobden, P.: 1994, 'The Role of Problems in the Teaching of Physical Science', in M. Glencross (ed.), *Proceedings of the Second Annual Meeting of Southern African Association for Research in Mathematics and Science Education*, University of Transkei, Umtata, South Africa, 153–162.
Hobden, P.: 1995a, 'Alternative Strategies for Teaching Problem Solving', Paper presented at KwaZulu-Natal Science Teachers Conference, University of Durban-Westville, South Africa.
Hobden, P.: 1995b, 'The Influence of High Stakes External Examinations on Classroom Environment', Paper presented at annual meeting of the American Educational Research Association, San Francisco, CA.
Johns, R.V., Ware, W.F. & Rees, A.I.: 1937, *Graded and Everyday Examples in Physics*, Macmillan, London.
Johsua, S. & Dupin, J.: 1991, 'In Physics Class, Exercises can also Cause Problems . . .', *International Journal of Science Education* 13, 291–301.
Lavoie, D.R. (ed.): 1995, *Toward a Cognitive-Science Perspective for Scientific Problem Solving* (Monograph of the National Association for Research in Science Teaching, Number Six), Ag Press, Manhattan, KS.
Lemke, J.L.: 1993, 'The Missing Context in Science Education: Science', Paper presented at the annual meeting of the American Educational Research Association, Atlanta, GA.
Maloney, D.P.: 1994, 'Research on Problem Solving: Physics', in D.L. Gabel (ed.), *Handbook of Research on Science Teaching and Learning*, Macmillan, New York, 327–354.
Noddings, N.: 1988, 'Preparing Teachers to Teach Mathematical Problem Solving', in R.I. Charles & E.A. Silver (eds.), *The Teaching and Assessing of Mathematical Problem Solving* (Volume 3), Lawrence Erlbaum, Hillsdale, NJ, 245–258.
Osborne, J.: 1990, 'Sacred Cows in Physics – Towards a Redefinition of Physics Education', *Physics Education* 25, 189–196.
Pickering, M.: 1990, 'Further Studies on Concept Learning versus Problem Solving', *Journal of Chemical Education* 67, 254–255.
Reif, F.: 1981, 'Teaching Problem Solving – A Scientific Approach', *The Physics Teacher* 19, 310–316.
Reusser, K.: 1988, 'Problem Solving Beyond the Logic of Things: Contextual Effects on Understanding and Solving Word Problems', *Instructional Science* 17, 309–338.
Sawrey, B.A.: 1990, 'Concept Learning versus Problem Solving: Revisited', *Journal of Chemical Education* 67, 253–254.
Schoenfeld, A.H.: 1988, 'When Good Teaching Leads to Bad Results: The Disasters of Well-Taught Mathematics Courses', *Educational Psychologist* 23, 145–166.
Schoenfeld, A.H.: 1992, 'Learning to Think Mathematically: Problem Solving, Metacognition and Sense Making in Mathematics', in D.A. Grouws (ed.), *Handbook of Research on Mathematics Teaching and Learning*, Macmillan, New York, 334–370.
Schuster, D.: 1993, 'Assessment as Curriculum in Science Education', Paper presented at the First International Conference on Science Education in Developing Countries, Jerusalem, Israel.
Stanic, G.M. & Kilpatrick, J.: 1988, 'Historical Perspectives on Problem Solving in the Mathematics Curriculum', in R.I. Charles & E.A. Silver (eds.), *The Teaching and Assessing of Mathematical Problem Solving* (Volume 3), Lawrence Erlbaum, Hillsdale, NJ, 1–22.
Sternberg, R.J. & Davidson, J.E.: 1994, 'Problem Solving', in M.C. Alkin (ed.), *Encyclopedia of Educational Research*, Macmillan, New York, 1037–1045.
Sweller, J.: 1988, 'Learning and Problem Solving: Disparate Goals?', in H. Burkhardt, S. Groves, A. Schoenfeld & K. Stacey (eds.), *Problem Solving: A World View* (Proceedings of the Problem Solving Theme Group, 5th International Congress on Mathematical Education), Shell Centre for Mathematical Education, University of Nottingham, Nottingham, UK, 187–191.
Sweller, J.: 1989, 'Cognitive Technology: Some Procedures for Facilitating Learning and Problem Solving in Mathematics and Science', *Journal of Educational Psychology* 81, 457–466.

Tobin, K. & Gallagher, J.J.: 1987, 'What Happens in High School Science Classrooms?', *Journal of Curriculum Studies* 19, 549–560.

Tobin, T., Tippins, D.J. & Gallard, A.J.: 1994, 'Research on Instructional Strategies for Teaching Science', in D.L. Gabel (ed.), *Handbook of Research on Science Teaching and Learning*, Macmillan, New York, 45–93.

van Heuvelen, A.: 1992, 'Models of Learning and Teaching', in D. Grayson (ed.), *Workshop on Research in Science and Mathematics Education*, South African Association for Research in Mathematics and Science Education, Cathedral Peak, South Africa, 56–67.

Watts, D.M. & Gilbert, J.K.: 1989, 'The "New Learning" Research, Development and the Reform of School Science Education', *Studies in Science Education* 16, 75–121.

Webb, J.: 1988, 'Problem Solving in South Africa', in H. Burkhardt, S. Groves, A. Schoenfeld & K. Stacey (eds.), *Problem Solving: A World View* (Proceedings of the Problem Solving Theme Group, 5th International Congress on Mathematical Education), Shell Centre for Mathematical Education, University of Nottingham, Nottingham, UK, 160–165.

## 2.7 The Complexity of Chemistry and Implications for Teaching

DOROTHY GABEL

*Indiana University, Bloomington, USA*

Chemistry teaching and learning are very complex human endeavours. One reason for this is the complexity of chemistry itself. In addition to many concepts being highly related to others, and therefore the learning of even relatively simple chemistry concepts being dependent on prerequisite knowledge, conceptual understanding requires the learner to link several modes of representing matter and the interactions that matter undergoes. Hence, students find chemistry one of their more difficult courses at the secondary and undergraduate levels.

In this review, studies that focus on students' conceptual understanding, non-scientific conceptions and problem-solving skills will not be examined. Excellent reviews of research studies in these areas already have been written by Andersson (1990), Gabel and Bunce (1994), Griffiths (1994), Krajcik (1991), Nakhleh (1992), Stavy (1995) and Wandersee, Mintzes and Novak (1994). Conclusions from these reviews indicate that students at all levels have many non-scientific conceptions about chemistry concepts that they have studied and that many students solve problems using algorithms rather than reasoning. Therefore, this chapter considers teaching strategies that hopefully promote conceptual understanding and problem solving in more meaningful ways.

An area of chemistry education research that has received considerable attention by chemistry education researchers in the past ten years is the understanding of the representation of matter at the *macroscopic*, *microscopic* and *symbolic* levels (both chemical and mathematical). Johnstone (1982, 1991a, 1991b, 1993) depicts the levels of thought about chemistry using a triangle. He maintains that, since the curriculum revision of the 1960s, the emphasis on chemistry teaching has focused on the symbolic level of representation rather than on the macroscopic level.

Many studies show that students do not understand chemistry at all three levels, that students do not easily shift from one level to another (Ben-Zvi, Eylon & Silberstein 1988), and that the lack of understanding exists even among students beginning graduate work in chemistry (Bodner 1991). This important area of chemical education research illustrates the complexity of chemistry teaching and

---

Chapter Consultant: Craig Bowen (University of Southern Mississippi, USA)

learning in which observations are made on the macroscopic level, yet explanations and theories which students are expected to understand depend on the atomic and molecular level which, in turn, are represented symbolically. It is no wonder that even chemistry teachers can be one cause of students' misconceptions because they also have been found to possess some of the same misconceptions (de Jong & Acampo 1993; Stromdahl, Tullberg & Lybeck 1994; Tullberg, Stromdahl & Lybeck 1994). In addition to students learning in limited contexts and generalising beyond these contexts (as well as arriving at misconceptions from observing natural events), it appears that language generalisations (Ringnes 1994; Schmidt 1994a, 1994b) and textbooks (de Berg & Treagust 1993; Staver & Lumpe 1993) contribute to the acquisition of misconceptions.

## EFFECTIVE TEACHING STRATEGIES

The complexity of examining effective teaching strategies is illustrated by the study by Garnett and Tobin (1989) in which they describe the chemistry teaching strategies of two effective but distinctly different teachers. One teacher used whole-group instruction, whereas the other used individualised instruction. The authors claim that both of these teachers were successful because they were efficient managers, had strong content knowledge, focused on instructional strategies that facilitated student understanding, asked appropriate questions and responded appropriately, and used effective monitoring techniques. In addition, both emphasised laboratory work and a Science-Technology-Society (STS) approach, and both had adequate pedagogical content knowledge.

Studies of this nature show that it is not one effective teaching strategy that leads to conceptual understanding, but the linking together of strategies in an appropriate fashion. Effective teachers frequently make use of the conceptual change teaching sequence described by Krajcik (1991), based on that of Driver and Oldham (1986), in which students describe their own understandings of a scientific concept, and restructure it via student-student and student-teacher exchange, and through conflict situations. New understandings are applied to another situation and new linkages between concepts occur. Students compare their new understandings with previous understandings. The process is repeated in different contexts until their conceptual understandings become more scientific.

Driver (1993) has described the role of the teacher from a constructivist viewpoint:

> ... [I]f students' understandings are to be changed toward those of accepted science, then intervention and negotiation with an authority, usually the teacher, is essential. Here, the critical feature is the nature of the dialogue process. The task of the authority figure has two important components. One is to introduce new ideas and to provide the support and guidance for students to make sense of these for themselves. The other is to listen and

diagnose the extent to which the instructional activities are being interpreted in the intended way in order to inform further action. (p. 73)

In addition to considering the conceptual change teaching sequence described above, effective teaching strategies also must be structured according to how students process information. The model used by Johnstone, Sleet and Vianna (1994) – shown in Figure 1 – reflects commonalties with other information-processing models. New information from external sources enters working memory (which has limited space), where it interacts with information stored in long-term memory and subsequently transforms it, and the new information then is stored permanently in long-term memory. Both this information-processing model of learning and the conceptual change teaching sequence must be considered when discussing effective teaching strategies.

*Social Interaction*

Student-student or student-teacher interactions are important components of learning from a constructivist viewpoint. Research in chemistry education explicitly in this area over the past six years has been relatively sparse. Studies by Ntho and Rollnick (1993) and Holme (1992) have examined teacher-student interactions. Ntho and Rollnick found that, when students worked with tutors, conceptual understanding was enhanced. Holme found that the use of the Socratic method of

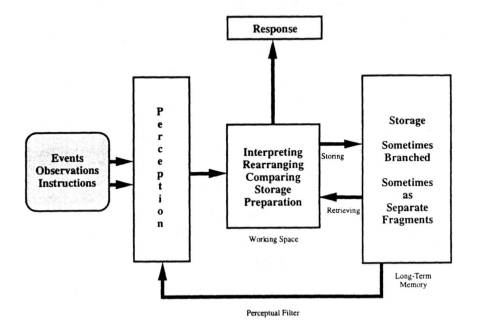

Figure 1: Information Processing Model (after Johnstone, Sleet & Vianna 1994)

questioning in an introductory university chemistry course that was delivered in a large lecture hall was well received by 73 percent of the students. Students sitting near one another collaborated on the answers and, although the lack of a control group makes comparisons of student understanding impossible, it appears that the collaboration created a better climate for learning.

Studies of students working together in collaborative groups showed positive effects for high-achieving Israeli secondary school students in learning about chemical energy (Cohen & Ben-Zvi 1992) and for secondary school general science students in learning about the particle nature of matter (Lonning 1993). Smith, Hinckley and Volk (1991) found that students who used the Jigsaw form of cooperative learning in an undergraduate chemistry laboratory situation achieved better on a combined test measuring understanding of acids and bases. The method was found to be particularly effective for low-achieving students. Basili and Sanford (1991) found that community college students who worked in cooperative groups using conceptual change techniques such as concept mapping decreased their nonscientific conceptions to a greater extent than did students receiving conventional instruction.

All of these methods give students opportunities to reflect on their own conceptions, modify them and thus restructure their conceptions. In terms of the information processing model, conceptions previously stored in long-term memory have been modified and new linkages have probably occurred.

*Use of Analogues, Models and Concept Maps*

Vosniadou and Brewer (1987) have suggested the use of analogues and physical models to facilitate the reconstruction of existing schema in long-term memory. Wilson (1994) suggests that concept mapping also could have this effect. Although many studies have examined the usefulness of using models and analogues in chemistry instruction, only two studies in addition to the one reported by Basili and Sanford (1991) mentioned previously, have reported having students construct concept maps using chemical concepts as an instructional strategy. Both used them in the laboratory. Stensvold and Wilson (1992) found no difference in achievement when concept maps were used as tools, whereas Wang (1993) found them useful in developing students' experimental skills.

The use of analogues and models as instructional tools for conceptual change could have certain limitations, and hence might not be as effective as creating a conflict situation. Duit (1990) speculates that, in order for an analogue to be effective, it must be familiar to the students and it must be in a domain in which students do not have any misconceptions. Lawson, Baker, Di Donato, Verdi and Johnson (1993) found that, if students have misconceptions in a given domain, the use of analogues are appropriate in producing conceptual change only if students have hypothetico-deductive reasoning ability. They were not found to be effective otherwise.

Stavy and Cohen (1989) compared the use of analogues and conflict training to overcome Israeli grade 5–7 students' misconceptions about the conservation of weight. They found that both treatments were effective in producing substantial conceptual change, and speculated that the use of analogues could be more appropriate because fewer students appear to regress to their previous conceptions when using this strategy rather than with conflict training.

Other studies of the use of analogues include Stavy and Tirosh's (1992, 1993a, 1993b) series of comparisons between geometric and chemical phenomena (line segment, copper wire and water), which showed that, when the same process was involved (successive division), grade 7–12 students did not differentiate among the content domains and overgeneralised. On the other hand, this might be expected because, unless students have received explicit instruction on ideas such as the way in which individual atoms do not have the properties of aggregates of atoms, students cannot be expected to have this nonintuitive knowledge (de Vos & Verdonk 1996). Pereira and Pestana (1991) found that students' use of space-filling models at the grade 8–12 levels revealed many misconceptions about the movement of the particles, expansion of individual particles in phase changes and bond angles.

Examples of using mental models to improve chemistry understanding of thermal equilibrium have been provided by Ben-Zvi, Silberstein and Mamlok (1989) and Ben-Zvi and Ramot (1994). Both studies used pictorial diagrams that showed heat distribution represented by shaded circles.

The use of pictorial diagrams also is being used to examine whether students make connections between the macroscopic, symbolic and microscopic levels of chemistry. The two-dimensional drawings and the three-dimensional models used to teach about the microscopic level might be considered to be analogues of actual atoms and molecules. Several studies have been reported that show promise for helping students to understand atoms and molecules and to link the three representations of matter. Lee, Eichinger, Anderson, Berkheimer and Blakeslee (1993) reported that middle school students had a better understanding of the particle nature of matter as a result of using modules that enlarged particles of matter undergoing physical changes than did students using a conventional text. Gabel (1993) found that, when teachers emphasised the particle nature of matter by having secondary school chemistry students make drawings of the particles representing physical and chemical changes over an extended period of time, students showed improved understanding of chemistry at all three levels. In a more extensive action research study involving 12 secondary school chemistry teachers, Bunce and Gabel (1995) found that integrating the macroscopic, particulate and symbolic levels was particularly effective for improving the chemistry achievement of girls. Williamson and Abraham (1995) showed the superiority of using dynamic computer simulations of atoms and molecules undergoing physical and chemical changes over the use of static pictures. Students who watched simulations in the lecture and in the computer laboratory achieved better than students who viewed them in the lecture only, and both of these groups achieved better than students who saw only static visuals.

Even though the research indicates that at least some students benefit from the use of analogues, Treagust, Duit, Joslin and Lindauer (1992) found that very few secondary school teachers in Australia actually use analogues as an instructional strategy. Theile and Treagust (1994) found that, when teachers did use analogues, they appeared to have a motivational effect on the students, that teachers rarely indicated the limitations of the analogues, and that there was little evidence that the analogues were a planned part of the instruction.

*Use of Technology*

Technology provides new ways of engaging students in learning, and hence increases conceptual understanding by providing opportunities for cognitive restructuring. The success of its use is probably dependent on the level of engagement and, even though chemistry educators might think that technology will never substitute for good teachers (Moore & Moore 1984), this might not be the case. A meta-analysis on the use of technology in chemistry teaching by Kulik (1983) indicated positive results on achievement and student ratings.

Technology can be used in a variety of ways in chemistry instruction, including: Computer Assisted Instruction (CAI) lessons for teaching conceptual understanding and for problem solving; and use in the laboratory as a substitute for performing experiments or in the actual gathering of data and then displaying, calculating and presenting the results.

The use of technology in CAI was reported by Dori and Barnea (1993) and Dori (1995) who engaged experts, teachers and students in the production of CAI modules, and found that students' attitudes towards the use of computers as learning tools improved significantly as a result of using the modules. Hameed, Hackling and Garnett (1993) reported that grade 12 students' conceptions of chemical equilibrium changed when CAI instruction was used, and that the effect was long-lasting. On the other hand, Wainwright (1989) found no significant difference in chemistry achievement for secondary school chemistry between students who used CAI instruction for teaching formula writing and equation balancing in comparison with those who used worksheets over a three-week period.

Two recent studies of the use of CAI in teaching students to solve chemistry problems do not provide convincing evidence for its use. Kramer-Pals (1994), who compared the effectiveness of a heuristic program to teach problem solving that was CAI-based versus one that was not, found no differences in achievement. Cracolice and Abraham (1996) compared the use of CAI, semi-programmed instruction and teacher assistant-led instruction on problem solving. Their data indicated that, for simple problems, all methods were equally effective but that, for more difficult problems, the semi-programmed instruction worksheets resulted in better achievement. Students' comments indicated that they liked the use of worksheets because they could work at their own pace. These results support those of Wainwright (1989) who found worksheets superior to CAI for problem-solving activities.

An area in which the use of technology appears more promising is in laboratory instruction. Jackman, Mollenberg and Brabson (1987), in a comparison of the effectiveness of using the learning cycle approach, a computer simulation and the traditional approach for teaching spectrophotometry, found that the computer simulation produced higher posttest scores. Bourque and Carlson (1987) compared the effectiveness of the use of simulation and hands-on activities in secondary school chemistry on acid-base titration, ionization constant and Avogadro's number. Although they found no significant differences when students used only one of the methods, they discovered that, for two of the three activities, there was a difference in achievement when students did the hands-on activity followed by simulation.

Russell, Staskun and Mitchell (1985) found that the use of videodisks in pre-laboratory instruction produced better laboratory performance at the undergraduate level than did the use of videotapes. Abraham and Williamson (1992) found that students had more positive attitudes about an introductory chemistry course when student data from experiments were shared in integrating the laboratory and the lecture using a computer-managed laboratory approach. Nakhleh and Krajcik (1993) compared students' verbalisations during an acid-base titration using indicators, pH meters or a microcomputer program. They found that students focused their attention on different aspects of the experiment according to the method used. Students using microcomputers focused more on graphing and used more concepts in their explanations.

In conclusion, it appears that the use of technology has beneficial effects on learning. Research indicates that it is more useful for teaching descriptive chemistry than for teaching problem solving, and that there are appropriate laboratory uses for it as well. It holds particular promise in helping students to visualise the motion and structure of molecules. Because of the usefulness of incorporating technology into science courses, it is important to include instruction in its use in teacher preparation courses. One method for doing this is to model its use with preservice and inservice teachers in science courses. Dori and Barnea (1994) found that teachers using a polymer CAI unit became more positive about the use of computers in their classrooms and that many actually used the unit in their own teaching.

*Laboratory Instruction*

In the previous section, the use of technology in laboratory instruction was shown to have some beneficial effects. Although laboratory instruction itself has not been particularly effective in enhancing chemistry achievement in the past – as shown in the review by Lunetta and Hofstein (1982) and more recently by Paik (1992) in Korea – the major goals of laboratory instruction appear to be changing, and this could affect its usefulness. Using data from a survey of university chemistry teachers in the USA, Abraham (1994) reported that the purpose of laboratory instruction has changed from being very fact-oriented and verification-oriented in the 1940s to the fostering of conceptual understanding today. In Jordan, Abdalla (1992,

1994) reports that, although chemistry instructors thought that the purpose of the laboratory was to promote conceptual understanding, students viewed it as a way to improve their practical skills. Perhaps this difference in view is one of the reasons why students, not only in Jordan but in many countries, do not perform as well as expected as a result of laboratory instruction, and why many researchers have explored ways to improve it.

Johnstone (1982, 1991a, 1991b) and Johnstone and Letton (1990, 1991) have researched the effectiveness of chemistry laboratory instruction for promoting conceptual understanding from the viewpoint of an information-processing approach over several years. They maintain that one reason for the lack of conceptual understanding resulting from laboratory instruction is that many concepts in chemistry are not directly observed and students do not link what they observe at the macroscopic level with what they are expected to learn at the atomic or molecular level. Learning is impeded because there is no way for students to pull out concepts from long-term memory or to anchor the new learning from the laboratory into long-term memory under traditional laboratory instruction. A major barrier is the limited space in short-term memory and the complexity of the directions for laboratory experiments as commonly written. In addition, the lack of familiarity of the students with the technical language used in chemistry adds an additional barrier to learning by stressing short-term memory. Roadruck (1993) makes a similar claim with regard to demonstrations. He indicates that care must be taken not to assume that students understand formal concepts that concrete demonstrations are supposed to illustrate. The complexity of the demonstration actually can obscure learning.

Johnstone and Letton (1991) suggest that the redesign of experimental instructions can help to improve the effectiveness of laboratory teaching in promoting conceptual understanding. Other modifications suggested by Johnstone (1993) include: (1) preparing of long-term memory to enable students to separate the 'signal' from 'noise'; (2) actively involving students in planning the experiment; (3) acquiring manual techniques needed for the experiment in advance to free space in working memory; and (4) providing an opportunity for problem solving to enable students to explore their long-term memory networks in lateral thinking. Johnstone, Sleet and Vianna (1994), who implemented the above principles into laboratory instruction over a three-year period, concluded that students' attitudes about how much they understood about laboratory learning were substantially modified as compared to those of a control group.

Research by others corroborates the usefulness of reducing the demand on short-term memory or working memory. Although Niaz and Logie (1993) distinguish between these two models of mental capacity, for the purposes of this review, no distinctions have been made. One way to reduce the demand is to prepare students sufficiently for doing the experiments. Wang and Horng (1992) found that, by using concept maps and mastery learning techniques, students not only performed the laboratory technique of distillation properly, but they maintained the technique learned over a period of time. Beasley and Heikkinen (1983) found that, although the use of written instructions (mental practice) combined with physical practice of skills needed in a laboratory experiment produced the most pronounced effect on posttest scores for another experiment that used the skills, mental practice alone was quite adequate. Isom and Rowsey (1989) found that providing students with separate pre-laboratory discussions prior to the day of the experiment was superior to using the first part of the laboratory period in preparing students for the experiment. Kempa and Ward (1988) found that, in multi-stimuli observational systems, detectability of one stimulus can seriously affect the presence of another, thus suggesting that simple experiments are preferable to complex ones.

Although conceptual understanding is an important objective of laboratory instruction and can be evaluated successfully in a variety of ways, such as by using network analysis (Paik & Pak 1993), the effect on students' attitudes is of equal importance because of its link to achievement (Germann 1988). Gabel (1994) found that the aspect of chemistry instruction which secondary school students liked most was laboratory work in chemistry, and Okebukola (1987) found that students' performance in practical chemistry was highly dependent on their participation in laboratory activities and their attitude towards chemistry. McGuire, Ealy and Pickering (1991) reported that, although expense and time to perform microscale experiments is reduced substantially, secondary school students showed a persistent preference for the macroscopic version of experiments over the microscopic version. This was particularly true for high-ability students.

*Science-Technology-Society Approaches*

Both theory and practice justify using a Science-Technology-Society (STS) approach to teaching science. Although research on using an STS approach does not indicate that students gain in conceptual understanding of the content taught over other approaches, STS has been found to improve students' attitudes towards science, process skills, ability to apply concepts in new situations and creativity (Yager & Tamir 1993). Relating science to the familiar, new concepts is likely to help students to link ideas to existing concepts and anchor them in long-term memory, and perhaps this is why students can apply what they have learned to new situations. However, at the secondary school level, textbooks in the USA – with the exception of *ChemCom: Chemistry in the Community* (American Chemical Society 1993) – do not use this approach (Chiappetta, Sethna & Fillman 1991), and STS is not valued by coordinators of introductory chemistry courses at the undergraduate level as a goal for secondary school chemistry (Shumba & Glass 1994).

However, in the mid-1980s, when success was being reported using the Personalized System of Instruction (PSI) as an alternative to more conventional approaches (Davis, Storch & Strawser 1987; Freeman 1984), evidence from an American Chemical Society symposium (Hostettler 1983) which reported chemistry program revisions (Chrostowski 1985) indicated that some undergraduate chemistry instructors were interested in making chemistry courses more relevant.

Research on the effectiveness of using STS approaches is difficult to analyse because comparisons are generally made between classes that incorporate other teaching strategies into the curriculum such as collaborative learning. Some components of STS teaching such as role playing (Reissman & Rollnick 1990) at the secondary level and employing a case-based curriculum for nursing students at the undergraduate level (Dori 1994) have been tested and found useful. Ben-Zvi and Gai (1994) found that students using an STS approach used micro explanations for chemical phenomena more frequently than students using a concept-building approach, and Garafalo and LoPresti (1993) reported increased achievement when students used an integrated biology/chemistry curriculum.

The use of STS approaches on a national level have been initiated in England with *Salters' Advanced Chemistry* (University of York Science Education Group 1994) and in the USA with *ChemCom: Chemistry in the Community* (American Chemical Society 1993) and *Chemistry in Context* (American Chemical Society 1994). Although the implementation of *Chemistry in Context* as an introductory undergraduate-level course for non-science majors is too recent for there to be complete evidence about its effectiveness, several studies have been reported for *ChemCom* and *Salters'* chemistry. Sutman and Bruce (1992) found that *ChemCom* students achieved better than students in more traditional undergraduate preparatory courses on test items designed by *ChemCom* writers to assess both the chemistry learned and applications, and Winther and Volk (1994) concluded that minority students learned more when using *ChemCom* compared with a more traditional approach. Results of both studies need to be interpreted with care because of the nature of the tests used for evaluation. More suitable methods of evaluation based on the work of Zoller (1990, 1993a, 1993b, 1994) might be considered. Zoller has developed a variety of examinations for which students actually compose questions and which require higher-order thinking. This is an important course outcome that should be considered in evaluating programs such as *ChemCom* and other programs based on a constructivist teaching philosophy.

*Salters'* chemistry, which is a much more comprehensive reform of the chemistry curriculum at the secondary level than *ChemCom*, consists of three programs: *Science: The Salters' Approach* (University of York Science Education Group 1990) for students of 11–14 years of age; *Chemistry: The Salters' Approach* (University of York Science Education Group 1992) for students in their last two years of compulsory education and aged 14–16 years; and *Salters' Advanced Chemistry* (University of York Science Education Group 1994) for students aged 16–18 years and leading to an Advanced Level examination. As indicated by Campbell, Lazonby, Millar, Nicolson, Ramsden and Waddington (1993), when one considers the numbers of schools adopting the programs, they can be considered a success. Although no research on the effectiveness of the programs on conceptual understanding has been reported, Ramsden (1992, 1994) has shown that, for the *Science: The Salters' Approach* course, University of York Science Education Group (1990), both students and teachers like the course which has resulted in positive student attitudes about what they are learning.

## CONCLUSIONS

Learning and teaching of chemistry can be approached in a variety of ways. These include teaching science from an interdisciplinary approach such as that advocated by *Project 2016* (Rutherford & Ahlgren 1990), or from a coordinated approach in which all sciences are taught every year. This is currently done in many countries and is being field tested in the USA in the National Science Teachers Association's (1993) *Scope, Sequence and Co-ordination of Secondary School Science* project, which includes the additional dimension of beginning with concrete instruction at grade 7 and making it more abstract at grade 12. Other approaches include making chemistry relevant to everyday life, such as in *Salters'* chemistry in the UK and *ChemCom* and *Chemistry in Context* in the USA. Research reported to date on the effectiveness of these large-scale programs supports the direction of reform that they espouse.

However, even within these programs, instruction can take many different forms. It will be necessary to sort out what strategies or combinations of strategies are particularly effective in promoting conceptual change for students at different cognitive and developmental levels. For example, the research on the learning cycle and modifications to it as described by Tobin, Tippins and Gallard (1994) needs to be considered when planning strategies within these programs. The evaluation of new programs that prescribe a more constructivist approach to learning, such as *Powerful Ideas in Physical Science* (American Association of Physics Teachers 1994), is likely to add to the research base on effective strategies for other national efforts to improve teaching and learning.

There also are other areas of research about the teaching of chemistry that need to be pursued. For example, the effectiveness of chemistry courses based on the coherent curriculum structure proposed by de Vos, van Berkel and Verdonk (1994), who gathered information in this area from chemistry educators around the world, needs to be explored. One area of chemistry education that appears to be sufficiently researched is that of chemistry misconceptions. Chemistry educators need to take the next step to examine teaching strategies that move students forward from their present ideas about scientific concepts to those that are more accepted by the scientific community (Treagust, Duit & Fraser 1996).

## REFERENCES

Abdalla, A.M.: 1992, 'Chemical Knowledge of Students Studying First Year Chemistry Practicals in Jordanian Universities', *Chemeda: Australian Journal of Chemistry Education* 34, 12–17.

Abdalla, A.M.: 1994, 'Students' Attitudes Towards Chemistry Practicals in Jordanian Universities', *Journal of Education and Science* 14, 11–35.

Abraham, M.R.: 1994, *General Chemistry Laboratory Survey*, Department of Chemistry and Biochemistry, University of Oklahoma, Norman, OK.

Abraham, M.R. & Williamson, V.M.: 1992, 'Integrating the Laboratory and Lecture with Computers', in W.J. McIntosh & M.W. Caprio (eds.), *Successful Approaches to Teaching Introductory Science Courses*, Society for College Science Teachers, Southern Utah University, Cedar City, UT, 21–28.

American Association of Physics Teachers: 1994, *Powerful Ideas in Physical Science*, Author, College Park, MD.
American Chemical Society: 1993, *ChemCom: Chemistry in the Community*, Kendall/Hunt, Dubuque, IA.
American Chemical Society: 1994, *Chemistry in Context*, Kendall/Hunt, Dubuque, IA.
Andersson, B.: 1990, 'Pupils' Conceptions of Matter and Its Transformations (Age 12–16)', *Studies in Science Education* 18, 53–85.
Basili, P.A. & Sanford, J.S.: 1991, 'Conceptual Change Strategies and Cooperative Group Work in Chemistry', *Journal of Research in Science Teaching* 28, 293–304.
Beasley, W.F. & Heikkinen, H.W.: 1983, 'Mental Practice as a Technique to Improve Laboratory Skill Development', *Journal of Chemical Education* 60, 488–489.
Ben-Zvi, N. & Gai, R.: 1994, 'Macro- and Micro-Chemical Comprehension of Real World Phenomena', *Journal of Chemical Education* 71, 730–732.
Ben-Zvi, R., Eylon, B. & Silberstein, J.: 1988, 'Theories, Principles and Laws', *Education in Chemistry* 25, 89–92.
Ben-Zvi, R. & Ramot, O.: 1994, 'Can Teachers Promote Concept Enchancement?', Paper presented at the 13th International Conference on Chemical Education, San Juan, Puerto Rico.
Ben-Zvi, R., Silberstein, J. & Mamlok, R.: 1989, 'Macro-Micro Relationships: A Key to the World of Chemistry', Paper presented at the Conference on Relating Macroscopic Phenomena to Microscopic Particles, Utrecht, The Netherlands.
Bodner, G.M.: 1991, 'I Have Found You an Argument: The Conceptual Knowledge of Beginning Chemistry Graduate Students', *Journal of Chemical Education* 68, 385–388.
Bourque, D.R. & Carlson, G.R.: 1987, 'Hands-on Versus Computer Simulation Methods in Chemistry', *Journal of Chemical Education* 64, 232–234.
Bunce, D.M. & Gabel, D.L.: 1995, 'The Classroom Teacher as Researcher: A University-High School Collaborative Effort', Paper presented at the meeting of the American Chemical Society, Chicago, IL.
Campbell, B., Lazonby, J., Millar, R., Nicolson, P., Ramsden, J. & Waddington, D.: 1993, *Science: The Salters' Approach: A Case Study of the Process of Large Scale Curriculum Development* (Science Education Research Paper 93/04), Science Education Group, University of York, UK.
Chiappetta, E.L., Sethna, G.H. & Fillman, D.A.: 1991, 'A Quantitative Analysis of High School Chemistry Textbooks for Scientific Literary Themes and Expository Learning Aids', *Journal of Research in Science Teaching* 28, 939–951.
Chrostowski, P.C.: 1985, 'The Environmental Chemistry Program at Vassar College', *Journal of Chemical Education* 62, 137–138.
Cohen, I. & Ben-Zvi, R.: 1992, 'Improving Student Achievement in the Topic of Chemical Energy by Implementing New Learning Materials and Strategies', *International Journal of Science Education* 14, 147–156.
Cracolice, M.S. & Abraham, M.R.: 1996, 'A Comparison of Computer-Assisted Instruction, Semi-Programmed Instruction and Teaching Assistant-Led Instruction in General Chemistry', *School Science and Mathematics* 96, 215–221.
Davis, L.P., Storch, D.M. & Strawser, L.D.: 1987, 'Physical Chemistry at USAFA: Personalized Instruction', *Journal of Chemical Education* 64, 784–787.
de Berg, K.C. & Treagust, D.F.: 1993, 'The Presentation of Gas Properties in Chemistry Textbooks and as Reported by Science Teachers', *Journal of Research in Science Teaching* 30, 871–880.
de Jong, O. & Acampo, J.: 1993, 'Creating Conditions of Learning Chemistry: Teachers' Actions and Reflections', in J.T. Voorbach (ed.), *Research and Developments on Teacher Education in the Netherlands*, Swets & Zeitlinger, Amsterdam, The Netherlands, 121–129.
de Vos, W., van Berkel, B. & Verdonk, A.H.: 1994, 'A Coherent Conceptual Structure of the Chemistry Curriculum', *Journal of Chemical Education* 71, 743–746.
de Vos, W. & Verdonk, A.H.: 1996, 'The Educational Nature of the Particulate Nature of Matter', *Journal of Research in Science Teaching* 33, 659–666.
Dori, Y.J.: 1994, 'Achievement and Attitude Evaluation of a Case-Based Chemistry Curriculum for Nursing Students', *Studies in Educational Evaluation* 20, 337–348.
Dori, Y.J.: 1995, 'Cooperative Studyware Development of an Organic Chemistry Module by Experts, Teachers and Students', *Journal of Science Education and Technology* 4, 163–170.
Dori, Y.J. & Barnea, N.: 1993, 'A Computer-Aided Instruction Module on Polymers', *Journal of Chemical Information in Computer Science* 33, 325–331.
Dori, Y.J. & Barnea, N.: 1994, *In-service Chemistry Teachers Training: The Impact of Introducing*

Computer Technology on Teachers' Attitudes and Classroom Implementation*, Israel Institute of Technology, Haifa, Israel.
Driver, R.: 1993, 'Constructivist Perspectives on Learning Science', in P.L. Lijnse (ed.), *European Research in Science Education: Proceedings of the First PhD Summerschool*, Utrecht University, The Netherlands, 65–74.
Driver, R. & Oldham, V.: 1986, 'A Constructionist Approach to Curriculum Development in Science', *Studies in Science Education* 13, 105–122.
Duit, R.: 1990, 'On the Role of Analogies, Similes and Metaphors in Learning Science', Paper presented at the annual meeting of the American Educational Research Association, Atlanta, GA.
Freeman, W.A.: 1984, 'Relative Long-Term Benefits of a PSI and a Traditional-Style Remedial Chemistry Course', *Journal of Chemical Education* 61, 617–619.
Gabel, D.: 1993, 'Use of the Particle Nature of Matter in Developing Conceptual Understanding', *Journal of Chemical Education* 70, 193–197.
Gabel, D.L.: 1994, 'Chemical Pedagogy', in M.V. Orna, J.O. Schreck & H. Heikkinen (eds.), *ChemSource* (Volume 1), American Chemical Society, New Rochelle, NY, 1–28.
Gabel, D.L. & Bunce, D.M.: 1994, 'Research on Problem Solving: Chemistry', in Dorothy L. Gabel (ed.), *Handbook of Research on Science Teaching and Learning*, Macmillan, New York, 301–326.
Garafalo, F. & LoPresti, V.: 1993, 'Evolution of an Integrated College Freshman Curriculum', *Journal of Chemical Education* 70, 352–359.
Garnett, P.J. & Tobin, K.: 1989, 'Teaching for Understanding: Exemplary Practice in High School Chemistry', *Journal of Research in Science Teaching* 26, 1–14.
Germann, P.J.: 1988, 'Development of the Attitude Toward Science in School Assessment and Its Use to Investigate the Relationship between Science Achievement and Attitude Toward Science in School', *Journal of Research in Science Teaching* 25, 689–703.
Griffiths, A.K.: 1994, 'A Critical Analysis and Synthesis of Research on Students' Chemistry Misconceptions', Paper presented at the International Seminar on Problem Solving and Misconceptions in Chemistry and Physics, Dortmund, Germany.
Hameed, H., Hackling, M.W. & Garnett, P.J.: 1993, 'Facilitating Conceptual Change in Chemical Equilibrium Using a CAI Strategy', *International Journal of Science Education* 15, 221–230.
Holme, T.A.: 1992, 'Using the Socratic Method in Large Lecture Courses: Increasing Student Interest and Involvement by Forming Instantaneous Groups', *Journal of Chemical Education* 69, 974–977.
Hostettler, J.D.: 1983, 'Introduction to the "Real World" Examples Symposium', *Journal of Chemical Education* 60, 1031–1032.
Isom, F.S. & Rowsey, R.E.: 1989, 'The Effect of a New Pre-laboratory Procedure on Students' Achievement in Chemistry', *Journal of Research in Science Teaching* 23, 231–236.
Jackman, L.E., Mollenberg, W.P. & Brabson, G.D.: 1987, 'Evaluation of Three Instructional Methods for Teaching General Chemistry', *Journal of Chemical Education* 64, 794–802.
Johnstone, A.H.: 1982, 'Macro- and Micro-Chemistry', *School Science Review* 64, 377–379.
Johnstone, A.H.: 1991a, 'Thinking about Thinking – A Practical Approach to Practical Work', in *Proceedings of the International Conference on Chemical Education*, University of Glasgow, Glasgow, UK, 69–76.
Johnstone, A.H.: 1991b, 'Why is Science Difficult to Learn? Things are Seldom What They Seem', *Journal of Computer Assisted Learning* 7, 75–83.
Johnstone, A.H.: 1993, 'The Development of Chemistry Teaching', *Journal of Chemical Education* 70, 701–703.
Johnstone, A.H. & Letton, K.M.: 1990, 'Investigating Undergraduate Lab Work', *Education in Chemistry* 27, 9–11.
Johnstone, A.H. & Letton, K.M.: 1991, 'Practical Measures for Practical Work', *Education in Chemistry* 28, 81–83.
Johnstone, A.H., Sleet, R.J. & Vianna, J.F.: 1994, 'An Information Processing Model of Learning: Its Application to an Undergraduate Laboratory Course in Chemistry', *Studies in Higher Education* 19, 77–87.
Kempa, R.F. & Ward, J.E.: 1988, 'Observational Thresholds in School Chemistry', *International Journal of Science Education* 10, 275–284.
Krajcik, J.S.: 1991, 'Developing Students' Understanding of Chemical Concepts', in S.M. Glynn, R.H. Yeany & B.K. Britton (eds.), *The Psychology of Learning Science*, Lawrence Erlbaum, Hillsdale, NJ, 117–147.
Kramer-Pals, H.: 1994, *Learning to Solve Explanation Problems in Chemical Education*, Unpublished doctoral dissertation, University of Twente, The Netherlands.

Kulik, J.A.: 1983, 'How Can Chemists Use Educational Technology Effectively?', *Journal of Chemical Education* 60, 957–959.

Lawson, A.E., Baker, W.P., Di Donato, L., Verdi, M.P. & Johnson, M.A.: 1993, 'The Role of Hypothetico-Deductive Reasoning and Physical Analogues of Molecular Interactions in Conceptual Change', *Journal of Research in Science Teaching* 30, 1073–1085.

Lee, O., Eichinger, D.C., Anderson, C.W., Berkheimer, G.D. & Blakeslee, T.D.: 1993, 'Changing Middle School Students' Conceptions of Matter and Molecules', *Journal of Research in Science Teaching* 30, 249–270.

Lonning, R.A.: 1993, 'Effect of Cooperative Learning Strategies on Student Verbal Interactions and Achievement During Conceptual Change of Instruction in 10th Grade General Science', *Journal of Research in Science Teaching* 30, 1087–1011.

Lunetta, V.N. & Hofstein, A.: 1982, 'The Role of the Laboratory in Science Teaching: Neglected Aspects of Research', *Review of Education Research* 52, 201–217.

McGuire, P., Ealy, J. & Pickering, M.: 1991, 'Microscale Laboratory at the High School Level: Time Efficiency and Student Response', *Journal of Chemical Education* 68, 869–871.

Moore, J.W. & Moore, E.A.: 1984, 'Computer Series, 48: Will Computers Replace TA's? Professors?', *Journal of Chemical Education* 61, 26–35.

Nakhleh, M.B.: 1992, 'Why Some Students Don't Learn Chemistry: Chemical Misconceptions', *Journal of Research in Chemical Education* 69, 191–196.

Nakhleh, M.B. & Krajcik, J.S.: 1993, 'A Protocol Analysis of the Influence of Technology on Students' Actions, Verbal Commentary and Thought Processes During the Performance of Acid-Base Titrations', *Journal of Research in Science Teaching* 30, 1149–1168.

National Science Teachers Association: 1993, *Scope, Sequence, and Coordination of Secondary School Science: Content Core* (Volume I), Author, Washington, DC.

Niaz, M. & Logie, R.H.: 1993, 'Working Memory, Mental Capacity and Science Education: Towards an Understanding of the Working Memory Overload Hypothesis', *Oxford Review of Education* 19, 511–525.

Ntho, T. & Rollnick, M.: 1993, 'Conceptual Change in Teaching of Concepts on Periodic Trends in a Physical Science Course', *South African Journal of Higher Education* 7, 110–119.

Okebukola, P.A.: 1987, 'Students' Performance in Practice Chemistry: A Study of Some Related Factors', *Journal of Research in Science Teaching* 24, 119–126.

Paik, S.H.: 1992, 'A Study of Students' Understanding and Teaching Method of Chemicals', *Chemical Education* 19, 124–135.

Paik, S.H. & Pak, S.J.: 1993, 'A Network Analysis Approach to the Evaluation of Students' Laboratory Reports', *Journal of Korean Science Education Society* 12, 93–101.

Pereira, M.P. & Pestana, M.E.: 1991, 'Pupils' Representations of Models of Water', *International Journal of Science Education* 13, 313–319.

Ramsden, J.M.: 1992, 'If It's Enjoyable, Is It Science?', *School Science Review* 73, 65–71.

Ramsden, J.M.: 1994, 'Context and Activity-Based Science in Action', *School Science Review* 75, 7–14.

Reissman, Q. & Rollnick, M.: 1990, 'The Use of Role Playing with School Students in Swaziland: A Case Study', *Science Education* 1, 15–30.

Ringnes, V.: 1994, 'Students' Understanding of Chemistry and Their Learning Difficulties', in H.J. Schmidt (ed.), *Proceedings of the International Seminar on Problem Solving and Misconceptions in Chemistry and Physics*, University of Dortmund, Dortmund, Germany, 100–112.

Roadruck, M.D.: 1993, 'Chemical Demonstrations: Learning Theories Suggest Caution', *Journal of Chemical Education* 70, 1025–1028.

Russell, A.A., Staskun, M.G. & Mitchell, B.L.: 1985, 'The Use and Evaluation of Videodiscs in the Chemistry Laboratory', *Journal of Chemical Education* 62, 420–422.

Rutherford, F.J. & Ahlgren, A.: 1990, *Science for All Americans*, Oxford University Press, New York.

Schmidt, H.: 1994a, *Students' Misconceptions in Chemistry – Looking for a Pattern*, University of Dortmund, Dortmund, Germany.

Schmidt, H.: 1994b, *Senior High School Students' Difficulties with the Oxidation Concept*, University of Dortmund, Dortmund, Germany.

Shumba, O. & Glass, L.W.: 1994, 'Perceptions of Coordinators of College Freshman Chemistry Regarding Selected Goals and Outcomes of High School Chemistry', *Journal of Research in Science Teaching* 31, 381–392.

Smith, M.E., Hinckley, C.C. & Volk, G.L.: 1991, 'Cooperative Learning in the Undergraduate Laboratory', *Journal of Chemical Education* 68, 413–415.

Staver, J.R. & Lumpe, A.T.: 1993, 'A Content Analysis of the Presentation of the Mole Conception in Chemistry Textbooks', *Journal of Research in Science Teaching* 30, 321–337.
Stavy, R.: 1995, 'Conceptual Development of Basic Ideas in Chemistry', in S. Glynn & R. Duit (eds.), *Learning Science in the Schools: Research Reforming Practice*, Lawrence Erlbaum, Mahwah, NJ, 131–154.
Stavy, R. & Cohen, M.: 1989, *Overcoming Students' Misconceptions about Conservation of Matter by Conflict Training and by Analogical Reasoning*, Tel Aviv University, Tel Aviv, Israel.
Stavy, R. & Tirosh, D.: 1992, 'Overgeneralization in Mathematics and Science: The Effect of External Similarity', *International Journal of Mathematical Education in Science and Technology* 23, 239–248.
Stavy, R. & Tirosh, D.: 1993a, 'When Analogy is Perceived as Such', *Journal of Research in Science Teaching* 30, 1229–1239.
Stavy, R. & Tirosh, D.: 1993b, 'Subdivision Processes in Mathematics and Science', *Journal of Research in Science Teaching* 30, 579–586.
Stensvold, M. & Wilson, J.T.: 1992, 'Using Concept Maps as a Tool to Apply Chemistry Concepts to Laboratory Activities', *Journal of Chemical Education* 69, 230–232.
Stromdahl, H., Tullberg, A. & Lybeck, L.: 1994, 'The Qualitative Different Conceptions of One Mole', *International Journal of Science Education* 16, 17–26.
Sutman, F.X. & Bruce, M.H.: 1992, 'Chemistry in the Community: A Five Year Evaluation', *Journal of Chemical Education* 69, 564–567.
Thiele, R.B. & Treagust, D.F.: 1994, 'An Interpretative Explanation of High School Chemistry Teachers' Analogical Explanations', *Journal of Research in Science Teaching* 31, 227–242.
Tobin, K., Tippins, D.J. & Gallard, A.J.: 1994, 'Research on Instructional Strategies for Teaching Science' in D.L. Gabel (ed.), *Handbook of Research on Science Teaching and Learning*, Macmillan, New York, 45–93.
Treagust, D.F., Duit, R. & Fraser, B.J. (eds.): 1996, *Improving Teaching and Learning in Science and Mathematics*, Teachers College Press, New York.
Treagust, D.F., Duit, R., Joslin, P. & Lindauer, I.: 1992, 'Science Teachers' Use of Analogies: Observation from Classroom Practical', *International Journal of Science Education* 14, 413–422.
Tullberg, A., Stromdahl, H. & Lybeck, L.: 1994, 'Students' Conceptions of One Mole and Educators' Conceptions of How They Teach the Mole', *International Journal of Science Education* 16, 145–151.
University of York Science Education Group: 1990, *Science: The Salters' Approach*, Heinemann Educational, Oxford, UK.
University of York Science Education Group: 1992, *Chemistry: The Salters' Approach*, Heinemann Educational, Oxford, UK.
University of York Science Education Group: 1994, *Salters' Advanced Chemistry*, Heinemann Educational, Oxford, UK.
Vosniadou, S. & Brewer, W.F.: 1987, 'Theories of Knowledge Restructuring in Development', *Review of Educational Research* 57, 51–67.
Wainwright, C.L.: 1989, 'The Effectiveness of a Computer-Assisted Instruction Package in High School Chemistry', *Journal of Research in Science Teaching* 26, 275–290.
Wandersee, J.H., Mintzes, J.J. & Novak, J.D.: 1994, 'Research on Alternative Conceptions in Science' in D.L. Gabel (ed.), *Handbook of Research on Science Teaching and Learning*, Macmillan, New York, 177–210.
Wang, C.H.: 1993, 'Improving Chemistry Experimental Learning through Process-Mapping Hydrolysis of p-nitroacetanilide', *Bulletin of National Taiwan Northern University* 38, 157–173.
Wang, C.H. & Horng, J.M.: 1992, *A Design of Group-Based Individual Instruction*, National Taiwan Normal University, Taipei, Taiwan.
Williamson, V.M. & Abraham, M.R.: 1995, 'The Effects of Computer Animation on the Particulate Mental Models of College Chemistry Students', *Journal of Research in Science Teaching* 32, 521–534.
Wilson, J.M.: 1994, 'Network Representations of Knowledge about Chemical Equilibrium: Variations with Achievement', *Journal of Research in Science Teaching* 31, 1133–1147.
Winther, A.A. & Volk, T.L.: 1994, 'Comparing Achievement of Inner-City High School Students in Traditional versus STS-Based Chemistry Courses', *Journal of Chemical Education* 71, 501–505.
Yager, R.E. & Tamir, P.: 1993, 'STS Approach: Reasons, Intentions, Accomplishments and Outcomes', *Science Education* 77, 637–658.
Zoller, U.: 1990, 'The IEE and STS Approach: A Course Format to Foster Problem Solving', *Journal of College Science Teaching* 19, 289–291.
Zoller, U.: 1993a, 'Are Lecture and Learning Compatible?', *Journal of Chemical Education* 70, 195–197.

Zoller, U.: 1993b, 'Scaling up of Higher-Order Cognitive Skills-Oriented College Chemistry Teaching and Evaluation', Paper presented at the 18th International Conference on Improving University Teaching, Schwabisch, Germany.

Zoller, U.: 1994, 'The Examination Where the Students Ask the Questions', *School Science & Mathematics* 94, 247–249.

## 2.8 The School Science Laboratory: Historical Perspectives and Contexts for Contemporary Teaching

VINCENT N. LUNETTA
*The Pennsylvania State University, University Park, USA*

The first of this chapter's three major sections provides a historical perspective on the use of laboratory activities in science education. The second section discusses the importance of achieving greater consistency between goals, theories and practices. Implications for teaching in school science laboratories form the focus of the third section.

### HISTORICAL PERSPECTIVE

Laboratory activities long have had a distinctive and central role in the science curriculum, and science educators have suggested that many benefits accrue from engaging students in science laboratory activities. From early in the 19th century, laboratory activities were reported to assist students in making observations about the natural world providing a basis for inferences based on that information (Edgeworth & Edgeworth 1811; Rosen 1954).

The progressive education movement in the early part of the 20th century advocated an investigative approach and, during that era, laboratory manuals acquired a more utilitarian, applied orientation. By the middle of the century, however, laboratory activities were used largely for illustrating and confirming information presented by the teacher and the textbook. In the 1960s, in both the USA and the UK, major science curriculum projects aimed to engage students in investigation and inquiry as a central part of their study of science. *Laboratory experiments* (terminology in the USA) and *practical activities* (terminology in the UK) continue to have a prominent place in the rhetoric of science educators today. These terms have embraced an array of activities, but normally they refer to experiences in school settings in which students interact with materials to observe and understand the natural world. Laboratory classes have ranged from activities in which data are gathered to *verify* a stated principle or relationship to *inductive* activities, in which students seek to identify patterns or relationships in data which

---

Chapter Consultants: Ed van den Berg (University of San Carlos, Philippines) and Joseph Krajcik (University of Michigan, USA)

they gather. Teacher guidance and instructions have ranged from highly structured to open inquiry. Laboratory classes have been conducted to engage students individually, in small groups and in large-group demonstration settings. Sometimes laboratory activities have incorporated a high level of instrumentation and at other times the use of any instrumentation has been meticulously avoided.

Laboratory activities have been said to offer important experiences in the learning of science that are unavailable in other school disciplines. For over a century, laboratory experiences have been purported to promote central science education goals including: understanding of scientific concepts; the development of scientific practical skills and problem-solving abilities; and interest and motivation.

A relevant instructional goal increasingly articulated in the latter part of the 20th century has been to enhance students' understanding of the nature of science (Duschl 1990).

## Mismatches Between Goals, Behaviour and Learning Outcomes

In recent decades, especially since the late 1970s, the role of the laboratory in science teaching and learning increasingly has been questioned (Bates 1978; Hofstein & Lunetta 1982). Questions arose from a variety of sources both within the science education community and beyond. New knowledge of children's development, new information about the learning of science concepts, and new perspectives about the nature of science fuelled many of the concerns about the ways introductory sciences should be taught to promote understanding. In addition, significant changes in computing and other technologies offered substantive new resources for teaching and learning science that can complement experiences in the school laboratory. Children were growing in a high technology culture that was increasingly perceived as disconnected with traditional school science.

Scholarly efforts have identified serious mismatches between goals for science education and learning outcomes visible in school graduates. Several researchers have reported that students regularly performed school science experiments with very different purposes in mind than those perceived by their teachers. Students tended to perceive either *following the instructions* or *getting the right answer* as the principal purpose for a school science task. In the laboratory especially, students can perceive that manipulating equipment and measuring are goals but they can fail to perceive conceptual or even procedural goals. Students often fail to understand the relationship between the purpose of the investigation and the design of the experiment which they had conducted, they do not connect the experiment with what they have done earlier, and they seldom note the discrepancies between their own concepts, the concepts of their peers, and those of the science community (Champagne, Gunstone & Klopfer 1985; Eylon & Linn 1988; Tasker 1981). To many students, a 'lab' means manipulating equipment but not manipulating ideas.

Comprehensive analyses of laboratory handbooks also provided evidence that there were major mismatches between goals espoused for science teaching and behaviours implicit in science laboratory activities associated with major curriculum projects. Tamir and Lunetta (1981) wrote that, in spite of attempts to reform curricula, students worked too often as technicians following 'cookbook' recipes in which they used lower-level skills; they were seldom given opportunities to discuss hypotheses, propose tests of those hypotheses, and engage in designing and then performing experimental procedures. Seldom, if ever, were students asked to formulate questions to be investigated or even to discuss sources of error and appropriate sample sizes. Students' performance in practical activities generally was not assessed nor were students asked to describe or explain their hypotheses, methodologies or the nature and results of their investigations (Hofstein & Lunetta 1982). Lunetta and Tamir (1979) were among those who recommended greater consistency between goals, theories and practices in the learning and teaching of science. Conditions were set for shifts in science education theory incorporating new kinds of scholarship with important implications for curriculum and practices in the laboratory classroom.

## TOWARDS GREATER CONSISTENCY BETWEEN GOALS, THEORIES AND PRACTICES

Over the years, science educators have sought theoretical organisers to inform curriculum development and instruction. Although the organisers and visions have changed with time, the predominant pattern of science instruction visible in schools has been based on a framework of *telling* the story of science. Within that framework, laboratory activities engaged students principally in following ritualistic procedures to verify the story that had been told. Students have had limited freedom to explore and discover. In the 1960s, major curriculum projects used the scholarship of Bruner and Piaget to justify a *hands-on* framework emphasising student inquiry and engagement. Projects including those of the Physical Science Study Committee and the Biological Sciences Curriculum Study in the USA and Nuffield in the UK developed inductive laboratory activities as a fundamental part of the science curriculum. The laboratory was to be a place for inquiring, for developing and testing assertions, and for practising 'the way of the scientist'.

A principal goal of the large science curriculum projects was to develop students' science concepts. Twenty years later, in the wake of evidence that students held many unscientific conceptions even after completing science courses, Nussbaum and Novick (1982) and others suggested that the school laboratory offered conditions and opportunities for conceptual conflict that could promote meaningful conceptual change for students. Several subsequent studies pointed out that students often held beliefs so intensely that even their observations in the laboratory were strongly influenced by those beliefs (Champagne, Gunstone & Klopfer 1985).

During the 1980s, a *constructivist* view of learning emerged that now serves as a theoretical organiser for many who are trying to understand cognition in science. Learners construct concepts and patterns to explain what they experience;

they continuously test these patterns with their experiences and modify them accordingly. There is a growing sense that learning is contextualised and that learners construct knowledge by solving meaningful problems (Brown, Collins & Duguid 1989). The evidence is increasingly clear that the learning of scientific *process* and *product* are interdependent and intertwined (Metz 1995).

A *social constructivist* framework has special potential for informing teaching in the laboratory. Millar and Driver (1987) were among those who recommended the use of extended, reflective investigations to promote the construction of more meaningful scientific concepts based upon the unique knowledge brought to the science classroom by individual learners. When students interact with problems that they perceive to be meaningful and connected to their experiences, they can develop more scientific concepts *in dialogue* with peer investigators. In the science education literature, there has been a growing commitment to move beyond a pedagogy of telling the story of science. The pedagogy that has emerged advocates that students engage intellectually with meaningful laboratory experiences and data with which they can construct shared understanding of scientific concepts in a community of learners in their classroom. Through interactions with representatives of the expert community (teachers, textbooks, etc.) and through discussions with groups of fellow student investigators seeking to find meaning in the laboratory-classroom, individual students can construct concepts more consistent with those of the scientific community. Strategies of this kind have been difficult for teachers to implement, but such experiences can help the learner to understand how the scientific community develops consensus and establishes the validity of scientific concepts.

Research suggests that, while laboratory investigations offer important opportunities to connect science concepts and theories discussed in the classroom and in textbooks with observations of phenomena and systems, laboratory inquiry alone is not sufficient to enable students to construct the complex conceptual understandings of the contemporary scientific community. If students' understandings are to be changed towards those of accepted science, then intervention and negotiation with an authority, usually a teacher, is essential (Driver 1995). Van den Berg, Katu and Lunetta (1994) reported that hands-on activities with introductory electricity materials in clinical studies with individual students facilitated their understanding of relationships among circuit elements and variables. The activities provided clear tests of the validity of the subject's ideas. 'Frequently they led to cognitive conflict . . . However, the carefully selected practical activities alone were not sufficient to enable the subject to develop a fully scientific model of a circuit system.' The findings suggested that greater engagement with conceptual organisers such as analogies and concept maps could have resulted in the development of more scientific concepts in basic electricity. Several researchers including Dupin and Joshua (1987) have reported similar findings. When laboratory experiences are integrated with other metacognitive learning experiences such as 'predict-explain-observe' demonstrations (White & Gunstone 1992) and when they incorporate the manipulation of ideas instead of simply materials and procedures, they can promote the learning of science.

## IMPLICATIONS FOR TEACHING IN THE SCHOOL LABORATORY

The laboratory offers practical experiences that, when properly developed, can enhance students' conceptual understandings and their understandings of the development and validation of scientific knowledge. Engaging in an authentic investigation in the school laboratory can enable students to compare and discuss the data that they have gathered. Individuals can come to understand that others have gathered different data and that some have different cause-effect explanations after pursuing similar questions and engaging in similar activities. Discussions within and among student teams can examine differences in data and in interpretation; students can consider the effects of sample size, alternative methodologies, error, etc. Discussing the interpretations of data generally can enable students to develop more sophisticated concepts and perceptions about the validity of data, about the generalisability of findings and about possible cause-effect relationships. Engaging in such discussions promises enhanced understanding of alternative interpretations and explanations, and an enhanced sense of the value and meaning of consensus in a scientific community.

*Engaging Student Collaboration*

The nature of the interactions among students and research teams in the school laboratory can enhance or inhibit science concept construction and related learning outcomes. Frequently, however, student groups do not function well (Kyle, Penick & Shymansky 1979). When teachers utilise cooperative learning strategies, researchers have reported better integration between classroom and laboratory activity and enhanced achievement, skill development and self esteem (Lunetta 1990; Quin, Johnson & Johnson 1995).

Because individual learners reorganise and construct their own scientific concepts and understanding, in part through dialogue with others, time should be provided for engaging students in generating driving questions, for team planning, for feedback and discussions about the nature and meaning of the data, and for discussing implications of the findings. In conventional laboratory classroom settings, well conducted *pre-labs* and *post-labs* must become more normal components of laboratory activity. Scientific knowledge is negotiated within a scientific community and, under the guidance of a skilled teacher, this negotiation can be simulated in a community of learners in the school laboratory (Krajcik, Blumenfeld, Marx & Solloway 1994). Collaborative planning and discussions of findings provide a forum in which implicit ideas can be made explicit; the ideas generated can become the substance for reflection and checking. Communication and reflection are encouraged through the preparation and discussion of laboratory *reports* or *journals*. Preparing a journal should provide opportunities for individual students to reflect upon and clarify their own observations, hypotheses, conceptions and theories. Dialogue also provides opportunities for individuals to build on the ideas of others to reach understanding and solutions.

## Levels of Inquiry

Student inquiry can range from independently conducted research on different questions to the investigation of one driving research question by the entire class. The students' freedom to explore and inquire can be very open or more focused and guided by teachers and laboratory handbooks.

On the basis of intensive research with open inquiry, Roth (1995) described how laboratory activities enable students to learn in small groups and share what they have learned with others at small-group and whole-class levels. He noted that laboratory activities are most productive when teachers view and conduct themselves as co-explorers rather than as disseminators of knowledge.

In contrast, extrapolating from their studies, Dekkers and Thijs (1993) suggested that, when a teacher's principal goal is to promote the understanding of specific scientific concepts, well-conducted (and well-controlled) teacher demonstrations can be more effective than student laboratory activities involving open inquiry. Consistent with what one would anticipate at a time of paradigm shift, these assertions appear to differ from those of Roth who examined the development of concepts and problem-solving skills more holistically and descriptively in a different setting. In Roth's 'open-inquiry' environment, student investigators shared at least five aspects in common with scientists' activities when engaging in meaningful learning experiences that incorporated: ill-defined problems and ambiguity; the social nature of scientific work and knowledge; learning that was predicated on their current knowledge; participation in an inquiring community; and access to the expertise of the more expert community.

For successful completion of many conventional secondary school laboratory activities, relatively sophisticated skills are needed to make measurements, perform tests, operate equipment and collect accurate data. However, just as research reveals that students have an abundance of less-than-scientific conceptions, studies also reveal that students often are deficient in simple skills prerequisite to performing conventional laboratory activities successfully (Bryce & Robertson 1985). Helping students to develop relevant instrumentation skills in controlled 'pre-lab' exercises can reduce the probability that important measurements in a laboratory experience will be compromised due to students' lack of expertise with the apparatus (Beasley 1985; Singer 1977). Variations in the openness of the classroom and laboratory environment are based, in part, upon the goals for learning outcomes and upon the teacher's perceptions of how students can learn in the context of unique school environments. These skills are consistent with goals of helping students to gather precise data and results with limited experimental error. Informed by a social constructivist model, however, the student's efforts to describe and reflect upon the investigation, the sources of error, and meaningful interpretations and applications of the findings in collaboration with peers are most highly valued. To that end, incorporating the open-inquiry model of Roth (1995), the teacher works alongside the student learners to model co-exploration and share technical skills of a more expert community. Others have suggested the importance of *scaffolding* instruction in response to the learner's developmental needs in constructing meaningful scientific concepts (see Linn's chapter in this *Handbook*).

*Phases of Investigation and 'Chunking' of Activities*

Students' behaviours and activities in laboratory investigations and related activities can be described in a variety of ways. While students' activities in developing understanding of the concepts and procedures of science (conceptual and procedural knowledge) are interrelated and cannot easily be isolated (decontextualised), teachers can find it helpful to identify four broad phases of laboratory activity. In a *planning and design phase,* students formulate questions to investigate, predict results, construct hypotheses to be tested, and design experimental procedures. In a *performance phase*, students conduct the investigation, manipulate materials, make decisions about investigative techniques, and observe and record data. In an *analysis and interpretation phase*, students organise and process the data, explain relationships, develop generalisations, examine the accuracy of data, outline assumptions and limitations, and formulate new questions based upon the data that they have collected. In an *application phase*, students formulate hypotheses on the basis of their investigation, make predictions about the application of the information to new situations, and apply to new problems the laboratory techniques which they have developed (Lunetta & Tamir 1979). As noted, these phases are interrelated and, in authentic inquiry, the investigator moves among the phases in a non-linear fashion probing and seeking to understand relationships.

If one looks carefully at what is possible within a single school laboratory session (normally from 40 minutes to no more than 110 minutes in length), it is clear that students can initiate, complete and understand few, if any, complete investigations in such short spans of time. Seldom do they have time to engage in the four phases of an authentic investigation in one class session. Alternatives that can result in meaningful learning need to be explored with care. Following a 'less is more' strategy (i.e., conducting a few investigations thoroughly and carefully) can result in more meaningful learning than engaging students in large numbers of conventional laboratory activities superficially. To this end, teachers can work with students or student teams to plan and conduct a small number of extended, authentic projects. Simultaneously, because it is clear that no single 40-minute experience in the school laboratory can promote all important learning outcomes, it is also useful to work with the students to identify a specific subset of goals or objectives that are especially appropriate for those students in laboratory experiences that might be conducted in one or more class sessions. To that end, van den Berg and Giddings (1992) advocated the delineation of distinct kinds of laboratory activities, each emphasising unique goals and teaching methodologies.

Teachers also can identify laboratory activities in which it is appropriate to focus on subsets of the four phases of investigation identified earlier. For example, a particular class activity might involve student teams in *planning and designing* a particular investigation. They might elaborate carefully the questions to be investigated, construct hypotheses to be tested, and discuss strategies and procedures for conducting the investigation. The activity, however, would not necessarily include involving the students in conducting the investigation. Another example is to present students with data gathered by 'other students' or by 'scientists' in an investigation that had been carefully described, perhaps including video clips of the people and the instruments involved in data collection. The investigation would be relevant to the students' development and interests. Students might then engage in an extended, intensive *analysis and interpretation* of the data using analytical tools including spreadsheets and interactive graphics as appropriate.

To reiterate, when a student conducts a few authentic investigations carefully, more meaningful learning generally results than when a larger number of laboratory activities is conducted superficially. Thus, while students might perform a variety of tasks in the laboratory-classroom, they generally should engage in a small number of extended, holistic projects.

*Intersections with the Context of Science Teaching*

With appropriate teaching strategies, relevant practical activities can help students to identify alternative conceptions, can motivate and can aid the search for more scientific conceptions. Practical activities can be important elements in model driven instruction to promote science learning. The *Generative Learning Model* (Osborne & Freyberg 1985) is a good example of one such model proposed on the basis of extensive research. Based upon earlier studies, Katu, Lunetta and van den Berg (1993) reported a study that included a concept development strategy in which students cycled through five interrelated phases of activity. The teacher worked with individual students to help them to (1) explicate their ideas about the topic under consideration, (2) design and engage in practical investigations identifying contrasts between events that occurred in the activities and the ideas that they articulated earlier, (3) find alternative explanations for observations and data that differed from their predictions, (4) apply and test their newly developed ideas and (5) contemplate which of their ideas and explanations were consistent with what they had observed and to note changes that had occurred in their ideas.

Laboratory activities are important elements in holistic science education. Yet, the van den Berg, Katu and Lunetta (1994) study referenced earlier reported that these relatively open investigations *per se* were insufficient to enable students to construct a complex and meaningful network of concepts in basic electricity consistent with those of the scientific community. Practical activities should be combined with discussions of competing explanatory models, analogies, diagrams, graphic representations or simulations to enable students to construct higher levels of scientific understanding. With appropriate complementary activities, well-conducted laboratory activities can play a role in enabling students to develop concepts and related skills. In the excellent science classroom, there is a spirit of inquiry, laboratory activities are integrated within a holistic science education experience, and students move back and forth between testing their conceptions and models in the laboratory, examining the implications of their findings, and examining the implications of concepts and models of the contemporary scientific community.

*Assessment*

Testing in school science often has been confined to paper-and-pencil tests that do not examine practical skills, understanding of the nature of scientific investigation, or the interpretation and justification of the findings of those investigations (Giddings, Hofstein & Lunetta 1991). Yet, if practical skills and abilities are important in the teaching and learning of science, they must be incorporated in the assessment of learning outcomes.

Tamir (1974) suggested that *practical tests* should confront students with a relevant and intrinsically valued problem that the student is not likely to have experienced. *Observational assessment* also has been advocated as a way in which the teacher, with input from the students, can rate students on skills and practices that can be observed, explicated and discussed with the students while they are conducting laboratory activities. While progress has been made in developing practical tests and observational classroom assessment, many educators now perceive these schemes to be difficult to employ in conventional schools. Nevertheless, any assessment system that incorporates the goals elaborated for contemporary science learning will contain elements of practical assessment.

In the holistic science classroom, students generate many products that reflect their practical skills and abilities and these can be included in classroom assessment. Certainly the student's laboratory reports or journals can be included with attention to the student's ability to describe the investigation including goals, questions investigated, relevant sources of error, interpretations and applications of the findings, and new questions that he or she has formulated based upon the results of the investigation. A comprehensive and valid assessment will examine students' understanding holistically and will assess skills and knowledge in the context of tasks for which the student can understand and share the purposes (Black 1995). It will assess students' behaviours in the planning and design, performance, analysis and interpretation, and application of laboratory activities.

*Engaging Technologies*

The school laboratory offers opportunities to engage students in using technologies to support the learning of science; these experiences can help students to understand relationships between science and technology. New technologies can complement and support student collaboration and engagement in school laboratory experiences. Yet, with the increasing availability of high technology resources and with growing sensitivity to the effects of experiences on students' concepts, it is increasingly important that teachers consider the appropriateness of specific technologies introduced in students' laboratory experiences. In the introductory study of electricity, for example, students can assess the quantity of current flowing in a circuit by observing the brightness of a light bulb or by reading an ammeter. Some learning conditions warrant the use of the low technology light bulb as a metering device while, in other conditions, it is more appropriate to introduce such 'black boxes' of higher technology as ammeters and microcomputer probes.

The potential of computer *spreadsheets*, coupled with interactive graphics, to assist students in visualising, organising, interpreting and *discussing* laboratory data is substantial. When students enter their collective data into a master spreadsheet for the class, the display can facilitate discussions of the effects of sample size, methodological and experimental error, and statistical distributions as well as comparisons of individual and group findings and norms (Pogge & Lunetta 1987). Students also can use spreadsheets to examine the effects of modifying variables or functional relationships as well as to make predictions. The graphics associated with contemporary spreadsheets offer visualisation that can have a positive impact on science learning outcomes. Following an extended series of studies, Heid and Zbiek (1995) reported that high school algebra instruction emphasising the understanding of functional relationships through the regular use of interactive computer-generated graphics has had positive effects on student achievement. Omasta and Lunetta (1988) reported a study with similar outcomes utilising graphics and calculators.

While spreadsheets and computer-generated graphics can help students in organising, interpreting and discussing laboratory data, contemporary calculators or *microcomputer-based labs (MBL)* 'probes', including digitised video, enable the gathering of data in the school laboratory in new ways with instantaneous graphic visualisation as the data are collected. The instantaneous and dynamic graphic display of data readily engages students and can help them to develop enhanced understanding of the relationships among variables. MBL systems enable students to conduct extensive, accurate and interesting investigations. The data can be merged with data gathered by other members of a laboratory team across multiple trials and across long time intervals (Friedler, Nachmias & Linn 1990; Mokros & Tinker 1987).

The science education literature also has reported the use of *electronic communication* (Krajcik, Blumenfeld, Marx & Soloway 1994) to enable students to share information about regional differences in data gathered on water or air quality, climate, populations, ecosystems, physical variables, etc. In addition, students can prepare and share their laboratory or project reports as electronic multimedia documents.

With the advent of CD-ROMs and the Internet, students in secondary schools, as well as in tertiary education, can access new sources of data. *Electronic databases* enable them to engage in original investigations using data relevant to problems in their world (e.g., population growth and demographics, public health and environmental data, distribution and consumption of energy and resources).

Just as laboratory activities have been a central part of school science, *simulations* also have been present throughout the history of science teaching. Simulations have been used to complement conventional laboratory activities in the 18th century orreries that simulated the relative motion of planets to two-dimensional and three-dimensional physical models that represent objects not easily observed directly (such as human organs or molecules), in analogue laboratory activities (such as the rolling of steel ball bearings to simulate rays of light) and in today's computer microworlds (simulating kinetic-molecular models, planetary systems, electric charges or currents, etc). Lunetta and Hofstein (1991) noted that interacting with an instructional simulation can help learners to understand a real system, process or phenomenon. Within purposefully contrived settings, both practical activities and instructional simulations enable students to confront and resolve problems, to make decisions and to observe effects. Laboratory practical activities in school science are themselves simulations of scientific practice. Good laboratory activities and good simulations engage students in dynamic problem solving. Laboratory activities have the distinct advantage of enabling students to work directly with materials and phenomena in the students' biological and physical environments. However, simulations can be designed to provide meaningful representations of laboratory experiences that are often not possible with *real* materials in introductory courses. Such simulations engage students in investigations that are too complex, dangerous, expensive, fast, slow, large, small, time-consuming or material-consuming to conduct in the school science laboratory.

Some studies have compared the effects of simulations with the effects of conventional laboratory activities. These studies have suggested that the instructional simulations examined were generally not as effective as hands-on experiences in promoting the development of manipulative skills. On the other hand, most of the studies found that computer-simulated laboratory activities were at least as effective as conventional laboratory work in promoting concept learning as measured by conventional tests. In some of the studies, specific simulations were found to be more effective than conventional laboratory activities in promoting concept learning. In general, engaging students in appropriate simulations takes considerably less time than engaging them in equivalent laboratory activities *per se* (Lunetta & Hofstein 1991). The literature has reported a large and growing number of computer-based simulation prototypes, and appropriate uses of computer simulations appear to have potential for promoting student learning in science. Yet, before definitive recommendations can be made about when and how to use simulations to complement investigations with materials, more focused pedagogical research is needed. In the interim, many of the pedagogical strategies appropriate for use with laboratory investigations seem also appropriate for use with simulated investigations.

## CONCLUSION

The teacher has many challenging roles to play in engaging students in appropriate laboratory activities, in serving as co-inquirer who models appropriate problem-solving strategies, in facilitating discussion of scientific practices, concepts and theories, in sensitively sharing strategies and explanations of the scientific community, and in engaging students in relevant, concept-building discussions. Making appropriate use of laboratory and simulation activities in the context of rich scientific experiences is no simple task. The scientific concepts and pedagogical knowledge needed for successful teaching in the school laboratory are complex and take time, education and experience to develop. Effective teachers incorporate what they know about the concepts and the nature of the science discipline to be taught, the ways in which students learn, and the knowledge and development of individual students in their teaching. Science education research in recent decades can inform more appropriate use of the school laboratory to promote important science learning outcomes. Yet, care, caution and a continuing scholarly approach to interpreting the data gathered in teaching are warranted. Classroom based studies reveal that the mismatches between goals, behaviour and learning outcomes outlined earlier in this chapter continue to limit the effectiveness of the laboratory in science teaching.

When laboratory activities are well integrated with other metacognitive learning experiences such as using analogies, concept maps and predict-explain-observe demonstrations and when they engage students in manipulating ideas (not just materials and procedures), laboratory activities can promote the learning of science. Supported by appropriate technologies such as *microcomputer based labs* and *spreadsheets*, well-conducted laboratory investigations can enable more effective science teaching and learning. Well-conducted laboratory experiences can be interesting and can help a student to develop self confidence in his or her ability to understand the natural world. Informed by a social constructivist framework, teachers can help students to plan, conduct and report their investigations. Such inquiry is one important component of holistic science education that develops students' thinking and communication skills, scientific concepts and understandings of the nature of science.

Progress has been made in our understanding of how individual learners construct meaningful scientific concepts. Yet, we are in the early stages of understanding how teachers can help to promote a student's conceptual development, and we still have much to learn about how experiences with laboratory materials influence that process. The need for pedagogical research and development on the laboratory and on related technologies that will inform the development of teaching models and practices continues to be an important challenge and opportunity.

# REFERENCES

Bates, G.R.: 1978, 'The Role of the Laboratory in Secondary School Science Programs', in M.B. Rowe (ed.), *What Research Says to the Science Teacher*, National Science Teachers Association, Washington DC, 55–82.

Beasley, W.F.: 1985, 'Improving Student Laboratory Performance: How Much Practice Makes Perfect?', *Science Education* 69, 567–576.

Black, P.: 1995, 'Assessment and Feedback in Science Education', in A. Hofstein, B.-S. Eylon & G. Giddings (eds.), *Science Education: From Theory to Practice*, Weizmann Institute of Science, Rehovot, Israel, 73–88.

Brown, J.S., Collins, A. & Duguid, P.: 1989, 'Situated Cognition and the Culture of Learning', *Educational Researcher* 18(1), 32–41.
Bryce, T.G.K. & Robertson, I.J.: 1985, 'What Can They Do? A Review of Practical Assessment in Science', *Studies in Science Education* 12, 1–24.
Champagne, A.B., Gunstone, R.F. & Klopfer, L.E.: 1985, 'Instructional Consequences of Students' Knowledge About Physical Phenomena', in L.H.T. West & A.L. Pines (eds.), *Cognitive Structure and Conceptual Change*, Academic Press, New York, 61–68.
Dekkers, P.J.J.M. & Thijs, G.D.: 1993, 'Effectiveness of Practical Work in the Remediation of Alternative Conceptions in Mechanics With Students in Botswana', Paper presented at the Third International Seminar on Misconceptions and Educational Strategies in Science and Mathematics, Cornell University, Ithaca, NY.
Driver, R.: 1995, 'Constructivist Approaches to Science Teaching', in L.P. Steffe & J. Gale (eds.), *Constructivism in Education*, Lawrence Erlbaum, Hillsdale, NJ, 385–400.
Dupin, J.J. & Joshua, S.: 1987, 'Analogies and "Modeling Analogies" in Teaching: Some Examples in Basic Electricity', *Science Education* 73, 791–806.
Duschl, R.A.: 1990, *Restructuring Science Education: The Importance of Theories and Their Development*, Teachers College Press, New York.
Edgeworth, R.L. & Edgeworth, M.: 1811, *Essays on Practical Education*, Johnson, London.
Eylon, B.-S. & Linn, M.C.: 1988, 'Learning and Instruction: An Examination of Four Research Perspectives in Science Education', *Review of Educational Research* 58, 251–301.
Friedler, Y., Nachmias, R. & Linn, M.C.: 1990, 'Learning Scientific Reasoning Skills in Microcomputer-Based Laboratories', *Journal of Research in Science Teaching* 27, 173–191.
Giddings, G.J., Hofstein, A. & Lunetta, V.N.: 1991, 'Assessment & Evaluation in the Science Laboratory', in B. Woolnough (ed.), *Practical Science: The Role and Reality of Practical Work in School Science*, Open University Press, Milton Keynes, UK, 166–177.
Heid, M.K. & Zbiek, R.M.: 1995, 'A Technology-Intensive Approach to Algebra', *The Mathematics Teacher* 88, 650–656.
Hofstein, A. & Lunetta, V.N.: 1982, 'The Role of the Laboratory in Science Teaching: Neglected Aspects of Research', *Review of Educational Research* 52, 201–217.
Katu, N., Lunetta, V.N. & van den Berg, E.: 1993, 'Teaching Experiment Methodology in the Study of Electricity Concepts', Paper presented at the Third International Seminar on Misconceptions and Educational Strategies in Science and Mathematics, Cornell University, Ithaca, NY.
Krajcik, J.S., Blumenfeld, P., Marx, R.W. & Soloway, E.: 1994, 'A Collaborative Model for Helping Middle Grade Science Teachers Learn Project-Based Instruction', *Elementary School Journal* 94, 483–497.
Kyle, W.C., Penick, J.E. & Shymansky, J.A.: 1979, 'Assessing and Analyzing the Performance of Students in College Science Laboratories', *Journal of Research in Science Teaching* 16, 545–551.
Lunetta, V.N.: 1990, 'Cooperative Learning in Science, Mathematics, and Computer Problem-Solving', in M. Gardner, J. Greeno, F. Reif, A. Schoenfeld, A. Disessa & E. Stage (eds.), *Toward a Scientific Practice of Science Education*, Lawrence Erlbaum, Hillsdale, NJ, 235–249.
Lunetta, V.N. & Hofstein, A.: 1991, 'Simulation and Laboratory Practical Activity', in B. Woolnough (ed.), *Practical Science: The Role and Reality of Practical Work in School Science*, Open University Press, Milton Keynes, UK, 125–137.
Lunetta, V.N. & Tamir, P.: 1979, 'Matching Lab Activities With Teaching Goals', *The Science Teacher* 46(5), 22–24.
Metz, K.E.: 1995, 'Reassessment of Developmental Constraints on Children's Science Instruction', *Review of Educational Research* 65, 93–127.
Millar, R. & Driver, R.: 1987, 'Beyond Processes', *Studies in Science Education* 14, 33–62.
Mokros, S.R. & Tinker, R.F.: 1987, 'The Impact of Micro-Computer Based Labs on Children's Ability to Interpret Graphs', *Journal of Research in Science Teaching* 24, 369–383.
Nussbaum, J. & Novick, S.: 1982, 'Alternative Frameworks, Conceptual Conflict and Accommodation: Toward a Principled Teaching Strategy', *Instructional Science* 11, 183–200.
Omasta, E. & Lunetta, V.N.: 1988, 'Exploring Functions: A Strategy for Teaching Physical Science Concepts and Problem Solving', *Science Education* 72, 652–636.
Osborne, R. & Freyberg, P.: 1985, *Learning in Science: The Implications of Children's Science*, Heinemann, London.
Pogge, A.F. & Lunetta, V.N.: 1987, 'Spreadsheets Answer "What If . . .?"', *The Science Teacher* 54(8), 46–49.

Quin, Z., Johnson, D.W. & Johnson, R.T.: 1995, 'Cooperative Versus Competitive Efforts and Problem-Solving', *Review of Educational Research* 65, 129–143.

Rosen, S.A.: 1954, 'History of the Physics Laboratory in the American Public Schools', *American Journal of Physics* 22, 194–204.

Roth, W.-M.: 1995, *Authentic School Science: Knowing and Learning in Open-Inquiry Science Laboratories*, Kluwer Academic Publishers, Dordrecht, The Netherlands.

Singer, R.N.: 1977, 'To Err or Not to Err: A Question for the Instruction of Psychomotor Skills', *Review of Educational Research* 47, 479–489.

Tamir, P.: 1974, 'An Inquiry Oriented Laboratory Examination', *Journal of Educational Measurement* 11, 25–33.

Tamir, P. & Lunetta, V.N.: 1981, 'Inquiry Related Tasks in High School Science Laboratory Handbooks', *Science Education* 65, 477–484.

Tasker, R.: 1981, 'Children's Views and Classroom Experiences', *Australian Science Teachers Journal* 27, 33–37.

van den Berg, E. & Giddings, G.: 1992, *Laboratory Practical Work: An Alternative View of Laboratory Teaching*, Curtin University of Technology, Perth, Australia.

van den Berg, E., Katu, N. & Lunetta, V.N.: 1994, 'The Role of "Experiments" in Conceptual Change', Paper presented at the annual meeting of the National Association for Research in Science Teaching, Anaheim, CA.

White, R.T. & Gunstone, R.F.: 1992, *Probing Understanding*, Falmer Press, London.

Section 3

Educational Technology

MARCIA C. LINN

# 3.1 The Impact of Technology on Science Instruction: Historical Trends and Current Opportunities

MARCIA C. LINN
*University of California, Berkeley, USA*

International assessments repeatedly report that students lack science understanding (International Association for the Evaluation of Educational Achievement 1988; LaPointe, Mead & Phillips 1989; National Assessment of Educational Progress 1978, 1988). Technology is often viewed as a catalyst, panacea or solution to limitations in students' science understanding. For example, in 1970, a government commission reported:

> Technology can make education more productive, individual and powerful, making learning more immediate; give instruction a more scientific base; and make access to education more equal. (Commission on Instructional Technology 1970, p. 7)

The report concludes that 'learning might be significantly improved if the so-called second industrial revolution – the revolution of information processing and communication – could be harnessed to the tasks of instruction' (Commission on Instructional Technology 1970, p. 6). This chapter explores the promise of technology for science education by examining historical, current and future contributions. The chapter's first section provides a definition of technology and a view of teaching and of science learning. The second section describes three instructional frameworks (explanation, hands-on, social-constructivist). The third section discusses the Scaffolded Knowledge Integration framework and the Computer as Learning Partner project. The chapter closes by identifying tensions between technological opportunities and instructional effectiveness that will challenge researchers for years to come.

## VIEW OF TECHNOLOGY AND SCIENCE LEARNING

In this chapter, technology refers to a wide array of tools used in science classes including (1) laboratory equipment such as measuring devices, (2) video materials

---

Chapter Consultant: Instead of using chapter consultants, chapter authors in the Educational Technology section provided editorial comments on each other's chapters.

such as movies, films, filmstrips, television, scientific visualisations and computer animations, (3) interactive media such as computer tutors, microworlds, programming environments and scaffolded learning environments and (4) electronic communication methods such as electronic mail, bulletin boards and discussion environments. As new technologies have become available to education, experts in science teaching, natural science and pedagogy have struggled to find ways to make them effective for all learners. A process of trial and refinement carried out by partnerships of experts in all the relevant disciplines has led to improved roles for technology in science learning and instruction and improved frameworks to guide future innovations.

Today, most would agree that understanding science involves developing and refining ideas about scientific phenomena into an integrated perspective. Many call for active learning but fail to distinguish activities that help students to develop powerful ideas and activities that entertain.

Scientific understanding requires analysing, linking, connecting, testing and reflecting on scientific ideas. Effective learners actively make sense of examples, abstractions and principles, distilling the information that they encounter into a cohesive, small set of what could be called *models*. Here, the term model refers to any pattern, idea, heuristic, explanation or rule-of-thumb that the learner reuses to explain another scientific event. Successful learners successively refine their models by distinguishing among them and assessing their effectiveness.

Both natural scientists and citizens need to develop an integrated set of scientific models and to refine them throughout their lives. The scientific ideas that citizens have the opportunity to refine apply to personally relevant problems such as home heating, nutrition, weather prediction and appliance repair. Natural scientists, in addition, tend to specialise in a specific topic and refine their ideas about that domain. Both groups need skills in learning autonomously after they complete their science courses, and effective courses should prepare students for lifelong learning.

Lifelong learners regularly apply their scientific ideas to new problems and revise, reformulate or refine their ideas based on these experiences. Evidence suggests that lifelong learners add new models of scientific phenomena and change the models that they prefer as the result of experience (e.g., Ben–Zvi, Eylon & Silberstein 1987, in press; Lewis & Linn 1994). However, even expert scientists tend to maintain a repertoire of models rather than 'replacing' or 'eliminating' models (e.g., Feynman 1995; Smith, diSessa & Roschelle 1994).

Successful science learners also develop knowledge and beliefs about models of scientific phenomena themselves. They learn that (1) new models result from scientific investigations, (2) models have limits and are regularly tested and (3) several alternative models can reasonably apply to the same scientific events (e.g., Feynman 1995).

Historically, tensions have arisen between the goal of preparing natural scientists and the goal of educating citizens. Although both researchers and citizens need spontaneously to seek explanations for puzzling phenomena, reflect on the relationships among alternative models, troubleshoot complex situations and regularly monitor their own understanding, the context for these activities is dramatically different. The models offered in science courses are best suited to research and cannot easily be applied to problems citizens encounter (e.g., Linn & Muilenburg 1995). For example, molecular kinetic theory is too abstract to help citizens design ways to keep food cold at a picnic.

In this chapter, the role of technology in fostering lifelong learning is analysed. In particular, technology can show students exciting displays or it can challenge learners to expand their scientific ideas. I argue that identifying effective roles for technology involves careful analysis of how students learn.

## INSTRUCTIONAL FRAMEWORKS

Instructional frameworks for achieving integrated understanding and motivating lifelong learning have shifted historically (Linn, Songer & Eylon 1996). This chapter examines four historical perspectives on instruction. The *explanation* framework emphasises a knowledge-telling view of instruction. The *hands-on* framework emphasises active learning informed by the work of Bruner (1960) and Piaget (1970). The *social-cultural* framework emphasises the social context of learning informed by Vygotsky (1962, 1978). The *scaffolded knowledge integration* framework (Linn 1995) attempts to integrate all the frameworks. Each of these frameworks suggests possible models for incorporating technology into science instruction.

*Explanation Framework*

A straightforward, intuitive perspective on science learning emphasises explanations. Texts, lectures, films and typical individual tutoring sessions present explanations of scientific phenomena. Students are expected to learn the models of science from this information. Successful scientists frequently point out that this approach worked for them. In contrast, extensive educational and psychological research suggests that, rather than understanding and utilising such explanations, many students isolate and forget them, preferring to retain their personally constructed, intuitive views (e.g., diSessa 1988). These explanations can become what Whitehead (1929) called 'inert knowledge', inaccessible to students in subsequent situations. Many studies have shown that students ignore, contort or dispute models which they encounter in science classes (e.g., Caramazza, McCloskey & Green 1981; Champagne, Klopfer & Gunstone 1982; Hewson & A'Becket Hewson 1984; Linn 1986). As a result, students often fail to add an explanation from science class to their repertoire, or to compare it to their existing views. Rather than gaining integrated understanding, students exposed only to explanations run the risk of either retaining existing intuitions, or adding disconnected ideas that remain inert.

Early textbook writers recognised the difficulty of making explanations effective and designed materials intended to encourage students to link new information to existing ideas such that models were refined. For example, the introduction to the *Elements of Physics* by Millikan and Gale (1927) explained:

> [The authors'] chief aim from the beginning has been 'to present elementary physics in such a way as to stimulate the pupil to do some thinking on his [sic] own account about the hows and whys of the physical world in which he lives.' Hence as to *subject matter* they have included in this book only such subjects as touch closely the everyday life of the average pupil. In a word, they have endeavored to make it represent the everyday physics which the average person needs to help him to adjust himself to his surroundings and to interpret his own experiences correctly. (p. iv)

Clearly, Millikan and Gale intended to help students develop integrated understanding and believed that such understanding would arise if scientific explanations were connected with students' existing ideas. In their book, Millikan and Gale used technology to illustrate models such as the heat flow model. For example, in the chapter on heat conductivity, one figure illustrates apparatus appropriate for measuring heat conductivity and the explanation describes how the apparatus can be used to determine which metals are the best conductors (Millikan & Gale 1927, p. 218). In another example, Millikan and Gale illustrate heat insulation of dwellings:

> It is estimated that an annual savings of at least $100,000,000 in fuel in the United States alone would be made if the walls of houses were properly constructed with respect to heat insulation. Such walls not only conserve heat in cold climates, but exclude it from houses in warm climates. (Millikan & Gale 1927, p. 227)

Finally, in questions for students at the end of the chapter, Millikan and Gale continue to emphasise connections to everyday phenomena asking, for example, 'Why do firemen wear flannel shirts in summer to keep cool and in winter to keep warm?', 'Why will a moistened finger or a tongue freeze instantly to a cold piece of metal on a cold winter's day but not to a piece of wood?' and 'Does clothing ever afford us heat in winter? How, then, does it keep us warm?' (Millikan & Gale 1927, p. 221). To make explanations effective, Millikan and Gale emphasised models that students could connect to their personal experiences and sought technological devices to connect science to everyday experience. In addition, besides providing explanations, Millikan and Gale also encouraged students to generate their own explanations for personally relevant phenomena and to distinguish their ideas from those presented in class.

During the reforms of the 1960s, explanations considered appropriate for science courses became more abstract. For example, the Physical Science Study Committee (PSSC), funded by a grant from the National Science Foundation, reviewed physics texts and called for a reform of the curriculum. The committee cited reasons for this reform: first, current science textbooks were out of date and no longer represented the views of the scientific community; second, the texts had a patchwork quality because new topics had been added but not fully integrated into the presentation; third, the sheer mass of material in the textbook no longer could be reasonably taught in the available time; fourth, increasing applications of science in everyday life had led to a course overloaded with technology and neglectful of basic scientific concepts (Physical Science Study Committee 1957, p. 3).

As a result of these findings, the PSSC called for major reforms meeting four important criteria: (1) emphasis on the major achievements of physics including conceptions such as the conservation of energy; (2) descriptions of how major ideas were developed; (3) interconnections between all of physics; and (4) presentation of physics as a 'human activity set within society and part of the history of mankind [sic]' (Physical Science Study Committee 1957, p. 4). The resulting PSSC textbook began with basic concepts including 'time, distance and matter; the structure of the Universe; the atomic structure of matter; the molecular interpretation of chemistry' (Physical Science Study Committee 1957, p. 4). Having covered this material in the first quarter of the course, the book goes on to optical phenomena, demonstrating the inadequacy of the particle theory and the need for a wave perspective. Thus, whereas Millikan and Gale emphasised technologies that explained everyday phenomena and connected to student experience, the PSSC emphasised abstract scientific principles, indicating that 'technology was eliminated with little or no regret' (Physical Science Study Committee 1957, p. 4).

To address potential difficulties in teaching the scientific concepts, the PSSC group sought to augment text explanations by using instructional technologies such as film, hands-on experiments, ripple tanks and air tracks (Haber-Schaim, Cross, Dodge & Walter 1976). Films were designed to present abstract ideas even when teachers had difficulty with them. Many believed that these approaches would provide the information that students needed while also ensuring that learners would be active investigators. The PSSC committee believed that its curriculum would 'restore the primacy of subject matter in the educational process' (Physical Science Study Committee 1957, p. 14). They recognised that teaching abstract concepts is difficult, but argued that, if students fail to learn, the responsibility lies either with the teacher or with the student and not with the choice of subject matter. Science 'must be learned by the student and the proper function of the teacher is to create conditions under which such learning is possible and likely' (Physical Science Study Committee 1957).

During the 1960s, the models emphasised in the science curriculum became more abstract and disconnected from personal experience, making knowledge integration more difficult. The natural scientists who jointly selected these goals concluded that those students who had the capability to learn science would learn these models, whereas those who did not succeed either had failed to work hard enough or had not been motivated by their instructors. Simply put, autonomous learners would succeed and others would fail.

Since 1960, most high school curriculum materials have retained the abstract models of the PSSC reformers, augmented by instructional technologies that illustrate the models. Extensive evaluation of the reformed curriculum confirms that students have difficulty learning these abstract models and that these abstract models fail to inform everyday problem solving. For example, students completing these courses retain flawed models of motion (McDermott 1984; Thornton & Sokoloff 1990), thermodynamics (Driver & Erickson 1983) and light. In addition, both students and research staff find these models difficult to apply to everyday problems (e.g., Lewis & Linn 1994; Linn & Muilenburg 1995).

For citizens to become lifelong learners, they need scientific models that they can continue to refine. Models that apply to personally relevant problems such as those found in the Millikan text offer promise for all citizens. Natural scientists will continue to refine the most abstract and comprehensive models in their area of specialisation. Research scientists, however, could neglect models from areas beyond their specialisation if they cannot apply them to problems which they encounter. Thus, to prepare lifelong learners, there are benefits in emphasising scientific models that apply to personally relevant problems. By eliminating such models and their accompanying technologies from reform textbooks, it is likely that opportunities to test, refine and link scientific ideas were also reduced.

To enhance explanations, both Millikan and Gale (1927) and the Physical Science Study Committee (PSSC) (1957) used instructional technologies such as scientific apparatus. The technologies were matched to the models studied in the curriculum, demonstrating the versatility of such tools. In addition, both groups expected students autonomously to analyse these explanations and link them to related ideas. As scientists' models became more abstract, the task of linking scientific explanations to personal views became more challenging.

*Hands-on Framework*

The reformers of the 1960s selected new goals for the curriculum and also reconsidered the role of hands-on activity in the science classroom. During the first half of the 20th century, science educators used scientific equipment to augment explanations of scientific models. Researchers compared lecture demonstrations to hands-on laboratory experience and found generally that lecture demonstrations were as good as hands-on methods (Cunningham 1946). For explaining scientific ideas, demonstrations were more efficient than hands-on experiments. Experiments performed by students took more time and did not improve performance on standard multiple-choice tests (Curtis 1926, 1931, 1939; Linn in press). Spurred by the research of Piaget (1970) and Bruner (1960), the reformers recognised that students needed insights into the reasoning processes of natural scientists and that those insights were best achieved if students engaged in scientific activities. Prior to this time, many believed that scientific reasoning could be learned by the study of Latin or geometry. However, research by Piaget and others convinced the reformers that students needed concrete experiences with scientific investigation to acquire the problem-solving skills of science (Inhelder & Piaget 1969). These reformers emphasised teaching models of scientific processes along with explanations of scientific phenomena. Thus, students explored the general process of hypothesis generation and experimental design. As argued elsewhere, students had difficulty learning these processes in the abstract and in applying them to concrete situations (Eylon & Linn 1988; Linn *et al.* 1996; Linn & Swiney 1981). As suggested by extensive research on students' intuitions, model development and models of scientific phenomena, refinement is closely tied to specific concepts. This view is consistent with calls for 'situated cognition' (Brown, Collins & Duguid 1989) and apprenticeship instruction (Collins, Brown & Holum 1991).

Many high school curriculum projects incorporated the concrete, hands-on framework, both for teaching scientific processes and for illustrating specific scientific phenomena, because it was consistent with techniques used by natural scientists. For example, the PSSC argued that apparatus such as the ripple tank and the air track illustrated scientific concepts in a way that other approaches could not imitate, thus making complex ideas more visible (Haber-Schaim *et al.* 1976; Linn 1996). These reformers pointed out that increased use of laboratory experience provided alternative paths towards understanding of ideas and also engaged students as scientists (Physical Science Study Committee 1957).

For primary school students, the hands-on framework was also embraced in the 1960s. For example, the Science Curriculum Improvement Study (SCIS) substituted laboratory equipment for the textbook (Karplus & Thier 1967). Based on the Piagetian view (Piaget 1970) that students were 'concrete' thinkers and could not learn abstractions before the age of 12 years, students conducted experiments using apparatus provided in a kit, primarily so they could learn scientific processes such as classification. The reliance on laboratory equipment, combined with the Piagetian emphasis on classification, meant that SCIS emphasised distinguishing such ideas as 'floating' from 'sinking' or 'living' from 'non-living.' In SCIS, students used concrete materials as the primary technology to develop and refine models for concepts such as density, and were likely to continue their explorations on their own.

The SCIS emphasis on hands-on experience bears resemblance to the Nature Study curriculum for primary science introduced at the turn of the century (Curtis 1932). Both were prompted by a belief that students learned best with concrete materials. The Nature Study approach, however, resulted mainly in students categorising plants and animals, whereas the SCIS included concepts such as buoyancy, energy and air that provide a good start on developing models for everyday events.

The hands-on tradition motivated advances in the development of cost-effective instructional technologies. The SCIS pioneered in the use of modern plastics to create inexpensive, lightweight, storable apparatus. For example, students used 'whirly-gigs' to explore the strength of various plastic propellers powered by rubber bands. One chemistry program developed inexpensive experiments using drops of chemicals. Other programs developed inexpensive versions of equipment found in research laboratories such as real-time data-collection (Linn 1989; Tinker 1987).

In SCIS laboratory activities, students also explored processes such as how scientists control variables to gather convincing evidence. They studied problems suited to the classroom rather than either everyday problems or famous scientific problems. Thus, the hands-on framework was endorsed to teach the thinking skills of scientists in a concrete way, yet these processes proved difficult to apply to new problems (Eylon & Linn 1988; Lawson 1985; Shulman & Tamir 1973).

Many comparison studies involving various types of laboratory equipment, films and computer-presented experiments were conducted to show benefits from these technologies. These studies were often inconclusive because the outcome measures were poorly linked to the potential benefits of hands-on science or because the students learned specific problem solutions rather than general processes (Linn 1996). For example, the reformers of the 1960s created student-appropriate laboratory equipment and designed materials to illustrate complex scientific ideas to students. However, as Walker and Shaffarzik (1974) pointed out, studies often used outcome measures inconsistent with the goals of the curriculum. Although some outcome measures show benefits of hands-on learning (Anderson, Kahl, Glass & Smith 1983; Hofstein & Lunetta 1982; Ramsey & Howe 1969), these studies also point out the need to examine the benefits of technological tools more carefully and, in particular, to assess their contributions to integrated understanding. For example, evaluation of the SCIS materials revealed that students who followed the curriculum for the primary grades gained deeper understanding of the science topics emphasised in the curriculum than did students who followed a text-based curriculum (Bowyer & Linn 1978; Shymansky, Kyle & Alport 1982), but also demonstrated that students rarely generalised the processes (e.g., Linn, Clement, Pulos & Sullivan 1989).

Thus, considerable research suggests that engaging in hands-on activity is insufficient to improve understanding of scientific processes. Instead, hands-on exploration seems more effective in fostering the construction of specific scientific topics (e.g., Driver & Easley 1978; Papert 1980; von Glaserfeld 1990). Many researchers have shown that hands-on learning affords the opportunity to construct scientific ideas personally in much the same way as scientists explore scientific phenomena. Rather than learning scientific processes, students refine their own scientific ideas. To succeed at learning from hands-on opportunities requires some discipline in reflecting on findings, planning new investigations and linking ideas. For hands-on science to become minds-on science, either a carefully orchestrated curriculum or an autonomous learner is required. The next section discusses a way to direct hands-on experiences.

*Social-Constructivist Framework*

Feminists (Keller 1985; Longino 1990; Rossiter 1982), anthropologists (Lave & Wenger 1991), philosophers (Kuhn 1962), psychologists (Vygotsky 1962; Vygotsky 1978), historians of science and natural scientists (Feynman 1963) have drawn attention to the social construction of science. Scientists share a goal of expanding knowledge and making sense of the natural world, but do not necessarily agree about what counts as scientific knowledge. In fact, the acceptance of methodologies, research paradigms and even research results depends on the social context in which they develop, as the story of Galileo and the church effectively illustrates. Recently, science education reformers have sought to capture the process of social construction of knowledge in instructional programs. Many draw on instructional technologies to improve collaboration (Brown & Campione 1994; Gordin, Polman & Pea 1994; Scardamalia & Bereiter 1992; Songer 1993).

Historically, communities of natural scientists have jointly developed research and ferreted out the most promising ideas in lively discourse. Natural scientists present explanations to their peers and negotiate the meaning of experiments at meetings, seminars and in written documents. Scientists jointly contribute to understanding complex phenomena, and science education reformers ought to capitalise on this process in precollege classrooms. From the standpoint of the science learner, discourse with peers has the potential of providing alternative models of scientific phenomena and of introducing criteria as well as evidence to help learners to distinguish among scientific models. The social-constructivist framework developed to extend the benefits of scholarly interaction to science learners. Recently, electronic communication technologies have expanded the opportunities for students to form a community of scholars.

This framework has been articulated in several ways. Observers noted that informal training in science included an apprenticeship and used this metaphor to define models for instruction (Collins *et al.* 1991; Collins, Brown & Newman 1989; Lave & Wenger 1991). Defining the nature of an apprenticeship in a science classroom proved difficult and several approaches have emerged. If teachers were the experts and students the apprentices, then how could teachers be trained to model ideas that students needed to learn? Could students learn from each other? A model frequently referred to as 'distributed cognition' argues that each student will become a master in some aspect of a scientific topic and help other students who wish to learn that material (Brown & Palincsar 1989; Pea 1993). Other programs capitalised on the social components of large collaborations and emphasised authentic experiences in projects (de Corte, Linn, Mandl & Verschaffel 1992). Still others noted the specific features of effective collaborations and taught students to emulate practices such as asking good questions (e.g., Palincsar & Brown 1984).

Close examination of programs based on the social-constructivist model has raised some questions about equity (Brown & Campione 1990; Burbules & Linn 1991; Nersessian 1991; Scardamalia & Bereiter 1985). Just as many have argued that natural scientists rely on others who are similar to themselves in gender, status and background, so too have educators worried that instruction incorporating this framework would involve such preferences. Research suggests that those welcomed to the informal learning of science are generally similar to those who have already succeeded in terms of race, gender and social status (Keller 1985; Rossiter 1982). Thus, emphasising the social context of learning might transmit not just the knowledge of science, but also the established expectations about the characteristics of scientists (e.g., Agogino & Linn 1992).

This point is illustrated by careful study of how newcomers succeed in science (Keller 1983; Rossiter 1982). Those who agree to adopt the cultural practices of the group independent of their relevance to the advancement of science succeed more readily. For example, women report constant struggles about how to incorporate family responsibilities into the 'work' of science. Established scientists might dismiss females who have children during the tenure track years. Complaints about the discourse community and its emphasis on arrogance and attack rather than equality and accommodation frequently revert to an implied view that those who cannot compete verbally should be excluded from science. Established scientists may expect every aspiring scientist to succeed in heated debate, yet at the same time criticise women who are combative and reward men who interrupt and ridicule others.

In designing educational programs that incorporate healthy debate and group learning, one must make sure that stereotyped norms do not become part of the informal science classroom community, reflected in comments like 'girls cannot do science' and in practices that silence some groups of students (Sadker & Sadker 1994; Sandler & Hall 1982). In following the social-cultural framework, reformers have begun to address the tension between maintaining essential aspects of scientific communities while overcoming stereotypes. Instructional technologies such as electronic communication offer promise for the social-constructivist perspective. They support science scholars who wish to combine geographically diverse data points (interpret data from current events including earthquakes and the weather) and access the views of individuals with diverse backgrounds (Gordin *et al.* 1994; Songer 1993). These programs often must help students learn to interpret the communications of individuals with varied expertise. By contrasting viewpoints, science courses can help students to make sense of news accounts and political debates relevant to science. Helping students to become autonomous learners who spontaneously critique ideas increases the effectiveness of this technology. Electronic media could provide a catalyst for breaking down stereotypes and contrasting opinions or could reinforce normative views.

The three frameworks discussed above (explanation, hands-on, social-constructivist) work best when students are autonomous learners. The explanation framework for science instruction is efficient and cost-effective for students who understand the explanations and reflect on them. Imparting knowledge in a lecture is less labour intensive (per student) than almost any educational alternative. Written explanations in textbooks are even more efficient. The framework also works best for autonomous learners who work hard to connect and link information from hands-on experiments. The social-cultural framework also benefits students who learn autonomously, because success depends on carefully critiquing the ideas of others as well as on soliciting alternative ideas, sorting out conflicting information and responding to other learners.

From the standpoint of making all citizens lifelong science learners, all three of these frameworks are incomplete. In the Computer as Learning Partner project, the Scaffolded Knowledge Integration framework was developed to combine the strength of these three frameworks, while addressing the need to help students become autonomous learners.

## SCAFFOLDED KNOWLEDGE INTEGRATION FRAMEWORK AND COMPUTER AS LEARNING PARTNER PROJECT

The Scaffolded Knowledge Integration (SKI) framework synthesises ten years of research developing the Computer as Learning Partner (CLP) curriculum (Linn 1992, 1995; Linn & Songer 1991; Linn et al. 1996; Linn, Songer, Lewis & Stern 1993; Songer & Linn 1991). CLP incorporates computer technology and real-time data collection into a middle-school science curriculum to impart integrated understanding and create lifelong learners. In creating this framework, the CLP group conducted a series of experiments and refined the curriculum based on the results. The framework reflects these results and also draws on substantial related work by other researchers.

The Scaffolded Knowledge Integration framework has four complementary elements, incorporating the tension between the frameworks described above while also capitalising on the role of technology. The first element concerns goals for science learning, including goals for lifelong learning. The second involves making the thinking of science visible, an amalgam of the explanation framework and the hands-on framework. To balance making thinking visible, the third element of the framework emphasises creating autonomous learners who eventually take responsibility for their own learning and for generating their own explanations. The fourth element of the framework concerns the social context of science learning and balances issues of relying on authoritative explanations versus careful criticism of evidence. This aspect of the framework also balances conflicting goals of supporting diversity while also shaping the nature of mainstream cultural transmission. The elements of the framework are described and illustrated below.

*Identifying New Goals*

To achieve the high-level goal of integrated understanding, science courses must motivate students to connect ideas, interpret explanations and link new information to existing views. To balance explanations and hands-on activities and to welcome under-represented groups, rather than preserve the current participants in science, courses need goals that are accessible to students. If ideas are accessible to students, then students can continue to refine their understandings long after they complete each science class. In spite of repeated arguments that 'less is more', curricular topics and goals continue to expand (American Association for the Advancement of Science 1993; Eylon & Linn 1988; National Committee on Science Education Standards and Assessment 1993). Reformers regularly suggest that more advanced, abstract topics be included in courses for younger students so that subsequent courses can go beyond these topics (Linn *et al.* 1996). Coming to grips with the information explosion is an essential component of setting goals for science instruction.

Sadly, the USA leads the world in the number of science textbook topics and pages assigned per year while trailing on all science achievement indicators, as shown in the Third International Mathematics and Science Study and other investigations (Jacobson & Doran 1985; National Assessment of Educational Progress 1988). Technological tools such as films and computer visualisations can expose students to increasingly more topics, encouraging educators to expand the curriculum rather than focus on integrated understanding. The Scaffolded Knowledge Integration framework emphasises goals that both foster integrated understanding and provide a firm foundation for subsequent science learning (Linn & Muilenburg 1995).

To establish goals that meet these dual criteria, the Scaffolded Knowledge Integration framework supports a 'repertoire of models' view of conceptual change (Linn, diSessa, Pea & Songer 1994). In this perspective, a learner holds multiple explanations for phenomena that, scientists would argue, have the same explanatory basis. For example, when explaining why a styrofoam cooler will keep drinks cold at a picnic, one could use an explanation based on molecular kinetic theory or one based on heat flow. Either of these models, it turns out, can generate predictions and provide interpretations of the results of an experiment using the cooler. Selecting the most promising model involves analysing the information needs of the situation. Choosing goals for science courses involves selecting models that students are likely to understand and distinguish from their existing perspectives. This is consistent with the curriculum of Millikan and Gale (1927) mentioned above.

For example, in middle-school science, the Computer as Learning Partner curriculum has advocated a heat-flow model rather than a molecular-kinetic model because the heat-flow model is accessible to far more students and provides a firm foundation for later instruction emphasising the molecular-kinetic model. Students often have many explanations for situations involving insulation and conduction, usually based on personal experiences. Lewis (1991) found that students often believe that aluminium foil is a good insulator because metals feel cold and therefore aluminium has the capability of imparting cold to objects. Similarly, Lewis found that students often believe that wool is a bad insulator for cold objects because wool sweaters cause people to warm up. Thus, students have heuristics, rules of thumb and descriptive explanations that they might access in trying to make sense of insulation and conduction experiments. Rather than attempt to eradicate these inadequate models, the repertoire of models perspective advocates helping students to add more comprehensive models and to distinguish among various models to identify the best explanation for a given situation. This approach maintains respect for ideas which students construct while it illustrates that the progress of science involves exploring alternative explanations. Skill in selecting among models is valuable for students, scientists and citizens alike, and therefore becomes an additional goal for science instruction.

Selecting models that students can understand supports the goal of helping students to become autonomous learners. If students understand the models presented in class, they are more likely to link new models to their existing ideas and to recognise the benefits of the new model. Thus, an aspect of the Scaffolded Knowledge Integration framework is to present intermediate models that connect to student ideas and provide a firm foundation for future science learning. Developing intermediate models has been a goal for several projects initiated by White (1993) (see also White's chapter in this *Handbook*).

In the Computer as Learning Partner project, a variety of mechanisms provide intermediate models for understanding scientific ideas and for helping students to expand their repertoire of models and to distinguish among them. First, the goals of the curriculum are 'pragmatic' models of thermodynamics and, more generally, energy. Rather than relying on microscopic and abstract models, the curriculum emphasises the heat-flow model for thermal events and a source-transmission-detection model for light and sound. Research shows that more students learn complex ideas with these models (Linn & Songer 1991). Also, students are encouraged to synthesise their ideas into patterns that apply widely. Specifically, in CLP, students conduct experiments, gather information and synthesise their understandings in patterns that are recorded in an electronic notebook (see Figure 1 for an example of a pattern note).

Students use results of their experiments to make predictions for subsequent experiments (Friedler, Nachmias & Linn 1990). They continuously refine their ideas by contrasting their predictions with experimental outcomes. Third, students integrate their knowledge using principles and prototypes, which are everyday situations in which students are certain of the outcome but could lack a mechanism to explain the phenomenon. By linking new situations to prototype situations, students can join events with similar outcomes and seek an explanation. For example, as previously mentioned, students have difficulty in understanding whether metals are insulators or conductors when thinking about metals keeping drinks cold. However, students are more clear about the nature of insulation and conduction when comparing metal and wood sticks for roasting marshmallows (Linn & Songer 1991). Thus, this prototype is linked to the cold drink situation and helps the students to distinguish among their perspectives on metal as a conductor (see Figure 2).

In summary, the Scaffolded Knowledge Integration framework features new goals for science courses that help students to link and connect ideas. To accomplish this, the CLP curriculum introduces intermediate models, encourages students to link everyday experiences and asks students to make predictions based on their best assessment of available evidence. As a result, students add models to their repertoire and increase their reliance on a heat-flow model (Linn & Songer 1991). The computer technology makes this curriculum richer because students can explore simulations of complex thermal events and experiment with everyday materials using real-time data collection.

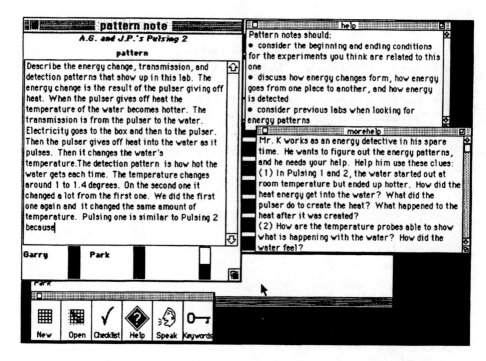

Figure 1: Examples of scaffolding provided in the Computer as Learning Partner curriculum. (Pattern notes spur students to identify connections among ideas. Two levels of help guide their thinking.)

*Make Thinking Visible*

In the Scaffolded Knowledge Integration framework, explanations provide opportunities to make scientific thinking visible. Two primary modes of making thinking visible characterise this component of the framework. First, instructors can make the problem-solving process – and not just the product of the problem-solving activity – visible using case studies and other instructional strategies (Linn & Clancy 1992b).

A second way to make thinking visible concerns the use of visualisations, films, models, simulations and other representations of scientific phenomena that help students to link and connect ideas. In the Computer as Learning Partner curriculum, real-time data collection and simulations are both used to help students understand the heat flow model and to make more visible aspects of heat flow, such as rate of heat flow, energy sources and specific heat. For example, the heat bar simulation allows students to contrast the rate of heat flow in bars of different materials (Foley 1994; Lewis 1991; see Figure 3).

In addition, the thermal model kit allows students to define situations such as large and small stove burners and to compare their effects when heating pots of similar and different mass. For further information on these simulations, see Linn

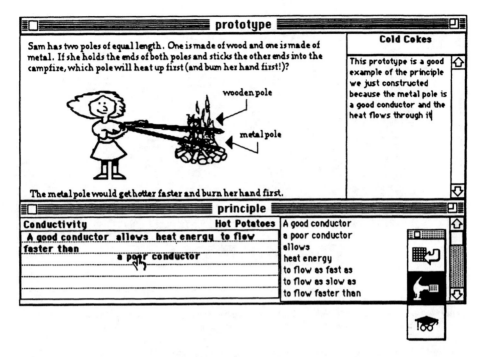

Figure 2: Examples of loci for knowledge integration in the Computer as Learning Partner Curriculum. (Prototypes help students integrate ideas across examples. Principles help students abstract their ideas into more general accounts of the phenomena they observe.)

(1995). These technological resources direct student attention to the important information while also increasing efficiency and reducing tedium. As a result, teachers are freed to spend more time tutoring individuals or small groups.

Perhaps the most compelling result of research on scientific explanations emphasises the diversity among learners (Gardner 1987). Providing different representations for explanations means that more students can access scientific ideas. Providing multiple representations of scientific ideas helps students to learn. Some students find visualisations helpful, others find verbal descriptions helpful, others find simulations and models useful, and others learn from kinaesthetic cues in hands-on experiments. Technological tools, such as films, computer models, visualisations and simulations, are particularly useful in this undertaking because they can increase the range of representations available in science classrooms. Creative uses of these approaches are described in several projects reported in this *Handbook* (see Schecker's chapter and White's chapter).

Making thinking visible suffers from some of the same drawbacks that were described for the explanation framework. Students could isolate explanations, rather than actively integrating them with their own perspectives and utilising them to explain new phenomena. They might misinterpret explanations. For example, they might not understand the limits of simulations. Technological tools including telescopes, microscopes, electron microscopes, lasers, computers and cyclotrons

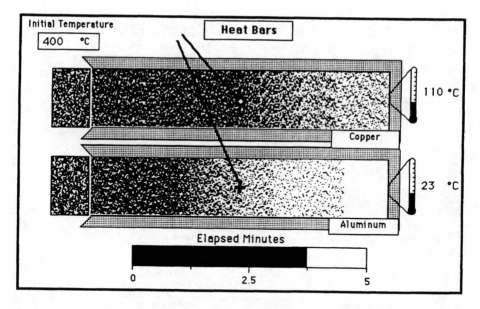

Figure 3: Example of a simulation for heat flow from the Computer as Learning Partner Curriculum. (Simulations make scientific ideas about thermal phenomena visible to students.)

can help to make ideas visible but also can confuse if links are not clear. For example, students interpreting information from a telescope in the Computer as Learning Partner classroom concluded that it brought the object close to them rather than increasing the accuracy of their view (Bell, Davis & Linn 1995). Both in teaching scientific concepts and in teaching about the history of scientific advances, these technologies offer promise for making thinking visible.

*Encourage Autonomy*

The Scaffolded Knowledge Integration framework balances explanations and knowledge integration activities to encourage autonomy. Autonomous learning involves self-monitoring and self-regulation. In the Scaffolded Knowledge Integration framework, students are encouraged to question, criticise, analyse, reflect upon and interpret the explanations that they encounter. In this sense learning is active, as emphasised in the hands-on framework, and knowledge integration is featured, because existing knowledge forms the basis for analysis and criticism.

A proportion of students behave autonomously without instruction. These students make self-explanations when reading texts or encountering scientific information (Chi & Bassok 1989). Students disposed towards autonomous learning also engage in self-monitoring and self-regulation (Corno 1992; Schoenfeld 1989). Students who reflect on their own understanding and identify inconsistencies and confusions also learn autonomously (Lewis & Linn 1993; Linn 1996; Songer 1989). How can more students learn these skills?

The Computer as Learning Partner project has documented that students vary in their beliefs about autonomous learning. Students' beliefs about the nature of their own learning were assessed in several studies (Linn & Songer 1991, 1993). Students who believe that learning is a passive process or that they cannot understand scientific ideas avoid autonomous learning. In contrast, students who believe that scientific ideas should make sense, that conundrums should be explained and that inconsistencies should be resolved are the ones who engage in autonomous learning. Autonomous learners independently take the role of critic in that they criticise each other, themselves and scientific experiments from the past. The CLP project has sought ways to increase the proportion of autonomous learners.

Champagne, Klopfer and Gunstone (1982), noting that students frequently observe scientific experiments without actively thinking about them, showed benefits for the predict-observe-explain approach to instruction. Students make predictions concerning outcomes of scientific experiments, observe the results and then explain their findings. A similar model called the Learning Cycle characterised the SCIS materials (Karplus & Thier 1967). CLP incorporated predictions to help students behave autonomously and showed the benefit of this approach (Linn & Songer 1991).

Another approach to encouraging autonomy involves engaging students as both investigators and critics. Students learn the strengths and limitations of their work by comparing their results to those of others. Often, however, students resist criticism and conclude that two conflicting answers to an experimental situation are both valid (Linn & Clark 1995). Helping students to expand their repertoire of models concerning the nature of scientific investigation, as well as realising that conflicting results require resolution at least in some situations, were emphasised in CLP as a result of these findings and remain challenging goals of the curriculum.

A third technique for encouraging autonomy involves prompting students to reflect, criticise and self-regulate while conducting scientific investigations. Computer technologies are effective at prompting and often lead students to reflect. Several levels of hints provide help for students. If these fail, the technology helps students decide to ask the teacher to tutor them. By prompting students to reflect, yet encouraging them to seek assistance when they reach impasses, the CLP computer laboratory notebook increases instructional efficiency. This approach fosters autonomy and frees teachers from mundane questions. The teacher can tutor students who need the most help.

Engaging students as tutors also offers promise when carefully implemented (Brown & Campione 1994). For example, Brown and her collaborators help students 'specialise' in topics so that they can become classroom experts. Encouraging all students to learn autonomously involves careful instructional planning. Observational studies demonstrate that instructors provide the best students and students who typify natural scientists more opportunities to become autonomous learners than others (Sadker & Sadker 1994). Thus, science teachers call on male students more often than on female students, ask male students harder questions and provide more prompts for female students. Teachers who monitor their behaviour and attempt to balance opportunities for all students in their classes will foster autonomy more broadly and increase access of under-represented groups to science learning. Engaging students as tutors also offers promise when carefully implemented.

Thus, to encourage autonomy, science courses can help students to gain insights into the practice of science, engage in self-monitoring and reflection, and act as both investigator and critic. Hands-on science alone, however, is insufficient to help learners become more autonomous. Changing laboratory experiences to prompt students to participate and allow students to engage in self-monitoring and reflection could increase student autonomy.

*Provide Social Support*

The cultural-social framework emphasises the social aspect of science learning, because social support might provide hints or prompts that allow students to go further in their reasoning than they would on their own (Vygotsky 1978). When teachers or students model appropriate autonomous scientific investigation strategies and prompt students to reflect, critique or monitor, they provide social support (Linn & Clancy 1992a). Students, behaving as a community of scholars, as implemented in several projects, can specialise in separate topics and provide each other with resources (Brown & Campione 1990; Scardamalia & Bereiter 1992). Thus, by drawing on the experience of other learners, a group of individuals might achieve more scientific understanding than each could achieve alone.

Instituting effective social support in science classrooms means dealing with social norms. Students could be marginalised, discriminated against or neglected. Students in science classrooms working in social groups can reinforce stereotypes (Hawkins 1991; Linn & Burbules 1993; Madhok 1992). In addition, creating collaborative environments, where students balance information provided by their peers against their own intuitions, requires careful instructional planning. Even in well-organised classrooms, the process can go awry and the students might reinforce each other's inadequate perceptions (Agogino & Linn 1992; Brown & Campione 1994; Linn & Burbules 1993).

Interactive technologies can foster effective social interactions. For example, the Computer as Learning Partner curriculum features what are called 'agreement bars' that ask group members to indicate independently their agreement with the synthesis provided by the group. This use of interactive technologies sometimes allows individuals and groups to get the attention of their peers (Linn 1992).

Many additional aspects of the Computer as Learning Partner curriculum contribute to an effective social environment in the classroom. Early research with group learning suggested that groups of two were more effective than larger groups both because of small computer screen sizes and because individuals in groups of two treated each other with more respect (Madhok 1992). Arranging classrooms to encourage social support requires an emphasis on respect as well as opportunities for all learners to develop expertise in some area (Cohen 1994; Cohen, Lotan & Leechor 1989). Students can more easily respect each other when they recognise the expertise of their peers.

Technological tools, especially network communication tools, expand the community of individuals working in science classrooms and permit greater opportunities for social support. Yet expanding the community to include natural scientists, accessible through the Internet, or parents and teachers in other classrooms could encourage students to rely on experts rather than actively to make sense of scientific phenomena or to monitor their own understanding. Helping students to learn to evaluate evidence from experts remains a challenge for science instruction.

Furthermore, economies of scale are necessary in order to support communication between students and scientists. There are far too many students for direct communication with natural scientists to be feasible. Instead, appropriate communication mechanisms for students and scientists are needed. One such approach is the 'Ask a Scientist' resource at Argonne National Laboratories (Figure 4). In this model, individuals wishing to find an answer to a science question first search all the available answers and then submit questions only if they have not yet been answered. Several scientists might respond to a new question and offer diverse views. In fact, scientists sometimes disagree with each other when responding to these questions. By reading a series of these responses, students gain insights into scientific discourse, alternative models for scientific phenomena and scientific issues. Guiding students to use such resources effectively remains an important challenge.

The Scaffolded Knowledge Integration framework emphasises balancing the four elements discussed above to encourage integrated understanding. New goals make science relevant to a broader range of students. Technologies that make thinking visible increase the variety of available explanations and, as a result, could meet the needs of a broader range of students. Technologies and classroom practices can help students become autonomous learners. The computer notebook prompts students and also frees the teacher for more comprehensive tutoring of students who need help to become autonomous. Classroom activities and technological resources allow learners to take advantage of the social nature of science learning. These four elements provide checks and balances, increasing the probability that more students will become autonomous learners and that diverse individuals will persist in science.

Keeping this framework in mind while analysing the impact of technology on science learning and instruction helps to capture the complexity of science education reform. In general, technological tools have served as catalysts for change in science education and as instigators of new approaches to science instruction. Often, however, technological tools are embraced uncritically even though they, in and of themselves, lack the power to transform science teaching and learning. The Scaffolded Knowledge Integration framework serves as a guide for incorporating modern technologies into the curriculum.

**QUESTION:**
From: STUDENT A
Message: Note 16 in AskAScientArc:General (There are 4 responses.)
Subject: Coldest Temperature?

I read that Antarctica had the coldest temperature on record. I was wondering why it was Antarctica and not the North Pole. Because they are both the same distance from the equator. Thanks for your help.
–STUDENT A
3rd grade
ANYTOWN Elementary School

**ANSWER(S):**
From: SCIENTIST A on Thu Jun 10 11:07:05 1993
Message: Note 16 in AskAScientArc:General (Response 1 of 4)
Subject: Coldest Temperature?

STUDENT A, this is an interesting question, and I'm afraid I don't have a good answer for it. Maybe it has something to do with how easy or hard it is to monitor temperatures in the two places. In other words, maybe there's a problem with getting to the cold places in the North Pole but not at the South Pole. It would be interesting to try and find out more information, like what is the average temperature at the North Pole? What's the average temperature at the South Pole? What's the lowest temperature ever recorded at the North Pole? And so forth. Finding out this information might help you understand the situation better.

Hope this helps! Let us know what you find out.
–SCIENTIST A

From: SCIENTIST B on Tue Jun 15 15:01:34 1993
Message: Note 16 in AskAScientArc:General (Response 2 of 4)
Subject: Coldest Temperature?

According to some books I consulted (e.g., 'The Ends of the Earth' by Isaac Asimov), it does get much colder in Antarctica than in the Arctic regions. Asimov quotes a record cold temperature of -127 degrees Fahrenheit measured at the Soviet station Vostok in 1960. (My World Almanac lists -128.6, again at Vostok, in 1983.) Here are some reasons why it gets so much colder there. Antarctica is actually a continent of much larger extent than the ice sheets of the Arctic. Thus, its interior is more isolated from the ocean waters (which moderate temperatures). Antarctica has the highest average altitude of the continents, much of it high plateau or mountainous; its thin, dry air allows intense cooling during the long winter.
–SCIENTIST B

From: SCIENTIST A on Fri Jun 25 14:53:32 1993
Message: Note 16 in AskAScientArc:General (Response 3 of 4)
Subject: Coldest Temperature?

All I can say is . . . cool.
–SCIENTIST A

From: SCIENTIST C on Wed Jul 21 18:02:15 1993
Message: Note 16 in AskAScientArc:General (Response 4 of 4)
Subject: Coldest Temperature?

Bingo. Continent not ocean, elevation of 3 km vs. sea level, and possibly less wind as well.
–SCIENTIST C

Figure 4: Example of discourse between scientists responding to student questions at the Ask-a-Scientist site maintained by Argonne National Laboratory. (Students get both scientific ideas and insights into scientific discussion.)

## CONCLUSIONS

Analysis of the four frameworks that historically have guided the design of science instruction and the incorporation of technology spotlights some important tensions for reformers. First, there is a tension between abstract and personally-relevant explanations and goals for science instruction. The Computer as Learning Partner (CLP) curriculum emphasises pragmatic principles that are personally relevant and argues that students gain a firm foundation for future scientific work from these principles. Using these principles sets students along a path of lifelong learning and prepares them to use network resources as their scientific ideas expand and improve. Others argue for better use of technology to illustrate abstract models of scientific phenomena (White & Frederiksen 1990; Wiser & Carey 1983).

A related tension arises between elitism and accessibility to science. From the elitist perspective, science is difficult and autonomous learners are born, not taught. In contrast, to increase equity in science, CLP attempts to teach autonomous learning so that more students gain access to scientific understanding. Because the benefits of expanding scientific literacy are far-reaching, greater effort could be required to attract and encourage diverse students to participate in science than would be necessary to engage traditional students in these activities.

Another tension arises between the goal of making thinking visible and the goal of encouraging autonomy. By making thinking visible, one provides effective explanations which increase the efficiency of learning. However, students might process explanations complacently rather than actively learning them. Technologies can help students to learn autonomously from explanations by prompting students to (1) reflect on information, (2) integrate information with familiar experiences or (3) make links between different aspects of a lesson. Interactive technologies can engage students as investigators and critics, thus helping them to become more autonomous and able to assess scientific ideas. This is an area ripe for further research.

A further tension arises between the goal of providing social support and the danger of reinforcing social stereotypes. When students work jointly to solve problems, individuals might behave stereotypically. For example, males might tell females that science is a male domain. Teachers could reinforce stereotypes inadvertently. Making these processes explicit can help learners to develop more respectful interaction patterns. At the same time, technology has both empowered those traditionally excluded from science and reinforced normative views concerning who should participate in science.

These tensions influence the design of science instruction both with and without technologies. However, some technologies contribute to these tensions. For example, multimedia explanations of abstract scientific concepts convince some to emphasise abstract ideas with increasingly younger students and such illustrations can frustrate learners. Ensuring an orderly progression of ideas, so that students continue to integrate rather than isolate or ignore scientific information, involves careful selection of goals of instruction. In another example, the opportunity to use interactive technologies to tutor and prompt students raises challenges concerning how these approaches can provide a balance between providing explanations and encouraging autonomy.

Network technologies provide access to a wide range of explanations from video servers to network sites. In addition, network technologies make connections possible between students and scientists and, as a result, again raise the tension between abstract and personally relevant explanations. Scientists following the views put forth by the PSSC might rely on abstract explanations when students need more pragmatic models. On the other hand, for those students who wish to understand the latest models, networked communication with natural scientists could be ideal.

*Scaffolded Knowledge Integration Framework*

Rapidly changing technological resources enable fundamental progress on the four aspects of the Scaffolded Knowledge Integration framework and also allow researchers to explore the tensions described above. Each aspect of the framework is discussed in turn below.

Incorporate Accessible Goals

As science advances, fostered by technological innovation, the difficulty of honing the curriculum to the available instructional time increases. Technological tools, however, offer promise for resolving some of these dilemmas. First, technological tools make qualitative science realistic by providing appropriate simulations (see other chapters in this section). In addition, network resources can move more science learning to the 'just in time' model. Thus, rather than teaching students everything that they need to know, the curriculum provides a firm foundation to continue to learn autonomously utilising network information. Just as the Computer as Learning Partner project emphasised the qualitative heat flow model rather than the quantitative molecular kinetic model, so subsequent courses might also emphasise qualitative principles that apply broadly rather than more abstract, mathematical and esoteric ideas. Rather than engaging all students in learning every scientific model, an alternative is to provide a subset of models for students and then allow students to conduct in-depth, sustained projects for which they choose to explore scientific problems at various levels of abstraction. Students with an engineering orientation could prefer to explore problems that rely more on pragmatic principles, while others might wish to utilise mathematical resources to explore more abstract representations of scientific phenomena. Models and simulations, including simulations of tools such as Working Model (Knowledge Revolution 1993) permit more powerful explorations in such projects than have been possible in the past.

Historically, new technologies have enabled science educators to modify the goals of the science curriculum, as noted for SCIS and PSSC. In the 1960s, laboratory equipment was used to emphasise the thinking skills of scientists. In the 1990s, laboratory equipment enabled reformers to emphasise integrated understanding. For example, these tools revitalise projects, a component of early recommendations from Dewey (1901). Projects allow students to learn to link and connect ideas (Linn & Clark 1995). Recently, computer technologies and networked technologies have expanded collaboration opportunities in projects and have supported increasingly complex scientific activities in classrooms (Goldstein 1992; Gordin & Pea 1995; Gordin et al. 1994). These technologies also help to support integrated understanding by providing prompts and help.

New, more advanced technological resources also motivate science education reformers to change their view of life-long learning. Access to more advanced scientific ideas becomes easier with networked resources. If students learn to use networked resources during pre-college instruction, they will be more likely to continue to use these tools throughout their lives. Students who learn to use modern technological tools are better prepared for the workplace and for opportunities to update and expand their scientific ideas.

Thus, new goals for science instruction need to respond to technological advance in two distinct ways: first, by carefully defining the goals of instruction so that they are both realistic and foundational; and, second, by including among those goals, an emphasis on technological understanding that supports workplace competence and life-long learning.

Design to Thinking More Visible

Technological tools, starting with laboratory equipment and including films, computer visualisations and other technological advances, have transformed our ability to visualise scientific phenomena. Qualitative models become far more realistic with the help of visualisation tools. In addition, interactive technologies allow students to access explanations on demand rather than as assigned. As other chapters in this section illustrate, technological tools have tremendous potential in making thinking visible. They also raise fundamental questions about how these resources should be designed.

Design to Autonomy

As technological tools become more widely available, the greatest challenge to fostering autonomous learning is the passive response of students to technological innovations such as film and video. Mindless interaction with technologies such as laboratory equipment or simulations creates similar drawbacks. Identifying instructional techniques that motivate learners to responsibly integrate, connect and distinguish scientific ideas remains a forefront issue. Of particular importance in promoting autonomy is promoting effective understanding of the nature of science and the nature of one's own learning. As information becomes more widely available in a vast array of media, students need more and more skill in determining the validity of scientific evidence, understanding the perspectives of those communicating the evidence, and interpreting news reports. Researchers, curriculum innovators and technology designers jointly need to address the issue of promoting autonomy in order to create citizens who can operate effectively in the 21st century.

Reachout to New Communities

Technological tools allow communication among a broader range of individuals while, at the same time, increasing the potential of hierarchical interactions that reinforce stereotypes. Creating communities that operate under the principle of mutual respect will improve science learning and understanding.

A challenge and opportunity in creating such communities involves a re-evaluation of the role of the science teacher. Previously, many viewed the science teacher as the fount of knowledge or the expert scientist spouting explanations. Redefining classrooms as communities of learners and connecting classrooms to electronic resources that provide diverse viewpoints concerning scientific phenomena calls for a redefinition of the goals of science teachers.

The role of teachers should be to increase their effectiveness by both creating effective learning communities and developing more sophisticated understandings of science teaching. It has become clear that effective science teaching involves designing instruction that balances explanations and autonomy. Teachers, as they re-teach topics year after year, have the opportunity to refine their model of students' understanding (Linn *et al*. 1994). Creating communities of teachers who use contrasting approaches to teaching concepts such as heat conduction could substantially improve science teaching and learning.

Many research studies emphasise the value of team teaching in science instruction (Little 1990). Electronic resources enable virtual teams to operate without requiring geographic proximity. Thus, teachers might divide responsibility for science topics and share curriculum materials, lesson plans and assessment devices. Members of a virtual team even might take responsibility for assessing groups of learners not in their classrooms, in order to improve learning outcomes. Exploring mechanisms for such virtual teams offers an exciting direction for new research.

*Future Challenges*

Technological innovation continues to outpace response in the educational community. Finding effective mechanisms for incorporating technology into science education remains challenging. Clearly the most promising techniques require trial and refinement to reach their full potential. Referred to as 'design experiments', this approach involves creating pilot materials and trying them in the classroom, and then refining them prior to the following trial (e.g., Brown *et al*. 1989). Virtual teams of teachers are particularly likely to be able to engage in such design experiments effectively. Furthermore, research programs involving partnerships among experts in pedagogy, classroom teaching, science disciplines and technology seem well-suited to incorporating technology into science instruction (diSessa 1993; Linn et al. 1996). An emerging group of educators with cross-training in a science discipline, technology and pedagogy are well positioned to advance instructional theory.

The driving forces in technological innovation are not the needs of education but the needs of corporations and scientific research groups. Often educational needs take second, third or fourth place to business applications. Thus we see educational innovators identifying ways to utilise business spreadsheets because creating applications particularly suited to science instruction would be too costly. Instead science educators must band together to create jointly the tools students need.

As the boundary between experts and novices becomes more fluid and as technological innovation makes the need for specialisation more and more necessary, demand for instructional provisions that create autonomous learners will increase. Current students will change careers several times during their working lives. Thus, equipping students to use modern technologies, as well as preparing them to engage in 'just in time' learning as they progress through careers, will become a greater necessity. Technological resources such as instructional materials, information bureaus and late breaking results should be accessible to all learners.

## OVERVIEW OF OTHER CHAPTERS

The section on educational technology consists of eight chapters. In this initial chapter, four historical frameworks of science instruction and the resulting roles for technology are explained. The Computer as Learning Partner project, a ten-year research program aimed at improving middle school science learning, is used to illustrate current directions. The chapter closes by identifying tensions between technological opportunities and instructional effectiveness that will challenge researchers for years to come.

Barbara White's chapter describes how the use of software tools enable students to use microworlds to interact with and create simulations as they engage in modelling activities that promote understandings of the subject matter and knowledge of the processes of scientific inquiry and modelling.

In the third chapter, Daniel Edelson presents a case study of the CoVis Project and a discussion of the common themes that underlie a number of projects that are investigating the use of technology to support more authentic science practices in science classrooms.

Nancy Butler Songer describes how The Kids as Global Scientists Project integrates the best features of the Internet and its resources into middle school science and facilitates the development of more personalised understandings in science.

Robert Sherwood and his colleagues from the Cognition and Technology Group at Vanderbilt describe Anchored Instruction and the implications of this model for science education. They provide examples of projects that have used the design principles, briefly summarise related research findings and indicate some of the opportunities for teachers as they implement these designs.

In the sixth chapter, Michele Wisnudel Spitulnik, Steve Stratford, Joseph Krajcik and Elliot Soloway explain how computer learning environments can promote scientific literacy by encouraging students to engage in scientific inquiry and to represent their understandings through dynamic models and hypermedia documents.

Horst Schecker explains how micro-based laboratory tools, a science spreadsheet and a dynamic model-building system can be used to alternate freely between measuring, mathematical modelling and conceptual modelling. Educational technology tools help to narrow the gap between conducting an experiment and developing a theoretical description by closely relating the processes of data acquisition and evaluation.

Finally, Angela McFarlane and Yael Friedler describe research in five countries on the use of portable computers in science teaching (with particular emphasis on an investigative approach to science education), compare the use of portables with the use of desk-top computers, consider learning gains, student attitudes and teacher development, and consider hardware and software.

## NOTE

This material is based upon research supported by the National Science Foundation under grants MDR–8954753 and MDR–9155744. Any opinions, findings and conclusions or recommendations expressed in this publication are those of the author and do not necessarily reflect the views of the National Science Foundation. Thanks to Dawn Davidson, Madeleine Bocaya, Patricia Kim and Jean Near for assistance with the production of this manuscript. The author gratefully acknowledges comments on an earlier draft of this manuscript provided by Brian Foley, Lawrence Muilenburg, Barbara White and Ellen Dill.

## REFERENCES

Agogino, A.M. & Linn, M.C.: 1992, 'Retaining Female Engineering Students: Will Early Design Experiences Help?', *NSF Directions* 5(2), 8–9.

American Association for the Advancement of Science: 1993, *Benchmarks for Science Literacy: Project 2061*, Oxford University Press, New York.

Anderson, R.D., Kahl, S.R., Glass, G.V. & Smith, M.L.: 1983, 'Science Education: A Meta-Analysis of Major Questions', *Journal of Research in Science Teaching* 20, 379–385.

Bell, P., Davis, E.A. & Linn, M.C.: 1995, 'The Knowledge Integration Environment: Theory and

Design', in *Proceedings of the Computer Supported Collaborative Learning Conference*, Lawrence Erlbaum, Mahwah, NJ, 14–21.
Ben-Zvi, R., Eylon, B. & Silberstein, J.: 1987, 'Students' Visualisation of a Chemical Reaction', *Education in Chemistry* 24, 117–120.
Ben-Zvi, R., Eylon, B. & Silberstein, J.: in press, 'Structure and Process in Chemistry: The Concepts and Their Conception', *Education in Chemistry*.
Bowyer, J.B. & Linn, M.C.: 1978, 'Effectiveness of the Science Curriculum Improvement Study in Teaching Scientific Literacy', *Journal of Research in Science Teaching* 15, 209–219.
Brown, A.L. & Campione, J.C.: 1990, 'Communities of Learning and Thinking, or A Context by Any Other Name', *Contributions to Human Development* 21, 108–126.
Brown, A.L. & Campione, J.C.: 1994, 'Guided Discovery in a Community of Learners', in K. McGilly (ed.), *Classroom Lessons: Integrating Cognitive Theory and Classroom Practice*, MIT Press/Bradford Books, Cambridge, MA, 229–270.
Brown, A.L. & Palincsar, A.S.: 1989, 'Guided, Cooperative Learning and Individual Knowledge Acquisition', in L.B. Resnick (ed.), *Knowing, Learning and Instruction: Essays in Honor of Robert Glaser*, Lawrence Erlbaum, Hillsdale, NJ, 393–451.
Brown, J.S., Collins, A. & Duguid, P.: 1989, 'Situated Cognition and the Culture of Learning', *Educational Researcher* 18(1), 32–41.
Bruner, J.S.: 1960, *The Process of Education*, Harvard University Press, Cambridge, MA.
Burbules, N.C. & Linn, M.C.: 1991, 'Science Education and the Philosophy of Science: Congruence or Contradiction?', *International Journal of Science Education* 13, 227–241.
Caramazza, A., McCloskey, M. & Green, B.: 1981, 'Naive Beliefs in "Sophisticated" Subjects: Misconceptions about Trajectories of Objects', *Cognition* 9, 117–123.
Champagne, A.B., Klopfer, L.E. & Gunstone, R.F.: 1982, 'Cognitive Research and the Design of Science Instruction', *Educational Psychologist* 17, 31–53.
Chi, M.T.H. & Bassok, M.: 1989, 'Learning from Examples via Self-Explanations', in L.B. Resnick (ed.), *Knowing, Learning and Instruction: Essays in Honor of Robert Glaser*, Lawrence Erlbaum, Hillsdale, NJ, 251–282.
Cohen, E.G.: 1994, 'Restructuring the Classroom: Conditions for Productive Small Groups', *Review of Educational Research* 64, 1–35.
Cohen, E.G., Lotan, R.A. & Leechor, C.: 1989, 'Can Classrooms Learn?', *Sociology of Education* 62, 75–94.
Collins, A., Brown, J.S. & Holum, A.: 1991, 'Cognitive Apprenticeship: Making Thinking Visible', *American Educator* 15(3), 6–11 & 38–39.
Collins, A., Brown, J.S. & Newman, S.E.: 1989, 'Cognitive Apprenticeship: Teaching the Craft of Reading, Writing, and Mathematics', in L.B. Resnick (ed.), *Cognition and Instruction: Issues and Agendas*, Lawrence Erlbaum, Hillsdale, NJ, 453–494.
Commission on Instructional Technology: 1970, *To Improve Learning* (A report to the President and the Congress of the United States), US Government Printing Office, Washington DC.
Corno, L.: 1992, 'Encouraging Students to Take Responsibility for Learning and Performance', *Elementary School Journal* 93, 69–83.
Cunningham, H.A.: 1946, 'Lecture Method Versus Individual Laboratory Method in Science Teaching: A Summary', *Science Education* 30, 70–82.
Curtis, F.: 1932, 'Some Contributions of Educational Research to the Solution of Teaching Problems in the Science Laboratory: Chapter VII', in G.M. Whipple (ed.), *A Program for Teaching Science: Part I of the Thirty-First Yearbook of the National Society for the Study of Education*, Public School Publishing, Bloomington, IN, 91–108.
Curtis, F.D.: 1926, *A Digest of Investigations in the Teaching of Science in the Elementary and Secondary Schools*, Maple Press, York, PA.
Curtis, F.D.: 1931, *Second Digest of Investigations in the Teaching of Science*, Maple Press, York, PA.
Curtis, F.D.: 1939, *Third Digest of Investigations in the Teaching of Science*, Maple Press, York, PA.
de Corte, E., Linn, M.C., Mandl, H. & Verschaffel, L. (eds.): 1992, *Computer-Based Learning Environments and Problem Solving* (NATO ASI Series F: Computer and System Series), Springer-Verlag, Berlin, Germany.
Dewey, J.: 1901, *Psychology and Social Practice* (Contributions to Education), University of Chicago Press, Chicago, IL.
diSessa, A.: 1988, 'Knowledge in Pieces', in G. Forman & P. Pufall (eds.), *Constructivism in the Computer Age*, Lawrence Erlbaum, Hillsdale, NJ, 49–70.

diSessa, A.: 1993, 'Toward an Epistemology of Physics', *Cognition and Instruction* 10(2–3), 105–225. (whole issues)
Driver, R. & Easley, J.: 1978, 'Pupils and Paradigms: A Review of Literature Related to Concept Development in Adolescent Science', *Studies in Science Education* 5, 61–84.
Driver, R. & Erickson, G.: 1983, 'Theories-in-Action: Some Theoretical and Empirical Issues in the Study of Students' Conceptual Frameworks in Science', *Studies in Science Education* 10, 37–60.
Eylon, B. & Linn, M.C.: 1988, 'Learning and Instruction: An Examination of Four Research Perspectives in Science Education', *Review of Educational Research* 58, 251–301.
Feynman, R.: 1963, *The Feynman Lectures on Physics*, Addison-Wesley, Reading, MA.
Feynman, R.P.: 1995, *Six Easy Pieces*, Addison-Wesley, New York.
Foley, B.: 1994, 'Teaching the Physics of Sound with an Animated Computer Simulation', Paper presented at the annual meeting of the American Educational Research Association, New Orleans, LA.
Friedler, Y., Nachmias, R. & Linn, M.C.: 1990, 'Learning Scientific Reasoning Skills in Microcomputer-Based Laboratories', *Journal of Research in Science Teaching* 27, 173–191.
Gardner, H.: 1987, *The Mind's New Science*, Basic Books, New York.
Goldstein, J.S.: 1992, *A Different Sort of Time: The Life of Jerrold R. Zacharias, Scientist, Engineer, Educator*, MIT Press, Cambridge, MA.
Gordin, D. & Pea, R.D.: 1995, 'Prospects for Scientific Visualization as an Educational Technology', *Journal of the Learning Sciences* 4, 249–279.
Gordin, D.N., Polman, J.L. & Pea, R.D.: 1994, 'The Climate Visualizer: Sense-Making Through Scientific Visualization', *Journal of Science Education and Technology* 3, 203–226.
Haber-Schaim, U., Cross, J.B., Dodge, J.H. & Walter, J.A.: 1976, *PSSC Physics: Teacher's Resource Book* (fourth edition), Heath, Lexington, MA.
Hawkins, J.: 1991, 'Technology-Mediated Communities for Learning: Designs and Consequences', in V.M. Horner & L.G. Roberts (eds.), *Electronic Links for Learning: The Annals of the American Academy of Political and Social Science* (Volume 514), Sage, Newbury Park, CA, 159–174.
Hewson, P.W. & A'Becket Hewson, M.G.: 1984, 'The Role of Conceptual Conflict in Conceptual Change and the Design of Science Instruction', *Instructional Science* 13, 1–13.
Hofstein, A. & Lunetta, V.N.: 1982, 'The Role of the Laboratory in Science Teaching: Neglected Aspects of Research', *Review of Educational Research* 52, 201–217.
Inhelder, B. & Piaget, J.: 1969, *The Early Growth of Logic in the Child*, Norton, New York.
International Association for the Evaluation of Educational Achievement: 1988, *Science Achievement in Seventeen Countries: A Preliminary Report*, Pergamon, Oxford, UK.
Jacobson, W.J. & Doran, R.L.: 1985, 'The Second International Science Study Results', *Phi Delta Kappan* 66, 414–417.
Karplus, R. & Thier, H.D.: 1967, *A New Look at Elementary School Science; Science Curriculum Improvement Study*, Rand McNally, Chicago, IL.
Keller, E.F.: 1983, *A Feeling for the Organism: The Life and Work of Barbara McClintock*, Freeman, San Francisco, CA.
Keller, E.F.: 1985, *Reflections on Gender and Science*, Yale University Press, New Haven, CT.
Knowledge Revolution: 1993, *Working Model* (Computer Simulation Program), San Francisco, CA.
Kuhn, T.S.: 1962, *The Structure of Scientific Revolutions* (first edition), University of Chicago Press, Chicago, IL.
LaPointe, A.E., Mead, N.A. & Phillips, G.W.: 1989, *A World of Differences: An International Assessment of Mathematics and Science* (No. 19–CAEP–01), Educational Testing Service, Princeton, NJ.
Lave, J. & Wenger, E.: 1991, *Situated Learning: Legitimate Peripheral Participation*, Cambridge University Press, Cambridge, MA.
Lawson, A.E.: 1985, 'A Review of Research on Formal Reasoning and Science Teaching', *Journal of Research in Science Teaching* 22, 569–617.
Lewis, E.L.: 1991, *The Process of Scientific Knowledge Acquisition among Middle School Students Learning Thermodynamics*, Unpublished doctoral dissertation, University of California, Berkeley, CA.
Lewis, E.L. & Linn, M.C.: 1993, *Conceptual Change in Middle School Science*, Computer as Laboratory Partner, University of California, Berkeley, CA.
Lewis, E.L. & Linn, M.C.: 1994, 'Heat Energy and Temperature Concepts of Adolescents, Adults, and Experts: Implications for Curricular Improvements', *Journal of Research in Science Teaching* 31, 657–677.
Linn, M.C.: 1986, 'Science', in R. Dillon & R.J. Sternberg (eds.), *Cognition and Instruction*, Academic Press, New York, 155–204.

Linn, M.C.: 1989, 'Science Education and the Challenge of Technology', in J. Ellis (ed.), *Informal Technologies and Science Education* (Association for the Education of Teachers in Science yearbook), ERIC Clearinghouse for Science, Mathematics and Environmental Education, Washington, DC, 119–144.

Linn, M.C.: 1992, 'The Computer as Learning Partner: Can Computer Tools Teach Science?', in K. Sheingold, L.G. Roberts & S.M. Malcom (eds.), *This Year in School Science 1991: Technology for Teaching and Learning*, American Association for the Advancement of Science, Washington, DC, 31–69.

Linn, M.C.: 1995, 'Designing Computer Learning Environments for Engineering and Computer Science: The Scaffolded Knowledge Integration Framework', *Journal of Science Education and Technology* 4, 103–126.

Linn, M.C.: 1996, 'From Separation to Partnership in Science Education: Students, Laboratories, and the Curriculum', in R. Tinker & T. Ellermeijer (eds.), *NATO Advanced Research Workshop on Microcomputer Based Labs: Educational Research and Standards*, Springer-Verlag, Brussels, Belgium, 13–46.

Linn, M.C. & Burbules, N.C.: 1993, 'Construction of Knowledge and Group Learning', in K. Tobin (ed.), *The Practice of Constructivism in Science Education*, American Association for the Advancement of Science (AAAS), Washington, DC, 91–119.

Linn, M.C. & Clancy, M.J.: 1992a, 'Can Experts' Explanations Help Students Develop Program Design Skills?', *International Journal of Man-Machine Studies* 36, 511–551.

Linn, M.C. & Clancy, M.J.: 1992b, 'The Case for Case Studies of Programming Problems', *Communications of the ACM* 35(3), 121–132.

Linn, M.C. & Clark, H.C.: 1995, 'How can Assessment Practices Foster Problem Solving?', in D.R. Lavoie (ed.), *Towards a Cognitive-Science Perspective for Scientific Problem Solving* (NARST Monograph Volume 6), National Association for Research in Science Teaching, Manhattan, KS, 142–180.

Linn, M.C., Clement, C., Pulos, S. & Sullivan, T.: 1989, 'Scientific Reasoning During Adolescence: The Influence of Instruction in Science Knowledge and Reasoning Strategies', *Journal of Research in Science Teaching* 26, 171–187.

Linn, M.C., diSessa, A., Pea, R.D. & Songer, N.B.: 1994, 'Can Research on Science Learning and Instruction Inform Standards for Science Education?', *Journal of Science Education and Technology* 3, 7–15.

Linn, M.C. & Muilenburg, L.: 1995, *How Can Less Be More?* (Technical Report), Computer as Learning Partner, University of California at Berkeley, CA.

Linn, M.C. & Songer, N.B.: 1991, 'Teaching Thermodynamics to Middle School Students: What are Appropriate Cognitive Demands?', *Journal of Research in Science Teaching* 28, 885–918.

Linn, M.C. & Songer, N.B.: 1993, 'How do Students Make Sense of Science?', *Merrill-Palmer Quarterly* 39(1), 47–73.

Linn, M.C., Songer, N.B. & Eylon, B.: 1996, 'Shifts and Convergences in Science Learning and Instruction: Alternative Views', in R. Calfee & D. Berliner (eds.), *Handbook of Educational Psychology*, Macmillan, Riverside, NJ, 438–490.

Linn, M.C., Songer, N.B., Lewis, E.L. & Stern, J.: 1993, 'Using Technology to Teach Thermodynamics: Achieving Integrated Understanding', in D.L. Ferguson (ed.), *Advanced Educational Technologies for Mathematics and Science* (Volume 107), Springer-Verlag, Berlin, Germany, 5–60.

Linn, M.C. & Swiney, J.J.: 1981, 'Individual Differences in Formal Thought: Role of Expectations and Aptitudes', *Journal of Educational Psychology* 73, 274–286.

Little, J.W.: 1990, 'The Mentor Phenomenon and the Social Organization of Teaching', in C.B. Cazden (ed.), *Review of Research in Education* (Volume 16), American Educational Research Association, Washington, DC, 297–351.

Longino, H.E.: 1990, *Science as Social Knowledge: Values and Objectivity in Scientific Inquiry*, Princeton University Press, Princeton, NJ.

Madhok, J.J.: 1992, 'Effect of Gender Composition on Group Interaction', in K. Hall, M. Bucholtz & B. Moonwomon (eds.), *Locating Power: Proceedings of the Second Berkeley Women and Language Conference*, Berkeley Women and Language Group, University of California at Berkeley, CA, 371–385.

McDermott, L.C.: 1984, 'Research on Conceptual Understanding in Mechanics', *Physics Today* 37, 24–32.

Millikan, R.A. & Gale, H.G.: 1927, *Elements of Physics*, Ginn, Boston, MA.

National Assessment of Educational Progress: 1978, *Three National Assessments of Science: Changes in Achievement 1969–77*, Education Commission of the States, Denver, CO.

National Assessment of Educational Progress: 1988, *The Science Report Card: Elements of Risk and Recovery: Trends and Achievement Based on the 1986 National Assessment*, Educational Testing Service, Princeton, NJ.

National Committee on Science Education Standards and Assessment: 1993, *National Science Education Standards: July '93 Progress Report*, National Research Council, Washington, DC.

Nersessian, N.J.: 1991, 'Conceptual Change in Science and in Science Education', in M.R. Matthews (ed.), *History, Philosophy, and Science Teaching*, Teachers College Press, New York, 133–148.

Palincsar, A.S. & Brown, A.C.: 1984, 'Reciprocal Teaching of Comprehension-Fostering and Comprehension-Monitoring Activities', *Cognition and Instruction* 1, 117–175.

Papert, S.A.: 1980, *Mindstorms: Children, Computers, and Powerful Ideas*, Basic Books, New York.

Pea, R.D.: 1993, 'The Collaborative Visualization Project', *Communications of the ACM* 36(5), 60–63.

Physical Science Study Committee: 1957, *First Annual Report of the Physical Science Study Committee*, Massachusetts Institute of Technology, Cambridge, MA.

Piaget, J.: 1970, *Science of Education and the Psychology of the Child*, Orion Press, New York.

Ramsey, G.A. & Howe, R.W.: 1969, 'An Analysis of Research on Instructional Procedures in Secondary School Science, Part II – Instructional Procedures', *The Science Teacher* 36, 62–67.

Rossiter, M.W.: 1982, *Women Scientists in America: Struggles and Strategies to 1940*, Johns Hopkins University Press, Baltimore, MD.

Sadker, M. & Sadker, D.: 1994, *Failing at Fairness: How America's Schools Cheat Girls*, Maxwell Macmillan International, New York.

Sandler, B. & Hall, R.: 1982, *The Classroom Climate: A Chilly One for Women?*, Association of American Colleagues, Project on the Status and Education of Women, Washington, DC.

Scardamalia, M. & Bereiter, C.: 1985, 'Fostering the Development of Self-Regulation in Children's Knowledge Processing', in S.F. Chipman, J.W. Segal & R. Glaser (eds.),*Thinking and Learning Skills: Research and Open Questions* (Volume 2), Lawrence Erlbaum, Hillsdale, NJ, 562–577.

Scardamalia, M. & Bereiter, C.: 1992, 'CSILE: A Knowledge-Building Environment for Today's Schools', Paper presented at the meeting of the Bay Area (Bay CHI) Computer Human Interaction group, Palo Alto, CA.

Schoenfeld, A.H.: 1989, 'Explorations of Students' Mathematical Beliefs and Behavior', *Journal for Research in Mathematics Education* 20, 338–355.

Shulman, L.S. & Tamir, P.: 1973, 'Research on Teaching in the Natural Sciences', in R.M.W. Travers (ed.), *Second Handbook of Research on Teaching*, Rand McNally, Chicago, IL, 1098–1148.

Shymansky, J.A., Kyle, W.C., Jr. & Alport, J.M.: 1982, 'How Effective Were the Hands-On Science Programs of Yesterday?', *Science and Children* 20(3), 14–15.

Smith, J.P., diSessa, A.A. & Roschelle, J.: 1994, 'Misconceptions Reconceived: A Constructivist Analysis of Knowledge in Transition', *Journal of the Learning Sciences*, 3(2), 115–163.

Songer, N.B.: 1989, *Promoting Integration of Instructed and Natural World Knowledge in Thermodynamics*, Unpublished doctoral dissertation, University of California, Berkeley, CA.

Songer, N.B.: 1993, 'Learning Science With a Child-Focused Resource: A Case Study of Kids as Global Scientists', in *Proceedings of the Fifteenth Annual Meeting of the Cognitive Science Society*, Lawrence Erlbaum, Hillsdale, NJ, 935–940.

Songer, N.B. & Linn, M.C.: 1991, 'How Do Students' Views of Science Influence Knowledge Integration?', *Journal of Research in Science Teaching* 28, 761–784.

Thornton, R.K. & Sokoloff, D.S.: 1990, 'Learning Motion Concepts Using Real-Time Microcomputer-Based Laboratory Tools', *American Journal of Physics* 58, 858–867.

Tinker, R.F.: 1987, 'In Our View: Real Science Education', *Hands On!* 10(1), 1.

von Glaserfeld, E.: 1990, 'An Exposition of Constructivism: Why Some Like it Radical', in R. Davis, C. Maher & N. Noddings (eds.), *Constructivist Views on the Teaching and Learning of Mathematics*, National Council of Teachers of Mathematics, Reston, VA, 19–29.

Vygotsky, L.S.: 1962, *Thought and Language*, MIT Press, Cambridge, MA.

Vygotsky, L.S.: 1978, *Mind in Society: The Development of Higher Psychological Processes*, Harvard University Press, Cambridge, MA.

Walker, D.F. & Schaffarzik, J.: 1974, 'Comparing Curricula', *Review of Educational Research* 44, 83–111.

White, B.Y.: 1993, 'ThinkerTools: Causal Models, Conceptual Change, and Science Education', *Cognition and Instruction* 10, 1–100.

White, B.Y. & Frederiksen, J.R.: 1990, 'Causal Model Progressions as a Foundation for Intelligent Learning Environments', *Artificial Intelligence* 24, 99–157.

Whitehead, A.N.: 1929, *The Aims of Education*, Macmillan, New York.

Wiser, M. & Carey, S.: 1983, 'When Heat and Temperature Were One', in D. Gentner & A.L. Stevens (eds.), *Mental Models*, Lawrence Erlbaum, Hillsdale, NJ, 267–298.

# 3.2 Computer Microworlds and Scientific Inquiry: An Alternative Approach to Science Education

BARBARA Y. WHITE
*University of California, Berkeley, USA*

The ability to create simulations and explore their behaviour is changing the way in which scientists and engineers do their research. Thirty years ago, computational models of physical systems were so novel that they were written about in magazines such as *Time* and *Newsweek*. Nowadays, they are commonplace. Technology is revolutionising the practices of science and engineering. It is enabling us to design artefacts in new ways and to create new types of theories about how the world works.

In this chapter, I argue for the importance and feasibility of undertaking a corresponding revolution in science education. We need to rethink the goals of science education, as well as our methods for achieving those goals. For instance, we need not only to develop new conceptions about expertise in understanding and doing science, but also to create new instructional approaches and conceptual tools that enable students to acquire this expertise.

The chapter falls into four major sections. In the first brief section, I argue that helping students to learn about the process of scientific modelling should be the main goal of science education. The second section considers three approaches to physics education (top-down, bottom-up and middle-out), and advocates and elaborates the middle-out approach. The third and fourth sections illustrate the use of Intermediate Causal Models and the middle-out approach by presenting an overview of research on two projects, namely, the QUEST Project and the ThinkerTools Project, respectively.

## THE IMPORTANCE OF SCIENTIFIC MODELS

Towards this reform agenda, I advocate in this chapter that the primary goal of science education should be to help students learn about the nature of scientific models, the process of constructing models, and the utility of models in predicting and explaining real-world phenomena (see also Collins & Ferguson 1993; Gilbert 1991; Glynn & Duit 1995; Halloun & Hestenes 1987; Mandinach 1989; Mellar,

---

Chapter Consultant: Instead of using chapter consultants, chapter authors in the Educational Technology section provided editorial comments on each other's chapters.

Bliss, Boohan, Ogborn & Tompsett 1994; Perkins, Schwartz, West & Wiske 1995; Stewart, Hafner, Johnson & Finkel 1992; Tinker 1993). Other chapters in this section of the *Handbook* by Linn, by Schecker and by Spitulnik and colleagues are relevant to the present chapter.

Further, I illustrate how models that enable both computers and humans to simulate the behaviour of a physical system have a key role to play in the development of such expertise. To elaborate, there are different types of computational models that simulate the behaviour of physical systems such as analytic and discrete-state models. There are also different types of mental models which humans could acquire for envisioning the behaviour of those systems (Gentner & Stevens 1983). There is an interesting intersection of the two containing models which not only simulate the behaviour of physical systems, but also serve as possible mental models for how people could conceptualise and reason about those systems. This set of 'mental simulation' models is turning out to be larger than one might have thought. This is partly because the field of artificial intelligence has introduced new types of models, such as qualitative models, that are good candidates for mental models (Bobrow 1985; Weld & deKleer 1990; Widman, Loparo & Nielsen 1989). It is also because some of the standard simulation techniques, such as iterating difference equations across time, produce models that are both computationally and psychologically tractable. That is, they are useful not only for simulation, but also for helping students to envision and understand a system's behaviour.

There is a particular type of mental simulation model, which I call Intermediate Causal Models (ICMs), that can be characterised as applying the laws of physics in causal form to predict what will happen as events occur. They also employ various visual representations to depict this sequence of behaviours at an intermediate level of abstraction (see Figures 3 and 8 later for examples of such representations). This type of model can be embedded in a computer simulation which can explain its reasoning verbally (using a speech output device) and can depict its behaviour visually and dynamically. They also can be internalised by students in the form of mental or conceptual models. Acquisition of such a mental model enables a student to step through time and/or events and to use the laws and representations to predict and explain what will occur.

ICMs, I argue, provide the best vehicle for reforming our conception of what it means to understand and teach science. They can change our view of scientific understanding by providing formal models of reasoning processes (such as causal, discrete-state reasoning) and representational forms (such as diagrams) that have always played a crucial role in scientific theorising, but that have not previously been given the respectability and prominence which they deserve (Larkin 1983; Nersessian 1992). ICMs also can change our view of science education because creating and experimenting with computer simulations of this type make possible an alternative approach to science education, which will make difficult subjects such as physics interesting and accessible to a wide range of students.

## ALTERNATIVE APPROACHES TO PHYSICS EDUCATION

To support this claim, I start by contrasting some alternative approaches to physics education.

*The Top-Down Approach*

The first is the traditional approach typically used at the high school and university levels. It can be characterised as a 'top-down' approach in which the laws of physics are presented as algebraic constraint equations, and in which the goal of physics education is for students to learn to solve well-defined quantitative problems by manipulating those equations.

There are many difficulties with this top-down approach. One is that the emphasis on constraint-based reasoning presents a very narrow view of what it means to understand and do physics and, as a consequence, many students lose interest in the subject. Another problem, revealed by research in both cognitive science and physics education, is that this approach does not work; while students might pass traditional physics courses, they frequently retain many of their misconceptions (McDermott 1984). In other words, there is an inconsistency between their formal knowledge and their experiential knowledge, probably because there is too big a gap between their real-world experiences and these highly formal and abstract algebraic ways of reasoning about physical systems.

*The Bottom-Up Approach*

These criticisms of the top-down approach are widely acknowledged (e.g., Clement 1981). Another approach that is gaining in popularity, however, is more difficult to criticise. It is being used at the primary and middle school levels, and could be characterised as a 'bottom-up' approach in which the laws of physics are gradually induced from experiences with real-world phenomena. There is theoretical support for this hands-on science movement from both situated-cognition theorists and many developmental psychologists who argue that children can only deal with concrete representations and simple forms of reasoning. In the USA, this view has been translated into curricular guidelines which assume that children are only capable of limited forms of inquiry (e.g., California Department of Education 1990). For example, in the early grades, it is recommended that students do simple tasks such as observing and classifying objects and then, at the middle-school level, they can progress to doing simple experiments in which they systematically manipulate a given variable.

There are many difficulties with this bottom-up approach. One is that it too presents a narrow view of scientific expertise and inquiry. The focus has simply changed from solving well-defined problems to doing well-defined experiments.

Unfortunately, scientific inquiry and modelling are not that simple. Learning science often involves a major change in a student's conception of the causality of the world. For example, does motion need a cause, or is it only a change in motion that needs a causal explanation?

Such a narrow view of scientific inquiry is not inherent in the bottom-up approach and could be corrected. However, there is a second, more serious objection to this approach that cannot be easily remedied: it advocates starting with the real world. Unfortunately, the real world is a confusing and complicated place. Because things happen quickly and many things happen simultaneously, it is difficult to make accurate observations. Because it also is hard to control and measure, it is difficult for students to do controlled experiments and obtain accurate data (Gunstone 1990). Furthermore, induction from real-world experiments is a slow and inefficient process that runs a risk of producing knowledge that is too situated, and this process reinforces a naive empiricist view of scientific inquiry. Thus, I contend that the real world is not a good starting point for deriving the powerful idealised abstractions that characterise scientific expertise.

*A Middle-Out Approach*

If I am arguing that you should not start with algebraic abstractions and acausal reasoning because it is too divorced from real-world experience, and if I am also arguing that you should not start with the real world and struggle to 'induce' the laws of physics, where should you start? I advocate starting at a level in between these two and pursuing what I characterise as a 'middle-out' approach to science education in which students work with and create causal models at an intermediate level of abstraction (White 1993b). These models are at an intermediate level of abstraction with regard to both the reasoning forms and the representations which they employ. The reasoning is causal where one typically steps through events using laws to predict what will happen next, and the representations are visual employing graphical symbols to portray this sequence of events in a diagrammatic form.

There are many advantages to starting with ICMs instead of algebraic abstractions or real-world experiments. The first is that even young children can make sense of such models because it is a small step from this form of representation and reasoning to how one naturally reasons about real-world phenomena. Furthermore, it is an efficient starting point because it is also a relatively small step from this form of representation and reasoning to algebraic abstractions and constraint-based reasoning which are useful for solving certain types of problems. Working with ICMs also potentially allows students to create a very coherent form of expertise in that these models provide a bridge between real-world phenomena and mathematical formalisms. Finally, ICMs are a useful knowledge form in their own right: They are particularly good for predicting and explaining real-world phenomena.

In this middle-out approach that I am advocating, the emphasis is on experimenting with computer models with the goal of creating conceptual models, and there is far less emphasis on doing real-world experiments or on solving quantitative problems (cf Richards, Barowy & Levin 1992; Sherin, diSessa & Hammer 1993; Snir, Smith & Grosslight 1993).

## ISSUES FOR THE DEVELOPMENT OF A MIDDLE-OUT APPROACH

There are two major issues that I want to address in this chapter. The first is to consider how ICMs can redefine our view of scientific expertise, both in terms of the conceptual models that I argue are the core of expertise, and in terms of the meta-knowledge that is needed to create and use such models effectively. The second is to consider how computer microworlds that embody and allow one to experiment with ICMs can be used to implement this alternative approach to science education in a way that makes science more interesting and accessible to students. I am going to address these two issues by presenting an overview of research on two projects, namely, the QUEST project and the ThinkerTools project.

In the QUEST Project, the domain of instruction is simple electrical circuits and my primary collaborator is John Frederiksen. (QUEST stands for Qualitative Understanding of Electrical Systems and their Troubleshooting and is the name given to a series of computer-based learning environments that we created.) In this work, we designed a range of different types of causal models, and then conducted a sequence of instructional experiments to see what impact they had on students' learning, understanding and problem solving. This research was done with students at the secondary school and undergraduate levels (Frederiksen & White 1992; White & Frederiksen 1990; White, Frederiksen & Spoehr 1993).

In the ThinkerTools Project, my primary collaborator was a physicist, Paul Horwitz. In this project, we worked with primary and middle school students and teachers to help them to learn how forces affect the motions of objects. We created a middle-out instructional approach that enables children to experiment with and internalise ICMs and, by so doing, to deal with relatively abstract and complex representations and reasoning forms. In this research, the focus was not only on helping children learn about physics, but also on helping them to learn about scientific inquiry in general. For instance, we wanted them to learn about the different forms that scientific laws and representations can take, about the process of constructing models that embody these laws and representations, and about the utility of such models in predicting and explaining real-world phenomena (Horwitz 1989; White 1993a; White & Horwitz 1988).

In addressing these two issues, I focus on ICMs and the nature of scientific expertise in my discussion of the QUEST Project, and on ICMs and the middle-out instructional approach in my discussion of the ThinkerTools Project, even though both projects address both issues.

## THE QUEST PROJECT

In the early stages of our research on the QUEST Project, we determined that the key ideas needed to understand electrical circuits are 'voltage' and 'voltage distribution'. These ideas are needed for reasoning about even very simple circuits, and they cause a great deal of difficulty for students (Cohen, Eylon & Ganiel 1983). Thus, we decided to focus on creating models that would help students to acquire these important but difficult ideas.

*Qualitative Models of Circuit Behaviour*

Our first hypothesis was that, if quantitative, constraint-based reasoning is a bad place to start instructionally, then qualitative, causal reasoning must be a good place to start. This hypothesis turned out to be too simplistic as the research presented here illustrates. In this research, we initially set about to create qualitative, causal models that reason about voltage distributions within a circuit at the macroscopic level (i.e., from the perspective of lights going on and off in a circuit as switches are opened and closed). This was the perspective taken by experts whom we studied as they reasoned about and engaged in troubleshooting simple electrical circuits (White & Frederiksen 1990).

The Qualitative Model (QM) that we created incorporates a cause-effect method for analysing and simulating circuit behaviour. It pays attention to events that might change the distribution of voltages within a circuit, which in turn might cause changes in the states of devices within the circuit. When something occurs that changes a device's conductivity, such as closing a switch, this event triggers a 'voltage redistribution process'. This process uses simple qualitative laws such as the following to determine the new distribution of voltages. 'If you have a circuit with an open in it, the only voltage drop will be across the open.' 'If you have a circuit that is a complete conductive loop, then there will be voltages across all resistive devices within that circuit.' Thus, when this process is triggered, the distribution of voltages can change which, in turn, can change the states of devices. For example, in the circuit shown in Figure 1, the light goes from being off to being on, because after the switch is closed, there is a voltage across the light, and the device model for a light reasons that 'a light is on when there is a voltage across it'.

In our instructional studies in which students interacted with these computational models in order to develop conceptual models, we found that students had no difficulty internalising such qualitative models. They enabled students to predict circuit behaviour and also supported the kind of troubleshooting that the QUEST intelligent learning environment teaches (White & Frederiksen 1990). In addition, studies conducted by our colleague, Kathy Spoehr, found that these models are particularly helpful for low-ability students (Spoehr 1989).

However, we discovered some problems. One is that having students simply internalise such models creates a shallow form of expertise. They have a very limited

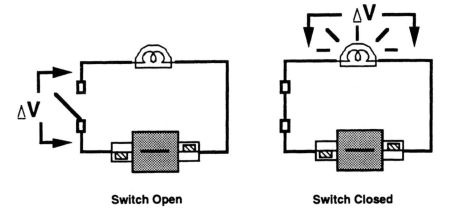

Figure 1: An illustration of the Qualitative Model as it reasons about the macroscopic behaviour of a simple electrical circuit containing a battery, a switch and a light bulb

conception of what voltage is and of how voltage determines current flow. In addition, when we asked students to give deeper explanations for what is going on within a circuit, we frequently got a very interesting form of reasoning. Many students think of electrical current as an intelligent agent that is making decisions locally as it moves through a circuit; it reasons about the relative resistance of alternative paths within a circuit and makes decisions about which way to flow based on the resistance. This type of reasoning, while very appealing, has serious limitations and makes predictions about circuit behaviour that are actually inconsistent with those of the macroscopic circuit models that the students had internalised.

*Linked ICMs: Unpacking the Origins of System Behaviour*

To remedy these problems, we hypothesised that we needed to create a sequence of mechanistic models that would unpack (or provide a theory for) the origins of macroscopic circuit behaviour. As a result, we set about to create flow models and particle models which would provide students with a deeper understanding of what goes on inside circuits. We believed that, if students worked with and internalised such a reductionistic sequence of models, they could overcome their lack of explanatory depth. For instance, they could understand the origins of the concept of voltage, the process of voltage distribution, and the macroscopic circuit laws. Working with the sequence also is likely to enable them to develop conceptual models that are consistent across both the macroscopic and the microscopic perspectives on circuit behaviour. We further conjectured that having this depth to their understanding might even improve their problem-solving performance on macroscopic circuit-behaviour problems (such as those shown later in Figure 5).

The first model that we created was the Flow Model (FM). In this model, to make it possible for students to think about what is happening locally within a circuit, we divided conductive materials into small slices. As these slices are uniform, one can talk about charge on a slice instead of charge density (which is a more complicated concept). Figure 2 illustrates two slices of a resistor, with the one on the right starting out with more negative charge than the one on the left. When you connect these two resistor slices together and step through time, you can see that charge flows from one to the other until the charges on the two slices are equal. You also can see that the rate of flow is proportional to the difference in charge. (This simple quantitative model uses the difference equation, flow($t$) = $k$ x $\Delta$charge($t$), where flow is the amount of charge transfer between two connected slices at time $t$, and $\Delta$charge is the difference in charge between the two slices at time $t$.)

The local-flow process shown in Figure 2 thus operates to equalise the charge across conductive materials. One can think about a battery as doing exactly the opposite. It tries to maintain a constant difference in charge across its two terminals. For example, if some negative charge is pulled off its negative terminal, it reacts to restore this difference in charge. So, if you put together a circuit, like the one shown in Figure 3 in which a five-slice resistor is connected directly across a battery, you can see how these two processes operate in opposition to one another. A gradient of charge builds up across the resistor and, as you step through time, current flows gradually equalise until the circuit reaches a steady state. Working with this model, you can step through time and query any slice to see (1) how the charge on that slice is going to change on the next time step, and (2) why it is going to change in that way. By creating and experimenting with more complicated circuits, students can see the mechanistic origins of the macroscopic circuit laws in both qualitative and quantitative form. That is, Ohm's Law and Kirchoff's Laws can be seen to be emergent properties of circuits operating according to these simple mechanisms.

We created an instructional sequence so that students could experiment with this flow model. In the early levels of the sequence, they work on understanding the local-flow and battery processes. They learn how these two processes operate in opposition to one another to produce charge gradients and constant current

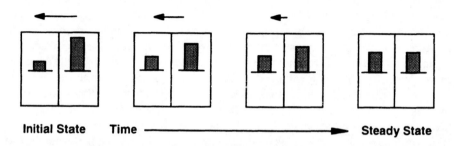

Figure 2: Using the Flow Model to illustrate the process of charge redistribution across time. (At steady state, charge densities are equal and no current flows.)

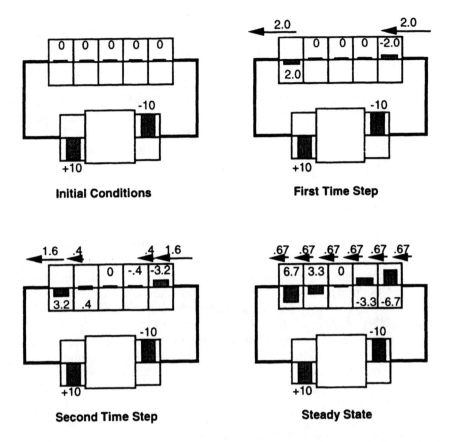

**Figure 3:** The Flow Model applied to a simple series circuit containing a resistor and a battery. (As the circuit approaches a steady state – going through more time steps than are shown – current flows equalise and charge densities form a stable gradient.)

flows within a circuit. Then, in the later levels, they build more complicated circuits and try to predict, before running the simulation, what will happen. They then experiment to see what actually happens, their task being to formulate laws which will enable them to determine more accurately what will happen the next time they see such a circuit.

We had two groups of students work with the flow model, with one group working exclusively with the flow model alone (the FM students) and another group, in the early levels, getting an unpacking of the local-flow process down to the behaviour of charged particles (the PM students). This was done by introducing them to a Particle Model (PM). In this particle model, the primitive process is the Coulomb interaction between particles (like charges repel, unlike charges attract). Thus, if one puts two slices next to one another as shown in Figure 4, and if there is a difference in their initial net charges, then there will be electrical forces exerted on the charged particles within the two adjacent slices. This can be thought of as due to the negative charges repelling one another and the positive charges attracting the negative charges. These forces accelerate the mobile charges (i.e., the electrons), causing them to be redistributed from the more negatively charged slice to the more positively charged slice until both slices have the same net charge. The model can be elaborated to explain resistance in terms of obstacles that affect the motion of charged particles. For the PM students who worked with this particle model, the flow model was just a more abstract representation of something that they had understood down at the level of charged particles. However, for the FM students who worked only with the flow model, the local-flow process was a mechanism in its own right that was not reduced further.

*Instructional Experiments with Linked ICMs*

What happened when we tried these two instructional sequences with undergraduate students who had little previous knowledge of electricity? We found that there were significant gains between pretest and posttest for both groups of students and that, on the posttest, both groups were performing at a high level on a wide range of circuit problems (White, Frederiksen & Spoehr 1993). These results support the viability of using ICMs to help students learn science.

When we analysed these results further, we found that there was a systematic difference in favour of students who got the unpacking of mechanism all the way down to the particle model. For example, the PM students were more likely to talk in terms of charge gradients causing current flow. The PM group also did better on problems (like the one shown in Figure 5, question A) in which one is asked to reason about the relative magnitudes of voltages and currents within a circuit (the PM group averaged 81 percent correct and the FM group averaged 66 percent). We found that this difference in performance was due primarily to the more difficult items in which the concept of a voltage divider is required. Thus, students who got the particle model unpacking were more successful in acquiring the difficult and key concepts of voltage and voltage divider. In addition, when we went on to look at performance on quantitative circuit-analysis problems (like the one shown in Figure 5, question B), we found again that the PM group did better (with a mean of 71 percent correct for the PM group compared with 51 percent for the FM group).

**Figure 4:** Using the Particle Model to illustrate how negative charges repel one another and become redistributed

**A. When this circuit reaches steady state, what can be said about the voltage across Resistor 1 vs. 2?**

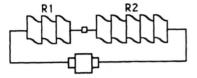

(a) It is larger across Resistor 1.

(b) It is larger across Resistor 2.

(c) It is the same across Resistors 1 and 2.

**B. The following circuit has two resistors and a 12 volt battery.**

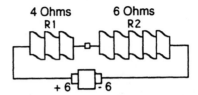

(a) What will be the voltage across Resistor 1 when the circuit reaches steady state? Resistor 2?

(b) What is the current through Resistor 1? Through Resistor 2?

Figure 5: Sample problems from the macroscopic circuit-behaviour tests: Question A is an example of a qualitative relative-magnitudes problem; and Question B is an example of a quantitative circuit-analysis problem.

*Educational Implications of Linked ICMs*

Why did students who received particle model explanations for the basic concepts and processes of the flow model do better on difficult problems that require an understanding of voltage and its distribution? Our hypothesis is that the particle model, in which individual objects act as causal agents, is more understandable and meaningful to students than the flow model, in which aggregate charge flows in response to pressure differences. The implication is that it is not enough to do what we did in creating the flow model, mainly to quantise the world (i.e. divide conductive materials into small slices and time into discrete steps) so that students can parse circuit behaviour into a sequence of local cause-effect events. Rather, there is a need to reduce phenomena further down to individual objects that are active agents. Because humans are individuals who act as causal agents in their world, models that have these characteristics can be more readily understandable to them. We think this finding has wide generality because models such as our particle and flow models can be used to explain the behaviours of many physical phenomena, not just those of electrical circuits. For example, they can be used to help students to understand heat flow, gas diffusion and how forces affect motion.

Our conclusion is that this sequence of particle and flow models grounds abstract laws via a mechanistic unpacking of the physical phenomena. In this way, circuit laws which reason about circuit behaviour at the macroscopic level can be understood to have mechanistic origins at the microscopic level. This reductionistic unpacking down to the particle level enabled students to develop consistent views of system behaviour, and to improve their performance in solving problems at the macroscopic level. The latter is a surprising and intriguing result. For example, if you were creating an AI expert system to solve problems like those in

Figure 5, it would only need to know what is necessary to solve these types of problems, namely, the macroscopic circuit laws. However, this study suggests that humans need to ground their understanding down to a more primitive level, such as the particle level, in order to make sense of, believe in, and use more abstract acausal laws. The unpacking is thus beneficial even though it requires students to learn more than one model and to understand the relationships among multiple models.

Working with ICMs also introduces students to some important epistemological ideas about the nature of scientific models and explanations. One is that phenomena can be explained from different perspectives, such as the microscopic, aggregate, and macroscopic perspectives. Another is that such alternative perspectives can cohere in that they can be derivationally linked. For instance, the emergent behaviours of our microscopic particle model become the primitive processes of our aggregate flow model; similarly, the emergent properties of our aggregate flow model become the primitive processes of our macroscopic qualitative model (Frederiksen, White & Gutwill 1996). This type of epistemological knowledge dealing with alternative perspectives and their relationships is not addressed in most science curricula.

Our research on the QUEST Project explored the space of conceptual models that I am calling ICMs, and conducted instructional experiments to see where to start in this space and how to progress to build coherent, in-depth expertise. We do not encounter much disagreement about starting with simulations like the particle model for helping students understand circuit behaviour, because it is difficult for students to do real-world experiments with atomic particles in which they easily can see the behaviour of those particles. However, for the next domain that I discuss (mainly force and motion), we encounter more disagreement because students can more easily perform real-world experiments. Also, because we are working with younger students, we encounter debates about the level of abstraction and the complexity of reasoning embodied in ICMs.

## THE THINKERTOOLS PROJECT

In my discussion of the ThinkerTools Project, I describe not only the form of ICMs which were used, but also the middle-out instructional approach which we developed and tried out in primary and middle school classrooms (Horwitz 1989; White 1993a; White & Horwitz 1988). Our goal in this project was to show that, with the right conceptual tools (ICMs) and the right instructional approach (middle-out), even young students can develop sophisticated conceptual models for reasoning about force and motion. Further, one can enable young students to learn about the properties of scientific laws and representations, the inquiry processes needed for creating them, and how conceptual models that embody these laws and representations can be used to predict and explain real-world phenomena. In general, we wanted to show how such a middle-out approach can make scientific abstraction, inquiry and modelling more interesting and accessible to students.

## The ThinkerTools Microworlds

In the ThinkerTools curriculum, the students' primary conceptual tools are a sequence of interactive microworlds that embody increasingly sophisticated models for how forces affect motion. These microworlds are all particle models which the QUEST research suggests should be helpful in enabling students to make sense of higher-order abstractions (like vector addition and **F = ma**). In working with the microworlds, children typically are engaged in game-like activities and experiments in which they try to control the motion of an object by applying forces to it. (This object is introduced as a generic object called a 'dot', which is the pictorial equivalent of a variable; students can map it onto different objects at different times, such as a spaceship or a billiard ball.) The activity shown in Figure 6, for example, is set in the context of a two-dimensional microworld in which there is no friction or gravity. In this game, students apply impulses to the dot so that it navigates the track and stops on the target. (Impulses are forces that act for a limited time, like a hit or a kick.)

The ThinkerTools microworlds all incorporate multiple levels of abstraction. For example, there is the dot itself which moves dynamically around on the screen. A more abstract representation of the dot's motion is provided by what we call

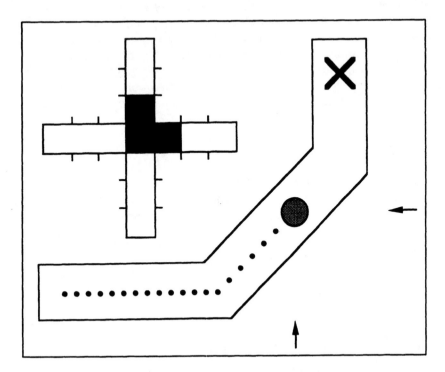

**Figure 6:** This screen image illustrating the representations of motion employed within the ICM microworlds (i.e., the moving dot and its wake, arrows and datacross).

wakes' or 'dotprints'. These are little dots that the big dot puts down at fixed time intervals. They show, by their position and spacing, where the dot has been and how fast it was going. At a higher level of abstraction, the students are encouraged to think in terms of the components of the dot's motion (i.e., its speed in the horizontal dimension and its speed in the vertical dimension). Components are useful for reasoning about complex situations such as when analysing trajectories. The components of the dot's velocity are illustrated dynamically by arrows which move so that they are constantly pointed at the dot. They are represented in more abstract form by a vector representation that we call a 'datacross' (shown at the top left of Figure 6) which indicates the amount of horizontal and the amount of vertical velocity.

In the early stages of working with these microworlds, children typically focus almost exclusively on the behaviour of the dot and ignore the other representations completely. The teacher then encourages them to understand and appreciate the utility of the more abstract representations by giving them activities in which, for example, the dot goes off the screen, so that the only way to understand and reason about its behaviour is to focus on the datacross vector representation.

The most recent version of the ThinkerTools software (White & Frederiksen 1995) also allows students to create their own activities easily. Students can use drawing tools to place dots, barriers and targets on the screen, and they can change a dot's properties such as its mass, position, velocity and elasticity. In addition, students can vary the magnitudes of forces like friction and gravity, and can apply impulses to any dot to make it accelerate. The software also makes available tools for measuring distances, times and velocities as well as providing the datacross and wake representations. These drawing, measurement and representation tools enable students to experiment with and analyse a microworld's behaviour in order to develop and refine their conceptual models of force and motion phenomena.

*The Middle-Out Instructional Approach*

In creating the ThinkerTools curriculum, we developed a four-phase inquiry cycle that is repeated with each new microworld in our sequence of increasingly complex microworlds. In the first phase, *the Motivation Phase*, children are asked to do thought experiments in which they imagine what would happen as forces are applied to an object. For instance, in conjunction with the first microworld (which is a one-dimensional world with no friction or gravity), they are asked to predict what will happen in the following situation:

> There is an object resting on a table. The table is very smooth, so there is no friction. A blast of air is applied to the object. Then, as it is moving along, a blast of air, the same size as the first, is applied in the opposite direction. What will the second blast of air do to the motion of the object?

The teacher simply listens to the children's reasoning and answers without commenting on their correctness or incorrectness. For instance, in response to the preceding question, some say that the second blast of air will cause the object to turn around and go in the opposite direction. Others say that the second blast of air cancels the first and the object will stop. Still others say that it simply slows the object down, but that the object keeps moving in the original direction. The children thus discover that different people have different hypotheses about what will happen in such situations. Such disagreements motivate them to find out who has the correct beliefs about how forces affect motion. In addition, because the questions are about the behaviour of real-world objects, they create the potential for linking what happens in the computer microworld with what happens in real-world situations.

The children then go on to the second phase of the inquiry cycle, *the Investigation Phase*, in which they work together in pairs with the computer games and inquiry activities. The games are designed to help them work with the different representations and to determine the laws that are governing the behaviour of the microworld (White 1984). The children are asked to predict what they think is going to happen within this world and then, as a result of playing the games and doing the experiments, they see what actually happens. Based on these experiences, they then create and write down the laws that they think are governing this microworld.

To facilitate this process, the children then proceed to the third phase of the inquiry cycle, *the Formalisation Phase*, in which they work together in larger groups to evaluate candidate laws for describing the behaviour of the microworld. These could include rules such as 'Whenever you apply an impulse to the dot, it changes speed' or 'Whenever you give the dot an impulse to the left, it slows down'. The laws vary in terms of their accuracy, precision and range of applicability. The children's task is to decide which of the laws are right and which of them are wrong. For the ones which they believe to be wrong, they have to give proofs to the rest of the class as to why they are wrong. For the ones which they believe to be correct, they have to decide which of them is the most useful and defend their choice. Through such activities, the children come to realise that a useful scientific law is something that makes precise predictions across a wide range of circumstances.

Then, in the fourth phase of the inquiry cycle, *the Transfer Phase*, the teacher finally introduces the real world. In this phase, the children do activities in which they try to determine whether their laws and representations are accurate and useful for explaining what happens in the everyday world. They inevitably discover limitations in their model. For example, their model might predict that an object will maintain a constant velocity after it has been hit but, in the real-world activity, they observe that the object slows down. To resolve such discrepancies, they carry out additional modelling activities such as going back to the microworld and putting in friction. They then can see that a microworld with friction behaves according to some of their original predictions and is a better model for many real-world situations. In this way, the children's conceptions of force and motion phenomena evolve.

This four-phase inquiry cycle is repeated with each new microworld. Each time it is repeated, more and more of the inquiry process is turned over to the students.

For example, in the early microworlds, they are given laws that we have created to evaluate but, in the later microworlds, they have to create and evaluate their own laws.

The children interact with a sequence of ICM microworlds that gradually increases in complexity. They begin with a simple one-dimensional world in which there is no friction or gravity. From working with this microworld, the children discover that they can use scalar arithmetic to model what happens. They then progress to a two-dimensional world in which they work towards creating a conceptual model that incorporates simple vector addition. Next, they go to a one-dimensional world in which gravity is acting. By doing a limit analysis, they learn to model continuous forces, like gravity, as a series of small impulses closely spaced in time. Finally, they progress to a two-dimensional world in which gravity is acting. Working with this microworld, the children use their conceptual model to analyse and explain the trajectories of objects. This involves stepping through time and applying their laws and representations to predict and encode what will occur.

*Instructional Trials of the Middle-Out Approach*

What happened when we tried this inquiry cycle, centred around these four microworlds, in public-school classes taught by a teacher who did not know any physics and never before had used computers in her classroom?

With regard to the development of the children's scientific inquiry skills (such as proving laws wrong, evaluating alternative laws and formulating their own), a brief example follows. One of the laws that the children evaluated after they had worked with the first one-dimensional microworld was: 'If the dot is stopped and you give an impulse to the right, and then you give an impulse to the left, the dot will stop again'. This law is correct, but of limited applicability. The children made different types of comments as to why this law is correct, ranging from an empirical proof ('because that's what happens'), to a more formal proof ('because an impulse to the right and an impulse to the left cancel out'), to a more mechanistic proof ('because there is pressure from both sides'). There were also interesting comments about the utility of the rule (e.g., 'it is bad, because you need more rules, it doesn't say enough' – in other words, it is not a general purpose rule). You can see from the types of comments that the children made that they are learning about the nature of scientific laws and about proving laws wrong. However, it takes a skilled teacher to make the most of such conversations, and creating activity structures with supporting materials to foster the development of such inquiry expertise is the focus of our recent work (White & Frederiksen 1995).

At the end of the ThinkerTools curriculum, the children did well on questions that assessed their understanding of the alternative representations of motion and their knowledge of the laws of the microworlds. For example, they were also asked to translate between the datacross and wake representations (e.g., 'draw the datacross that goes with this wake'), to choose which wake corresponds to statements such as 'the dot is accelerating', and to predict the wake that the dot would make if it were given certain impulses in microworlds with and without friction or with and without gravity. For a test consisting of such questions, the children averaged 77 percent correct with a standard deviation of 19 percent. The evidence thus indicates that most of them understood the laws and representations used in the microworld.

Could the children take their conceptual model and apply it to real-world situations? To assess this, we constructed a test consisting of problems, such as the one shown in Figure 7, that involve predicting what will happen in simple, real-world situations. For these real-world transfer questions, children who had taken the ThinkerTools curriculum did considerably better than a control group of their peers (65 percent compared with 44 percent). They also did significantly better than high-school physics students in the same school system who were taught force and motion using a traditional physics text (and who averaged only 58 percent correct). We were surprised at this finding because not only were the high-school physics students older, but they were also a much more select group because few students choose to take physics in high school.

We went on to look at how the children reasoned when they attempted to answer such problems by interviewing some of them at the end of the curriculum. For example, consider what occurred when I asked one student about throwing a ball upward: 'Imagine that we throw a ball straight up into the air. Describe what happens to the motion of the ball, and then explain why that happens.' First of all, he described the motion of the ball in a step-by-step manner: 'It will start going up at the speed you threw it up, and then it will gradually slow down, and there will be a second when it is stopped in the air, and then it will start coming down, and it will gradually speed up'. Next, he gave a causal account of those events using

**Suppose that we have two identical rivers with two identical boats trying to cross those rivers. The only difference is that one river has a current flowing and the other does not. Both boats have the same motors and leave at the same time. Which boat gets to the other side first?**

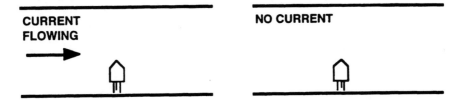

(a) The one crossing the river without a current flowing;

(b) The one crossing the river with a current flowing; or

(c) Both boats get to the other side at the same time.

Figure 7: A sample problem from the transfer test which asks students to apply their conceptual model to real-world situations.

the conceptual model that he had developed: 'Because going up, the gravity keeps pulling, adding another impulse down, and that eventually will stop the ball, and then going down it keeps adding another impulse down which makes it go faster and faster'. So, he was able to do a causal, step-by-step analysis of what would happen, and showed no evidence of any of the common force and motion misconceptions.

We found that many students also could engage in a more precise and abstract form of reasoning. For example, I asked the students to draw and explain what would happen if the following sequence of impulses were applied to an object in a frictionless environment: right, down, left and up. One student drew the diagram shown in Figure 8. Each time that an impulse was applied, she drew a datacross to indicate the object's new velocity. Based on the velocity shown in the datacross, she determined the position of the object's next wake point. In this way, she stepped through time and analysed events to determine the object's velocity and position.

Many students who worked through the ThinkerTools curriculum were able to use these intermediate abstractions and causal laws (in both qualitative and quantitative form) to step through time and to thereby simulate what would happen (as illustrated in Figure 8). They also had acquired some knowledge of scientific inquiry with regard to the form and properties of scientific models, the inquiry

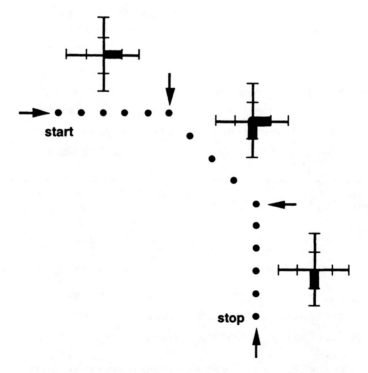

Figure 8: A student's ICM diagram portraying how a sequence of impulses affects an object's motion.

process needed for creating them, and how to apply conceptual models to predict and explain real-world phenomena.

## CONCLUSIONS

The first of two major issues addressed in this chapter concerned the nature of scientific expertise. I argued that the key to developing scientific expertise is to have students work with and create causal models that enable them to envision sequences of events at an intermediate level of abstraction. These models are useful for explanation and understanding, as well as for linking formal physics to real-world phenomena.

Such ICMs have the potential to create a coherent body of knowledge. They can provide consistency between students' intuitive knowledge and their formal knowledge. Also, as the QUEST research illustrates, they can create consistency across models that reason about system behaviour from different perspectives, such as the microscopic and the macroscopic perspective. This kind of cohesiveness is important, because a major goal of scientific theorising is to create models with the kind of coherence embodied in the set of QUEST models (i.e., microscopic [PM], aggregate [FM], macroscopic [QM]). These models are not only consistent with one another, they also are derivationally linked.

To create this kind of expertise, ICMs by themselves are not enough. One needs a new instructional approach, like the middle-out approach described in this chapter, that enables students to believe in and internalise these models. In this approach, there are powerful conceptual tools in the form of microworlds that let students interact with and create simulations that embody ICMs. There also are modelling activities, set in the context of an inquiry cycle, which include experimenting with these simulations and creating explicit, written-down laws and representations to characterise their behaviour. These activities enable students' understandings of the subject matter to evolve. In addition, they enable students to develop knowledge of the processes of scientific inquiry and modelling (White 1993a; White & Frederiksen 1995).

There is a major challenge, however, in disseminating this view of science education. For example, teachers often use the software tools in a manner that fits their existing view of science education. If they are high school teachers, they use the software to create simulated laboratory experiments, but they do not incorporate the idea of a model. If they are primary school teachers, they create computer games which they view as entertaining and motivating for students, but they mainly focus on creating real-world activities during which they believe all of the important learning is going to occur. Thus, in the beginning, they do not accept the idea of using the software as a tool for enabling students to experiment with and create conceptual models. As they work with the ThinkerTools curriculum materials and computer-based tools, they gradually come to appreciate the potential of this inquiry-oriented, model-based approach to science education.

The conclusion is that such new approaches to science education are going to meet with resistance. Arguments like those I have attempted to present in this chapter need to be presented for their value (see also White 1993a, 1993b). In addition, curriculum materials need to be developed that make it easier for teachers to adopt them (e.g., White & Frederiksen 1995). Finally, research has to be done to show that they can be successful with regular students and teachers, as in the research of the ThinkerTools project. Otherwise, such software tools – that are the foundation for new approaches to scientific theorising and inquiry – will get used in trivial ways, if indeed they get used at all, and they will not reach their potential to change and improve dramatically the way in which science is taught.

## REFERENCES

Bobrow, D.: 1985, *Qualitative Reasoning About Physical Systems*, MIT Press, Cambridge, MA.
California Department of Education: 1990, *Science Framework for California Public Schools: Kindergarten Through Grade Twelve*, California Department of Education, Sacramento, CA.
Clement, J.: 1981, 'Solving Problems With Formulas: Some Limitations', *Engineering Education* 72, 158–162.
Cohen, R., Eylon, B. & Ganiel, U.: 1983, 'Potential Difference and Current in Simple Electric Circuits: A Study of Students' Concepts', *American Journal of Physics* 51, 407–412.
Collins, A. & Ferguson, W.: 1993, 'Epistemic Forms and Epistemic Games: Structures and Strategies to Guide Inquiry', *Educational Psychologist* 28, 25–42.
Frederiksen, J. & White, B.: 1992, 'Mental Models and Understanding: A Problem for Science Education', in E. Scanlon & T. O'Shea (eds.), *New Directions in Educational Technology*, Springer Verlag, New York, 211–226.
Frederiksen, J., White, B. & Gutwill, J.: 1996, *Dynamic Mental Models in Learning Science: The Importance of Constructing Derivational Linkages Among Models* (Technical Report CM-96-01), School of Education, University of California at Berkeley, CA.
Gentner, D. & Stevens, A.: 1983, *Mental Models*, Lawrence Erlbaum, Hillsdale, NJ.
Gilbert, S.: 1991, 'Model Building and a Definition of Science', *Journal of Research in Science Teaching* 28, 73–79.
Glynn, S. & Duit, R.: 1995, *Learning Science in the Schools: Research Reforming Practice*, Lawrence Erlbaum, Hillsdale, NJ.
Gunstone, R.: 1990, 'Reconstructing Theory from Practical Experience', in B. Woolnough (ed.), *Practical Science*, Open University Press, Milton Keynes, UK, 67–77.
Halloun, I. & Hestenes, D.: 1987, 'Modeling Instruction in Mechanics', *American Journal of Physics* 55, 455–462.
Horwitz, P.: 1989, 'Interactive Simulations and their Implications for Science Teaching', in J. Ellis (ed.), *The 1988 AETS Yearbook*, Association for the Education of Teachers of Science, Auburn, AL, 59–70.
Larkin, J.: 1983, 'The Role of Problem Representation in Physics', in D. Gentner & A. Stevens (eds.), *Mental Models*, Lawrence Erlbaum, Hillsdale, NJ, 75–98.
Mandinach, E.: 1989, 'Model-Building and the Use of Computer Simulation of Dynamic Systems', *Educational Computing Research* 5, 221–243.
McDermott, L.: 1984, 'Research on Conceptual Understanding in Mechanics', *Physics Today* 37, 24–32.
Mellar, H., Bliss, J., Boohan, R., Ogborn, J. & Tompsett, C. (eds.): 1994, *Learning with Artificial Worlds: Computer Based Modelling in the Curriculum*, Falmer Press, London.
Nersessian, N.: 1992, 'How Do Scientists Think? Capturing the Dynamics of Conceptual Change in Science', in R. Giere (ed.), *Cognitive Models of Science*, University of Minnesota Press, Minneapolis, MN, 3–44.
Perkins, D., Schwartz, J., West, M. & Wiske, M. (eds.): 1995, *Software Goes to School*, Oxford University Press, New York.
Richards, J., Barowy, W. & Levin, D.: 1992, 'Computer Simulation in the Science Classroom', *Journal of Science Education and Technology* 1, 67–79.
Sherin, B., diSessa, A. & Hammer, D.: 1993, 'Dynaturtle Revisited: Learning Physics Through Collaborative Design of a Computer Model', *Interactive Learning Environments* 3(3), 91–118.

Snir, J., Smith, C. & Grosslight, L.: 1993, 'Conceptually Enhanced Simulations: A Computer Tool for Science Teaching', *Journal of Science Education and Technology* 2, 373–388.
Spoehr, K.: 1989, 'Conventional versus Model-Based Instruction', Paper presented at the ARI contractors meeting, Princeton, NJ.
Stewart, J., Hafner, R., Johnson, S. & Finkel, E.: 1992, 'Science as Model Building: Computers and High School Genetics', *Educational Psychologist* 27, 317–336.
Tinker, R.: 1993, 'Modelling and Theory Building: Technology in Support of Students' Theorizing', in D. Ferguson (ed.), *Advanced Educational Technologies for Mathematics and Science*, Springer, Berlin, Germany, 91–113.
Weld, D. & deKleer, J. (eds.): 1990, *Readings in Qualitative Reasoning About Physical Systems*, Morgan Kaufmann, Palo Alto, CA.
White, B.: 1984, 'Designing Computer Activities to Help Physics Students Understand Newton's Laws of Motion', *Cognition and Instruction* 1, 69–108.
White, B.: 1993a, 'ThinkerTools: Causal Models, Conceptual Change, and Science Education', *Cognition and Instruction* 10, 1–100.
White, B.: 1993b, 'Intermediate Causal Models: A Missing Link for Successful Science Education?', in R. Glaser (ed.), *Advances in Instructional Psychology, Volume 4*, Lawrence Erlbaum, Hillsdale, NJ, 177–252.
White, B. & Frederiksen, J.: 1990, 'Causal Model Progressions as a Foundation for Intelligent Learning Environments', *Artificial Intelligence* 24, 99–157.
White, B. & Frederiksen, J.: 1995, *The ThinkerTools Inquiry Project: Making Scientific Inquiry Accessible to Students and Teachers* (Technical Report CM-95-02), School of Education, University of California at Berkeley, CA.
White, B., Frederiksen, J. & Spoehr, K.: 1993, 'Conceptual Models for Understanding the Behavior of Electrical Circuits', in M. Caillot (ed.), *Learning Electricity and Electronics with Advanced Educational Technology*, Springer Verlag, New York, 77–95.
White, B. & Horwitz, P.: 1988, 'Computer Microworlds and Conceptual Change: A New Approach to Science Education', in P. Ramsden (ed.), *Improving Learning: New Perspectives*, Kogan Page, London, 69–80.
Widman, L., Loparo, K. & Nielsen, N.: 1989, *Artificial Intelligence, Simulation & Modeling*, Wiley, New York.

# 3.3 Realising Authentic Science Learning through the Adaptation of Scientific Practice

DANIEL C. EDELSON
*Northwestern University, Evanston, Illinois, USA*

Making science learning better resemble science practice has been a common goal among education reformers at least since Dewey (1964). The potential benefits are clear. Students become active learners, they acquire scientific knowledge in a meaningful context, and they develop styles of inquiry and communication that will help them to be effective life-long learners. As appealing as this goal might be, unfortunately it has remained difficult to achieve in practice. However, the increasing availability of computer and networking technologies for the classroom, as well as the growing role of these technologies in science practice itself, offers new opportunities for the successful adaptation of scientific practice for learning environments. Specifically, technology can place a greater range of tools and resources at the disposal of teachers and students, and it can increase the opportunities for the social interchange that is at the heart of authentic science practice.

In this chapter, I discuss the adaptation of scientific practices for the classroom to achieve the benefits of authenticity. The first section contains a discussion of authentic science practice and the major challenges of adapting science practice for education. The second section presents a case study of the CoVis Project, an ongoing research and development effort using technology to assist the adaptation and implementation of scientific practice for learning. The final section describes a number of closely related projects that are exploring a variety of ways in which to use technology to bring authenticity to the classroom.

## AUTHENTICITY

In a modern echo of Dewey's (1964) recommendations, authenticity has become a rallying cry for innovation in education. Following Brown, Collins and Duguid (1989), who argued that '[a]uthentic activity . . . is important for learners, because it is the only way they gain access to the standpoint that enables practitioners to

---

Chapter Consultant: Instead of using chapter consultants, chapter authors in the Educational Technology section provided editorial comments on each other's chapters.

act meaningfully and purposefully' (p. 36), numerous educational researchers have adopted authenticity as a crucial objective for learning. The reason why authentic settings are important, and why authenticity is receiving so much attention, is the recognition that the knowledge and skills that learning activities produce are tied to the situation in which they are learned. For many topics taught in schools and universities, researchers have uncovered a great deal of discouraging evidence showing that students are often unable to meaningfully apply the knowledge which they acquire in school (e.g., Caramazza, McCloskey & Green 1981; Halloun & Hestenes 1985). This inability to apply knowledge in real-world settings has been attributed to the *situated* nature of knowledge (Brown, Collins & Duguid 1989). When individuals learn, they use the features of the situation in which they learn something to index the knowledge in their memories so that they can access it in the future (Schank 1982). If the indices in the learner's memory are too specific, then knowledge becomes situated too strongly within the setting in which it was learned, and it will not be applied in other settings. If the learning context accurately reflects the context in which the new understanding will be useful, however, then the situated nature of the learning will benefit learners and enable them to recognise opportunities to apply the learning.

*Characteristics of Authentic Science Practice*

In adapting science practice to the classroom, it is seductively easy to focus on scientific knowledge, tools and techniques at the expense of other elements of scientific practice. However, scientists' attitudes and their social interactions are also defining features of scientific practice. For students, understanding these attitudes and interactions is essential in order to understand the scientific process and to interpret the products of science. The key features of scientific practice fall into the three categories of attitudes, tools and techniques, and social interaction.

*Attitudes.* Scientific practice is characterised by the attitudes of uncertainty and commitment. Science is uncertain in that it is the pursuit of unanswered questions. By nature, both the techniques and results of scientific inquiry are subject to continual re-examination. Although scientists pursue issues that are important to them for a wide variety of reasons, effective science practice is always characterised by a commitment on the part of the scientists to the questions which they are attempting to resolve.

Any translation of scientific practice into educational settings must encourage these attitudes in order to provide authenticity. Therefore, students must have the opportunity to adopt questions that represent true uncertainty in their world. To foster commitment among students, the questions that they pursue must have ramifications that are meaningful within the value systems of these students.

*Tools and Techniques.* The practice of science in any modern discipline includes a set of tools and techniques that have been developed and refined over the history of the field. These tools and techniques permit scientists to pose and investigate a range of questions. The fact that these practices are shared across a community of scientists establishes a shared context that facilitates communication within the community.

*Social Interaction.* Science is not just investigation. It includes the sharing of results, concerns and questions among a community of scientists. This interaction has the same mix of cooperation and competition, agreement and argumentation that accompanies all human social activity.

For scientists, their attitudes, tools, techniques and social interaction are all supported by a body of knowledge that provides a meaningful context for scientific activity. The goal of providing students with the means to engage in adapted scientific practice enables them to acquire a body of scientific knowledge that is integrated with an understanding of science knowledge, attitudes, tools, techniques and social interaction. The successful adaptation of scientific practice for learning will place the tools and techniques of scientists into the hands of students in a context that reflects the characteristics of science practice outlined above. A vision of learning that integrates these features of scientific practice has students investigating open questions about which they are genuinely concerned, by using methods that parallel those of scientists. Throughout the process, they are engaged in active interchange with others who share their interest. Just as scientists accumulate knowledge and understanding through the course of posing and investigating research questions, students will too. Before engaging in any investigations, students will need to accumulate enough knowledge to pose well-framed questions. In the course of conducting investigations, they will need to master the tools and techniques that allow them to generate and analyse meaningful data. Finally, to be able to work with others, they must develop a vocabulary and framework for their understanding that allows them to communicate clearly about the knowledge that they acquire. The result of these learning activities will be student knowledge that is firmly situated in a context that reinforces both the applicability and value of that knowledge. While emphasising the importance of having students engage in activities that *resemble* those of scientists, it is important to acknowledge the significant differences that necessarily distinguish the practices of learners from those of scientists. The adapted scientific activities for students must reflect the vastly different interests, background knowledge and motivations of the two dramatically different populations.

*Adapting Scientific Practice to Learning*

The vision of students learning science by engaging in science is hardly new. However, it has been notoriously difficult to implement. Often the problem has been an incomplete implementation of the features described above. For example, traditional laboratory activities are intended to give students the opportunity to employ authentic scientific tools and techniques. However, the design of laboratory experiments usually removes any uncertainty, does little to obtain student commitment, and places minimal importance on social interaction.

Any complete adaptation of scientific practice will need to address three primary issues: curriculum structure; teacher preparation; and learner-appropriate resources, tools and techniques. The challenge in addressing these issues is achieving authenticity within the practical constraints of the classroom environment. Current, fixed curricula present a significant obstacle to the use of authentic scientific practice in the classroom, because of the flexibility in time and topic required for students to wrestle with uncertainty and pursue issues to which they are personally committed. Traditional training for teachers has not prepared them for new roles in which they must engage students in uncertain science, help them to formulate and refine research questions, identify resources and tools that will allow them to expand their understanding, and foster authentic scientific debate. Providing scientific resources, tools and techniques for use by students requires the modification of facilities designed for expert scientific practitioners to allow students to ask questions and pursue them in ways that are similar to those of scientists.

## THE ROLE OF TECHNOLOGY: A CASE STUDY OF THE COVIS PROJECT

Technology has an important role to play in addressing these challenges, particularly in the adaptation of resources, tools and techniques for students. In this section, I discuss The Learning Through Collaborative Visualization (CoVis) Project, a research project at Northwestern University that is capitalising on advanced technologies in an effort to provide characteristics of authentic science practice for students (Edelson, Pea & Gomez 1996a; Pea 1993). In addition to focusing on technology development, this project and others discussed in the next section, are confronting the issues of curriculum structure, teacher preparation and the adaptation of resources, tools and techniques.

The CoVis Project's objective is to use advanced computing and communications technologies to support innovative high school science education. In its technological design and development, CoVis has focused on two types of tools for students: scientific visualisation tools; and communication and collaboration tools. In its approach to pedagogy, the project has focused on project-enhanced (Ruopp, Gal, Drayton & Pfister 1993) and project-based (Blumenfeld, Soloway, Marx, Krajcik, Guzdial & Palincsar 1991) science learning. The project has specialised in earth and environmental science. In the 1993–1994 and 1994–1995 school years, the first two years of implementation, two high schools with six teachers and nearly 300 students participated in CoVis each year. In the 1995–1996 school year, approximately 40 middle and high schools throughout the USA participated in the project.

*Curriculum Structures*

The CoVis Project does not provide teachers with a fixed curriculum or a required set of activities. Instead it presents them with a set of resources and technologies. Participating teachers are encouraged to organise their courses around project cycles, but the duration, structures and topics of these projects are left to the discretion of the individual teachers. For example, project cycles in CoVis classes have varied from a few days to a half a year. In some classes, the structure and topic of projects have been specified in advance by the teacher and, in other classes, they are left to the choice of the students. The final product and the standards of evaluation are also left to the discretion of the teacher. The CoVis project's role is to assist teachers to set these parameters for their classes in an informed way and to provide them with access to supporting resources.

Beginning with the 1995-1996 school year, the CoVis project is providing teachers with a resource guide containing a collection of inquiry-based activities. This guide does not constitute a curriculum *per se*, but rather provides a starting point for teachers in their development of classroom activities. The resource guide is distributed to teachers in a print form and is also available through the CoVis Geosciences Server, a World Wide Web site maintained by the CoVis project. The activities included in the resource guide range from highly-structured, traditional laboratories to suggested topics for open-ended investigations. The goal is to provide teachers and students with a range of activities that will allow them to migrate from the practices with which they are familiar and comfortable, to less-structured, more authentic activities. Some of these activities are CoVis Inter-school Activities (CIAs) which, are scheduled, coordinated activities involving communities of schools. These CIAs follow the model of the Global Lab (Berenfeld 1993; TERC [Technical Education Research Center] 1991) and Kids as Global Scientist (see Songer's chapter in this *Handbook*) projects discussed in the next section.

*Teacher Preparation*

The teacher development activities in the CoVis project are conducted with the goal of helping teachers to improve their confidence and abilities in facilitating the inquiry process on the part of their students. This requires flexibility, comfort with non-traditional teacher roles, familiarity with available resources and tools, and an ability to monitor and assess the needs of their students. Because teachers are the richest source of available expertise on how to manage inquiry-oriented classrooms, most of the teacher development activities are aimed at the establishment of a community of teachers for the exchange of ideas, experiences and strategies. Communications technologies play a central role in creating and maintaining this community of teachers. Through a combination of face-to-face workshops and electronic communications, CoVis teachers engage in discussions of pedagogy and practice.

## Adaptation of Resources, Tools and Techniques

In its adaptation of scientific resources, tools and techniques, the CoVis project has focused on computational tools, especially tools for scientific visualisation and collaboration. Scientific visualisation refers to a set of display techniques used by scientists for the analysis of data (Brodie, Carpenter, Earnshaw, Gallop, Hubbold, Mumford, Osland & Quarendon 1992). A scientific visualisation is an image or animation in which numeric values in data sets are represented visually as colours, shapes or symbols. By exploiting the mapping between visual representations and underlying data values, scientists are able to discover important properties of the data through the observation of visual patterns in images. Because visualisation makes it possible to interpret data through visual perception rather than numerical analysis, it offers the potential for an impact on science education that is comparable to the enormous impact it has had on many scientific disciplines in the last decade.

The primary challenge in adapting scientific visualisation for learning stems from the fact that current scientific visualisation tools have been designed for use by researchers. When scientists use a scientific visualisation tool to investigate a body of data, they draw upon a rich store of knowledge about scientific phenomena, the source and limitations of the data, and the central questions of the field. In order to take advantage of visualisation as a means for learning through inquiry, students must be able to use scientific visualisation tools without possessing the background knowledge of a scientist. On the other hand, their use of the tool should enable them to acquire a meaningful portion of that scientific knowledge. In short, the goal of the adaptation process is to take a tool that scientists use to extend human knowledge about a subject and transform it into a tool that a learner can use in a similar way in order to extend his or her personal knowledge about that subject.

Three visualisation environments, called *visualizers*, have been developed using this process. They cover three aspects of atmospheric science and are called the Weather Visualizer, the Climate Visualizer, and the Greenhouse Effect Visualizer. Each of these is built on top of a visualisation tool used by researchers and provides learners with a more structured and more supportive user interface than the scientists' tool.

The Weather Visualizer (Fishman & D'Amico 1994) is an interface to current weather data for the USA. It enables students to view satellite images, weather maps displaying graphical symbols describing local atmospheric conditions, and false-colour images showing meteorological variables such as temperature, wind speed and direction, atmospheric pressure and dew point. Students use the Weather Visualizer to conduct 'nowcasts' and forecasts in ways that are similar to the daily activities of researchers in meteorology and to conduct research into weather-related topics. For example, one student used archived weather maps as part of an investigation of the conditions that led to disastrous wild fires in the Los Angeles area in 1993.

Like the Weather Visualizer, the Climate Visualizer (Gordin & Pea 1995; Gordin, Polman & Pea 1994) provides students with access to weather data. However, the Climate Visualizer draws from a data set that contains 25 years of twice-daily temperature, wind and atmospheric pressure values from the early 1960s to the late 1980s for most of the northern hemisphere. The Climate Visualizer allows students to display temperature as colour, wind as vectors, and barometric pressure gradients as contours. Through operations on data, the Climate Visualizer enables students to track trends over diurnal, seasonal and annual time periods. Students have used the Climate Visualizer to conduct investigations into topics of their own choosing that include the effect of coastlines on local temperatures, the impact of volcanoes on weather, and projections of future climate.

The Greenhouse Effect Visualizer allows students to visualise and manipulate data having to do with the balance of incoming and outgoing solar radiation in the earth's atmosphere (Gordin, Edelson & Pea 1995). The Greenhouse Effect Visualizer combines measured data with derived data and provides students with access to variables such as incoming solar radiation (insolation), reflectance of the earth's surface (albedo) and surface temperatures of the earth. Students have used the Greenhouse Effect Visualizer to investigate the possibility of global warming.

The second set of tools that the CoVis project has developed for high school students supports communication and collaboration, both within and beyond the classroom. The goal of these tools is to foster the same sort of dialogue about scientific topics that the scientific community engages in. Recognising that a classroom is often not a large enough community to allow a student with a particular interest to find others who share that interest, the CoVis project is using networking technologies to create a larger community for students and teachers. This larger community includes – in addition to students and teachers – scientists in relevant content areas, museum staff at the Exploratorium in San Francisco, and educational researchers. In addition to these committed participants, students have access to the resources and members of the Internet community at large.

To create and support this community, the CoVis project has appropriated and adapted tools developed for use in scientific and corporate workplaces. These technologies enable students and others to work together within classrooms and across the country, at the same time (synchronously) or at different times (asynchronously). To engage in synchronous collaboration, several individuals can sit together at the same computer, or work together at a distance as if they were sitting at the same computer. This is achieved through desktop video teleconferencing coupled with remote screensharing.

Asynchronous collaboration in CoVis classrooms is supported both by conventional communication applications such as e-mail and newsgroup discussions, and by a collaboration environment developed by the CoVis project called the Collaboratory Notebook (Edelson, Pea & Gomez 1996b; O'Neill & Gomez 1994). Students use e-mail and newsgroups to contact remote experts and to post queries to both the CoVis community and the Internet community at large. They use the Collaboratory Notebook as a collaboration environment for extended research projects.

The Collaboratory Notebook is a networked, multimedia database that provides learners with a mechanism for working cooperatively with others and is structured to support the inquiry process. Some of its features resemble those of other collaboration and shared database environments, such as CSILE (Scardamalia, Bereiter, Brett, Burtis, Calhoun & Lea 1992) and INQUIRE (Hawkins & Pea 1987). It is also designed to provide teachers and other mentors with a window into the thinking processes and activities of students. In a prototypical use of the Collaboratory Notebook, a group of students would develop an idea for an investigation and begin by recording some questions and hypotheses using the Collaboratory Notebook. These can be followed by a plan for how to pursue these issues. A teacher could read the students' questions, hypotheses and plans and add comments to help them to focus their efforts or to alert them to resources that they might find useful. In the next stage, individual students might engage in separate research activities that they could record individually for the others to view. In doing so, they might store both data and analyses within the Notebook. Without needing to meet in person, students could exchange questions and comments on their findings. Once they have conducted their investigations, they could get further guidance from an instructor or a scientist mentor, and then use the information which they have recorded to draw conclusions or initiate further investigations.

Loosely modelled on the metaphor of a scientists' notebook, the Collaboratory Notebook provides users with the ability to author pages individually or in groups and to read the pages authored by others. Pages are labelled according to the role that they play in the inquiry process (e.g., question, plan, conjecture, evidence-for, evidence-against, commentary) and can be linked via a hypermedia interface to other pages. The Collaboratory Notebook database is divided into individual workspaces, called *notebooks,* that students and teachers can create to serve specific purposes. In addition to text, students are able to store images, graphs, sound, animation or any other computer-generated media in their notebooks. Taking advantage of the capability to use the software from anywhere on the Internet, one teacher at a Chicago-area school used the Collaboratory Notebook in an activity in which students in his class entered information about topics in mineralogy, while scientist mentors at the University of Illinois at Urbana-Champaign and the Exploratorium in San Francisco interceded with questions designed to impel students to probe more deeply.

*Authenticity*

The CoVis project has adapted investigation tools used by scientists in order to provide students with the opportunity to engage in authentic scientific practice. The adaptation methodology used for the tools is designed to ensure that the tools are suitable for the knowledge and abilities of high school students. The project supplements these tools with communication and collaboration tools that enable students to engage in dialogue with a community of science practice that extends far beyond the boundaries of their classroom. In its work with teachers to develop project-enhanced science classrooms, the CoVis project encourages them to make uncertainty, commitment and social interaction over science a central part of students' activities.

## A BROADER PERSPECTIVE ON SUPPORTING AUTHENTICITY WITH TECHNOLOGY

The CoVis project is one of many that use technology to facilitate authentic scientific practices in classrooms. In this section, I use the framework of curriculum structures, teacher support and adaptation of scientific tools and techniques to describe some underlying themes in this research and to highlight some distinguishing characteristics of a number of these projects.

*Curriculum Structures*

Researchers who aim to foster authentic science practice in classrooms tend to describe their approach to science learning as project-based (Blumenfeld *et al.* 1991), project-enhanced (Ruopp *et al.* 1993) or problem-based (Barrows & Tamblyn 1980) learning. Their curriculum structures typically take the form of activities that are oriented around one or a cluster of open-ended questions. In the case of the projects described in this section, they all take advantage of technology to support these activities. Two important themes in the curriculum activities of current projects are (1) a focus on local phenomena and (2) activities conducted in multi-school communities.

Three projects that focus on the study of local phenomena are Global Laboratory, Kids as Global Scientists, and ScienceWare (see Spitulnik *et al.*'s chapter in this *Handbook*). In the Global Lab project, schools around the world each conduct an environmental study of a local site on or near their school's grounds. In the Kids as Global Scientists project, students study the local weather patterns in their area and the particular geographic features that lead to these patterns. In the ScienceWare project, students at a school in Michigan study the ecology of a stream that passes close to the school. The benefit of studying these local phenomena is the motivation that comes from understanding one's personal environment better and the recognition that science is relevant to one's personal lives. These benefits pay off through the sense of commitment, a hallmark of authentic science, that students feel when studying their local environment.

Like CoVis, two of these projects – Global Lab and Kids as Global Scientists – also take advantage of networking technologies to provide a multi-school community in which to conduct the activities. These multi-school communities replicate in a fashion the collection of diverse perspectives and experiences that make up the scientific community. While CoVis spans a broad diversity of settings within the USA, the Global Lab and Kids as Global Scientists community span the globe. In all three projects, students have the opportunity to share their personal experience and local perspectives in the pursuit of questions of a scientific nature.

*Teacher Support and Development*

For many science teachers, providing students with the opportunity to pursue open-ended inquiry is not a part of their current practice. This shift in approach requires a significant amount of support. Several research projects are providing this support through the development of specialised resource materials and the use of networking technologies to conduct on-line professional development activities.

Projects involving large numbers of schools, such as CoVis, Global Lab and Kids as Global Scientists, have found it valuable to develop resource materials for distribution to participating teachers. In the case of all three of these projects, the resource materials include careful descriptions of both small-scale, well-structured activities resembling traditional laboratories and larger-scale extended activities which provide a great deal of room for teacher and student specification of the structure. In addition, both the Global Lab and CoVis resource guides include writings on approaches to inquiry-based science pedagogy. In the case of the CoVis project, these materials are available to teachers on-line via the CoVis Geosciences Server.

These projects also have taken advantage of communications technologies to enable moderated discussions of pedagogy and practice over networks. These professional development activities allow teachers to form their own communities of inquiry and practice. These communities provide them both with moral support and an opportunity to share experience and expertise. In a profession for which isolation is a very common complaint, especially among science teachers, these electronic communities can be an important key to enabling teachers to transform their practice to support authentic activities.

*Adaptation of Scientific Tools, Techniques and Resources*

The most fertile use of technology in adapting scientific practice for the purposes of science learning has been in the adaptation of scientific tools, techniques and resources. Four strands are evident in current research that all facilitate the adoption of the attitudes, techniques and social interactions that characterise the scientific community by the science education community. These strands are (1) collection and sharing of data, (2) analysis of data through modelling and visualisation, (3) evidence gathering and evaluation and (4) communication and collaboration. Each strand is discussed in turn below.

## Collection and Sharing of Data

The collection of data is an essential element of scientific practice. In a large number of current educational research projects, students use scientific measurement devices that have been adapted from scientists' devices to be both suitable and affordable for educational environments. In an analogy to the way in which the CoVis project has adapted scientists' visualisation tools for learners, projects such as Global Lab, ScienceWare, the Princeton Earth Physics Project (Nolet 1994) and the MicroObservatory Project (Gingerich 1994) have developed student-appropriate tools and techniques for the collection and sharing of data. As described above, in both the Global Lab project and ScienceWare, students collect data in their local environment for analysis. The Global Lab project has devoted its resources to developing inexpensive devices that provide reliable and accurate measurements for quantities such as atmospheric ozone and ultraviolet radiation, but are inexpensive and robust enough for classroom use. Many of these devices are constructed by the teachers and students using materials that are common worldwide. The collection of data by Global Lab schools is coordinated to maximise comparability across sites and then entered into a shared, centralised database to which all the schools have access. Similarly, the ScienceWare project has developed interfaces to hand-held Newton computers for inexpensive probes that are useful in tracking stream ecology. In the geological sciences, the Princeton Earth Physics Project (PEPP) has been developing low-cost accurate seismographs to enable schools to monitor seismic activity. This project links schools together using electronic networking in order to share data. In this case, differential measurements from sites at different locations provide important information about the nature of the seismic activity observed. The MicroObservatory Project at the Harvard College Observatory is engaged in a similar effort to develop and distribute a low-cost, automated-imaging telescope for use in schools, similar to those used by astronomers (Gingerich 1994). These computer-controlled telescopes will be used by students who then will share their data with others via the Internet. It is no coincidence that all of these projects are in the environmental and earth sciences. The professional communities in these fields have been at the forefront in terms of the use of advanced technologies for the global collection and sharing of scientific data.

## Evidence Gathering and Evaluation

In a similar vein, two projects have focused on the identification of evidence by students to support scientific reasoning and argumentation. The Knowledge Integration and Environment (KIE) Project (Bell, Davis & Linn 1995) at Berkeley has developed a learning environment that provides students with the opportunity to search a database for relevant evidence and construct a scientific argument using that evidence. Using a network browser based on one that scientists use to conduct similar searches on the World Wide Web, students work in an educationally supportive environment in order to accumulate evidence from a database designed by researchers for its pedagogical value. In the Science-in-Action series being created by the Cognition and Technology Group at Vanderbilt, students view video-recorded scenarios containing problems that students need to identify and solve. An important element of their challenge is to identify the evidence embedded in the video scenario and accompanying print materials that will allow them to construct a well-supported argument (see Sherwood et al.'s chapter in this *Handbook*). An important feature of both of these projects in terms of providing an authentic experience of science is the attitude of uncertainty that accompanies the central questions that these learning environments raise for students.

Analysis of Data Through Modelling and Visualisation

A large number of recent projects has focused on the creation of tools to support the modelling and visualisation of data. Educational researchers have not failed to notice the impact of computational tools for modelling and visualisation on scientific practice and have been quick to explore ways to adapt these tools for learners. The Blue-Skies project (Samson, Steremberg, Ferguson & Kamprath 1994) and Project Globe (National Science Foundation 1994) are two educational projects that, like CoVis, are exploring the development of learner-appropriate visualisation tools for atmospheric data sets. Hands On Universe (Friedman 1995) has developed analogous data manipulation and display tools for astronomy. Science-Ware, through its Model-It environment (see Spitulnik et al.'s chapter in this *Handbook*), is providing students with a tool for constructing models to fit the data that they gather about stream ecology. In all of these projects, the goal of the development effort has been to provide students with tools that enable them to learn about a subject matter through the investigation meaningful questions in ways that resemble the methods of practising scientists. The challenge is creating tools that can scaffold students in inquiry despite the limitations of their scientific background and understanding.

Communication and Collaboration

The importance of tools for collaboration and communication in science practice is evidenced by the fact that the explosive growth of the Internet, Usenet and the World Wide Web were all ignited by the needs of the scientific research community. These networking environments support the social interaction that is at the heart of scientific advancement. Collaboration tools that are adapted and structured for learners, such as the CoVis Collaboratory Notebook, Global Lab's Alice software (Hunter 1993), KIE's speakeasy (Bell, Davis & Linn 1995) and the

CSILE (Scardamalia *et al.* 1992) software being used in conjunction with the Scientist-in-Action series, all reflect efforts by researchers to facilitate a similar process of social interaction among learners over open-ended scientific questions. In their own way, all of these collaboration environments attempt to support and gently guide learners as they engage in complex activities.

## CONCLUSION

In this chapter, I have argued for an increased role for authentic science practice in science learning. However, it would be a mistake to view this argument as one that only has room for extremes in that a class must be either traditional and didactic or entirely practice-oriented. Effective science education will always consist of an appropriate balance between didactic instruction and hands-on activity. Meaningful science practice at any level requires an understanding of relevant fundamental science principles. It is a mistake to believe that the right activities will allow students to discover those principles entirely on their own, just as it is wrong to believe that they will understand them as a result of memorising them. Students should have the opportunity to acquire and apply knowledge through scientific inquiry, as well as to expand and structure their knowledge through lectures and readings. In the end, the responsibility for striking the balance lies with the individual teacher who is able to make judgements about what is suitable for his or her environment and students.

The ability of teachers to exercise this judgement, however, can be enhanced greatly by the increased focus of researchers and scientists on the adaptation of science practice for educational settings. The examples in this chapter demonstrate valuable roles for technology in this adaptation. In particular, they show how the computational resources, tools and techniques used by scientists in their practice can be adapted for use in the classroom. In addition, they exploit networking technologies to allow both students and teachers to become a part of a larger community of science practitioners. However, the research described here reflects a second lesson that researchers have learned: focusing on the issues of technology and adaptation of scientific tools alone is insufficient to achieve authentic, suitable science learning. Achieving this goal requires strategies, both technological and non-technological, for establishing appropriate curriculum structures and preparing teachers.

In the complexity of the real world, reaching these objectives is extremely difficult. The research projects described here are addressing the constraints of working in real classrooms with teachers and students and are experiencing challenges that belie the rosy tone of this chapter. Nevertheless, the broad involvement in these efforts indicates an enthusiasm for incorporating authentic practices into science education that goes beyond researchers to include the teachers and students themselves.

## ACKNOWLEDGEMENT

This material is based in part on work supported by the National Science Foundation under grant numbers MDR–9253462 and RED–9454729. The author would like to acknowledge the critical role of the members of the CoVis research team, Roy Pea, Louis Gomez, Barry Fishman, Eileen Lento, Laura D'Amico, Doug Gordin, Steven McGee, Kevin O'Neill and Joe Polman, in defining and refining the ideas in this chapter.

## REFERENCES

Barrows, H.S. & Tamblyn, R.: 1980, *Problem-Based Learning: An Approach to Medical Education*, Springer, New York.
Bell, P., Davis, E.A. & Linn, M.C.: 1995, 'The Knowledge Integration Environment: Theory and Design', in J.L. Schnase & E.L. Cunnius (eds.), *Proceedings of CSCL 95: Computer-Support for Collaborative Learning*, Lawrence Erlbaum, Mahwah, NJ, 14–21.
Berenfeld, B.: 1993, 'A Moment of Glory in San Antonio: A Global Lab Story', *Hands On!* 16(2), 1 & 19–21.
Blumenfeld, P.C., Soloway, E., Marx, R., Krajcik, J.S., Guzdial, M. & Palincsar, A.: 1991, 'Motivating Project-Based Learning: Sustaining the Doing, Supporting the Learning', *Educational Psychologist* 26, 369–398.
Brodie, K.W., Carpenter, L.A., Earnshaw, R.A., Gallop, J.R., Hubbold, R.J., Mumford, A.M., Osland, C.D. & Quarendon, P.: 1992, *Scientific Visualization*, Springer-Verlag, Berlin, Germany.
Brown, J.S., Collins, A. & Duguid, P.: 1989, 'Situated Cognition and the Culture of Learning', *Educational Researcher* 18(1), 32–42.
Caramazza, A., McCloskey, M. & Green, B.: 1981, 'Naive Beliefs in "Sophisticated" Subjects: Misconceptions about Trajectories of Objects', *Cognition* 9, 117–123.
Dewey, J.: 1964, 'Science as Subject Matter and as Method', in R.D. Archambault (ed.), *John Dewey On Education: Selected Writings*, University of Chicago Press, Chicago, IL, 121–127.
Edelson, D.C., Pea, R.D. & Gomez, L.: 1996a, 'Constructivism in the Collaboratory', in B. Wilson (ed.), *Constructivist Learning Environments: Case Studies in Instructional Design*, Educational Technology Publications, Englewood Cliffs, NJ, 151–164.
Edelson, D.C., Pea, R.D. & Gomez, L.M.: 1996b, 'The Collaboratory Notebook: Support for Collaborative Inquiry', *Communications of the ACM* 39, 32–33.
Fishman, B. & D'Amico, L.: 1994, 'Which Way Will the Wind Blow? Networked Tools for Studying the Weather', in T. Ottmann & I. Tomek (eds.), *Educational Multimedia and Hypermedia, 1994 (Ed-Media '94)*, Association for the Advancement of Computing in Education, Charlottesville, VA, 209–216.
Friedman, J.: 1995, 'Image Processing in a Science Classroom: Students' Interpretations of Images', Paper presented at the annual meeting of the National Association for Research on Science Teaching, San Francisco, CA.
Gingerich, O.: 1994, '*MicroObservatory II*', in *Descriptions of Current Projects: Applications of Advanced Technologies Program*, National Science Foundation, Washington, DC.
Gordin, D.N., Edelson, D.C. & Pea, R.D.: 1995, 'The Greenhouse Effect Visualizer: A Tool for the Science Classroom', in D.R. Smith & L. Bastiaans (eds.), *Proceedings of the Fourth American Meteorological Society Education Symposium* (Section J6), American Meteorological Society, Boston, MA, 47–52.
Gordin, D.N. & Pea, R.D.: 1995, 'Prospects for Scientific Visualization as an Educational Technology', *Journal of the Learning Sciences* 4, 249–279.
Gordin, D.N., Polman, J.L. & Pea, R.D.: 1994, 'The Climate Visualizer: Sense-Making through Scientific Visualization', *Journal of Science Education and Technology* 3, 203–226.
Halloun, I.A. & Hestenes, D.: 1985, 'The Initial Knowledge State of College Physics Students', *American Journal of Physics* 53, 1043–1055.
Hawkins, J. & Pea, R.D.: 1987, 'Tools for Bridging the Cultures of Everyday and Scientific Thinking', *Journal of Research in Science Teaching* 24, 291–307.
Hunter, B.: 1993, 'Collaborative Inquiry in Networked Communities', *Hands On!* 16(2), 16–18.

National Science Foundation: 1994, *Global Learning and Observations to Benefit the Environment (GLOBE)* (Announcement No. NSF-94-152), Author, Washington, DC.

Nolet, G.: 1994, 'The Princeton Earth Physics Project (PEPP) – A Short History', *PEPP Newsletter* 1(1), 1 & 7–8.

O'Neill, D.K. & Gomez, L.M.: 1994, 'The Collaboratory Notebook: A Distributed Knowledge-Building Environment for Project-Enhanced Learning', in T. Ottmann & I. Tomek (eds.), *Educational Multimedia and Hypermedia, 1994 (Ed-Media '94)*, Association for the Advancement of Computing in Education, Charlottesville, VA, 416–423.

Pea, R.D.: 1993, 'Distributed Multimedia Learning Environments: The Collaborative Visualization Project', *Communications of the ACM* 36(5), 60–63.

Ruopp, R., Gal, S., Drayton, B. & Pfister, M.: 1993, *LabNet: Toward a Community of Practice*, Lawrence Erlbaum, Hillsdale, NJ.

Samson, P.J., Steremberg, A., Ferguson, J. & Kamprath, M.: 1994, 'Blue-Skies: A New Interactive Teaching Tool for K–12 Education', in *Proceedings of the 10th International Conference on Interactive Information and Processing Systems for Meteorology, Oceanography, and Hydrology* (Section J), American Meteorological Society, Boston, MA, 9–14.

Scardamalia, M., Bereiter, C., Brett, C., Burtis, P.J., Calhoun, C. & Lea, N.S.: 1992, 'Educational Applications of a Networked Communal Database', *Interactive Learning Environments* 2, 45–71.

Schank, R.C.: 1982, *Dynamic Memory*, Cambridge University Press, Cambridge, UK.

TERC (Technical Education Research Center): 1991, *Global Laboratory Notebook*, Author, Cambridge, MA.

# 3.4 Can Technology Bring Students Closer to Science?

NANCY BUTLER SONGER

*University of Michigan, Ann Arbor, USA*

Across scientific disciplines, professional scientists are embracing emerging technologies and incorporating new technological features into their practice in remarkable ways. Telecommunications technologies are increasingly a part of most professional scientists' lives as they utilize the Internet as an important resource for data, data analysis, professional collaboration and data exchange. In one recent example, astronomers located in several locations worldwide use the Internet for real-time control of telescopes in distant locations, such as Antarctica, so that they can analyze the chemical composition of stars from several sites and in collaboration with each other. Users can control both the selection of instruments and a range of other sophisticated controls, such as the choice of filters and settings prior to receiving and storing the digitized image on their machine at the control site. While it is doubtful that this technology will remove the need for scientists to be in residence at the remote locations, it is clear that providing a larger community of scientists with distant control of the sophisticated technology, and the ability to collaborate from several distinct locations, will change both the nature of the data collected and the analysis and collaboration available for those data.

In this and other examples, scientists themselves testify that network technologies are beginning to affect profoundly the way in which they do their work, although the changes are not always in expected ways. Dr Doyal Harper, chief administrator of the consortium of organizations which supports the South Pole telescopes believes that:

> ... the human benefits are at least as important to South Pole astronomy as is the Internet's remote-control capability. It's absolutely essential to tie together the men and women who work here and the scientists outside Antarctica who use our telescopes, into effective working teams. The people in the States must remain engaged in the problems and realities of life down here, and the winter-over people who work here must not feel cut off from the other world. For the first time, e-mail will let them communicate not only with the professional people outside Antarctica, but with their families and friends. We have begun a new era ... [T]he Internet is changing the flavor of science in Antarctica. (Browne 1995, B5 & B9)

---

Chapter Consultant: Instead of using chapter consultants, chapter authors in the Educational Technology section provided editorial comments on each other's chapters.

In essence, the technology provides a combined interactive communications tool with new real-time or near-time data and data manipulation features. This combination helps networks to be a communication tool unlike the next best communications tool available, the telephone, in its ability to help professionals simultaneously get 'closer to the science'. Scientists are able to build new relationships with the science and each other through: the combination of the interactive remote control of instruments by a wider range of users; the collaborative and iterative analysis of data and hypotheses; and the first-hand, real-time or near-time iterative correspondence with colleagues and interested others.

Similarly, telecommunications technologies have begun to transform the work of classroom teachers in many ways (Putnam & Borko in press; Edelson's chapter in this *Handbook*). Already, many teachers acknowledge the importance of the Internet as a professional development tool, especially because of its ability to provide up-to-date communication on the dynamic aspects of teachers' professional lives (Black, Klingenstein & Songer 1995). Researchers discuss the many role changes that become necessary for teachers in integrating the technological tools successfully into their classrooms. Such role changes include the willingness of teachers to recognize and feel comfortable not knowing everything about the new resource, a willingness to learn alongside and in collaboration with their students, and a willingness to seek out help and support from other teachers and more knowledgeable resources in many areas, including technology, pedagogy and subject matter content. Teachers using telecommunications-based curricula testify that, as a result of the new content resources available for their classrooms, using real-time, Internet-available resources has allowed them to question some of their own assumptions about science content, and therefore has motivated them to continue to learn and become 'closer' to the science that they are helping their students to understand (Songer 1995).

This chapter outlines one example of how telecommunications technologies such as the Internet are providing opportunities for middle school students to become 'closer to science' in ways similar to that described for professional scientists and teachers above. The first section outlines the development of an Internet-based science education program called Kids as Global Scientists (KGS). Recent research results from this project are discussed in the second section. The third section presents case study information from one KGS site, emphasising first-hand views from the teacher in this location. The final section summarizes general strengths and weaknesses of this approach to middle school science education, and concludes with implications for improving science education.

## THE KIDS AS GLOBAL SCIENTISTS PROJECT

Recognizing the need to understand the educational potential of telecommunications technologies for pre-college students, the Kids as Global Scientists project began with a desire to transform current thinking about student learning into innovative curricula which capitalise on Internet-available resources. In particular, the design of KGS activities incorporated current thinking in the learning sciences which:

a) considered the active nature of student learning,
b) recognized the importance of a sequence of explanatory models, including the value of intermediate-level, qualitative understandings which bridge familiar and unfamiliar representations (Lewis & Linn 1994; Linn, diSessa, Pea & Songer 1994; White 1989),
c) helped students to build increasingly integrated understandings which work with prior knowledge including natural world experiences (Linn, Songer & Eylon in press; Songer 1996),
d) capitalized on the social construction of knowledge and collaboration in appropriate ways, and
e) worked to achieve a learning environment which allowed a wide range of learners to interact, both as contributors and consumers (Brown et al. 1993; Songer 1996).

In particular, KGS worked to design ways in which middle school students' science could be made more authentic and personalized through the use of the Internet and its resources, including real-time and near-time data and visualization resources, and peers distributed worldwide. While many recent research projects define and interpret authentic learning in different ways (Collins, Brown & Holum 1991), we interpret the authentic nature of our activities as the opportunity for students both to work directly with professional-quality, real-time data in much the same way professional scientists do, and to communicate directly with individuals who have first-hand experiences with the phenomena which they are studying. Particularly distinctive features of our project include:

- classroom learning that includes peer coaching and information exchange worldwide;
- learning which can capitalize on today's science developments;
- a coordinated, yet flexible, curricular shell which allows adaptation to a variety of connectivity levels and local customization by teachers and students at each site.

*A Rationale for Global Exchange*

To achieve the more personalized learning approach desired, it was necessary to customize and develop our own curriculum and learning approach. In particular, the project recognized early on that the Internet was a strong communication tool, and that capitalizing on this communication feature might allow the nature of student learning to change in distinct and interesting ways. In addition, the Internet provided an opportunity for students to experience first-hand science imagery in real-time or near-time, and to ask new questions and propose hypotheses about real-time events. Students could download and map the development of a hurricane as it approached the Florida coast, make predictions about the path of a developing storm, or compare the forecast simulations from experts with what they saw outside their classroom windows.

In recognition of the potential of Internet access to grow within many schools in the USA and beyond, the curriculum needed to capitalize on students distributed in many different locations worldwide. Because the content information gathered in each distributed location is distinct, the curriculum needed to emphasise a means for middle school students to use each other as sources of information, much as professional scientists use each other to share information and data in areas beyond their own specialty. Distributed individuals also needed to find an opportunity to share what they had learned with those who might find it useful, and to gain from the knowledge of others.

In general, using peers as information resources and professional-quality real-time visualizations as information sources allow middle school students opportunities to experience aspects of authentic practice which are common among professional scientists. Most importantly, however, the research team continuously questions when and where authentic practice is applicable to middle school science education. Clearly, providing students with raw data, complex interfaces, or activities which have not been carefully scaffolded to allow rich and integrated understandings might be authentic in nature, but do not support the development of integrated understandings. In addition, many opportunities for the social construction of knowledge do not benefit all individuals equally.

Recent research on gender issues and the social construction of knowledge indicates that care must be taken to ensure that a diverse range of views are respected (Wellesley College Center for Research on Women 1992). In favorable circumstances, however, the opportunity to support students' developing understandings in science, by building first on familiar instances and then using these experiences as a foundation for developing an understanding of these same issues on a larger and more complex scale, is a developmental model of learning which supports increasingly rich, interdisciplinary understandings for the application of more formal and complex understandings at a later age. In addition, the correspondence, real-time imagery and group questioning adds aspects of authenticity that often lead to higher levels of motivation to learn for many students, including even some who previously have shown little interest in this or many other science topics. These results are discussed in greater detail in subsequent sections.

The curriculum that was developed is called Global Exchange. The learning approach uses three curricular phases to help students identify and find explanations for the knowledge that becomes available at each location, and to find an effective means for sharing and incorporating this knowledge into their thinking. The activities are organized into three phases: an introductory phase; a development of expertise phase; and a sharing phase. While the details of the Global Exchange curricula are beyond the scope of this chapter, general features are outlined below (also see Songer 1995).

*The Global Exchange Curricula*

Based on our observations that students and teachers needed an opportunity to establish social connections prior to becoming science resources for each other, the eight-week curriculum begins with a 1.5-week introductory phase. Students and teachers are matched to a cluster of about 10–12 schools which are geographically distributed all over the world. Preliminary interactions allow students to describe distinguishing features of their own geography and culture.

After introducing themselves, students spend the next three weeks collecting data and information leading to the development of local expertise in one of five topics in atmospheric science: clouds and humidity; winds; precipitation; severe weather; or environmental issues. In this phase, students collect at least two weeks of their own weather data, such as wind speed and direction, cloud observations, precipitation or tests for acid rain or snow. Local data are reviewed for patterns, and reviewed by peers studying the same phenomena in different locations. Electronic dialogue with local mentors in atmospheric science encourages questioning and further explanations of students' understandings. Further activities in the expertise phase include hands-on experiments or demonstrations for each topic, such as opportunities to estimate the percentage of cloud coverage or cloud speed, and the development of pictorial and textual explanations for the local phenomena. In addition, students work on activities which help them bridge from concrete representations of data (their own data) to more abstract views, such as the real-time satellite imagery or weather maps available via the Internet, the Weather Channel or other resources.

During the next four weeks (the sharing phase), students are encouraged to build on their local expertise through the comparison of information with others studying the same phenomena in different locations. Students are instructed to send comparison questions and data to a few other groups, and to critique the ideas and data sent to them within 48 hours. Students check their predictions and dialogue through the use of real-time and archival satellite and weather imagery, focusing especially on the areas where their correspondents live.

Often, particularly interesting or sensational scientific events occur sometime during the eight-week period. This provides a unique opportunity for all students within the same Global Exchange to take advantage of the power of first-hand resources, whether local or across the globe, as they further investigate the science behind the occurrence.

## RESEARCH RESULTS FROM THE SECOND GLOBAL EXCHANGE

The second Global Exchange occurred during Spring 1994. The following sections outline the project participants, sites, connectivity and research results.

*Participants in the Second Global Exchange*

A total of 800 students and 100 adults from 26 schools and 7 countries participated in the second Global Exchange. KGS locations were diverse along many criteria including very remote locations in northern Alaska, schools in urban and diverse American locations such as Harlem, Sacramento, Albuquerque and Denver, and other overseas locations in Finland, Israel, Eastern Australia, Mexico and South Africa.

KGS sites were located in a variety of diverse locations, including rural (36 percent), suburban (39 percent) and urban (25 percent) sites. Schools were also chosen for their ethnic and cultural diversity, with 55 percent of KGS schools reporting greater than 30 percent diverse students. While about 90 percent of students were middle school age, a small percentage of sites were located in primary or high schools.

Science Mentors

A total of 24 mentors were involved in the second Global Exchange, ranging from professional meteorologists to graduate students in atmospheric science, physics, astronomy and science education. Several of the international sites initiated their own contact with local meteorologists, and involved them in helping the students with their research. The mentors assisted students in finding data for their location, as well as answering basic questions on local weather phenomena. Mentor training packets were developed by the KGS team to help mentors distinguish between appropriate and inappropriate correspondence.

Connectivity

To allow participation by as wide a range of schools as possible, the KGS curricular activities were designed for compatibility with diverse levels of connectivity. Essential components of the Global Exchange, such as student correspondence and exchange of data files, were conducted as text exchanges to allow the easiest possible send-and-receive operations by all participants. Real-time or near-time imagery could be accessed through a variety of means. Classrooms with direct connectivity (ISDN or T1 lines) could utilise FTP (file transfer protocol), gopher or World Wide Web (WWW) clients such as Blue-Skies or Mosaic to download imagery. Classrooms with slower rates of connectivity often used other vehicles for real-time imagery, such as *The Weather Channel* or the *USA Today* weather page. Despite these differences, all schools followed the same curricular outline and progressed through all phases of the exchange.

With regard to 'Internet protocol', students' access to e-mail and the Internet ranged from the use of a single account with dial-in modem access, to individual accounts with direct connectivity for each student. While connectivity levels were extremely varied among project sites, a majority of sites (over 85 percent) corresponded via some form of modem connectivity. Teachers by and large found their connectivity to be 'very reliable' (75 percent), which therefore allowed many of them to be effective participants.

*Results of Teacher Surveys*

A variety of data were obtained from the detailed teacher surveys which were sent to instructors at each project site. Surveys contained many short-answer and selected-answer questions, and were completed by teachers at three prescribed intervals during the project.

Throughout the project, many teachers reported that their students were sending and receiving between 5–15 notes a week, and that two areas of the project provided the greatest success, namely, the contact with students worldwide and high student motivation. The greatest disappointment was that the other sites were not sending as many notes back as some sites expected. Teachers in general felt that students were learning a great deal about the use of the technology, with 100 percent of the teachers commenting at the end of the project that the students were comfortable using the technology, and 86 percent of the teachers reporting that the project was a very good learning experience for their students. These results appear to demonstrate a quite dramatic growth of comfort with technology for these teachers in that, while a majority of the teachers (88 percent) had used computers in teaching before, almost no teachers previously had used the Internet in their classroom or participated in Internet-based curricular projects such as this one. Concerning continued participation, 100 percent of the teachers expressed that they would be interested in participating in the KGS project in the future.

In summary, a majority of the teachers felt the best project experiences were student contact with the world and high student motivation. The weakest project areas were not enough responses (25 percent), technical problems (25 percent) and organization and time problems (25 percent).

*Results on Student Learning*

All teachers were given pre- and post-written assessments to administer to their participating students. Student assessments consisted of 20 short-answer and short-essay questions focusing on a variety of topics, including students' understandings of weather both locally and in distant locations, students' explanatory models, information on what makes a good question, and student attitudes about learning science, both within this project and compared to previous learning experiences. The project staff received complete pre- and post-assessment sets from a total of nine sites. Five complete sets of student assessments were coded and analyzed.

In general, students who participated in the Global Exchange learned a great deal about local weather phenomena, the students in the other locations, the new technology, and the weather in their specialty area but in different locations.

In particular, data analysis focused on how the learning of the student experts in a given topic was similar or different from the learning of that topic by the individuals who had not specialized in that topic. This kind of analysis helped to illustrate how effective the Global Exchange curriculum might be in helping students to gain a rich understanding in one topic area which goes beyond local understandings. Results indicate that student experts often outperform their peers in their specialty topic areas. A two-treatment Group x Time of Testing analysis of covariance showed that wind experts, for example, were able to give significantly more complete explanations on what causes wind in their area than students who had focused on other topic areas [$F(1, 67) = 80$, $p<.023$]. Similarly, Colorado precipitation experts were able to explain more effectively than their peers how precipitation in a nearby location is similar to and different from their own patterns [paired $t$-test, $t(1, 29) = 2.31$, $p<.028$]. While other topic area differences followed this trend, the results were not always significant. In addition, many students learned other important inquiry skills such as the development of criteria for what makes a good question to ask an on-line mentor or peer [paired $t$-test, $t(1, 29) = 2.11$, $p<.043$], and the ability to determine and justify their choice of an appropriate question to send to others.

Another interesting set of learning differences occurred when the performance on content area questions was compared for boys and girls. While in general the significant differences between genders were subtle, girls outperformed boys on 64 percent (9) of the content area questions, while boys outperformed girls on 36 percent (5) of these questions. Considering that the topic area of atmospheric science often is associated with some of the greatest gender differences in science favouring males, the indication of a possible trend in the opposite direction is intriguing. Presently, the KGS team is developing and conducting a gender study to investigate further the nature and possible stability of these gender differences.

In summary, by the end of the eight-week exchange, many students had both developed a locally-based understanding of weather and atmospheric science, in addition to extending that understanding through questioning and answering with knowledgeable peers. Students also developed a proficiency in the use of Internet technology, and a greater awareness of inquiry skills such as appropriate questioning.

*Problem Areas*

In addition to the cataloging of project successes, research staff investigated several sites which, for a variety of reasons, were not so successful. Researchers performed detailed telephone interviews and discovered locations with time-management problems, staffing/weather difficulties and technical problems.

Participants at two locations discussed problems with KGS that could have been experienced with any new curricular innovation. At an urban and diverse site in the western USA, participants commented that they experienced frustration with the KGS project because their local curricular requirements mandated that they continue with another conflicting curriculum at the same time as they attempted KGS. Despite this setback, the teacher commented that aspects of the program were successful: 'The students did have a good time communicating with students around the world. They picked up some information about the weather around them.'

Similarly, in a location in southeastern Australia, participants commented that changes in the teaching staff in the middle of the project caused a major interruption in the progress of the project. The region also experienced no precipitation for 47 consecutive days, and therefore shifted precipitation experts' responsibilities to ulraviolet radiation level readings and daily sunlight time information. This teacher also commented on some project successes: 'The students have been very keen to read the replies and letters sent. I have seven rain gauges made by the students which have not been used – dusty but fine.'

Participants at a third site in South Carolina mentioned that they experienced technical and time problems. 'I was unable to arrange a working modem connection at school. I was also unable to arrange for my students to participate in the activities. We had the old problem of "time". I know that the students were excited about the project.'

In summary, participation in any complex and demanding curricular project requires a great deal of expertise in a diverse set of areas including time management and new pedagogical approaches. In addition, KGS participation adds complicated issues of technology, and of becoming comfortable with new content and resources. Many sites experienced setbacks in at least one of these areas. Despite some disappointments, nearly every project participant had something positive to say about the unique learning opportunities which the worldwide, science-based project provides. Our work continues to iron out the problem areas so that the highest possible levels of success can be experienced at a greater number of locations.

## CASE STUDY OF A SUCCESSFUL KGS SITE: UPSTATE NEW YORK

New York proved to be an interesting location to study, in part because of its tremendous success with the project, despite many obstacles. First, the school was located in a primarily rural, middle-class community. The teacher helped all 32 of her sixth-graders to be active participants in the project, despite a level of connectivity which was slower and more difficult than most locations. Students accessed the Internet via one Apple IIe computer and a 2400-baud modem, using one account on Learning Link, a public television-supported educational network. Despite these obstacles, the participation of the New York school was exemplary on many criteria, with all students actively participating and corresponding with their meteorologists on-line, peers worldwide and each other. In addition, the New

York teacher entered a contest sponsored by her local public broadcasting station which encompasses 14 New York counties. This teacher was awarded first prize for 'Most Innovative Use of Technology in the Classroom' as a result of her participation in the KGS project.

*Vignette 1: General Teacher Comments On The Project*

In general, the New York site appeared to have a very positive experience with the KGS project. The teacher commented on many aspects which appeared to contribute to project success, including the wide range of new learning for her students, the cohesiveness of the curriculum, and the worldwide forum for science learning. The students appeared very motivated. The project encompassed a reasonably large amount of time, and resulted in students discussing weather and other project issues often. The teacher expresses these views:

Teacher 1: It was a wonderful project. I enjoyed it immensely. And the kids really enjoyed it too . . . We would spend at least an hour a day working on the project. And, most of the time, we spent much more than an hour. There was so much that could have been done and that we did do with the project that we could have spent all day, every day on it . . . We managed to get just about every subject area involved in the project. Lots of writing. Mathematics. Science of course. Spelling.

The teacher expresses that the motivation level of the students was considerably higher than she normally sees for interdisciplinary science projects:

Teacher 1: One of the best things that I liked about the project was just the whole motivation; the kids were really psyched up about the whole, the whole thing. They were excited about it. We'd talk about the weather constantly. And they worked in their groups. And it was just kind of a cohesiveness to the whole class while we were doing the project.

Interviewer: . . . It sounds like . . . they were pretty motivated to do it. Is that similar to or different from other projects that you usually do for science or other topics?

Teacher 1: . . . This one, I think, . . . was especially good because of the involvement with the other schools. And the kids would talk about the kids in the other schools. And they just felt that there was a specialness, a special quality to it because they were doing the same things as all these kids in the different schools . . . And geography . . . we had a world map set up the whole time and we would just constantly check things out. When we got a message from the other schools, we would always check the map to make sure we knew where the kids were from. And it just opened up so many different areas of conversation and exploration.

The success levels and high motivation at this site at least partially can be attributed to the large number of notes sent and received at this site. The New York site participants testified that they had successful correspondence with about 25 other schools – virtually every project participant. Clearly, at this site, there were no problems with the lack of return messages as experienced at other schools where this was the greatest project disappointment. Research results lead us to believe that the New York teacher worked hard to ensure that a much larger number of notes went out successfully than was guaranteed at many of the other locations.

In addition to receiving many responses from others, the New York site appeared to obtain valuable information from others, both from local and from distant locations:

Interviewer: Do you think that your students learned something from the other sites, from the kids and also from the meteorologists?

Teacher 1: Yes. They definitely learned things. . . . cultural things . . . a message from the school in Scotland telling about how they had some treacherous weather and the school bus almost didn't make it to school. They could really identify because we have much the same weather here. But then Scotland's flowers started blooming and we were still covered under three feet of snow. So that was interesting too. Another thing that they learned from kids in Colorado was that business about the red days and blue days, about burning wood stoves . . . We were totally fascinated with that. So there was a great deal of learning that went on . . . Our meteorologist was great . . . There were some questions the kids could not find answers to, and he would always have the answer back the next day. . . . It turned out that he and I had some mutual friends . . . I had my student . . . who was very, kind of, . . . really into higher mathematics. And David hooked us up with another student at the college who is majoring in mathematics and then he and my student kept a mathematics relationship going on after the project ended. So that was another bonus.

Therefore, it appears as if the New York site was able to implement a successful dialogue and image-based weather curriculum which was an especially motivating learning opportunity for students, despite a lack of state-of-the-art technological equipment. The following example illustrates another aspect of the New York experience which contributed to its success.

*Vignette 2: An Example of the Project Really Making a Difference – Sharing Your Storm in Real-Time With Those Less Fortunate*

In this vignette, the teacher describes one particular weather occurrence and the corresponding KGS activities which helped capitalize on that event for participation in many other locations:

> Teacher 1: Oh, it was fantastic. Because we had our biggest blizzard of the year happen right in the middle of the project. And it was so great because we had just learned about the barometer and all that business. And, just that week the barometer was taking a real nose dive. And one of the kids said, 'Look at the barometer. We're going to have a big storm.' And the next day was when we got that huge storm. We got 26 inches of snow. That was really one of the highlights of the project. And then, of course, the communication with the other schools went really thick and fast because we were telling everybody about the blizzard. And, you know kids were writing back from St. Petersburg, Florida saying, 'Oh, you guys are so lucky'. And it was ... that part that was really, really neat.

A teacher from Florida also comments on this storm from her perspective:

> Teacher 2: We have very mild weather during this project. We didn't even get any rainfall. . . . We did have fun watching the fronts that were going across [the continental USA] 'cause there was some pretty crazy weather further north. And we could watch that. And what we'd do is take the page out of the newspaper that shows, you know, that map with the fronts and stuff, and we'd put it next to the computer where our satellite pictures come in. And then, when a picture would come in, we'd be able to see that those clouds are often exactly where those little jagged lines on the newspaper map were. And so that was kind of neat for them to really see that the map is really showing the real thing.

In previous Global Exchanges, particularly interesting or sensational weather events occurred at one or more participating sites during the project duration. In a unique way each time, the power of multiple views of these real-time events, whether they were satellite pictures, newspaper maps, observations of the sky outside or peer dialogue on the topic, served as particularly interesting group learning opportunities. Called Real-Time Knowledge (Songer 1996), these events allowed students to be closer to dramatic scientific events which they could not experience first hand. In general, this ability to couple real-time or near-time data with personalized information from students in that location has continuously proven to be an important learning opportunity for all involved. In many cases, teachers testify after the exchange that these events were the learning highlights of the project.

*Vignette 3: The Most Important Learning*

Finally, the New York teacher commented on what she believed to be the most important learning the students experienced:

> Interviewer: ... [D]id you think that they learned a lot about weather . . .?
> Teacher 1: Yeah. I think one thing that they really learned was why we get so much rain around here . . . They definitely learned about the barometer. They learned about humidity. They learned about clouds. . . . Although they did learn . . . much content, I think probably the learning about the technology, the working together, probably were more valuable than the actual content. Although there are still kids who are watching the Weather Channel faithfully at home every day because they got so into it. And they're talking about the barometer. So, I think some things will stick with them.

It is interesting how similar this quote is to that from Dr Harper, the scientist heading the Antarctica telescope project: 'The human benefits are at least as important to South Pole astronomy as is the Internet's remote-control capability . . .'.

Clearly, these technologies are allowing students and professional scientists to interact with scientific data in new ways. Perhaps, more importantly, the technology is allowing individuals to work more closely together, both to the science itself and to the other individuals involved. Thus, the science education experience takes on more of a personalized or social collaborative experience, rather than a solitary one.

## SUMMARY AND IMPLICATIONS

Despite considerable challenges in implementing this complicated curriculum, Global Exchange appears to offer new opportunities for middle school students to become closer to science and to those experiencing science first-hand. We believe that the successes are due to:

- transforming students from consumers of information to reporters and participants;
- using communication as 'the hook' to get students to engage in meaningful scientific dialogue and explanation-building;
- building new relationships with the subject matter through greater personalization of information.

Each of these areas is discussed in greater detail below.

## Transformation of Students from Consumers of Information to Reporters and Participants

Students in New York and other locations appeared to benefit from the Global Exchange approach as they experienced the excitement and empowerment of sharing their storm 'with those less fortunate'. This ability to capitalize on the responsibility to share what they have learned as a part of their own learning appears similar to the benefits of student tutoring. It also provides a somewhat unique means of adding authenticity to student learning.

Other educational researchers utilizing the Internet for correspondence also mention this benefit: 'My students moved from reading about current events into reporting them as they exchanged e-mail with students in Moscow during last year's coup' (Serim 1994).

Clearly, these electronic communities do not always work effectively. Recent research demonstrates that several key components of successful electronic collaborations include (1) an exchange facilitator who monitors dialogue and assists when problems arise, (2) users who work with concrete tasks in which all have a shared interest, (3) exchanges that encourage analysis and (4) technical problems that are minimal (Honey, Tally & McMilan 1995).

## Communication as 'The Hook' to Motivate Deeper Science Learning

When the New York site participants experienced their biggest blizzard of the year during the Exchange, students were able to capitalize on the communication aspects of the Internet to share their storm with others. The teacher discussed how important that communication was to help encourage her students to learn more about the science of the storm. Having not only a real audience which was interested in their understandings, but also an audience of peers who were simultaneously focused on the topics which they were investigating, contributed greatly to the higher motivation experienced by the New York students. The chance to tell others what they knew of the blizzard, and to share what they learned, appear to have deepened students' interest in learning the science.

## Building New Relationships with the Science Content

Finally, the opportunity to experience real-time events in multiple representations appears to provide students with an opportunity to develop new relationships with the content which they are studying. Students collect their own data, visualize professional imagery or weather maps, obtain explanations for weather phenomena from scientists or peers and send their own questions to others who might have more information than themselves. When utilized successfully and integrated appropriately, the power of these multiple, more personal representations of the study phenomena appeared to strengthen student understandings and motivation.

In summary, the KGS project provides new opportunities for students to recognize and build on the social nature of learning as students and teachers learn to value the intertwining of social and personal constructs as a part of knowledge development and construction. Students utilize others and first-hand data for personal communication and expertise. Therefore, we speculate that the key to education with new telecommunications technologies is the power of capitalizing on the human factor, the individuals on either ends of the wires, for the achievement of complex educational goals.

## REFERENCES

Black, L., Klingenstein, K. & Songer, N.B.: 1995, 'Part One: Observations from the Boulder Valley Internet Project', *T.H.E. Journal* 22(10), 75-80.

Brown, A.L., Ash, D., Rutherford, M., Nakagawa, K., Gordon, A. & Campione, J.C.: 1993, 'Distributed Expertise in the Classroom', in G. Salomon (ed.), *Distributed Cognitions: Psychological and Educational Considerations*, Cambridge University Press, New York, 188-228.

Browne, M.W.: 1995, 'South Pole Ready for Internet Revolution', *The New York Times*, January 10, B5 & B9.

Collins, A., Brown, J.S. & Holum, A.: 1991, 'Cognitive Apprenticeship: Making Thinking Visible', *American Educator* 15(3), 6-11 & 38-39.

Honey, M., Tally, B. & McMilan, K.: 1995, 'On-line Communities: They Can't Happen Without Thought and Hard Work', *Electronic Learning* 14(4), 12-13.

Lewis, E.L. & Linn, M.C.: 1994, 'Heat Energy and Temperature Concepts of Adolescents, Naive Adults and Experts: Implications for Curricular Improvements', *Journal of Research in Science Teaching* 31, 657-677.

Linn, M.C., diSessa, A., Pea, R.D. & Songer, N.B.: 1994, 'Can Research on Science Learning and Instruction Inform Standards for Science Education?', *Journal of Science Education and Technology* 3(1), 7-15.

Linn, M.C., Songer, N.B. & Eylon, B.: in press, 'Shifts and Convergences in Science Learning and Instruction', in D. Berliner & R. Calfee (eds.), *The Handbook of Educational Psychology*, Macmillan, New York.

Putnam, R.T. & Borko, H.: in press, 'Teacher Learning: Implications of New Views of Cognition', in B. Biddle, T.L. Good & I.F. Goodson (eds.), *The International Handbook of Teachers and Teaching*, Kluwer, Dordrecht, The Netherlands.

Serim, F.: 1994, 'Notes From A Virtual Panelist', *The Consortium for School Networking (CoSN) Discussion List.*, cosndisc@yukon.cren.org., October 12.

Songer, N.B.: 1995, 'Ways of Seeing: Exploring the Educational Potential of Real-Time Imagery and Communication in K-12 Science Classrooms', Paper presented at the annual meeting of the Geological Society of America, New Orleans, LA.

Songer, N.B.: 1996, 'Exploring Learning Opportunities in Coordinated Network-Enhanced Classrooms: A Case of Kids as Global Scientists', *The Journal of the Learning Sciences* 5, 297-327.

Wellesley College Center for Research on Women: 1992, *How Schools Shortchange Girls*, American Association of University Women Educational Foundation, Washington, DC.

White, B.Y.: 1989, 'The Role of Intermediate Abstractions in Understanding Science and Mathematics', *The Proceedings of the Eleventh Annual Meeting of the Cognitive Science Society*, Lawrence Erlbaum, Hillsdale, NJ, 972-979.

# 3.5 Problem-Based Macro Contexts in Science Instruction: Design Issues and Applications

ROBERT D. SHERWOOD, ANTHONY J. PETROSINO, XIAODONG LIN and COGNITION AND TECHNOLOGY GROUP AT VANDERBILT
*Vanderbilt University, Nashville, USA*

This chapter considers the development and application of a constructivist model of instruction which we call 'anchored instruction' and the implications that this model has for science education. In the first section, we describe the evolving theory of anchored instruction and associated design principles. The second section of the chapter provides examples of projects that have used these design principles, and briefly summarise related research findings. Some of the implications and opportunities for teachers as they implement these designs are considered in the third section.

## ANCHORED INSTRUCTION

*Anchoring Instruction in Meaningful Contents*

We propose that a major goal of instruction is to allow students and teachers to experience the kinds of problems and opportunities that experts in various areas encounter (Bransford, Sherwood, Vye & Rieser 1986; Perfetto, Bransford & Franks 1983). Experts in an area are immersed in phenomena and are more familiar with how they have been thinking about them than are novices (e.g., Chi, Glaser & Farr 1991). When introduced to new theories, concepts and principles that are relevant to their areas of interest, experts experience the changes in their own thinking that these ideas afford (e.g., Dewey 1933; Hanson 1970; Schwab 1960). For novices, however, the introduction of concepts and theories often seems like the mere addition of new facts or mechanical procedures to be memorised rather than major components of the discipline. Because novices have not been immersed in the phenomena being investigated, they are unable to experience the effects of the new information on their own noticing and understanding.

The anchored instruction approach represents an attempt to help students

---

Chapter Consultant: Instead of using chapter consultants, chapter authors in the Educational Technology section provided editorial comments on each other's chapters.

become actively engaged in learning by situating or anchoring instruction in interesting and realistic problem-solving environments. These environments are designed to invite the kinds of thinking that help students to develop *general* skills and attitudes that contribute to effective problem solving, in addition to acquiring *specific* concepts and principles that allow them to think effectively about particular domains (Cognition and Technology Group at Vanderbilt 1990, 1992a). Anchored instruction environments, like inquiry environments, do not 'directly' instruct students but provide a situation in which learning can take place (Schwab 1978).

Anchored instruction also reflects constructivist theories (e.g., Bransford & Vye 1989; Clement 1982; Duffy & Bednar 1991; Minstrell 1989; Perkins 1991; Scardamalia & Bereiter 1991; Schoenfeld 1989). According to a constructivist perspective, knowledge is actively constructed by learners through interaction with their physical and social environments and through the reorganisation of their own mental structures (Brown, Collins & Duguid 1989; Cobb, Yackel & Wood 1992; Wheatley 1991). Instead of having teachers 'transmit' information that students then 'receive', these theorists place a great amount of importance on having students become actively involved in the construction of knowledge. For instance, constructivist theorists assist students to construct and coordinate effective problem representations through the use of physical and symbolic models, through reasoning and argumentation, and through deliberate application of problem-solving strategies (e.g., Bransford & Stein 1993; Brown, Collins & Duguid 1989; Clement 1982; Minstrell 1989; Palincsar & Brown 1984; Scardamalia & Bereiter 1991; Schoenfeld 1989). A fundamental assumption of the constructivist position is that students cannot learn to engage in successful knowledge-building activities simply by passively being told new information (Bransford, Franks, Vye & Sherwood 1989). Rather, students need repeated opportunities to engage in sustained exploration, assessment and revision of their ideas over extended periods of time and through a series of reiterations (Cognition and Technology Group at Vanderbilt 1992b). Anchored instruction engages students in realistic problem-solving environments that allow them to make modifications to their current understanding.

More extended discussions of anchored instruction can be found in publications by the Cognition and Technology Group at Vanderbilt (1990, 1992c). There we note that the general idea of anchored instruction has a long history and is related to ideas about project-based learning (Dewey 1933), case-based learning (Gragg 1940) and problem-based learning (e.g., Barrows 1985; Williams 1992). We also connect our use of anchors to the 'situated cognition' arguments of Brown, Collins & Duguid (1989) (see Moore, Lin, Schwartz, Petrosino, Hickey, Campbell, Hmelo & Cognition and Technology Group at Vanderbilt 1996). We have studied the uses of anchored instruction in domains that focus on literacy (Bransford, Kinzer, Risko, Rowe & Vye 1989; Kinzer, Risko, Goodman, McLarty & Carson 1990; Kinzer, Risko, Vye & Sherwood 1988), mathematics (Cognition and Technology Group at Vanderbilt 1992a, 1993a, 1993b, 1993c) and science (Cognition and Technology Group at Vanderbilt 1992c; Goldman, Petrosino, Sherwood, Garrison, Hickey, Bransford & Pellegrino 1996; Sherwood, Petrosino, Goldman, Garrison, Hickey, Bransford & Pellegrino 1993). We discuss some of the lessons learned about ways to increase the power of anchored instruction to produce flexible transfer in Barron, Vye, Zech, Schwartz, Bransford, Goldman, Pellegrino, Morris, Garrison & Kantor 1995; Cognition and Technology Group at Vanderbilt 1993a, 1994.

*Design Principles for Anchored Instruction*

There are seven design principles that guide our work in the variety of domains mentioned previously (Cognition and Technology Group at Vanderbilt 1993a). These design principles mutually influence one another and operate as a Gestalt rather than as a set of independent features of the materials. For example, the *narrative format*, the *generative design of the stories* and the fact that the adventures include *embedded data* make it possible for students to learn to generate problem-solving goals, find relevant information and engage in reasoned decision making. The *complexity of the problems* helps students deal with this important aspect of problem solving and the *use of video* helps make the complexity manageable. The video format also makes it easier to embed the kinds of information that provide opportunities for *links across the curricula*. It is important for *pairs of episodes* to be developed to afford discussions about transfer of problem-solving skills.

The design principles of anchored instruction are also consistent with a constructivist cognitive science perspective. If students are actively going to make changes in their cognitive structures, we need to create environments that afford them the opportunities to do this. Using the design principles of anchored instruction, we have begun to provide these types of environments. Students need interesting and challenging problem situations that are embedded in familiar and engaging scenarios in order to facilitate the utilisation of the knowledge which they already have available. We believe that this assists the learner in the creation of new and modified knowledge structures.

## APPLICATIONS OF THE ANCHORED INSTRUCTION DESIGN PRINCIPLES

*The Scientists-in-Action Series*

There are two features of the instantiation of the design principles for anchored instruction that are unique to our video-based middle school science project, the *Scientist-in-Action* series. First, challenges are posed several times during the course of the story rather than at the end as in our earlier work in mathematics, *The Adventures of Jasper Woodbury* (Cognition and Technology Group at Vanderbilt 1993b). These 'interruptions' to deal with a problem enable students to be a part of the problem-solving process and to have multiple opportunities to work on the same problems on which the scientists in the video are working. When the video resumes, students can compare and contrast their solutions to what the scientists actually did. There is a tradeoff in this design in contrast with the design of the

Jasper adventures: the modification made in the *Scientist in Action* episodes provides fewer opportunities for students to gain experience in formulating solutions to complex, ill-defined problems due to the fact that the problems at the interruption points might not be as complex or as ill-defined as problems that would come at the end of a video story.

A second modification of the original anchored instruction design principles is that much of the data needed to solve the problem occur in ancillary materials rather than being embedded in the video, which was the design used in the Jasper series. These ancillary materials are authentic (topographic maps, actual data from experiments on water quality and lead in paint, etc.). Further, teachers are encouraged to help students to conduct laboratory experiences in the classroom.

While our initial studies of a pilot episode were encouraging (Goldman *et al.* 1996; Sherwood *et al.* 1993), they also tended to show that our pilot episode was somewhat limited in the depth to which students became engaged in the problem-solving process. This was due to the fact that the problems encountered by the students were identified in the video and they were asked to solve them. This lack of *generativity* in the design was a concern to the development team. Other issues, such as the authenticity of the situation and the need to bring students into situations with more extended data analysis and hypothesis testing, drew the development team to reconsider some of the design features mentioned previously and to modify them further.

The three completed episodes still reflect many of the design principles that were used in the pilot version, but with some modifications. The episodes all use a narrative with a *video format*, although the video format is being supplemented with more materials such as maps, data sets, simulations and reference materials. While the problems posed in the adventure offer students a chance to work within a problem space that is *complex*, they do not have answers that are based simply on one or two scientific principles. Although data are still *embedded* in the video, more data are available to the students from sources not directly in the video but referred to by the characters and utilised by the students to solve the overall problem. The data are currently accessible via paper-and-pencil handouts but we anticipate using CD-ROM, commercial database applications and electronic networks in the final version. We are also experimenting with the inclusion of anomalous data (Chinn & Brewer 1993) in order to facilitate the development of various critical thinking skills among middle school students. *Links across the curricula*, especially social issues related to science, have been established.

In order to discuss some of the design principle modifications and initial research conducted in the project, some indication of the nature of the first episode is needed. The first episode deals with the issue of water quality and the degradation of it by some action. The story line starts with a team of scientists and students at a local school conducting an extensive water quality survey of a local river system. Pairs of students and scientists conduct 'electronic field trips' to various sites on the river and collect such data as chemical tests for dissolved oxygen, ammonia levels, pH and temperature. In addition, biological surveys such as benthic macro invertebrate sampling are performed to give a picture of overall water quality. Data, which could be in the form of raw numbers, video clips of various sections of the river and pictures of invertebrates, are sent back to the 'base' classroom through electronic means. The technology to do this in real life is not completely developed but could be available by the time the series is completed (1998).

Students who are using the program have an opportunity to work in small groups, summarise raw data, develop graphs and charts for classroom discussion and research the science behind the particular tests used in the video. We believe that, in this manner, students will develop a stronger understanding of the environmental system under study and be prepared to tackle the 'incident' problem that will occur next in the video.

In the episode, a noticeable change in the data collected from a new sample is an indication to the students that a change in the river system has occurred, and that they need to collect additional data to compare to their baseline data to help isolate what has happened and to determine the source of the contamination. The story is designed so that the expected hypothesis generated by the students will be incorrect in terms of the source of the pollution. The video then provides additional data that allow students to rethink their initial ideas to determine the actual source and cause of the problem. In this manner, we allow students to experience some of the difficulties of doing science in natural systems (i.e., that first hypotheses often need rejection or refinement).

As indicated, the use of both *embedded and external* data, the *electronic field trip* component, substantial work with *baseline data* and more *hypothesis testing* are all substantial modifications of the pilot episode. These design changes along with the use of CD-ROM technology to replace the videodisc technology of our pilot and earlier Jasper project has allowed us to develop a product that appears to be both interesting and challenging to middle school science students.

*The Role of Field Testing in Design and Development*

We would like to illustrate how we integrate design and development processes with field experiments and testing in order to modify design decisions effectively. The script for the first episode, *The Stones River Mystery*, not only reflected the design principles discussed in the previous sections, but also took into consideration our understanding of students' domain knowledge through field testing and review of existing literature. The goal is always to build on students' current understanding rather than simply prescribing activities designed to help students reach pre-set objectives by particular points in time. To do so, the research and design of the video were concurrent, so that our most timely understanding of students' domain knowledge and conceptual understanding could be incorporated in the design processes. In this way, repetition of a scripted version of what worked someplace else was avoided, while retaining the flexibility to make the design and development more situational and reflective of classroom needs (Lin, Bransford, Hmelo, Kantor, Hickey, Secules, Petrosino, Goldman & Cognitive and Technology Group at Vanderbilt 1995).

There were three major phases of field testing during the design and development of the *Stones River Mystery*. The first phase focused on (1) students' conceptual understanding of water pollution (e.g., How do they understand water pollution? How do they think that pollution should be tested? What are the variables that should be considered while assessing water pollution? How should pollution be cleaned?); (2) whether the topic (water pollution in this case) selected for the video was motivating to students and teachers and whether the topic could be merged with existing curricula and goals of science learning; and (3) what kinds of conceptual connections should be made explicit in this video-based science learning environment and what kinds of connections should be left for students to discover.

As an example, throughout the design process in the first phase of this research design, we did several experiments with sixth and seventh graders on their knowledge about pollution. We designed various scenarios around concepts, such as dissolved oxygen testing and cleaning pollution, to study how students understand those concepts. Questions were asked, such as the following. 'Imagine you have an opened bottle of soda. If you don't open the soda, you cannot see bubbles. However, when you open the bottle, we see a lot of bubbles trying to come out. Can you explain where these bubbles come from?' We found that most students did not understand the notion of a gas dissolved in a liquid. Additionally, if we asked them from where fish get their oxygen, they had little idea. When these findings were reported to the design team, we immediately revised the existing episode to add more situations in which dissolved oxygen could be further explored and discussed. For instance, we added to the script one incident in which one of the students in the *Scientists In Action* team accidentally slipped into the river. He asks the scientists in the team what algae are good for except making people slip. We took this opportunity to explain how the dissolved oxygen can be produced by having the field biologist explain that, 'when the sun shines on the algae, they produce oxygen that fish and invertebrates need'. Further discussions around the notion were evoked by the incident among the students and the scientists.

In addition to our concerns over the domain knowledge acquisition in the design process, we also gave considerable attention to how students reacted to the materials, acting and the learning environments created in the video. We continued the field experiments even after the video was produced so that further revisions could be made according to the suggestions and reactions from teachers and students. These experiments helped to guide our post-production editing. Additionally, we synthesised the results from our field experiments, our review of existing literature and the suggestions from teachers, our national review panels and students to guide further improvements in the design of the hands-on laboratory experiments and teacher guidelines to go with the episode.

The second sets of field studies extended the first developmental set by investigating whether the video, *The Stones River Mystery*, served as a good problem-based macro-context in which to situate further learning and instruction. Two classes of seventh grade students, with a total of 25 students in each class, were involved in the study. The treatment class watched *The Stones River Mystery* and subsequently read a chapter that had content related to the science and problem solving involved in the video. The control class followed the same protocol except that they watched the video called, *Return to Rochester* (the second episode in the series), instead of *The Stones River Mystery*. Thus, the treatment class saw a video that situated their readings, whereas the control class did not. This minor study, in which neither group of students carried out ancillary activities associated with *The Stones River Mystery* (i.e., hands-on experiments, computer simulations, or solving the problems in the video), provided a way to establish the baseline effects of a video-based macro-context.

We looked at three primary sets of dependent measures: motivational, problem-solving and noticing measures. With respect to motivation, we asked students to rate their on-line 'intellectual energy' using a scale from 1 to 7. As they watched the video, read the chapter and worked on the pretest and posttest, we interrupted them every few minutes to mark down their intellectual energy level. The treatment group reported a higher level of intellectual energy when reading the chapter than the control group. We infer that the context created by *The Stones River Mystery* made the reading more 'interesting'. The treatment group also showed a higher level of energy when working on the posttest items. This might have been because they experienced an increased sense of efficacy when working on the problems.

On the pretest and posttest, we asked questions that required the students to solve problems that they had not seen or read about directly. For example, one question asked why putting too much food in a fish tank could kill the fish, even if they did not over eat. This question was targeted at the idea that pollution often has indirect effects and is not simply a poison (a misconception that we discovered in our field work prior to producing the video). The results revealed that the control group did not improve on these types of questions (although sufficient information was in the chapter), whereas the treatment group improved substantially. Finally, we examined the children's ability to notice actions or situations that were relevant to pollution and its measurement. One of the potential benefits of a video-based macro-context, in addition to situating subsequent instruction, is that it provides students with a virtual experience whereby they can make contact with the phenomena of interest. Consequently, we believe that students who watch the SIA adventures will be more likely to notice and identify scientifically relevant things in everyday life. In the current study, we could not take the students to a river to see if they would notice situations relevant to pollution. Instead, we made a five-minute video of two college students who spent the day monitoring water quality. The video included many events that could be considered relevant to pollution, as well as events that reflected mistakes in procedure and interpretation. For example, they have lunch at a 'burger joint' at one point in the video, they drive past run-off barriers at a construction area at another point, and they interpret a lack of macro-invertebrates as a sign that the water is clear and unpolluted at yet another. We simply showed the children in the study the 'noticing video' at posttest and asked them to write down what they noticed. The results indicated that the treatment group was more attuned both to pollution relevant situations and to indicators of pollution.

The results from these initial field studies suggest that the video-based, macro-context format of the SIA adventures has beneficial effects, even without extensive wrap-around instruction.

*Other Projects Related to Science Using a Similar Design*

Two other projects are being undertaken by teams of researchers at Vanderbilt and elsewhere using anchored instruction principles. These include *Mission to Mars* and *Schools for Thought*. The *Mission to Mars* learning environment (Hickey, Petrosino, Pellegrino, Bransford, Goldman & Sherwood 1994) is designed to lead students to generate problems about the scientific challenge of planning a human expedition to the planet Mars, and then support student inquiry into solving these problems. As with several of the projects using anchored instruction, a short video has been developed as the 'anchor' around which instruction is developed. Students then work in cooperative groups to plan the mission and report their designs to their peers.

In *The Schools For Thought Project*, three groups of researchers are cooperating with sites in three diverse geographic locations to test an integration of projects that have showed potential for instruction. The three projects being integrated are:

- *Fostering Communities of Learners*, developed by a team at the University of California at Berkeley (Brown, Ash, Rutherford, Nakagawa, Gordon & Campione 1993; Brown & Campione 1990; Brown & Campione 1994).
- The *CSILE* project (Computer Supported Intentional Learning Environments) developed at the Ontario Institute for Studies in Education (Scardamalia, Bereiter, Brett, Burtis, Calhoun & Smith 1992; Scardamalia, Bereiter & Lamon 1994; Scardamalia, Bereiter, McLean, Swallow & Woodruff 1989).
- The *Jasper Woodbury Project* previously mentioned in this chapter (Cognition and Technology Group at Vanderbilt 1990; 1992b; 1993a, 1993b).

A more extensive report on the integration project can be found in Lamon (1993) but, in summary, the project is using the instructional designs outlined by *Fostering Communities of Learners*, with the computer-based learning environment of *CSILE* and the anchored instruction materials from the *Jasper Woodbury Project* to change radically the school-day experiences for children in Nashville (Tennessee), Ontario (Canada) and Oakland (California) field sites.

*Projects at Other Institutions*

In this section, we would like to direct readers of this *Handbook* to chapters by our colleagues who are currently implementing long-term projects at institutions around the USA. These projects, such as the ThinkerTools (White 1993) described by Barbara White in a chapter in this section of the *Handbook*, the Kids as Global Scientists project under the leadership of Nancy Songer (Songer 1993), the CoVis (Collaborative Visualization) project discussed by Daniel Edelson (Edelson & O'Neill 1994) and the Computers as Learning Partners project by Marcia Linn's group (see Linn's chapter in this *Handbook*), have attempted to integrate advances in educational technology and cognitive psychology into the field of science education.

Four key design principles span across these research programs. First, all groups place importance on collaboration and inquiry activities (Linn 1986; White 1993). According to Linn, diSessa, Pea and Songer (1994), large collaborations such as the Human Genome Project and the high energy physics groups, are necessary for the advancement of many fields. Secondly, all groups utilise everyday problems (also known as 'real world' problems) in making content and instruction both relevant and motivating for the students. Thirdly, we believe that these projects have been committed to the incorporation of technology to facilitate such aspects of learning as mental model construction (Computers as Learning Partners, ThinkerTools, CoVis), multiple representations (ThinkerTools, Computers as Learning Partners, CoVis) and innovative use of the Internet and telecomputing applications (CoVis, Kids as Global Scientists, CSILE), to bridge the distance between everyday discourse and that of scientific debate and vocabulary building (CoVis, CSILE). Finally, these groups have attempted to merge technology, pedagogy and epistemology as they seek to create classrooms which model the scientific enterprise of a research community. To this end, these projects represent long-term commitments between the research institution and the local school system. Staff of these projects have done their best to develop and sustain active communities of learners (Rogoff 1990) and practitioners (Lave & Wenger 1991) in which teachers, students, researchers and content area specialists pool their respective expertise to create classrooms in which thinking becomes visually, literally and orally apparent.

While each project certainly has its own distinguishing and unique characteristics and points of emphasis – such as Linn's advocacy for respect for diversity of opinion as well as representation issues of females in science classrooms (Agogino & Linn 1992; Linn & Songer 1991) or Edelson and his group's visionary use of the extraordinary resources and power of the Internet in their CoVis project – it is in the shared characteristics mentioned previously that we begin to see the exciting actualisation of theory and research findings in applied settings.

## IMPLICATIONS FOR TEACHERS USING ANCHORED INSTRUCTION DESIGNS

Scardamalia and Bereiter (1991), in discussing knowledge building in children, postulate that there are three different types of teachers in current instructional settings. The type 'A' teacher holds a 'task' model which fits the current style of many classrooms and focuses on seatwork (tasks) to be completed by the student. The type 'B' teacher, in contrast, holds a 'knowledge-based' model. Scardamalia and Bereiter claim that, in this model, the 'focus tends to be on understanding, and the teacher's role includes setting cognitive goals, activating prior knowledge, asking stimulating and leading questions, directing inquiry, and monitoring comprehension' (p. 39). They note that this model is one that is often taught as a model of exemplary teaching and is often found in teachers' guides. Their reservations about the model centre on the fact that the teacher is very much in 'control' of the learning process. Their model 'C' teacher exhibits the characteristics of the model 'B' teacher, but with the addition of trying to turn over to the students the higher-level processes that the teacher would control in model 'B'. Scardamalia and Bereiter note that 'there is a concern with helping student to formulate their own goals, do their own activating of prior knowledge, ask their own questions, direct their own inquiry, and do their own monitoring of comprehension' (p. 39). They note that this model is followed in the reciprocal teaching model they have developed (Brown & Palincsar 1989; Palincsar & Brown 1984). This type 'C' teacher is also consistent with Wheatley's (1991) call for the teacher to view the learner 'more as a growing tree than a sponge' (p. 19).

In order for the teacher to use effectively the materials proposed in this chapter, several things need to occur. First, teachers' views of instruction must undergo some modification if they are from type 'A' or, to some extent, from the type 'B'. For some teachers, this will be a very large step while, for others, it will seem a natural progression. Secondly, our work with the Jasper series has indicated that there were several 'special challenges' that needed to be attended to in working with the Jasper materials:

- allowing students to pursue 'wrong' pathways that lead to alternative solutions that might not be optimum;
- knowing when to assist students versus letting them 'struggle' a little longer;
- where to put alternative materials such as Jasper (and the ones currently under development) in the 'regular' curriculum (see also Hofwolt 1992).

We also realise that many of our teachers might need additional subject knowledge support. Our experiences indicate that this support is particularly beneficial at the lower grade levels (fifth and sixth) of our target grades (fifth through eighth). In some Jasper episodes, teachers have reported difficulty with some of the statistics and geometry concepts. This is not surprising, given that they might not have taught these particular concepts and their instruction in them could have been several years ago. We are attempting to address this issue both through video and print materials to assist teachers to become more comfortable with the mathematics and science that are involved in the episodes.

## CONCLUDING REMARKS

While the projects described in this chapter are all relatively new and evaluative studies are only at the preliminary stages, it would appear that they offer promise for science instruction. They require school environments to be structured differently with both the teacher and student taking on roles that, in many ways, are not traditional. Teachers will need to be more guiders of instruction who help students to develop their understandings. Students will need to be able to act as active pursuers of knowledge and not just passive acceptors of information. It will require cooperation of parents and school systems to allow trials of various programs, and it will place demands on the research and development communities to create and research innovative and theoretically-sound instructional materials.

## NOTE

The Cognition and Technology Group at Vanderbilt (CTGV) is an interdisciplinary team associated with the Vanderbilt Learning Technology Center, Box 45, Peabody College, Vanderbilt University, Nashville, TN 37203, USA. Members of the CTGV who contributed to this article are John Bransford, Susan Goldman, Dan Hickey, Clifford Hofwolt, Mary Lamon, Teresa Secules and Dan Schwartz. Partial funding for various aspects of the research and development described in this chapter were provided by the James S. McDonnell Foundation, NASA, and The National Science Foundation (Grant # ESI–9350510). Opinions expressed are those of the authors and not necessarily those of the funding agencies.

## REFERENCES

Agogino, A.M. & Linn, M.C.: 1992, 'Retaining Female Engineering Students: Will Early Design Experiences Help?', *NSF Directions* 5(2), 8–9.
Barron, B., Vye, N.J., Zech, L., Schwartz, D., Bransford, J.D., Goldman, S.R., Pellegrino, J., Morris, J., Garrison, S. & Kantor, R.: 1995, 'Creating Contexts for Community Based Problem Solving: The Jasper Challenge Series', in C. Hedley, P. Antonacci & M. Rabinowitz (eds.), *Thinking and Literacy: The Mind at Work*, Lawrence Erlbaum, Hillsdale, NJ, 47–71.
Barrows, H.S.: 1985, *How to Design a Problem-Based Curriculum for the Preclinical Years*, Springer, New York.
Bransford, J., Franks, J.J., Vye, N.J. & Sherwood, R.D.: 1989, 'New Approaches to Instruction: Because Wisdom Can't Be Told', in S. Vosniadou & A. Ortony (eds.), *Similarity and Analogical Reasoning*, Cambridge University Press, New York, 470–497.
Bransford, J., Kinzer, C., Risko, V., Rowe, D. & Vye, N.: 1989, 'Designing Invitations to Thinking: Some Initial Thoughts', in S. McCormick, J. Zutrell, P. Scharer & P.O. Keefe (eds.), *Cognitive and Social Perspectives for Literacy Research and Instruction*, National Reading Conference, Chicago, IL, 35–54.
Bransford, J.D., Sherwood, R.D., Vye, N. & Rieser, J.: 1986, 'Teaching Thinking and Problem Solving', *American Psychologist* 41, 1078–1089.
Bransford, J. & Stein, B.S.: 1993, *The IDEAL Problem Solver* (second edition), Freeman, New York.
Bransford, J.D. & Vye, N.J.: 1989, 'A Perspective on Cognitive Research and Its Implications for Instruction', in L. Resnick & L.E. Klopfer (eds.), *Toward the Thinking Curriculum: Current Cognitive Research*, Association for Supervision and Curriculum Development, Alexandria, VA, 173–205.
Brown, A.L., Ash, D., Rutherford, M., Nakagawa, K., Gordon, A. & Campione, J.C.: 1993, 'Distributed Expertise in the Classroom', in G. Salomon (ed.), *Distributed Cognitions: Psychological and Educational Considerations*, Cambridge University Press, New York, 188–228.

Brown, A.L. & Campione, J.C.: 1990, 'Communities of Learning and Thinking or a Context by Any Other Name', *Human Development* 21, 108–125.

Brown, A.L. & Campione, J.C.: 1994, 'Guided Discovery in a Community of Learners', in K. McGilly (ed.), *Classroom Lessons: Integrating Cognitive Theory and Classroom Practice*, MIT Press/Bradford Books, Cambridge, MA, 229–272.

Brown, A.L. & Palincsar, A.S.: 1989, 'Guided Cooperative Learning and Individual Knowledge Acquisition', in L.B. Resnick (ed.), *Knowing, Learning and Instruction: Essays in Honor of Robert Glaser*, Lawrence Erlbaum, Hillsdale, NJ, 393–451.

Brown, J.S., Collins, A. & Duguid, P.: 1989, 'Situated Cognition and the Culture of Learning', *Educational Researcher* 18(1), 32–42.

Chi, M.T.H., Glaser, R. & Farr, M.: 1991, *The Nature of Expertise*, Lawrence Erlbaum, Hillsdale, NJ.

Chinn, C.A. & Brewer, W.F.: 1993, 'The Role of Anomalous Data in Knowledge Acquisition: A Theoretical Framework and Implications for Science Instruction', *Review of Educational Research* 63, 1–49.

Clement, J.: 1982, 'Algebra Word Problem Solutions: Thought Processes Underlying a Common Misconception', *Journal of Research in Mathematics Education* 13, 16–30.

Cobb, P., Yackel, E. & Wood, T.:1992, 'A Constructivist Alternative to the Representational View of Mind in Mathematics Education', *Journal for Research in Mathematics Education* 19, 99–114.

Cognition and Technology Group at Vanderbilt: 1990, 'Anchored Instruction and Its Relationship to Situated Cognition', *Educational Researcher* 19(6), 2–10.

Cognition and Technology Group at Vanderbilt: 1992a, 'The Jasper Experiment: An Exploration of Issues in Learning and Instructional Design', *Educational Technology Research and Development* 40(1), 65–80.

Cognition and Technology Group at Vanderbilt: 1992b, 'The Jasper Series as an Example of Anchored Instruction: Theory, Program Description, and Assessment Data', *Educational Psychologist* 27, 291–315.

Cognition and Technology Group at Vanderbilt: 1992c, 'Anchored Instruction in Science and Mathematics: Theoretical Basis, Developmental Projects, and Initial Research Findings', in R. Duschl & R. Hamilton (eds.), *Philosophy of Science, Cognitive Psychology and Educational Theory and Practice*, SUNY Press, New York, 244–273.

Cognition and Technology Group at Vanderbilt: 1993a, 'Anchored Instruction and Situated Cognition Revisited', *Educational Technology* 33, 52–70.

Cognition and Technology Group at Vanderbilt: 1993b, 'The Jasper Experiment: Using Video to Furnish Real-World Problem-Solving Contexts' [Special issue], *The Arithmetic Teacher* 4, 474–478.

Cognition and Technology Group at Vanderbilt: 1993c, 'Toward Integrated Curricula: Possibilities from Anchored Instruction', in M. Rabinowitz (ed.), *Cognitive Science Foundations of Instruction*, Lawrence Erlbaum, Hillsdale, NJ, 33–55.

Cognition and Technology Group at Vanderbilt: 1994, 'From Visual Word Problems to Learning Communities: Changing Conceptions of Cognitive Research', in K. McGilly (ed.), *Classroom Lessons: Integrating Cognitive Theory and Classroom Practice*, MIT Press/Bradford Books, Cambridge, MA, 157–200.

Dewey, S.: 1933, *How We Think: Restatement of the Relation of Reflective Thinking to the Educative Process*, Heath, Boston, MA.

Duffy, T.M. & Bednar, A.K.: 1991, 'Attempting to Come to Grips with Alternative Perspectives', *Educational Technology* 31(9), 12–15.

Edelson, D.C. & O'Neill, D.K.: 1994, 'The CoVis Collaboratory Notebook: Supporting Collaborative Scientific Enquiry', in A. Best (ed.), *Proceedings of the 1994 National Educational Computing Conference*, International Society for Technology in Education in cooperation with the National Education Computing Association, Eugene, OR, 146–152.

Goldman, S.R., Petrosino, A.J., Sherwood, R.D., Garrison, S., Hickey, D., Bransford, J.D. & Pellegrino, J.W.: 1996, 'Anchoring Science Instruction in Multimedia Learning Environments', in S. Vosniadou, E. De Corte, R. Glaser & H. Mandl (eds.), *International Perspectives on the Psychological Foundations of Technology-Based Learning Environments*, Lawrence Erlbaum, Hillsdale, NJ, 257–284.

Gragg, C.I.: 1940, 'Because Wisdom Can't Be Told', *Harvard Alumni Bulletin*, October 19, 78–84.

Hanson, N.R.: 1970, 'A Picture Theory of Theory Meaning', in R.G. Colodny (ed.), *The Nature and Function of Scientific Theories*, University of Pittsburgh Press, Pittsburgh, PA, 233–274.

Hickey, D.T., Petrosino, A.J., Pellegrino, J.W., Bransford, J.D., Goldman, S.R. & Sherwood, R.D.: 1994, 'The Mars Mission Challenge: A Generative Problem Solving School Science Environment', in S. Vosniadou, E. De Corte & H. Mandl (eds.), *Technology-Based Learning Environments: Psychological and Educational Foundations* (NATO ASI Series), Springer-Verlag, Berlin, Germany, 97–103.

Hofwolt, C.: 1992, 'HyperScience for Middle School Students', Paper presented at the annual convention of the National Science Teachers Association, Boston, MA.

Kinzer, C.K., Risko, V.J., Goodman, J., McLarty, K. & Carson, J.: 1990, 'A Study of Teachers Using Videodisc Anchors in Literacy Instruction', Paper presented at the annual meeting of the National Reading Association, Miami, FL.

Kinzer, C.K., Risko, V., Vye, N.J. & Sherwood, R.D.: 1988, 'Macrocontexts for Enhancing Instruction', Paper presented at the annual meeting of the American Educational Research Association, New Orleans, LA.

Lamon, M.: 1993, *Cognitive Studies for Restructuring Middle School Education* (Report of the St Louis Science Center/St Louis Public Schools Middle School Curriculum Collaborative), St Louis Public Schools, St Louis, MO.

Lave, J. & Wenger, E.: 1991, *Situated Learning: Legitimate Peripheral Participation*, Cambridge University Press, New York.

Lin, X.L., Bransford, J.D., Hmelo, C.E., Kantor, R.J., Hickey, D.T., Secules, T., Petrosino, A., Goldman, S.R. & Cognition and Technology Group at Vanderbilt: 1995, 'Instructional Design and Development of Learning Communities: An Invitation to a Dialogue', *Educational Technology* 35(5), 53–63.

Linn, M.C.: 1986, 'Science', in R.F. Dillon & R.J. Sternberg (eds.), *Cognition and Instruction*, Academic Press, Orlando, FL, 155–204.

Linn, M.C., diSessa, A., Pea, R.D. & Songer, N.B.: 1994, 'Can Research on Science Learning and Instruction Inform Standards for Science Education?', *Journal of Science Education and Technology* 3(1), 7–15.

Linn, M.C. & Songer, N.B.: 1991, 'Cognitive and Conceptual Change in Adolescence', *American Journal of Education* 99, 379–417.

Minstrell, J.A.: 1989, 'Teaching Science for Understanding', in L.B. Resnick & L.E. Klopfer (eds.), *Toward the Thinking Curriculum: Current Cognitive Research*, Association for Supervision and Curriculum Development, Alexandria, VA, 129–149.

Moore, J.L., Lin, X., Schwartz, D.L., Petrosino, A.J., Hickey, D.T., Campbell, O., Hmelo, C. & Cognition and Technology Group at Vanderbilt: 1996, 'The Relationship Between Situated Cognition and Anchored Instruction: A Response to Tripp', in H. McLellan (ed.), *Perspectives on Situated Learning*, Educational Technology Publishers, Englewood Cliffs, NJ, 123–154.

Palincsar, A.S. & Brown, A.L.: 1984, 'Reciprocal Teaching of Comprehension-Fostering and Comprehension Monitoring Activities', *Cognition and Instruction* 1, 117–175.

Perfetto, G.A., Bransford, J.D. & Franks, J.J.: 1983, 'Constraints on Access in a Problem Solving Context', *Memory and Cognition* 11, 24–31.

Perkins, D.N.: 1991, 'What Constructivism Demands of the Learner', *Educational Technology* 31(9), 19–21.

Rogoff, B.: 1990, *Apprenticeship in Thinking*, Oxford University Press, New York.

Scardamalia, M. & Bereiter, C.: 1991, 'Higher Levels of Agency for Children in Knowledge Building: A Challenge for the Design of New Knowledge Media, *Journal of the Learning Sciences* 1(1), 37–68.

Scardamalia, M., Bereiter, C., Brett, C., Burtis, P.J., Calhoun, C. & Smith L.N.: 1992, 'Educational Applications of a Networked Communal Database', *Interactive Learning Environments* 2(1), 45–71.

Scardamalia, M., Bereiter, C. & Lamon, M.: 1994, 'The CSILE Project: Trying to Bring the Classroom into the World 3', in K. McGilly (ed.), *Classroom Lessons: Integrating Cognitive Theory and Classroom Practice*, MIT Press/Bradford Books, Cambridge, MA, 201–228.

Scardamalia, M., Bereiter, C., McLean, R.S., Swallow, J. & Woodruff, E.: 1989, 'Computer Supported Intentional Learning Environments', *Journal of Educational Computing Research* 5, 51–68.

Schoenfeld, A.H.: 1989, 'Teaching Mathematical Thinking and Problem Solving', in L.B. Resnick & L.E. Klopfer (eds.), *Toward the Thinking Curriculum: Current Cognitive Research*, Association for Supervision and Curriculum Development, Alexandria, VA, 83–103.

Schwab, J.J.: 1960, 'What Do Scientists Do?' *Behavioral Science* 5, 1–27.

Schwab, J.J.: 1978, *Science, Curriculum, and Liberal Education*, University of Chicago Press, Chicago, IL.

Sherwood, R.D., Petrosino, A.J., Goldman, S.R., Garrison, S., Hickey, D., Bransford, J.D. & Pellegrino, J.W.: 1993, 'An Experimental Study of a Multimedia Instructional Environment in a Science Classroom', Paper presented at the annual meeting of the American Educational Research Association, Atlanta, GA.

Songer, N.B.: 1993, 'Learning Science with a Child-Focused Resource: A Case Study of Kids as Global Scientists', in *Proceedings of the Fifteenth Annual Meeting of the Cognitive Science Society*, Lawrence Erlbaum, Hillsdale, NJ, 935–940.

Wheatley, G.: 1991, 'Constructivist Perspectives on Science and Mathematics Learning', *Science Education* 75, 9–21.

White, B.Y.: 1993, 'ThinkerTools: Causal Models, Conceptual Change, and Science Education', *Cognition and Instruction* 10(1), 1–100. (whole issue)

Williams, S.M.: 1992, 'Putting Case-Based Instruction into Context: Examples from Legal and Medical Education', *Journal of the Learning Sciences* 2, 367–427.

# 3.6 Using Technology to Support Students' Artefact Construction in Science

MICHELE WISNUDEL SPITULNIK
*University of Michigan, Ann Arbor, USA*

STEVE STRATFORD
*Maranatha Baptist Bible College, Watertown, USA*

JOSEPH KRAJCIK and ELLIOT SOLOWAY
*University of Michigan, Ann Arbor, USA*

A new vision of science education is emerging from a number of national efforts (American Association for the Advancement of Science 1993; National Research Council 1996). These efforts call for scientific literacy which involves students engaging in scientific inquiry activities and using scientific understanding to make informed decisions regarding personal and social issues. Promoting scientific literacy requires schools and teachers to adopt constructivist methods of instruction. Teachers need to provide environments which support students inquiring, collaborating and communicating. Such environments can help students to generate scientific understanding. Technological tools can help facilitate this effort.

Various research groups are creating tools to promote student inquiry and investigation of authentic problems (see Edelson's, Linn's and White's chapters in the *Handbook*). These environments support student construction, rather than transmission of knowledge. Although we also create and explore learning environments and technological tools that support inquiry activities, we focus on environments which emphasise the collaborative construction of technological artefacts, that is, student-designed and student-built intellectual products. There is growing evidence which suggests that artefacts represent students' cognitive processes and students' emerging understandings of science concepts and processes (Lehrer 1993; Spitulnik 1995). There are endless possibilities for student-produced artefacts; we focus here on two types of technological artefacts; dynamic models and hypermedia documents.

In this chapter's first section, we present a theoretical rationale for engaging students in technological artefact construction which includes a description of why

---

Chapter Consultant: Instead of using chapter consultants, chapter authors in the Educational Technology section provided editorial comments on each other's chapters.

we incorporate technology into the building of artefacts, why we engage students in building artefacts and, more specifically, why we engage students in dynamic model construction and hypermedia document construction. Finally, the cases that we present examine how construction of technological artefacts facilitates students' explanations of scientific phenomena and development of conceptual understanding in science. The first case involves students constructing dynamic (time-based) models of complex ecological systems, whereas the second case involves high school chemistry students in building hypermedia documents around general chemistry concepts and kinetic molecular theory.

## THEORETICAL RATIONALE

Today's educational researchers are likely to view learners as active constructors of knowledge rather than as passive recipients, and to emphasise the 'active, reflective, and social nature of learning' (Brown & Campione in press). Learning occurs as students make use of prior knowledge and new experiences to build more complex understandings. Learning in communities is believed to foster development of deep understandings better than learning in isolation (Brown & Campione 1994). Teachers should provide engaging and motivating learning opportunities for students (Blumenfeld, Soloway, Marx, Krajcik, Guzdial & Palincsar 1991) – opportunities fostered by personally meaningful activities situated in authentic real-world contexts (Brown, Collins & Duguid 1989; Cognition and Technology Group at Vanderbilt 1992). Of particular interest in this chapter are computer-based 'interactive learning environments', that is, computer applications that allow student-directed learning activities rather than computer-driven tutoring. Such computer-based environments have been called 'cognitive tools' (Salomon 1993), because they can amplify and extend the cognitive abilities of learners. They also can provide scaffolding to support the creation of technological artefacts of understanding, as described in the following section.

*Why Use Technology?*

Dynamic models and hypermedia documents arguably do not exist without modern high-speed multimedia computers. Technology, particularly computer technology, provides unique information-processing characteristics: multiple representations; interactive real-time testing; and document revision. These attributes can facilitate learners in constructing understanding.

Technological tools allow users to represent phenomena and concepts using different representations and to make connections between these representations. It is relatively easy to incorporate diagrams, drawings, graphs, animations and video into computer-generated artefacts. Because scientists use numerous representations to describe phenomena, the incorporation of multiple representations of concepts within dynamic models and hypermedia documents is especially important in helping students to develop conceptual understanding of scientific concepts (Lemke 1990; Kozma 1991). Moreover, by building thoughtful connections and constructing meaningful relationships between representations, students form more integrated understandings (Novak & Gowin 1984; Linn's chapter in this *Handbook*).

Technology allows students to test their constructions more interactively in real time. With appropriate technological tools, students can generate and test hypotheses and predictions in order to help them build explanations of scientific phenomena. Interactive testing, therefore, allows students to make connections between their ideas and the information gained through testing.

Technology allows students more readily to revise their work and develop understanding over time (Harel 1991; Spitulnik 1995). The digital nature of the dynamic models and hypermedia documents allows students to revisit their productions. Students save their work in the current version and continue constructing newer versions by further developing ideas, reworking representations and building connections between concepts. The construction of several versions of artefacts allows students to reflect on their process and conceptual development and provides a continuum of students' demonstrated understandings for evaluation by teachers and educational researchers.

*Why Artefacts?*

Constructing artefacts not only provides students with an opportunity to develop understanding, it also provides a context in which students can demonstrate their understanding (Perkins 1986). The process of producing an artefact requires students to engage in many elements of design – for example, gathering data from multiple sources, decomposing topics into subtopics, organising diverse and contradictory information, formulating questions and presenting information (Lehrer 1993; Perkins 1986). As students engage in these design elements, they enhance their conceptual understanding, which can be described as the richness of interconnections and relationships made between concepts and the structure which organises those concepts (Novak & Gowin 1984). Students construct and reconstruct their understanding as they synthesise information and work with ideas to form them into a coherent structure (Papert 1993; Perkins 1986). Students also construct understanding as they design their artefacts collaboratively so that, during the process, students communicate and defend their ideas (Brown, Collins & Duguid 1989).

As students build artefacts, they integrate new information into their conceptual understanding, they build connections between concepts and they demonstrate these connections within their artefacts. The degree to which students make connections and draw relationships between concepts within their artefacts provides insight into students' understanding of concepts. Artefacts are external representations of students' understandings. Analysis of student artefacts suggests that differing degrees of understanding are evident within artefacts (Spitulnik 1995).

## Why Dynamic Models?

The construction of dynamic computer models by students is valuable for several reasons. First, many complex scientific systems are difficult to explore or understand adequately without a computerised model; (Forrester 1968). Constructing dynamic models can help students to understand complex phenomena as interconnected systems (Miller, Ogborn, Briggs, Brough, Bliss, Boohan, Brosnan, Mellar & Sakonidis 1993), develop understandings of how scientists use models (Richards, Barowy & Levin 1992) and understand and apply concepts of systems thinking to scientific problems (Mandinach & Cline 1989). Second, new software modelling tools (along with the increasing presence of computers in classrooms) are opening the doors to dynamic model-building at the pre-college level (High Performance Systems 1992; Jackson, Stratford, Krajcik & Soloway 1995). One of the primary reasons why students in pre-college classrooms historically have not been able to engage in dynamic model-building in their science classrooms is the lack of age-appropriate and ability-appropriate tools. Third, allowing students actually to design and create models helps them to understand the nature of models and their possible uses. Reform efforts are suggesting that students should gain a better understanding of models in science (e.g., American Association for the Advancement of Science 1993). Finally, student creation and exploration of simulations of real-world phenomena enable different approaches to science education that can make science accessible to a more diverse population of students (see White's chapter in this *Handbook*).

The construction of dynamic models encourages analysing, synthesising, reasoning and explaining. These processes help students to develop understandings of complex systems. Modelling requires careful analysis of a system – for example, breaking it down into component parts, suggesting and testing cause-and-effect relationships between those parts, explaining and justifying those relationships, and testing and debugging the model. Building a dynamic model is a concrete way in which to help students construct mental connections between real-world concepts by providing an environment in which they can formulate and test their mental models of a phenomenon. Student models contain representations of how and to what extent they understand the complex system in question.

## Why Hypermedia Documents?

Hypermedia document construction involves the same design elements as artefact construction. In order for students to create a meaningful hypermedia artefact, they need to organise a corpus of material. This process of organisation can encourage learners to think, not only about ideas, but also about how the ideas and concepts are interrelated (Salomon 1988). Salomon suggests that students can develop an understanding of the structure of the domain of information which they are trying to represent by getting 'hands-on' experience with the information as they organise and build their artefacts. Lehrer describes this process of organising and structuring as students developed artefacts on the Civil War using many sources of information (Lehrer 1993).

Students construct understanding out of information by designing and creating appropriate representations of concepts. Daiute (1992) observed that, as students' incorporate and create multiple representations such as images, text and sounds, their cultures, values and interests are expressed in their artefacts. The self-constructed artefacts create familiar contexts and symbols on which students can focus their academic work. Finally, as students construct the hyper-links between text and other symbolic expressions within their artefacts, they begin to define and develop relationships between representations and concepts.

## LEARNING ENVIRONMENTS WHICH SUPPORT ARTEFACT CONSTRUCTION

We have worked with science teachers to design and utilise project-based learning environments that allow students to build authentic artefacts. These environments encourage students to take responsibility for their learning by providing an investigatory framework in which they find solutions to authentic problems through collaboration and the use of technological tools (Krajcik, Blumenfeld, Marx & Soloway 1994).

Other efforts are underway to create similar learning environments in which students engage in investigations, including the CoVis (see Edelson's chapter in this *Handbook*) and TERC (Technical Education Research Center) projects. The CoVis project involves students collecting, analysing and visualising weather data and collaborating with telecommunications. TERC is leading in developing student inquiry projects, one of which is project GLOBE (Berenfeld 1993), for which students, scientists, teachers and others collaborate globally with telecommunications.

Although we attempt to create environments that foster student inquiry, we also focus on students constructing artefacts during these projects in order to support students' developing understanding. As students engage in these projects, they simultaneously construct artefacts with representations of their understandings. These artefacts develop over time throughout the projects and often undergo several revisions, with peer and teacher evaluations, until they are presented to the larger classroom (and community) audience. Our work also differs because we have attempted to provide support to a large number of teachers as they are in transition to project-based science (see Marx, Freeman, Krajcik & Blumenfeld's chapter in this *Handbook*).

## CASE STUDY 1: BUILDING MODELS OF COMPLEX ECOLOGICAL SYSTEMS

This chapter provides two case studies in which students constructed technological artefacts while they engaged in projects. In the first case study, students built ecological systems models using a dynamic modelling tool. In a second case study described later in this chapter, students built hypermedia documents with authoring tools. For each case study, there is a major section devoted to project purposes, background and characteristics, and a second major section devoted to illustrating the instruction of the specific artefacts.

In a project-based science curriculum integration effort entitled Foundations of Science, 22 ninth and tenth grade science students at a public high school in Michigan used a dynamic modelling tool called Model-It to create artefacts of complex stream ecosystems. Model-It was developed by the Highly Interactive Computing group at the University of Michigan (Jackson et al. 1995).

*Description of Dynamic Modelling Tools*

Dynamic computer modelling tools allow users to construct time-based models of complex systems. In a dynamic model, interconnected time-dependent constructs (usually mathematical) represent relationships between components of a system. Models are created by recognising, identifying and operationalising patterns of causes and effects. Modern dynamic modelling tools allow visual representation of static components and relationships (line drawings, icons, pictures) and model output (meters, graphs, animated graphics). They support an iterative, interactive process of building, testing and debugging. They are generative, placing few constraints upon actual content, and are wholly dependent upon the modeller for their design, structure and rationale. In sum, they provide an environment conducive to constructing connections between causally-related concepts, representing understanding through design, rationale and representation, and revising understandings through iterative revision.

Model-It designed by the University of Michigan's Highly-Interactive Computing Group, provides a dynamic modelling environment designed especially for learners who are unfamiliar with dynamic modelling and lack mathematical or symbol manipulation skills. Models consist of *objects* ('things' in the system being modelled), *factors* (measurable attributes of objects) and *relationships* (usually causal) between factors. These modelling terms are concrete and familiar enough to many students to break down any terminology barrier that could exist. Objects are represented visually with photorealistic or graphical images. Factors are defined with text (e.g., 'As food supply increases, rate of growth increases'; see Figure 1a) or graphically (see Figure 1b).

The interactivity and speed of computers allow real-time model testing and swift (but distinct) transitions between building and testing. Output is presented with sliding scale meters or with graphs (Figure 2). Closely linking design and testing can allow students to connect mentally their model with its behaviour and facilitate exploration and experimentation.

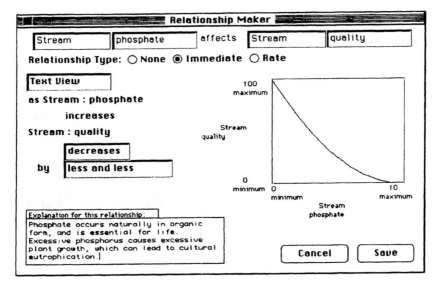

Figure 1a: Defining relationships: textually

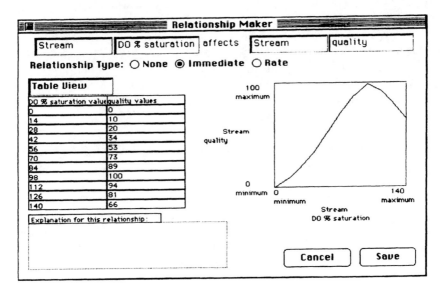

Figure 1b: Defining relationships: graphically

*Using Model-It in the Classroom*

In project-based science, learning is structured around authentic investigations of meaningful questions such as 'Is Traver Creek safe?' This project involves collecting chemical, biological and habitat data from a local stream. Students measure

370  *Spitulnik*

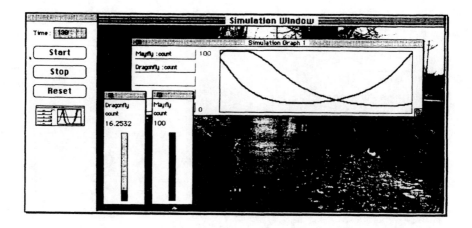

**Figure 2: Running a model**

levels of phosphates, nitrates and pH (among others) and record plant life, macroinvertebrate organism and physical data. To develop further their understanding of ecosystems, students constructed dynamic models, working together in pairs. After working with a prepared guide for a few days, they designed their own models based on suggested scenarios. Modelling allowed them to investigate the stream in a unique way: they could *cause* a change in a system and *see* the effects immediately. They also could quickly formulate and test questions by modifying their model. Engaging in such in-depth scientific investigations can enable students to construct situated understandings of complex relationships in the stream.

## CASE STUDY 1: ILLUSTRATING THE CONSTRUCTION OF DYNAMIC MODEL ARTEFACTS

In this section we summarise evidence supporting three themes related to dynamic modelling artefacts: (1) making connections by modelling cause-and-effect relationships; (2) representing knowledge in and constructing knowledge about models; and (3) constructing knowledge by building and testing models. Our illustrations are drawn from two studies of modelling in project-based science classrooms, both using the same students.

*Making Connections by Building Cause-and-Effect Relationships*

Sample Model With Factor Map

The sample model in Figure 3 is a typical model from the second study. Two students worked on this model for about a week both in and out of class. It illustrates the connections and interconnections possible in a model. It contains 11 factors (miniature pictures) and three populations (three groups of three factors at the right). There are 22 relationships (lines between factors), with some defined as text and others as graphs (e.g., Figures 1a and 1b). Note that populations are predefined in Model-It, with factors and relationships required for growth and decay.

Students also entered explanations for relationships (not visible). For example, the explanation for the relationship between 'Total Phosphates' and 'BOD' (biochemical oxygen demand) reads as follows. 'As the number of phosphates in the water increases, the plant growth increases, which makes more organic material for the aerobic bacteria to decompose, increasing the BOD.' Articulating explanations can help students to construct deeper understandings of the modelled relationships.

Reflecting Upon Models and Model-Building

In interviews, students reflected upon models in both simplistic and sophisticated terms. Some said that models were useful for organising data, showing 'what might have made the changes between 1993 and 1994'. Others said that it was a tool for forecasting future events and trends. One said that creating the model helped her to 'see the process of understanding' the relationships in the ecosystem. We think, therefore, that dynamic modelling helps students to understand how models can be used as a tool for understanding science.

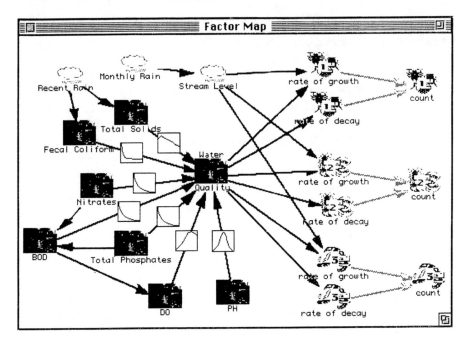

Figure 3: Sample model (our representation)

## Representing Knowledge in and Constructing Knowledge about Models

### Knowledge and Understanding in Models

In our studies, students constructed models related to water quality, including algae blooms, land use and the impact of quality on macroinvertebrates, among others. We found that they can construct interesting (though occasionally flawed) models relatively quickly. We also found that, given longer time to work (our second study), their artefacts were more fully developed and contained fewer problems.

Models in the second study revealed a range of understandings of how an environmental factor, rain, might affect an ecosystem. Some expressed the relationship unidimensionally (one rainfall factor affecting one to three stream factors), whereas others were multidimensional (such as Figure 3), representing other aspects of rainfall (e.g., monthly rain) and showing effects on macroinvertebrates directly (e.g., washing more food down the stream for them). Notice the web of interrelationships between the three populations of pollution-sensitive and -insensitive macroinvertebrates at the right and the two stream factors called 'Stream Level' and 'Water Quality'. Closer examination of relationships (combined with written explanation) reveals that they believe that all three populations will be positively affected by rainfall, but in varying ways depending upon their pollution sensitivity. This observation is supported by observing the behaviour of the model as it is run.

### Knowledge and Understanding about Models

Responses to interview questions about the relationship between the stream and their model revealed a range of understandings. One student asserted: 'The model *is* Traver Creek', although later comments indicate a more realistic view. Others noted that they were 'similar in a lot of areas' or that they 'sort of start out the same' but later 'branch apart'. Another stated that they were very different because their model did not account for other factors. In general, students viewed models as representations of reality; most seemed able to distinguish between acceptable and poor representations and to articulate reasons why.

## Building and Revising Knowledge by Building and Testing Models

### An Example of Model Building and Revision

To illustrate model planning, building and revising, we present a short excerpt from one modelling session. In the scenario, two students (J and P) first plan, then construct and test, and then revise, extend and retest their model iteratively. First they plan their model, expressing ideas about what their model could include:

J: Let's use that one.
P: The golf course?
J: Yeah, we haven't used that one yet.
P: How the golf course affects what, though?
J: How the golf course affects, um, bacteria.
P: Too hard.
J: It's easy. Because, a lot of geese are on the golf course, and the geese faeces go in the water, . . .
P: . . . oh, and it affects faecal coliform, . . .
J: . . . which in turn affects the bacteria, and the faecal coliform grows on bacteria.

First, students discussed how to put the golf course in their model (which already contained some water-quality relationships and factors). After some discussion, they decided to use 'size' and to include 'geese' as a golf course factor:

P: Here, 'edit', it should say, . . . it should say 'size' I think, so the object is a 'golf course', . . .
J: Uh huh, . . .
P: For the factor of the 'golf course', see, we can have different factors, like number of 'geese'. Do you want to do that?
J: Yeah.
P: So 'size' is the acres, then, . . . um, 'new', um. How do you spell 'geese'?

Students proceed to reason that golf course size and number of geese would impact on the amount of fertiliser (phosphates and nitrates) and faecal coliform entering the stream, respectively. So, after creating 'size' and 'geese', they created the corresponding relationships ('geese' to 'faecal coliform' and 'size' to the fertiliser factors). In the next conversation, the students are testing their model and verifying that their relationships work as they expect:

P: So if you have, . . . like a ton of geese. If you have a hundred geese [changing number of geese to 100], faecal coliform is 100 percent, and if you have . . .
J: . . . a golf course, . . .
P: The golf course is a hundred acres [changing golf course size to 100], our stream quality is around, . . . like really bad.

After realising that larger golf courses probably have more geese, students created a relationship connecting 'size' to 'geese', and tested again to verify that it worked as expected. Finally, they discussed how stream quality might affect a macroinvertebrate population. So, they inserted a population object into their model, created the appropriate relationship, and tested it over a range of golf course sizes to verify that it worked.

Not all students followed the same interactive process. Some preferred to plan on paper and enter factors and relationships all at once, followed by testing. Typically, though, we have seen students building as in the example above by planning, creating, testing and revising.

## Student Reflection on the Model-Building Process

In interviews prompting students to reflect upon their own model-building process, students described incremental strategies involving: creating some factors and relationships; connecting them in some populations; and finally identifying additional influential factors. Some said that they identified relationships by looking for ways to explain differences in data. One group felt that identifying external factors which could influence the whole system and account for differences in data was a 'pointless idea' because of the large number of potentially influential factors.

In this section, we have presented some data from varied sources, organised around the themes of making connections, representing knowledge, and constructing and revising knowledge. While we don't attempt to generalise far beyond our formative studies, we feel that these data support the notion that creating dynamic model-artefacts of real phenomena can help students to understand modelling better in general and complex phenomena in particular.

## CASE STUDY 2: CREATING HYPERMEDIA ARTEFACTS

Creating artefacts using technology is likely to enable students to construct complex understandings of science concepts. Hypermedia authoring tools, including HyperCard and HyperStudio (Hypercard, Apple Computer, Inc.; HyperStudio, Roger Wagner, Inc.), were used by students at a public magnet high school in south central Michigan, as an integral part of their tenth grade chemistry class.

Authoring tools like HyperCard and HyperStudio allow users to construct presentations or 'stacks'. A stack is composed of 'cards' or computer screens, and the cards are 'linked' together with buttons. The stack designer defines the links so that, when a button is clicked, a new card appears. The design of hypermedia authoring tools facilitates the linking of ideas, constructing relationships or connection building. Scanned images, sounds and video can readily be incorporated into these stacks and this allows the construction of personally meaningful artefacts.

Students constructed hypermedia artefacts to learn and represent their understanding of some fundamental chemistry concepts. The teacher presented the purpose of the hypermedia artefacts as instructional software from which other high school chemistry students could learn. In building their artefacts, students decided how to organise and teach self-selected information. Complex skills, including cognitive strategies, became explicit because students had the opportunity to make design decisions about how to teach certain material (Harel 1991). For example, students grappled with giving content material structure and making it interactive.

During construction of the artefacts, the teacher led discussions and students reflected in writing about what it means to learn and teach. The students decided how best to structure their ideas and concepts, how to represent concepts, and how to include instructional strategies (e.g., motivating their potential student users and encouraging their users to evaluate their knowledge while using the instructional tool).

## CASE STUDY 2: ILLUSTRATING THE CONSTRUCTION OF HYPERMEDIA ARTEFACTS

Below, we describe two projects in which hypermedia artefact construction was incorporated into the chemistry class. The first occurred in the 1992–1993 school year and the second occurred during the 1993–1994 school year. The first project involved students building hypermedia artefacts around a chemistry element of their choice. The second project involved building hypermedia artefacts to explain kinetic molecular theory.

### The First Project: The Elements

The first hypermedia construction project occurred as an adjunct to the regular chemistry class; in other words, students worked on their artefacts as a part of their chemistry class, but the content of the artefacts was not fully integrated into the chemistry class curriculum. Students selected a chemical element and built artefacts to explain the chemistry associated with their elements. Students were encouraged to research and incorporate content on the following topics: chemical structure; properties; discovery; and practical or real-world applications or uses.

The research focus for this project centred around four questions. First, what content do students include in their stacks? Second, how do students represent science concepts? Third, how do representations reflect student understanding? Fourth, what kinds of relationships do students make between concepts, real-world applications of concepts, and popular culture? Below we report the analysis of the students' final artefacts.

### The Range of Content and Its Representation

Students presented a variety of chemistry concepts and topics to a varying degree of depth, breadth and accuracy. The topics presented in the stacks can be categorised as periodic table information, atomic structure, molecular structure, lattice structure, states of matter, physical properties, reactions, uses, sources, production, discovery and other. No stacks included all of this content, but most stacks included content from at least five of these categories.

Students portrayed concepts with many representations and with several types of media, including text, drawings, graphics, sounds, animations and video. We describe two examples of how students used representations to portray concepts and we explore the relationship between the depth of student understanding and the representations which they created.

In constructing their hypermedia artefacts, students explored a variety of concepts, with the depth of understanding represented in the artefacts varying. For instance, one group of students represented and described (in text) their element, carbon, as a monatomic element – clearly a simplistic understanding of the nature of carbon. This can be compared to the thulium group of students who presented the lattice structure of their element with multiple representations. They described the structure of their element as hexagonal close packed with text, represented this structure with a drawing, and incorporated a student-produced video of a styrofoam model rotating in space. These students exhibited a more well-developed understanding of the structure of their element (Figure 4).

Constructing Relationships

Students constructed relationships within their artefacts by (1) forming connections between chemistry concepts, (2) generating links to practical, real-world applications and (3) creating ties between science concepts and students' popular culture.

Students constructed relationships between chemistry concepts by making explicit connections between an element's atomic or molecular structure and its chemical properties. The plutonium stack students made this connection as they described the colour of plutonium compounds when plutonium exists with different valences. The carbon stack students also created this type of link between

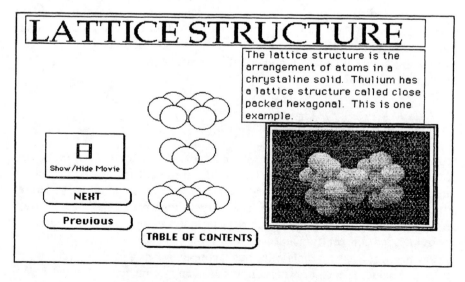

Figure 4: The Thulium Stack – Lattice structure

chemistry concepts as they explained the properties of diamonds; they identified the connection between high density and high melting point.

Students generated links between chemistry concepts and real-world applications. The thulium stack students described x-ray photography as an application of the radioactive property, gamma radiation production. The zirconium stack students described zirconium as an industrial substitute for diamonds because they have similar molecular structures, and they described zirconium as a good substitute for steel and titanium because zirconium has a high corrosion factor. The plutonium students relate radioactivity to the discovery and application of the destructive atomic bomb. These students incorporate a video clip of a detonated atomic bomb, as well as information about the bomb target, Nagasaki, Japan.

Students created ties between their chemistry elements and concepts, their popular culture and their world outside of the classroom. Students incorporated examples of their world into their stacks by including sound bites from popular television, recorded music ranging from rap music to the Star Wars theme song, hand-drawn cartoon characters and video clips from movies. The students expressed themselves, with many forms of media, throughout the stacks and usually this expression was related to the chemistry content that the students presented. For example, the plutonium stack students described plutonium as an energy source and they coupled text which describes the isotopes and reaction of plutonium with two film clips from the movie, Back to the Future. These film clips show a scientist and a younger student discussing plutonium as an energy source for a futuristic vehicle.

*The Second Project: Kinetic Molecular Theory*

The second hypermedia construction project occurred as a more integrated project within the chemistry class curriculum. The first semester of this chemistry course was structured around the question, 'What is in our air?' This question provided the framework for basic chemistry concepts including: matter as composed of atoms; conservation of matter and balancing reactions; states of matter; composition of gases and their properties; and air quality/air pollution. During this project, students constructed hypermedia artefacts on kinetic molecular theory. Students were asked to include content on states of matter, changing states of matter, and applications of states of matter. Students were encouraged to select a chemical element or molecule and focus their stacks around that substance. Students were also encouraged to construct multiple representations of concepts.

This research focused on (1) how students represent and model the concepts associated with kinetic molecular theory, (2) how students' representations reflect understanding of concepts and (3) how students revise their conceptions of kinetic molecular theory through revising and editing their artefacts. We analysed successive versions of students' artefacts to show how students' representations and understanding changed as their artefacts were developed.

In early versions of students' artefacts, students represented concepts associated with kinetic molecular theory with only a few representations of concepts and connections were not drawn between these representations. For example, two students working on an artefact about water represented the melting of ice (Figure 5). These students described with text what happens to particles, or molecules, as they change phase and melt, and they included a temperature vs. time graph of the phenomena. However, the students also included a simple animation with a temperature gauge increasing as a cube of ice melts. Although these representations all depicted the concept of melting, the representations of melting were conflicting and not connected. These students in this early artefact did not demonstrate an understanding of the concept of melting. Also, in early versions of the artefacts, students included few real-world examples of the concepts which they represented.

As students evaluated each other's artefacts, they revised the artefacts and the conceptions developed. It is apparent from analysing later and final versions of artefacts that students included more types of representations, including graphs and more sophisticated animations. They made more explicit connections between the various representations of concepts, either by referring to the other representations or by comparing representations. Likewise, in final versions of the artefacts, students had fewer conflicting representations (e.g., students' representations of molecular animations correspond with graphs and text explaining the animation). An example here (Figure 6) shows a later version of an artefact constructed by the students discussed above. They continued to develop the concept of melting by adding a more detailed description of melting and a molecular-level animation

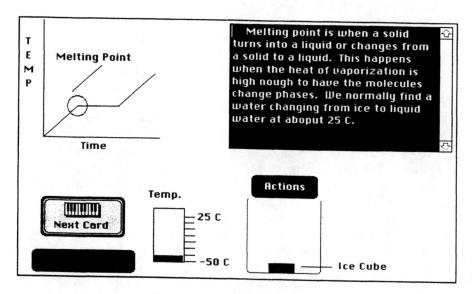

Figure 5: An early version of an artefact

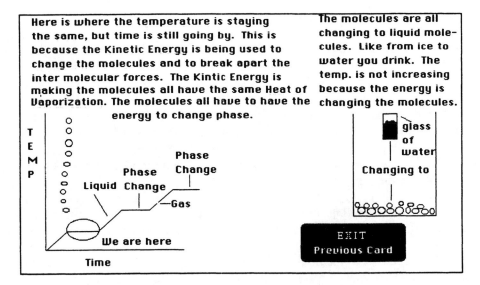

Figure 6: A later version of an artefact

of melting. Their different representations of melting were more integrated and communicated a better understanding of the phenomenon. Finally, in later versions, students incorporated more real-world examples into their artefacts.

Although we have been successful with students creating hypermedia artefacts within a chemistry class context, we are just beginning to learn how students represent different concepts and how these representations reflect student understanding. As we observed in the construction of dynamic models, hypermedia document construction also provides a means by which students can make connections between concepts, represent knowledge using different media, and iteratively construct and revise their understandings. We continue to analyse how students revise their understanding over time by revising their artefacts.

## CONCLUSIONS

Educational environments that support students designing and constructing technological artefacts respond to the call of national reform efforts in science education to have students use scientific knowledge to predict, explain and model phenomena. Our thesis is that environments which provide opportunities for students to design, create and revise technological artefacts are particularly valuable in helping them to construct integrated understandings of those concepts and phenomena. The two cases presented illustrate that, as students design and construct technological artefacts, they re-present their understanding of the concepts and phenomena. First, artefact construction allowed students to make connections between concepts. In creating dynamic models, students made connections by stipulating the relationships between factors. In the construction of the hypermedia documents, students made connections by building hyperlinks between various parts of their documents. Second, the creation of technological artefacts allowed students to build multiple representations of scientific knowledge. Both in model building and in document construction, students represented and explained science concepts. Finally, the artefact construction occurred in various stages. With more time, students developed more elaborate artefacts. This revision and elaboration of the artefacts represented the development of students' understandings. As students revised their artefacts, they built more integrated understandings of the concepts.

Although designing and creating technological artefacts promotes students' integrated understanding, we still have much to learn about technological artefact construction. For instance, how well do the technological artefacts represent students' conceptual understanding? How is the understanding that students exhibit in artefacts related to the understanding that students exhibit in other forms of assessment? How can we better design technological tools to support the design and construction of artefacts? Further research is also needed to identify: the processes in which students engage as they work on and revise their thinking and their artefacts; and how students collaboratively negotiate meaning while they create these artefacts and construct understanding. We are beginning to research and analyse the many influences on students as they construct their artefacts in classrooms. These influences range from external factors, including State, district and school curricula and policies and popular culture, to internal factors, including curricula, teachers, discourse, textbooks, software, and structure and focus of projects.

## REFERENCES

American Association for the Advancement of Science: 1993, *Benchmarks for Science Literacy: Project 2061*, Oxford University Press, New York.

Berenfeld, B.: 1993, 'A Moment of Glory in San Antonio', *Hands On!* 16(2), 1 & 19–21.

Blumenfeld, P.C., Soloway, E., Marx, R.W., Krajcik, J.S., Guzdial, M. & Palincsar, A.: 1991, 'Motivating Project-Based Learning: Sustaining the Doing, Supporting the Learning', *Educational Psychologist* 26, 369–398.

Brown, A. & Campione, J.: 1994, 'Guided Discovery in a Community of Learners', in K. McGilly (ed.), *Classroom Lessons: Integrating Cognitive Theory and Classroom Practice*, MIT Press/Bradford Books, Cambridge, MA, 229–270.

Brown, A. & Campione, J.: in press, 'Psychological Theory and the Design of Innovative Learning Environments: On Procedures, Principles and Systems', in L. Schauble & R. Glaser (eds.), *Contributions of Instructional Innovation to Understanding Learning*, Lawrence Erlbaum, Hillsdale, NJ.

Brown, J.S., Collins, A. & Duguid, P.: 1989, 'Situated Cognition and the Culture of Learning', *Educational Researcher* 18(1), 32–42.

Cognition and Technology Group at Vanderbilt: 1992, 'Jasper Series as an Example of Anchored Instruction: Theory, Program, Description and Assessment Data', *Educational Psychologist* 27, 291–315.

Daiute, C.: 1992, 'Multimedia Composing: Extending the Resources of Kindergarten to Writers Across the Grades', *Language Arts* 69, 250–260.

Forrester, J.W.: 1968, *Principles of Systems*, Wright-Allen Press, Cambridge, MA.
Harel, I.: 1991, *Children Designers*, Ablex, Norwood, NJ.
High Performance Systems.: 1992, *Stella II: An Introduction to Systems Thinking*, Author, Hanover, NH.
Jackson, S., Stratford, S.J., Krajcik, J. & Soloway, E.: 1995, 'Making System Dynamics Modeling Accessible to Pre-College Science Students', Paper presented at the annual meeting of the American Educational Research Association, San Francisco, CA.
Kozma, R.: 1991, 'Learning With Media', *Review of Educational Research* 61, 179–211.
Krajcik, J.S., Blumenfeld, P.C., Marx, R.W. & Soloway, E.: 1994, 'A Collaborative Model for Helping Teachers Learn Project-Based Instruction', *Elementary School Journal* 94, 483–497.
Lehrer, R.: 1993, 'Authors of Knowledge: Patterns of Hypermedia Design', in S.P. Lajoie & S.J. Derry (eds.), *Computers as Cognitive Tools*, Lawrence Erlbaum, Hillsdale, NJ, 197–228.
Lemke, J.: 1990, *Talking Science: Language, Learning and Values*, Ablex, Norwood, NJ.
Mandinach, E. & Cline, H.: 1989, 'Applications of Simulation and Modeling in Precollege Instruction', *Machine-Mediated Learning* 3, 189–205.
Miller. R., Ogborn, J., Briggs, J., Brough, D., Bliss, J., Boohan, R., Brosnan, T., Mellar, H. & Sakonidis, B.: 1993, 'Educational Tools for Computational Modeling', *Computers in Education* 21, 205–261.
National Research Council: 1996, *National Science Education Standards*, National Academy Press, Washington, DC.
Novak, J.D. & Gowin, D.B.: 1984, *Learning How to Learn*, Cambridge University Press, Cambridge, UK.
Papert, S.: 1993, *The Children's Machine: Rethinking School in the Age of the Computer*, Basic Books, New York.
Perkins, D.N.: 1986, *Knowledge as Design*, Lawrence Erlbaum, Hillsdale, NJ.
Richards, J., Barowy, W. & Levin, D.: 1992, 'Computer Simulations in the Science Classroom', *Journal of Science Education and Technology* 1(1), 67–79.
Salomon, G.: 1988, 'AI in Reverse: Computer Tools that Turn Cognitive', *Journal of Educational Computing Research* 4, 123–134.
Salomon, G.: 1993, 'On the Nature of Pedagogic Computer Tools: The Case of the Writing Partner', in S. Lajoie & S. Derry (eds.), *Computers as Cognitive Tools*, Lawrence Erlbaum, Hillsdale, NJ, 179–196.
Spitulnik, M.W.: 1995, 'Students' Modeling Concepts and Conceptions', Paper presented at the annual meeting of the National Research in Science Teaching, San Francisco, CA.

## 3.7 Integration of Experimenting and Modelling by Advanced Educational Technology: Examples from Nuclear Physics

HORST P. SCHECKER
*University of Bremen, Germany*

Experimenting and modelling are the most prominent activities in science education. While experimenting can be understood as purposively probing the behaviour of systems prepared in laboratory settings, modelling has a broader meaning. Modelling can mean *mathematical modelling* (i.e., forming and solving a set of differential equations to predict the behaviour of a system). Modelling also refers to mental or *conceptual modelling* (i.e., developing some *qualitative* understanding of the concepts and principles involved in a scientific line of argumentation).

Students tend to make a distinction between physical models or 'theories' and 'practice' (cf Meyling 1990; Schecker 1985). In the students' view, theory tells us how things should work under ideal conditions, while practice in the laboratory hardly ever behaves according to the rules – at least if the apparatus is not artificially tailored to educational purposes, like air tracks or vacuum tubes. To explain the discrepancies, teachers use conditional clauses such as: 'If there was no friction, the cart would . . .' or 'If insulation was perfect, then the temperature after mixing the liquids would . . .'. Students interpret this as: 'Theory is good for calculations in the next examination, but you can't use it in practice'. One way of dealing with this problem is to include epistemological aspects of scientific theory construction explicitly into science courses (cf Meyling 1990). Another is to relate more closely the activities of experimenting and modelling. If hands-on sessions are to provide 'flesh' to the theoretical 'skeleton', students must experience that 'abstract' principles really help to describe and explain in detail – not just in principle – phenomena investigated in and outside the laboratory. We will show how educational technology can serve as a means to tie experimentation and theorising closer together.

A second contribution of an integrated measuring and modelling approach to knowledge integration lies in prompting students to shift perspectives between theorising and experimentation. The interplay between theoretical predictions,

---

Chapter Consultant: Instead of using chapter consultants, chapter authors in the Educational Technology section provided editorial comments on each other's chapters.

empirical investigations, explanations, reformulations of the theory and new predictions forms the basis of the scientific enterprise. Experiments can serve as empirical starting points for modelling, while modelling results can stimulate new experiments or new ways of evaluating experimental data. When students alternate between theory construction and experimentation, they look at the same problem from different angles. A tentative conceptual model – first ideas formulated in words after observing an experiment – has to be formalised, either in equation form (mathematical model) or with a dynamic modelling system as a network of interacting quantities (conceptual model). Theoretical predictions have to be tested against empirical data by designing and performing relevant experiments. Verbal descriptions, mathematical formalisations, concept maps, the apparatus of a related experiment and the evaluation of a data table aim at different input channels and different cognitive preferences of students.

This chapter falls into two major sections. The first section considers the role of technology in supporting learning in science, including integrating tools and open-ended and targeted tools. The second section provides a specific example involving the measuring and modelling of radioactive decay.

## THE ROLE OF TECHNOLOGY

How can educational technology support meaningful learning and knowledge integration? This chapter concentrates on two categories of tools: micro-based laboratories (MBL); and dynamic model building systems (MBS). Some of the ways in which these two tools support learning and knowledge integration are given below:

- MBL narrow the time gap between measuring and evaluating the data. While it takes several minutes to produce a displacement-time-graph from a list of s-t-data with paper and pencil, the computer displays a graphical representation parallel to the experiment. Brasell (1987) found that real-time graphing is crucial for students to relate graphical representations of velocity to observed motion.
- MBL-probes extend experimental possibilities beyond standard laboratory apparatus. With sonar range meters, students can investigate their own body motion (Thornton 1987). Contact-free pliers-ammeters measure the current through any point of a circuit, including batteries or neon tubes (Girwidz 1994). Learners thus gain more flexible opportunities to plan and realise their own investigations.
- MBS reduce restrictions for school science topics that arise from *mathematical* boundaries. Quantitative investigations of real-life phenomena, like the motion of a parachutist, thus can be included when normally one would be restrained by the students' insufficient *mathematical* competence. The students can concentrate on the *physical* aspects of the description – that is, conceptualising and applying

principles like Newton's laws – while the computer numerically solves the differential equations.
- An important component of practical activities in the laboratory is *experimenting with things*. Students always have had opportunities to modify settings and design their own experiments. MBS open up new opportunities for the students to *experiment with ideas*. Modifying a theory and constructing their own approaches is encouraged because MBS are based on the system dynamics approach (Forrester 1968) which presents the model in a flow diagram similar to a concept map on the screen ('making thinking visible'; cf Linn's chapter in this *Handbook*). New elements easily can be added and relationships between quantities reshaped. The consequences can immediately be tested in simulation runs. The learner does not depend on the teacher's feedback but can work autonomously.

Research results encourage the assumption that computer-aided experimentation and the use of modelling systems foster scientific understanding (e.g., Schecker 1993; Thornton 1992). One step further in educational technology is the use of MBL and MBS in an *integrated approach*. Tinker (1986) suggests that there be a 'symbiosis' between MBL and modelling. Although Tinker (1993) found flow diagrams used by dynamic modelling systems inappropriate for younger students, he still favours the combination of modelling and MBL.

*Integrating Tools*

The goal of integrating MBL and MBS tools is to enable students to change freely from measuring to modelling and vice versa so that the two activities mutually profit from each other. Alternating should not be understood as performing an experiment on phenomenon A and then constructing a theory for phenomenon B, but as measuring *and* modelling A, and applying the model to predict a new phenomenon B, which subsequently also becomes the subject of experimental investigations. Each step in the interplay between measuring and modelling can stimulate deeper probing into the topic domain and thus scaffolds the path to a robust understanding.

There have been several projects to develop powerful software and hardware packages for multipurpose use in science education, covering applications such as:

- data acquisition: interface systems with probes and software;
- data analysis tools: spreadsheets with tables and graphs;
- modelling tools: dynamic modelling systems, programming platforms;
- mathematics tools: function plotters, computer algebra systems.

Integration can be achieved either by developing software modules for a master program or by interconnecting existing applications under a common user interface. The Dutch package *IP-Coach* (cf Dorenbos & Dulfer 1992) is a prototype for the first category. *IP-Coach* provides a hardware adapter board and probes plus a comprehensive software package including a user interface for operating the board, a calculating sheet, a dynamic modelling program and a video evaluation tool. The German product *Multilab* (Brandenburg 1993) also offers data analysis and modelling facilities together with a universal driver software for different MBL hardware.

Graphical operating systems with standard formats for exchanging data (e.g., tables, graphs) between applications make it possible to join together distinct tools for scientific explorations, so that they mould into a comprehensive environment. Available applications like spreadsheets thus can be linked with new programs developed for special instructional purposes. The most prominent example for this line of work is given by the *Comprehensive Unified Physics Learning Environment* (*CUPLE*, Wilson & Redish 1992). The *CUPLE* toolbox provides various applications ranging from word processors and spreadsheets, to programming, video analysis, databases and mathematics tools. A similar approach is taken by *BremLab* (Schecker 1996), which combines a dynamic modelling system, a science spreadsheet, mathematics tools and a universal hardware adapter. The *Tools for Scientific Thinking* materials (Thornton 1987), with their major contributions to MBL-technology and their facilities for mathematical modelling, also can be used as an open toolbox.

*Open-Ended and Targeted Tools*

The *CUPLE* toolbox and *BremLab* are meant for undergraduate and upper secondary school students who have specifically chosen physics as a course of study. Autonomous usage of these complex packages demands a high degree of experience, expertise and motivation. It takes a number of examples until students learn to apply the given opportunities creatively to open-ended problems. This effort is worth being invested because the materials can be employed over a longer period of time in different contexts. *BremLab* for example is adaptable to large parts of German high (secondary) school physics courses.

Open-endedness and universality are a *chance* for the learners and at the same time a *challenge* – learners *can* formulate a modelling or a measurement task on their own, but they also *have to* do so. The learners have to work out for themselves which quantities are to be considered, which parameters seem to be appropriate, and how the data can be processed and displayed. The packages support autonomous learners but lack inherent facilities to assist learners in building the necessary competencies. It is the teacher's task to create appropriate learning environments.

The fact that many students, particularly younger ones, need assistance to become autonomous learners contributed to the development of educational packages that are more content specific and which prompt students to perform certain activities in the course of knowledge acquisition in a domain. Without narrowly guiding the students, the materials assist learners to target key steps in the learning process. In the *Computer as Learning Partner* project for middle school thermodynamics (Linn 1992), the prompted activities comprise predicting, real-time data collection, simulating, observing, explaining and conceptual modelling – but also reflecting, critiquing and exchanging ideas with others. Linn's chapter in this *Handbook* calls this approach 'scaffolded knowledge integration'. A content-specific curriculum developed by Laws (1991) replaces traditional lectures by collaborative work in laboratory settings supported by advanced MBL-tools.

An important source for the development of specific, in contrast to universal, tools is research into students' alternative conceptions. Two decades of research (Duit 1994) has led to a set of consensual findings about students' ideas concerning force, light and heat. Educational technology has been used to address directly the resulting learning obstacles. Hennessy, Twigger, Driver, O'Shea, O'Malley, Byard, Draper, Hartley, Mohamed & Scanlon (1995) designed special simulation environments to trigger conceptual change concerning the intuitive force concept. Goldberg and Bendall (1995) have worked on a computer-based curriculum in the domain of geometrical optics where experiments with laboratory apparatus, digitised pictures of experiments and modelling (e.g., in the form of students drawing ray diagrams into these pictures by direct manipulation) are connected.

Both approaches, open-ended universal systems and specific, more supportive environments have their merits. Their appropriate use largely depends on the students' age and level of autonomy. This chapter concentrates on the first category of tools with a combination of MBL with dynamic modelling systems.

*Methodological Aspects*

Open-ended tools do not have an inherent pedagogical impetus. It is the teacher who decides how to make use of them – either in a teacher-oriented lecture style, or in a student-oriented pedagogical framework, with extensive group work on experiments and models. The educational potential of integrated measuring and modelling systems can be fully exploited if they are used in a *constructivist* teaching and learning strategy (see section entitled 'Learning' in this *Handbook*). It is not technology in itself that causes knowledge acquisition but the active engagement of students in tackling scientific problems. In constructivist teaching, the teacher becomes a counsellor and a stimulator of thinking rather than a transmitter of knowledge. The students have to use the MBS and MBL tools themselves. They must be given time to develop and test their own theoretical approaches as well as creating their own experiments. Formal mathematical operations, like solving equation systems, have to be reduced in favour of speaking about physical approaches (e.g., discussing and testing different assumptions for a conceptual model structure).

If students are encouraged to conduct their own experiments and express their own ideas, they will often arrive at questions and results differing from the scientist's or the teacher's view. Teachers tend to ignore deviating ideas and wait for the 'right' answer. From a constructivist perspective, the teacher has to refrain from immediate corrections of 'wrong' answers, from ignoring deviating proposals, or from blocking alternative ways to tackle a problem. By temporarily withholding expert knowledge, the teacher stimulates students to explicate their own ideas. Often, too much time in class is wasted on formal operations that stay largely meaningless for the students. If students are asked what they remember about radioactive decay, they mainly remember the mathematical model: 'Well, there was something about an exponential function'. Few students can argue about qualitative relationships between parent and daughter nuclei and radioactive activities. This calls for shifting the focus of instruction towards more qualitative aspects.

## MEASURING AND MODELLING RADIOACTIVE DECAY

This section exemplifies a multi-stage interplay between measuring, mathematical description and conceptual modelling in the domain of nuclear physics. *BremLab* is used as the software platform, but the investigations can be carried out in a similar way with any of the open-ended systems mentioned above. Specific features of the *BremLab* tools are described only so far as they are necessary to follow the process of investigations. The unit has been tested in teacher training courses and with university students. Major parts of it are contained in a physics and computer science curriculum for German upper secondary schools (Reger & Freiberger 1989).

The unit is presented in a prototypical form. Actual realisations will deviate from the straightforward path presented below. Reger and Freiberger (1989) suggest that different groups of students work on different parts of the project in parallel. If one aims at helping students to integrate experimental and theoretical knowledge, it is important for them to see the steps as a sequence of interrelated activities. Sharing the work between groups thus requires intense exchanges of results in several intermediate class forums. These forums can serve to develop ideas about further questions to examine. In other phases, particularly steps 1 and 6 outlined below, the teacher will have to initiate a new phase of the project.

Students taking part in the project should have experiences with the dynamic modelling system and the spreadsheet from earlier examples. As mentioned before, open-ended, universal systems have to be used over a longer period of time, in order to develop their role as tools. In this cycle, nuclear physics is situated towards the end of a high school physics course.

The sequence of investigations starts off with measuring the decay rate of a barium isotope. MBL hardware and software tools are used to acquire the data which are then exported to a science spreadsheet for evaluation. The decay rate can be *mathematically* modelled by a simple exponential function. A closer look at the *physical* assumptions is taken by employing a dynamic modelling system. The structural relationships between the relevant concepts are developed in the form of a concept map and tested against the experimental data in a simulation run. The model can be extended for the prediction of the production of the barium isotope in a new experiment. Performing the experiment confirms the theoretical assumptions. The decay model is then applied to a third experiment, this time with activated silver. A comparison of data yields a discrepancy between measurement and model prediction. The decay model has to be revised to fit the empirical data.

The focus of reception should not lie on the physical aspects of the phenomena but on the method of alternating between experimental and theoretical (i.e., mathematical and conceptual) activities, supported by a common software environment. The transfer of data between the applications is closely connected to changing viewpoints in the course of the investigations.

*Step 1: Measuring the Decay of Barium 137$^m$*

In the first teacher-experiment, a meta-stable isotope of barium ($^{137}Ba^m$) is 'milked' from a 'radionuclide cow' by pressing some millilitres of a hydrochloric acid-saline solution through a miniature radioisotope generator containing a minute quantity of caesium 137. $^{137}Cs$ decays by beta-ray emission to $^{137}Ba^m$, which in turn decays to stable $^{137}Ba$. While $^{137}Cs$ is a long-lived nuclide (half-life about 30 years), $^{137}Ba^m$ is short-lived.

The probe is placed in front of a Geiger-Müller tube that is connected to an interface adapter which transfers the number of counts per 10 seconds to a computer. The count rate can be taken as proportional to the activity of the $^{137}Ba^m$ preparation. The incoming data are displayed in real-time as a count-time graph. First ideas about the shape of the graph can be made during the experiment. As opposed to conventional ways of registering, there is no time gap between reading numbers from a standard digital counter, writing them down, drawing a graph and reflecting the outcome.

*Step 2: Developing a Mathematical Model*

The MBL software saves the data in a table that can be imported by student groups for evaluation with a science spreadsheet. In the spreadsheet table, the ground rate is subtracted and the rate (=[counts/10s]-ground) can be displayed over time. The science spreadsheet *MatheLab* (Weißgerber, Andraschko & Lorbeer 1994) offers direct-manipulation tools for evaluating the graph together with plotting facilities. The half-life of about 160s can be scanned from the graph (see Figure 1).

A mathematical model is gained from the assumption that the half-life is a constant:

$$Rate = initialRate \cdot \frac{1}{2}^{\frac{Time}{HalfLife}}$$

This mathematical description can be tested by plotting a function graph over the data (see line in Figure 1).

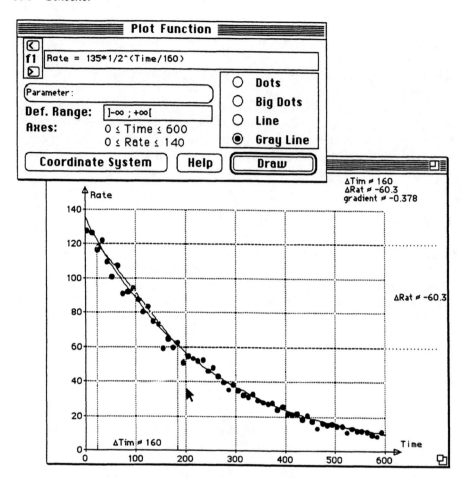

Figure 1: Evaluation of the experimental data by direct manipulation (ΔTime, ΔRate) and plot of an exponential function

*Step 3: Developing a Conceptual Model*

The algebraic formula works quite well to make predictions for the decay of $^{137}Ba^m$. However, at this point, there is no conceptual basis for this prediction. The students are prompted to construct a conceptual model for the decay process. Model building packages help learners to work out the quantities and the relationships to be included in the physical description of a process. With symbolic modelling packages, like *Stella* (High Performance Systems 1994) or *Powersim* (Gonzales 1994), the conceptual features of the model are graphically designed to be similar to a concept map on the screen. This 'simulation diagram' visualises the students' assumptions about the decay process. The computer translates the iconic representation into a corresponding mathematical representation. This results in a set of differential, or rather, difference equations for a numerical simulation of the system's behaviour over time. The user adds initial values and constants and specifies functional relationships. System dynamics modelling requires only limited mathematical skills, not exceeding linear functions. *Stella* is the prototype of this software category. It follows Forrester's (1968) theory of dynamic systems, which focuses on state variables and their rates of change. The top of Figure 2 shows the four types of structural elements of which all *Stella*-models are composed. These building blocks are placed in the model window and connected by mouse operations similar to those using a drawing program.

The basic physical structure of a decay model is shown in Figure 2. There is a certain quantity of radioactive $^{137}Ba^m$. Its rate of change is given by the activity. The activity determines how many $^{137}Ba^m$ nuclei decay per time unit. The decrease of $^{137}Ba^m$ leads to an increase of stable $^{137}Ba$. The most important feature of the model lies in the assumption that the activity is proportional to the number of $^{137}Ba^m$ nuclei still in stock. This means that the activity depends on the quantity 'Ba_127_meta-stable' and on a proportional constant called 'decay_constant'. The relationships represented graphically in the simulation diagram of Figure 2 can be expressed mathematically as:

$$Ba\_137\_meta - stable(t) = Ba\_137\_meta - stable(t - \Delta t) - Activity(t) \cdot \Delta t$$
$$Activity = Ba\_137\_meta - stable * decay\_constant$$

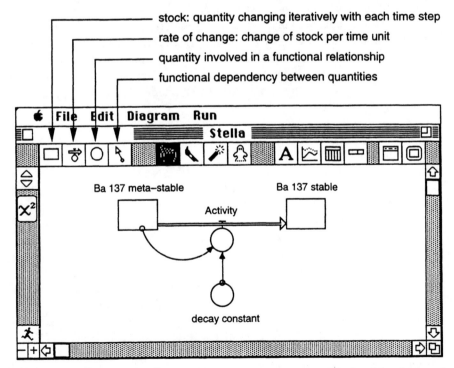

**Figure 2: Stella-model of the decay process (The structural elements are explained, and the other icons in the bar are used to edit the model and to create graphs and tables.)**

In order to test the model in a simulation run, the relationships have to be quantified. For the initial values, it is arbitrarily assumed that in the beginning there is a certain number of meta-stable barium nuclei and no stable nuclei. The simplest mathematical assumption for the activity is a linear function. The crucial point is to determine the decay constant, k. The decay constant can be understood as a certain *percentage* of parent nuclei decaying *per time unit*: $k=(\Delta N/N)/\Delta t$. We can estimate a value for k from the experimental data. Given that the rate acquired in the experiment is a measure for the actual activity of the preparation, and that the activity is proportional to the number of meta-stable barium, we can say that:

$$\text{decay constant } k = \frac{\Delta Rate / mean Rate}{\Delta Time}$$

The necessary data for calculating k already have been gained from Step 2 of our investigations (see Figure 2). A rough estimate leads to:

$$k = \frac{60.3 / 90}{160 \text{ s}} = 0.0042 \text{ s}^{-1}$$

This value can be used in the model. A simulation run can now be started. The data predicted by the *Stella*-model are exported to the spreadsheet table in new columns parallel to the experimental data. The simulation graph looks very similar to the exponential function displayed in Figure 1 (which is not surprising, because k was calculated from the same data). However, there is a major difference between the two approaches. In the first case, purely *mathematical* assumptions about the type of function describing the decay process were made. The second approach is based on *physical* reflection.

## Step 4: Extending and Applying the Model

The *Stella*-model can easily be extended to include the *production* of $^{137}Ba^m$. In our trials, this idea was proposed by the learners themselves. The basic structure shown in Figure 2 can be duplicated and added to the existing model. There is still no need for calculating sums of exponential functions as required in a mathematical approach. Figure 3 shows a refined model that can assist in understanding the concept of radioactive equilibrium.

Due to a half-life about 30 years, the decay constant of $^{137}Cs$ is very small. As only a comparatively small number of $^{137}Cs$ nuclei decay in the time span investigated, the activity remains almost constant. The resulting $^{137}Ba^m$ nuclei do not accumulate constantly because of their rather high decay constant. Their number increases until the activity of the $^{137}Ba^m$ preparation reaches the value of the production rate (= activity of $^{137}Cs$). The model predicts a duration of about 12 to 15 minutes.

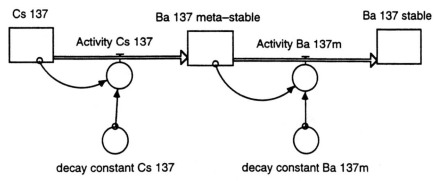

Figure 3: Decay model extended to include the production of $^{137}Ba^m$

*Step 5: An Empirical Test*

If the extended model is appropriate, there should be some empirical equivalent to the predicted time span until equilibrium. The students have to design an appropriate experiment. The perspective thus changes back to experimental issues. After 'milking' the radionuclide generator from $^{137}Ba^m$, radioactive equilibrium is disturbed. It takes some time until the activity measured at the surface of the generator again reaches a constant value. School equipment does not allow for quantitative differentiation between the activities of $^{137}Cs$ and $^{137}Ba^m$ as in the model (the beta-activity of $^{137}Cs$ cannot be measured at the surface of the generator anyway), but the $^{137}Ba^m$ activity outside the generator is a good enough test for the model prediction. Performing the experiment results in a time interval of about 14 minutes until a fairly constant rate is measured. This value is in accordance with the model prediction.

In Steps 4 and 5, the model prediction precedes a corresponding experiment while, in Steps 1 and 2, the experiment was performed first. Both approaches are consistent with accepted scientific methods of inquiry. Going back and forth between experiment and theory strengthens the ties between the two approaches.

*Step 6: New Decay Experiment*

For the next stage, the teacher introduces a new decay experiment. A small silver plate is inserted into a neutron source for several minutes. This exposure to thermic neutral radiation makes the plate radioactive. It is then put in front of a Geiger-Müller tube. The rates are acquired, displayed and saved by the MBL-tools. The students can try the mathematical description developed in Step 2 and/or the decay model from Step 3. In both cases, there are discrepancies. If the decay constant is calculated from data of the first phase of the experiment, the predicted curve approaches zero too soon. With data from the second phase, the predicted decay is too slow.

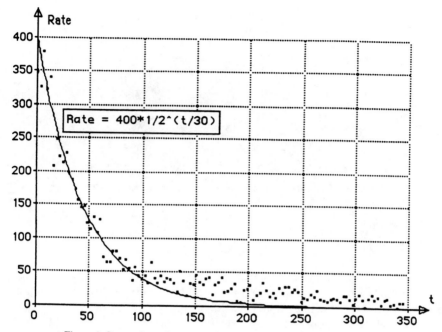

Figure 4: Comparison of measured rates and exponential function for Rate

*Step 7: Model Revision and New Simulation Runs*

The theory has to be improved. If purely mathematical considerations are employed, curve fitting could help us to find an appropriate function. Using predominantly physical aspects, it is preferable to work on the dynamic model and revise the physical assumptions. One of the main assumptions thus far was that, in the decay experiments, only *one* isotope was involved and its rate measured. A look into the nuclide-chart shows that, in the case of silver, *two* isotopes are produced by neutron radiation. The processes are:

$$^{109}_{47}Ag\,(n,\gamma)\; ^{110}_{47}Ag \xrightarrow{\beta,\gamma(24\,s)} \; ^{110}_{48}Cd$$

$$^{109}_{47}Ag\,(n,\gamma)\; ^{108}_{47}Ag \xrightarrow{\beta,\gamma(144\,s)} \; ^{108}_{48}Cd$$

These findings are used to modify the decay model so that two silver isotopes with different decay constants are considered. The basic model structure remains. It just appears twice for two separate isotopes. The decay constants are chosen according to the nuclide-chart data. The initial values for 'Ag_108' and 'Ag_110' are equal. In the beginning, the activity of $^{110}Ag$ exceeds that of $^{108}Ag$. $^{110}Ag$ decays rapidly so that, in the second phase, the rate is dominated by the $^{108}Ag$ decay. This explains why a single half-life or decay constant in the mathematical model did not suffice. Comparison in *MatheLab* yields a satisfactory accordance of the simulated data for the total activity (sum of the two single ones) and the measured rates. One now could enter into a deeper mathematical analysis of the decay process, like calculating the logarithms of the count rates and linearising the logarithmic curves. However, this would exceed the scope of this chapter.

## CONCLUSION

Radioactive decay was chosen as a prototype for connecting empirical and theoretical explorations of a scientific phenomenon, aided by information technology tools. More examples are contained in Niedderer, Schecker and Bethge (1991), Schecker (1996) and Tinker (1986).

Figures 5 and 6 look at the course of investigations, but emphasise structural aspects and leave out the particular contents. The arrows show that there is a rapid flow of ideas between the different approaches – experimental, mathematical and conceptual. There is a bidirectional relationship between experimentation and modelling. The initial experiment triggers mathematical and conceptual modelling. Vice versa, extending the basic model to a more complex case stimulates the design of empirical tests.

The consequent steps – designing a new experiment, requantifying the model for new simulation runs or working on the model structure – foster a deeper understanding of the scientific topic and of the scientific method. Experimentation, mathematical modelling and conceptual modelling contribute in different ways to understanding:

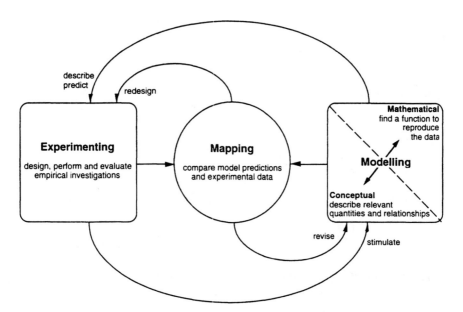

Figure 5: The experimenting and modelling interplay

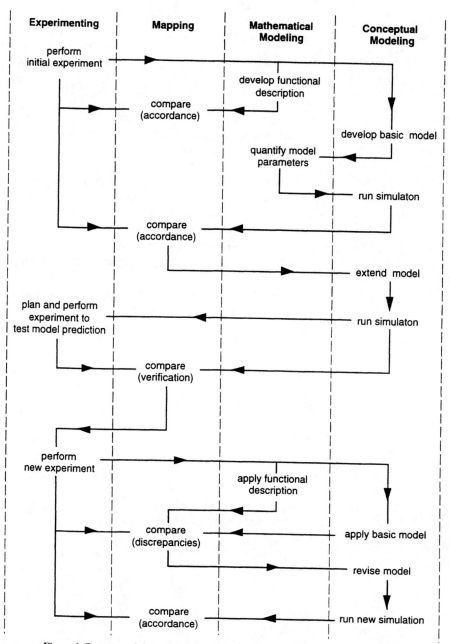

Figure 6: Structure of the measuring and modelling process about radioactive decay

- Experimentation provides the phenomenological basis from which first ideas about proper descriptions can emerge and 'solid data' against which a theoretical model later can be tested.
- Conceptual modelling helps to clarify qualitative assumptions about the process, such as 'the more nuclei there are, the more of them decay', but it does not result in an analytic formula for calculating decay curves.
- Mathematical modelling results in universal solutions that are very useful for quantitative predictions. However, the final formula no longer shows which assumptions were made to develop it.

A crucial role is played by mapping experimental and model data. Discrepancies are often more productive than agreement, because they cause further theoretical reflection or a revision of the experiment. There are at least three possible reasons for discrepancies:

- The theoretical predictions make sense; there could have been some mistake in performing the experiment or in handling the acquired data.
- The model is structurally correct, but the parameters (e.g., initial values, constants, time parameters) are not properly chosen.
- The physical assumptions of the model are not appropriate for the investigated phenomenon. The model *structure* has to be altered.

Dynamic modelling systems primarily support conceptual modelling. Quantitative predictions can be reached by launching numerical simulations. Mathematics tools support experimentation by processing measured data and also mathematical modelling by laying function graphs over the data. MBL-systems tie observation and data representation closer together. The technical integration of these tools opens up new opportunities for educational learning environments supporting integrated knowledge acquisition.

REFERENCES

Brandenburg, R.J.: 1993, 'MultiLab – Integrierte Software zum Messen, Auswerten und Modell-Bilden' [MultiLab – Integrated Software for Measuring, Evaluation and Model-Building], *Praxis der Naturwissenschaften – Physik* 42(5), 12–15.
Brasell, H.: 1987, 'The Effect of Real-Time Laboratory Graphing on Learning Graphing Representations of Distance and Velocity', *Journal of Research in Science Teaching* 24, 385–395.
Dorenbos, V. & Dulfer, G.H.: 1992, 'IP-Coach – An Open Science Interfacing Package for IBM-Compatible Computers', Paper presented at the NATO Advanced Research Workshop 'Microcomputer Based Labs: Educational Research and Standards', Amsterdam, The Netherlands.
Duit, R.: 1994, 'Research on Students' Conceptions – Developments and Trends', in H. Pfundt & R. Duit (eds.), *Bibliography: Students' Alternative Frameworks and Science Education* (fourth edition), Institute for Science Education at the University of Kiel, Kiel, Germany, xxii–xlii.
Forrester, J.W.: 1968, *Principles of Systems*, Wright-Allen, Cambridge, MA.
Girwidz, R.: 1994, *Beiträge und Analysen zum Lernen in Physik mit dem Computer als Hilfsmittel* [Contributions to and Analyses of Computer-Aided Physics Learning], Doctoral dissertation, University of Würzburg, Würzburg, Germany.
Goldberg, F. & Bendall, S.: 1995, 'Making the Invisible Visible: A Teaching and Learning Environment that Builds on a New View of the Physics Learner', *American Journal of Physics* 63, 978–991.

Gonzalez, J.J.: 1994, *Powersim*, Powersim AS, Bergen, Norway.
Hennessy, S., Twigger, D., Driver, R., O'Shea, T., O'Malley, C.E., Byard, M., Draper, S., Hartley, R., Mohamed, R. & Scanlon, E.: 1995, 'Design of a Computer-Augmented Curriculum for Mechanics', *International Journal of Science Education* 17, 75–92.
High Performance Systems: 1994, *Stella*, Author, Dartmouth, NH.
Laws, P.W.: 1991, 'Calculus-Based Physics Without Lectures', *Physics Today* 44(12), 24–31.
Linn, M.C.: 1992, 'The Computer as Learning Partner: Can Computer Tools Teach Science?', in K. Sheingold, L.G. Roberts & S.M. Malcolm (eds.), *This Year in School Science 1991: Technology for Teaching and Learning*, American Association for the Advancement of Science, Washington, DC, 31–69.
Meyling, H.: 1990, *Wissenschaftstheorie im Physikunterricht der gymnasialen Oberstufe* [Science Philosophy in Upper Secondary Physics Courses], Doctoral dissertation, University of Bremen, Bremen, Germany.
Niedderer, H., Schecker, H. & Bethge, T.: 1991, 'The Role of Computer-Aided Modelling in Learning Physics', *Journal of Computer Assisted Learning* 7(2), 84–95.
Reger, R. & Freiberger, U. (eds.): 1989, *Informatik-Themen im Grundkurs Physik (Informatik)* [Computer-Based Physics Curriculum], Bayer Schulbuch-Verlag, Munich, Germany.
Schecker, H.: 1985, *Das Schülervorverständnis zur Mechanik* [Students' Alternative Conceptions in Mechanics], Doctoral dissertation, University of Bremen, Bremen, Germany.
Schecker, H.: 1993, 'The Didactic Potential of Computer Aided Modelling for Physics Education', in D.L. Ferguson (ed.), *Advanced Educational Technologies for Mathematics and Science*, Springer, Berlin, Germany 165–208.
Schecker, H.: 1996, 'Bremer Interface-System: Didactic Guidelines for a Universal, Open, and User-Friendly MBL-System', in R.F. Tinker (ed.), *Microcomputer-Based Labs: Educational Research and Standards*, Springer, Berlin, Germany 351–367.
Thornton, R.K.: 1987, 'Tools for Scientific Thinking – Microcomputer-Based Laboratories for Teaching Physics', *Physics Education* 22, 230–238.
Thornton, R.K.: 1992, 'Enhancing and Evaluating Students' Learning of Motion Concepts', in A. Tiberghien & H. Mandl (eds.), *Physics and Learning Environments*, Springer, Berlin, Germany, 265–283.
Tinker, R.F.: 1986, 'Modelling and MBL: Software Tools for Science', Paper presented at the National Educational Computer Conference 1986, San Diego, CA.
Tinker, R.F.: 1993, 'Modelling and Theory Building: Technology in Support of Student Theorizing', in D.L. Ferguson (ed.), *Advanced Educational Technologies for Mathematics and Science*, Springer, Berlin, Germany 91–113.
Weißgerber, H., Andraschko, H. & Lorbeer, W.: 1994, *MatheLab – Funktionale Tabelle, Graphik, Iteration* [MatheLab – Spreadsheet, Graphs, Iteration], Landesbildstelle, Bremen, Germany.
Wilson, J.M. & Redish, E.F.: 1992, 'The Comprehensive Unified Physics Learning Environment', *Computers in Physics* 6, 202–209 & 282–286.

## 3.8 Where You Want IT, When You Want IT: The Role of Portable Computers in Science Education

ANGELA E. McFARLANE
*Homerton College, University of Cambridge, UK*

YAEL FRIEDLER
*The Hebrew University of Jerusalem, Israel*

The world of science beyond schools has been transformed by the use of computer-based technologies. No contemporary professional scientist could imagine working without access to data capture, manipulation and display systems, electronic databases including on-line and CD ROM systems, or word processing and desktop publishing for the preparation of papers and reports. There is a growing body of research evidence suggesting that using such systems can assist school students in the acquisition of the process skills central to carrying out successful science investigation (Brasell 1987; Friedler, Nachmias & Linn 1990; Friedler, Nachmias & Songer 1989; Heutinck 1992; Linn, Layman & Nachmias 1987; Linn & Songer 1991; Mokros & Tinker 1987; Stuessy & Rowland 1989; Thornton 1987; for further discussion of the area see other chapters in this section of the *Handbook*, particularly the overview provided by Marcia Linn.) Yet the use of such systems remains rare in schools. In the UK, for example, a recent survey (National Council for Educational Technology 1993) showed that, although many schools use some Information Technology (IT) in science, the amount of use is very small, with only 11 percent of schools using it for more than 3 hours per year. Surveys by the Department for Education (1989, 1991) in England and Wales suggest that this is not solely due to a lack of computers or teacher expertise in IT. Even if the schools were restricted to one computer in the science department, the level of use is very low. Given such low exposure, it is therefore not surprising that quantifiable learning gains in mainstream classrooms related to the use of IT have been somewhat elusive (Johnson, Cox & Watson 1994; Watson 1993).

Factors limiting the implementation of IT in science can include inadequate teacher development, lack of hardware and appropriate software, inappropriate curriculum and poor integration of computer-based activities into lesson plans. In fact, there is still a general lack of recognition of the implications of IT for science education in the science education community. Major science education

---

Chapter Consultant: Instead of using chapter consultants, chapter authors in the Educational Technology section provided editorial comments on each other's chapters.

journals rarely carry papers relating to the use of IT; even when the central findings are of greater relevance to science education than IT, such work is still largely to be found in the educational technology journals.

Several major projects have attempted to address some of the issues limiting the implementation of IT, by providing access to IT in the school science laboratory through the medium of portable computers. The experiences of those involved help to identify which factors inhibit the use of IT in mainstream science education and the learning gains to be expected if those factors can be overcome.

The specifications of portable computers are advancing rapidly, and none of the particular machines used (even in the most recent evaluations) are still in production. However, this technological advance should not be seen as making what we have learned so far about portable computers in science obsolete. Observations in many schools in five countries have provided valuable insight into the effective implementation of information technology in science education using portable computers. What we will focus on here is not the technical detail of the machines used, but the generic features of these machines which distinguish their use from that of desk-top computers.

This chapter draws heavily on the experiences of teachers and students using portable computers during four major evaluation projects in five countries. They are the Sunrise project in Queensland, Australia, the PULSE project in New Jersey, USA, the PLAIT project in Northern Ireland and the Portables Evaluation Pilot in England and Wales (see Table 1). The last of these projects was an order of magnitude larger than any of the others, involving a total of 118 separate projects, over 250 schools and 2,500 machines. Each of these projects has published at least one report which summarises its findings (Bowell, France & Redfern 1994; Gardner, Morrison, Jarman, Reilly & McNally 1992, 1993; McMillan & Honey 1993; Rowe 1993; Stradling, Sims & Jamison 1994). In preparing this chapter, the authors also had access to individual project reports from the Portables Evaluation Pilot, one of which was managed by the authors. This chapter concentrates on an analysis of common (and therefore presumably) reproducible findings from the various summary reports and the individual reports from the Portables Evaluation Pilot. Material has been drawn exclusively from those reports which make direct reference to the teaching of science unless otherwise indicated.

The first section briefly considers the potential uses of computers in science teaching, with particular emphasis on an investigative approach to science education. In the second section, the advantages of portable computers are outlined (availability, spatial impact, time management, ownership, etc.). What follows is a summary based on evidence drawn from all of the portable evaluation projects. We compare the use of portables to the use of desk-top computers, considering learning gains (the third section of the chapter), student attitudes (the fourth section), teacher development (the fifth section), and appropriate hardware and software (the sixth section). Finally, key points for the successful implementation of portable computers in science are given in the seventh section, concluding with a comment on the likely future of IT in science education in the eighth section.

Table 1: A summary of four evaluations of projects involving portable computers

| Project | Sunrise, Queensland, Australia | PULSE, New Jersey, USA | PLAIT-N. Ireland | Portables Pilot, England and Wales |
|---|---|---|---|---|
| Date | 1990/1991 | 1991/1992 | 1991/1992 | 1992/1993 |
| Duration | 1 year | 2 years | 1 year | 1 year |
| Population | 115 students | 25 students | 235 students | Thousands of students (250 schools) |
| Age group | 11 years | 13–14 years | 10–15 years | 5–17 years |
| Objectives | Investigate the potential of portable computers on teaching and learning. Emphasis on attitudes, computer literacy, gender differences | Investigate the potential of portable computers on teaching and learning. Emphasis on classroom interactions, teachers' goals and practices | Investigate the potential of portable computers on teaching and learning. Emphasis on IT needs and classroom processes | Investigate the potential of portable computers on teaching and learning. Emphasis on policy issues, process skills, special need students |
| Sources of data | Questionnaires and interviews (teachers and students), observations, tests | Interviews (teachers), observation, tests | Diaries (teachers and students), interviews (teachers and parents), observations, tests | Questionnaires (teachers, students, parents), interviews (teachers, students), observations, tests |
| Computer allocation | One computer/student | One computer/student | One computer/student | Varied among projects |
| Software | Logo Writer | MS Works | MS Works, Data-logging software | Varied among projects. Mainly wordprocessors, spreadsheets and data-logging software |
| Areas of use reported | Typing, Writing procedures, Mathematics | English, Science | English, Science, Mathematics | English, Mathematics, Science, Geography |

## USE OF INFORMATION TECHNOLOGY IN SCIENCE

Information Technology (IT) is a term increasingly used to describe the application of computers and appropriate software to any information-handling task. It encompasses the use of computers to collect, analyse, interpret, present and communicate information and therefore covers everything from word processing to computer modelling. IT has many possible roles in the teaching and learning of science, particularly skill-based science, which often is called process science (Frost, McFarlane, Hemsley, Wardle & Wellington 1994). There is a range of possible applications of IT in science. Scientific activity in the classroom has been broken down into a series of processes, or skills, and each one can be paired with possible supporting IT applications (see Table 2).

This list is intended to be indicative and is not exhaustive. The effectiveness of these uses of IT in science skill development has been widely studied in a variety of research projects (e.g., Brasell 1987; Friedler *et al.* 1989, 1990; Heutinck 1992; Linn, Layman & Nachmias 1987; Linn & Songer 1991; Mokros & Tinker 1987; Stuessy & Rowland 1989; Thornton 1987). However, a feature which makes the portables studies unique is that these projects used small computers to make these IT tools available in routine school science settings, integrated into standard curricula under the management of regular school teachers. A major factor which distinguishes the portables studies from the ImpacT report (Johnson *et al.* 1994; Watson 1993) is that of access; students in the portables studies were able to use the computers more frequently, *at the point of task*, and in greater numbers. It was not uncommon for portables projects to provide individual or small-group access, which is a situation rarely found in ImpacT schools where no additional hardware was provided and where schools worked only with installed desktop technology.

Table 2: Possible supporting IT applications for science processes or skills

| Process or Skill | IT Applications |
| --- | --- |
| Planning an investigation, including hypothesis construction and collaborative writing | Word processing |
| Researching a topic, including on-line services | CD ROM, databases |
| Taking measurements | Data logging (MBL) |
| Making results tables | Data logging, spreadsheets |
| Drawing graphs | Data logging, spreadsheets, databases |
| Doing calculations | Data logging software, spreadsheets |
| Searching for patterns, including analysing and interpreting data | Data logging software, spreadsheets, simulations, modelling software |
| Asking what if . . .? questions | Simulations, modelling software |
| Comparing results with others | On-line services, e-mail, CD ROM |
| Presenting information in a report | Word processing, desk-top publishing, spreadsheets |

## WHY PORTABLE COMPUTERS?

Here we examine the particular impact of portable computers on the implementation of IT in science as outlined in the previous section. What can portable computers offer that desk-top computers cannot? The evaluation projects have shown that the range of uses of IT in science identified above can be accomplished using a portable, notebook computer provided that it has appropriate software and an adequate hardware specification with which it can be run. However, it should be noted that, at the time of the projects, none of the portable computers used could read CD-ROM disks. We consider the role of the portable in terms of availability, spatial impact, issues of time and curriculum coverage, personal and group use, and the range of activities supported.

*Availability – The Time, The Place*

A common model of computer distribution in secondary schools in most countries is to site desk-top machines in a designated computer room. Such rooms rarely provide sufficient space to set up science apparatus. Furthermore, the designation of computer rooms as dry zones means that any wet labs are immediately off limits. Even where computers are available in a laboratory, the problem is not entirely solved. Few school laboratories can accommodate a desk-top computer for a group of students and still have sufficient bench space for apparatus. One computer per student is out of the question. Desk-top machines are very difficult to move and, even when mounted on trolleys, considerable effort is required to move several between rooms. This means that one laboratory must be designated as the science computer laboratory. Any teacher wishing to use the computers must book this room for her class and any class scheduled in that room has to work around the computers, whether or not they are needed for that lesson.

This whole situation changes dramatically when notebook computers are offered as an alternative. The small size and weight of the notebooks mean they can be stored centrally and moved easily between laboratories as and when required, as with any other class set of apparatus. In many projects, this process was facilitated by the storage of the computers not in the protective carry cases provided with the machines, but in open trays stored on a trolley. The trays made it slightly quicker to set up and store the computer at the beginning and end of lessons but, more importantly, it meant that the teacher could easily do a visual check to ensure that all the equipment was present when the computers were collected.

The portable computers were also used outside the school laboratory. Students took them home to work on assignments or to finish writing up laboratory work, or on field study visits to collect data.

During tasks which do not involve data-logging, portables offer the advantage that they can be used as and when required. There is no need to go off to a special room, or even a different area of the same room to work. When a student wants to write something or use a spreadsheet to make a graph, she can turn to the portable on the bench beside her. Like a calculator or a pencil case, the portable can be there on the table, in use or ignored as the current task requires. Arguably, the writing or data manipulation aspects of a science activity can be carried out in the computer room. This would mean dedicating a lesson to this one element whereas, with the portables, it becomes part of the laboratory process which naturally leads into, or grows from, the experimental phase as each group becomes ready. The ability to use a portable – or set it aside – as the task in hand demanded, was undoubtedly one of the strongest features of this technology in the classroom.

*Spatial Impact – Small Indeed Can Be Beautiful*

Portable computers have shown themselves to be especially effective in the process of data logging, or in microcomputer-based laboratories (MBL). Here, students use probes attached to the computer to monitor physical variables, and software is used which displays the values of these variables over time on the screen, usually as a line graph. The bulk of desk-top computers means that they are not easily incorporated into a practical science investigation. Desk-top computers dominate the work space by nature of their size and their large, visually-distracting screens. The actual experiment, to which the computer is a useful adjunct, can become a secondary concern. The computer operator could be unable to see the data generation process, and those manipulating the apparatus are unlikely to be able to watch the screen simultaneously. The desk-top machine provides a visual obstruction which disrupts communication between the members of a group, and between student and teacher.

When using the notebook computer during a practical laboratory, the dynamics of the work space are altered radically. The machine takes its rightful place as another tool in the experiment. It neither dominates the apparatus, blocks the view of the process under investigation, nor prevents dialogue among students or between teacher and student by obscuring the student operating the machine. Everyone can see the practical activity which is generating data, whilst simultaneously having sight of the screen displaying that data in an interpretable format.

The ergonomics of the writing or analysis phases of the science process are also changed. Again, the machine could be passed across the table whenever each student wished to make a contribution, and even the teacher could join in with a comment without disturbing the dynamics of the group. Discussion about the task could continue across the top of the machine uninterrupted.

*Time Management*

It is common for the use of computers to be associated with time saving. Certainly a computer can collect data and plot a graph far faster than could be accomplished manually. For text entry, older students with poor keyboard skills are often slowed down by the use of a word processor, although very young children, even with no keyboard skills, often produce text faster as their immature motor skills make writing manually very laborious.

It was noticeable that, in projects where classes with access to portables were compared to parallel groups without them, classes with access always ended up slipping behind in terms of curriculum coverage (Gardner *et al.* 1993; National Council for Educational Technology 1994) by as much as three weeks over one year was reported. On examination, this proved to be due to classes taking longer to complete practical work when using the portables. This was not, as might be imagined, caused solely by the time taken to set up the computers or in learning to use software. Rather, it was because students with instant access to data in an interpretable form became aware of the shortcomings of their practical techniques and experiment design. Naturally, they wanted to refine their procedure and collect additional or alternative data, which took longer. Students without the portables did not usually become aware of any inadequacies until they plotted graphs, often for homework, and by this time it was too late for the problem to be remedied, even if it was recognised.

*Ownership*

Different evaluation projects employed different models of computer distribution. The two most common scenarios in projects involving science were either that each student under study had a machine for personal use at school and home in any subject area, or the science department had a group of eight computers for use when required in the department. Clearly, in the second model, the computers were most commonly used by groups of students working collaboratively. However, some individual use was possible especially with older students who had the opportunity of booking a machine for use during study periods. In addition, the presence of one computer per child did not preclude collaborative working, with some machines being set aside for a period or machines being exchanged by students working together.

Surprisingly, problems rarely arose when students needed to proceed with a task in a subsequent lesson when small numbers of machines were being used by many groups. They easily identified which hard disk-based machine contained their work, retrieved it and continued. This was the case even with very young children, for whom colour coding of the machines proved very effective.

Many students appreciated the privacy offered by a portable. When trying something new, or concentrating on a difficult task, many students appreciated the way in which they could turn down the screen so that they alone could see their work. In contrast, when working collaboratively, they had no problem sharing their work by using a portable with partners or by exchanging machines.

*Portable Computer as a Part of the Kit*

Once established as an off-the-shelf tool to assist students in their work, the use of the notebooks in science classes quickly expanded to include word processing and desk-top publishing of laboratory reports, and the use of spread sheets as well as data logging. This was greatly facilitated where students could employ IT skills which they already possessed in the use of the application software on the notebooks. Demand for these facilities was often student led. For example, the pasting of graphs from a data logging application into a laboratory report was common, but rarely was it a skill taught by the teacher. Often teachers learned new techniques, or about facilities which they had not yet discovered, from students. Students might have had greater knowledge of the technology in some cases, but teachers recognised the applications most likely to lead to enhanced learning. For example, teachers were usually responsible for the introduction of word processing to the planning phase of an investigation, whereas students might only have recognised their use for report writing.

Table 3 shows data collected from eight classes, with varying student ages ranging from 12 to 17 years, in one secondary school. The students were asked to give their perceptions of what they had used the portable computers for during a science lesson.

There was no indication that the students' description of a task actually reflected the nature of that task, or how tasks were classified (e.g., would writing up an experiment also score as word processing?). However the data indicates that the portables indeed were used for a wide range of activities, with data logging being the foremost.

The experience of students' use of the computer as a tool in science is, of course,

Table 3: Students' perceptions of what they used portable computers for during a science lesson.

| Process Skill | Students (in %) |
| --- | --- |
| Planning investigation | 8.0 |
| Word processing | 8.8 |
| Write up experiments | 3.6 |
| Check spelling | 1.5 |
| Log data | 59.9 |
| Display data | 8.0 |
| Key in data | 2.2 |
| Do calculations | 4.4 |
| Run a program | 2.2 |
| A control application | 1.5 |

Adapted from National Council for Educational Technology (1994)

not unique to portables (for example, see Nancy Butler Songer's chapter in this *Handbook*). However, the use of portables does afford widespread access by all students being taught in a science department, on a demand led basis. Additionally, the portables projects did not require a novel curriculum, but rather they facilitated the integration of computer use into the existing curriculum. Because changing the whole curriculum is not an option in many schools, or even countries, this becomes a very significant issue.

## LEARNING OUTCOMES

The portable projects were able to offer copious qualitative data on student learning gains, in addition to some quantitative data. Teachers reported improved learning in three main categories: the ability to use computers, data loggers and various software in a science context; improved investigative skills, including data analysis; and better understanding of content. It is clear that portables are having a positive impact on curriculum delivery (teacher quoted in Stradling *et al*. 1994).

As with the use of desk-top computers in science, learning gains were most apparent where inquiry-led, investigative, independent learning was an objective (Watson 1993). Furthermore, where the process skills enhanced by the use of IT were not included in the assessment process, students using IT in science, especially during the early parts of the project, did not perform so well on tests. In an example of students investigating motion, students performing data logging did significantly less well than students using conventional apparatus in a test on the laws of motion dealing only with subject knowledge (Rogers 1994). This also links to the issue of time spent and content covered (see time management section above).

Where students use data-logging software on desk-top computers to plot line graphs as data are collected, the contiguous nature of these events has been shown to have remarkable impact on students' understanding, both of the processes under investigation and of the nature of graphs representing the relationship between variables (Brasell 1987). This learning gain is equally evident where students use portable computers (Friedler & McFarlane in press; McFarlane, Friedler, Warwick & Chaplain 1995). Furthermore, if portable computers are used, these activities can be undertaken as part of normal class activities and therefore integrated into the science curriculum. Lessons involving such computer use can be managed by regular teachers, under normal school conditions, for all students whether working individually or in groups. The smaller, monochrome screens of the portables do not distract from this process. Additionally, for all the reasons elucidated earlier, data logging as part of a science investigation is better facilitated through the medium of portable computers than desk-top machines.

Tables 4 and 5 show some quantitative data collected during one of the Portable Computers Pilot Evaluations carried out by the authors. The project took place over the summer and autumn terms of 1993. In the summer, two parallel classes in each of two secondary schools and two primary schools were identified and one class in each school was designated as the experimental group. Four sessions, each of two hours in duration, introduced both control and experimental group teachers to the project, its aims and scope, and the new resources which they would be using. These new resources included portable computers, data loggers and probes, data-logging software and a word processor. Once the teachers had some understanding of the process of data logging, the teachers and researchers together designed activities which would suit the curriculum planned for the following two terms.

The skills sought during this project were to develop reading of patterns (i.e., the relationship between the variables temperature and time) on line graphs, representation of patterns on a line graph by students and the interpretation of events described on line graphs. In addition, with the secondary students, content knowledge and process skill development were monitored. These skills were assessed via pretest and posttest questionnaires and interviews.

The posttest results for the two groups of 14-year-old students after one term of science are shown in Table 4. The control group carried out two practical investigations using non-computer-based techniques throughout. The experimental group used portable computers and data loggers for the practical data collection and analysis phases of the same investigations. Pretests had already established that the two groups were not significantly different in their mastery of the skills

Table 4: Posttest comparison of experimental and control groups in Year 9 for individual question types.

| Question type | Control group N=46 | | Experimental group N=40 | | |
|---|---|---|---|---|---|
| | Mean % | SD | Mean % | SD | t |
| Drawing graphs (subject matter independent) | 45.6 | 5.0 | 65.0 | 4.8 | 1.81* |
| Accuracy of cooling curves | 54.3 | 5.0 | 73.7 | 4.4 | 1.91* |
| Sketching predictive heating curves | 52.2 | 5.1 | 75.0 | 4.4 | 2.24** |
| Accuracy of heating curves | 11.9 | 2.4 | 33.7 | 4.0 | 3.01** |
| Content specific knowledge (heating different volumes) | 71.7 | 4.4 | 87.5 | 3.3 | 1.84* |
| Process skills (fair testing) | 71.7 | 3.7 | 83.7 | 2.5 | 1.77* |
| Explanation of heating curves for two different volumes drawn by the students | 39.1 | 5.4 | 61.2 | 4.8 | 2.00* |
| Transformation from text to graph (moving between two different symbol systems) | 22.8 | 4.2 | 42.5 | 5.0 | 1.96* |

*$p \leq 0.05$
**$p \leq 0.01$
Adapted from Friedler & McFarlane (in press)

prior to the project. It is evident from the posttest results that the experimental group showed better development of skills both in the context of their investigations, heating and cooling curves, and also in contexts independent of subject matter. There was evidence of transfer of developed skills relating to the drawing of graphs more accurately and transformation of data between text and graphs in novel contexts (Friedler & McFarlane in press).

The posttest results for the control and experimental groups in the two primary schools are shown in Table 5. In this project, the children in the experimental classes had their first introduction to line graphs not as a Cartesian plot (for which the ability to identify positions correctly on a grid was the objective), but rather they were introduced to graphs as a representation of the relationship between two variables, namely, temperature and time. They saw a system which they had the power to influence directly, depicted in terms of a line graph. Their ability to read and interpret temperature/time graphs was greatly enhanced as a result, and it is particularly significant that their ability to sketch temperature/time curves to predict the behaviour of a novel system also improved.

The children in the control group showed no significant gains in the same skill set, despite being taught by more experienced, highly-skilled teachers using exemplary methodology, but restricted to traditional resources. Even though these children plotted points manually onto a grid as they took measurements (a process which might be thought to mimic the real-time data logging very closely), there was no evidence of any improvement in their understanding of graphs as a representation of changes in the variables which they were studying, according to the test instruments used. This further supports the findings of other researchers in the field (Linn, Layman & Nachmias 1987; Linn & Songer 1991; Mokros & Tinker 1987; Stuessy & Rowland 1989). However, what distinguishes the portables study reported here is that these data were collected in full-size, mainstream classes, and taught by normal teachers as part of the normal curriculum in standard science laboratories or classrooms. This situation was only practically supportable in a real school setting using portable computers. We are discussing a situation in

Table 5: Pretest and posttest comparisons for control and experimental classes in two primary schools.

| School testing | Control | | | | Experimental | | | |
| --- | --- | --- | --- | --- | --- | --- | --- | --- |
| | N | Mean % | SD | $t$ | N | Mean % | SD | $t$ |
| School A Pretest | 13 | 35.29 | 22.0 | 1.29 | 13 | 34.84 | 22.46 | 1.96* |
| School A Posttest | 10 | 48.53 | 25.91 | | 13 | 53.85 | 26.83 | |
| School B Pretest | 12 | 24.02 | 16.72 | 1.35 | 11 | 32.08 | 7.61 | 2.64** |
| School B Posttest | 12 | 35.78 | 25.01 | | 11 | 51.34 | 22.95 | |

* $p \leq 0.05$
** $p \leq 0.01$
Adapted from McFarlane et al. (1995)

which all students in a large number of schools with more than 800 students have access to a computer in science laboratories, and in which the learning benefits previously only found in research projects are realised.

Traditional teaching methods are quite effective for teaching students the skills required to handle graphs manually by the time they are 11 years of age (Swatton & Taylor 1994; Taylor & Swatton 1990). However, the same sources show that, even by the age of 13 or 15 years, the vast majority of children are not able to appreciate their graph's power as models of underlying variable relationships presented symbolically in this way. In traditional curricula, children are taught to draw graphs before they spend time interpreting them and, in many secondary science curricula, the skills of interpretation, linking the events depicted and the symbolic representation of those events are never explicitly taught. Rather, this appreciation is expected to develop by osmosis.

The data collected in this project suggest not only that the use of data logging using portable computers with young children (aged 8 to 9 years) is manageable within the average primary classroom, but that to do so has very positive learning outcomes which are not achievable in other ways. Furthermore it suggests that introducing the children to the graph holistically – as a model of the relationship between changing variables – before introducing the manual skills of plotting might be a way to bridge the conceptual gulf which otherwise develops. With older children aged 13 or 14 years, we found that even quite able students go through the motions of manual plotting of graphs in science automatically. They cannot articulate why they use a graph, or why they use a bar or line graph to represent data. 'Because we always do in science' was a much volunteered response. Gains towards bridging this conceptual gulf perhaps could be attained by shifting the emphasis away from the manual plotting skills towards the higher-order interpretative skills, and by reversing the order in which these skills are introduced in the curriculum. This might be achieved by using real-time plotting by computer to remove the initial need for manual plotting skills. Removing the need for manual plotting permits the introduction of graphs at a much earlier stage of intellectual development, and allows the students to meet graphs first as a representation of the relationship between changing variables, not as an exercise in Cartesian coordinate mapping.

In addition to the use of data logging, the use of computerised spreadsheets and databases to give students rapid access to many different representations of the same data set is a useful strategy in improving the understanding of the complexities of the graph symbol system. The use of spreadsheets to present and graph data on portable computers was shown to have a positive effect on students' graphing skills in the PLAIT project (Gardner *et al.* 1993). They also proved effective for helping students understand the links between variables in scientific formulae (Stradling *et al.* 1994). However, it seems that the contiguous manipulation of physical variables and dynamic plotting could be a more powerful way to introduce even very young children to the links between graphing and the ability to handle variables and their interactions, which is at the heart of process science.

Computer-generated graphs from data-logging or data-handling software lead students to engage in quantitative analysis of data earlier in a lesson. The result seen in many projects was a greater understanding of the underlying content.

It was noted that the use of collaborative writing techniques, facilitated by the group use of a small computer, led to a general improvement in the planning process and therefore a better outcome from the practical. Students working together with a portable computer collaborate in terms of sharing ideas, rather than merely the division of labour, whilst developing scientific observation skills (Bowell *et al.* 1994).

There were no reported learning gains identified specifically with the use of e-mail in science in the PULSE project or elsewhere. However, the collection and sharing of data with students in other schools proved highly motivating (McMillan & Honey 1993).

Learning gains suffered where the software used proved inappropriate due to the difficulty in using it, particularly for less able students (Gardner *et al.* 1992). In contrast, learning gains were greatest where the use of appropriate software was integrated into a task by a teacher who understood the potential of the technology.

## STUDENT ATTITUDES TO PORTABLES IN SCIENCE

Perhaps the most universal finding from all the evaluation projects was the enthusiasm of students for the portable computers. The use of these machines proved highly motivational. This manifested itself not only in suggestions and requests to use the computers in lessons, but as increased student time on task. It was common for students to return in order to complete tasks out of lesson time, which included taking the machines home with them. Questionnaire data show that students felt that using the portable increased their interest in science. Using the machines even made some children feel more scientific. The more frequently that secondary students used a portable, the more likely that they were to claim that what they liked most about it was its impact on their learning (teacher quoted in Stradling *et al.* 1994). These findings are interesting in that they also apply in situations that were previously generously resourced with desk-top computers, but students (and teachers) did not show the same degree of enthusiasm prior to the introduction of the portables. It is not simply a computer-as-novelty phenomenon, but seems to be related to issues of ownership and the less intimidating physical characteristics of a portable.

Interestingly, where portables were used, there was no evidence of the gender bias found in conventional IT situations in schools (Nickell 1987). Girls were equally, or even slightly more, likely to use these small computers than boys wherever this parameter was monitored. In one school, where 14–16 year-olds worked in groups for practical investigations, the computer was most frequently operated by one of the girls during the data-logging stage. This involved setting up software and designing a virtual representation of the investigation, a key role rather than a supportive one in the group. The reasons for the lack of gender bias are unclear and more research on this particular finding might provide useful indicators to strategies to improve uptake of science generally by girls. It is possible that, in a practical situation, the girls took to the computers as a strategy for avoiding an interaction with practical science apparatus, preferring to leave this to the boys. However, their enthusiasm for the computers was equally clear in single-gender groups, for which girls were just as likely as the boys to spend more time repeating experiments and thereby improving their method with each iteration.

Positive student reaction was not entirely universal. There were instances of younger students apparently deliberately leaving computers at home, although this was attributed by their teachers to the fact that they found the machines heavy and awkward to carry. Some students, who were already computer-phobic, were not any happier with portables. However, these negative reactions were exceptional and there were many more instances of students reacting with enthusiasm when they had opportunities to use portables, even when they had shown little interest in science or school generally prior to the project. The smaller size of these machines generally seemed to make them less intimidating for the computer-phobic, and particularly for young children who found them more user-friendly than desk-top computers.

This predominantly positive attitude is also evidenced by the absence of any reports of wilful damage or misappropriation of computers, although there were some instances of accidental damage and some deliberate mischief. It was not a good idea, for example, to use password-protected machines, as this proved too great a provocation to the highly computer-literate students. In one case, a student changed a password and then forgot the new one. The repair to the machine proved ridiculously costly!

## TEACHER PREPARATION AND DEVELOPMENT

It was rarely the case that teachers involved in any of the portable evaluation projects had prior expertise in the use of information technology in their teaching. Most projects built an element of staff development into the model. However, this was inevitably given limited time. In one model, the staff were not trained at all, and instead 'experts' went to the school and taught the students the computer-related skills needed for the project. Not surprisingly, this led to an alienation of the teachers of the classes involved and the failure of the project to bring about lasting change in the school. In stark contrast, projects for which innovation proved effective and apparently sustainable used collaborative models of staff development. Teachers and researchers planned the integration of the portables into the science curriculum together.

Successful staff development models used in projects usually included some designated sessions at which teachers could learn the mechanics of setting up the computers and peripherals, and try out the software. This could have been supplemented by occasional expert support in the form of a team teacher joining classes from time to time. This was particularly effective at the beginning of the projects when students were all learning how to set up equipment and when there was simultaneous demand for teacher input to several groups.

This model, of course, is no different from the desk-top based experience which many teachers have without subsequently making any change to their classroom practice. A crucial difference here was that teachers could take home the small computers and practice in privacy. In fact, it became common in one project for teachers to take a machine with them when they were called to substitute teach. They could make mistakes and find solutions in their own time without the pressures of losing face in front of colleagues or students and having a lesson disrupted by such failure. They also had the additional incentive that appropriate equipment would be available in the laboratory or classroom when they came to teach the planned lesson.

The resulting growth in confidence was remarkable. Even the small minority of teachers who experienced serious problems with their machines were not put off by the idea of integrating portables into their teaching in the future (teacher quoted in Stradling *et al*. 1994). There is considerable evidence of teachers becoming so comfortable with the computers that they were prepared for students to experiment with new applications of IT on a spontaneous basis, and responded positively to students when they made legitimate requests to use the technology in a context for which the teacher had not previously planned. Students 'came up with good ideas for using sensors I had not even thought of' (teacher in interview with authors).

Another important strategy in the growth of teacher comfort with the technology was the provision of a portable as an aid in administrative tasks. Although there is good evidence that the personal use of IT by teachers does not automatically lead to its use in the classroom (Department for Education 1989, 1991), clearly this is a shorter step for the IT literate. There was good evidence in the evaluation projects that the teachers who made the greatest advances in terms of innovative teaching and enhanced student learning were not those with IT development skills prior to the project, but those with a proven record of successful curriculum development.

The most effective projects were those for which the staff development included not only an opportunity to come to grips with the technology, but also an element of curriculum planning and integration. Lessons for which the technology was applied to a traditional scenario as an extra layer, without consideration of the implications for the learning outcomes, were generally less successful (Rogers 1994). Those lessons which recognised the potential impact of the technology on the lesson outcomes and planned accordingly were generally more effective (Friedler & McFarlane in press; McFarlane *et al.* 1995).

## HARDWARE AND SOFTWARE

The broadest definition of a portable computer is simply one which is designed so that it can be used away from an AC power supply. At the time of writing, portable computers can be divided into the two distinct categories of notebooks and sub-notebooks. Notebook computers are those which can perform all the functions of a desk-top computer and which can run conventional desk-top computer software. Here, the term sub-notebooks is used to describe all of those portable computers which do not run desk-top computer software, including designated word processors and hand-held palm-top computers including personal digital assistants (PDAs). Some sub-notebook models have performed well in the field for manual data collection and note making. The portables facilitated collecting data in the field, with children being able to work effectively both independently and cooperatively in groups (teacher referring to palm-tops in Stradling *et al.* 1994). Some UK manufacturers, and some USA developers, have considered these machines as suitable adjuncts to data loggers, and so they could have a future in school science. Unless otherwise indicated, all portable computer use in this chapter refers to notebook computers (notebooks).

The notebook computers used in the England and Wales evaluation used pointer-driven software, such as that used in the Macintosh or Microsoft Windows environments. These machines therefore needed a hard disk drive and mouse or other pointing device, usually a tracker ball. The machines in the earlier studies usually had no hard disk drive or pointing device and therefore used software which did not require them.

One anticipated advantage of notebook computers was the added mobility provided by battery power. Most projects planned initially to use these machines widely outside the classroom. However one universal fact to emerge was that battery life, between charging, never matched that claimed by the manufacturers. A functional battery life of around two hours was common. This was long enough for most laboratories, but hopeless for long-term monitoring of the environment, for example.

In a science laboratory, unlike other classrooms, reliance on AC power is not an insurmountable problem because power outlets are usually plentiful. This does not entirely preclude easy sharing of the machine by moving it around the table. However, battery characteristics are one aspect of this technology which is lagging behind in development and major advances are required before the real freedom that portability should afford is realised.

The software used throughout the projects varied enormously. However, there was good evidence that the greatest benefit resulted when the software used in science complemented that used in the rest of the school. Science teachers had to spend less time teaching students how to use the software, which was something which they were reluctant to do, particularly because they already had the additional tasks of teaching students how to set up the portable computers and data logging equipment where this was used.

The main use of the notebook computers in science in the Pilot Evaluation and PLAIT projects was for data logging. This did require the use of novel software, but again this was easier to initiate in the classroom when the data-logging application operated in the same user environment as the rest of the school-based applications (e.g., where Microsoft Windows had been adopted as a standard). Failure to do this, such as when the notebooks were not powerful enough to run Windows software, often had dire consequences. The use of lower specification computers, which could not run mouse driven software proved more difficult. As frequently reported in the PLAIT project, significant numbers of students and staff could not reliably save and retrieve their work at the end of the first year (Gardner *et al.* 1993). This was a devastating blow, preventing any measurable learning gains from taking place in those cases.

Interestingly, when considering the problem of science software and the notebook computer hardware platform, departure from the school policy was not evident where the notebooks used one form of mouse-driven user interface and the rest of the school computers used another. In this case, even very young children moved between the two environments quite happily, often recognising that a new skill in data manipulation learned on one computer could be transferred to the other.

Low availability of good modelling and simulation software for science education developed for use with the current more powerful computers, was widely noted. Some modelling with spreadsheets was recorded.

## SUCCESSFUL IMPLEMENTATION

Factors which have proved important in the successful integration of the use of portable computers into the science curriculum are summarised below:

The Teacher
> Teacher motivation (Teachers are key innovators.)
> Teacher development (training in use of computers and software, and curriculum integration)
> Possession of computer literacy skills
> Awareness of the potential of portables in the science curriculum
> Previous use of open learning and inquiry approach to teaching
> Management support (especially for teacher release for staff development and planning)

The Student
> All students get the chance to be involved in the whole process of using portables – from setting up the computers to collecting, processing and presenting data. Portables are used frequently so they become a natural part of school and, ideally, home life.

Computer Provision
> Compact portable computers
> One or two machines for staff use
> Some spare machines in reserve
> Enough power supplies and access to AC power
> Suitable software
> Technical support

The Curriculum
> Greatest effects where curriculum uses a process model and encourages independent student investigation
> Correct balance between technology and subject matter
> Planning for investigative skill development
> Recognition of investigative skill development (e.g., in assessment).

Providing schools with access to portable computers and appropriate software alone will not secure the integration of IT into the science curriculum. Adequate teacher support, both in teaching them how to operate the technology and how to integrate its use into the curriculum, is crucial. Finally, unless the curriculum, particularly the process of student assessment, recognises the value of investigative skill development in science, the whole innovation will fail to produce effective outcomes in terms of recognised learning gains.

## THE FUTURE

Portable computers will continue to become more powerful, lighter and cheaper in the future. Machines with colour screens, which support CD ROM and full multimedia, will be widely available before the millennium. There is likely to be convergence between today's gaming consoles and such computers, which is a declared intention of some of the games manufacturers. Ownership of powerful portable computers by children is therefore set to become very widespread. Central provision of IT equipment by schools, limited by available funding, will no longer be the main source of computers in schools. Rather, schools will assume the role of coordinating the use of students' own machines, provision of spare machines for those who do not have a working machine on a given day, or who do not own one, and the setting of standards (e.g., for software). This will be greatly assisted by the convergence of major hardware and software standards which is currently underway.

The lack of access which currently holds up the use of IT in science education will no longer be an issue. It is therefore essential that strategies can be identified for overcoming other constraints, such as those imposed by lack of teacher development, lack of appropriate software, content-driven curricula and the resulting assessment frameworks. Failure to do this will prevent students benefiting from the skills enhancement which the use of their portables can afford in science.

Further research in this area is needed, particularly into the long-term effects of the presence of portables in the science classroom. None of the projects referred to here reported on use in science beyond the first year. Additionally, the use of on-line services, access to multimedia and widespread use of simulations and models were either absent entirely or referred to very briefly.

It seems that the use of portable computers can provide the science student with access to IT where and when required, and that this can provide enhanced and extended learning. In ten years or less, portables will be part of the furniture and we will wonder what the fuss was about.

ACKNOWLEDGEMENTS

The individual project reports from the Portable Computers Pilot Evaluation were made available by the National Council for Educational Technology (NCET) who managed the overall project.

REFERENCES

Bowell, B., France, S. & Redfern, S.: 1994, *Portable Computers in Action*, National Council for Educational Technology, Coventry, UK.
Brasell, H.: 1987, 'The Effect of Real-Time Laboratory Graphing on Learning Graphic Representations of Distance and Velocity', *Journal of Research in Science Teaching* 24, 385–395.
Department for Education: 1989, *Statistical Bulletin: Survey of Information Technology in Schools*, Her Majesty's Stationery Office, London.
Department for Education: 1991, *Statistical Bulletin: Survey of Information Technology in Schools*, Her Majesty's Stationery Office, London.
Friedler, Y. & McFarlane, A.E.: in press 'Data Logging With Portable Computers', *Journal of Computers in Mathematics and Science Teaching*.
Friedler, Y., Nachmias, R. & Linn, M.C.: 1990, 'Learning Scientific Reasoning Skills in Microcomputer-Based Laboratories', *Journal of Research in Science Teaching* 27, 173–191.
Friedler, Y., Nachmias, R. & Songer, N.B.: 1989, 'Teaching Scientific Reasoning Skills. A Case Study of a Microcomputer-Based Curriculum', *School Science and Mathematics* 89, 58–67.
Frost, R., McFarlane, A., Hemsley, K., Wardle, J. & Wellington, J.: 1994, *Planning for IT*, National Council for Educational Technology, Coventry, UK.
Gardner, J., Morrison, H., Jarman, R., Reilly, C. & McNally, H.: 1992, *Pupils Learning and Access to Information Technology*, School of Education, Queens University of Belfast, Ireland.
Gardner, J., Morrison, H., Jarman, R., Reilly, C. & McNally, H.: 1993, *Personal Portable Computers and the Curriculum*, The Scottish Council for Research in Education, Edinburgh, Scotland.
Heutinck, L.: 1992, 'Laboratory Connections: Understanding Graphing Through Microcomputer-Based Laboratories', *Journal of Computers in Mathematics and Science Teaching* 11, 95–100.
Johnson, D.C., Cox, M.J. & Watson, D.M.: 1994, 'Evaluating the Impact of IT on Pupils' Achievements', *Journal of Computer Assisted Learning* 10, 138–156.
Linn, M.C, Layman, J. & Nachmias, R: 1987, 'Cognitive Consequences of Microcomputer-Based Laboratories Graphing skills Development', *Contemporary Educational Psychology* 12, 244–253.
Linn, M.C. & Songer, N.B.: 1991, 'Teaching Thermodynamics to Middle School Students: What are Appropriate Cognitive Demands?', *Journal of Research in Science Teaching* 28, 885–918.
McFarlane, A.E., Friedler, Y., Warwick, P. & Chaplain, R.: 1995, 'Developing an Understanding of the Meaning of Line Graphs in Primary Science Investigations Using Portable Computers and Data Logging Software', *Journal of Computers in Mathematics and Science Teaching* 14, 461–480.
McMillan, K. & Honey, M.: 1993, *Year One of Project Pulse: Pupils Using Laptops in Science and English: A Final Report* (Technical Report No 26), Education Development Center, New York.
Mokros, J.R. & Tinker, R.F.: 1987, 'The Impact of Microcomputer-Based Labs on Children's Ability to Interpret Graphs', *Journal of Research in Science Teaching* 24, 369–383.
National Council for Educational Technology: 1993, *Evaluation of IT in Science*, Author, Coventry, UK.
National Council for Educational Technology: 1994, *Project Number 239 – University of Bath School of Education, Final Report*, Author, Coventry, UK.
Nickell, G.S.: 1987, *Gender and Sex-Role Differences in Computer Attitudes, Performance and Experiences*, ERIC Document Reproduction Service. (ED 284 114)

Rogers, L.: 1994, 'The Use of IT in Practical Science – A Practical Study in Three Schools', *School Science Review* 75, 21–28.
Rowe, H.A.H.: 1993, *Learning with Portable Computers*, Australian Council for Educational Research, Melbourne, Australia.
Stradling, B., Sims, D. & Jamison, J.: 1994, *Portable Computers Pilot Evaluation Summary*, National Council for Educational Technology, Coventry, UK.
Stuessy, C.L. & Rowland, P.M.: 1989, 'Advantages of Micro-Based Labs: Electronic Data Acquisition, Computerized Graphing, or Both?', *Journal of Computers in Mathematics and Science Teaching* 8, 18–21.
Swatton, P. & Taylor, R.M.: 1994, 'Pupil Performance in Graphical Tasks and its Relationship to the Ability to Handle Variables', *British Educational Research Journal* 20, 227–245.
Taylor, R.M. & Swatton, P.: 1990, *Assessment Matters, No 1: Graph Work in School Science*, Her Majesty's Stationery Office, London.
Thornton, R.K.: 1987, 'Access to College Science: Microcomputer-Based Laboratories for the Naive Science Learner', *Collegiate Microcomputer* 5, 100–106.
Watson, D.M.: 1993, *The ImpacT Report: An Evaluation of the Impact of Information Technology on Children's Achievements in Primary and Secondary Schools*, King's College, London.

Section 4

Curriculum

JAN VAN DEN AKKER

# 4.1 The Science Curriculum: Between Ideals and Outcomes

JAN VAN DEN AKKER
*University of Twente, Enschede, and Free University of Amsterdam, The Netherlands*

DEFINITIONS OF CURRICULUM

Among the many abstract concepts in educational literature, 'curriculum' probably belongs to the most elusive ones (Jackson 1992). To understand a publication on curriculum issues, it is usually helpful for readers if authors clarify the intended meaning, scope and context of the term. In this chapter, many potential curriculum representations and levels are considered.

As a concise and general definition, the notion of curriculum as a 'plan for learning' (Taba 1962) is helpful. This definition refers to the core of a wide variety of other, usually much longer, definitions. Moreover, this wording resembles traditional definitions of curriculum in many languages including Dutch, German and Swedish. The major advantage, however, is that this definition allows specification for many educational levels, representations and contexts. Another definition of curriculum that is still generic, but more informative about the nature of a plan for learning, is offered by Walker (1990, p. 5): 'The curriculum refers to the *content* and *purpose* of an educational program together with their *organization*'.

The conceptual work of John Goodlad (1979, 1994) is excellent for differentiating between various curriculum interpretations. Although a more refined differentiation is possible, the following threefold distinction has proven to be clear and useful worldwide when talking about many curricular activities (policy-making, design and development, evaluation and implementation): *societal* or system level ('macro'); *institutional* or school level ('meso'); and *classroom* level ('micro').

It is often illuminating to distinguish between various curriculum representations, such as:

- the *ideal* curriculum: the original vision underlying a curriculum (basic philosophy, rationale or mission);
- the *formal* curriculum: the vision elaborated in a curriculum document (with either a prescribed/obligatory or exemplary/voluntary status);

---

Chapter Consultant: Geoffrey Giddings (Curtin University of Technology, Australia)

- the *perceived* curriculum: the curriculum as interpreted by its users (especially teachers);
- the *operational* curriculum: the actual instructional process in the classroom, as guided by previous representations (also often referred to as the curriculum-in-action or the enacted curriculum);
- the *experiential* curriculum: the actual learning experiences of the students;
- the *attained* curriculum: the resulting learning outcomes of the students.

This typology of curriculum representations is particularly instructive for analysing the roots and fruits of numerous curriculum innovation efforts in science education over the previous decades. Curriculum reform typically is aimed at reducing the incongruence between 'new' ideals and current student learning. Many previous reviewers note the all-too-common failure of science curriculum reform projects to affect 'average' classroom processes (operational curriculum) and student outcomes (attained curriculum). Often, past reviews lack an appropriate interpretation of the findings in terms of curriculum level or representation, such as directly comparing student outcomes at the micro level (attained curriculum) with policy intentions at macro level (ideal curriculum) without taking notice of all intermediate processes. However, if the curriculum concept is used inaccurately or too narrowly, efforts to produce any explanation for these disappointing outcomes incur the risk of simplistic criticism and one-sided blaming of specific stakeholders.

Proponents of recent initiatives for science curriculum improvement – and the last decade of this century seems to be characterised by a new wave of such reform programs worldwide – can benefit from a more comprehensive, differential and systemic view of the multiple realities of science curricula and the many factors that influence those different representations (cf Anderson 1992). A major theme running through this chapter is the recognition that science curriculum change is complex but not impossible. Easy and general recipes are not available, but a lot can be learned from analysing experiences worldwide (e.g., Black & Atkin 1996; Fraser & Walberg 1995), using a broad orientation on conceptual issues and empirical information.

This chapter aims to provide such a broad orientation, taking into account all three types of phenomena and problems that are most characteristic of the curriculum field: *substantive*; *socio-political*; and *technical-professional* (Goodlad 1994). Two classic curricular questions frequently raised in curriculum literature refer to: (1) what knowledge is of most worth for the science curriculum; and (2) how science curriculum programs and materials should be developed and implemented. Whereas the first question is closely related to substantive matters, the second one relates to technical-professional perspectives. Socio-political aspects often play a role in answering both questions, because values and interests of many individuals, groups and organisations are usually at stake in making curricular decisions.

A comprehensive and systemic analysis of curriculum problems gives rise to many links with other sections in this *Handbook*. The wide range of relevant perspectives (e.g., from macro-policy to micro-instructional), activities (e.g., from philosophical deliberation about curriculum rationale, to materials design, to assessment of student achievement), representations (e.g., from mission statement, to lesson unit, to student experiences) and stakeholders (e.g., from politicians, to educational publishers, to teachers, to parents, to students), makes connections to other areas inevitable and desirable.

This chapter is structured into four major sections. First, we offer a historical and international overview of debates and policies related to the goals and content of science education, including a description of related curriculum development activities and their impact. Second, current school and classroom practices are considered in international perspective. Third, recent science curriculum development and reform initiatives worldwide are considered and common patterns and constraints are identified. Fourth, we consider possibilities for changing the science curriculum and propose the guidelines for various participants and stakeholders wishing to improve the science curriculum.

## HISTORICAL OVERVIEW OF SCIENCE CURRICULUM DEVELOPMENT

Surprisingly, previous efforts to sketch the history of science curriculum development show considerable variation in their distinctions between subsequent 'waves' of activities. It seems that the time frame chosen plays an important role. For example, whereas Bybee and DeBoer (1994), Jenkins (1994) and Matthews (1994) considered the last two centuries (and occasionally even more), Coble and Koballa (1996) started their coverage of science education reform in the 1980s. Obviously, such different timelines lead to identification of different waves. Moreover, the extent to which authors pay attention to an international perspective, instead of limiting themselves to an Anglo-American stance, creates different stories.

Like many reviews (e.g., Fensham 1992; Tamir 1991), this chapter's account of science curriculum development history starts in the 1950s when science education began to be a subject of serious policy making and when the first large-scale science curriculum development activities were initiated. Of course, it would be a mistake to suggest that the history of science education only began at the middle of this century, because influences of previous times played a significant role in those developments. Insightful information about these matters is offered in historically-oriented publications on science education such as DeBoer (1991), Matthews (1994) and Jenkins (1994).

Although the start of the first reform period is easy to identify (i.e., the late 1950s in the USA and UK), its endpoint varies across countries. For the Anglo-American context, the mid-1970s often are mentioned as the end of the first, intensive wave of curriculum reform. But, for many countries in other continents, reform lasted through the 1970s (in fact, often just starting then) and ended only in the 1980s. (However, it is sometimes rather difficult to trace relevant activities in non-English speaking countries because they are reported less often in the international literature.) The second wave of science curriculum reform (with different time limits, but focusing on the 1980s) gave less emphasis to large-scale development projects and more emphasis to debates about the most desirable reform directions to pursue. These discussions were influenced by both the disappointing impact of previous projects and emerging societal concerns and scientific insights.

Although the reforms in different countries were neither at the same pace (with different waves in different nations overlapping each other) nor in the same direction, currently we are on the verge of a new intensive period of science curriculum reform. This new impetus is built on a stronger consensus about worthwhile approaches as a result of the debates, and on a stronger awareness of the necessity for a systemic approach towards such reform endeavours. These first two waves, characterised as 'development projects' and 'debates on priorities', are described in the following two subsections. The most recent trend, systemic reform, is discussed later in this chapter

*The Era of Development Projects*

The origin of modern curriculum development for school science was in the late 1950s. The most prominent and extensively documented projects (which strongly influenced later development work in other countries) were initiated in the USA and the UK. Development work in the rest of the world shows a more scattered pattern.

The USA

In the USA, dissatisfaction grew about the poor output of the education system in terms of meeting the increasing demand of the labour market for personnel with appropriate training in science and technology. The content of science programs and materials had not kept up with rapid scientific and technological developments, and this did not match the optimistic view and high expectations that American society had regarding the applicability of science and technology at that time. The launching of the Sputnik by the Soviets in 1957 – a serious blow for American pride – helped to create the general spirit for action and the readiness for big investments in large-scale development initiatives. During the next two decades (later called the Golden Age of science curriculum development), approximately two billion American dollars were spent on these programs (and associated teacher training and school support components), usually via projects sponsored by the National Science Foundation (NSF).

The initial emphases in these projects (especially those for secondary education) was the modernisation of the curriculum content and objectives of the various science subjects and an emphasis on the structure of the disciplines. Also, the importance of science processes, especially scientific inquiry, was reflected in the development efforts. Both of these features were intended to improve the relevance of science education for students entering science-based university courses. Attention to practical and technological applications of science usually were limited. Among the many projects for secondary science (Hurd 1969) were Physical Science Study Committee (PSSC), Biological Sciences Curriculum Study (BSCS) and Chemical Educational Materials Study (CHEMS). In primary science curriculum projects, pedagogical aspects such as learning-by-doing and hands-on activities received more attention. Well-known primary projects (Hurd & Gallagher 1968) were Elementary School Science (ESS), Science Curriculum Improvement Study (SCIS) and Science – A Process Approach (SAPA).

These 'alphabet-soup' curricula during the 1960s typically were characterised by:

- broad development teams, with prominent science scholars in leading positions and educational theorists in advisory roles;
- frequent use of writing conferences for designing and revising materials;
- involvement of educational publishers in the later stages;
- products comprising a wide range of components including textbooks, worksheets, teacher guides and audiovisuals;
- frequent inservice training for teachers through workshops and summer institutes.

After 1970, the number of development initiatives decreased, partly because content modernisation more or less had been accomplished and sufficient materials had been produced. At the same time, societal awareness of the results of science and technology grew, especially in terms of undesirable side-effects such as environmental problems and aggressive military applications in Vietnam. Although the National Science Foundation ceased its sponsoring of new programs after 1976, it initiated a series of large-scale evaluation studies to gather information about the impact of the curriculum reform activities. In particular, Helgeson, Blosser and Howe (1977) reviewed research on the impact of reform activities on educational practices, Weiss (1978) surveyed teachers and other participants on curriculum use, course offering and instructional practices, and Stake and Easley (1978) conducted a series of 11 in-depth case studies on science education practices. Other data, especially about the knowledge, skills, attitudes and experiences of students became available through the Third National Assessment of Science (National Assessment of Educational Progress 1978).

In the subsequent Project Synthesis, Harms and Yager (1981) attempted to summarise and interpret the results from these different studies by contrasting the actual findings with desirable images, and deriving recommendations from the discrepancies. The conclusions (see also Welch 1979) were sobering:

- Preparation for the next academic level seemed to be the almost exclusive goal of most teachers.
- Science instruction appeared to be overly dependent on textbook use.
- Direct experience, inquiry approaches and other forms of intellectual stimulation were uncommon.

This picture also was confirmed in the comprehensive study of American school practice by Goodlad (1984, p. 215), who concluded that 'the gap between the expectations and the teaching practices . . . was formidable'. These findings contributed to a new 'crisis in science education' which are discussed in the later section on 'Debates on Priorities'.

### The UK

In England and Wales, a wave of science curriculum development occurred at about the same period (1960–1980) as in the USA. Major examples are the Nuffield Science Courses (for both primary and secondary education), funded by the Nuffield Foundation, and the Science 5–13 project of the Schools Council. In many respects, they resembled the National Science Foundation projects in the USA in terms of their emphasis on academic upgrading at the secondary level and the attention given to process skills in primary school education. The latter feature corresponded with the plea in the UK for an open, child-oriented teaching approach, with much room for discovery learning, as advocated in the Plowden (1967) report. Evaluation studies (Harlen 1978; Ingle & Jennings 1981; Kerr & Engel 1983) revealed that teachers experienced great difficulties in using the materials that had been developed. In particular, selecting and structuring activities appeared to be complicated. Although much inservice education was organised to assist teachers in such tasks, the implementation remained troublesome. The English educational inspectorate (Department of Education and Science 1978) concluded that the impact of the ideas and materials of the primary science projects on classroom practice was limited. According to the inspectorate, the quality of science education was satisfactory in only ten percent of the schools. In general, it was felt that the proposed science approach required too much of the average teacher in terms of science content knowledge, change of teaching role and personal commitment to science education (Black 1983; Harlen 1985).

### Rest of the World

Since the 1960s, science curriculum development has rapidly grown as a worldwide phenomenon. Many countries (including the developing countries discussed by Ware 1992) in all continents started science projects (see Lockhard 1963–1977 for regular international inventories). The large amounts of money, time and expertise spent in science education were not spent in other subject areas. In many countries, deliberate use was made of the American and British examples, initially by adoption (after translation) and later through intensive adaptations (see Blum 1979 for an analysis of activities and problems with those adaptation efforts).

Some examples of well-known first-wave curricula in the rest of the world are discussed below. However, it should be noted that many of these early projects differ in several respects from the American and British forerunner projects, because important lessons had been learned from evaluation findings and were influenced by emerging insights and concerns at the time of their development. Therefore, sometimes they already incorporated elements that seem more characteristic of the later second-wave proposals in the USA, including more attention to pedagogical aspects in secondary science projects, and more attention to the relation of science to other fields and to societal problems.

Examples of curriculum projects that have been the focus of retrospective accounts, evaluations or related research are the Australian Science Education Project (Cohen & Fraser 1987; Fraser & Cohen 1989), Learning in Science Project in New Zealand (Osborne & Freyberg 1985), Israel High School Biology Project (Tamir 1985) and Science Education Project in South Africa (Rogan & McDonald 1985).

An implementation perspective on experiences with and reflections on science curriculum development in many countries is reported in Tamir, Blum, Hofstein and Sabar (1979). A review of the impact of science reform projects in both advanced and developing countries is offered by Walberg (1991). Although there are substantial differences between countries in approach and timing, a general conclusion concerning the impact of those projects on instructional practices has been sobering: the diffusion and uptake of the project materials was limited; the actual classroom use hardly corresponded with the intentions of the developers; and effects on student outcomes were unclear.

Nevertheless, the first-wave curriculum reform movement had some profound influences on the teaching of science across the world, including (1) modernisation of content in chemistry and physics, (2) emergence of biology as a coherent field, (3) renewed emphasis on investigation and laboratory work, (4) introduction of primary school science and (5) reduced emphasis on the applications of science (Keeves & Aikenhead 1995). All of these seem to be lasting changes, as Keeves and Aikenhead suggest, although the last one is more questionable. Attention to applications of science and technology definitely has returned to science reform agenda, but probably with more productive intentions than in the past.

*The Era of Debates on Priorities*

The second wave of curriculum reform is more difficult to portray in unequivocal terms than the first wave. The landscape is clearest for the USA because the second wave can be interpreted in contrast to the first wave in some respects. The situation in the rest of the world is more diffuse and is more extended over time. Moreover, the second wave is more characterised by reflection, debates and proposals for specific problems than by large-scale development work and practical realisation. Thus, generalisations over many different approaches are hard to make. Nevertheless, some major themes in that era of debate can be identified.

A major drive for renewed reform efforts arose from the awareness of poor student learning outcomes in science. Although a meta-analysis of studies on the new curricula showed several positive effects on student performance (Shymansky, Kyle & Alport 1983), the broader impact was limited because of the poor spread and implementation of the programs. In the USA, discomfort grew with the low achievements of American students in international comparisons, such as those conducted by the International Association for the Evaluation of Educational Achievement (IEA; Keeves & Aikenhead 1995). The *Nation at Risk* report in 1983, by pointing to the poor results of American schools and creating a sense of crisis comparable with the Sputnik shock, helped to mobilise the American spirit for new reform efforts (National Commission on Excellence in Education 1983).

Another impulse for change came from the evaluation and implementation studies mentioned previously which led to the conclusion that the programs and materials of first-wave science curricula were considered too difficult for most students and teachers (Hurd 1986; Walberg 1991). More specifically, primary school teachers appeared to struggle with four implementation problems (Harlen 1985; van den Akker 1988). First, lesson preparation was seen as a complex and time-consuming chore. Second, there was a lack of background knowledge and confidence in science subject matter and skills. Third, there was considerable difficulty in changing the teacher's role, especially with regard to inquiry learning. Fourth, there was little realisation of the learning effects with students.

Secondary school science teachers, also facing difficulties with changing their role as information transmitters and reducing their chalk-and-talk approach, seemed especially hindered by their persistent 'preparation ethic' associated with their traditional view about the main purpose of science teaching.

A strong stimulus for change originated from emerging societal concerns in the 1970s. Environmental issues were an important driving force calling for more attention to promoting an understanding of how science, technology and society influence each other (usually referred to as the Science, Technology, Society approach; see Solomon & Aikenhead 1994). Under this approach, students not only should learn to appreciate the value of science and technology for society, but also should have an open eye for their limitations. Such knowledge also should enable learners to become better informed decision makers in everyday life situations. Related claims were made for including historical and cultural aspects of science and technology in the curriculum, and for familiarising students more with the creative practices and critical attitudes of scientists. A common label for these related goals is 'scientific literacy'.

This plea for scientific literacy also grew out of changes in the target group of science education. With rapidly increasing proportions of student cohorts entering and remaining longer in the school system, not only did the nature of the student population became more heterogeneous, but also the relevance of goals had to be re-examined. As preparation for continued science education was no longer the exclusive aim, scientific literacy for all students came more to the foreground (Fensham 1985), especially at the lower secondary education level. This trend also implied more attention to the perspectives of traditionally under-represented groups in science education, namely, girls and minorities.

Discussion started about the relative merit of more integrated (thematic or topic-based) and modular approaches compared to the traditional single-subject programs. Although this discussion has not been resolved yet, it resulted in a lot of modular materials around specific science themes (see Blum 1994).

Major pressures for rethinking the science curriculum also have come from new theoretical insights into the most appropriate ways of learning and teaching in relation to the complex goals emphasised in second-wave science proposals. As Linn and Eylon (1994, p. 5338) state:

> ... this new consensus incorporates greater emphasis on the social context of learning, increased understanding of contextual influences on problem-solving, and growing respect for the learner's struggle to make sense of scientific phenomena.

Anderson, Anderson, Varanka-Martin, Romagnamo, Bielenberg, Flory, Mieras, and Whitworth (1994, p. 7) summarised the main trends in the literature on science and mathematics education in the USA as follows:

(1) All students need to develop higher-order thinking skills.
(2) Student learning is not a passive activity in which teachers disseminate knowledge to students, but rather an activity in which students must actively construct their own knowledge through a complex process of interaction with their own knowledge structures, engagement with the materials and attention to the dialogue through which they are developing meaning.
(3) Learning less information in greater depth is preferable to covering a large number of facts and concepts with little or no understanding (i.e., 'less is more').

These trends strongly reflect notions of constructivism discussed in Fensham, Gunstone and White (1994) and Tobin (1993). Authors such as Driver and Scott (1996) have formulated implications for constructivist curriculum development in science, drawing upon extensive research on how students interpret science phenomena, how they form concepts about them, and how teachers can facilitate such learning. However, the amount of knowledge about which teaching approaches are effective in stimulating and supporting these widely proclaimed constructivist learning activities is still modest, especially when this is applied in average or less favourable classroom conditions.

Some further characteristics of the second-wave reforms relate to the development approach itself. They can be described as reactions to features of the first-wave reforms. Second-wave reforms were characterised by:

- less large-scale development projects at the national level, and more macro-curricular frameworks at the national level (e.g., the National Curriculum in the UK) or the State level (e.g., California) supplemented by more local materials development;

- more emphasis on teacher involvement and initiative, sometimes even aimed at making teachers 'curriculum-proof' instead of making the curriculum 'teacher-proof';
- teacher training not only focusing on content, but also on pedagogy (with more linkages with preservice education and attention to comprehensive professional development programs);
- different role distributions in development work, with less scientists and more science educators and specialists from other educational fields (psychology, curriculum, technology).

## CURRENT INSTRUCTIONAL PRACTICES IN SCIENCE EDUCATION

In this section, information about current school and classroom practices in science education is summarised from an international perspective. This is not easy because of the problems in providing a description that is valid across many countries, and because the available research knowledge is rather limited. Most information stems from cross-national surveys that are somewhat out of date. Moreover, those surveys tend to rely heavily on self reports by teachers via questionnaires which attempt to avoid superficiality and inaccuracy, despite the fact that they have to be usable across many countries. Additional information comes from case studies that often produce richer, more specific information but which are difficult to generalise because of limited sampling.

### IEA Studies

Valuable survey information on instructional practices in science education comes from IEA (International Association for the Evaluation of Educational Achievement) studies. These studies produce more data on the intended curriculum and the attained curriculum (student achievement) than on actual classroom practices (the implemented curriculum). Because data from the ongoing Third International Mathematics and Science Study (TIMSS) are not yet published, the latest available IEA information comes from the Second International Science Study (SISS) conducted in 1983–1986 in 23 countries. Information about the science curriculum in those countries is summarised by Rosier and Keeves (1991), while Postlethwaite and Wiley (1991) report student achievement for three student populations (10-year-olds, 14-year-olds and the terminal secondary level). Further summaries and commentaries are offered by Keeves and Aikenhead (1995) and Keeves (1995). Some of these findings from the SISS are summarised below:

- Especially in upper secondary education, there is considerable consensus about the content of science in the major subjects of physics, chemistry and biology.
- Attention to the teaching of science as an integrated subject in the primary

school has increased in the last two decades. There has been increasing use of textbooks and experiments in science lessons.
- Differences exist between countries in the extent that the four fields of science (biology, chemistry, earth science, biology) are taught separately or integrated in lower and secondary education. The nature of training and the position of teachers are very influential in these matters.
- With regard to the acquisition of various inquiry skills (Klopfer 1971), most emphasis is on knowledge, observation, measurement, problem solving, interpretation of data, manual skills and attitudes, interests and values. Little attention is paid to the formulation of generalisations, model building, and the limitations of science and scientific methods.
- Achievement in science at all age levels appears to be influenced primarily by (1) the status and educational resources of the home, (2) the aptitude, prior learning and gender of the student, (3) the attitudes and values towards science held by the student and (4) the motivation of the student.
- The achievement level of classroom groups is influenced by school and teacher factors, namely, (1) the competence of the teacher, (2) emphasis on inquiry and practical work in the classroom, (3) the amount of time allocated and the opportunity to learn and (4) the average level of attitudes and values of the school and classroom group.
- Further studies of science was found to be influenced by (1) science values (career interest and beneficial aspects of science), (2) aspirations (expected occupation and postsecondary education), (3) amount of science studied (class time and number of science courses taken) and (4) science attitudes (interest in science and ease of learning science).

*Example of a National Study*

With regard to instructional classroom processes in science education, much less information is available, particularly on a cross-national scale, but also at the national level. A highly informative study of science classroom practices in a single country has been reported by Kuiper (1993, 1995) for The Netherlands. The study involved a comprehensive curriculum reform effort in Dutch lower secondary education which emphasised context-based and activity-based approaches to teaching and learning in an attempt to promote the formation of meaningful concepts, to facilitate application of concepts to realistic settings, and to make instruction more attractive and motivating. The main part of that study consisted of a representative survey of science teaching practices in lower secondary education. Data were collected by administering a precoded written questionnaire to a stratified sample of 944 physics, chemistry and biology teachers. The response rate was 73 percent. The study indicated that:

- textbooks play a dominant role in all three subjects;
- context-based modes of dealing with subject matter are rare;

- in higher grades, teachers place little emphasis on real-life situations and applications and pay more attention to subject-specific formal concepts, principles and rules;
- experiments are conducted most frequently in chemistry and less in physics and biology;
- teachers need adequate facilities for practical work (especially support from laboratory assistants);
- whole-class instruction (sometimes supplemented by teacher questioning of students) is the dominant mode of instruction.

The general conclusion from the survey was that there is a considerable gap between the intended (ideal plus formal) curriculum and the operational curriculum.

The second part of the research consisted of case studies which provided in-depth description and analysis of the teaching practices of eight teachers (four in physics, four in biology) selected from respondents to the survey. These teachers were exemplary in terms of their obvious efforts to apply the reform proposals. The data collected included: lesson observations in various classes (about nine per teacher) focusing on the operational curriculum; textbook analysis (the formal curriculum); interviews with the teachers concerning their curriculum interpretation and opinions (the perceived curriculum); reflective journals assembled by all teachers (the perceived curriculum); analysis of tests used by the teachers (the tested curriculum); and a pre-coded student questionnaire, containing items about students' perceptions of the operational curriculum (the experiential curriculum), administered to 937 students.

The case study findings pointed to varied teaching practices that reflect some, but far from all, elements of the intended science curriculum reform. All teachers expressed clear limits as to what they perceived to be desirable and feasible within the practical constraints (limited laboratory facilities, group size, content overload, pressure from final examinations). Kuiper concluded that these findings of limited implementation even among exemplary teachers should warn all participants in curriculum reform about the complexity of the intended changes. The findings of Goodlad (1984) about the persistency of traditional teaching practices seem to be rather universal.

*Exemplary Cases*

More information about the perceived and operational science curriculum can be obtained from case studies of exemplary teaching (e.g., Tobin & Fraser 1990). Usually, these case studies are limited in the number of teachers, schools and systems involved, with the exceptions of the well-known but by now somewhat outdated Case Studies in Science Education (Stake & Easley 1978). Although small-scale case studies can reveal a lot about specific phenomena and problems, and help us to understand better (e.g., implementation problems), they are not adequate sources of information about common instructional practices. However, recent interesting findings from more comprehensive sets of case studies of science education practices have become available. We have selected three of those publications to summarise here because they provide useful baseline information about actual classroom practices from various perspectives. Such knowledge could help in making realistic judgements of promises and pitfalls regarding the kind of science improvement initiatives to be discussed later.

Changing the Subject

The first rich source of information on international trends in science policies and practices is a book entitled *Changing the Subject: Innovations in Science, Mathematics and Technology Education* (Black & Atkin 1996). This book is based on 23 case studies of educational innovations in 13 OECD (Organisation for Economic Cooperation and Development) countries and involves many stakeholders (school teachers, inspectors, academics in both subject matter and educational research, policy makers and advisers).

An emphasis on student practical work was common to most change endeavours (covering both primary and secondary education). Another characteristic was the effort to make connections, both between the sciences and between the sciences and other fields of study. There was a general tendency to pursue science as a way of knowing about how the world works or manifests itself.

Although contexts differ, the following common forces that drive reform activities across countries were identified: concerns about economic competitiveness; preparing future citizens; improving inclusiveness and equity; enhancing student learning; and empowering teachers. The studies also made clear that systemic reform, which addresses all elements of a whole education system and includes all the people and institutions which have any stake in the quality of education, is essential for success.

These findings do not refer to average practices, but to real-life exemplary practices in the context of innovative programs. However, most of these approaches share the notion that the following five traditional assumptions of learning should be challenged (Black & Atkin 1996, p. 62). First, 'knowing that' must come before 'knowing how'. Second, the effective sequence for learning is first to receive and memorise, then to use in routine exercises so as to develop familiarity and understanding, and finally to attempt to apply. Third, it is better to teach at the abstract level first and to leave application in many different contexts to a later stage. Fourth, motivation should be achieved by external pressure on the learner, not by changes in the mode of learning or the presentation of the subject. Fifth, difficulty or failure in learning by the traditional route arises from an innate lack of ability, or inadequate effort, rather than from any mismatch between the teacher's and the student's preferred learning styles. Black and Atkin (1996, p. 63) also comment that the new approaches to learning are:

> . . . a formidable challenge to teachers. Ideally teachers will be trying to offer their students some new learning process by which to achieve a deeper mutual

transaction of learning. But to achieve it, teachers will have to change almost every aspect of their professional equipment. They will have to reconsider themselves entirely: not only the structures of their material and their classroom techniques, but even their fundamental beliefs and attitudes concerning learning.

Many cases analysed in this book underline the serious difficulties that teachers experience in changing their roles. Some of the features that are characteristic for science innovation endeavours across countries are: inclusion of more 'hands-on' activity; broadening of the range of resources for learning (e.g., video and computer materials); connecting students' work to the real world; promoting collaborative activities; and more responsibility to students for their own learning. The cases also make it clear that this latter ambitious aim of greater responsibility for the student often raises difficult questions: 'How much can teachers dare to risk of their professional persona, that sense of their own identity in which both institutional authority and expertise as physicist, mathematician or whatever play so large a part? How can they be sure that what is to be learned is indeed learned? For what can they be held accountable? Who authorizes these practices? Where will they find a satisfactory alternative to the security of the text? If not in direct leadership in the classroom, then wherein lies their expertise?' (Black & Atkin 1996, p. 121).

Study of Curriculum Reform

Anderson (1995a, 1995b) reports a four-year research project on curriculum reform, with specific attention to the three areas of science education, mathematics education and higher-order thinking across the disciplines. Following a comprehensive literature review (Anderson *et al.* 1994), nine case studies of middle and high schools across the USA were conducted. Schools were selected as examples of sites where successful engagement in the process of curriculum reform was ongoing. The major aim of the case studies was to acquire an understanding of both the substance of the reforms and the means by which they were put into practice. To gather information, a researcher spent a minimum of 20 days in a given school for observing classes, interviewing students, teachers and others, attending teacher meetings, viewing student work products, and collecting documents and artifacts. Cross-site analysis examined (within both technical as well as political and cultural dimensions) the nature of the reforms, barriers to change and essentials for attaining the reforms.

The following common themes across the different reform efforts emerged: integrating themes in the subject matter; teaching for understanding by focusing in depth on major concepts rather than covering lots of details; making connections between subject matter and its applications; and reaching all students – not just the elite – with rigorous content and attention to critical thinking.

The case studies made it clear that these reforms are of major proportions, involve radical differences from traditional schooling, and connect with all aspects of teaching and learning. Reorientation from traditional to reform approaches is evident in the student's role (from passive receiver to self-directed learner), in student work (from teacher-prescribed activities to student-directed learning) and in the teacher's role. Table 1 summarises changes needed as the teacher's role changes from a dispenser of knowledge to a coach and facilitator (Anderson 1995a).

The research findings also underline that putting reforms into practice is a difficult and demanding task that takes a lot of energy and time. In their efforts to internalise the new approaches, teachers appeared to experience tensions concerning their beliefs and values about their classroom role. Probably the most difficult dilemma refers to coverage of content versus depth of understanding:

> The notion of preparing for the next level of schooling . . . is so deeply ingrained in the culture of schools that to omit any topic they know will be encountered later makes teachers feel inadequate. But teaching for understanding means that some topics must be left uncovered. (Anderson 1995b, p. 34)

If teachers agree with a constructivist approach to learning, they don't automatically know how to teach accordingly. They often experience a strong need for help with learning 'how to do it'. Teacher learning (and support for that) is essential not only as a foundation for acquiring a new role, but also in its own right in view of continuous professional development. Anderson (1995a) claims that such learning benefits most from everyday collaborative work between teachers.

## Portraits of Productive Schools

The third example of an informative set of case studies comes from international research on the institutionalisation of activity-based teaching and learning in primary science (Hameyer, van den Akker, Anderson & Ekholm 1995). Based on a common conceptual framework, set of questions and methodology, 15 schools (spread over Germany, the Netherlands, the USA and Sweden) were studied. These schools were seen as productive because they had succeeded not only in putting activity-based programs into practice, but also in maintaining them over the years.

The research approach was reconstructive. After a careful analysis of how

**Table 1: Contrasting roles for a teacher who is a dispenser of knowledge and a coach/facilitator (Anderson 1995a)**

| Role for Teacher as Dispenser of Knowledge | Role for Teacher as Coach and Facilitator |
|---|---|
| Transmits information | Helps students process information |
| Directs student actions | Coaches student actions |
| Explains conceptual relationships | Facilitates student thinking |
| Emphasises static knowledge | Models the learning process |
| Directs use of textbook | Uses materials flexibly |

activity-based science instruction looked in the specific sites, attempts were made to reconstruct the process of change that the schools had experienced, with specific attention to the conditions that made this sustained improvement possible. Data collection predominantly involved a combination of classroom observations and interviews with teachers, principals and other key personnel who were part of the process of practising or supporting activity-based science learning. Cross-site and cross-country analyses of instructional practices revealed: a high level of student involvement and enthusiasm; increased student initiative in the learning process; a lot of group work and interaction; teaching involving stimulation and facilitation; increased variety of resources (materials and objects) and experiences; extensive integration of science topics with project-oriented activities over a long period; and a lot of emphasis on process skills for exploration, learning to learn, and attitudinal goals such as curiosity, precision and perseverance.

Notwithstanding these general trends, a wide variation in the beliefs and practices of teachers was noticed, both among and within individual schools. Also, the implementation of activity-based learning had been characterised by the following typical problems (Hameyer et al. 1995, p. 118): the conceptual and organisational complexity of the teacher's role (e.g., dealing with unanticipated student questions, diagnosing learning problems in matters for which the teachers sometimes do not feel confident, and maintaining a balance between autonomous student initiatives, small-group activities, whole-class events and structured lessons); time-consuming lesson preparation; and difficulties in assessing and evaluating learning outcomes.

Reconstruction of the change processes showed that it usually took a long time before the activity-based learning approaches were firmly established in schools. Periods of five to seven years were most common. Although factors and conditions for change differed considerably, the following consistently beneficial factors were identified: social pressure; flexibility in use of time; availability of appropriate learning materials; collective ownership of learning principles; inservice education; staff stability; active dissemination by schools themselves; outside legitimisation; and few conflicts. Based on their findings, the authors made recommendations to school personnel (such as teachers and principals) and to those outside the school, such as support personnel, administrators and policy-makers. Recommendations were made separately for the three main phases of innovation processes, namely, initiation (starting up), implementation (going further) and institutionalisation (anchoring the new).

*Developing Countries*

All three sets of case studies described above involve science education practices and innovation initiatives in developed countries. Although Walberg's (1991) comprehensive review of research on the effectiveness of primary and secondary science education focuses on developing countries, as does the book of proceedings from a recent conference in Israel (Hofstein, Eylon & Giddings 1995), only limited attention in this chapter is given to the impact of science curriculum reform movements in developing countries because relevant research findings are scarce.

Information on patterns of teaching and learning (the implemented curriculum) in science classrooms is mostly anecdotal. Only few empirical studies of limited scope are available (cf Fuller & Clarke 1994). Nevertheless, the information about science classroom practices is consistent and unfavourable. De Feiter, Vonk and van den Akker (1995), in their review of science education in Southern Africa, conclude that science education has deteriorated in the last decade in most African countries only because of the poor socioeconomic situation and its negative consequences for the conditions of teaching. Many obstacles (e.g., shortage of qualified teachers, inadequate textbooks and facilities, weak communications network, conflicting policies, slow rate of information diffusion) still stand in the way of effective science education programs (Ogunniyi 1995). Also, 'the developing world is littered with educational reform proposals that either have not been seriously implemented or that were overtaken by events before their effects were transparent' (Lewin, 1995, p. 3).

A main characteristic of typical science classroom practice, as portrayed by both empirical studies and anecdotal information, is the overdependence on the lecture method ('chalk and talk') in contrast to the intended student-centredness of instruction in many curriculum reform movements in Southern Africa (de Feiter et al. 1995). A summary of typical problems was offered by Prophet (1990), who studied the effectiveness of a reform project in Botswana, but whose observations seem valid for many African practices. First, the most striking aspect of the science lessons observed was the passive nature of the learners. The majority of work involved teacher talk using either a lecture technique or a simple question-and-answer routine that demands only basic recall from the learners, often as single words or short sentences. Second, generally only correct answers were accepted by the teacher and acted upon. Incorrect answers tended simply to be ignored and all the lessons were characterised by an almost total lack of learner questioning. Third, the other major learner activity was the whole-class activity of writing which generally meant filling in correct answers on worksheets.

Stuart's (1991) research on classroom practices in Lesotho led to a comparable picture. First, most teaching was formal and didactic, involving questions, answers and teacher exposition (with 85 percent teacher talk and 15 percent involving students answering questions). Second, students seldom asked questions or initiated an exchange of thoughts. Third, teacher questions were pitched at the recall and comprehension levels and seldom involved comparing, inferring, reasoning and evaluating. Fourth, no models of higher cognitive skills were present in the teachers' lectures. Fifth, very little practical work was carried out, often because of lack of time or equipment.

The typical reasons for these patterns in developing countries (cf Lockheed 1993) are not encouraging for proponents of more meaningful science education. Some explanations that frequently come up in the literature (de Feiter et al. 1995) include: insufficient confidence and mastery by many teachers of both subject content and basic teaching skills (like questioning); language problems for both teachers and learners, with English being their second or even third language; the inconsistency between school science often involving Western-based curricula and textbooks (if available) and African life outside the school environment; tension between African culture (e.g., values about the relation between adults and children) and the spirit of inquiry and critical questioning required in school science; poor facilities (e.g., equipment) in schools and classrooms; and a weak alignment of 'innovative' curricular aims and typical assessment and examination practices.

A major conclusion from this brief overview is the huge gap between improvement ideals in various reform efforts and current classroom processes. Moreover, many basic conditions for improvement appear to be lacking. This situation creates such serious difficulties for the improvement efforts that a reconsideration of those aims might seem wise. One wonders whether the innovation gap is too big to bridge. This question seems justified in view of the fact that, also in more advanced countries, these kinds of ideals hardly have been implemented.

What is a realistic (ambitious, but feasible) perspective for improving science teaching in developing countries in the short and long range? In view of the many problems facing teachers, some researchers (e.g., Walberg 1991) have suggested that it might be wise to give priority to the improvement of conventional teaching approaches instead of aiming at more complex, student-centred approaches. Conventional teaching could be improved, for example, by making it more interactive and concrete through questioning and demonstrations. No doubt, it also makes sense to pay ample attention to the kinds of basic teaching skills (such as regular supervision and control of students' seat work) that are involved in conventional teaching approaches. These kinds of improvements seem necessary as a sort of foundation for more complex (constructivist) types of instruction. Probably, it is best not to think in terms of conflicting approaches, but to strive towards conventional teaching practices in which teachers are competent and confident but which also help teachers to deal with more complex goals and student-centred aspects of instruction (de Feiter *et al.* 1995).

## RECENT INITIATIVES AND TRENDS IN IMPROVING THE SCIENCE CURRICULUM

Initiatives in improving the science curriculum can be found at many levels varying from individual teacher efforts for site-specific changes to system-wide programs for generic reform. This section summarises seven international trends that can be recognised in initiatives worldwide. Later, the recently-published *National Science Education Standards* in the USA will be discussed briefly as an exemplary elaboration of many of those trends.

*Common Characteristics*

National Guidelines

One trend is the development of national guidelines (in whatever form) for science education. The aim, scope, status and degree of specification can vary considerably across countries, but the trend is unmistakable: there is a tendency towards curricular convergence. Those guidelines either can be highly prescriptive and have an obligatory status, or they can have a more exemplary and visionary status and aim to inspire and involve many stakeholders. However, differences in status might have less differential impact than perhaps expected. Compulsory curricula are often formulated in rather general terms in view of political feasibility – too much specification would elicit strong opposition – while exemplary frameworks, presented and discussed at a national level, can have the same effect of bringing coherence in the support of teaching and learning.

Meaningful Content

Another trend across many initiatives (especially in secondary science) is the emphasis on teaching key conceptual issues in depth instead of covering ever-increasing amounts of information. Continuous expansion of scientific content, as a consequence of rapid growth in scientific and technological knowledge, is no longer a primary goal of reform. Teaching for understanding is seen as more important (Tobin, Kahle & Fraser 1990). With that aim in mind, there is also some tendency (although with hesitations) to soften rigid disciplinary boundaries. Also, greater emphasis is placed on making connections among the sciences and among other disciplines (technology, mathematics and social studies) and with societal phenomena.

Scientific Literacy for All

Science education increasingly is seen as essential for all students, from their years in nursery schools until the end of secondary education. Science is important not only for those who will specialise in continued science studies or who will prepare for science-related professions, but also for all students as a contribution to their intellectual and cultural education and their preparation for informed citizenship. Because students in many countries are remaining longer in the school system, larger and more heterogeneous proportions of age cohorts need to be served. Special attention is given to traditionally under-represented groups in science education, such as girls and ethnic minorities.

Learning to Learn

The goals for science learning are increasingly put in the perspective of lifelong learning. The aforementioned focus on foundational knowledge is combined with more emphasis on skills in problem solving, inquiry, information and communication, and a preference for active, investigatory and independent forms of learning. The most common label for the approach characterised by all of these inter-related aspects is 'learning to learn' (Hurd 1993).

Alignment of Curriculum and Assessment

Urged by previous failures of science curriculum reform, more efforts are being made to achieve correspondence between curricular guidelines, frameworks and materials and the approaches and emphases in assessing and evaluating student learning. Those efforts refer to assessment as an integrated component of the instructional process as well as to examination criteria at the system level. This trend is closely related to the pursuit of coherence (which was mentioned in relation to the previous trend). It also reflects the intention to make assessment a more productive component of instructional processes. However, we face many technical and practical challenges in these efforts (cf Welch 1995).

Teacher Development

Another important trend also deals with the lack of success in previous reforms: the essential connection of curriculum improvement and the professional development of teachers. From political, cultural and technical viewpoints, it is highly desirable to encourage, facilitate and support teacher development. Without the commitment and continuous professional growth of teachers (starting with pre-service education, but also seen as a lifelong process), curriculum change is doomed to remain superficial and volatile. Although consensus on this essential link is becoming more evident, this is also an area in which numerous problems and obstacles prevent us from identifying empirically proven and easy-to-implement 'solutions'.

Information and Communication Technology

A final trend that is hard to miss is the rapidly-growing influence of information and communication technology on science education. This technology not only implies some content changes, but manifests itself especially by providing numerous instructional tools for teachers and learners. Although the extent of utilisation of technology in instructional practices still could be modest, the potential and the external pressure for using it are so overwhelming that a strongly increasing impact is to be expected. Obviously, this will require major learning on the part of teachers (Plomp & Voogt 1995).

*National Science Education Standards*

A document that will be influential for science curriculum reform in the forthcoming years is the *National Science Education Standards* (National Research Council 1995). Although focusing on science education in the USA, it probably also will have an international impact because of its ambitious nature, comprehensive scope and wide publicity. The report seems to include all major trends in the debates on science education reform in the last decade. It builds on several previous reports, such as *Science for All Americans* (American Association for the Advancement of Science 1989) and *Benchmarks for Science Literacy* (American Association for the Advancement of Science 1993), both stemming from Project 2061. The *Standards* seems to be based on nation-wide consensus by all stakeholders. It is a very ambitious document in that it calls for dramatic changes throughout school systems.

The *Standards* document, covering general schooling from kindergarten to grade 12, formulates quality criteria in the six components of science teaching, professional development of science teachers, assessment in science education, science content, science education programs and science education systems. For each of these components a list of areas is elaborated. The central goals for school science that underlie the *Standards* are to educate students who are able to:

- experience the richness and excitement of knowing about and understanding the natural world;
- use appropriate scientific processes and principles in making personal decisions;
- engage intelligently in public discourse and debate about matters of scientific and technological concern;
- increase their economic productivity through the use of the knowledge, understanding, and skills of the scientifically literate person in their careers. (National Research Council 1995, p. 13)

The *National Science Education Standards* summarises some of the essential changing emphases that are crucial for curricular change. These changing emphases in content are shown in Figure 1, while the changing emphases in teaching are shown in Figure 2.

The intended characteristics in Figures 1 and 2 are very ambitious. A comparison with the review of current practices in the previous section reveals large discrepancies even if we look at exemplary sites. The standards concerning professional development of science teachers recognise that the substantive changes in content and teaching at issue require equally important changes in professional development practices. Some of the catchwords used to indicate preferable approaches include collegial and collaborative learning, long-term coherent plans, mix of internal and external expertise, staff developers as facilitators, consultants and planners, teacher as intellectual, reflective practitioner, teacher as member of a collegial professional community, and teacher as source of and facilitator of change (National Research Council 1995, p. 72).

The conditions suggested for these practices (strongly emphasising collaboration) make sense, yet their ambitious nature should not be underestimated. It will be very difficult to put the new standards into full practice. The most fundamental problem might be that these kinds of reforms require significant changes in teachers' values and beliefs about appropriate science education practices and their own role in that practice (Anderson 1996; Tobin 1995). The frequent suggestion about

| Less emphasis on: | More emphasis on: |
|---|---|
| Knowing scientific facts and information | Understanding scientific concepts and developing abilities of inquiry |
| Studying subject matter disciplines (physical, life, earth sciences) for their own sake | Learning subject matter disciplines in the context of inquiry, technology, science in personal and social perspectives, and history and nature of science |
| Separating science knowledge and science process | Integrating all aspects of science content |
| Covering many science topics | Studying a few fundamental science concepts |
| Implementing inquiry as a set of processes | Implementing inquiry as instructional strategies, abilities and ideas to be learned |

**Figure 1: Changing emphases in content** (From the *National Science Education Standards*, National Research Council 1995)

| Less emphasis on: | More emphasis on: |
|---|---|
| Treating all students alike and responding to the group as a whole | Understanding and responding to individual student's interests, strengths, experiences and needs |
| Rigidly following the curriculum | Selecting and adapting the curriculum |
| Focusing on student acquisition of information | Focusing on student understanding and use of scientific knowledge, ideas and inquiry processes |
| Presenting scientific knowledge through lecture, text and demonstration | Guiding students in active and extended scientific inquiry |
| Asking for recitation of acquired knowledge | Providing opportunities for scientific discussion and debate among students |
| Testing students for factual information at the end of the unit or chapter | Continuously assessing student understanding |
| Maintaining responsibility and authority | Sharing responsibility for learning with students |
| Supporting competition | Supporting a classroom community with cooperation, shared responsibility and respect |
| Working alone | Working with other teachers to enhance the science program |

**Figure 2: Changing emphases in teaching** (From the *National Science Education Standards*, National Research Council 1995)

stimulating teacher collaboration as a vehicle for teacher learning seems to have some merit, but this remains vague as well as difficult to realise within the typical school context.

## CONCLUSION AND GUIDELINES FOR IMPROVEMENT

'It is easier to get a man to the moon than to reform schools', wrote Cuban (1992, p. 216) when introducing the theme of curricular stability and change. Although this statement is too negative for portraying the history of science curriculum reform, it raises our awareness of the continuity and durability of science instructional practices. As long as proposed changes in the science curriculum are of a first-order (incremental) nature, aimed at making the existing curriculum more efficient and effective, signs of progress can be traced (although our information base about what is actually taught and learned remains very limited). However, second-order (fundamental) changes, aimed at making essential alterations in goals and roles, are much more difficult to realise. Yet, current improvement initiatives in science education appear to imply the importance of that very type of restructuring (cf Fullan 1991). The fate of previous improvement efforts should prevent us from being overly optimistic about the chances for success, especially expecting quick results. Cuban (1992) concludes that most changes hardly ever pass the intended curriculum stage; the implemented curriculum tends to be very durable. Although the different waves in science curriculum reform have left their residue, there is far more continuity in what teachers actually do in their classroom than is suggested by changes in the intended curriculum.

Although many pressures for curricular change come from outside schools, it is naive to expect much from top-down strategies of curriculum reform. We need more comprehensive approaches which take into account the many perspectives that belong to a systemic view, and which start with a realistic time frame in mind. Although there is nothing wrong with great aspirations – they are desirable for mobilising resources and bringing direction to reform initiatives – new curriculum ideas will not affect classroom processes until teachers have had sufficient opportunity and support to internalise the teaching repertoire, particularly beliefs associated with those actions.

The overview of desirable trends in professional development for science teachers discussed in the previous sections seems promising, but how can such trends be brought to life? Many suggestions can be found in literature on professional development, both in general and for science education in particular. The following fruitful set of guidelines is based on lessons learned in recent science improvement endeavours internationally (Black & Atkin 1996, pp. 146–147):

- Change begins with 'disequilibrium' or a realisation that current practices cannot achieve current educational goals.
- Exposure to other ideas, resources and opportunities broadens teachers' awareness of possibilities for change and fosters a sense that alternatives are available.
- 'Existence proof' of new methods under normal classroom conditions gives moral support to teachers and challenges them.
- Demonstration of actions, reflecting the new ideas, in a real context deepens teachers' understanding. Also, such modelling strengthens the proof of existence.
- Personal support (both knowledgeable and close at hand) is essential for facing the innovation risks and overcoming the usual performance dip during initial implementation. Teaming up teachers is valuable.

- Encouragement of experimentation through the whole-school environment reduces perceived risks.
- Experimentation should be intertwined with reflection to increase understanding and competence.

Many groups (e.g., policy-makers, school administrators, supporting agencies, principals and teachers themselves) have their responsibilities in creating positive conditions for such teacher development. The major challenge is probably how schools can become more productive learning environments for teachers. Only then, real and sustainable curriculum changes can be expected.

The type of research that seems most promising for contributing to science curriculum improvement can be characterised as formative inquiry, design studies or developmental research (Walker 1992). Such approaches can help to build a bridge between innovative theoretical ideas and regular classroom practices by providing empirically-based exemplary elaborations for teaching scripts (Lijnse 1995). Also, the feasibility and effectiveness of improvement procedures themselves (e.g., successful peer coaching scenarios) can be explored using a cyclic approach of systematic design and evaluation. Last, but not least, this kind of action-research approach can contribute to the professional development of all participants, with teachers being major partners in curriculum design (van den Akker 1994).

The other chapters in this section provide five diverse perspectives on curriculum in science. First, Reuven Lazarowitz and Rachel Hertz-Lazarowitz overview numerous cooperative learning methods, their implementation in science classrooms, and research on their effectiveness. Second, John Wallace and William Louden examine three waves of science curriculum reform over the past 40 years and derive lessons about how best to proceed with curriculum reform in the future. Third, Rodger Bybee and Nava Ben-Zvi's 'Science Curriculum: Transforming Goals to Practices' synthesises diverse ideas on science curriculum reform, especially how to transform goals for scientific literacy into policy, programs and practices. Fourth, Donna Berlin and Arthur White propose a theoretical framework for integrating science and mathematics education and identify research and development activities which could employ this model in order to realise the potential of integrated teaching and learning of science and mathematics. Fifth, Michael Abraham describes an inquiry-oriented laboratory-based instructional strategy called the 'learning cycle' in terms of its historical development, its basic characteristics and the research evidence supporting its efficacy.

## REFERENCES

American Association for the Advancement of Science: 1989, *Science for All Americans: A Project 2061 Report on Literacy Goals in Science, Mathematics, and Technology*, Oxford University Press, New York.

American Association for the Advancement of Science: 1993, *Benchmarks for Science Literacy*, Oxford University Press, New York.

Anderson, R.D.: 1992, 'Perspectives on Complexity: An Essay on Curricular Reform', *Journal of Research in Science Teaching* 29, 861–876.
Anderson, R.D., Anderson, B.L., Varanka-Martin, M., Romagnamo, L., Bielenberg, J., Flory, M., Mieras, B. & Whitworth, J.: 1994, *Issues of Curriculum Reform in Science, Mathematics and Higher Order Thinking Across the Disciplines*, US Government Printing Office, Washington, DC.
Anderson, R.D.: 1995a, *Final Technical Research Report: Study of Curriculum Reform*, University of Colorado, Boulder, CO.
Anderson, R.D.: 1995b, 'Curriculum Reform: Dilemmas and Promise', *Phi Delta Kappan* 77(1), 33–36.
Anderson, R.D.: 1996, 'Putting the National Science Education Standards into Practice: Needed Research', Paper presented at the annual meeting of the National Association for Research in Science Teaching, St. Louis, MO.
Black, P.: 1983, 'Why Hasn't It Worked?', in C. Richards & D. Holford (eds.), *The Teaching of Primary Science*, Falmer Press, London, 29–32.
Black, P. & Atkin, J.M. (eds.): 1996, *Changing the Subject: Innovations in Science, Mathematics and Technology Education*, Routledge, London.
Blum, A.: 1979, 'Curriculum Adaptation in Science Education: Why and How?', *Science Education* 63, 693–704.
Blum, A.: 1994, 'Integrated and General Science', in T. Husén & T.N. Postlethwaite (eds.), *The International Encyclopedia of Education*, Pergamon Press, Oxford, UK, 2897-2903.
Bybee, R.W. & DeBoer, G.E.: 1994, 'Research on Goals for the Science Curriculum', in D.L. Gabel (ed.), *Handbook of Research on Science Teaching and Learning*, Macmillan, New York, 357–387.
Coble, C. & Koballa, T.: 1996, 'Science Education', in J. Sikula (ed.), *Handbook of Research on Teacher Education*, Macmillan, New York, 459–484.
Cohen, D. & Fraser, B.J.: 1987, *The Processes of Curriculum Development and Evaluation: A Retrospective Account of the Processes of the Australian Science Education Project*, Curriculum Development Centre, Canberra, Australia.
Cuban, L.: 1992, 'Curriculum Stability and Change', in P.W. Jackson (ed.), *Handbook of Research on Curriculum*, Macmillan, New York, 216–247.
DeBoer, G.E.: 1991, *A History of Ideas in Science Education*, Teachers College Press, New York.
de Feiter, L., Vonk, H. & van den Akker, J.: 1995, *Towards More Effective Science Teacher Development in Southern Africa*, Free University Press, Amsterdam, The Netherlands.
Department of Education and Science: 1978, *Primary Education in England: A Survey by HM Inspectors of Schools*, Her Majesty's Stationery Office, London.
Driver, R. & Scott, P.: 1996, 'Curriculum Development as Research: A Constructivist Approach to Science Curriculum Development and Teaching', in D.F. Treagust, R. Duit & B.J. Fraser (eds.), *Improving Teaching and Learning in Science and Mathematics*, Teachers College Press, New York, 94–108.
Fensham, P.: 1985, 'Science for All', *Journal of Curriculum Studies* 17, 415–435.
Fensham, P.: 1992, 'Science and Technology', in P.W. Jackson (ed.), *Handbook of Research on Curriculum*, Macmillan, New York, 789–829.
Fensham, P., Gunstone, R. & White, R. (eds.): 1994, *The Content of Science: A Constructivist Approach to its Teaching and Learning*, Falmer Press, London.
Fraser, B. & Cohen, D.: 1989, 'A Retrospective Account of the Development and Evaluation Processes of a Science Curriculum Project', *Science Education* 73, 25–44.
Fraser, B. & Walberg, H. (eds.): 1995, *Improving Science Education*, National Society for the Study of Education, Chicago, IL.
Fullan, M.: 1991, *The New Meaning of Educational Change*, Teachers College Press, New York.
Fuller, B. & Clarke, P.: 1994, 'Raising School Effects While Ignoring Culture? Local Conditions and the Influence of Classroom Tools, Rules and Pedagogy', *Review of Educational Research* 64, 119–158.
Goodlad, J.: 1979, *Curriculum Inquiry: The Study of Curriculum Practice*, McGraw-Hill, New York.
Goodlad, J.: 1984, *A Place Called School*, McGraw-Hill, New York.
Goodlad, J.: 1994, 'Curriculum as a Field of Study' in T. Husén & T.N. Postlethwaite (eds.), *The International Encyclopedia of Education*, Pergamon Press, Oxford, UK, 1262–1267.
Hameyer, U., van den Akker, J., Anderson, R.D. & Ekholm, M.: 1995, *Portraits of Productive Schools. An International Study of Institutionalizing Activity-Based Practices in Elementary Science*, State University of New York Press, Albany, NY.
Harlen, W.: 1978, 'Does Content Matter in Primary Science?', *School Science Review* 59, 614–625.

Harlen, W.: 1985, 'Science Education: Primary-School Programmes', in T. Husén & T.N. Postlethwaite (eds.), *The International Encyclopedia of Education*, Pergamon Press, Oxford, 4456–4461.
Harms, N. & Yager, R.: 1981, *What Research Says to the Science Teacher (Vol. II)*, National Science Teachers Association, Washington, DC.
Helgeson, S., Blosser, P. & Howe, R.: 1977, *The Status of Pre-College Science, Mathematics and Social Studies Education*, Center for Science and Mathematics Education, Ohio State University, Columbus, OH.
Hofstein, A., Eylon, B.-S. & Giddings, G.G. (eds.): 1995, *Science Education: From Theory to Practice*, Department of Science Teaching, Weizmann Institute of Science, Rehovot, Israel.
Hurd, P.: 1969, *New Directions in Teaching Secondary School Science*, Rand McNally, Chicago, IL.
Hurd, P.: 1986, 'Perspectives for the Reform of Science Education', *Phi Delta Kappan* 67, 353–357.
Hurd, P. & Gallagher, J.J.: 1968, *New Directions in Elementary Science Teaching*, Wadsworth, Belmont, CA.
Hurd, P.: 1993, 'Comment on Science Education: A Crisis of Confidence', *Journal of Research on Science Teaching* 30, 1009–1011.
Ingle, R. & Jennings, A.: 1981, *Science in Schools: Which Way Now?* Institute of Education, University of London, London.
Jackson, P.: 1992, 'Conceptions of Curriculum and Curriculum Specialists', in P.W. Jackson (ed.), *Handbook of Research on Curriculum*, Macmillan, New York, 3–40.
Jenkins, E.W.: 1994, 'History of Science Education', in T. Husén & T.N. Postlethwaite (eds.), *The International Encyclopedia of Education*, Pergamon Press, Oxford, UK, 5324–5328.
Keeves, J.P.: 1995, 'Cross-National Comparisons of Outcomes in Science Education', in B.J. Fraser & H.J. Walberg (eds.), *Improving Science Education*, National Society for the Study of Education, Chicago, IL, 211–233.
Keeves, J.P. & Aikenhead, G.S.: 1995, 'Science Curricula in a Changing World', in B.J. Fraser & H.J. Walberg (eds.), *Improving Science Education*, National Society for the Study of Education, Chicago, IL, 13–45.
Kerr, J. & Engel, E.: 1983, 'Can Science Be Taught in Primary Schools?', in C. Richards & D. Holford (eds.), *The Teaching of Primary Science: Policy and Practice*, Falmer Press, London, 45–52.
Klopfer, L.: 1971, 'Evaluation of Learning of Science', in B. Bloom, T. Hastings & G. Madaus (eds.), *Handbook of Formative and Summative Evaluation of Student Learning*, McGraw-Hill, New York, 33–47.
Kuiper, W.: 1993, *Curriculumvernieuwing en Lespraktijk* [Curriculum Renewal and Classroom Practice], University of Twente, Enschede, The Netherlands.
Kuiper, W.: 1995, 'The Implementation of Context- and Activity-Based Science Education: Intention and Reality', Paper presented at the annual meeting of the National Association for Research in Science Teaching, San Francisco, CA.
Lewin, K.: 1995, 'Development Policy, Planning and Practice for Science and Technology', Paper presented at the Annual Conference of the Southern African Association of Research in Mathematics and Science Education, Capetown, South Africa.
Lijnse, P.: 1995, ' "Developmental Research" as a Way to an Empirically Based "Didactical Structure" of Science', *Science Education* 79, 189–199.
Linn, M. & Eylon, B.S.: 1994, 'Learning and Instruction of Science', in T. Husén & T.N. Postlethwaite (eds.), *The International Encyclopedia of Education*, Pergamon Press, Oxford, UK, 5338–5342.
Lockard, D.: 1963–1977, *Reports of the International Clearinghouse on Science and Mathematics Curricular Developments*, University of Maryland, College Park, MD.
Lockheed, M.: 1993, 'The Condition of Primary Education in Developing Countries', in M. Lockheed & H. Levin (eds.), *Effective Schools in Developing Countries*, Falmer Press, London, 20–40.
Matthews, M.R.: 1994, *Science Teaching: The Role of History and Philosophy in Science*, Routledge, New York.
National Assessment of Educational Progress: 1978, *The Third Assessment of Science, 1976–1977*, Author, Denver, CO.
National Commission on Excellence in Education: 1983, *A Nation at Risk: The Imperative for Educational Reform*, US Government Printing Office, Washington, DC.
National Research Council: 1995, *National Science Education Standards*, National Academy Press, Washington, DC.
Ogunniyi, M.: 1995, 'The Development of Science Education in Botswana', *International Journal of Science Education* 79, 95–109.

Osborne, R. & Freyberg, P. (eds.): 1985, *Learning in Science: The Implications of Children's Science*, Heinemann, Auckland, New Zealand.
Plomp, T. & Voogt, J.: 1995, 'Use of Computers', in B.J. Fraser & H.J. Walberg (eds.), *Improving Science Education*, National Society for the Study of Education, Chicago, IL, 171–185.
Plowden, B.H.: 1967, *Children and Their Primary Schools: A Report of the Central Advisory Council for Education (England)*, Central Advisory Council for Education, Her Majesty's Stationery Office, London.
Postlethwaite, T.N. & Wiley, D.: 1991, *Science Achievement in Twenty-Three Countries*, Pergamon Press, Oxford, UK.
Prophet, R.: 1990, 'Rhetoric and Realty in Science Curriculum Development in Botswana', *International Journal of Science Education* 12, 13–24.
Rogan, J.M. & McDonald, M.A.: 1985, 'The In-Service Education Component of an Innovation: Case study in an African Setting', *Journal of Curriculum Studies* 17, 63–85.
Rosier, M. & Keeves, J.: 1991, Science Education and Curricula in Twenty-Three Countries, Pergamon Press, Oxford, UK.
Shymansky, J.A., Kyle, W.C. & Alport, J.M.: 1983, 'The Effects of New Science Curricula on Student Performance', *Journal of Research on Science Teaching* 20, 387–404.
Solomon, J. & Aikenhead, G. (eds.): 1994, *STS Education: International Perspectives on Reform*, Teachers College Press, New York.
Stake, R. & Easley, J.: 1978, *Case Studies in Science Education, Volumes I and II*, Center for Instructional Research and Curriculum Evaluation, University of Illinois, Champaign, IL.
Stuart, J.: 1991, 'Classroom Interaction Research in Africa: A Study of Curriculum and Professional Development', in K. Lewin & J. Stuart (eds.), *Educational Innovation in Developing Countries: Case Studies of Change Makers*, Macmillan, London, 127–153.
Taba, H.: 1962, *Curriculum Development: Theory and Practice*, Harcourt Brace Jovanovitch, New York.
Tamir, P.: 1985, 'The Evaluation of the Israeli High School Biology Project', in P. Tamir (ed.), *The Role of Evaluators in Curriculum Development*, Croom Helm, London, 162–183.
Tamir, P.: 1991, 'Reforms in Science Education', in A. Lewy (ed.), *The International Encyclopedia of Education*, Pergamon Press, Oxford, UK, 899–901.
Tamir, P., Blum, A., Hofstein, A. & Sabar, N. (eds.): 1979, *Curriculum Implementation and its Relationship to Curriculum Development in Science*, Israeli Science Teaching Center, Hebrew University of Jerusalem, Jerusalem, Israel.
Tobin, K. (ed.): 1993, *The Practice of Constructivism in Science Education*, Lawrence Erlbaum, Hillsdale, NJ.
Tobin, K.: 1995, 'Teacher Change and the Assessment of Teacher Performance', in B.J. Fraser & H.J. Walberg (eds.), *Improving Science Education*, National Society for the Study of Education, Chicago, IL, 145–170.
Tobin, K. & Fraser, B.J.: 1990, 'What Does it Mean to be an Exemplary Science Teacher?', *Journal of Research in Science Teaching* 27, 3–25.
Tobin, K., Kahle, J.B. & Fraser, B.J. (eds.): 1990, *Windows into Science Classrooms: Problems Associated with Higher-Level Learning*, Falmer Press, London.
van den Akker, J.: 1988, 'The Teacher as Learner in Curriculum Implementation', *Journal of Curriculum Studies* 20, 47–55.
van den Akker, J.: 1994, 'Designing Innovations from an Implementation Perspective', in T.Husén & T.N. Postlethwaite (eds.), *The International Encyclopedia of Education*, Pergamon Press, Oxford, UK, 1491–1494.
Walberg, H.J.: 1991, 'Improving School Science in Advanced and Developing Countries', *Review of Educational Research* 61, 25–69.
Walker, D.: 1990, *Fundamentals of Curriculum*, Harcourt Brace Jovanovich, San Diego, CA.
Walker, D.: 1992, 'Methodological Issues in Curriculum Research', in P.W. Jackson (ed.), *Handbook of Research on Curriculum*, Macmillan, New York, 98–118
Ware, S.: 1992, *Secondary School Science in Developing Countries: Status and Issues*, The World Bank, Washington, DC.
Weiss, I.: 1978, *National Survey of Science, Mathematics and Social Studies Education*, Institute of Education, Educational Resources Information Center, Washington, DC.
Welch, W.: 1979, 'Twenty Years of Science Curriculum Development: A LookBack', in D. Berliner (ed.), *Review of Research in Education*, American Educational Research Association, Washington, DC, 282–306.
Welch, W.: 1995, 'Student Assessment and Curriculum Evaluation', in B.J. Fraser & H.J. Walberg (eds.), *Improving Science Education*, National Society for the Study of Education, Chicago, IL, 90–116.

# 4.2 Cooperative Learning in the Science Curriculum

REUVEN LAZAROWITZ
*Technion – Israel Institute of Technology, Haifa, Israel*

RACHEL HERTZ-LAZAROWITZ
*University of Haifa, Israel*

Science is one of the leading curriculum areas in that it has formed the focus for developing, researching and implementing innovative curricula and instructional methods at all levels of education, from the kindergarten to the university level (Gabel 1994; National Assessment of Educational Progress 1988). Since the 1960s, science curricula have emphasised the inquiry approach by claiming that doing science like real scientists is the most promising method by which students will master inquiry skills and become literate in science and thus literate citizens of today's society (Mid-Continent Regional Educational Laboratory and Biological Sciences Curriculum Study 1969; Project 2061 1989; Schwab 1963; Yager & Lutz 1994).

After three decades of inquiry-driven curricula, only a small percentage of students chose to study science in high school or at college level. Academic achievement in science is decreasing, and science literacy is far from what was hoped by educators and the academic community. Yager and Lutz (1994) mentioned that there is no evidence that science courses were successful in enhancing students' science literacy. This agonising state might be attributed to the over-academic approach of learning and teaching science involving either the inquiry approach or other traditional approaches. The inquiry approach emphasised academic, intellectual and cognitive aspects of science and neglected the human, cultural, social and affective dimensions of science within a rapidly changing society. The ideal model of an expert researcher overlooked the novice students within their social contexts. Students' cultural, racial, social and ethnic backgrounds and their needs, preferences and interests could not be met by most methods utilised in either the inquiry, the traditional or the individual method of teaching-learning (Yager & Lutz 1994). Only in the late 1970s, following the school integration reform (Miller & Brewer 1984), cooperative learning methods were integrated into science classrooms and laboratories in an attempt to enhance students' learning within a peer context.

---

Chapter Consultant: James Barufaldi (University of Texas-Austin, USA)

The overall goal of the chapter is to discuss the potential of cooperative learning for science education. The chapter presents in its first section two theoretical claims about science as a human and a constructivist endeavour. The second section provides a conceptual overview and a summary table for cooperative learning methods, followed by a description of cooperative learning instructional methods used in the science classroom. The third part of the chapter reviews and analyses research based on the use of cooperative learning methods in teaching science, with a special focus on curricular issues. The literature review, encompassing various levels of schooling, leads to the fourth section involving an analysis of the implementation of cooperative learning methods in science classrooms taking a curricular perspective.

## SCIENCE: A SOCIAL-HUMAN AND CONSTRUCTIVIST ENDEAVOUR

Human nature and society are characterised by cooperation and competition (Deutsch 1949). Competition exists in society between individuals, social groups, organisations and countries. But cooperation is vital and important in sharing resources, taking mutual responsibility, and relating to the 'public good', with all of these being necessary conditions for the continuation of the human race. This balance between cooperation and competition is missing in the socialisation process of students and adults.

Research in science includes cooperative activities in addition to individualistic and competitive behaviours. Scientific breakthroughs were always products of individual and teamwork, which interact and work with each other. The accomplishments in reaching the moon, producing lasers, silicon chips and television sets in the electronic industry, and genetic engineering based in the decoding of the DNA structure 'are products of teams who worked cooperatively in order to advance society, scientific knowledge and technology, and to improve peoples health' (Biological Sciences Curriculum Study 1989).

The political infrastructure of the world leads to the creation of organisations which are based on large-scale cooperation such as the World Bank, the European Economic Market and the North America Free Trade Agreement (NAFTA). We can expect that the social skills needed in this post-industrial world can be acquired by using cooperative learning in the classroom. Interaction between social settings and instruction can be a factor which affects thinking and learning. It is desirable to provide students with cooperative-interactive learning experiences in which knowledge is acquired.

### Inquiry, Constructivist Theory and Cooperative Learning

The inquiry method was an exciting innovation, yet it was only partially suited to the complex and diverse context of the schools. Unfortunately, most of the student population did not obtain higher academic achievement, mastery of inquiry skills, development of understanding and positive attitudes towards science (Yager & Lutz 1994).

Sensitivity to variation in students' background and its relationship to ways of thinking and learning led to the introduction of the concept of 'construction of knowledge' in science teaching. The constructivist approach in science education emphasises the way in which learners construct their knowledge by gradually progressing within their Zone of Proximal Development, with the assistance of a peer or an adult (Piaget 1926; Vygotsky 1978). The role of constructivism in science education was elaborated by Driver and Oldham (1986) and Driver and Bell (1986), who claimed that learners accomplish understanding through the social interaction which occurs in the classroom. They identified six characteristics of the constructivist approach in science education:

(1) Academic outcomes depend on the knowledge, purposes and motivation that students bring to the classroom.
(2) Personal construction of meaning is involved in learning.
(3) Learners construct meaning through an active continuous process.
(4) Learners evaluate constructed meanings through the process of rejection or acceptance.
(5) The final responsibility for learning lies with the student.
(6) Students construct types of meanings by experience with real objects and through natural language.

The construction of meaning takes place when students negotiate their understanding through class discussion and by exchanging thoughts and ideas (Prawat 1989).

Johnston and Driver (1990) suggested that cognitive construction is facilitated through the following activities, all of which are based on peer-interaction: students present their own ideas by explaining them to other group members; they think and talk about their experiences; they suggest and try out new ideas; they reflect on changes in their ideas; they negotiate and aid other students to clarify their thoughts; and they move ideas forward by making sense of new ones. Indeed, constructivist theory brings to light the significance of social-cognitive interaction, cooperation and collaboration to the science teaching-learning context.

## COOPERATIVE LEARNING METHODS

Cooperative learning developed from social-psychological studies of cooperation and competition in human behaviour (Deutsch 1949). Since the early 1970s, cooperative learning has been one of the most often implemented and researched instructional movements across various subject matters (Hertz-Lazarowitz & Miller 1992; Sharan 1994; Slavin 1990). Cooperative learning is the antithesis of the expository competitive classroom teaching approach. Cooperative learning brings to the school a different learning organisation in which the classroom is structured into cooperative teams of learners, thus making learning together a way of life.

Students tutor each other, conduct group projects, practice mutual assistance by sharing and exchanging information, and create a collaborative-cooperative learning environment (Hertz-Lazarowitz 1993). At its best implementation, cooperative learning interweaves cognitive academic behaviours with social skills such as active listening, responsibility, dependability, mutual respect, helping and sharing behaviours, positive-social interaction, respect for others, emphatic sensitivity to people and concerns towards the environment. It can be assumed that, when competition and individualism took priority in schools, many of the skills mentioned above were neglected.

The development of cooperative learning instructional methods followed the reform of school desegregation in the late 1960s and early 1970s. Teaching and learning in the heterogeneous classroom led to a questioning of the practices used in former segregated classrooms. Science teachers were challenged by students' diversity in academic ability, motivation, needs, interests, future careers and values. They were ready to experiment with new instructional methods, to meet students' multidimensional diversity, and to help them to learn, because achievement in high school is a key factor in determining students' future status and mobility.

Since the early development of cooperative learning methods (Aronson, Stephan, Sikes, Blaney & Snapps 1978; Johnson & Johnson 1975), many cooperative learning methods have been developed, and many of these have been implemented and researched within the science classroom.

*An Overview of Cooperative Learning Methods*

Four conceptual guidelines can be presented for the various cooperative learning methods:

(1) Cooperative learning is defined as an *educational movement* because it is based on concepts that stem from philosophical, psychological, cognitive and instructional theories. The combination of these concepts builds a *vision* that usually includes beliefs, values and ideology about education.
(2) Several schools exist within the cooperative learning movement and these are associated with distinct theories.
(3) Schools of cooperative learning generate and develop different and unique *methods*. The methods are identified by names, structures, sequence of stages, content specification and learning materials, and they often are described in manuals. New methods are being developed constantly within the cooperative learning schools by individuals and research and development centres.
(4) *Cooperative structures/techniques* can be used in the classroom as independent activities, outside of a cooperative learning method. Many cooperative learning techniques such as 'pair and share', 'peer discussion' and 'round table' are implemented in the classroom.

Table 1 presents a summary of the main features of various cooperative learning methods, organised by each distinctive theory that generated various methods within its theory/school. The researchers who are most identified with each method

Table 1: Cooperative learning method – A summary table

| Theory | Methods | Developers | Emphasised elements |
|---|---|---|---|
| Behaviourism | STAD – Student team achievement division<br>TGT – Team game tournament<br>TAI – Team assisted instruction | Slavin (1983) | Teams, enhancing motivation by rewards<br>Cooperation within teams<br>Competition between teams |
| Social Psychology | Learning together<br>Groups of four<br>Cooperative structure<br>Jigsaw<br>Expert jigsaw | Johnson & Johnson (1975)<br>Barnes & Todd (1977)<br>Kagan (1985)<br>Aronson et al. (1978) | Division of tasks<br>Specialisation, interdependence<br>Individual and group product |
| Progressive Education | Group – investigation<br>Jigsaw – investigative group<br>Complex instruction | Sharan & Hertz-Lazarowitz (1980)<br>Lazarowitz & Karsenty (1990)<br>Cohen (1986) | Students choice of task<br>Interest and inquiry<br>Coordination between groups/curricula topics<br>Classroom as group of groups |
| Cognitive/learning | Scripted cooperation<br>CIRC – Cooperative integrated reading and composition | O'Donnell & Dansereau (1992)<br>Stevens, Madden, Slavin & Farnish (1987) | Role exchange teacher/learner<br>Text comprehension<br>Reading/writing literacy<br>Writing as a learning process |
| Cognitive/developmental | Collaboration<br>Reciprocal teaching, Tutoring | Damon (1984)<br>Brown & Palincsar (1989)<br>Allen (1976) | Non-structured interaction<br>Teachers' role as an expert<br>Individual products<br>Scaffolding and construction of students' learning |
| Instructional | Cooperative classroom context:<br>Six mirrors of the classroom | Hertz-Lazarowitz (1992) | Contextualising six dimensions in the classroom, gradual implementation of cooperative learning across methods, coordinating individual, group and whole-class instruction |

are listed and the elements that are emphasised in each method are highlighted by the names of the methods. Hopefully this table will clarify the variations in terms and methods currently found in the literature. Further description of these methods can be found in Sharan (1994), Slavin (1990) and Hertz-Lazarowitz and Miller (1992). Below, more detail is provided about some of the cooperative methods of instruction which have been used in science education.

*Learning Together*

In the learning together method (Johnson & Johnson 1975), students are assigned to groups of four or five. They are assigned to learn together in order to accomplish a common goal. Each individual is accountable within the group and has to show that he or she masters the learning material. In order to cooperate, students have to acquire interpersonal and group skills. Those skills relate to the ability of students to learn together, discuss, share ideas and prepare as a group for achieving the common goal. The group is heterogeneous and everyone is expected to help and be helped. The group's tasks include group discussions and group projects which can be performed in the science laboratory.

*The Jigsaw Method*

In the jigsaw method (Aronson, Stephan, Sikes, Blaney & Snapp 1978), the class is divided into small heterogeneous groups of five students who can treat each other as resources. The learning goals and materials are structured by the teacher and are divided into independent sub-units which can be learned separately so that one sub-unit does not depend on the mastery of the others. The jigsaw method is based on two cooperative structures: the jigsaw group (five students a to e); and the experts group (five students with the same part 5a, 5b, etc.). In the expert group, students master their part and prepare for peer-tutoring, and then they return to the jigsaw group to tutor their team-mates and prepare for a test. The original jigsaw has been extended to jigsaw II, involving expert-jigsaw and jigsaw-investigative groups.

*Student Teams and Achievement Divisions (STAD) and Teams Games Tournaments (TGT)*

The student teams and achievement divisions (Slavin 1978) has five components:

(1) *Class presentation.* Each week new material is first presented by the teacher to the whole class in a lecture, discussion or video technology format.

(2) *Teams.* Students are assigned to four-member or five-member heterogeneous learning teams. Team members work together on worksheets written by the teacher. They can work in pairs to master the material. Teams are also given answer sheets.
(3) *Quizzes.* After the team practises, each student takes a quiz on the material.
(4) *Individual improvement scores.* A scoring system allows students to earn points for their team based on individual improvement over past performance.
(5) *Team recognition.* Teams are recognised for high individual performance and high team scores (Slavin 1978).

The teams-games-tournament (De Vries & Slavin 1978) is the same as STAD except that it replaces quizzes with a game or tournament. Students compete as representatives of their team with others who are judged to be similar to them in academic standing.

*Group Investigation (GI)*

Group investigation (Sharan & Hertz-Lazarowitz 1980) is rooted in Dewey's (1970) philosophy of education. Group investigation integrates the four basic features of investigation, interaction, interpretation and intrinsic motivation. These features are combined into the six stages of the model: (1) the class determines subtopics and is organised into research groups; (2) the groups plan their investigations; (3) the groups carry out their investigations; (4) the groups plan their presentations; (5) the groups make their presentations; and (6) the teacher and students evaluate their projects.

In group investigation, the investigation process is presented in each stage; groups select topics for investigation based on their interest and curiosity. Thus, in the group investigation classroom, groups work on *different* but *related* topics of investigation. They use a variety of resources to generate questions, gather information in the investigation and become active in constructing their knowledge. The teacher is a facilitator, a mentor and a collaborator in the students' inquiry process (Hertz-Lazarowitz & Calderón 1992; Sharan & Sharan 1992).

*Peer Tutoring in Small Investigative Groups (PTSIG)*

The method was developed by Lazarowitz and Karsenty (1990) as a combination of the jigsaw method (Aronson *et al.* 1978) and group investigation (Sharan & Hertz-Lazarowitz 1980). In this chapter, we present the method as either PTSIG or jigsaw-investigative group (JIG).

The Jigsaw-Investigative Group for the high school level consists of the jigsaw structure for peer-tutoring and the group investigation structure for the expert countergroup. The teacher, as a curriculum developer, designs the science-related learning tasks for each sub-unit as an inquiry-investigative sequence of activities.

Therefore, students work, especially in their expert group, on complex and rich learning tasks. In their expert group, students are reading, making observations related to the objects being studied and generating questions for laboratory investigative experiments. The tasks include open questions and problems which could be solved only, for example, by using microscopes, preparing slides or performing experiments with other group members. After they finish their learning tasks in the expert group, students return to their jigsaw group for peer tutoring. Usually, the different sub-topics which were investigated are presented and discussed within the original jigsaw group so that students acquire a general understanding and knowledge of the topic.

The evaluation is based on students' academic products in their expert groups and on their grades in a test based on all the units. The students prepare for the final test with further reading. The teacher occasionally leads the discussion with the whole class in order to organise and conceptualise significant scientific concepts. Topics such as cells, animal physiology, photosynthesis in higher plants, and evolution are topics which can be divided naturally into five independent sub-units that can be learned in the jigsaw investigative method. Teachers/researchers in Israel have implemented jigsaw-investigative group developed elaborate curricula to be used in high school biology classrooms (Lazarowitz 1991; Lazarowitz & Karsenty 1990).

## REVIEW OF RESEARCH ON COOPERATIVE LEARNING IN SCIENCE EDUCATION

Tables 2–5 present salient features of research on the implementation of cooperative learning methods which has been undertaken since the mid-1970s and which has been reported in refereed journals in science education. The research is divided by school level, from primary school (Table 2), to junior high school (Table 3), to high school (Table 4) and to college (Table 5). The focus of the tables is implementation of cooperative learning in science classes. The tables provide, for each study, information about the sample, the type of cooperative learning method of instruction, the subject matter and curriculum topics, and the variables investigated. Research findings related to the studies are reported in Lazarowitz (1995).

*Analysis of Cooperative Learning Studies*

Tables 2–5 encompass 36 studies of the implementation of cooperative learning methods in science education. Most of the studies were carried out at the secondary school level (11 at the junior high level and 15 at the senior high level). Five studies were reported at the primary level, and six at college level. Studies of the college level were mostly in courses for preservice science teachers. Five cooperative learning methods were used in science classrooms, namely, learning together, jigsaw, jigsaw-investigative group, teams games tournaments, and student teams

Table 2: Research on cooperative learning in science – Primary level

| Authors | Sample | Method of instruction | Subject | Topics | Variables investigated |
|---|---|---|---|---|---|
| Jones & Steinbrink (1989) | 50 inservice science teachers | Two level small groups developed on jigsaw II | Energy | Energy from fossil fuels | Academic achievement Learning behaviour Helping behaviour |
| Snyder* (1987) | Grade 5 (physically, academically and emotionally handicapped students) | TGT | Energy | Energy from fossil fuels | Academic achievement Attitudes to cooperative learning Increased support for peers |
| Soderland* (1987) | Grade 6 (low ability students) | Two level small group | Energy | Energy from fossil fuels | Achievement |
| Jones* (1987) | Grade 5 | Two level small group | Energy | Energy from fossil fuels | Reduced failures Achievement |
| Guidos-Henderson* (1988) | Grade 5 (low ability students) | Two level small group | Energy | Energy from fossil fuels | Academic achievement |

* Unpublished Master's projects at the University of Houston, Clear Lake, 1987–1989; reported by Jones and Steinbrink (1989)

**Table 3: Research on cooperative learning in science – Junior high school level**

| Authors | Sample | Method of instruction | Subject | Topics | Variables investigated |
|---|---|---|---|---|---|
| Humphreys, Johnson & Johnson (1982) | Grade 9 N = 44 | Learning together | Earth science | Health Light & sound Nuclear energy | Academic achievement Students' attitudes Students' self-esteem |
| Okebukola (1986a) | Grade 9 N = 221 | Learning together Laboratory work | Biology | Nutrition in animals and plants | Students' attitude to laboratory work |
| Okebukola (1986b) | Age 14 N = 493 | Learning together | Biology | Diversity of organisms | Students' academic achievement Students' preference for methods of instruction |
| Walters (1988) | 13+ age group | Working in groups | Biology | Human body – cells and the organisation of cells into organs and systems Plants – transport and reproductive systems Animals – respiratory systems, digestive, skeletal and nervous systems | Active learning Learning skills Research skills Collaboration and strengths within the group Skills of reporting Enjoyment Students' motivation |
| Lazarowitz & Galon (1990) Lazarowitz (1991) | Grade 9 N = 201 | Jigsaw-investigative group | Biology | Mitosis and meiosis | Students' academic achievement Gender differences Cognitive stages Classroom learning environment |
| Rogg & Kahle (1992) | Grade 9 N = 44 | Learning together Continuum of homogeneous to heterogeneous groups | Biology | Evolution Taxonomy of organisms | Students' on task behaviours Students' interactions Student-teacher interactions |

**Table 3:** Contined

| Authors | Sample | Method of instruction | Subject | Topics | Variables investigated |
|---|---|---|---|---|---|
| Okebukola (1985) | Age 12.7<br>N = 630 | Learning together<br>Jigsaw<br>TGT and STAD | Integrated Science | – | Academic achievement<br>Low and high cognitive test items |
| Okebukola & Ogunniyi (1984) | Age 13.4<br>N = 1025 | Learning together<br>Mixed ability groups | Biology, Chemistry<br>Physics | Laboratory work | Academic achievement<br>Practical skills |
| Chang & Lederman (1994) | Grade 7<br>N = 141 | Groups with and without assigned roles | Physics | Laboratory work | Academic achievement<br>Investigative skills<br>Social skills<br>Non-learning behaviours |
| Yeroslavski, Dori & Lazarowitz (1995) | Grade 8<br>N = 112 | Jigsaw-investigative group | Biology | The cell theory<br>Structures and functions | Academic achievement<br>Learning activity<br>Attitudes towards the method |

Table 4: Research on cooperative learning in science – Senior high school level

| Authors | Sample | Method of instruction | Subject | Topics | Variables investigated |
|---|---|---|---|---|---|
| Lazarowitz, Baird, Hertz-Lazarowitz & Jenkins (1985) | Grade 10 N = 69 | Jigsaw investigative group | Biology | Cell theory | Academic achievement Low and high students' ability Students' self-esteem Attitudes towards biology Involvement in learning Classroom learning environment |
| Lazarowitz, Hertz-Lazarowitz, Baird & Bowlden (1988) | Grade 10 N = 113 | Jigsaw investigative group | Biology | Cell structure and function Plant morphology and physiology | Academic achievement On-task behaviour |
| Sherman (1988) | Senior High N = 46 | Group investigation | Biology | Ecology | Effectiveness for learning in an investigative approach Training of teachers in the method |
| Lazarowitz & Karsenty (1990) | Grade 10 N = 108 | Peer-tutoring in small investigative groups Jigsaw investigative groups | Biology | Plants Photosynthesis | Academic achievement Inquiry skills Classroom learning environment Students' self-esteem |
| Watson (1991) | Age 14-17 N = 715 | Group education modules Learning together | Biology | Biological topics undefined | Academic achievement |
| Jegede, Okebukola & Ajewole (1991) | Age 16.4 N = 64 | Groups working with computers | Biology | Life cells Mammals and plants Photosynthesis and enzymes Respiration and excretion Transport and osmosis Sensitivity and coordination Reproduction, cycles and micro-organisms | Academic achievement Attitudes towards the use of computers |

**Table 4:** Continued

| Authors | Sample | Method of instruction | Subject | Topics | Variables investigated |
|---|---|---|---|---|---|
| Welicker & Lazarowitz (1995) | Grade 10 N = 192 | Peer tutoring in small investigative groups | Biology | Human body Cardiovascular system and diseases | Academic achievement Inquiry skills Attitudes towards the topics Attitudes towards the method Learning behaviours |
| Ron & Lazarowitz (1995) | Grades 11–12 N = 152 | Peer tutoring in small investigative groups | Biology | Evolution | Academic achievement Peer tutors' achievement Classroom learning environment |
| Lazarowitz, Hertz-Lazarowitz & Baird (1994) | Grades 10–12 N = 73 | Jigsaw investigative groups | Geology | Energy sources | Academic achievement Essay measuring number of words and ideas Students' self esteem Number of friends Classroom learning environment |
| Hertz-Lazarowitz, Baird & Lazarowitz (1994) | Grades 10–11 N = 200 | Jigsaw investigative groups | Biology Geology | Cell structure and function Plant life Energy resources | Affective outcomes: listening, peer tutoring, exchanging ideas, student-teacher interactions, student learning behaviours |
| Zadock (1983) | Grade 11 N = 385 | Inquiry groups | Chemistry | Chemical energy | Academic achievement Inquiry questionnaire Problem solving Attitudes towards chemistry Social aspects: interactions |

**Table 4:** Continued

| Authors | Sample | Method of instruction | Subject | Topics | Variables investigated |
|---|---|---|---|---|---|
| Cohen (1987) | Grades 11–12 N = 1355 | Group activities | Chemistry | Chemical energy Chemical equilibrium | Academic achievement Inquiry skills Low and high ability students Affective domain group setting Attitudes towards the learning strategies and topics Attitudes and achievement |
| Tingle & Good (1990) | Grade 11–12 N = 178 | Solving problems in groups | Chemistry | Stoichiometric problems The mole concept Empirical equations | Academic achievement Solving problems Supportive climate Organisation and division of work, responsibility |
| Lonning (1993) | Grade 10 N = 36 | Learning together | General science | Particle model Solids, liquids and gases | Academic achievement Use of verbal patterns Conceptual change |
| Scott & Heller (1991) | High school, emphasis on minorities and females | Jigsaw and group work | Physics | Candle in water demonstration | Interdependence and individual accountability Learning activities Helping behaviour Students' interactions |

Table 5: Research on cooperative learning in science – College level

| Authors | Sample | Method of instruction | Subject | Topics | Variables investigated |
|---|---|---|---|---|---|
| Okebukola & Jegede (1988) | High school and university Age 15.3–21.2 N = 145 | Learning in groups | Biology | Photosynthesis Concept mapping | Academic achievement Cognitive preferences |
| Basili & Sanford (1991) | Community college N = 62 | Learning in groups | Chemistry | Energy, material preservation Gases, liquids and solids | Changing misconceptions Verbal behaviour Previous concepts held in relation to learning and the leader role in the group |
| Smith, Hinckley & Volk (1991) | College non-majors in chemistry N = 52 | Jigsaw | Chemistry laboratory | Acids and bases | Academic achievement Students with different academic status |
| Heller & Huann-Shyang (1992) | Introductory college physics N = 207 | Learning together | Physics | Context-rich problems | Problem-solving strategy Conceptual understanding Supportive environment Gender |
| Watson & Marshall (1992) | Primary education majors Ages 19–32 N = 116 | Cooperative groups Heterogeneous and homogeneous groups | General biology | Different topics | Academic achievement |
| Burron, Lynn James & Ambrosio (1993) | College preservice primary and middle level teachers in science | Learning together | Physics Chemistry | Laboratory work Velocity and acceleration Newton's laws of motion Atomic structures, antimatter, states of matter, changes in state | Academic achievement Contribution of ideas and suggestions Participation On-task behaviour Learning behaviours |

and achievement divisions. More studies used jigsaw and jigsaw-investigative group than the other three cooperative learning methods.

The subject matter involved most in the implementation of cooperative learning in the research literature is biology, and this is followed by chemistry, physics and earth science. It seems that the structure of biological science knowledge can be restructured to jigsaw and jigsaw-investigative group in a generic manner. Biology is a subject matter which uses classification as one of its most important tools. In the past, as in the present, the structure of knowledge is hierarchical. Organisms are classified from single cell organisms to mutli-cellular ones. The structures and functions of organelles in cells, tissues, organs and organisms are well defined and arranged around concepts and principles. Science curricula of the 1960s and 1970s were assembled also in a hierarchical mode based on seven biological organisation levels, namely, molecules, cells, tissues and organs, organisms, population, society and biome. This systematic order of biology content knowledge permits the task of identifying and selecting specific topics to be taught and studied in a cooperative learning approach.

Two curricular examples of cooperative learning methods are discussed below. At the cell level, the cell division process, mitosis, includes the five distinctive steps of interphase, prophase, metaphase, anaphase and telophase. These five steps can be studied as five sub-learning units. Each step can be studied separately and put together by peer-tutoring to form an entity of one physiological process which occurs in every cell whenever it divides itself. Another curricular example is the photosynthesis process which occurs in any green plant. This topic can be divided in five sub-learning units: the importance of the green plants for the living world (production of organic material and energy cycles); morphology and anatomy of the leaves; the role of chlorophyll, light and water in photosynthesis; gas exchange between plants and the atmosphere; and the biochemistry of the photosynthesis process.

Both topics include classroom and laboratory work. The mitosis topic is built of short learning tasks which can be concluded in two weeks. The photosynthesis topic consists of a sequence of extended learning tasks which can be accomplished in two months. It is suggested that teachers willing to implement cooperative learning should start first with short topics and later proceed to longer ones.

The sub-learning units of both topics demand different cognitive levels and can accommodate heterogeneous group members. Therefore, the structure of knowledge in biology fits the requirements of the jigsaw methods, which use highly structured curricular material.

The teacher, as a jigsaw curriculum developer, is in charge of students' learning, because the jigsaw and jigsaw-investigative group learning material is structured by the teacher and each student is assigned a specific role. Five learners are given five concepts or principles, which belong to the same topic, to study cooperatively using various organisations. The teacher supervises the process and knows what material was learned, what was the responsibility of every student and what was accomplished. Science teachers who are asked to change their instruction typically accommodate more easily to jigsaw, because they still have a feeling of being in charge of the learning pace. The transition from individual inquiry to group investigation, using a structured jigsaw-investigative group, is becoming more common in biology teaching.

In other science subjects, such as chemistry and physics, few studies were conducted. The topics studied in cooperative learning were related to energy, which is a challenging topic for students to learn. Perhaps educators turned to the cooperative learning method in order to overcome difficulties in teaching such topics. Because physics and chemistry represent the 'hard sciences', it is important to identify 'difficult' learning topics and design curricula to be learned through cooperative learning.

Integrating jigsaw-investigative group, or other group inquiry methods, into the science curriculum proved to help all students and particularly weak students to achieve more academically. Cooperative learning has been found to enhance girls' academic achievement in science, as well as self-esteem, attitudes and social skills for all students (Heller & Huann-Shyang 1992; Lazarowitz 1991).

*Results of Cooperative Learning Research Studies*

At the primary school level, overall cooperative learning was linked with increased students' academic achievement. Gains for cooperative learning in the social-affective areas included increased helping behaviour and peer support (Table 2). At the junior high school level (Table 3), more variables were studied within each of the cognitive, affective and social domains. The general picture which emerged is that, in the cognitive area, students' academic achievement increased and research, investigative, practical, learning and reporting skills were enhanced. In the affective domain, cooperative learning students' attitudes towards learning and laboratory work and their self-esteem became more positive, and they were more engaged in the active learning process and in on-task behaviour. Students' motivation and enjoyment increased, they displayed more collaboration within the group and student-student and student-teacher interactions were more intense.

At the senior high school level, the variables investigated included students' learning ability and academic achievement, inquiry and other cognitive skills (Table 4). Affective measures included student self-esteem, attitudes towards subject matter and cooperative learning, and on-task behaviour. On all cognitive and affective measures, gains and positive results related to cooperative learning were observed. An additional variable investigated in both junior and senior high schools was the classroom learning environment (Fraser 1989); results suggested that cooperative learning promoted a positive classroom environment.

Another variable studied is the number of friends which students made during their learning with the cooperative learning method. Using the number of words and ideas generated by students in cooperative learning settings is a new trend in research into possible relationships between learning in cooperative groups and students' creativity (Lazarowitz, Hertz-Lazarowitz & Baird 1994).

Ron and Lazarowitz (1995) reported no differences in academic achievement between peer-tutors and their tutees on the sub-topics taught and learned by the group members.

The studies carried out at the college level (Table 5) show that students' cognitive preferences, concept learning and gender differences were investigated in addition to cognitive and affective measures. The results show similar patterns to findings at other educational levels.

## CONCLUSION

The research reviewed in this chapter indicates that cooperative learning is a promising innovation for the academic, social and cognitive development of students. The greatest challenge facing teachers and researchers is to develop relevant, rich and challenging curricula. In the future, cooperative learning will involve interacting with peers and computer data bases. Only cooperative teams will meet the challenge to access information and transfer it to meaningful knowledge.

Some implications for improving science education based on this chapter's material on cooperative learning are provided below:

(1) Because most students like to study science using cooperative learning methods, science teachers should strive to develop learning units and tasks that utilise cooperative learning methods to enhance student academic achievement and perceptions of classroom learning environment.

(2) Because cooperative learning methods in science, especially in biology and chemistry, can be implemented at all grade levels and in various cultures, science teachers should use cooperative learning methods with heterogeneous groups of students in different countries and from varied cultures.

(3) The topics in science that are more suitable to be taught using cooperative learning methods are those that can be divided into sub-topics or which stimulate inquiry. Topics which emphasise memorisation or basic information are not as suitable.

(4) Teachers should choose, from the many cooperative learning methods that are available, the methods with which they feel comfortable and that fit their particular student population. Teachers can be creative and innovative with cooperative learning methods by choosing and planning structures that suit them best.

(5) Because cooperative learning emphasises social, affective and academic domains, teachers should implement cooperative learning for at least 30 percent of the teaching/learning time in their classroom in order to contribute to students' potential to interact and cooperate with other people in school and society.

# REFERENCES

Allen, V. (ed.): 1976, *Children at Teachers*, Academic Press, New York.
Aronson, E., Stephan, C., Sikes, J., Blaney, N. & Snapp, M.: 1978, *The Jigsaw Classroom*, Sage Publication, Beverly Hills, CA.
Barnes, D. & Todd, F.: 1977, *Communication and Learning in Small Groups*, Routledge & Kegan Paul, London.
Basili, P.A. & Sanford, J.P.: 1991, 'Conceptual Change Strategies and Cooperative Group Work in Chemistry', *Journal of Research in Science Teaching* 28, 293–304.
Biological Sciences Curriculum Study: 1989, *The Natural Selection* (Newsletter), Author, Colorado Springs, CO.
Brown, A. & Palincsar, A.S.: 1989, 'Guided, Cooperative Learning and Individual Knowledge Acquisition', in L. Resnick (ed.), *Knowing, Learning and Instruction*, Earlbaum, Hillsdale, NJ, 393–451.
Burron, B., Lynn James, M. & Ambrosio, A.L.: 1993, 'The Effects of Cooperative Learning in a Physical Science Course for Elementary/Middle Level Preservice Teachers', *Journal of Research in Science Teaching* 30, 697–707.
Chang, H.P. & Lederman, N.G.: 1994, 'The Effect of Levels of Cooperation Within Physical Science Laboratory Groups on Physical Science Achievement', *Journal of Research in Science Teaching* 31, 167–181.
Cohen, E.: 1986, *Designing Groupwork: Strategies for the Heterogeneous Classroom*, Teachers College Press, New York.
Cohen, I.: 1987, *Various Strategies for the Teaching of the Topics of Chemical Energy and Chemical Equilibrium: Development, Implementation and Evaluation*, Unpublished PhD thesis, Weizmann Institute of Science, Rehovot, Israel.
Damon, W.: 1984, 'Peer Education: The Untapped Potential', *Journal of Applied Developmental Psychology* 5, 331–343.
Deutsch, M.: 1949, 'A Theory of Cooperation and Competition', *Human Relations* 2, 129–152.
De Vries, D.L. & Slavin, R.E.: 1978, 'Teams-Games-Tournament (TGT): Review of Ten Classroom Experiments', *Journal of Research and Development in Education* 12, 28–38.
Dewey, J.: 1970, *Experience and Education*, Collier, New York.
Driver, R.G. & Bell, B.: 1986, 'Students' Thinking and the Learning of Science: A Constructivist View', *School Science Review* 67, 443–456.
Driver, R.G. & Oldham, V.: 1986, 'A Constructivist Approach to Curriculum Development in Science', *Studies in Science Education* 12, 105–122.
Fraser, B.J.: 1989, 'Twenty Years of Classroom Climate Work: Progress and Prospect', *Journal of Curriculum Studies* 21, 207–327.
Gabel, D.L. (ed.): 1994, *Handbook of Research on Science Teaching and Learning*, Macmillan, New York.
Heller, P. & Huann-Shyang, L.: 1992, 'Teaching Physics Problem Solving through Cooperative Grouping: Do Men Perform Better than Women?', Paper presented at the annual meeting of the National Association for Research in Science Teaching, Boston, MA.
Hertz-Lazarowitz, R.: 1992, 'Understanding Interactive Behaviors: Looking at Six Mirrors of the Classroom', in R. Hertz-Lazarowitz & N. Miller (eds.), *Interaction in Cooperative Groups: The Theoretical Anatomy of Group Learning*, Cambridge University Press, New York, 71–101.
Hertz-Lazarowitz, R.: 1993, 'On Becoming a Cooperative Teacher: Using the Model of the Six Mirrors of the Classroom to Document the Transitional Phase of Two Teachers', *Texas Researcher* 4, 97–110.
Hertz-Lazarowitz, R., Baird, H.J. & Lazarowitz, R.: 1994, 'Affective Measures on High School Students Who Learned Science in a Cooperative Mode', *Australian Science Teachers Journal* 40(2), 67–71.
Hertz-Lazarowitz, R. & Calderón, M.: 1992, *Group Investigation Magnifies Cooperative Learning*, University of Texas, El-Paso, TX.
Hertz-Lazarowitz, R. & Miller, N. (eds.): 1992, *Interaction in Cooperative Groups: The Theoretical Anatomy of Group Learning*, Cambridge University Press, New York.
Humphreys, B., Johnson, R.T. & Johnson, D.W.: 1982, 'Effects of Cooperative, Competitive and Individualistic Learning on Students' Achievement in Science Class', *Journal of Research in Science Teaching* 19, 351–356.
Jegede, O.J., Okebukola, P.A. & Ajewole, G.A.: 1991, 'Computers and the Learning of Biological Concepts: Attitudes and Achievements of Nigerian Students', *Science Education* 75, 701–706.

Johnson, D.W. & Johnson, R.T.: 1975, *Learning Together and Alone*, Prentice Hall, Englewood Cliffs, NJ.

Johnston, K. & Driver, R.G.: 1990, *Children's Learning in Science Project: Interactive Teaching in Science-Workshop for Training Courses*, Centre for Studies in Science and Mathematics Education, University of Leeds, Leeds, UK.

Jones, R.M. & Steinbrink, J.E.: 1989, 'Using Cooperative Groups in Science Teaching', *School, Science and Mathematics* 89, 541–551.

Kagan, S.: 1985, 'Dimensions of Cooperative Classroom Structure', in R. Slavin, S. Sharan, S. Kagan, R. Hertz-Lazarowitz, W. Webb & R. Schmuck (eds.), *Learning to Cooperate, Cooperating to Learn*, Plenum Press, New York, 67–96.

Lazarowitz, R.: 1991, 'Learning Biology Cooperatively: An Israeli Junior High School Study', *Cooperative Learning: The Magazine for Cooperation in Education* 11, 19–21.

Lazarowitz, R.: 1995, 'Learning Science in Cooperative Modes in Junior- and Senior-High School: Cognitive and Affective Outcomes', in E.J. Pedersen & D.A. Digby (eds.), *Cooperative Learning and Secondary Schools: Theory, Models and Strategies*, Garland Press, New York, 185–227.

Lazarowitz, R., Baird, J.H., Hertz-Lazarowitz, R. & Jenkins, J.: 1985, 'The Effects of Modified Jigsaw on Achievement, Classroom Social Climate, and Self-Esteem in High School Science Classes', in R. Slavin, S. Sharan, S. Kagan, R. Hertz-Lazarowitz, W. Webb & R. Schmuck (eds.), *Learning to Cooperate, Cooperating to Learn*, Plenum Press, New York, 231–253.

Lazarowitz, R. & Galon, M.: 1990, 'Learning Biology in a Cooperative Setting: Ninth Grade Students' Achievement and Cognitive Reasoning States', Paper presented at the 1990 International Convention on Cooperative Learning, Baltimore Convention Center, Baltimore, MD.

Lazarowitz, R., Hertz-Lazarowitz, R. & Baird, J.H.: 1994, 'Learning Science in a Cooperative Setting: Academic Achievement and Affective Outcomes', *Journal of Research in Science Teaching* 31, 1121–1131.

Lazarowitz, R., Hertz-Lazarowitz, R., Baird, J.H. & Bowlden, V.: 1988, 'Academic Achievement and On-Task Behavior of High School Biology Students Instructed in a Cooperative Small Investigative Group', *Science Education* 72, 475–487.

Lazarowitz, R. & Karsenty, G.: 1990, 'Cooperative Learning and Students' Self-Esteem in Tenth Grade Biology Classrooms', in S. Sharan (ed.), *Cooperative Learning, Theory and Research*, Praeger Publishers, New York, 123–149.

Lonning, R.A.: 1993, 'Effect of Cooperative Learning Strategies on Student Verbal Interactions and Achievement During Conceptual Change Instruction in Tenth Grade General Science', *Journal of Research in Science Teaching* 30, 1087–1101.

Mid-Continent Regional Educational Laboratory and Biological Sciences Curriculum Study: 1969, *Inquiry Objectives for the Teaching of Biology*, Mid-Continental Educational Laboratory, Kansas City, KS.

Miller, N. & Brewer, B. (eds.): 1984, *Groups in Contact: The Psychology of Desegregation*, Academic Press, Orlando, FL.

National Assessment of Educational Progress: 1988, *The Science Report Card: Elements of Risk and Recovery*, Educational Testing Service, Princeton NJ.

O'Donnell, A.M. & Dansereau, D.F.: 1992, 'Scripted Cooperation in Student Dyads: A Method for Analyzing and Enhancing Academic Learning and Performance', in R. Hertz-Lazarowitz & N. Miller (eds.), *Interactions in Cooperative Groups*, Cambridge University Press, New York, 120–141.

Okebukola, P.A.O.: 1985, 'The Relative Effectiveness of Cooperative, Competitive Interaction Techniques in Strengthening Students' Performance in Science Classes', *Science Education* 69, 501–509.

Okebukola, P.A.O.: 1986a, 'Cooperative Learning and Students' Attitudes to Laboratory Work', *School Science and Mathematics* 86, 582–590.

Okebukola, P.A.O.: 1986b, 'The Influence of Preferred Learning Styles on Cooperative Learning in Science', *Science Education* 70, 509–517.

Okebukola, P.A.O. & Jegede, O.J.: 1988, 'Cognitive Preference and Learning Mode as Determinants of Meaningful Learning Through Concept Mapping', *Science Education* 72, 489–500.

Okebukola, P.A.O. & Ogunniyi, M.B.: 1984, 'Cooperative, Competitive and Individualistic Laboratory Interaction Patterns: Effects on Students' Performance and Acquisition of Practical Skills', *Journal of Research in Science Teaching* 21, 875–884.

Piaget, J.: 1926, *Language and Thought of the Child*, Harcourt Brace, New York.

Prawat, R.S.: 1989, 'Teaching for Understanding: Three Key Attributes', *Teaching and Teacher Education* 5, 315–328.

Project 2061: 1989, *Science for All Americans*, American Association for Advancement of Science, Washington, DC.

Rogg, S.R. & Kahle, J.B.: 1992, 'The Characterization of Small Instructional Work Groups in Ninth-Grade Biology', Paper presented at the annual meeting of the National Association for Research in Science Teaching, Boston, MA.

Ron, S. & Lazarowitz, R.: 1995, 'Learning Environment and Academic Achievement of High School Students Who Learned Evolution in a Cooperative Mode', Paper presented at the annual meeting of the National Association for Research in Science Teaching, San Francisco, CA.

Schwab, J.H.: 1963, *Biology Teachers' Handbook*, Wiley, New York.

Scott, L.W. & Heller, P.: 1991, 'Team Work Strategies for Integrating Women and Minorities into the Physical Sciences', *The Science Teacher* 58, 24–28.

Sharan, S. (ed.): 1994, *Handbook of Cooperative Learning Method*, Greenwood Press, Westport, CT.

Sharan, S. & Hertz-Lazarowitz, R.: 1980, 'A Group Investigation Method of Cooperative Learning in the Classroom', in S. Sharan, P. Hare, C. Webb & R. Hertz-Lazarowitz (eds.), *Cooperation in Education*, Brigham Young University Press, Provo, UT, 14–46.

Sharan, Y. & Sharan, S.: 1992, *Expanding Cooperative Learning Through Group Investigation*, Teachers College Press, New York.

Sherman, L.W.: 1988, 'A Comparative Study of Cooperative and Competitive Achievement in Two Secondary Biology Classrooms: The Group Investigative Model Versus an Individually Competitive Goal Structure', *Journal of Research in Science Teaching* 26, 55–64.

Slavin, R.E.: 1978, 'Student Teams and Achievement Divisions', *Journal of Research and Development in Education* 12, 39–49.

Slavin, R.W.: 1983, *Cooperative Learning*, Longman, New York.

Slavin, R.E.: 1990, *Cooperative Learning: Theory, Research and Practice*, Allyn and Bowen, Boston, MA.

Smith, M., Hinkley, C.C. & Volk, G.L.: 1991, 'Cooperative Learning in the Undergraduate Laboratory', *Journal of Chemical Education* 68, 413–416.

Stevens, R., Madden, N., Slavin, R. & Farnish, A.: 1987, 'Cooperative Integrated Reading and Composition: Two Field Experiments', *Reading Research Quarterly* 22, 433–454.

Tingle, J.B. & Good, R.: 1990, 'Effects of Cooperative Grouping on Stoichiometric Problem Solving in High School Chemistry', *Journal of Research in Science Teaching* 27, 671–683.

Vygotsky, L.S.: 1978, *Mind in Society: The Development of Higher Psychological Process*, Harvard University Press, Cambridge, MA.

Walters, J.: 1988, 'Teaching Biological Systems', *Journal of Biological Education* 22, 87.

Watson, S.B.: 1991, 'Cooperative Learning and Group Education Modules: Effects on Cognitive Achievement of High School Biology Students', *Journal of Research in Science Teaching* 28, 141–146.

Watson, S.B. & Marshall, J.E.: 1992, 'Cooperative Incentives and Heterogeneous Arrangement of Cooperative Learning Groups: Effects on Achievement of Elementary Education Majors in an Introductory Life Science Course', Paper presented at the annual meeting of the National Association for Research in Science Teaching, Boston, MA.

Welicker, M. & Lazarowitz, R.: 1995, 'Performance Tasks and Performance Assessment of High School Students Studying Primary Prevention of Cardiovascular Diseases', Paper presented at the annual meeting of the National Association for Research in Science Teaching, Boston, MA.

Yager, R.E. & Lutz, M.V.: 1994, 'Integrated Science: The Importance of "How" Versus "What"', *School Science and Mathematics* 94, 338–346.

Yeroslavski, O., Dori, J. & Lazarowitz, R.: 1995, 'The Effect of Teaching the Cell Topic Using the Jigsaw Method on Students' Achievement and Learning Activity', Paper presented at the annual meeting of the National Association for Research in Science Teaching, Boston, MA.

Zadock, N.: 1983, *The Learning of the Concept 'Chemical Energy' in Inquiry Groups: Evaluation of Effectiveness* [in Hebrew], Unpublished MSc thesis, Weizmann Institute of Science, Rehovot, Israel.

# 4.3 Curriculum Change in Science: Riding the Waves of Reform

JOHN WALLACE
*Curtin University of Technology, Perth, Australia*

WILLIAM LOUDEN
*Edith Cowan University, Perth, Australia*

This chapter examines the multiple understandings of curriculum reform in science. Curriculum activity over the past four decades has been characterised by different understandings of the role of science in the curriculum. In the 1950s and 1960s, science was understood as *discipline knowledge,* with the goal being to develop a base in schools from which new scientific discoveries could be made. In the 1970s and 1980s, science was understood as *relevant knowledge,* and the goal was to use science as a tool for personal and societal improvement. In the 1980s and 1990s, we have seen a recognition of science as *imperfect knowledge,* with a focus on the individual, social and cultural influences on the formation of scientific knowledge. While these periods overlap and remnants of earlier goals persist to the present day, what they have in common is the promotion of *science* – as distinct from other forms of human endeavour – as a means of improving society. It is this belief in the pre-eminence of science which has animated much of the activity in science curriculum development over the past 40 years (Hurd 1969; Welch 1979).

The aim of this chapter is to extract from the reform experience of the past 40 years some lessons about how to proceed with curriculum change in science. Three waves of reform are analysed in terms of their attention to the discipline of science, teachers, students and the milieu of schooling. The chapter outlines a set of key considerations that should guide curriculum reform in science and proposes a balance between these considerations and the moral purpose of teaching and of science itself.

## SCIENCE AS DISCIPLINE KNOWLEDGE

The post-war reformation of science education, particularly in Western countries, was driven by a belief in the value of science as means of ensuring national security

---

Chapter Consultant: John Olson (Queen's University, Canada)

and competitiveness. Jerome Bruner's influential book *The Process of Education* (1960) outlined a set of understandings about children and scientific knowledge. Bruner's vision – that children should be exposed to the formal concepts and structure of the discipline – became the philosophical basis for the science curriculum projects of the next decade:

> The curriculum of a subject should be determined by the most fundamental understanding that can be achieved of the underlying principles that give structure to that subject. (p. 31)

Most of the early curriculum reforms of the 1960s employed scientists as the project directors and curriculum development was structured by questions about the structure of the discipline. Chem Study, Physical Sciences Study Committee (PSSC) and Biological Sciences Curriculum Study (BSCS) in the USA, as well as Nuffield in the UK, adopted scientific knowledge as the primary goal and scientific methods as the means of achieving this goal.

The early efforts at science curriculum development in the 1960s, particularly in the USA, were accompanied by essentially positive reports. Teachers in trial schools were engaged with the materials and attendance at summer training institutes was high (Crane 1976). Most of the early studies of students' experiences with the science curriculum focused directly or indirectly on Bruner's goal of understanding the discipline. Of 90 studies published between 1965 and 1975 examining science courses in the USA, 82 used student effects as the main criteria (Welch 1979). Measures of achievement, attitude, laboratory skills, critical thinking, problem solving and logical reasoning were used to describe and compare classrooms (see White & Tisher 1986).

Over time, it became clear that the hopes and good intentions of the reformers of the 1960s were not being achieved in schools. Three major studies conducted for the National Science Foundation in the USA (Hegelson, Blosser & Howe 1977; Research Triangle Institute 1977; Stake & Easley 1978) documented the shortfall in a particularly thorough way. Stake and Easley's (1978) research involved a series of case studies that pointed to the conservative nature of schools and the central role of the teacher in curriculum reform. These case studies also set a trend for new ways of researching and evaluating the curriculum of schools and classrooms. The studies provided glimpses of why curriculum materials were adopted in some settings and not others. Importantly, these studies reached conclusions which were very different from the essentially positive reports of the projects in trial schools (Welch 1979). The prestige associated with high-profile curriculum-development projects helped systems or schools to make decisions to adopt particular ideas and materials. However, the profile of the project had much less effect on how the materials were implemented in schools. In his 1979 review of 20 years of science curriculum development in the USA, Welch said:

> In spite of the expenditures of millions of dollars and the involvement of some of the most brilliant scientific minds, the science classroom of today is little different from one of 20 years ago. While there may be new books on

the shelves and clever gadgets in the storage cabinets, the day-to-day operation of the class remains largely unchanged. A teacher tells his or her students what is important to learn and so the class progresses. (Welch 1979, p. 303)

The prevailing view was that these science curriculum reforms were unsuccessful because the intentions of the curriculum developers were not reflected in the teachers' actions. This was seen as a problem of adoption; teachers needed to overcome barriers to change such as lack of time and money, poor textbooks, inappropriate pedagogy and poor community or school support. An alternative view was offered by Connelly (1972) who argued that it was wrong to talk of a deficit on the part of teachers. Teachers' responsibility to the classroom means that they interpret and translate the curriculum-as-offered into a curriculum-in-use. Often, teachers find creative but unintended ways of releasing the 'curriculum potential' (Ben-Peretz 1975) of project materials.

Many other countries followed the lead of the USA and the UK in adopting big-budget, discipline-knowledge curricula, sometimes by slavishly importing these curricula under a new form of educational imperialism (Fensham 1992). With the exception of these late starters to the reformation, the impetus for the first wave of reform had slowed considerably by the early 1970s. The curriculum materials from the various projects had been disseminated widely and many teachers had been initiated into their intended use with varying degrees of thoroughness (Fensham 1992).

## SCIENCE AS RELEVANT KNOWLEDGE

As the first wave of science curriculum reform slowed, many commentators argued that the emphasis of the 1950s and 1960s on the discipline of science had been adopted at the expense of the personal, historical and applied aspects of science. In the 1970s, science educators began calling for science curricula which emphasised an understanding of the importance of scientific literacy. Hurd, for example, argued:

> The goal of science teaching is to foster an enlightened citizenry, capable of using the intellectual resources of science to create a favorable environment that will promote the development of man as a human being. (Hurd 1970, p. 14)

The term 'scientific literacy' described a wide range of progressive educational goals through the 1970s and 1980s. It also was associated with other major reform movements concerned with scientific relevance including science, technology and society (Gallagher 1971), the environment movement (Bybee 1979) and science for all (Fensham 1985). Educators began to advocate curricula which drew upon the social and cultural interests of the community to prepare students to participate in that community. Curriculum developers set out to produce materials which were accessible to all students and, at the same time, provided a sound basis for those who wished to pursue their scientific studies to a higher level. The developers often employed a thematic approach in an attempt to tap into young people's interests in science. Examples included the Individualised Science Instructional System (ISIS) from the USA, the Nuffield Science Teaching Project from the UK (Waring 1979) and the Australian Science Education Project (ASEP) from Australia (Owen 1978). By 1979 Haggis and Adey were able to identify 130 such integrated curriculum programs worldwide. More recent variations on this approach include the PLON project from the Netherlands (Eijkelhof & Kortland 1988) and the Salter's Project from the UK (Campbell, Lazonby, Millar, Nicolson, Ramsden & Waddington 1994).

The shift towards a more socially relevant base for science education was taking place at the same time as researchers such as Connelly (1972) were advocating an enhanced role for teachers in the curriculum process. Some projects recognised the need for increased teacher involvement in curriculum development (White & Tisher 1986) but, for the most part, little had changed in the practice of curriculum implementation. Materials were developed from a view of the role of science in transforming the knowledge and skills of young people. Arguments raged about the kind of science being promoted in these curricula (Good, Herron, Lawson & Renner 1985), but most of the resources were devoted to development rather than implementation. Evaluations pointed to the numbers of students using a particular program as well as teacher and student receptivity to the use of the materials as indicators of the success of the program (Campbell et al. 1994). However, increases in student numbers implied nothing about the intervening processes, or about the experiences of those teachers and students who were using the material. Evaluators found that it was not easy to get teachers or students to articulate their views on the effects of particular learning activities, on learning of specific concepts or on the nature of science (Ramsden 1994). Responses from teachers often were couched in the most general terms suggesting that the main criteria against which teachers judged the success of the lessons was the overall classroom atmosphere. Students found it difficult to reconcile their enjoyment of an activity with a view of proper science as being hard and boring. Importantly, it was almost impossible to compare the learning outcomes or the management effects of using particular materials with those which would have been achieved by using some other materials or simply by modifying a previous teaching program (Campbell et al. 1994).

When evaluators did make forays into classrooms to observe the teaching and learning process, they discovered that teachers were powerful mediators of the intended learning environment. For example, in studies of the Australian Science Education Project (ASEP) classrooms, it was found that teachers organised their classes in diverse ways to retain control over the self-paced learning environment (Power & Tisher 1976). The influential work of Fullan (1982) on educational change emphasised that teachers were likely to adopt a new set of materials only when they had became dissatisfied, for some reason, with their previous practice. According to Fullan (1991), the grounds for teacher dissatisfaction could be driven either internally or externally. Teachers might wish to take up a new project if it enabled them to make a change which they wished to make or because they recognised that this was a better way to teach the subject. Alternatively, because of external pressures requiring them to change their teaching program, teachers might take up a new project if the new materials seemed the most suitable to their context and needs. In all these cases, the new curriculum materials offer a 'solution' to the user's 'problem'. Variations in the scientific basis or content of a curriculum appeared to exert little influence on a teacher's decision to adopt particular materials (Crocker & Banfield 1986).

## SCIENCE AS IMPERFECT KNOWLEDGE

The most recent wave of science curriculum reform, commencing in the early 1980s, is based on the notion that individual students are constructors of their own scientific knowledge. This wave drew upon theoretical work in the philosophy of science and empirical work in cognitive science.

One source of the pressure for change in the science curriculum has emerged from an epistemological shift. School textbooks have tended to describe the practice of science as a linear, rational and objective process, discovered and guaranteed by the certainty of an objective experimental method. However, philosophers of science such as Lakatos (1970), Popper (1972) and Feyerabend (1975) have argued that knowledge is not discovered but rather constructed within communities of like-minded people. The consequence of this line of argument is that scientific knowledge is seen as imperfect and imperfectible, depending for its warrants and truth claims on the socially constructed knowledge of particular communities of scientists. The implications for school science of this conception of the discipline knowledge of science have been explored by writers such as von Glasersfeld (1987), Tobin, Kahle & Fraser (1990) and Matthews (1990). Curriculum applications of these new understandings about the history and philosophy of science include Wandersee's (1990) use of historical vignettes, Hodson's (1990) model for curriculum planning and Shortland and Warwick's (1989) suggestions for teaching the history of science.

Cognitive science has been a second source of thinking about imperfect knowledge in the science curriculum. Building on the observation that students' common-sense practical knowledge of physical phenomena is not necessarily contradicted by their formal science knowledge, science educators and cognitive scientists began to study students' misconceptions and the process of conceptual change (West & Pines 1985). Studies in the 1980s focusing on the scientific understandings that children bring to the classroom (Champagne, Gunstone & Klopfer 1985; Osborne & Freyberg 1985) led to various theories about student engagement and conceptual change (Pines & West 1985). Under the more familiar title of 'constructivism', these ideas have now been elaborated in considerable detail. They are based on the idea that learners are the constructors of their own knowledge. From the constructivist perspective, knowledge resides in individuals and cannot be transferred intact from 'the head of the teacher to the heads of students' (Lorsbach & Tobin 1992, p. 9). At the heart of the theory is the negotiation of meaning, at a personal level and also at a social level (Tobin 1990).

Much of the curriculum development emerging from these ideas centred on ways of bridging the gap between children's common-sense knowledge and the formal knowledge of science. Based on models developed in New Zealand (Osborne & Wittrock 1985), the UK (Driver & Bell 1986) and Australia (Fensham 1989), considerable work has been carried out in developing teaching materials which promote a 'constructivist' approach to learning. Examples include the Children's Learning in Science Project (1987) from the UK, the Atlantic Science Curriculum Project in Canada (McFadden, Morrison, Armour, Hammond, Haysom, Moore, Nicoll & Smyth 1986), and the Chemistry in a Thousand Questions text from the Netherlands (de Vos & Verdonk 1987). In Australia, the Project to Enhance Effective Learning (PEEL) utilised a collaborative action research model to help teachers identify students' science conceptions and develop cognitive change teaching strategies (Baird 1986).

While cognitive scientists and philosophers of science share some common territory, there are important differences in their approach to science education. While the first group sees student knowledge as imperfect – and often in need of perfection – the second group views science itself as imperfect and imperfectible. These different perspectives on science knowledge have raised questions about the viability of 'constructivist' approaches to teaching science. In her critique of constructivism, Solomon (1994) points out that the theory does not account for the processes used by students as they struggle to comprehend the formal language of science used in textbooks. 'No amount of recollection of their own remembered territory with shut eyes', says Solomon (1994, p.16), will help them with the foreign task of understanding the canons of science. This view is amplified by Martin (1993) who argues that science cannot be understood in the student's 'own words', because science 'has evolved a special use of language in order to interpret the world in its own, not in common sense, terms' (Martin 1993, p. 200). O'Loughlin (1992) argues that constructivism is flawed because it fails to come to grips with issues such as culture and power in the classroom. In our own work (Wildy & Wallace 1995), we have suggested that constructivist teaching has been conceived narrowly because it does not take into account the cultural context of children's learning.

At about the same time as science education researchers became interested in children's constructions of scientific knowledge, a parallel strand of research on teachers' knowledge emerged. Unlike the psychologically-based information processing models of teachers' thinking which were common in the 1970s, new descriptions of teachers' knowledge problematised the kinds of knowledge, skills and meaning structures which teachers have and use. Elbaz (1983) elaborated the structures of a teacher's practical knowledge and explored how that knowledge was used. Similarly, Connelly and Clandinin (1988) explored the relationship between teachers' personal practical knowledge and their work in enacting the curriculum in individual classrooms. In a related strand of research, Shulman (1986) and his colleagues developed a series of accounts of teachers' 'pedagogical content knowledge', a kind of knowledge that is specific to teaching particular areas of subject content. Hashweh (1987), for example, elaborated some of the pedagogical content knowledge used by teachers in particular content areas in biology and physics. Other complementary strands of research on teachers' knowledge focused on the influence of teachers' lives and careers (Goodson 1992) and craft knowledge (Grimmett & MacKinnon 1992).

In the context of science teaching specifically, our own work on science has focused on the connections between teachers' knowledge and curriculum reform (Louden 1991; Louden & Wallace 1994; Wallace & Louden 1992). We argue that reforms such as constructivist science syllabuses only can be implemented through programs which acknowledge the central place of teachers' knowledge in determining what actually happens in the classroom. The process of becoming a constructivist science teacher, in our view, involves teachers reconstructing their own incomplete and imperfect knowledge of science and science teaching. Unfortunately, science curriculum reformers fall too often into what we call 'the constructivist paradox': anxious to see teachers pay respect to students' constructions of science, they fail to pay comparable respect to teachers' current constructions of teaching. Like students, teachers need opportunities to rework their current understandings and practices in the light of new ideas contained in the science curriculum materials which they are expected to implement.

As one group of science education researchers focused on students' and teachers' constructions of science knowledge, others were recognising the importance of school culture (Cobern 1993) and social justice. Several new curriculum questions were asked of science about the growing disparity between racial, ethnic and gender demographics of the school population and these demographics within the science and science education communities. This disparity has raised larger issues relating to authority and power, conceptions of truth and professionalism in science education. A major issue of the 1990s in science education is how modern science – which is largely the product of a European masculine culture and world view – can provide a framework for an inclusive and liberating science curriculum for all cultures. In recent years, there has been burgeoning literature on these issues of curriculum implementation as they relate to the education of girls (Kahle & Meece 1994), cultural minorities (Pomeroy 1994), non-Western science (Cobern 1996), language and science (Sutton 1989), school culture and science (Wildy & Wallace 1995) and policy making (Miller 1993).

Various approaches to these issues have been proposed and implemented, including removing bias in curriculum materials, increasing access to science for minority groups and introducing more inclusive pedagogies. The critics of these approaches suggest that they are merely a means to the end of inducting minorities and different cultural groups into a science that is essentially unchanged. O'Loughlin (1992), for example, proposes that it is not enough to teach science in a student-centred way:

> Who decides on the pedagogy and curriculum of the constructivist classroom and in whose interests? ... Are students better off in the apparently more

amenable social and intellectual milieu of a constructivist, student-centred classroom if the epistemological messages they receive about themselves and their world are identical to the messages they would have received in a traditional didactic classroom? (pp. 806–808)

As these commentators were highlighting the power relationships between students and school science, others were examining the relationship between teachers and their moral purposes and authority structures. Avenues of research into teachers' milieu and the curriculum have included the role of peers (Little 1984), external influences (Fullan 1991), social organisation (Rosenholtz 1989), subject departments (Siskin 1994), teacher involvement in curriculum development (Ruddock 1991; Wallace, Parker & Wildy 1995), school leadership (Hallinger & Murphy 1985), knowledge structures (Apple 1993) and the moral dimensions of teaching (Olson 1992; Sockett 1993). This corpus of research reinforces the view that teaching sits within the wider context of schooling which exerts a powerful overt and covert influence on the way in which curriculum is enacted.

## MULTIPLE UNDERSTANDINGS AND CURRICULUM CHANGE

A key lesson of the last 40 years is that curriculum change requires attention to multiple sites of understanding: the contested understandings of what counts as essential knowledge in scientific disciplines; students' prior understandings and misunderstandings of science; teachers' prior understandings of science and of teaching and learning; and the complex local understandings required to teach and learn science in a wide range of particular contexts characterised by differences of gender, culture, language background and ethnicity. These four forms of understandings of discipline, students, teachers and milieu – called curriculum 'commonplaces' by Joseph Schwab (1973) – received different emphases in each of the three periods of curriculum reform.

In the first wave of science curriculum reform, which focused on *discipline knowledge*, the understanding which received most attention was the curriculum developers' understanding of their subjects. From their point of view, the most important curriculum question seemed to be: 'What shall we include and exclude in order to provide students with the most rigorous possible introduction to understanding the disciplines of science?' The consequences for curriculum implementation are clear. Many curriculum packages were produced and many teachers attended inservice courses. Teachers adopted the curriculum materials, but they did not implement the understanding of discipline knowledge intended by the curriculum developers. Implementation plans paid little attention to students' prior understanding of science. Even less attention was paid to the impact of the milieu on students' and teachers' capacity to realise the discipline knowledge of the curriculum designers.

The second wave of science curriculum reform – science as *relevant knowledge* –

focused less on refining the knowledge required in preparation for advanced studies of science disciplines, and more on understanding the relationship between science and society. Curriculum developers produced materials which attended to the relationships between science, technology and society, to environmental concerns, to scientific literacy, and to young peoples' interests in science. These new curriculum approaches were implemented by schools and school systems with the same enthusiasm as the first wave of curriculum change. With a few exceptions, most of the resources were directed to curriculum development and few resources to implementation strategies. As with the first wave of change, little attention was given to students' understandings and even less importance to the impact of differences in school contexts on teaching and learning. The role of teachers in implementation, however, was explored more fully in evaluation studies conducted during the second wave of reform. Instead of assuming that teachers' failure to use curriculum materials as intended was the result of wilful resistance, researchers began to see that teachers' intentions and understandings are crucial to curriculum implementation. Teachers' perceptions of the need to change became the crucial implementation issue, rather than the quality of the materials or the slickness of the implementation strategy.

The third wave of curriculum reform in science shifted the focus from social relevance to students' own understandings and misunderstandings of science – from science as relevant knowledge to science as *imperfect knowledge*. Implementation of the many curriculum projects developed in the third wave paid more attention to students' understandings of science and the relationship between students' understandings and the formal discipline knowledge of science. Less frequently, teachers' understandings of science and teaching were included as part of the process of curriculum implementation in constructivist science curriculum projects. During the third wave, issues of school context or milieu also received more active consideration in curriculum implementation. Burgeoning research involved the effect of context on the curriculum, including the impact of students' gender, ethnicity and language background, the impact of school and science department cultures, the morality of teaching, and the impact of non-Western conceptions of science.

What do these three waves of curriculum reform teach about the possibility of change in science education? On the credit side of the balance sheet, the three waves demonstrate teachers' and researchers' capacity to learn from experience. Whereas the first wave attended only to the discipline of science, evaluations of second-wave curriculum change paid more attention to teachers' prior understandings of science and teaching. The third wave focused primarily on students, but also was accompanied by considerations of impact of the teachers and the milieu on curriculum reform.

On the debit side of the curriculum reform balance sheet, however, awareness of a wider range of curriculum commonplaces is incomplete. Curriculum reformers have learned to pay more attention to students' understandings, but have focused primarily on how the curriculum can develop students' naive understandings of science into the understandings authorised in the appropriate scientific discipline. Similarly, curriculum reformers have learned that teachers' intentions are important, but they show less patience with teachers' imperfect understandings than students' imperfect understandings. Curriculum developers also have learned that differences among students, such as gender, have an impact, but science educators still are working within masculine and Eurocentric understandings of science.

In our view, curriculum reform needs to begin by making problematic the issue of understanding. We think that there are four key considerations that should guide curriculum reform in science.

*Consideration 1: Whose Understanding of Science Will be Operationalised in the Curriculum?*

Effective curriculum change requires consideration of a range of stakeholders' understandings of discipline knowledge. Science educators' views of what counts as most important to teach and learn in science should be considered alongside professional scientists' views, and so too should the views of curriculum generalists who could advise about the impact of discipline-based decisions on the implementability of curriculum programs. For example, curriculum programs based on students' constructing their own science understandings are likely to conflict with the preconceptions about science held by the general public as well as by some students and teachers. Consequently, the desire for bold, constructivist curriculum change is likely to increase the complexity of the implementation stage of the curriculum cycle. Finally and importantly, curriculum developers need to listen to their own moral voice and to ask themselves: 'For what good and for whose good is this curriculum being developed?' Developers need to recognise and attend to the voices of these multiple stakeholders in the curriculum process.

*Consideration 2: How Does the Curriculum Relate to the Understandings of School and Science Held by Students?*

Students rarely have a voice in curriculum implementation, but they play a powerful part in determining which changes teachers are able to make. Students familiar with the security of chalk and talk science, for example, might resist teachers' efforts to implement approaches to teaching that require more high-level thinking, teamwork or problem solving. Teachers' difficulty in resisting the steering effect of students' expectations about what counts as school science has consequences for curriculum change. The more radical the departure intended by a curriculum change, the more important it is to involve students explicitly in understanding the change. A more student-centred science curriculum easily could founder on students' unwillingness to renegotiate their expectation that teachers will teach and test a static body of canonical knowledge.

*Consideration 3: What Do Teachers Already Know About Teaching and Learning in Science?*

Although research conducted during the second and third waves of change in science increasingly has acknowledged the importance of teachers' understandings in the success or failure of curriculum reforms, this insight has not been translated always into the practice of curriculum implementation. It is easier to secure funding for the development stage of curriculum projects than for the implementation stage. Too often, funds and energy are exhausted by the time the early adopters have implemented a curriculum change, and long before the more cautious and traditional teachers have had an opportunity to consider and experiment with new approaches and materials. We argue that it is important to secure sufficient funds to support the long chain of implementation.

The mind-set of curriculum implementors is as important as the availability of implementation funds. Because teachers' prior understandings are so important, curriculum implementors need to approach teachers with humility, tact and respect for tradition. Equally, they need to provide practical support, encourage collaboration among teachers and provide opportunities for involvement in adaptation of the approaches championed by the leaders of curriculum change. Without the enthusiasm of researchers, school board officials and the opinion leaders among teachers, there is likely to be little change. Equally, few changes are likely to be sustained without respect for teachers' knowledge and moral purpose.

*Consideration 4: What are the Range of Contexts in Which the Curriculum Will be Implemented?*

Finally, the bewildering range of particular contexts in which a curriculum change must be achieved remains a serious constraint. With a shrinking globe and rapidly changing school demographics and societal expectations, no longer can we act as though all students were the same. Neither can we assume that students and teachers share the same cultural and socioeconomic background and goals. Equally, consideration must be given to the impact of the different contexts in which curriculum will be implemented. For example, implementation strategies must be sufficiently flexible to be useful in school contexts where there is strong leadership and a tradition of collaboration, and in contexts where teachers have learned to work in isolation. In short, curriculum reform needs to move away from generalisations about implementation and attend more to the particulars of individuals, classrooms and school contexts.

## CONCLUSION

This chapter has described three major waves of reform which have dominated science curriculum change over the past 40 years. Our analysis of the impact of each wave of reform is based on the curriculum commonplaces of discipline, teachers, students and the milieu. We have proposed that future curriculum developers attend to four considerations to ensure an appropriate balance between these commonplaces. Taken together, these four considerations demonstrate the complexity of achieving change in science classrooms. As the curriculum buzz-words of today are replaced imperceptibly by a new set of imperatives for change, the complexity of achieving worthwhile curriculum change will remain. Each of us involved in curriculum reform is likely to attend most closely to those forms of understandings which are held most closely. For professional scientists, the temptation must be to struggle to refine understandings of the underlying structure of the disciplines; others might find themselves drawn more powerfully towards consideration of the impact of students, teachers or the milieu on curriculum implementation. Underpinning decisions about what to attend to in curriculum reform is the individual's sense of the morality or virtue of teaching and of science itself. Criteria such as fairness, inclusiveness, appropriateness, caring, honesty and equity are important. Moral purpose helps professionals to construct and reconstruct the science curriculum and provides a frame for making decisions about the balance between competing demands. The key message of the past 40 years is that curriculum change is a complex mixture of the facts of the change process and the values which underpin the change. Understanding and respecting the different dimensions of this complex relationship provide a way forward, albeit tentatively, for teachers and others involved in curriculum reform in science.

## REFERENCES

Apple, M.W.: 1993, *Official Knowledge: Democratic Education in a Conservative Age*, Routledge, New York.
Baird, J.: 1986, 'Improving Learning Through Enhanced Metacognition: A Classroom Study', *European Journal of Science Education* 8, 263–282.
Ben-Peretz, M.: 1975, 'The Concept of Curriculum Potential', *Curriculum Theory Network* 5, 151–159.
Bruner, J.S.: 1960, *The Process of Education*, Vintage, New York.
Bybee, R.W.: 1979, 'Science Education and the Emerging Ecological Society', *Science Education* 63, 95–109.
Campbell, B., Lazonby, J., Millar, R., Nicolson, P., Ramsden, J. & Waddington, D.: 1994, 'Science: The Salters' Approach – A Case Study of the Process of Large Scale Curriculum Development', *Science Education* 78, 415–447.
Champagne, A.B., Gunstone, R.F. & Klopfer, L.E.: 1985, 'Effecting Changes in Cognitive Structures Among Physics Students', in L. West & L. Pines (eds.), *Cognitive Structure and Conceptual Change*, Academic Press, Orlando, FL, 163–186.
Children's Learning in Science Project: 1987, *Approaches to Teaching the Particulate Theory of Matter*, Centre for Studies in Science and Mathematics Education, University of Leeds, Leeds, UK.
Cobern, W.W.: 1993, 'Contextual Constructivism: The Impact of Culture on the Learning and Teaching of Science', in K. Tobin (ed.), *The Practice of Constructivism in Science Education*, AAAS Press, Washington, DC, 51–69.
Cobern, W.W.: 1996, 'Constructivism and Non-Western Science Education Research', *International Journal of Science Education* 18, 295–310.
Connelly, M.F.: 1972, 'The Functions of Curriculum Development', *Interchange* 3(2–3), 161–177.
Connelly, F.M. & Clandinin, D.J.: 1988, *Teachers as Curriculum Planners*, Teachers College Press, New York.

Crane, T.: 1976, *The National Science Foundation and Pre-College Science Education: 1950–1975*, US Government Printing Office, Washington, DC.
Crocker, R.K. & Banfield, H.: 1986, 'Factors Influencing Teacher Decisions on School Classroom and Curriculum', *Journal of Research in Science Teaching* 3, 805–816.
de Vos, W. & Verdonk, A.H.: 1987, 'A New Road to Reactions; Part 4: Substance and its Molecules', *Journal of Chemical Education* 64, 692–694.
Driver, R. & Bell, B.: 1986, 'Students' Thinking and the Learning of Science: A Constructivist View', *School Science Review* 67(240), 443–456.
Eijkelhof, H.M.C. & Kortland, K.: 1988, 'Broadening the Aims of Physics Education', in P. Fensham (ed.), *Development and Dilemmas in Science Education*, Falmer Press, London, 282–305.
Elbaz, F.L.: 1983, *Teacher Thinking: A Study of Practical Knowledge*, Croom Helm, London.
Fensham, P.J.: 1985, 'Science for All: A Reflective Essay', *Journal of Curriculum Studies* 17, 415–435.
Fensham, P.J.: 1989, 'Theory in Practice: How to Assist Teachers to Teach Constructively', in P. Adey (ed.), *Adolescent Development and School Science*, Falmer Press, London, 61–77.
Fensham, P.J.: 1992, 'Science and Technology', in P. Jackson (ed.), *Handbook of Research on Curriculum*, Macmillan, New York, 789–829.
Feyerabend, P.K.: 1975, *Against Method: Outline of an Anarchistic Theory of Knowledge*, Verso, London.
Fullan, M.: 1982, *The Meaning of Educational Change*, Teachers College Press, New York.
Fullan, M.: 1991, *The New Meaning of Educational Change*, Teachers College Press, New York.
Gallagher, J.J.: 1971, 'A Broader Base for Science Teaching', *Science Education* 55, 329–338.
Good, R., Herron, J., Lawson, A. & Renner, J.: 1985, 'The Domain of Science Education', *Science Education* 69, 139–141.
Goodson, I. (ed.): 1992, *Studying Teachers' Lives*, Teachers College Press, New York.
Grimmet, P.P. & MacKinnon, A.M.: 1992, 'Craft Knowledge and the Education of Teachers', in G. Grant (ed.), *Review of Research in Education,* 18, American Educational Research Association, Washington, DC, 385–456.
Haggis, S. & Adey, P.: 1979, 'A Review of Integrated Science Education Worldwide', *Studies in Science Education* 6, 69–89.
Hallinger, P. & Murphy, J.: 1985, 'Assessing the Instructional Management Behavior of Principals', *The Elementary School Journal* 86, 217–247.
Hashweh, M.Z.: 1987, 'Effects of Subject Matter Knowledge in Teaching Biology and Physics', *Teaching and Teacher Education* 3, 109–120.
Hegelson, S.L., Blosser, P.E. & Howe, R.: 1977, *The Status of Pre-College Science, Mathematics and Social Studies Education 1955–1975, Vol 1. Science Education*, National Science Foundation, Washington, DC.
Hodson, D.: 1990, 'Making the Implicit Explicit: A Curriculum Planning Model for Enhancing Children's Understanding of Science', in D. Hegert (ed.), *More History and Philosophy of Science in Science Teaching*, Florida State University, Tallahassee, FL, 277–283.
Hurd, P.D.: 1969, *New Directions in Teaching Secondary School Science*, Rand McNally, Chicago, IL.
Hurd, P.D.: 1970, 'Scientific Enlightenment in an Age of Science', *The Science Teacher* 37, 13.
Kahle, J.B. & Meece, J.: 1994, 'Research on Gender Issues in the Classroom', in D. Gabel (ed.), *Handbook of Research on Science Teaching and Learning*, Macmillan, New York, 542–557.
Lakatos, I.: 1970, 'Falsification and the Methodology of Scientific Research Programs', in I. Lakatos & A. Musgrave (eds.), *Criticism and the Growth of Knowledge*, Cambridge University Press, Cambridge, UK, 91–181.
Little, J.W.: 1984, 'Seductive Images and Organisational Realities in Professional Development', *Teachers College Record* 86, 84–102.
Lorsbash, A. & Tobin, K.: 1992, Constructivism as a Referent for Science Teaching, *NARST News* 30, 9–11.
Louden, W.: 1991, *Understanding Teaching: Continuity and Change in Teachers' Knowledge*, Cassell, London and Teachers College Press, New York.
Louden, W. & Wallace, J.: 1994, Knowing and Teaching Science: The Constructivist Paradox, *International Journal of Science Education* 16, 649–657.
Martin, J.R.: 1993, 'Literacy in Science: Learning to Handle Text as a Technology', in M.A.K. Halliday & J.R. Martin, *Writing Science: Literacy and Discursive Power*, Falmer Press, London, 106–202.
Matthews, M.: 1990, 'Galileo and Pendulum Motion: A Case for History and Philosophy in the Science Classroom', *Australian Science Teachers Journal* 36(1), 7–13.

McFadden, C.P., Morrison, E.S., Armour, N., Hammond, A.R., Haysom, J., Moore, A., Nicoll, E.M. & Smyth, M.M.: 1986, *Science Plus, Vols 1, 2, 3, & Teachers' Guide of the Atlantic Science Curriculum Project*, Academic Press, Toronto, Canada.

Miller, S.: 1993, 'Minorities Move from Lab Rats to Policy Wonks', *Science* 262(5136), 1101–1102.

O'Loughlin, M.: 1992, 'Rethinking Science Education: Beyond Piagetian Constructivism Towards a Sociocultural Model of Teaching and Learning', *Journal of Research in Science Teaching* 29, 791–820.

Olson, J.: 1992, *Understanding Teaching: Beyond Expertise*, Open University Press, Milton Keynes, UK.

Osborne, R. & Freyberg, P.: 1985, *Learning in Science: The Implications of Children's Science*, Heinemann, Auckland, New Zealand.

Osborne, R.J. & Wittrock, M.C.: 1985, 'The Generative Learning Model and its Implications for Science Education', *Studies in Science Education* 12, 59–87.

Owen, J.M.: 1978, *The Impact of the Australian Science Education Project on Schools*, Curriculum Development Centre, Canberra, Australia.

Pines, A.L. & West, L.H.T.: 1985, 'Conceptual Understanding and Science Learning: An Interpretation of Research Within a Sources-of-Knowledge Framework', *Science Education* 70, 583–604.

Pomeroy, D.: 1994, 'Science Education and Cultural Diversity: Mapping the Field', *Studies in Science Education* 24, 49–73.

Popper, K.R.: 1972, *Conjectures and Refutations: The Growth of Scientific Knowledge*, Routledge & Kegan Paul, London.

Power, C.N. & Tisher, R.P.: 1976, 'Relationships Between Classroom Behavior and Instructional Outcomes in an Individualised Science Program', *Journal of Research in Science Teaching* 13, 489–497.

Ramsden, J.: 1994, 'Context and Activity-Based Science in Action: Some Teachers' Views of the Effects on Pupils', *School Science Review* 75(272), 7–14.

Research Triangle Institute: 1977, *National Survey of Science Education Curriculum Usage*, Research Triangle Institute, Research Triangle Park, NY.

Rosenholtz, S.J.: 1989, *Teachers' Workplace: The Social Organisation of Schools*, Longman, New York.

Ruddock, J.: 1991, *Innovation and Understanding*, Open University Press, Milton Keynes, UK.

Schwab, J.J.: 1973, 'The Practical 3: Translation into Curriculum', *School Review* 81, 501–522.

Shortland, M. & Warwick, A. (eds.): 1989, *Teaching the History of Science*, Basil Blackwell, Oxford, UK.

Shulman, L.S.: 1986, 'Those Who Understand: Knowledge Growth in Teaching', *Educational Researcher* 15(2), 4–14.

Siskin, L.S.: 1994: *Realms of Knowledge: Academic Departments in Secondary Schools*, Falmer Press, Washington DC.

Sockett, H.: 1993, *The Moral Base for Teacher Professionalism*, Teachers College Press, New York.

Solomon, J.: 1994, 'The Rise and Fall of Constructivism', *Studies in Science Education* 23, 1–19.

Stake, R.E. & Easley, J.A.: 1978, *Case Studies in Science Education*, US Government Printing Office, Washington, DC.

Sutton, C.R.: 1989, 'Reading and Writing in Science: The Hidden Messages', in R. Millar (ed.), *Doing Science: Images of Science in Science Education*, Falmer Press, London, 137–159.

Tobin, K.: 1990, 'Social Constructivist Perspectives on the Reform of Science Education', *Australian Science Teachers Journal* 36(4), 29–35.

Tobin, K.G., Kahle, J.B. & Fraser, B.J. (eds.): 1990, *Windows into Science Classrooms: Problems Associated with Higher-Level Cognitive Learning*, Falmer Press, London.

von Glasersfeld, E.: 1987, *Thought and Language* (translated by E. Hanfmann & G. Vaker), MIT Press, Cambridge, MA.

Wallace, J. & Louden, W.: 1992, 'Science Teaching and Teachers' Knowledge: Prospects for Reform of Elementary Classrooms', *Science Education* 76, 501–521.

Wallace, J., Parker, L. & Wildy, H.: 1995, 'Curriculum Reform and the Case of the Disappearing Agents', *Educational Studies* 21(1), 41–54.

Wandersee, J.: 1990, 'On the Value and Use of the History of Science in Teaching Today's Science: Constructing Historical Vignettes', in D. Hegert (ed.), *More History and Philosophy of Science in Science Teaching*, Florida State University, Tallahassee, FL, 277–283.

Waring, M.: 1979, *Social Pressures and Curriculum Innovation: A Study of the Nuffield Foundation Science Teaching Project*, Methuen, London.

Welch, W.W.: 1979, 'Twenty Years of Science Curriculum Development: A Look Back', *Review of Research in Education* 7, 282–306.
West, L.H.T. & Pines, A.L.: 1985, *Cognitive Structure and Conceptual Change*, Academic Press, Orlando, FL.
White, R.T. & Tisher, R.P.: 1986, 'Research on Natural Sciences', in M. Wittrock (ed.), *Handbook of Research on Teaching* (third edition), Macmillan, New York, 874–905.
Wildy, H. & Wallace, J.: 1995, 'Understanding Teaching or Teaching for Understanding: Alternative Frameworks for Science Classrooms', *Journal of Research in Science Teaching* 32, 143–156.

# 4.4 Science Curriculum: Transforming Goals to Practices

RODGER W. BYBEE
*National Research Council, Washington D.C., USA*

NAVA BEN-ZVI
*The Hebrew University of Jerusalem, Israel*

Within a short period of 25 years, science educators developed perspectives that have expanded from national to global dimensions. We are indeed in a global revolution in science, mathematics and technology education (Organisation for Economic Cooperation and Development 1996). No doubt this developmental change relates to advances in science and technology, especially those involving transportation and communication. But, like most changes, new issues and problems emerge. In science education, one of those new issues is a need to reform the curriculum.

Although reform must include changing multiple facets of science education, such as teaching and assessment, clearly the curriculum emerges as a central focus of any reform effort. Considering global perspectives and science curriculum reform engages the problem that forms the theme of this chapter: 'How can educators both reform the science curriculum to achieve the common goal of scientific literacy and accommodate the unique needs and aspirations of their region or nation?' The subtitle of the chapter anticipates the answer to the question. Once science educators have identified and clarified the curriculum's goals for scientific literacy, the critical issue becomes the transformation of those goals to policy, programs, and practices. Associated with any answer to the question is the use of resources, which include those human and material aspects that can be used for support or help when needed. What are the available resources for designing and developing a science curriculum? Answers to this question define the possibilities and limits of curriculum reform and they are certainly unique to a region or nation.

The perspectives developed in this chapter are based on presentations from the curriculum strand at an international conference entitled *International Conference on Science Education in Developing Countries: From Theory to Practice* held in Israel in 1993. To provide coordination and coherence among the presentations at that conference, we developed an organisation that generally parallels (with critical additions) the conference sub-theme of 'Transforming Goals to Practices'.

---

Chapter Consultant: Winston King (University of the West Indies, West Indies)

## SCIENTIFIC LITERACY AND THE SCIENCE CURRICULUM

The process of curriculum reform must begin with, and be informed by, the larger and more comprehensive purposes on which science educators from diverse countries can find common agreement. During the conference, the goal of scientific and technological literacy transcended the unique requirements of more-developed and less-developed countries. Presentations by Haggis (France), Cobern (USA), Whittle (Malawi), D'Ambrosio (Brazil) and Roseman (USA) developed the important first step of curriculum reform, namely, the reappraisal and reformulation of goals so that they represent the contemporary perspective of scientific literacy.

David Layton, Edgar Jenkins and James Donnelly, from the University of Leeds, Great Britain, completed a comprehensive, international review of scientific and technological literacy for UNESCO (1993). A careful review of this thorough summary of the literature clarifies the meaning of scientific literacy. In this chapter, instead of attempting a thorough review of scientific literacy, we describe some important aspects of scientific literacy for the science curriculum. We largely based this discussion on the aforementioned review by Layton, Jenkins and Donnelly (1993) and *Achieving Scientific Literacy: From Purpose to Practice* (Bybee 1997).

Using the term 'scientific literacy' implies a general education approach for the science curriculum. General education suggests that part of a student's education that emphasises an orientation towards personal development and citizenship. A general education orientation contrasts with a specialised education that emphasises competence in a specific career or occupation, such as science. Assuming a general education orientation for the science curriculum suggests that one should begin the design of a program by asking what it is that a student ought to know, value, and do as a citizen.

We contrast this approach to designing a science curriculum with an initial effort in which individuals ask what it is about physics, chemistry, biology and the earth sciences that students should learn. We view the science-technology-society (S-T-S) approach to science curriculum as a means of introducing the general education idea within the general purpose of achieving scientific literacy.

Closely associated with scientific literacy is the recommendation that educators design science curriculum materials for *all* students. Peter Fensham, an Australian science educator, has written several thoughtful essays on the topic of science education for all students (Fensham 1985, 1986/87, 1988). Paralleling the discussion on general education, Fensham clearly differentiates the demands of teaching science for future scientists from the goal of a scientifically literate citizenry. The 'scientific literacy' and 'science for *all*' themes are complementary. Fensham (1985) provides a useful metaphor to help developers realise the difference between a science curriculum for future scientists and one for all students. The former views science internally, presenting science as scientists see it; the latter views science externally, from the perspective of someone in society. From the perspective described here, science should be organised and taught in a context that continually affirms a personal and social perspective.

It is more beneficial to think of the aforementioned components of scientific literacy as goals towards which all individuals can develop for a lifetime. Indeed, one can define goals or thresholds for purposes of curriculum and instruction. The business of science education ought to be the continual development of individuals' understandings and abilities within the components described above.

Within the components of scientific literacy, one also has to represent a variety of expressions of understandings and abilities. For example, individuals probably can use scientific terms correctly, but most agree that equating scientific literacy with vocabulary represents a lack of understanding of the idea of scientific literacy. Figure 1, based on the work of Bybee (Biological Sciences Curriculum Study 1993; Bybee 1995, 1997; Uno & Bybee 1994), presents different aspects of scientific literacy that could be helpful to those designing science curricula.

## RETHINKING GOALS FOR THE SCIENCE CURRICULUM

The history of science teaching reveals several common goals for students and that historical changes are primarily shifts in emphasis among the goals rather than the creation of entirely new goals. Thus, studying the history of science education, in particular the changing structure and function of goals, provides insights for curriculum developers today. Extended historical reviews of the goals of science teaching include *A History of Ideas in Science Education* (DeBoer 1991), *Reforming Science Education* (Bybee 1993) and the chapter entitled 'Goals for the Science Curriculum' in *Handbook of Research on Science Teaching and Learning* (Bybee & DeBoer 1994).

Throughout the history of science education, three major goals for students have been (1) to acquire scientific knowledge, (2) to learn the procedures or methodologies of science and (3) to understand the applications of science, especially the relationship between science and society. The terms used to express these goals have changed throughout history. Scientific knowledge, for example, has been called facts, principles, conceptual schemes and major themes. Scientific procedures have been referred to as the scientific method, problem solving, scientific inquiry and the nature of science. Also, for a long time, there has been confusion between an emphasis on *knowing* about the procedures of science and *doing* scientific investigations. The applications of science have found expression as life adjustment, science personpower shortage and the contemporary science-technology-society (S-T-S) movement. In the following discussion, we use the terms 'scientific method', 'knowledge' and 'applications' broadly and generically to encompass the variety of terms used by science educators.

Sometimes the goals of scientific knowledge, method and applications are accompanied by clearly articulated justifications, but at other times they are advanced and accepted less critically. Science educators periodically should examine goals and their representation in science programs. Science programs will represent a curricular emphasis (Roberts 1982) – the structure and function of goals. Science educators should decide why they hold the views which they do and if they

This framework presents scientific and technological literacy as a continuum in which an individual develops greater and more sophisticated understanding of science and technology. This framework functions as a taxonomy for extant programs and practices and as a guide for curriculum and instruction.

**Nominal Scientific and Technological Literacy**

Individuals demonstrating *nominal* literacy associate vocabulary with the general area of science and technology. However, the association represents a misconception, naive theory or inaccurate concept. Using the basic definition of nominal, the relationship between science and technology terms and acceptable definitions is small and significant. There is, at best, only a token understanding that bears little or no relationship to real understanding.

**Functional Scientific and Technological Literacy**

Individuals demonstrating *functional* literacy respond adequately and appropriately to vocabulary associated with science and technology. They meet minimum standards of literacy. That is, they can read and write passages with simple scientific and technological vocabulary. Individuals might associate vocabulary with larger conceptual schemes, for example, that genetics is associated with variation within a species and variation is associated with evolution, but with token understanding of associations.

**Conceptual and Procedural Scientific and Technological Literacy**

Individuals demonstrate *conceptual and procedural* literacy by an understanding of both the parts (e.g., facts and information) and the whole (e.g., concepts, structure of a discipline) of science and technology as disciplines. Further, the individual can identify the way the parts form a whole *vis a vis* major conceptual schemes and the way new explanations and inventions develop *vis a vis* the processes of scientific inquiry and technological design. These individuals understand and can use the structure of scientific disciplines and the procedures of inquiry and design for developing new knowledge and techniques.

**Multidimensional Scientific and Technological Literacy**

*Multidimensional* literacy consists of understanding the essential conceptual structures of science and technology plus understanding features that make that understanding of the disciplines more complete, for example, the history and nature of science. In addition, individuals at this level understand the relationship of disciplines to the whole of science and technology, to society and to contemporary science-related and technology-related social issues.

Figure 1: A framework for scientific and technological literacy

can justify the particular emphasis in light of contemporary societal demands. This examination enables them to justify their focus and determine specifically what they mean by each of these major goals in curriculum design.

Reasons for teaching science knowledge, methods and applications have included (1) enhancing personal development, which includes aesthetic appreciation, intellectual development and career awareness, (2) maintaining and improving society, which includes the maintenance of a stable social order, economic productivity and the preparation of citizens who feel comfortable in a scientific and technological world, and (3) sustaining and developing the scientific enterprise itself. This enterprise involves the transmission of scientific knowledge from one generation to the next (so that each subsequent generation has a knowledge base from which new scientific discoveries can be made) and the formation of a scientifically enlightened citizenry sympathetic to the importance of science as a field of inquiry.

The challenge for science educators now, as in the past, is establishing an appropriate balance among competing goals given today's social needs. Recently, more than ever before, there is recognition of the potential relationship between the three major goal areas and this allows us to balance our curriculum focus on scientific knowledge, method, and applications without viewing the goals as mutually exclusive and thus diminishing our support for any one of them. In the following paragraphs, we briefly examine the goals of scientific knowledge, methods, and applications for the curriculum.

In primary and secondary schools, the main reason for teaching science today is the same as it has been in the past – to give students an understanding of the natural world and the abilities to reason and think critically as they explain their world. Students should begin early with observing and describing the world around them and moving towards progressively more elaborated scientific explanations of phenomena. By the end of high school, students should be able to provide comprehensive explanations for the most obvious and compelling events that they experience, such as the seasons, day and night, disease, heredity and species variation, and dangers of hazardous substances.

With respect to the methods of science, students should learn a disciplined way of asking questions, making investigations and constructing explanations of a scientific and technological nature. The latter certainly can be developed in a personal/societal context. Students should learn that scientific inquiry is a powerful, but not the only, route to progress in our world. Inquiry should not be taught in isolation but as a tool for finding answers to questions about the world in which students live. Science curricula and teaching consistently should emphasise students' conceptual development of scientific explanations, as opposed to step-by-step methods that too often characterise the nature of scientific inquiry.

Concerning the applications of science, students should confront contemporary and historical examples of how scientific knowledge is related to social advances and how society influences scientific advances. Once again, the focus should not be on learning about science and society for their own sakes, but to bring students to an appreciation of the complexity of the scientific/technological enterprise and to provide contexts and explanations for important science-related and technology-related societal challenges which they confront.

Scientific knowledge, method, and applications should not be taught in isolation. Each needs to be taught in connection with the other, with the aim of enlarging students' understanding of their world in meaningful ways.

Scientific literacy expresses the configuration and balance of goals for science education and thus the design or review of curriculum should assess the degree to which the curriculum incorporates the acquisition of scientific knowledge, development of inquiry abilities and understanding of the applications of science. To what degree and in what form are the goals expressed? Does the curriculum suggest one orientation for the structuring of the goals or does it suggest variations in the structuring of goals? Are there guidelines or suggestions for the use of goals in the design of curriculum materials, teaching strategies and assessment practices? If the curriculum were achieved, what levels and types of scientific literacy would be developed for the constellation of components described earlier? Would individuals continue into careers associated with science, engineering and related work?

Science curriculum developers should continue to work towards an integration of the three major goals of acquiring scientific knowledge, developing the abilities of scientific inquiry, and applying the understandings and abilities of science to personal decisions and societal challenges. If developers do, students' lives will be enriched, the levels of scientific literacy will be heightened, and the sympathy towards science as a way of knowing will be enlarged. More students will pursue careers in science and engineering, and we should continue to develop the understanding and skills required to solve our most vexing problems.

## TRANSFORMING GOALS TO A FRAMEWORK FOR CURRICULUM AND INSTRUCTION

An essential step, and one that is neglected too often, consists of the translation of goals into a curriculum framework and the need to address the critical issues in design and development of new science curricula. The framework specifies and explains the basic components used to design the science program. In the past, it was common for a curriculum framework to specify only criteria for content selection. Little, if any, attention was paid to a learning-teaching model and few curriculum developers attended to assessment practices. A complete framework provides information needed to make decisions about the content, the scope and sequence of activities, the selection of effective instructional strategies and techniques, appropriate assessment practices and other specifics of the curriculum. At a minimum, a framework defines enough of the science curriculum to differentiate it from other science programs.

Using a technological metaphor, a framework provides the requirements and specifications for a design project. The framework has to fulfil certain criteria and acknowledge constraints. At the same time, the more specific details must be left to those who actually will develop the science curriculum. Although there will be modification as the curriculum framework actually is developed and implemented, there should be fidelity to the original intentions, specifications, constraints and overall design.

A framework has advantages and disadvantages. An advantage of the framework is that program developers at local, state and national levels have opportunities to provide specific ideas. One assumes those decisions would be made in terms of the unique characteristics of students, schools and states, yet still fulfil the curriculum developer's requirements. A disadvantage is that it is incomplete. It lacks a full scope and sequence of lessons, the precise placement of concepts and skills, the selection of topics and learning activities, and the strategies for assessment, management of materials and other practical matters. In a sense, the incomplete nature of a framework is a necessity based on the understanding that curriculum developers and school personnel must make the final adjustments to the curriculum in terms of the unique characteristics of the school, community and students.

In Figure 2, we list some general characteristics that should facilitate the translation of goals to a curriculum framework. The list of characteristics is by no means complete or intended to suggest a particular curriculum. We intend only to provide an orientation or characteristics that individuals can use to develop a curriculum framework. We have found it quite helpful to use this list also to describe the current curriculum, thus creating a discrepancy model for current and proposed science curriculum.

---

Goals (e.g., learn science knowledge and processes)
Rationale (e.g., social change, scientific advances)
Grade Levels (e.g., K–6, 10–12)
Time Requirements (e.g., 20 Min./Day at K–2; 55 Min./Day at 10–12)
Student Population (e.g., all students, at-risk)
Type of Schools (e.g., urban, rural)
Academic Subjects (e.g., life science, integrated)
Curriculum Emphasis (e.g., STS, fewer topics in greater depth)
Relationship to Other Subjects (e.g., complements health, supports reading)
Curriculum Materials (e.g., textbook, student modules)
Instructional Emphasis (e.g., reading, active learning)
Instructional Strategies (e.g., hands-on, cooperative learning)
Instructional Model (e.g., BSCS 5Es Model, Learning Cycle)
Educational Equipment (e.g., kits, local equipment)
Educational Courseware (e.g., MBL, telecommunications)
Assessment (e.g., built into instruction, portfolios, end-of-unit tests, performance-based)
Implementation (e.g., concerns-based adoption model, staff development)

---

**Figure 2: Some characteristics of a curriculum framework for science**

This discussion of the development of curriculum frameworks makes the general and abstract nature of the goals of developing scientific literacy more specific and concrete, thus taking an initial step towards the problem of achieving a common goal and accommodating the unique requirements of different regions or countries.

## TRANSFORMING FRAMEWORKS TO SCIENCE CURRICULUM

In this section, we discuss some issues associated with the development of materials for school science programs. Our discussion might be general due to the international audience and the broad range of criteria and constraints that individual science educators must consider. We begin with the premise that transforming criteria described in curriculum frameworks for science programs represents the best efforts of individuals within the constraints of time and budget and the requirements of their countries or regions within their countries.

In some situations, individual science teachers or teams will develop new materials based on their curriculum framework. We only can imagine that these materials can range from single lessons used in rural schools by teachers who have little science background, to complete science programs used in regions or entire countries. The curriculum frameworks should be invaluable resources in the design of science curriculum; that is why we placed significant emphasis on that section. However, when it comes to the actual development of materials, we have several recommendations. First, teams of scientists, science educators, and science teachers should be used to design and develop the materials. Second, individuals who are not associated directly with development should review the materials for accuracy of science content, usability of the materials by science teachers, and alignment of curricular components such as teaching methods and assessment strategies. Third, materials should be field tested. If possible, curriculum developers should include at least one round of field testing and revision during program development.

Some schools, regions or countries will adapt curriculum materials for their unique setting. In the 1960s and 1970s, for example, over 20 different countries adapted materials from the Biological Sciences Curriculum Study (BSCS). Adaptation of science curriculum should include more than translation of the program from one language to another. The curriculum framework can establish the 'closeness of fit' or 'compatibility' of a current science curriculum with the specifications of the school, region, or country. The probability of identifying a program that is perfectly compatible with specifications is quite low, which is the reason for our recommendation to adapt a program. Using similar characteristics and comparing the extant program with the curriculum framework will identify those aspects of the program that require modifications and perhaps additions. Depending on the degree of adaptation, the new curriculum might need review, field testing and evaluation.

## TRANSFORMING CLASSROOM PRACTICES

Changing the science curriculum implies changes in teaching science. In fact, consistent with the theme of this chapter, the final translation of a curriculum into actual classroom practices must be considered an essential aspect of any curriculum development effort. Science teachers must make the final decisions about the use of science materials in their classrooms. Of necessity, they will make decisions based on unique aspects of the classroom. Some of those unique aspects include the basic needs of students, students' conceptual levels and developmental stages, available resources, background of the science teacher, time available for instruction, the teacher's understanding of science, and assumptions about students' learning.

We associate terms such as 'implementation' and 'staff development' with this final step of curriculum reform. Based on research (Fullan 1982; Hall & Hord 1987; James & Hord 1988), we recommend that implementation and staff development should be a significant consideration in the development of any science curriculum. We recognise the complicated nature, time, and constraints when considering the diversity of schools and science curricula. But, teachers must understand science content and pedagogy. They should understand the philosophy and materials of the program, and the expected outcomes and assessment strategies of the program. Some of the responsibility for full implementation of a science curriculum belongs to the curriculum developers; the curriculum must be complete, accurate and provide the appropriate resources and support for teachers. Others will have to assume responsibility for administrative support, staff development, and the provision of materials and equipment.

## SOME PRINCIPLES FOR DESIGNING AND DEVELOPING SCIENCE CURRICULUM

This section presents some fundamental principles that transcend differences in economic, developmental, political ideology and religious preference. Core ideas about the science curriculum that could facilitate constructive adaptation and cooperative development are identified. These principles are presented from the perspective of science curriculum, but our general view is that they represent a synthesis that incorporates many issues confronting the international community in the reform of science education.

*Design and Development of Science Curriculum Should Recognise that Student Learning Neither Begins Nor Ends With the Science Curriculum.*

Constructivist literature on learning informs us that students already have conceptions of the natural and designed world and that they have developed explanations for many phenomena. In addition, an individual's science education occurs outside the school and in other settings, such as the family and peer culture, and through other social influences, such as the media and organised religion. Although the form and accuracy of the science education which a student receives from these sources can be questioned, its occurrence and influence cannot be questioned. This suggests the need for recognition of students' current understandings, and the meanings of those understandings for the students, as science programs are designed, developed and implemented.

*Design and Development of Science Curriculum Should Include Strategies, Resources and Materials for Science Content, Teaching, Assessment and Implementation.*

In the past, science educators primarily have focused on the content and secondarily on instruction, leaving assessment and implementation to others or ignoring them completely. The design and development of science curriculum must incorporate assessment as a part of the curriculum and instruction. The development of science curricula must account for who is going to use the science program, where they are going to use the program, and how they are going to learn about the program. This recommendation places an additional burden on those responsible for new science curriculum, but it provides a complete, coherent and consistent program that actually is used. The burden is a challenge but it is worth the effort.

*Design and Development Should Recognise and Incorporate the Human Resources, Especially Science Teachers, in the Transformation of Goals to Practices in Order to Optimise the Classroom Learning Environment and Enhance the Development of the Learner.*

Any science curriculum must meet general criteria, such as accurately representing science concepts and methods of inquiry and appropriately accommodating students' learning and development. Curriculum developers have some responsibility to recognise science teachers' needs, such as classroom management and program effectiveness, as it is the science teacher who has to establish the connections between the curriculum and students and use those connections to develop meaning and enhance learning. Development of science curriculum must be done with a sensitivity to the environments and situations in which teachers use the materials and a realisation that any new program probably requires more change for those teaching the program than for those designing and developing the program.

Science teachers have the responsibility to adapt materials for their unique situations, to be clear and to modify their teaching to both new science content and educational approaches. The changes in teaching science suggested in this discussion are likely to result in personal and professional stress among the teachers responsible for implementing the science curriculum. On the part of those developing science programs, we must recognise the changes that will occur and design programs that best meet the science teachers' needs, such as management, effectiveness and efficiency. Others in the educational system must assume responsibility for supporting the implementation process through adequate materials, supplies, equipment, and staff development. We cannot avoid the personal and professional stress associated with change, but we can recognise it and provide adequate support for the science teacher who must assume responsibility for implementing the program.

*Design and Development of Science Curricula Should Consider the Culture and Educational Context From Which, and Into Which, One Plans to Implement the Curriculum.*

Science educators must strive to maintain the integrity of science, education, and cultural diversity. We can avoid the pitfalls of one region or nation dictating curriculum, even unintentionally. To do this, we must be sensitive to educational needs and requirements and focus on helping the classroom teacher to adapt the curriculum framework to usable materials and adjust current teaching practices so they are more effective. Science education personnel and programs can be thought of as resources. Some nations, regions, territories and schools have more resources than others. The critical issue is not the existence of resources, but the way in which the resources are used, the way in which they are exported and imported, and the way in which they are modified to meet the needs of science teachers.

## CONCLUSION

In this chapter, we have attempted to synthesise many excellent ideas from diverse individuals who, over the years, have addressed the general theme of science curriculum. Collectively, they convey a message of hope that we can help students progress towards a goal of scientific literacy. Individually, the work of reforming science programs will take the creativity, insight, knowledge and skills of those who understand the unique requirements of their nations, schools and teachers.

## REFERENCES

Biological Sciences Curriculum Study: 1993, *Developing Biological Literacy*, Author. Colorado Springs, CO.
Bybee, R.W.: 1993, *Reforming Science Education: Social Perspectives and Personal Reflections*, Teachers College Press, New York.
Bybee, R.W.: 1995, 'Achieving Scientific Literacy', *The Science Teacher* 63, 28–33.
Bybee, R.W.: 1997, *Achieving Scientific Literacy. From Purpose to Practice*. Portsmouth, NH: Heinemann.
Bybee, R.W. & DeBoer, G.E.: 1994, 'Goals for the Science Curriculum', in D. Gabel (ed.), *Handbook of Research on Science Teaching and Learning*. National Science Teachers Association, Washington, DC, 357–387.

DeBoer, G.E.: 1991, *A History of Ideas in Science Education*, Teachers College Press, New York.
Fensham, P.: 1985, 'Science for All', *Journal of Curriculum Studies* 17, 415–435.
Fensham, P.: 1986/87, 'Science for All', *Educational Leadership* 44, 18–23.
Fensham, P.: 1988, 'Science for All: A Vision Splendid', *Proceedings of the Royal Society of New Zealand* 116, 191–197.
Fullan, M.: 1982, *The Meaning of Educational Change*, Teachers College Press, Columbia University, New York.
Hall, G. & Hord, S.: 1987, *Change in Schools: Facilitating the Process*, State University of New York Press, Albany, NY.
James, R. & Hord, S.: 1988, 'Implementing Elementary School Science Programs', *School Science and Mathematics* 88, 315–334.
Layton, D., Jenkins, E. & Donnelly, J.: 1993, *Scientific and Technological Literacy, Meanings and Rationales: An Annotated Bibliography*, Unpublished manuscript developed for UNESCO, University of Leeds, Leeds, UK.
Organisation for Economic Cooperation and Development 1996: 1996, *Changing the Subject: Innovation in Science, Mathematics, and Technology Education*, Routledge, New York.
Roberts, D.A. (ed.): 1982, *Science and Society Teaching Units*, Toronto Institute for Studies in Education Press, Toronto, Canada.
Uno, G.E. & Bybee, R.W.: 1994, 'Understanding the Dimensions of Biological Literacy', *BioScience* 44, 553–557.

# 4.5 Integrated Science and Mathematics Education: Evolution and Implications of a Theoretical Model

DONNA F. BERLIN and ARTHUR L. WHITE
*The Ohio State University, Columbus, USA*

Post-Sputnik curriculum reform efforts have focused on science and mathematics education for a variety of reasons, including low levels of achievement, international comparisons, economic competitiveness, workplace preparation and the need for scientifically and mathematically informed citizens. These political, economic and quality of life concerns become more critical as the public and the profession become more vocal in calling for improved teaching and learning of science and mathematics.

In this climate of educational change, the integration of science and mathematics has been suggested as a promising path toward improved student understanding, performance and attitudes related to science and mathematics. Science and mathematics are related naturally and logically in the real world. The integration of science and mathematics teaching and learning has the potential to help students perceive the relevance and power of this union.

The chapter begins with a discussion of general integration models from which emerges the need for a model specific to the integration of school science and mathematics. A short historical review of the conceptual development of the Berlin-White Integrated Science and Mathematics (BWISM) Model is followed by a comprehensive discussion of the components of the model. Two strategies to facilitate the infusion of integrated science and mathematics into classroom practice are discussed. The challenge of developing assessment strategies to measure student outcomes associated with integrated science and mathematics experiences is then explored. The chapter closes with recommendations for research and development efforts which are grounded in a theoretical framework and which aim to realise the potential of integrated science and mathematics teaching and learning.

## THEORETICAL MODELS

Today, one has little difficulty finding proponents of science and mathematics integration. Current USA reform documents recommend integration of content

---

Chapter Consultant: Mary Budd Rowe (Stanford University, USA)

and instruction in a changing curriculum. Documents such as *Project 2061* (American Association for the Advancement of Science 1989), *Science for All Americans* (Rutherford & Ahlgren 1990), *Benchmarks for Science Literacy* (American Association for the Advancement of Science 1993), *National Science Education Standards* (National Research Council 1996), *Curriculum and Evaluation Standards for School Mathematics* (National Council of Teachers of Mathematics 1989) and *Reshaping School Mathematics: A Philosophy and Framework for Curriculum* (National Research Council 1990) recognise the integration of science and mathematics as a necessary component of reform:

> The alliance between science and mathematics has a long history, dating back centuries. Science provides mathematics with interesting problems to investigate, and mathematics provides science with powerful tools to use in analyzing data. . . . Science and mathematics are both trying to discover general patterns and relationships, and in this sense they are part of the same endeavor. (Rutherford & Ahlgren 1990, pp. 16–17)

> Since mathematics is both the language of science and a science of patterns, the special links between mathematics and science are far more than just those between theory and applications. The methodology of mathematical inquiry shares with the scientific method a focus on exploration, investigation, conjecture, evidence, and reasoning. Firmer school ties between science and mathematics should especially help students' grasp of both fields. (National Research Council 1990, pp. 44–45)

The National Curriculum in England and Wales, although conceived in terms of traditional subject areas, recognises the importance of cross-curriculum themes and the place of interdisciplinary education. A number of writers have voiced fundamental concern for the subject-matter orientation of the National Curriculum and have argued for integrated subject experiences that are both participatory and experiential (Buck & Inman 1993; Hargreaves 1994; Kerry & Eggleston 1994).

In general, there is rhetorical agreement for integration of school science and mathematics and connecting both subjects to the real world. However, the literature is vague on what constitutes 'integration' as indicated by the plethora of terms used to describe it (e.g., connected, coordinated, correlated, cross-disciplinary, interrelated, multidisciplinary and unified). This vagueness indicates the absence of an agreed-upon theoretical view, which is needed to provide a language and conceptualisation of integrated science and mathematics education to direct research, further curricular discussion and contribute to the improvement of science and mathematics education. Given this goal, below we first describe some general integration models followed by a model specific to integrated school science and mathematics.

*General Models*

Fogarty (1991a, 1991b) assumes a continuum of models for integrating curricula. At one end of the continuum are models of integration within single disciplines; the other end includes models that integrate within learners and across networks of learners. The middle of the continuum includes models that integrate across disciplines and is most relevant to the present discussion. These models are defined as follows (Fogarty 1991a, p. xv):

- *Sequenced model.* Topics or units are rearranged and sequenced to coincide with one another. Similar ideas are taught in concert while the subjects remain separate.
- *Shared model.* Shared planning and teaching take place in two disciplines in which overlapping concepts emerge as organising elements.
- *Webbed model.* A fertile theme is webbed to curriculum contents and disciplines; subjects use the theme to sift out appropriate concepts, topics and ideas.
- *Threaded model.* The metacurricular approach threads thinking skills, social skills, multiple intelligences, technology and study skills through the various disciplines.
- *Integrated model.* This interdisciplinary approach matches subjects for overlaps in topics and concepts with some team teaching in an authentic integrated model.

Both Ost (1975) and Dossey (1994) offer general models of integrated curricula with a focus on science and mathematics education.

Ost operationally defines six models related to science, mathematics and social studies:

- *Interdisciplinary model.* 'Interdisciplinary is generally applied to courses of study which merge for purposes of instructional expediency two or more bodies of knowledge' (p. 50). Themes often are in response to student demand and reflect current social issues and concerns (e.g., environmental issues). The result is often two or more mini-courses that are team taught with little attempt to integrate the content conceptually. This model requires students to make connections between disciplines.
- *Unified model.* 'The term unified is usually applied only to courses or programs within which unifying themes or concepts can be developed' (p. 50). For example, the concept of energy could serve to unify the sciences (life, earth and physical). Often there is a sound theoretical basis for this connection grounded in the structure of the discipline.
- *Integrated model.* 'Integrated most often is used in connection with mathematics and science, although in recent years other disciplines are being included under the umbrella of integration' (p. 51). Integrated is 'really interdisciplinary, but at a more sophisticated level' (p. 51). However, the mathematics is often applied mathematics. Although conceptually integrated into the curriculum, it could be taught separately.

- *Correlated model.* 'Correlated is a term used to describe attempts made to relate skills or concepts from one discipline to the other' (p. 51). In this approach, mathematics retains its integrity while providing appropriate skills to be applied in other disciplines (e.g., physics). It also can provide connections between courses for students, enhancing both relevance and meaning.
- *Coordinated model.* 'Coordinated programs are probably attempts to remove redundancy' (p. 52). Programs, textbooks or individual course content can be coordinated to increase efficiency and focus on materials in a conceptual, developmental or systematic (e.g., spiral) way.
- *Comprehensive problem-solving model.* Using the unit concept, 'the student defines a comprehensive problem and is asked to apply various skills and knowledge from the sciences, mathematics, and social sciences in an attempt to optimize some solution' (p. 52).

Dossey (1994) describes five models related to integrated curricula. In the *simultaneous model*, students take courses from the selected disciplines at the same time and attempts are made to connect the content in both courses. Similar to the spiral curriculum, the *braided model* involves strands derived from the content of the disciplines that are taught in cyclical patterns. The *topical model* involves the development of curriculum consisting of specific topics taught during specific years. The *unified model* identifies unifying ideas (e.g., functions, structure) that relate concepts, principles and skills from the disciplines (e.g., science and mathematics). The fifth model, a full *interdisciplinary model*, completely merges the content of the disciplines based on a topic or unit.

*Integrated Science and Mathematics Model*

A review of the literature reveals general models of integrated curricula, but none specific to science and mathematics education. As a first step in creating an infrastructure for an integrated science and mathematics model, a comprehensive bibliography of theoretical, research, curricular and instructional literature was compiled (Berlin 1991). A cursory look at the bibliography reveals an assortment of terms used to refer to 'integration'. The definition of integration is not simply a question of semantics; it points to fundamental questions for teachers and researchers. Can a model be devised to define the critical components and issues for integrated science and mathematics? Can such a model be translated into curricula, implemented in practice and evaluated through systematic research?

To this comprehensive literature review, we added the results of our empirical research related to brain processing, the development of verbal and visuo-spatial ability, decision-making skills, the integration of technology with other instructional resources and the assessment of pattern recognition ability. This research, along with reviews of current integrated science and mathematics programs and projects, further contributed to the delineation of the aspects of the model.

In 1991, a conference entitled 'A Network for Integrated Science and

Mathematics Teaching and Learning' was held at The Johnson Foundation Wingspread facility in Wisconsin with sponsorship from the National Science Foundation, the School Science and Mathematics Association and The Johnson Foundation. Attendees included leaders from education, professional associations, curriculum development and the scientific community. This diverse group focused on three tasks: (1) defining and developing a rationale for integrated science and mathematics teaching and learning; (2) listing guidelines for the infusion of integrated science and mathematics teaching and learning into school practice; and (3) identifying high-priority research questions related to integrated science and mathematics teaching and learning. A consensus was not reached related to a definition of integrated science and mathematics teaching and learning although infusion guidelines and a research agenda were generated (Berlin & White 1992). A second working conference was held at The National Center for Science Teaching and Learning, Columbus, Ohio in 1992. A subset of the participants at the Wingspread Conference and others gathered to continue grappling with the definition of integrated science and mathematics teaching and learning.

Based upon these endeavours, the *Berlin-White Integrated Science and Mathematics (BWISM) Model* was developed to include six aspects (Berlin & White 1994, 1995a, 1995b):

- *Ways of learning.* Integration can be based on how students experience, organise and think about science and mathematics. Based on a constructivist/ neuropsychological perspective or rationale, students must do science and mathematics and be actively involved in the learning process.
- *Ways of knowing.* Integrated school science and mathematics can reinforce the cyclical relationships between inductive-deductive and qualitative-quantitative views of the world. In science and mathematics, new knowledge is often produced through a combination of induction and deduction. For this discussion, induction means looking at numerous examples to find a pattern (qualitative) that can be translated into a rule (quantitative). The application of this rule in a new context is deduction.
- *Process and thinking skills.* Integrated science and mathematics can develop processes and skills related to inquiry, problem-solving and higher-order thinking skills. Integration of science and mathematics can focus on ways of collecting and using information gathered by investigation, exploration, experimentation and problem solving. Skills such as classifying, collecting and organising data, communicating, controlling variables, developing models, estimating, experimenting, graphing, inferring, interpreting data, making hypotheses, measuring, observing, recognising patterns and predicting are representative of this aspect.
- *Content knowledge.* Science and mathematics can be integrated in terms of content that is overlapping or analogous. The examination of concepts, principles, laws and theories of science and mathematics reveal ideas that are unique to each discipline and ideas that overlap or are analogous (e.g., the fulcrum of a lever and the mean of a distribution).

- *Attitudes and perceptions.* Integration can be viewed from what children believe about science and mathematics, their involvement and their confidence in their ability to do science and mathematics. Similarities and differences related to scientific and mathematical attitudes/perceptions or 'habits of mind' can be identified. The values, attitudes and ways of thinking shared between science and mathematics education include accepting the changing nature of science and mathematics, basing decisions and actions on data, a desire for knowledge, a healthy degree of scepticism, honesty and objectivity, relying on logical reasoning, willingness to consider other explanations and working together to achieve better understanding.
- *Teaching strategies.* Integration can be viewed from the teaching methods valued by both science and mathematics educators. Integrated science and mathematics teaching should include a broad range of content, provide time for inquiry-based learning, provide opportunities to use laboratory instruments and other tools, provide appropriate uses of technology (e.g., calculators and computers), embed assessment within instruction and maximise opportunities for successful connections between science and mathematics.

The identification and elaboration of these aspects is meant to clarify the characteristics in constant interplay in defining integration. It is expected that the real value will be in identifying the links and overlap among the aspects rather than attending to them in isolation. The BWISM Model is designed to provide a conceptual base and a common language that advances the research agenda, to serve as a template for characterising current resources and to guide in the development of new materials related to integrated science and mathematics teaching and learning.

## STRATEGIES FOR IMPLEMENTATION

This section explores two promising strategies for integrating science and mathematics teaching and learning, namely, the *project approach* and *the infusion of technology*. Individuals simultaneously make use of scientific and mathematical thinking and dispositions in their daily personal/social interactions and endeavours. The capacity to solve problems and make informed decisions is dependent upon a basic understanding of science, mathematics and the relationships between them. To provide knowledge and skills to deal with real-world issues and concerns, it seems logical to contextualise school learning experiences through integrated experiences that mirror everyday situations and environments.

### The Project Approach

The project approach provides students with real-world problems in which the conceptual and procedural knowledge of science and mathematics are functionally interrelated. As reported by Kerry and Eggleston (1994), the historical roots of the project approach date back to the 1700s and Rousseau's philosophy of discovery learning and child-centred experiences. At the turn of the century, both Dewey (1910) and Kilpatrick (cited in Kerry & Eggleston 1994) promoted the 'project method', which is characterised by interdisciplinary activities, real-world experiences, students actively working on their own problems at their own pace, the interplay of content and process and the role of the teacher as facilitator and guide.

The British primary school system, influenced by these philosophers and the Plowden Report of 1967, has favoured use of the project method or integrated topic work to give 'coherence to what is studied' (Kerry & Eggleston 1994, p. 192; see also Ryan 1994). Although the Plowden Report (Department of Education and Science 1967) discussed the curriculum under traditional subject headings (e.g., mathematics, science), it advocated a more subject-integrated approach to teaching. 'A school is not merely a teaching shop, it must transmit values and attitudes. ... The school sets out deliberately to devise the right environment for children, to allow them to be themselves and to develop in the way and at the pace appropriate to them. ... It lays special stress on individual discovery, on first-hand experience and on opportunities for creative work. It insists that knowledge does not fall into neatly separate compartments and that work and play are not opposite but complementary' (Department of Education and Science 1967, Vol. 1, pp. 187–188). Implicit in this quote is support for both integrated curricula (e.g., science and mathematics) and integrated instruction (e.g., the project approach).

Although there seems to be a paradox between current trends towards defining the curriculum in terms of subject matter, as in the National Curriculum in England and Wales (Her Majesty's Stationery Office 1988), and separate 'Standards' for science and mathematics in the USA (National Council of Teachers of Mathematics 1989; National Research Council 1996), and integrated curricula and instruction (Skilbeck 1994), we perceive these positions as compatible and complementary. Discipline-specific conceptual and procedural knowledge are cornerstones for developing integrated curricula and instruction. It is the interplay between discipline-specific and general conceptual and procedural knowledge within an integrated instructional environment that enables students to develop connected, coherent knowledge that is both school-based and real-world based.

Compatible with our view is the work at TERC 'oriented to three areas of innovative student project activities, better curricula to support a project environment, and alternative assessment' (Tinker 1994, p. 51). A curriculum designed to support a project-oriented learning environment will 'empower students to undertake original investigations, to do mathematics and science' (1994, p. 50). There is need for basic knowledge and skills (e.g., reading, writing and numeracy) as well as project-essential, subject-specific knowledge and skills taught as separate, compartmentalised subjects. However, it is imperative to provide an integrated, relevant context in which such knowledge and skills can be used, applied and practised in order to facilitate the development of cognitive links that are meaningful and coherent.

It is our belief that current constructivist views of teaching and learning provide support for integrated curricula and the project approach. Constructivist principles have influenced much of the work of the Cognition and Technology Group at

Vanderbilt (1990, 1993) (Bransford & Cognition and Technology Group at Vanderbilt 1994). This group, with goals and methods similar to those we recommend, has been developing and exploring the potential of 'anchored instruction as an approach to integrative collaborative inquiry' (Cognition and Technology Group at Vanderbilt 1993, p. 43). Specially-designed video environments provide a realistic context for student and teacher inquiry, problem solving and reasoning. Mathematics, science, geography, history and literature are integrated in this format. The intent of this integration is to (1) optimise instructional time across subject areas, (2) help students acquire knowledge that is less inert, (3) help students appreciate the power of common concepts and methods of inquiry and (4) help students to adopt multiple perspectives when solving problems (Cognition and Technology Group at Vanderbilt 1993).

The anchors or cases being developed by the Cognition and Technology Group at Vanderbilt are similar to the inquiry-based problems that would be embedded within our project approach and support a project-oriented learning environment. However, unlike the Cognition and Technology Group at Vanderbilt problems, project-based problems would be student-generated and situated within and emerging from the project. The project approach involves students in the identification of a project question that is of special academic interest or of immediate or 'emerging' personal relevance (Brooks & Brooks 1993).

Both problem-based and project-based experiences are not independent of the core or standards-based curricula. Subject matter content and processes should be taught before, during and after these experiences and be responsive to the demands of the curricula and specific problems/projects. As such, the general expectations for student projects include:

- helping students to establish, in their own minds, the need to acquire discipline-specific (science and mathematics) conceptual and procedural knowledge;
- helping students link discipline-specific (science and mathematics) conceptual and procedural knowledge while engaged in real-world problem-solving and decision-making situations;
- providing students with opportunity and motivation for synthesis of knowledge within and across discipline boundaries.

It is politic to identify topics within the science and mathematics core or standards-based curricula that enable student exploration using the project approach. Selecting from criteria generated by Ryan (1994, p. 197), chosen topics should: stimulate work both inside and outside of school; provide for active, concrete experiences; suggest a variety of learning strategies; develop skills in a variety of fields; have immediate relevance to students; embrace the priorities (conceptual and procedural) of the national curriculum; enable process and content to merge; and have potential for discipline progression, continuity and differentiation. To this list, we would add the potential for integrating within and across disciplines. Topics that facilitate the blending of instructional methods (i.e., employ logical, quantitative methods in science and investigative methods in mathematics) optimise integration across disciplines (Steen 1994).

To maximise the success of the project approach, we suggest surveying students to identify potential projects that would be of interest and relevance to students and their community. Such projects enable students to experience 'anchored instruction' or authentic and meaningful application of discipline knowledge (Bransford & Cognition and Technology Group at Vanderbilt 1994). We recommend that students be provided with the opportunity to choose projects as well as strategies to 'do' the projects, and that members of the community with experience and expertise in day-to-day learning and application of science and mathematics be involved as collaborative partners.

*Technology Infusion*

In this age of high technology and the information superhighway, the impact of new technologies upon the teaching and learning of science and mathematics presents a challenge and a promise. 'We are now at a stage where teachers and students must move from seeing technology as a source of knowledge (coach, drill) to viewing it as a medium or forum for communication and intelligent adventure' (Hamm 1992, p. 7). Videodiscs, modelling tools, analytic tools, calculator-based and microcomputer-based laboratories, multimedia publishing software and teleconferencing are technological tools that can support a problem-based or project-oriented learning environment in which science and mathematics can be integrated conceptually and procedurally in a meaningful and authentic format (Bransford & Cognition and Technology Group at Vanderbilt 1994; Tinker 1994).

The use of new technologies can break down barriers between science and mathematics (Berger 1994). In a five-year project, Berger and his associates found that student graphical problem-solving ability was enhanced by the use of a computerised data-gathering and data-analysis program (Jackson, Berger & Edwards 1990). The use of technology (computer and calculator) to collect, analyse and display data in a variety of graphics formats can strengthen the connections between science and mathematics and between school and real-world science and mathematics.

Most of the support for integrated curricula enhanced by the project approach and technologies is theoretical and intuitive. Roth (1992) investigated elements that are compatible with those posited in this chapter, focusing upon an innovative physics course integrating mathematics, science and technology. Students engaged in experiments coupled with two heuristic devices to promote student reflection: (1) the concept map, which is a graphical way to represent knowledge; and (2) the epistemological vee, which is a method for attending to both conceptual and methodological sides of an investigation. Students framed their own questions and used graphics calculators, computer statistical analysis packages and advanced science and engineering software packages to process the information they gathered.

The study involved three, grade 11 physics classrooms in an urban, Canadian school of college-bound students. Data included videotapes and student feedback in the form of (1) essays about student attitudes toward the course, the structure of the course and their collaboration in research teams, (2) discussions of positive and negative aspects of the course in the presence of the instructor and (3) responses to the short form of the Individualised Classroom Environment Questionnaire (Fraser 1990) to evaluate differences between preferred and actual classroom environments. A significant difference was found related to personalisation, which reflects students' desire for more individual attention. Essay-type evaluations revealed that students reacted positively towards having the opportunity to choose research topics and the independence to do their own experiments.

## ASSESSMENT

Despite nearly a century of philosophical and intuitive support for integrated school science and mathematics, there is little empirical research documenting its benefits (Berlin 1991, 1995a). There is some evidence of achievement gains and positive attitudinal changes (Friend 1985; Goldberg & Wagreich 1991; Kolb 1967–1968; Kren & Huntsberger 1977; Scarborough 1993a, 1993b; Shann 1977). However, research that is available is based on various definitions of integration and generally relies on traditional assessment measures specific to either science or mathematics. Future research requires development of instruments and procedures specific to integrated science and mathematics experiences.

The goal of an ongoing, collaborative research project with the AIMS Education Foundation is to develop an assessment package that measures the specific understandings, skills and attitudes that students acquire through integrated science and mathematics experiences. Initial efforts identified outcomes for students who had experienced the Activities Integrating Mathematics and Science (AIMS) program. Data were gathered by seven research team leaders and 45 teacher-researchers of grades 4–6 from six States in the USA. There were 2,025 students representing diverse backgrounds. The data consisted of classroom observation journals, teacher responses to standardised, open-ended focus questions and classroom audio and video recordings. A reiterative method of analytic induction was employed. On the basis of this analysis, we enumerated 423 cognitive, 234 affective and 188 social outcomes (Berlin & Hillen 1994).

To explore further the nature of these outcomes, two independent researchers were given data from four sites – two predominantly Anglo and two predominantly Hispanic-American. They were instructed to 'code, sort and classify the outcomes according to category, frequency, intensity and source of evidence' and to report their categorisation process, rationale, results and possible implications for assessment. Log linear and chi square analyses were used to detect relationships between the categories and ethnicity. The following conclusions applied equally, with one exception, at Hispanic-American and Anglo sites (Berlin 1995b):

(1) Teachers tended to identify more cognitive than affective outcomes. However, while able to recognise, describe and provide evidence for affective outcomes,

cognitive outcomes often were identified incorrectly, stated vaguely, of a trivial nature or unrelated to the educational experience.
(2) Few outcomes were linked to underlying science or mathematics concepts. The teacher-researchers might not have attended to underlying concepts due to the traditional proclivity to 'observe' cognitive outcomes through paper-and-pencil tests. There could have been insufficient emphasis on the underlying concepts in the integrated activities, or these primary and middle school teachers might have had inadequate mathematics and/or science backgrounds.
(3) Approximately 75 percent of the process-skills outcomes reported dealt with lower-level skills (e.g., observing, measuring and using numbers). Few outcomes were generated for such higher-level process skills as inferring, predicting, formulating hypotheses, identifying and controlling variables, collecting and interpreting data, experimenting, applying and constructing models. Perhaps, the teacher-researchers were more familiar with the lower-level process-skills or the objectives of the AIMS activities focused more on those skills.
(4) Of the cognitive-skills outcomes identified, more than 85 percent were lower-level, such as knowledge and comprehension, rather than higher-level cognitive skills of application, analysis, synthesis and evaluation.
(5) Of the affective outcomes, more that 85 percent were lower-level skills, such as receiving and responding, rather than higher-level affective skills of valuing, organising and characterising.
(6) In contrast to the similarities previously observed between predominantly Hispanic-American and Anglo sites, personal motivation/emotion outcomes were more frequently identified at the Hispanic-American sites and cooperation outcomes more frequently at the Anglo sites.

From these conclusions, we make the following suggestions for developing an assessment package for integrated science and mathematics experiences. Appropriate assessment should:

- include definitions and examples of mathematics and science concepts and processes;
- deliberately and explicitly attend to science and mathematics concepts;
- include affective outcomes;
- provide opportunities for students to use the three most frequently identified process skills and expand and extend the repertoire to include higher-level skills;
- attend to higher-level cognitive and affective skills;
- include cooperative group activities, with opportunities for both group and individual assessment.

CONCLUSIONS

In this chapter, we have traced the roots of the Berlin-White Integrated Science and Mathematics (BWISM) Model. BWISM can contribute to the conceptualisation and organisation of needed research and development aimed at realising the teaching and learning potential of integrated science and mathematics. In designing and implementing this agenda, we recommend that research and development efforts should:

- be grounded in a theoretical framework such as the Berlin-White Integrated Science and Mathematics (BWISM) Model;
- promote the development of discipline-specific conceptual and procedural knowledge identified in core or standards-based curricula;
- realise the power and coherence of overlapping or analogous conceptual and procedural knowledge;
- explore integrated, collaborative, inquiry-based learning environments such as problem-based learning and the project approach;
- harness the potential of information and communication technologies to collect, organise, analyse and communicate data and advance student and teacher knowledge;
- encourage the development and use of alternative assessments for student outcomes specific to integrated science and mathematics experiences;
- determine when and how best to integrate science and mathematics teaching and learning.

REFERENCES

American Association for the Advancement of Science: 1989, *Project 2061: Science for All Americans. Summary*, Author, Washington, DC.
American Association for the Advancement of Science: 1993, *Benchmarks for Science Literacy*, Author, Oxford University Press, New York.
Berger, C.: 1994, 'Breaking What Barriers Between Science and Mathematics? Six Myths from a Technological Perspective', in D.F. Berlin (ed.), *NSF/SSMA Wingspread Conference: A Network for Integrated Science and Mathematics Teaching and Learning. Conference Plenary Papers* (School Science and Mathematics Association Topics for Teachers Series Number 7), National Center for Science Teaching and Learning, Ohio State University, Columbus, OH, 23–27.
Berlin, D.F.: 1991, *Integrating Science and Mathematics in Teaching and Learning: A Bibliography* (School Science and Mathematics Association Topics for Teachers Series Number 6), ERIC Clearinghouse for Science, Mathematics and Environmental Education, Columbus, OH.
Berlin, D.F.: 1995a, *A Bibliography of Integrated Science and Mathematics Teaching and Learning Literature: 1991–1994* (second edition), Unpublished manuscript, National Center for Science Teaching and Learning, Ohio State University, Columbus, OH.
Berlin, D.F.: 1995b, 'Practitioners and Researchers Identifying and Analyzing Student Outcomes: Toward the Development of a Culturally Appropriate Assessment Package for Integrated Science and Mathematics', Paper presented at the annual meeting of the National Association for Research in Science Teaching, San Francisco, CA.
Berlin, D.F. & Hillen, J.A.: 1994, 'Making Connections in Mathematics and Science: Identifying Student Outcomes', *School Science and Mathematics* 94, 283–290.
Berlin, D.F. & White, A.L.: 1992, 'Report from the NSF/SSMA Wingspread Conference: A Network for Integrated Science and Mathematics Teaching and Learning', *School Science and Mathematics* 92, 340–342.
Berlin, D.F. & White, A.L.: 1994, 'The Berlin-White Integrated Science and Mathematics Model', *School Science and Mathematics* 94, 2–4.
Berlin, D.F. & White, A.L.: 1995a, 'Connecting School Science and Mathematics', in P.A. House & A.F. Coxford (eds.), *Connecting Mathematics Across the Curriculum. NCTM 1995 Yearbook*, National Council of Teachers of Mathematics, Reston, VA, 22–33.

Berlin, D.F. & White, A.L.: 1995b, 'Using Technology in Assessing Integrated Science and Mathematics Learning', *Journal of Science Education and Technology* 4, 47–56.

Bransford, J.D. & Cognition and Technology Group at Vanderbilt: 1994, 'Video Environments for Connecting Mathematics, Science and Other Disciplines', in D.F. Berlin (ed.), *NSF/SSMA Wingspread Conference: A Network for Integrated Science and Mathematics Teaching and Learning. Conference Plenary Papers* (School Science and Mathematics Association Topics for Teachers Series Number 7), National Center for Science Teaching and Learning, Ohio State University, Columbus, OH, 29–48.

Brooks, J.G. & Brooks, M.G.: 1993, *In Search of Understanding: The Case for Constructivist Classrooms*, Association for Supervision and Curriculum Development, Alexandria, VA.

Buck, M. & Inman, S.: 1993, 'Provision for Personal and Social Development Through the Cross-Curricular Themes: A Framework', in P. O'Hear & J. White (eds.), *Assessing the National Curriculum*, Paul Chapman Publishing, London, 103–111.

Cognition and Technology Group at Vanderbilt: 1990, 'Anchored Instruction and its Relationship to Situated Cognition', *Educational Researcher* 19(6), 2–10.

Cognition and Technology Group at Vanderbilt: 1993, 'Toward Integrated Curricula: Possibilities from Anchored Instruction', in M. Rabinowitz (ed.), *Cognitive Science Foundations of Instruction*, Lawrence Erlbaum, Hillsdale, NJ, 33–55.

Department of Education and Science: 1967, *Children and their Primary Schools, A Report of Central Advisory Council for Education* (Plowden Report; two volumes), Her Majesty's Stationery Office, London.

Dewey, S.: 1910, *How We Think*, Heath, Boston, MA.

Dossey, J.: 1994, 'Mathematics and Science Education: Convergence or Divergence?', in D.F. Berlin (ed.), *NSF/SSMA Wingspread Conference: A Network for Integrated Science and Mathematics Teaching and Learning. Conference Plenary Papers* (School Science and Mathematics Association Topics for Teachers Series Number 7), National Center for Science Teaching and Learning, Ohio State University, Columbus, OH, 13–21.

Fogarty, R.: 1991a, *The Mindful School: How to Integrate the Curricula*, Skylight Publishing, Palatine, IL.

Fogarty, R.: 1991b, 'Ten Ways to Integrate Curriculum', *Educational Leadership* 49, 61–65.

Fraser, B.J.: 1990, *Individualised Classroom Environment Questionnaire*, Australian Council for Educational Research, Melbourne, Australia.

Friend, H.: 1985, 'The Effect of Science and Mathematics Integration on Selected Seventh Grade Students' Attitudes Toward and Achievement in Science', *School Science and Mathematics* 85, 453–461.

Goldberg, H. & Wagreich, P.: 1991, 'A Model Integrated Mathematics Science Program for the Elementary School', *International Journal of Educational Research* 14, 193–214.

Hamm, M.: 1992, 'Achieving Scientific Literacy Through a Curriculum Connected with Mathematics and Technology', *School Science and Mathematics* 92, 6–9.

Hargreaves, D.: 1994, 'Coherence and Manageability: Reflections on the National Curriculum and Cross-Curricular Provision', in A. Pollard & J. Bourne (eds.), *Teaching and Learning in the Primary School*, Routledge, London, 184–187.

Her Majesty's Stationery Office: 1988, *Education Reform Act*, Author, London.

Jackson, D., Berger, C.F. & Edwards, B.: 1990, 'The Use of Computers to Aid in Teaching Graphics: Analysis, Software Design and Student Response Analysis', Paper presented at the annual meeting of the National Association for Research in Science Teaching, Atlanta, GA.

Kerry, T. & Eggleston, J.: 1994, 'The Evolution of the Topic', in A. Pollard & J. Bourne (eds.), *Teaching and Learning in the Primary School*, Routledge, London, 188–193.

Kolb, J.R.: 1967–1968, 'Effects of Relating Mathematics to Science Instruction on the Acquisition of Quantitative Science Behaviors', *Journal of Research in Science Teaching* 5, 174–182.

Kren, S.R. & Huntsberger, J.P.: 1977, 'Should Science Teaching be used to Teach Mathematical Skills?', *Journal of Research in Science Teaching* 14, 557–561.

National Council of Teachers of Mathematics: 1989, *Curriculum and Evaluation Standards for School Mathematics*, Author, Reston, VA.

National Research Council: 1990, *Reshaping School Mathematics: A Philosophy and Framework for Curriculum*, National Academy Press, Washington, DC.

National Research Council: 1996, *National Science Education Standards*, Author, Washington, DC.

Ost, D.H.: 1975, 'Changing Curriculum Patterns in Science, Mathematics and Social Studies', *School Science and Mathematics* 75, 48–52.

Roth, W.M.: 1992, 'Bridging the Gap Between School and Real Life: Toward an Integration of Science, Mathematics, and Technology in the Context of Authentic Practice', *School Science and Mathematics* 92, 307–317.

Rutherford, F.J. & Ahlgren, A.: 1990, *Science for All Americans*, Oxford University Press, New York.

Ryan, A.: 1994, 'Preserving Integration with the National Curriculum in Primary Schools: Approaching a School Development Plan', in A. Pollard & J. Bourne (eds.), *Teaching and Learning in the Primary School*, Routledge, London, 194–205.

Scarborough, J.D.: 1993a, 'PHYS-MA-TECH: Integrated Models for Teachers', *The Technology Teacher* 52, 26–30.

Scarborough, J.D.: 1993b, 'PHYS-MA-TECH: Operating Strategies, Barriers, and Attitudes', *The Technology Teacher* 52, 35–38.

Shann, M.H.: 1977, 'Evaluation of an Interdisciplinary, Problem Solving Curriculum in Elementary Science and Mathematics', *Science Education* 61, 491–502.

Skilbeck, M.: 1994, 'The Core Curriculum: An International Perspective', in A. Pollard & J. Bourne (eds.), *Teaching and Learning in the Primary School*, Routledge, London, 167–171.

Steen, L.A.: 1994, 'Integrating School Science and Mathematics: Fad or Folly?', in D.F. Berlin (ed.), *NSF/SSMA Wingspread Conference: A Network for Integrated Science and Mathematics Teaching and Learning. Conference Plenary Papers* (School Science and Mathematics Association Topics for Teachers Series Number 7), National Center for Science Teaching and Learning, Ohio State University, Columbus, OH, 7–12.

Tinker, R.: 1994, 'Integrating Mathematics and Science', in D.F. Berlin (ed.), *NSF/SSMA Wingspread Conference: A Network for Integrated Science and Mathematics Teaching and Learning. Conference Plenary Papers* (School Science and Mathematics Association Topics for Teachers Series Number 7), National Center for Science Teaching and Learning, Ohio State University, Columbus, OH, 49–51.

# 4.6 The Learning Cycle Approach as a Strategy for Instruction in Science

MICHAEL R. ABRAHAM
*University of Oklahoma, Norman, USA*

The focus of this chapter is instructional strategies designed to teach science concepts, particularly an instructional strategy called the Learning Cycle Approach. A wide variety of instructional strategies can be characterised as consisting of one or more of three phases: (1) identification of a concept; (2) demonstration of the concept; and (3) application of the concept. Many science educators have recommended instructional strategies consisting of these phases, although some writers have subdivided the phases into more components (Bybee & Landes 1990; Hewson 1981; Karplus & Thier 1967; Renner 1982; Torrance 1979). Instructional strategies differ in how these three phases are arranged, the formats of the activities in each of the phases, and the number of the different phases utilised in instruction.

In order to illustrate these ideas in this chapter, two instructional strategies are compared with regard to phases of instruction (Renner 1982). In the first strategy, called the traditional approach, students first are informed of what they are expected to know. The informing is accomplished via a textbook, a lecture or other media. Next, the concept is verified or confirmed for or by the student by demonstrating the truth of the concept. In science, this often is accomplished by using laboratory activities. Finally, the student answers questions, solves problems or engages in some form of practice with the new information. The 'inform-verify-practice' (I–>V–>P) sequence corresponds to the three phases previously discussed.

## THE LEARNING CYCLE

The learning cycle approach is an inquiry-based instructional strategy which also divides instruction into three phases. First, in the exploration phase (E), students are given experience with the concept to be developed, often involving a laboratory experiment. Second, in the conceptual invention phase (I), the student and/or teacher derives the concept from the data, with this usually being carried out during a classroom discussion. Third, the application phase (A) gives the student the

---

Chapter Consultant: George Glasson (Virginia Polytechnic Institute & State University, USA)

opportunity to explore the usefulness and application of the concept. Each of the three phases of the learning cycle approach corresponds to the three phases of the traditional approach, although the sequence and format vary.

There are several characteristics which, when used in combination, establish the learning cycle approach as a distinct instructional strategy. The most important of these is the presence of all the phases in a specific sequence, E->I->A. The exploration phase comes first, implying that the information exposed will be used inductively during the invention phase. This is not to deny the role that prior knowledge and deductive thought play during exploration. Indeed Lawson (1988) discusses various kinds of thinking processes active during exploration. Because the exploration in science instruction is most commonly a laboratory activity, the data generated by the laboratory activity will be generalised to a concept. Laboratory work takes on a more central role in instruction because it is used as an introduction to a concept in the learning cycle approach. Traditional laboratory activities used to confirm concepts are more peripheral to the main focus of instruction.

An example of a simple learning cycle lesson is an instructional unit designed to teach the concept of acids and bases. Students are provided with a number of unknown solutions and asked to perform tests on them (e.g., acid/base indicators, conductivity, etc.). During this exploration phase of instruction, the students might be asked to identify patterns which would enable the unknown solutions to be grouped. During a subsequent class discussion, students compare data and notice that the solutions could be classified into three groups. The instructor then might define these groups operationally and label them as acids, bases and neutral substances. Depending on the age and scientific sophistication of the students, the instructor might continue the discussion by introducing the chemical formulae of the solutions and ask what the formulae in each category have in common. This information could be used to invent a theoretical definition of acids and bases. Using these ideas, students might be given application activities involving additional solutions.

## HISTORICAL AND THEORETICAL PERSPECTIVES

The learning cycle approach as a recognised instructional strategy can be traced to the Science Curriculum Improvement Study (SCIS), a primary school science curriculum project initiated during the late 1950s (Atkin & Karplus 1962). During the development and modification of instructional materials, a theoretical model for instruction began to take shape and the idea of an instructional strategy consisting of phases was proposed (Karplus & Thier 1967). The apparent first use of the term 'learning cycle' appears in early teacher's guides for the SCIS instructional units.

The names of the phases in the learning cycle have changed as their roles have become refined and understood in greater detail, and as the theory has been introduced to various audiences. Originally, the three phases were named 'preliminary exploration, invention, and discovery' (Karplus & Thier 1967). This was modified to 'exploration, concept introduction, and concept application' (Karplus, Lawson, Wollman, Appel, Bernoff, Howe, Rusch & Sullivan 1977).

Since then, many authors have modified the names of the phases (Abraham & Renner 1986; Glasson & Lalik 1993; Good & Lavoie 1986; Lawson 1988; Renner 1985).

The theoretical justification for the learning cycle approach can be found in the history and philosophy of science and in the psychology of learning, especially the developmental psychology of Jean Piaget. Although Piaget's theories are too complex to discuss in detail here, a brief consideration of one aspect of his ideas is provided to clarify how the learning cycle approach is consistent with these ideas.

According to Piaget (1970), human beings have mental structures which interact with the environment. We assimilate or transform information from our environment into our existing mental structures. Our mental structures operate on the assimilated information and transform it in a process of accommodating to it. Thus, information from the environment transforms our mental structures, while at the same time our mental structures transform the information. This change is driven and controlled by the process of disequilibration. When our mental structures have accommodated to the assimilated information, we are in a state of equilibrium and have reached an 'accord of thought with things' (Piaget 1963, p. 8). In accommodating the information, however, the altered mental structure can become disequilibrated with related existing mental structures. The new structure must be organised with respect to the old structures to develop a new equilibrated organisation. In other words, we must bring the 'accord of thought with itself' (Piaget 1963, p. 8).

If learning occurs spontaneously through a process of assimilation accommodation and organisation, then instruction could take advantage by sequencing instructional activities to be compatible. In order to facilitate assimilation, instructional activities should expose the learner to a segment of the environment that demonstrates the information to be accommodated. This should be followed by activities which help the learner to accommodate to the information. Finally, in order to organise the accommodated information, activities should be developed that help the learner to see the relation between the new information and other previously learned information.

## RESEARCH ON THE LEARNING CYCLE

Considerable research concerning the learning cycle approach has been conducted since its origins in the 1960s. Most of the research discussed here is abstracted from Lawson, Abraham and Renner (1989) and updated with studies done since that monograph was published.

### Research on SCIS

A large amount of literature is related to the SCIS program and with Piaget's theory of intellectual development as a theoretical focus. The effect of SCIS instructional materials on the attitudes, achievement and reasoning ability of students is discussed below.

Most studies comparing SCIS to non-SCIS programs found that SCIS was superior in developing attitudes towards science (Brown 1973; Krockover & Malcolm 1978; Lowery, Bowyer & Padilla 1980), better motivation towards learning (Allen 1973a), higher levels of self-concept in the areas of intellect and school status (Malcolm 1976) and more positive attitudes towards experimentation (Lowery, Bowyer & Padilla 1980). When Brown, Weber and Renner (1975) used a measure of attitudes towards science and scientists, no significant difference was found between the attitudes of the SCIS students and those of professional scientists.

Many of the studies involving the SCIS program assessed student achievement in intellectual outcomes such as content and process learning. Bowyer's (1976) study of scientific literacy among 521 rural sixth grade students showed significant gains in basic process skills and content knowledge associated with exposure to the SCIS program. Brown, Weber and Renner (1975) found that SCIS students had superior attainment of scientific process skills relative to non-SCIS students.

Other researchers found that, compared to traditionally-taught students, SCIS students exhibited superior inquiry skills (TaFoya 1976), figural creativity abilities (Brown 1973), ability to isolate and control variables (Linn & Thier 1975), ability to describe objects by their properties, ability to describe similarities and differences between different forms of the same substance, and ability to observe an experiment and use observations to describe what happened in the experiment (Thier 1965). Finally, Nussbaum (1979) found that the SCIS was effective in teaching concepts in a way that was lasting and generalisable, as well as in promoting minor changes in Piagetian developmental level.

Two large-scale longitudinal studies of the effect of the SCIS program on primary school children confirmed the superiority of the SCIS program to non-SCIS approaches in a large number of cognitive factors (Allen 1971, 1972, 1973a, 1973b; Renner, Stafford, Coffia, Kellogg & Weber 1973).

As was the case with many of the curriculum projects produced in the 1960s, teacher education programs accompanied the development of the SCIS curriculum materials and formed the focus for research and evaluation. Some of this research was associated with how teachers utilise the learning cycle approach. Relative to teachers who were not trained to teach SCIS, SCIS teachers spent more time teaching science (Kyle 1985), focused more on higher-order, open-ended questions rather than fact-oriented questions (Moon 1969; Porterfield 1969; Wilson 1969), and were more open-minded, asked more higher-level questions and had students with superior science process achievement (Eaton 1974). Lawlor (1974) found that students of SCIS-trained teachers had better attitudes towards science. Using interaction analysis, Simmons (1974) found that SCIS teachers were more student oriented than non-SCIS teachers.

Because much of the previously cited research compared the SCIS program with a textbook or 'non-SCIS' program, it is possible that some other aspect of the program besides the learning cycle could be responsible for its success. For example, the effectiveness of the SCIS program might be due to its hands-on or laboratory approach.

*Learning Cycle Research*

As a result of the success of the SCIS program, many science educators consider the learning cycle to be a potentially useful model for instruction and curriculum development. Therefore, many groups developed curricula using the learning cycle approach and did research to test the effectiveness of these programs. As with the SCIS program, using the learning cycle in other curricula often provided more positive student attitudes towards science and science instruction relative to other more 'traditional approaches'. This was true at varying levels from primary school to college (Bishop 1980; Campbell 1977; Cumo 1992; Davis 1978; Klindienst 1993; Shadburn 1990; Veath 1989).

It is more difficult to find unambiguous positive results in content achievement using the learning cycle approach. A group of studies either found no difference or mixed results when they compared the learning cycle and traditional approaches (Bishop 1980; Champion 1993; Davis 1978; Vermont 1985; Ward & Herron 1980).

In contrast, a wide range of studies has shown unambiguous improvements among students using the learning cycle approach in terms of process skill development (Berndt 1994; Cumo 1992; Davidson 1989; Veath 1989), content learning attainment (Berndt 1994; Campbell 1977; Kurey 1991; Purser & Renner 1983; Saunders & Shepardson 1987; Schneider & Renner 1980; Shadburn 1990) and in reducing misconceptions (Lawson & Thompson 1988; Marek, Cowan & Cavallo 1994; Scharmann 1991).

Research on the learning cycle approach also has focused on student reasoning abilities. In most cases, 'reasoning ability' was based on the ideas of concrete and formal operational reasoning as defined by Piagetian type tasks. Numerous studies involving science majors and non-science majors and preservice and inservice teachers from a variety of backgrounds and different educational levels have shown that the learning cycle approach can bring about growth in reasoning abilities greater than those of other instructional strategies or those expected from maturation effects alone ( Lawson & Snitgen 1982; Marek & Methven 1991; McKinnon & Renner 1971; Purser & Renner 1983; Renner & Lawson 1975; Renner & Paske 1977; Rubin & Norman 1992; Saunders & Shepardson 1987; Schneider & Renner 1980).

This research, together with the research reported previously on the SCIS program, seems to indicate that the learning cycle approach has real promise in terms of attitudes towards science and science instruction, content achievement and reasoning ability. However, the question remains open as to whether the use of the laboratory within the learning cycle approach is different from and more effective than other instructional approaches that use laboratory or hands-on activities.

*Research on Aspects of the Learning Cycle*

Much of the research reported in the previous sections is subject to the criticism that comparisons of global instructional strategies, such as the learning cycle, with even well-defined 'traditional' approaches do not identify the specific nature or causes of any advantages or disadvantages of instructional strategies. It might be claimed that the approaches are so different that one cannot isolate the critical variables that explain the results. What is needed is research into the aspects of instruction that characterise or define the success of the learning cycle approach. Two of the most critical distinguishing characteristics of the learning cycle approach discussed below are the *inductive use of the laboratory* and *defined phases of instruction*.

Inductive Use of the Laboratory

In a meta-analysis of 39 studies comparing inductive and deductive teaching approaches, Lott (1983) found no overall differences between the two approaches for the whole group of studies, but several interaction effects were apparent. First, inductive approaches had more positive effects on intermediate level students and were superior when higher levels of thought and outcome demands were involved. Second, students in smaller classes containing 17–26 students performed better under inductive approaches. As the size of the class increased, performance differences decreased relative to deductive approaches. Finally, inductive approaches functioned better when they were part of a complete program rather than when they were used in isolated units of instruction.

Phases of Instruction

Two large-scale multi-experiment studies investigated the learning cycle approach in terms of the three phases of instruction. The three research questions involved whether (1) all three phases were necessary, (2) whether there was an optimum order in which the phases were utilised, and (3) whether some formats of instruction (i.e., readings, lectures, laboratories) were more effective than others. Learning cycle lessons in high school physics and chemistry were modified in order to investigate these questions. A large proportion of the physics students had high abstract-reasoning abilities, while the chemistry students were an even mix of abstract and concrete reasoners. The overall results of these experiments suggest that abstract reasoning physics students are not dependent on the sequence and form of instruction for optimum learning as long as all three phases of instruction are present. Concrete physics students and all chemistry students, however, achieve better with instruction consisting of the learning cycle sequence of phases, laboratory-centred formats, and all three phases. Both abstract and concrete reasoning students prefer instruction consisting of the learning cycle sequence of phases, laboratory-centred formats, and all three phases (Abraham 1988–89; Abraham & Renner 1983, 1986; Renner, Abraham & Birnie 1983, 1985, 1988). The apparent discrepancies between the physics and chemistry samples might be explained by the higher proportion of abstract reasoning students in the physics sample and/or differences in the concrete/abstract nature of the disciplines.

*Fine Tuning the Learning Cycle Approach*

In 1988, a symposium at the annual conference of the National Association for Research in Science Teaching provided a retrospective look on the learning cycle approach and suggestions for revising the instructional strategy (Good, Renner, Lawson & Abraham 1988). Since then, there has been added to the literature research findings concerning the learning cycle approach and suggested refinements involving combining the learning cycle with other instructional techniques and concerns.

Two recent surveys of the learning cycle approach assessed its usefulness as an instructional strategy. Gabel and Bunce (1994) concluded that instructional strategies, like the learning cycle, that are compatible with the learning characteristics of students can be expected to have positive effects on achievement. Tobin, Tippins and Gallard (1994) are more critical, claiming that research on the learning cycle has focused on the approach as a model for organising instruction rather than on its effects on teachers' actions. As a consequence, these researchers argue, we don't know enough about the appropriate uses and adaptations for the model under different instructional variables.

Lawson (1988) has examined a wide variety of learning cycle lessons and proposed that learning cycles can be classified as *descriptive, empirical-abductive* or *hypothetical-deductive*. For descriptive learning cycles, students discover and describe an empirical pattern within a specific context. Descriptive learning cycles answer the 'what' question, but do not raise the causal 'why' question. In empirical-abductive learning cycles, observations are made initially in a descriptive fashion, but then generate and initially test one or more causes. Hypothetical-deductive learning cycles are initiated with the statement of a causal question to which the students are asked to generate alternative explanations. Students try to deduce the logical consequences of these explanations and design and conduct experiments to test them.

Westbrook and Rogers (1994) studied different types of learning cycles by modifying the expansion phase of the learning cycle and giving students opportunities for hypothesis testing and/or designing experiments. They found that this modification enhanced process skills and logical thinking abilities among ninth grade physical science students.

It has been suggested that additional phases of instruction might improve the learning cycle. Cognitive science and artificial intelligence theories suggest the need for a 'prediction' phase to be added to the learning cycle as the first phase of instruction (Good 1987). Bybee and Landes (1990) have suggested adding 'engagement' and 'evaluation' phases. The engagement phase is an introductory activity designed to involve the student in the learning cycle lesson.

Nicolo (1994) found that adding cooperative learning activities to learning cycle lessons produced a greater sense of student control and therefore enhanced self-esteem. Rogers (1993) has used concept mapping activities (Novak & Wandersee 1990) in learning cycle lessons with ninth grade physical science students.

Glasson and Lalik (1993) investigated whether the learning cycle approach could be used by teachers to engage students in social constructivist learning. These researchers suggested several modifications designed to make the learning cycle approach more congruent with social constructivist teaching and to avoid the tendency for teachers to treat present knowledge as absolute truth: (1) changing the label for the invention phase to 'clarification' to emphasise the learner's primary role in knowledge construction, (2) using the term 'elaboration' for application phase activities to emphasise divergent problem solving and (3) renaming the learning cycle to Language-Oriented Learning Cycle to emphasise the critical role of discourse and social interaction. Also concerned with constructivist issues, Gallagher (1992) has proposed an instructional strategy (consisting of acquisition of scientific ideas, integration and application) which is compatible with the learning cycle approach and which he believes focuses more on student learning.

Additional research is needed, however, before the effectiveness of these and other suggested modifications can be fully judged.

## SUMMARY

In considering an inquiry-oriented laboratory-based instructional strategy called the learning cycle, this chapter has focused on its basic characteristics, its historical development, and the research rationale for its effectiveness. A considerable body of research has confirmed that the learning cycle is an effective instructional strategy with many advantages over more traditional approaches in terms of student attitudes, motivation, process learning and concept learning. This research suggests that science teachers should utilise instructional materials that have some of the key characteristics of the Learning Cycle Approach. Perhaps the most important conclusion from this consideration of instructional strategies is that merely providing students with hands-on laboratory experiences is not by itself enough. Laboratory activities should be used to introduce concepts so that students are given the opportunity to construct knowledge from their own experience and apply that knowledge to new situations.

## REFERENCES

Abraham, M.R.: 1988–89, 'Research on Instruction Strategies', *Journal of College Science Teaching* 18, 185–187, 200.
Abraham, M.R. & Renner, J.W.: 1983, *Sequencing Language and Activities in Teaching High School*

Chemistry: A Report to the National Science Foundation, Science Education Centre, University of Oklahoma, Norman, OK. (ERIC Document Reproduction Service No. ED 241 267)

Abraham, M.R. & Renner, J.W.: 1986, 'The Sequence of Learning Cycle Activities in High School Chemistry', *Journal of Research in Science Teaching* 23, 121–143.

Allen, L.R.: 1971, 'An Examination of the Ability of First Graders from the Science Curriculum Study Program to Describe an Object by its Properties', *Science Education* 55, 61–67.

Allen, L.R.: 1972, 'An Evaluation of Children's Performance on Certain Cognitive, Affective, and Motivational Aspects of the Interaction Unit of the Science Curriculum Improvement Study', *Journal of Research in Science Teaching* 9, 167–173.

Allen, L.R.: 1973a, 'An Evaluation of Children's Performance in Certain Cognitive, Affective, and Motivational Aspects of the Systems and Subsystems Unit of the Science Curriculum Improvement Study Elementary Science Program', *Journal of Research in Science Teaching* 10, 125–134.

Allen, L.R.: 1973b, 'An Examination of the Ability of Third-Grade Children from the Science Curriculum Improvement Study Elementary Science Program to Identify Experimental Variables and to Recognize Change', *Science Education* 57, 135–151.

Atkin, J.M. & Karplus, R.: 1962, 'Discovery or Invention?', *The Science Teacher* 29, 45–51.

Berndt, J.A.: 1994, 'The Effect of the Learning Cycle in Teaching Natural Resource Sciences in the Elementary School Classroom' (Doctoral Dissertation, West Virginia University, 1994), *Dissertation Abstracts International* 54(11), 4052A.

Bishop, J.E.: 1980, 'The Development and Testing of a Participatory Planetarium Unit Employing Projective Astronomy, Concepts and Utilizing the Karplus Learning Cycle, Student Model Manipulation and Student Drawing with Eighth-Grade Students' (Doctoral Dissertation, University of Akron, 1980), *Dissertation Abstracts International* 41(3), 1010A.

Bowyer, J.A.B.: 1976, 'Science Curriculum Improvement Study and the Development of Scientific Literacy' (Doctoral Dissertation, University of California, Berkeley, 1975), *Dissertation Abstracts International* 37(1), 107A.

Brown, T.W. 1973, 'The Influence of the Science Curriculum Improvement Study on Affective Process Development and Creative Thinking' (Doctoral Dissertation, University of Oklahoma, 1973), *Dissertation Abstracts International* 34(6), 3175A.

Brown, T.W., Weber, M.C. & Renner, J.W.: 1975, 'Research on the Development of Scientific Literacy', *Science and Children* 12, 13–15.

Bybee, R.W. & Landes, N.M.: 1990, 'Science for Life and Living', *The American Biology Teacher* 52, 92–98.

Campbell, T.C.: 1977, 'An Evaluation of a Learning Cycle Intervention Strategy for Enhancing the Use of Formal Operational Thought by Beginning Physics Students' (Doctoral Dissertation, University of Nebraska-Lincoln, 1977), *Dissertation Abstracts International* 38(7), 3903.

Champion, T.D.: 1993, 'A Comparison of Learning Cycle and Expository Laboratory Instruction in Human Biochemistry' (Doctoral Dissertation, University of Northern Colorado, 1993), *Dissertation Abstracts International* 54(4), 1308A.

Cumo, J.M.: 1992, 'Effects of the Learning Cycle Instructional Method on Cognitive Development, Process, and Attitude Toward Science in Seventh Graders' (Doctoral Dissertation, Kent State University, 1991), *Dissertation Abstracts International* 53(2), 387A.

Davidson, M.A.: 1989, 'Use of the Learning Cycle to Promote Cognitive Development' (Doctoral Dissertation, Purdue University, 1988), *Dissertation Abstracts International* 49(11), 3320A.

Davis, J.O.: 1978, 'The Effects of Three Approaches to Science Instruction on the Science Achievement, Understanding, and Attitudes of Selected Fifth and Sixth Grade Students' (Doctoral Dissertation, University of North Carolina, 1977), *Dissertation Abstracts International* 39(1), 211A.

Eaton, D.: 1974, 'An Investigation of the Effects of an In-Service Workshop Designed to Implement the Science Curriculum Improvement Study Upon Selected Teacher-Pupil Behavior and Perceptions' (Doctoral Dissertation, University of West Virginia, 1974), *Dissertation Abstracts International* 35(4), 2096A.

Gabel, D.L. & Bunce, D.M.: 1994, 'Research on Problem Solving: Chemistry', in D.L. Gabel (ed.), *Handbook of Research on Science Teaching and Learning*, Macmillan, New York, 301–326.

Gallagher, J.J.: 1992, 'Secondary Science Teachers and Constructivist Practice', in K. Tobin (ed.), *The Practice of Constructivism in Science Education*, American Association for the Advancement of Science Press, Washington, DC, 181–191.

Glasson, G.E. & Lalik, R.V.: 1993, 'Reinterpreting the Learning Cycle from a Social Constructivist Perspective: A Qualitative Study of Teachers' Beliefs and Practices', *Journal of Research in Science Teaching* 30, 187–207.

Good, R.: 1987, 'Artificial Intelligence and Science Education', *Journal of Research in Science Teaching* 24, 325–342.

Good, R. & Lavoie, D.: 1986, 'The Importance of Prediction in Science Learning Cycles', *Pioneer Journal of the Florida Association of Science Teachers* 1, 24–35.

Good, R., Renner, J.W., Lawson, A.E. & Abraham, M.R.: 1988, 'Research and Conceptual Bases for Revising the Karplus-Renner Learning Cycle . . .', Paper presented at the annual meeting of the National Association for Research in Science Teaching, Lake Ozark, MO.

Hewson, P.W.: 1981, 'A Conceptual Change Approach to Learning Science', *European Journal of Science Education* 3, 383–396.

Karplus, R., Lawson, A.E., Wollman, W., Appel, M., Bernoff, R., Howe, A., Rusch, J.J. & Sullivan, F.: 1977, *Science Teaching and the Development of Reasoning: A Workshop*, Regents of the University of California, Berkeley, CA.

Karplus, R. & Thier, H.D.: 1967, *A New Look at Elementary School Science*, Rand McNally, Chicago, IL.

Klindienst, D.B.: 1993, 'The Effects of the Learning Cycle Lessons Dealing with Electricity on the Cognitive Structures, Attitudes Toward Science and Achievement of Urban Middle School Students' (Doctoral Dissertation, Pennsylvania State University, 1993), *Dissertation Abstracts International* 54(5), 1748A.

Krockover, G.H. & Malcolm, M.D.: 1978, 'The Effects of the Science Curriculum Improvement Study on a Child's Self-Concept', *Journal of Research in Science Teaching* 14, 295–299.

Kurey, M.M.: 1991, 'The Traditional and Learning Cycle Approaches to Performance in High School Chemistry Topics by Students Tested for Piagetian Cognitive Development' (Doctoral Dissertation, Temple University, 1991), *Dissertation Abstracts International* 52(2), 411A.

Kyle, W.C., Jr.: 1985, 'What Research Says: Science Through Discovery: Students Love It', *Science and Children* 23, 39–41.

Lawlor, F.X.: 1974, 'A Study of the Effects of a CCSSP Teacher Training Program on the Attitudes of Children Toward Science', Paper presented at the annual meeting of the National Association for Research in Science Teaching, Chicago, IL.

Lawson, A.E.: 1988, 'A Better Way to Teach Biology', *American Biology Teacher* 50, 266–289.

Lawson, A.E., Abraham, M.R. & Renner, J.W.: 1989, *A Theory of Instruction: Using the Learning Cycle to Teach Science Concepts and Thinking Skills* (Monograph, Number One), National Association for Research in Science Teaching, Kansas State University, Manhattan, KS.

Lawson, A.E. & Snitgen, D.: 1982, 'Teaching Formal Reasoning in a College Biology Course for Pre-Service Teachers', *Journal of Research in Science Teaching* 19, 233–248.

Lawson, A.E. & Thompson, L.D.: 1988, 'Formal Reasoning Ability and Misconceptions Concerning Genetics and Natural Selection', *Journal of Research in Science Teaching* 25, 733–746.

Linn, M.C. & Thier H.D.: 1975, 'The Effect of Experimental Science on Development of Logical Thinking in Children', *Journal of Research in Science Teaching* 12, 49–62.

Lott, G.W.: 1983, 'The Effect of Inquiry Teaching and Advanced Organizers Upon Student Outcomes in Science Education', *Journal of Research in Science Teaching* 20, 437–451.

Lowery, L.F., Bowyer, J. & Padilla, M.J.: 1980, 'The Science Curriculum Improvement Study and Student Attitudes', *Journal of Research in Science Teaching* 17, 327–355.

Malcolm, M.D.: 1976, 'The Effect of the Science Curriculum Improvement Study on a Child's Self-Concept and Attitude Toward Science' (Doctoral Dissertation, Purdue University, 1975), *Dissertation Abstracts International* 36(10), 6617A.

Marek, E.A., Cowan, C.C. & Cavallo, A.M.L.: 1994, 'Students' Misconceptions About Diffusion: How Can They be Eliminated?', *The American Biology Teacher* 56, 74–77.

Marek, E.A. & Methven, S.B.: 1991, 'Effects of the Learning Cycle Upon Student and Classroom Teacher Performance', *Journal of Research in Science Teaching* 28, 41–53.

McKinnon, J.W. & Renner, J.W.: 1971, 'Are Colleges Concerned with Intellectual Development?', *American Journal of Physics* 39, 1047–1052.

Moon, T.C.: 1969, 'A Study of Verbal Behavior Patterns in Primary Grade Classrooms During Science Activities' (Doctoral Dissertation, Michigan State University, 1969), *Dissertation Abstracts International* 30(12), 5325A.

Nicolo, E.: 1994, 'The Effects of Cooperative Learning and the Learning Cycle on Students' Locus-of-Control' (Doctoral Dissertation, Temple University, 1993), *Dissertation Abstracts International* 54(7), 2527A.

Novak, J.D. & Wandersee, J.H. (guest eds.): 1990, 'Special Issue: Perspectives on Concept Mapping', *Journal of Research in Science Teaching* 27(10), 921–1075.

Nussbaum, J.: 1979, 'The Effect of the SCIS's Relativity Unit on the Child's Conception of Space', *Journal of Research in Science Teaching* 16, 45–51.
Piaget, J.: 1963, *The Origins of Intelligence in Children*, Norton, New York.
Piaget, J.: 1970, *Structuralism*, Harper and Row, New York.
Porterfield, D.R.: 1969, 'Influence of Preparation in the Science Curriculum Improvement Study on Questioning Behavior of Selected Second and Fourth-Grade Reading Teachers' (Doctoral Dissertation, University of Oklahoma, 1969), *Dissertation Abstracts International* 30(4), 1341A.
Purser, R.K. & Renner, J.W.: 1983, 'Results of Two Tenth-Grade Biology Teaching Procedures', *Science Education* 67, 85–98.
Renner, J.W.: 1982, 'The Power of Purpose', *Science Education* 66, 709–716.
Renner, J.W.: 1985, *The Learning Cycle and Secondary School Science Teaching*, University of Oklahoma Press, Norman, OK.
Renner, J.W., Abraham, M.R. & Birnie, H.H.: 1983, *Sequencing Language and Activities in Teaching High School Physics: A Report to the National Science Foundation*, Science Education Center, University of Oklahoma, Norman, OK. (ERIC Document Reproduction Service No. ED 238 732)
Renner, J.W., Abraham, M.R. & Birnie, H.H.: 1985, 'The Importance of the Form of Student Acquisition of Data in Physics Learning Cycles', *Journal of Research in Science Teaching* 22, 303–325.
Renner, J.W., Abraham, M.R. & Birnie H.H.: 1988, 'The Necessity of Each Phase of the Learning Cycle in Teaching High School Physics', *Journal of Research in Science Teaching* 25, 39–58.
Renner, J.W. & Lawson, A.E.: 1975, 'Intellectual Development in Pre-Service Elementary School Teachers: An Evaluation', *Journal of College Science Teaching* 5, 89–92.
Renner, J.W. & Paske, W.C.: 1977, 'Comparing Two Forms of Instruction in College Physics', *American Journal of Physics* 45, 851–859.
Renner, J.W., Stafford, D.G., Coffia, W.J., Kellogg, D.H. & Weber, M.C.: 1973, 'An Evaluation of the Science Curriculum Improvement Study', *School Science and Mathematics* 73, 291–318.
Rogers, L.N.: 1993, 'Conceptual Organization in a Learning Cycle Classroom' (Doctoral Dissertation, University of Oklahoma, 1993), *Dissertation Abstracts International* 54(4), 1309A.
Rubin, R.L. & Norman, J.T.: 1992, 'Systematic Modeling Versus the Learning Cycle: Comparative Effects on Integrated Science Process Skill Achievement', *Journal of Research in Science Teaching* 29, 715–727.
Saunders, W.L. & Shepardson, D.: 1987, 'A Comparison of Concrete and Formal Science Instruction Upon Science Achievement and Reasoning Ability of Sixth-Grade Students', *Journal of Research in Science Teaching* 24, 39–51.
Scharmann, L.C.: 1991, 'Teaching Angiosperm Reproduction by Means of the Learning Cycle', *School Science and Mathematics* 91, 100–104.
Schneider, L.S. & Renner, J.W.: 1980, 'Concrete and Formal Teaching', *Journal of Research in Science Teaching* 17, 503–517.
Shadburn, R.G., 1990, 'An Evaluation of a Learning Cycle Intervention Method in Introductory Physical Science Laboratories in Order to Promote Formal Operational Thought Process' (Doctoral Dissertation, University of Mississippi, 1989), *Dissertation Abstracts International* 51(6), 1897A.
Simmons, H.N.: 1974, 'An Evaluation of Attitudinal Changes and Changes in Teaching Behavior of Elementary Teachers Enrolled in Eleven SCIS Workshops Directed by Leadership Teams Trained in a SCIS Leader's Workshop' (Doctoral Dissertation, University of Kansas, 1973), *Dissertation Abstracts International* 34(7), 4068A.
TaFoya, M.E.: 1976, 'Assessing Inquiry Potential in Elementary Science Curriculum Materials' (Doctoral Dissertation, University of Maryland, 1976), *Dissertation Abstracts International* 37(6), 3401A.
Thier, H.D.: 1965, 'A Look at a First-Grader's Understanding of Matter', *Journal of Research in Science Teaching* 3, 84–89.
Tobin, K., Tippins, D.J. & Gallard, A.J.: 1994, 'Research on Instructional Strategies for Teaching Science', in D.L. Gabel (ed.), *Handbook of Research on Science Teaching and Learning*, Macmillan, New York, 45–93.
Torrance, E.P.: 1979, 'A Three-Stage Model for Teaching Creative Thinking', in A.E. Lawson (ed.), *The Psychology of Teaching for Thinking and Creativity: 1980 AETS Yearbook*, ERIC/SMEAC and the Association for the Education of Teachers in Science, Columbus, OH, 226–253.
Veath, M.L.: 1989, 'Comparing the Effects of Different Laboratory Approaches in Bringing About a Conceptual Change in the Understanding of Physics by University Students' (Doctoral Dissertation, University of Wyoming, 1988), *Dissertation Abstracts International* 49(10), 2987A.

Vermont, D.R.: 1985, 'Comparative/Effectiveness of Instructional Strategies on Developing the Chemical Mole Concept', (Doctoral Dissertation, University of Missouri, St. Louis, 1984), *Dissertation Abstracts International* 45(8), 2473.

Ward, C.R. & Herron, J.D.: 1980, 'Helping Students Understand Formal Chemical Concepts', *Journal of Research in Science Teaching* 17, 387–400.

Westbrook, S.L. & Rogers, L.N.: 1994, 'Examining the Development of Scientific Reasoning in Ninth-Grade Physical Science Students', *Journal of Research in Science Teaching* 31, 65–76.

Wilson, J.H.: 1969, 'The "New" Science Teachers are Asking More and Better Questions', *Journal of Research in Science Teaching* 6, 49–53.

Section 5

Learning Environments

BARRY J. FRASER

# 5.1 Science Learning Environments: Assessment, Effects and Determinants

BARRY J. FRASER

*Curtin University of Technology, Perth, Australia*

Although research and evaluation in science education have relied heavily on the assessment of academic achievement and other valued learning outcomes, these measures cannot give a complete picture of the educational process. Because students spend up to 15,000 hours at school by the time they finish senior high school (Rutter, Maughan, Mortimore, Ouston & Smith 1979), students have a large stake in what happens to them at school and their reactions to and perceptions of their school experiences are significant. This chapter reviews the remarkable progress over the past 30 years in conceptualising, assessing and investigating the determinants and effects of social and psychological aspects of the learning environments of classrooms and schools.

This chapter falls into seven main parts. First, an introductory section provides background information about the field of learning environment (including alternative assessment approaches, historical perspectives on past work, the distinction between school and classroom environment, and the unit-of-analysis question). Second, a section is devoted to specific instruments for assessing perceptions of classroom environment. Third, some important developments with learning environment instruments are outlined (preferred forms, short versions, hand scoring, the distinction between Personal and Class forms). Fourth, the validation of learning environment scales is discussed. Fifth, assessment instruments for school environment are considered. Sixth, an overview is given of several lines of past research involving environment assessments in science classrooms (including associations between outcomes and environment, use of environment dimensions as criterion variables, and person-environment fit studies of whether students achieve better in their preferred environment). Seventh, consideration is given to teachers' use of classroom and school environment instruments in practical attempts to improve their own classrooms and schools. Eighth, current trends and future desirable directions in research on educational environments are identified (e.g., combining quantitative and qualitative methods, school-level environments, school psychology, links between educational environments, cross-national studies, transition between primary and secondary schooling, teacher education and teacher assessment).

---

Chapter Consultant: Peter Okebukola (Lagos State University, Nigeria)

## BACKGROUND

*Approaches to Studying Educational Environments*

Using students' and teachers' perceptions to study educational environments can be contrasted with the external observer's direct observation and systematic coding of classroom communication and events (Brophy & Good 1986). Murray (1938) introduced the term *alpha press* to describe the environment as assessed by a detached observer and the term *beta press* to describe the environment as perceived by milieu inhabitants. Another approach to studying educational environments involves application of the techniques of naturalistic inquiry, ethnography, case study or interpretive research (see Erickson's chapter in this *Handbook*). Defining the classroom or school environment in terms of the shared perceptions of the students and teachers has the dual advantage of characterising the setting through the eyes of the participants themselves and capturing data which the observer could miss or consider unimportant. Students are at a good vantage point to make judgements about classrooms because they have encountered many different learning environments and have enough time in a class to form accurate impressions. Also, even if teachers are inconsistent in their day-to-day behaviour, they usually project a consistent image of the long-standing attributes of classroom environment. Later in this chapter, discussion focuses on the merits of combining quantitative and qualitative methods when studying educational environments (Fraser & Tobin 1991).

*Historical Perspectives*

Thirty years ago, Herbert Walberg and Rudolf Moos began seminal independent programs of research which form the starting points for the work reviewed in this chapter. Walberg developed the widely-used *Learning Environment Inventory* (LEI) as part of the research and evaluation activities of Harvard Project Physics (Walberg & Anderson 1968). Moos began developing the first of his social climate scales, including those for use in psychiatric hospitals and correctional institutions, which ultimately resulted in the development of the *Classroom Environment Scale* (CES) (Moos 1979; Moos & Trickett 1987). The way in which the important pioneering work of Walberg and Moos on perceptions of classroom environment developed into major research programs and spawned a lot of other research is reflected in books (Fraser 1986; Fraser & Walberg 1991; Moos 1979; Walberg 1979), literature reviews (Fraser 1994; MacAuley 1990; von Saldern 1992) and monographs sponsored by the American Educational Research Association's Special Interest Group (SIG) on the Study of Learning Environments (e.g., Fisher 1994).

The work on educational environments over the previous 30 years builds upon the earlier ideas of Lewin and Murray and their followers (such as Pace and Stern). Lewin's (1936) seminal work on field theory recognised that both the environment and its interaction with personal characteristics of the individual are potent determinants of human behaviour. The familiar Lewinian formula, $B = f(P, E)$, was first enunciated to stress the need for new research strategies in which behaviour is considered to be a function of the person and the environment. Murray (1938) was first to follow Lewin's approach by proposing a needs-press model which allows the analogous representation of person and environment in common terms. Personal needs refer to motivational personality characteristics representing tendencies to move in the direction of certain goals, while environmental press provides an external situational counterpart which supports or frustrates the expression of internalised personality needs. Needs-press theory has been popularised and elucidated by Pace and Stern (e.g., Stern 1970).

*School-Level vs. Classroom-Level Environment*

It is useful to distinguish classroom or classroom-level environment from school or school-level environment, which involves psychosocial aspects of the climate of whole schools (Fraser & Rentoul 1982). School climate research owes much in theory, instrumentation and methodology to earlier work on organisational climate in business contexts. Two widely-used instruments in school environment research, namely, Halpin and Croft's (1963) *Organizational Climate Description Questionnaire* (OCDQ) and Stern's (1970) *College Characteristics Index* (CCI), relied heavily on previous work in business organisations. Two features of school-level environment work which distinguishes it from classroom-level environment research are that the former has tended to be associated with the field of educational administration and to involve the climate of higher education institutions. Despite their simultaneous development and logical linkages, the fields of classroom-level and school-level environment have remained remarkably independent. Although the focus of past research in science education has been primarily upon classroom-level environment, it would be desirable to break away from the existing tradition of independence of the two fields of school and classroom environment and for there to be a confluence of the two areas.

*Level of Analysis: Private and Consensual Press*

Murray's distinction between alpha press (the environment as observed by an external observer) and beta press (the environment as perceived by milieu inhabitants) has been extended by Stern, Stein and Bloom (1956) who distinguish between the idiosyncratic view that each person has of the environment (*private* beta press) and the shared view that members of a group hold about the environment (*consensual* beta press). Private and consensual beta press could differ from each other, and both could differ from the detached view of alpha press of a trained non-participant observer. In designing classroom environment studies, researchers must decide whether their analyses will involve the perception scores obtained from individual students (private press) or whether these will be combined to obtain the average of the environment scores of all students within the same class (consensual press).

A growing body of literature acknowledges the importance and consequences of the choice of level or unit of statistical analysis and considers the hierarchical analysis and multilevel analysis of data (Bock 1989; Bryk & Raudenbush 1992; Goldstein 1987). The choice of unit of analysis is important because: measures having the same operational definition can have different substantive interpretations with different levels of aggregation; relationships obtained using one unit of analysis could differ in magnitude and even in sign from relationships obtained using another unit; the use of certain units of analysis (e.g., individuals when classes are the primary sampling units) violates the requirement of independence of observations and calls into question the results of any statistical significance tests because an unjustifiably small estimate of the sampling error is used; and the use of different units of analysis involves the testing of conceptually different hypotheses. Although the unit of analysis problem has received considerable attention in the context of testing hypotheses using already-developed learning environment instruments, Sirotnik (1980) considers it ironic that concerns about analytic units have been virtually nonexistent at the stage of developing and empirically investigating the dimensionality of new instruments.

## INSTRUMENTS FOR ASSESSING CLASSROOM ENVIRONMENT

This section describes the following historically important and contemporary instruments: Learning Environment Inventory (LEI); Classroom Environment Scale (CES); Individualised Classroom Environment Questionnaire (ICEQ); My Class Inventory (MCI); College and University Classroom Environment Inventory (CUCEI); Questionnaire on Teacher Interaction (QTI); Science Laboratory Environment Inventory (SLEI); Constructivist Learning Environment Survey (CLES); and What Is Happening In This Class (WIHIC) questionnaire. Table 1 shows the name of each scale in each instrument, the level (primary, secondary, higher education) for which each instrument is suited, the number of items contained in each scale, and the classification of each scale according to Moos's (1974) scheme for classifying human environments. Moos's three basic types of dimension are Relationship Dimensions (which identify the nature and intensity of personal relationships within the environment and assess the extent to which people are involved in the environment and support and help each other), Personal Development Dimensions (which assess basic directions along which personal growth and self-enhancement tend to occur) and System Maintenance and System Change Dimensions (which involve the extent to which the environment is orderly, clear in expectations, maintains control and is responsive to change).

**Table 1: Overview of scales contained in nine classroom environment instruments (LEI, CES, ICEQ, MCI, CUCEI, QTI, SLEI, CLES and WIHIC)**

| Instrument | Level | Items per scale | Scales Classified According to Moos's Scheme | | |
|---|---|---|---|---|---|
| | | | Relationship dimensions | Personal development dimensions | System maintenance and change dimensions |
| Learning Environment Inventory (LEI) | Secondary | 7 | Cohesiveness Friction Favouritism Cliqueness Satisfaction Apathy | Speed Difficulty Competitiveness | Diversity Formality Material Environment Goal Direction Disorganisation Democracy |
| Classroom Environment Scale (CES) | Secondary | 10 | Involvement Affiliation Teacher Support | Task Orientation Competition | Order and Organisation Rule Clarity Teacher Control Innovation |
| Individualised Classroom Environment Questionnaire (ICEQ) | Secondary | 10 | Personalisation Participation | Independence Investigation | Differentiation |
| My Class Inventory (MCI) | Elementary | 6–9 | Cohesiveness Friction Satisfaction | Difficulty Competitiveness | |
| College and University Classroom Environment Inventory (CUCEI) | Higher Education | 7 | Personalisation Involvement Student Cohesiveness Satisfaction | Task Orientation | Innovation Individualisation |
| Questionnaire on Teacher Interaction (QTI) | Secondary/ Primary | 8–10 | Helpful/Friendly Understanding Dissatisfied Admonishing | | Leadership Student Responsibility and Freedom Uncertain Strict |
| Science Laboratory Environment Inventory (SLEI) | Upper Secondary/ Higher Education | 7 | Student Cohesiveness | Open-Endedness Integration | Rule Clarity Material Environment |
| Constructivist Learning Environment Survey (CLES) | Secondary | 7 | Personal Relevance Uncertainty | Critical Voice Shared Control | Student Negotiation |
| What Is Happening In This Classroom (WIHIC) | Secondary | 8 | Student Cohesiveness Teacher Support Involvement | Investigation Task Orientation Cooperation | Equity |

## Learning Environment Inventory (LEI)

The initial development and validation of a preliminary version of the LEI began in the late 1960s in conjunction with the evaluation and research related to Harvard Project Physics (Fraser, Anderson & Walberg 1982; Walberg & Anderson 1968). The final version contains a total of 105 statements (or seven per scale) descriptive of typical school classes. The respondent expresses degree of agreement or disagreement with each statement using the four response alternatives of Strongly Disagree, Disagree, Agree and Strongly Agree. The scoring direction (or polarity) is reversed for some items. A typical item in the Cohesiveness scale is: 'All students know each other very well' and in the Speed scale is: 'The pace of the class is rushed'.

## Classroom Environment Scale (CES)

The CES was developed by Rudolf Moos at Stanford University (Fisher & Fraser 1983c; Moos 1979; Moos & Trickett 1987) and grew out of a comprehensive program of research involving perceptual measures of a variety of human environments including psychiatric hospitals, prisons, university residences and work milieus (Moos 1974). The final published version contains nine scales with 10 items of True-False response format in each scale. Published materials include a test manual, a questionnaire, an answer sheet and a transparent hand scoring key. Typical items in the CES are: 'The teacher takes a personal interest in the students' (Teacher Support) and 'There is a clear set of rules for students to follow' (Rule Clarity).

## Individualised Classroom Environment Questionnaire (ICEQ)

The ICEQ assesses those dimensions which distinguish individualised classrooms from conventional ones. The initial development of the ICEQ (Rentoul & Fraser 1979) was guided by: the literature on individualised open and inquiry-based education; extensive interviewing of teachers and secondary school students; and reactions to draft versions sought from selected experts, teachers and junior high school students. The final published version of the ICEQ (Fraser 1990) contains 50 items altogether, with an equal number of items belonging to each of the five scales. Each item is responded to on a five-point scale with the alternatives of Almost Never, Seldom, Sometimes, Often and Very Often. The scoring direction is reversed for many of the items. Typical items are: 'The teacher considers students' feelings' (Personalisation) and 'Different students use different books, equipment and materials' (Differentiation). The published version has a progressive copyright arrangement which gives permission to purchasers to make an unlimited number of copies of the questionnaires and response sheets.

*My Class Inventory (MCI)*

The LEI has been simplified to form the MCI for use among children aged 8–12 years (Fisher & Fraser 1981; Fraser, Anderson & Walberg 1982; Fraser & O'Brien 1985). Although the MCI was developed originally for use at the primary school level, it also has been found to be very useful with students in the junior high school, especially those who might experience reading difficulties with other instruments. The MCI differs from the LEI in four important ways. First, in order to minimise fatigue among younger children, the MCI contains only five of the LEI's original 15 scales. Second, item wording has been simplified to enhance readability. Third, the LEI's four-point response format has been reduced to a two-point (Yes-No) response format. Fourth, students answer on the questionnaire itself instead of on a separate response sheet to avoid errors in transferring responses from one place to another. The final form of the MCI contains 38 items altogether, with typical items being: 'Children are always fighting with each other' (Friction) and 'Children seem to like the class' (Satisfaction). Although the MCI traditionally has been used with a Yes-No response format, Goh, Young and Fraser (1995) have successfully used a three-point response format (Seldom, Sometimes and Most of the Time) with a modified version of the MCI which includes a Task Orientation scale.

*College and University Classroom Environment Inventory (CUCEI)*

Although some notable prior work has focused on the institutional-level or school-level environment in colleges and universities (e.g., Halpin & Croft 1963; Stern 1970), surprisingly little work has been done in higher education classrooms which is parallel to the traditions of classroom environment research at the secondary and primary school levels. Consequently, the CUCEI was developed for use in small classes (say up to 30 students) sometimes referred to as 'seminars' (Fraser & Treagust 1986; Fraser, Treagust & Dennis 1986). The final form of the CUCEI contains seven seven-item scales. Each item has four responses (Strongly Agree, Agree, Disagree, Strongly Disagree) and the polarity is reversed for approximately half of the items. Typical items are: 'Activities in this class are clearly and carefully planned' (Task Orientation) and 'Teaching approaches allow students to proceed at their own pace' (Individualisation).

*Questionnaire on Teacher Interaction (QTI)*

Research which originated in The Netherlands focuses on the nature and quality of interpersonal relationships between teachers and students (Créton, Hermans & Wubbels 1990; Wubbels, Brekelmans & Hooymayers 1991; Wubbels & Levy 1993). Drawing upon a theoretical model of proximity (cooperation-opposition) and influence (dominance-submission), the QTI was developed to assess student perceptions of eight behaviour aspects. Each item has a five-point response scale ranging from Never to Always. Typical items are 'She/he gives us a lot of free time' (Student Responsibility and Freedom behaviour) and 'She/he gets angry' (Admonishing behaviour).

Although research with the QTI began at the senior high school level in The Netherlands, cross-validation and comparative work has been completed at various grade levels in the USA (Wubbels & Levy 1993), Australia (Fisher, Henderson & Fraser 1995), Singapore (Goh & Fraser 1996) and Brunei (Riah, Fraser & Rickards 1997), and a more economical 48-item version has been developed and validated (Goh & Fraser 1996). Also, Cresswell and Fisher (1997) modified the QTI to form the *Principal Interaction Questionnaire* (PIQ) which assesses teachers' or principals' perceptions of the same eight dimensions of a principal's interaction with teachers. Further information about research involving the QTI can be found in Wubbels and Brekelmans (1997) and Wubbels and Brekelmans' chapter in this *Handbook*.

*Science Laboratory Environment Inventory (SLEI)*

Because of the critical importance and uniqueness of laboratory settings in science education, an instrument specifically suited to assessing the environment of science laboratory classes at the senior high school or higher education levels was developed (Fraser, Giddings & McRobbie 1995; Fraser & McRobbie 1995; Fraser, McRobbie & Giddings 1993). The SLEI has five scales (each with seven items) and the five response alternatives are Almost Never, Seldom, Sometimes, Often and Very Often. Typical items are 'I use the theory from my regular science class sessions during laboratory activities' (Integration) and 'We know the results that we are supposed to get before we commence a laboratory activity' (Open-Endedness). The Open-Endedness scale was included because of the importance of open-ended laboratory activities often claimed in the literature (e.g., Hodson 1988). The SLEI was field tested and validated simultaneously with a sample of over 5,447 students in 269 classes in six different countries (the USA, Canada, England, Israel, Australia and Nigeria), and cross-validated with 1,594 Australian students in 92 classes (Fraser & McRobbie 1995), 489 senior high school biology students in Australia (Fisher, Henderson & Fraser 1997) and 1,592 grade 10 chemistry students in Singapore (Wong & Fraser 1995).

*Constructivist Learning Environment Survey (CLES)*

According to the constructivist view, meaningful learning is a cognitive process in which individuals make sense of the world in relation to the knowledge which they already have constructed, and this sense-making process involves active negotiation and consensus building. The CLES (Taylor, Dawson & Fraser 1995; Taylor, Fraser & Fisher 1997) was developed to assist researchers and teachers to assess the degree to which a particular classroom's environment is consistent with a constructivist epistemology, and to assist teachers to reflect on their epistemological assumptions and reshape their teaching practice. Appendix A contains a complete copy of the CLES's 'actual' form (see a later section for a clarification of the distinction between 'actual' and 'preferred' forms). Recent studies that have used the CLES include Dryden and Fraser (1996) and Roth and Roychoudhury (1994).

*What Is Happening In This Class (WIHIC) Questionnaire*

The WIHIC questionnaire brings parsimony to the field of learning environment by combining modified versions of the most salient scales from a wide range of existing questionnaires with additional scales that accommodate contemporary educational concerns (e.g., equity and constructivism). Also, the WIHIC has a separate Class form (which assesses a student's perceptions of the class as a whole) and Personal form (which assesses a student's personal perceptions of his or her role in a classroom), as discussed in detail later in this chapter.

The original 90-item nine-scale version was refined by both statistical analysis of data from 355 junior high school science students, and extensive interviewing of students about their views of their classroom environments in general, the wording and salience of individual items and their questionnaire responses (Fraser, Fisher & McRobbie 1996). Only 54 items in seven scales survived these procedures, although this set of items was expanded to 80 items in eight scales for the field testing of the second version of the WIHIC, which involved junior high school science classes in Australia and Taiwan. Whereas the Australian sample of 1,081 students in 50 classes responded to the original English version, a Taiwanese sample of 1,879 students in 50 classes responded to a Chinese version that had undergone careful procedures of translation and back translation (Huang & Fraser 1997). This led to a final form of the WIHIC containing the seven eight-item scales. The WIHIC has been used successfully in its original form or in modified form in studies involving 250 adult learners in Singapore (Khoo & Fraser 1997) and 2,310 high school students in Singapore (Chionh & Fraser 1998).

*Other Instruments*

Many studies have drawn on scales and items in existing questionnaires to develop modified instruments which better suit particular research purposes and research contexts. For a study of the classroom environment of Catholic schools, Dorman, Fraser and McRobbie (1997) developed a 66-item instrument which drew on the CES, CUCEI and ICEQ but made important modifications. The seven scales in this study (Student Application, Interactions, Cooperation, Task Orientation, Order and Organisation, Individualisation and Teacher Control) were validated using a sample of 2,211 grade 9 and 12 students in 104 classes.

Because a limited number of classroom environment instruments have a reading level suitable for the primary school level, Sinclair and Fraser (1997) developed a questionnaire based on the MCI and WIHIC for use in teachers' action research attempts to improve their primary classroom environments in an urban school district. The instrument has the four scales of Cooperation, Teacher Empathy/Equity, Task Orientation and Involvement, and it was validated with a sample of 745 students in 43 grade 6–8 classes.

In evaluations of computer-assisted learning, Maor and Fraser (1996) and Teh and Fraser (1994, 1995b) drew on existing scales in developing specific-purpose instruments. Maor and Fraser developed a five-scale classroom environment instrument (assessing Investigation, Open-Endedness, Organisation, Material Environment and Satisfaction) based on the LEI, ICEQ and SLEI and validated it with a sample of 120 grade 11 students in Australia. Teh and Fraser developed a four-scale instrument to assess Gender Equity, Investigation, Innovation and Resource Adequacy, and validated it among 671 high school geography students in Singapore.

In the first learning environment study worldwide specifically in agricultural science classes, Idiris and Fraser (1997) selected and adapted scales from CLES and ICEQ in developing a five-scale instrument to assess Negotiation, Autonomy, Student Centredness, Investigation and Differentiation. This instrument was validated with a sample of 1,175 students in 50 high school agricultural science classes in eight States of Nigeria.

Influenced partly by the CES, Wong (1993) developed a questionnaire to assess the actual and preferred environment of classes in Hong Kong along the dimensions of Enjoyable, Order, Involvement, Achievement Orientation, Teacher Led, Teacher Involvement, Teacher Support and Collaborativeness.

Whereas most classroom environment instruments focus on general psychosocial characteristics, Woods and Fraser (1995) developed a questionnaire to assess student perceptions of specific teacher behaviours. The *Classroom Interaction Patterns Questionnaire* (CIPQ) assesses teaching style with the scales of Praise and Encouragement, Open Questioning, Lecture and Direction, Individual Work, Discipline and Management, and Group Work. Successive versions were field tested with a total of 1,470 grade 8–10 students in 62 classes in Western Australia.

Based partly on existing instruments, Fisher and Waldrip (1997) developed a questionnaire to assess culturally sensitive factors of learning environments. The 40-item *Cultural Learning Environment Questionnaire* (CLEQ) assesses students perceptions of Equity, Collaboration, Risk Involvement, Competition, Teacher Authority, Modelling, Congruence and Communication. Administration of the new questionnaire to 3,031 secondary science students in 135 classes in Australia provided support for the internal consistency reliability and factorial validity of the CLEQ.

Jegede, Fraser and Fisher (1995) developed the *Distance and Open Learning Environment Scale* (DOLES) for use among university students studying by distance education. The DOLES has the five core scales of Student Cohesiveness,

Teacher Support, Personal Involvement and Flexibility, Task Orientation and Material Environment, and Home Environment, as well as the two optional scales of Study Centre Environment and Information Technology Resources. Administration of the DOLES to 660 university students provided support for its internal consistency reliability and factor structure.

## IMPORTANT DEVELOPMENTS WITH LEARNING ENVIRONMENT INSTRUMENTS

### Preferred Forms of Scales

A distinctive feature of most of the instruments in Table 1 is that they have, not only a form to measure perceptions of 'actual' or experienced classroom environment, but also another form to measure perceptions of 'preferred' or ideal classroom environment. The preferred forms are concerned with goals and value orientations and measure perceptions of the classroom environment ideally liked or preferred. Although item wording is similar for actual and preferred forms, slightly different instructions for answering each are used. For example, an item in the actual form such as 'There *is* a clear set of rules for students to follow' would be changed in the preferred form to 'There *would be* a clear set of rules for students to follow'.

### Short Forms of ICEQ, MCI and CES

Although the long forms of classroom environment instruments have been used successfully for a variety of purposes, some researchers and teachers have reported that they would like instruments to take less time to administer and score. Consequently, short forms of the ICEQ, MCI and CES were developed (Fraser 1982a; Fraser & Fisher 1983a) to satisfy three main criteria. First, the total number of items in each instrument was reduced to approximately 25 to provide greater economy in testing and scoring time. Second, the short forms were designed to be amenable to easy hand scoring. Third, although long forms of instruments might be needed to provide adequate reliability for the assessment of the perceptions of individual students, short forms are likely to have adequate reliability for the many applications which involve averaging the perceptions of students within a class to obtain class means. The development of the short form was based largely on the results of several item analyses performed on data obtained by administering the long forms of each instrument to a large sample. The short form of the ICEQ and the MCI each consist of 25 items divided equally among the five scales comprising the long form. Because the long form of the CES consisted of 90 items, this was reduced to a short version with 24 items divided equally among six of the original nine scales.

## Hand Scoring Procedures

Appendix A illustrates typical hand scoring procedures for the CLES. Items are arranged in blocks so that all items from the same scale are found together. All items in Appendix A except Item 6 are scored by allocating the circled number (i.e., 1 for Almost Never, 2 for Seldom, etc.). Item 6 is scored in the reverse manner. Omitted or invalidly answered items are scored 3. To obtain scale totals, the six item scores for each scale are added. For example, the total score for the first scale (Personal Relevance) is obtained by adding scores for items 1 to 6.

## Personal Forms of Scales

Fraser and Tobin (1991) point out that there is potentially a major problem with nearly all existing classroom environment instruments when they are used to identify differences between subgroups within a classroom (e.g., males and females) or in the construction of case studies of individual students. The problem is that items are worded in such a way that they elicit an individual student's perceptions of the class as a whole, as distinct from a student's perceptions of his/her own role within the classroom. For example, items in the traditional class form might seek students' opinions about whether '*the work of the class is difficult*' or whether '*the teacher is friendly towards the class*'. In contrast, a personal form of the same items would seek opinions about whether '*I find the work of the class difficult*' or whether '*the teacher is friendly towards me*'. Confounding could have arisen in past studies which employed the class form because, for example, males could find a class less difficult than females, yet males and females still could agree when asked for their opinions about the class as a whole. The distinction between personal and class forms is consistent with Stern, Stein and Bloom's (1956) terms of 'private' beta press, the idiosyncratic view that each person has of the environment, and 'consensual' beta press, the shared view that members of a group hold of the environment.

When Fraser, Giddings and McRobbie (1995) developed and validated parallel class and personal forms of both an actual and preferred version of the SLEI, item and factor analyses confirmed that the personal form had a similar factor structure and comparable statistical characteristics (e.g., internal consistency, discriminant validity) to the class form when either the individual student or the class mean was used as the unit of analysis. Also students' scores on the class form were found to be systematically more favourable than their scores on the personal form, perhaps suggesting that students have a more detached view of the environment as it applies to the class as a whole. As hypothesised, gender differences in perceptions were somewhat larger on the personal form than on the class form. Although a study of associations between student outcomes and their perceptions of the science laboratory environment revealed that the magnitudes of associations were comparable for class and personal forms of the SLEI, commonality analyses showed that each form accounted for appreciable amounts of outcome variance which was independent of that explained by the other form (Fraser & McRobbie 1995). This finding justifies the decision to evolve separate class and personal forms because they appear to measure different, albeit overlapping, aspects of the science laboratory classroom environment.

Administration of the WIHIC questionnaire followed by interviews with 45 students showed that many students have perceptions from the perspective of the class as a whole that differ from their perceptions of their personal role within the classroom (Fraser, Fisher & McRobbie 1996). Underlying many of the responses was the idea that, because the individual student is only part of the class, interactions with an individual student (Personal form) are less frequent than the interactions with the class as a whole (Class form). Further discussion of the distinction between Personal and Class forms can be found in McRobbie, Fisher and Wong's chapter in this *Handbook*.

## VALIDATION OF SCALES

This section reports typical validation data for selected classroom environment scales. Table 2 provides a summary of a limited amount of statistical information for the nine instruments (LEI, CES, ICEQ, MCI, CUCEI, QTI, SLEI, CLES and WIHIC) considered previously. Attention is restricted to the student actual form and to the use of the individual student as the unit of analysis. Table 2 provides information about each scale's internal consistency reliability (alpha coefficient) and discriminant validity (using the mean correlation of a scale with the other scales in the same instrument as a convenient index), and the ability of a scale to differentiate between the perceptions of students in different classrooms (significance level and $eta^2$ statistic from ANOVAs). Statistics are based on 1,048 students for the LEI, except for discriminant validity data which are based on 149 class means (Fraser, Anderson & Walberg 1982), 1,083 students for the CES (Fisher & Fraser 1983c), 1,849 students for the ICEQ (Fraser 1990), 2,305 students for the MCI (Fisher & Fraser 1981), 372 students for the CUCEI (Fraser & Treagust 1986), 3,994 high school science and mathematics students for the QTI (Fisher, Fraser & Rickards 1997), 3,727 senior high school students for the SLEI (Fraser, Giddings & McRobbie 1995) and 1,081 high school science students for both the CLES and WIHIC (previously unpublished results).

## INSTRUMENTS FOR ASSESSING SCHOOL ENVIRONMENT

In contrast to work on classroom-level environment, relatively little research has been directed towards helping teachers assess and improve the environments of their own schools. Earlier instruments include Stern's (1970) College Characteristics Index (CCI) and Halpin and Croft's (1963) Organizational Climate Description Questionnaire (OCDQ). The *Work Environment Scale* (WES; Moos 1981) was designed for use in any work milieu rather than for use specifically in schools. To

Table 2: Internal consistency (alpha reliability), discriminant validity (mean correlation of a scale with other scales), and ANOVA results for class membership differences (eta$^2$ statistic and significance level) for student actual form of nine instruments using individual as unit of analysis

| Scale | Alpha rel. | Mean correl. with other scales | ANOVA results eta$^2$ | Scale | Alpha rel. | Mean correl. with other scales | ANOVA results eta$^2$ |
|---|---|---|---|---|---|---|---|
| **Learning Environment Inventory (LEI)** | | | | **College and University Classroom Environment Inventory (CUCEI)** | | | |
| (N=1,048 (N=149 students) classes) | | | | (N=372 students) | | | |
| Cohesiveness | 0.69 | 0.14 | –[a] | Personalisation | 0.75 | 0.46 | 0.35* |
| Diversity | 0.54 | 0.16 | – | Involvement | 0.70 | 0.47 | 0.40* |
| Formality | 0.76 | 0.18 | – | Student Cohesiveness | 0.90 | 0.45 | 0.47* |
| Speed | 0.70 | 0.17 | – | Satisfaction | 0.88 | 0.45 | 0.32* |
| Material Environment | 0.56 | 0.24 | – | Task Orientation | 0.75 | 0.38 | 0.43* |
| Friction | 0.72 | 0.36 | – | Innovation | 0.81 | 0.46 | 0.41* |
| Goal Direction | 0.85 | 0.37 | – | Individualisation | 0.78 | 0.34 | 0.46* |
| Favouritism | 0.78 | 0.32 | – | **Questionnaire on Teacher Interaction (QTI)** | | | |
| Difficulty | 0.64 | 0.16 | – | (N=3,994 students) | | | |
| Apathy | 0.82 | 0.39 | – | Leadership | 0.82 | –[b] | 0.33* |
| Democracy | 0.67 | 0.34 | – | Helping/Friendly | 0.88 | – | 0.35* |
| Cliqueness | 0.65 | 0.33 | – | Understanding | 0.85 | – | 0.32* |
| Satisfaction | 0.79 | 0.39 | – | Student Responsibility/Freedom | 0.66 | – | 0.26* |
| Disorganisation | 0.82 | 0.40 | – | Uncertain | 0.72 | – | 0.22* |
| Competitiveness | 0.78 | 0.08 | – | Dissatisfied | 0.80 | – | 0.23* |
| **Classroom Environment Scale (CES)** | | | | Admonishing | 0.76 | – | 0.31* |
| (N=1,083 students) | | | | Strict | 0.63 | – | 0.23* |
| Involvement | 0.70 | 0.40 | 0.29* | **Science Laboratory Environment Inventory (SLEI)** | | | |
| Affiliation | 0.60 | 0.24 | 0.21* | (N=3,727 students) | | | |
| Teacher Support | 0.72 | 0.29 | 0.34* | Student Cohesiveness | 0.77 | 0.34 | 0.21* |
| Task Orientation | 0.58 | 0.23 | 0.25* | Open-Endedness | 0.70 | 0.07 | 0.19* |
| Competition | 0.51 | 0.09 | 0.18* | Integration | 0.83 | 0.37 | 0.23* |
| Order and Organisation | 0.75 | 0.29 | 0.43* | Rule Clarity | 0.75 | 0.33 | 0.21* |
| Rule Clarity | 0.63 | 0.29 | 0.21* | Material Environment | 0.75 | 0.37 | 0.21* |
| Teacher Control | 0.60 | 0.16 | 0.27* | **Constructivist Learning Environment Survey (CLES)** | | | |
| Innovation | 0.52 | 0.19 | 0.26* | (N=1,081 students) | | | |
| **Individualised Classroom Environment Questionnaire (ICEQ)** | | | | Personal Relevance | 0.88 | 0.43 | 0.16* |
| (N=1,849 students) | | | | Uncertainty | 0.76 | 0.44 | 0.14* |
| Personalisation | 0.79 | 0.28 | 0.31* | Critical View | 0.85 | 0.31 | 0.14* |
| Participation | 0.70 | 0.27 | 0.21* | Shared Control | 0.91 | 0.41 | 0.17* |
| Independence | 0.68 | 0.07 | 0.30* | Student Negotiation | 0.89 | 0.40 | 0.14* |
| Investigation | 0.71 | 0.21 | 0.20* | **What Is Happening In This Classroom (WIHIC)** | | | |
| Differentiation | 0.76 | 0.10 | 0.43* | (N=1,081 students) | | | |
| **My Class Inventory (MCI)** | | | | Student Cohesiveness | 0.81 | 0.37 | 0.09* |
| (N=2,305 students) | | | | Teacher Support | 0.88 | 0.43 | 0.15* |
| Cohesiveness | 0.67 | 0.20 | 0.21* | Involvement | 0.84 | 0.45 | 0.10* |
| Friction | 0.67 | 0.26 | 0.31* | Investigation | 0.88 | 0.41 | 0.15* |
| Difficulty | 0.62 | 0.14 | 0.18* | Task Orientation | 0.88 | 0.42 | 0.15* |
| Satisfaction | 0.78 | 0.23 | 0.30* | Cooperation | 0.89 | 0.45 | 0.12* |
| Competitiveness | 0.71 | 0.10 | 0.19* | Equity | 0.93 | 0.46 | 0.13* |

a This statistic is not available for the LEI.
b This statistic is not relevant for the QTI.

improve the WES's face validity for use in schools, the word 'people' was changed to 'teachers', 'supervisor' was changed to 'senior staff', and 'employee' was changed to 'teacher' (Fisher & Fraser 1983b; Fraser, Docker & Fisher 1988). Of the WES's 10 scales, three measure Relationship Dimensions (Involvement, Peer Cohesion, Staff Support), two measure Personal Development Dimensions (Autonomy, Task Orientation) and five measure System Maintenance and System Change Dimensions (Work Pressure, Clarity, Control, Innovation, Physical Comfort). The WES consists of 90 items of True/False response format, with an equal number of items in each scale. Validation data for the WES were generated in a study of 34 primary and secondary schools in Tasmania (Docker, Fraser & Fisher 1989).

The *School-Level Environment Questionnaire* (SLEQ) was designed especially to assess school teachers' perceptions of psychosocial dimensions of the environment of the school. A review of potential strengths and problems associated with existing school environment instruments suggested that the SLEQ should contain eight scales (Fisher & Fraser 1991; Rentoul & Fraser 1983). Two scales measure Relationship Dimensions (Student Support, Affiliation), one measures the Personal Development Dimension (Professional Interest) and five measure System Maintenance and System Change Dimensions (Staff Freedom, Participatory Decision Making, Innovation, Resource Adequacy and Work Pressure). The SLEQ consists of 56 items, with each of the eight scales being assessed by seven items. Each item is scored on a five-point scale with the responses of Strongly Agree, Agree, Not Sure, Disagree and Strongly Disagree. In addition to an actual form which assesses perceptions of what a school's work environment is actually like, the SLEQ also has a preferred form. In a study of the school-level environment of Catholic schools, Dorman, Fraser and McRobbie (1997) developed a 57-item school environment instrument which includes modified versions of five SLEQ scales (Student Support, Affiliation, Professional Interest, Resource Adequacy and Work Pressure), but which adds the two new scales of Empowerment (the extent to which teachers are empowered and encouraged to be involved in decision-making processes) and Mission Consensus (the extent to which consensus exists within the staff with regard to the overarching goals of the school). This instrument was used in studies of differences in the school environment of Catholic and government schools (Dorman & Fraser 1996) and of associations between school environment and classroom environment (Dorman, Fraser & McRobbie 1997).

## RESEARCH INVOLVING EDUCATIONAL ENVIRONMENT INSTRUMENTS

Three types of past research considered in this section involve (1) associations between student outcomes and environment, (2) use of environment dimensions as criterion variables (including the evaluation of educational innovations and investigations of differences between students' and teachers' perceptions of the same classrooms) and (3) investigations of whether students achieve better when in their preferred environments. A separate section later focuses on teachers' practical attempts to improve their classroom and school climates.

*Associations Between Student Outcomes and Environment*

The strongest tradition in past classroom environment research has involved investigation of associations between students' cognitive and affective learning outcomes and their perceptions of psychosocial characteristics of their classrooms (Fraser & Fisher 1982; Haertel, Walberg & Haertel 1981; McRobbie & Fraser 1993). Numerous research programs have shown that student perceptions account for appreciable amounts of variance in learning outcomes, often beyond that attributable to background student characteristics. For example, Fraser's (1994) tabulation of 40 past studies in science education shows that associations between outcome measures and classroom environment perceptions have been replicated for a variety of cognitive and affective outcome measures, a variety of classroom environment instruments and a variety of samples (ranging across numerous countries and grade levels).

Using the SLEI, associations with students' cognitive and affective outcomes have been established for a sample of approximately 80 senior high school chemistry classes in Australia (Fraser & McRobbie 1995; McRobbie & Fraser 1993), 489 senior high school biology students in Australia (Fisher, Henderson & Fraser 1997) and 1,592 grade 10 chemistry students in Singapore (Wong & Fraser 1996). Using an instrument suited for computer-assisted instruction classrooms, Teh and Fraser (1995a) established associations between classroom environment, achievement and attitudes among a sample of 671 high school geography students in 24 classes in Singapore. Using the QTI, associations between student outcomes and perceived patterns of teacher-student interaction were reported for samples of 489 senior high school biology students in Australia (Fisher, Henderson & Fraser 1995), 3,994 high school science and mathematics students in Australia (Fisher, Fraser & Rickards 1997) and 1,512 primary school mathematics student in Singapore (Goh, Young & Fraser 1995).

Multilevel Analysis

While many past learning environment studies have employed techniques such as multiple regression analysis, few have used the multilevel analysis (Bock 1989; Bryk & Raudenbush 1992; Goldstein 1987), which takes cognisance of the hierarchical nature of classroom settings. Because classroom environment data typically are derived from students in intact classes, they are inherently hierarchical. Ignoring this nested structure can give rise to problems of aggregation bias (within-group homogeneity) and imprecision.

Two studies of outcome-environment associations compared the results obtained from multiple regression analysis with those obtained from an analysis involving the hierarchical linear model. The multiple regression analyses were performed separately at the individual student level and the class mean level. In the HLM analyses, the environment variables were investigated at the individual level, and were aggregated at the class level. In Wong, Young and Fraser's (1997) study involving 1,592 grade 10 students in 56 chemistry classes in Singapore, associations were investigated between three student attitude measures and a modified version of the SLEI. In Goh, Young and Fraser's (1995) study with 1,512 grade 5 mathematics students in 39 classes in Singapore, scores on a modified version of the MCI were related to student achievement and attitude. Most of the significant results from the multiple regression analyses were replicated in the HLM analyses, as well as being consistent in direction.

Meta-Analysis of Studies

The findings from prior research are highlighted in the results of a meta-analysis involving 734 correlations from 12 studies involving 823 classes, eight subject areas, 17,805 students and four nations (Haertel, Walberg & Haertel 1981). Learning posttest scores and regression-adjusted gains were found to be consistently and strongly associated with cognitive and affective learning outcomes, although correlations were generally higher in samples of older students and in studies employing collectivities such as classes and schools (in contrast to individual students) as the units of statistical analysis. In particular, better achievement on a variety of outcome measures was found consistently in classes perceived as having greater Cohesiveness, Satisfaction and Goal Direction and less Disorganisation and Friction. Other meta-analyses synthesised by Fraser, Walberg, Welch and Hattie (1987) provide further evidence supporting the link between educational environments and student outcomes.

Cooperative Learning

Among the various lines of programmatic research on classroom environment, the work on the relative effectiveness of cooperative, competitive and individualistic goal structures stands out because of the volume of studies completed (Johnson & Johnson 1991). Although many past studies of student achievement have found that cooperative learning is more successful than either competitive or individualistic learning, the evidence is not always consistent. The generally positive effect of cooperative learning approaches on student achievement is illustrated by the findings of a comprehensive meta-analysis involving 122 studies (Johnson, Maruyama, Johnson, Nelson & Skon 1981), but this synthesis is not totally conclusive and generalisable. For instance, a large proportion of these studies involved group outcomes (e.g., the group's ability to solve problems) rather than the conventional student individual outcome which is so important in primary and secondary schooling.

Educational Productivity Research

Psychosocial learning environment has been incorporated as one factor in a multi-factor psychological model of educational productivity (Walberg 1981). This theory, which is based on an economic model of agricultural, industrial and national productivity, holds that learning is a multiplicative, diminishing-returns function of student age, ability and motivation; of quality and quantity of instruction; and of the psychosocial environments of the home, the classroom, the peer group and the mass media. Because the function is multiplicative, it can be argued in principle that any factor at a zero-point will result in zero learning; thus either zero motivation or zero time for instruction will result in zero learning. Moreover, it will do less good to raise a factor that already is high than to improve a factor that currently is the main constraint to learning. Empirical probes of the educational productivity model were made by carrying out extensive research syntheses involving the correlations of learning with the factors in the model (Fraser, Walberg, Welch & Hattie 1987; Walberg 1986) and secondary analyses of large data bases collected as part of the National Assessment of Educational Achievement (Walberg 1986) and National Assessment of Educational Progress (Fraser, Welch & Walberg 1986; Walberg, Fraser & Welch 1986). Classroom and school environment was found to be a strong predictor of both achievement and attitudes even when a comprehensive set of other factors was held constant.

*Use of Environment Perceptions as Criterion Variables*

Evaluation of Educational Innovations

Classroom environment instruments can be used as a source of process criteria in the evaluation of educational innovations (Fraser, Williamson & Tobin 1987). An evaluation of the Australian Science Education Project (ASEP) revealed that, in comparison with a control group, ASEP students perceived their classrooms as being more satisfying and individualised and having a better material environment (Fraser 1979). The significance of this evaluation is that classroom environment variables differentiated revealingly between curricula, even when various outcome measures showed negligible differences. Recently, the incorporation of a classroom environment instrument within an evaluation of the use of a computerised database revealed that students perceived that their classes became more inquiry oriented during the use of the innovation (Maor & Fraser 1996). Similarly, in two studies in Singapore, classroom environment measures were used as dependent variables in evaluations of computer-assisted learning (Teh & Fraser 1994) and computer application courses for adults (Khoo & Fraser 1997). In an evaluation of an urban systemic reform initiative in the USA, use of the CLES painted a disappointing picture in terms of a lack of success in achieving constructivist-oriented reform of science education (Dryden & Fraser 1996).

## Differences Between Student and Teacher Perceptions of Actual and Preferred Environment

An investigation of differences between students and teachers in their perceptions of the same actual classroom environment and of differences between the actual environment and that preferred by students or teachers was reported by Fisher and Fraser (1983a) using the ICEQ with a sample of 116 classes for the comparisons of student actual with student preferred scores and a subsample of 56 of the teachers of these classes for contrasting teachers' and students' scores. Students preferred a more positive classroom environment than was actually present for all five ICEQ dimensions. Also, teachers perceived a more positive classroom environment than did their students in the same classrooms on four of the ICEQ's dimensions. These results replicate patterns emerging in other studies in school classrooms in the USA (Moos 1979), Israel (Hofstein & Lazarowitz 1986), The Netherlands (Wubbels, Brekelmans & Hooymayers 1991) and Australia (Fraser 1982b; Fraser & McRobbie 1995), and in other settings such as hospital wards and work milieus (e.g., Moos 1974).

## Studies Involving Other Independent Variables

Classroom environment dimensions have been used as criterion variables in research aimed at identifying how the classroom environment varies with such factors as teacher personality, class size, grade level, subject matter, the nature of the school-level environment and the type of school (Fraser 1994). For example, larger class sizes were found to be associated with greater classroom Formality and less Cohesiveness (Anderson & Walberg 1972). Kent and Fisher (1997) established associations between teacher personality and classroom environment (e.g., extravert teachers' classes having high levels of Student Cohesiveness). Knight (1992) reported differences in the classroom environment perceptions of African American and Hispanic students, and Levy, Wubbels, Brekelmans and Morganfield (1994) reported cultural differences (based on place of birth and primary language spoken at home) in student perceptions of teacher-student interaction.

Several studies have attempted to bring the fields of classroom environment and school environment together by investigating links between classroom and school environment (Fisher, Fraser & Wubbels 1993; Fisher, Grady & Fraser 1995; Fraser & Rentoul 1982). When Dorman, Fraser and McRobbie (1997) administered a classroom environment instrument to 2,211 students in 104 classes and a school environment instrument to 208 teachers of these classes, only weak associations between classroom environment and school environment were found. Although school rhetoric often would suggest that the school ethos would be transmitted to the classroom level, it appears that classrooms are somewhat insulated from the school as a whole.

In a study of students' preferences for different types of classroom environments, girls were found to prefer cooperation more than boys, but boys preferred both competition and individualisation more than girls (Owens & Straton 1980). Similarly, Byrne, Hattie and Fraser (1986) found that boys preferred friction, competitiveness and differentiation more than girls, whereas girls preferred teacher structure, personalisation and participation more than boys. Several studies have revealed that females generally hold perceptions of their classroom environments that are somewhat more favourable than the perceptions of males in the same classes (Fisher, Fraser & Rickards 1997; Fraser, Giddings & McRobbie 1995; Henderson, Fisher & Fraser 1995).

*Person-Environment Fit Studies of Whether Students Achieve Better in Their Preferred Environment*

Using both actual and preferred forms of educational environment instruments permits exploration of whether students achieve better when there is a higher similarity between the actual classroom environment and that preferred by students. By using a person-environment interaction framework, it is possible to investigate whether student outcomes depend, not only on the nature of the actual classroom environment, but also on the match between students' preferences and the actual environment (Fraser & Fisher 1983b, 1983c; Wong & Watkins 1996). Using the ICEQ with a sample of 116 class means, Fraser and Fisher's study involved the prediction of posttest achievement from pretest performance, general ability, the five actual individualisation variables and five variables indicating actual-preferred interaction. Overall, the findings suggested that actual-preferred congruence (or person-environment fit) could be as important as individualisation *per se* in predicting student achievement of important affective and cognitive aims. The practical implication of these findings is that class achievement of certain outcomes might be enhanced by attempting to change the actual classroom environment in ways which make it more congruent with that preferred by the class.

## TEACHERS' ATTEMPTS TO IMPROVE CLASSROOM AND SCHOOL ENVIRONMENTS

Although much research has been conducted on educational environments, less has been done to help teachers to improve the environments of their own classrooms or schools. This section reports how feedback information based on student or teacher perceptions can be employed as a basis for reflection upon, discussion of, and systematic attempts to improve classroom and school environments (Fraser 1981). The proposed methods have been applied successfully in studies at the early childhood (Fisher, Fraser & Bassett 1995), primary (Fraser & Deer 1983; Fraser, Docker & Fisher 1988), secondary (Fraser, Seddon & Eagleson 1982; Thorp, Burden & Fraser 1994; Woods & Fraser 1996) and higher education levels (Fisher & Parkinson in press; Yarrow & Millwater 1995; Yarrow, Millwater & Fraser 1997).

The attempt at improving classroom environments described below (Fraser & Fisher 1986) made use of the short 24-item version of the CES discussed previously. The class involved in the study consisted of 22 grade 9 boys and girls of mixed ability studying science at a government school in Tasmania. The procedure followed by the teacher of this class incorporated the following five steps:

(1) *Assessment*. All students in the class responded to the preferred form of the CES first, while the actual form was administered in the same time slot one week later.

(2) *Feedback*. The teacher was provided with feedback information derived from student responses in the form of the profiles shown in Figure 1 representing the class means of students' actual and preferred environment scores. These profiles permitted ready identification of the changes in classroom environment needed to reduce major differences between the nature of the actual environment and the preferred environment as currently perceived by students. Figure 1 shows that the interpretation of the larger differences was that students would prefer less Friction, less Competitiveness and more Cohesiveness.

(3) *Reflection and discussion*. The teacher engaged in private reflection and informal discussion about the profiles in order to provide a basis for a decision about whether an attempt would be made to change the environment in terms of some of the dimensions. The main criteria used for selection of dimensions for change were, first, that there should exist a sizeable actual-preferred difference on that variable and, second, that the teacher should feel concerned about this difference and want to make an effort to reduce it. These considerations led the teacher to decide to introduce an intervention aimed at increasing the levels of Teacher Support and Order and Organisation in the class.

(4) *Intervention*. The teacher introduced an intervention of approximately two months' duration in an attempt to change the classroom environment. This intervention consisted of a variety of strategies, some of which originated during discussions between teachers, and others of which were suggested by examining ideas contained in individual CES items. For example, strategies used to enhance Teacher Support involved the teacher moving around the class more to mix with students, providing assistance to students and talking with them more than previously. Strategies used to increase Order and Organisation involved taking considerable care with distribution and collection of materials during activities and ensuring that students worked more quietly.

(5) *Reassessment*. The student actual form of the scales was re-administered at the end of the intervention to see whether students were perceiving their classroom environments differently from before.

The results summarised graphically in Figure 1 show that some change in actual environment occurred during the time of the intervention. When tests of statistical significance were performed, it was found that pretest-posttest differences were

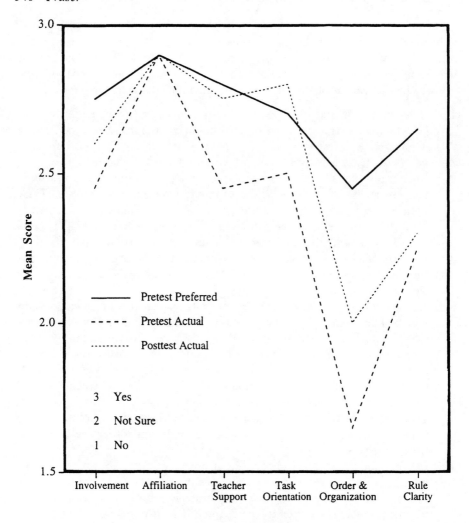

**Figure 1: Pretest actual, pretest preferred and posttest actual means.**

significant ($p<0.05$) only for Teacher Support, Task Orientation and Order and Organisation. These findings are noteworthy because two of the dimensions on which appreciable changes were recorded were those on which the teacher had attempted to promote change. (Note also that there appears to be a side effect in that the intervention could have resulted in the classroom becoming more task oriented than the students would have preferred.) Although the second administration of the environment scales marked the end of this teacher's attempt at changing a classroom, it might have been thought of as simply the beginning of another cycle.

Woods and Fraser (1995) used this basic approach to improving classroom environments with 16 teachers who used the actual and preferred forms of the Classroom Interaction Patterns Questionnaire to assess student perceptions of six dimensions of teacher behaviour (Praise and Encouragement, Open Questioning, Lecture and Direction, Individual Work, Discipline and Management, and Group Work). Whereas half of the teachers received feedback and attempted changes in their classrooms, the other half only administered the questionnaires. The study showed that the teachers who received feedback, compared with the teachers who didn't receive feedback, were able to achieve more reductions in actual-preferred discrepancies on most classroom environment dimensions.

Yarrow, Millwater and Fraser (1997) reported a study in which 117 preservice education teachers were introduced to the field of learning environment through being involved in action research aimed at improving their university teacher education classes and their 117 primary school classes during teaching practice. The CUCEI was used at the university level and the MCI was used at the primary level. Improvements in classroom environment were observed, and the preservice teachers generally valued both the inclusion of the topic of learning environment in their preservice programs and the opportunity to be involved in action research aimed at improving classroom environments.

The methods described previously for improving classroom environments have been adapted for use by teachers wishing to improve their school-level environments. Fraser, Docker and Fisher (1988) used the WES as part of teacher development activities and reported a case study of a successful school change attempt in a primary school with a staff of 24 teachers. The SLEQ (Fisher & Fraser 1991) was used in similar school improvement studies using the same basic strategy in a primary school of 15 teachers. After an intervention had been implemented for approximately 10 weeks, it was found that sizeable changes had occurred in two of the targeted areas (of about two-thirds of a standard deviation and about half a standard deviation, respectively).

## RECENT TRENDS AND DESIRABLE FUTURE DIRECTIONS

### Combining Quantitative and Qualitative Methods

Educational researchers claim that there are merits in moving beyond choosing between quantitative *or* qualitative methods, to combining quantitative *and* qualitative methods. In recent years, significant progress has been made towards the desirable goal of combining quantitative and qualitative methods within the same study in research on classroom learning environments (see Fraser & Tobin 1991 and Tobin & Fraser's chapter in this *Handbook*).

For example, a team of 13 researchers was involved in over 500 hours of intensive classroom observation of 22 exemplary teachers and a comparison group of nonexemplary teachers (Fraser & Tobin 1989). The main data collection methods were based on interpretive research methods and involved classroom observation, interviewing of students and teachers, and the construction of case studies. But, a distinctive feature was that the qualitative information was complemented by quantitative information obtained from questionnaires assessing student perceptions of classroom psychosocial environment. These instruments furnished a picture of life in exemplary teachers' classrooms as seen through the students' eyes. The study suggested that, first, exemplary and non-exemplary teachers could be differentiated in terms of the psychosocial environments of their classrooms as seen through their students' eyes and, second, that exemplary teachers typically create and maintain environments that are markedly more favourable than those of non-exemplary teachers (Fraser & Tobin 1989).

In a study which focused on the elusive goal of higher-level cognitive learning, a team of six researchers intensively studied the grade 10 science classes of two teachers (Peter and Sandra) over a ten-week period (Tobin, Kahle & Fraser 1990). Each class was observed by several researchers, interviewing of students and teachers took place on a daily basis, and students' written work was examined. The study also involved quantitative information from questionnaires assessing student perceptions of classroom psychosocial environment. Students' perceptions of the learning environment within each class were consistent with the observers' field records of the patterns of learning activities and engagement in each classroom. For example, the high level of Personalisation perceived in Sandra's classroom matched the large proportion of time that she spent in small-group activities during which she constantly moved about the classroom interacting with students. The lower level of Personalisation perceived in Peter's class was associated partly with the larger amount of time spent in the whole-class mode and the generally public nature of his interactions with students.

Fraser's (1996) multilevel study of the learning environment of a science class in Australia incorporated a teacher-researcher perspective as well as the perspective of six university-based researchers. The research commenced with an interpretive study of a grade 10 science teacher's classroom learning environment at one school, which provided a challenging learning environment in that many students were from working class backgrounds, some were experiencing problems at home, and others had English as a second language. Qualitative methods involved several of the researchers visiting this class each time it met over five weeks, using student diaries, and interviewing the teacher-researcher, students, school administrators and parents. A video camera recorded activities during each lesson for later analysis. Field notes were written during and soon after each observation, and team meetings took place three times per week. The qualitative component of the study was complemented by a quantitative component involving the use of a questionnaire which linked three levels: the class in which the interpretive study was undertaken; selected classes from within the school; and classes distributed throughout the same State. This enabled a judgement to be made about whether this teacher was typical of other teachers at her school, and whether the school was typical of other schools within the State. Some of the features identified as salient in this teacher's classroom environment were peer pressure and an emphasis on laboratory activities. (For another example of a multilevel classroom environment study which combined quantitative and qualitative methods, see Waxman, Huang & Wang 1996.)

*School-Level Environments*

Although science education researchers have paid more attention to classroom environment research than to school environment research, desirable future directions include a greater emphasis on the school-level environment and the integration of classroom and school climate variables within the same study. Docker, Fraser and Fisher (1989) reported the use of the WES with a sample of 599 teachers in investigating differences between the environment of various school types. Reasonable similarity was found for preferred environment scales, but teachers' perceptions of their actual school environments varied markedly in that the climate in primary schools was more favourable than the environment of high schools on most scales. For example, primary schools were viewed as having greater Involvement, Staff Support, Autonomy, Task Orientation, Clarity, Innovation and Physical Comfort and less Work Pressure. Similarly, when the SLEQ was used in a study of differences between the climates of primary and high schools for a sample of 109 teachers in 10 schools (Fisher & Fraser 1991), the most striking finding was that the climate in primary schools emerged as more favourable than the environment of high schools on most SLEQ scales. Dorman and Fraser (1996) used a school environment questionnaire based on the SLEQ in a comparison of Catholic and government schools. Data from 208 science and religion teachers from 32 schools showed significant differences of approximately one standard deviation between the two school types on teacher-perceived Mission Consensus and Empowerment. Catholic school teachers saw their schools as more empowering and higher on Mission Consensus than government school teachers.

*School Psychology*

Given the school psychologist's changing role, the field of psychosocial learning environment provides a good example of an area which furnishes a number of ideas, techniques and research findings which could be valuable in school psychology (Fraser 1987; Hertz-Lazarowitz & Od-Cohen 1992). Traditionally, school psychologists have tended to concentrate heavily and sometimes exclusively on their roles in assessing and enhancing academic achievement and other valued learning outcomes. The field of classroom environment provides an opportunity for school psychologists and teachers to become sensitised to subtle but important aspects of classroom life, and to use discrepancies between students' perceptions of actual and preferred environment as a basis to guide improvements in classrooms (Burden

& Fraser 1993). Similarly, expertise in assessing and improving school environment can be considered important in the work of educational psychologists (Burden & Fraser 1994).

*Links Between Educational Environments*

Although most individual studies of educational environments in the past have tended to focus on a single environment, there is potential in simultaneously considering the links between and joint influence of two or more environments. For example, Marjoribanks (1991) shows how the environments of the home and school interact and codetermine school achievement, and Moos (1991) illustrates the links between school, home and parents' work environments. Several studies have established associations between school-level and classroom-level environment (Dorman, Fraser & McRobbie 1997; Fraser & Rentoul 1982). In order to investigate whether the socio-cultural environment influences Nigerian students' learning of science, Jegede, Fraser and Okebukola (1994) developed and validated the *Socio-Cultural Environment Scale* to assess students' perceptions of Authoritarianism, Goal Structure, African World-View, Societal Expectations and Sacredness of Science with 600 senior secondary students. Apparently, students' socio-cultural environment in non-Western societies can create a wedge between what is taught and what is learned.

*Cross-National Studies*

Science education research which crosses national boundaries offers much promise for generating new insights for at least two reasons (Fraser 1997). First, there usually is greater variation in variables of interest (e.g., teaching methods, student attitudes) in a sample drawn from multiple countries than from a one-country sample. Second, the taken-for-granted familiar educational practices, beliefs and attitudes in one country can be exposed, made 'strange' and questioned when research involves two countries.

Huang and Fraser (1997) reported one of the few cross-national studies undertaken in science education. It involved six Australian and seven Taiwanese science education researchers in working together on a cross-national study of learning environments. The WIHIC was administered to 50 junior high school science classes in each of Taiwan (1,879 students) and Australia (1,081 students). An English version of the questionnaire was translated into Chinese, followed by an independent back translation of the Chinese version into English again by team members who were not involved in the original translation. Qualitative data, involving interviews with teachers and students and classroom observations, were collected to complement the quantitative information and to clarify reasons for patterns and differences in the means in each country.

The scales of Involvement and Equity had the largest differences in means between the two countries, with Australian students perceiving each scale more positively than students from Taiwan. Data from the questionnaires were used to guide the collection of qualitative data. Student responses to individual items were used to form an interview schedule which was used to clarify whether items had been interpreted consistently by students and to help to explain differences in questionnaire scale means between countries. Classrooms were selected for observations on the basis of the questionnaire data, and specific scales formed the focus for observations in these classrooms. The qualitative data provided valuable insights into the perceptions of students in each of the countries, helped to explain some of the differences in the means between countries, and highlighted the need for caution when interpreting differences between the questionnaire results from two countries with cultural differences.

*Transition from Primary to High School*

There is considerable interest in the effects on early adolescents of the transition from primary school to the larger, less personal environment of the junior high school at this time of life. Midgley, Eccles and Feldlaufer (1991) reported a deterioration in the classroom environment when students moved from generally smaller primary schools to larger, departmentally-organised lower secondary schools, perhaps because of less positive student relations with teachers and reduced student opportunities for decision making in the classroom. Ferguson and Fraser's (1996) study of 1,040 students from 47 feeder primary schools and 16 linked high schools in Australia also indicated that students perceived their high school classroom environments less favourably than their primary school classroom environments, but the transition experience was different for boys and girls and for different school size 'pathways'.

*Teacher Education*

Although the field of psychosocial learning environment provides a number of potentially valuable ideas and techniques for inclusion in teacher education programs, little progress has been made in incorporating these ideas into teacher education. Fraser (1993) reported some case studies of how classroom and school environment work has been used within preservice and inservice teacher education to (1) sensitise teachers to subtle but important aspects of classroom life, (2) illustrate the usefulness of including classroom and school environment assessments as part of a teacher's overall evaluation/monitoring activities, (3) show how assessment of classroom and school environment can be used to facilitate practical improvements in classrooms and schools and (4) provide a valuable source of feedback about teaching performance for the formative and summative evaluation of student teaching. It appears that information on student perceptions of the classroom learning environment during preservice teachers' field experience adds usefully to the information obtained from university supervisors, school-based cooperating teachers and student teacher self-evaluation (Duschl & Waxman 1991). Créton, Hermans and Wubbels (1990) have used a systems communication perspective to provide guidance on how teacher education programs can be changed to improve interpersonal teacher behaviour in the classroom (e.g., changing escalating spirals of breakdown in communication).

*Teacher Assessment*

An innovative teacher assessment system called the Louisiana STAR (System for Teaching and Learning Assessment and Review) specifically includes learning environment dimensions among a set of four performance dimensions (Ellett, Loup & Chauvin 1989). The other three performance dimensions are Preparation, Planning and Evaluation (e.g., teaching methods, homework, assessment), Classroom and Behaviour Management (e.g., student engagement, monitoring student behaviour) and Enhancement of Learning (e.g., content accuracy, thinking skills, pace, feedback). With the STAR, multiple observers complete an assessment in 45 minutes by focusing on preparation and planning in addition to in-class performance, on student learning as well as teaching behaviour, on higher-level as well as lower-level student learning, and on differential provision for different children. Teachers who were effective in terms of the psychosocial learning environment dimension were found to encourage positive interpersonal relationships within a classroom environment in which students felt comfortable and accepted. The teacher, through verbal and non-verbal behaviours, modelled enthusiasm and interest in learning, included all students in learning activities and encouraged active involvement.

## DISCUSSION AND CONCLUSION

The major purpose of this chapter devoted to perceptions of psychosocial characteristics of classroom and school environments has been to make this exciting research tradition in science education more accessible to wider audiences. In its attempt to portray prior work, attention has been given to instruments for assessing classroom and school environments (including some interesting new instruments and the distinction between Personal and Class forms), several lines of previous research (e.g., associations between outcomes and environment, use of environment dimensions as dependent variables, person-environment fit studies of whether students achieve better in their preferred environment), and teachers' use of learning environment perceptions in guiding practical attempts to improve their own classrooms and schools. Also new lines of research which suggest desirable future directions for the field were discussed, including the desirability of combining quantitative and qualitative methods, school-level environment, links between different educational environments, cross-national studies, changes in

environment during the transition from primary to high school, and incorporating educational environment ideas into school psychology, teacher education and teacher assessment.

This section of the *Handbook* devoted to learning environments has five other chapters which are summarised below. In 'The Teacher Factor in the Social Climate of the Classroom', Theo Wubbels and Mieke Brekelmans review research on teachers' contributions to a positive social climate in science classes, particularly through their interaction or communication with students. The way in which a teacher interacts with students is important because it is a predictor of student learning and discipline problems and of teacher job satisfaction and burnout. The chapter considers data-gathering methods (such as observation and questionnaires, including the Questionnaire on Teacher Interaction, associations between student outcomes and teacher-student interactions, and correlates of teacher-student interactions (e.g., teacher age, experience and cognition, student gender and setting).

McRobbie, Fisher and Wong's chapter on 'Personal and Class Forms of Classroom Environment Instruments' differentiates between a learning environment questionnaire which assesses the whole-class environment (Class form) and one which assesses a student's perception of his or her role within a classroom (Personal form). Personal forms are better suited for investigating within-class subgroups and for case studies of individual students. Differences were found in the means obtained on Personal and Class forms, and these were illuminated through student interviews. The Personal form and the Class form each accounted for unique variance in student outcomes that could not be explained by the other form.

Hanna Arzi's chapter entitled 'Enhancing Science Education Through Laboratory Environments: More than Walls, Benches and Widgets' assumes that laboratory work is both a means and an end in science education and that some of school science teaching should be carried out in a flexibly-designed laboratory. The chapter considers the goal of laboratory work, the structure and function of laboratories, and the physical design of laboratories. A case study of designing science environments is reported.

In 'Reading the Furniture: The Semiotic Interpretation of Science Learning Environments', Bonnie Shapiro broadens the term 'learning environment' to include signs, symbols and rule sets as powerful features that influence learning and teaching. The chapter considers the historical development of semiotics as a research approach, and provides five case studies of the use of a semiotic interpretive perspective. Finally, implications of the semiotic perspective for teaching, learning and curriculum organisation are explored, and the value of a semiotic awareness of school learning environments is discussed.

Kenneth Tobin and Barry Fraser, in 'Qualitative and Quantitative Landscapes of Classroom Learning Environments', consider multiple theoretical perspectives for framing learning environment research and its methods, and they advocate combining qualitative and quantitative methods to maximise the potential of research. A major contribution of the chapter is a case study involving an analysis of a learning environment at multiple levels or 'grain sizes'. The credibility of assertions from this study was enhanced by the use of qualitative and quantitative information from multiple data sources and grain sizes.

Based on research on learning environments, several practical implications for policy-makers and practitioners can be drawn (see, Fraser & Wubbels 1995). First, learning environment assessments should be used in addition to student learning outcome measures to provide information about subtle but important aspects of classroom life. Second, because teachers and students have systematically different perceptions of the same classrooms, student feedback about classrooms should be collected. Third, teachers should strive to create 'productive' classroom learning environments as identified by research (e.g., classroom environments with greater organisation, cohesiveness and goal direction and less friction). Fourth, in order to improve student outcomes, classroom environments should be changed to make them more similar to those preferred by the students. Fifth, the evaluation of innovations, new curricula and reform efforts should include classroom environment assessments to provide process measures of effectiveness. Sixth, teachers should use assessments of actual and the preferred learning environments to monitor and guide attempts to improve classrooms and schools. Seventh, learning environment assessments should be used by school psychologists in helping teachers change their styles of interacting with students and improve their classroom and school environments.

## REFERENCES

Anderson, G.J. & Walberg, H.J.: 1972, 'Class Size and the Social Environment of Learning: A Mixed Replication and Extension', *Alberta Journal of Educational Research* 18, 277–286.

Bock, R.D. (ed.): 1989, *Multilevel Analysis of Educational Data*, Academic Press, San Diego, CA.

Brophy, J. & Good, T.L.: 1986, 'Teacher Behavior and Student Achievement', in M.C. Wittrock (ed.), *Handbook of Research on Teaching* (third edition), Macmillan, New York, 328–375.

Bryk, A.S. & Raudenbush, S.W.: 1992, *Hierarchical Linear Models: Applications and Data Analysis Methods*, Sage, Newbury Park, CA.

Burden, R. & Fraser, B.J.: 1993, 'Use of Classroom Environment Assessments in School Psychology: A British Perspective', *Psychology in the Schools* 30 232–240.

Burden, R.L. & Fraser, B.J.: 1994, 'Examining Teachers' Perceptions of Their Working Environments: Introducing the School Level Environment Questionnaire', *Educational Psychology and Practice* 10, 67–73.

Byrne, D.B., Hattie, J.A. & Fraser, B.J.: 1986, 'Student Perceptions of Preferred Classroom Learning Environment', *Journal of Educational Research* 81, 10–18.

Chionh, Y.H. & Fraser, B.J.: 1998, 'Validation of the "What Is Happening In This Class" Questionnaire', Paper presented at the annual meeting of the National Association for Research in Science Teaching, San Diego, CA.

Cresswell, J. & Fisher, D.L.: 1997, 'A Comparison of Actual and Preferred Principal Interpersonal Behavior', Paper presented at the annual meeting of the American Educational Research Association, Chicago, IL.

Créton, H., Hermans, J. & Wubbels, Th.: 1990, 'Improving Interpersonal Teacher Behaviour in the Classroom: A Systems Communication Perspective', *South Pacific Journal of Teacher Education* 18, 85–94.

Docker, J.G., Fraser, B.J. & Fisher, D.L.: 1989, 'Differences in the Psychosocial Work Environment of Different Types of Schools', *Journal of Research in Childhood Education* 4, 5–17.

Dorman, J.P. & Fraser, B.J.: 1996, 'Teachers' Perceptions of School Environment in Australian Catholic and Government Secondary Schools', *International Studies in Educational Administration* 24(1), 78–87.

Dorman, J.P., Fraser, B.J. & McRobbie, C.J.: 1997, 'Relationship Between School-Level and Classroom-Level Environments in Secondary Schools', *Journal of Educational Administration* 35, 74–91.

Dryden, M. & Fraser, B.J.: 1996, 'Evaluating Urban Systemic Reform Using Classroom Learning Environment Instruments', Paper presented at the annual meeting of the American Educational Research Association, New York.

Duschl, R.A. & Waxman, H.C.: 1991, 'Influencing the Learning Environment of Student Teaching', in B.J. Fraser & H.J. Walberg (eds.), *Educational Environments: Evaluation, Antecedents and Consequences*, Pergamon, London, 255–270.

Ellett, C.D., Loup, K.S. & Chauvin, S.W.: 1989, *System for Teaching and Learning Assessment and Review (STAR): Annotated Guide to Teaching and Learning*, Louisiana Teaching Internship and Teacher Evaluation Projects, College of Education, Louisiana State University, Barton Rouge, LA.

Ferguson, P.D. & Fraser, B.J.: 1996, 'School Size, Gender, and Changes in Learning Environment Perceptions During the Transition from Elementary to High School', Paper presented at the annual meeting of the American Educational Research Association, New York.

Fisher, D.L. (ed.): 1994, *The Study of Learning Environments, Volume 6*, Curtin University of Technology, Perth, Australia.

Fisher, D.L. & Fraser, B.J.: 1981, 'Validity and Use of My Class Inventory', *Science Education* 65, 145–156.

Fisher, D.L. & Fraser, B.J.: 1983a, 'A Comparison of Actual and Preferred Classroom Environment as Perceived by Science Teachers and Students', *Journal of Research in Science Teaching* 20, 55–61.

Fisher, D.L. & Fraser, B.J.: 1983b, 'Use of WES to Assess Science Teachers' Perceptions of School Environment', *European Journal of Science Education* 5, 231–233.

Fisher, D.L. & Fraser, B.J.: 1983c, 'Validity and Use of Classroom Environment Scale', *Educational Evaluation and Policy Analysis* 5, 261–271.

Fisher, D.L. & Fraser, B.J.: 1991, 'School Climate and Teacher Professional Development', *South Pacific Journal of Teacher Education* 19(1), 17–32.

Fisher, D.L., Fraser. B.J. & Bassett, J.: 1995, 'Using a Classroom Environment Instrument in an Early Childhood Classroom', *Australian Journal of Early Childhood* 20(3), 10–15.

Fisher, D.L., Fraser, B.J. & Rickards, T.: 1997, 'Gender and Cultural Differences in Teacher-Student Interpersonal Behavior', Paper presented at the annual meeting of the American Educational Research Association, Chicago, IL.

Fisher, D.L., Fraser, B.J. & Wubbels, Th.: 1993, 'Interpersonal Teacher Behavior and School Environment', in Th. Wubbels & J. Levy (eds.), *Do You Know What You Look Like: Interpersonal Relationships in Education*, Falmer Press, London, 103–112.

Fisher, D.L., Grady, N. & Fraser, B.: 1995, 'Associations Between School-Level and Classroom-Level Environment', *International Studies in Educational Administration* 23, 1–15.

Fisher, D.L., Henderson, D. & Fraser, B.J.: 1995, 'Interpersonal Behaviour in Senior High School Biology Classes', *Research in Science Education* 25, 125–133.

Fisher, D., Henderson, D. & Fraser, B.: 1997, 'Laboratory Environments & Student Outcomes in Senior High School Biology', *American Biology Teacher* 59, 214–219.

Fisher, D.L. & Parkinson, A.: in press, 'Improving Nursing Education Classroom Environment', *Journal of Nursing Education*.

Fisher, D.L. & Waldrip, B.G.: 1997, 'Assessing Culturally Sensitive Factors in the Learning Environment of Science Classrooms', *Research in Science Education* 27, 41–49.

Fraser, B.J.: 1979, 'Evaluation of a Science-Based Curriculum', in H.J. Walberg (ed.), *Educational Environments and Effects: Evaluation, Policy, and Productivity*, McCutchan, Berkeley, CA, 218–234.

Fraser, B.J.: 1981, 'Using Environmental Assessments to Make Better Classrooms', *Journal of Curriculum Studies* 13, 131–144.

Fraser, B.J.: 1982a, 'Development of Short Forms of Several Classroom Environment Scales', *Journal of Educational Measurement* 19, 221–227.

Fraser, B.J.: 1982b, 'Differences Between Student and Teacher Perceptions of Actual and Preferred Classroom Learning Environment', *Educational Evaluation and Policy Analysis* 4, 511–519.

Fraser, B.J.: 1986, *Classroom Environment*, Croom Helm, London.

Fraser, B.J.: 1987, 'Use of Classroom Environment Assessments in School Psychology', *School Psychology International* 8, 205–219.

Fraser, B.J.: 1990, *Individualised Classroom Environment Questionnaire*, Australian Council for Educational Research, Melbourne, Australia.

Fraser, B.J.: 1993, 'Incorporating Classroom and School Environment Ideas into Teacher Education Programs', in T.A. Simpson (ed.), *Teacher Educators' Annual Handbook*, Queensland University of Technology, Brisbane, Australia, 135–152.
Fraser, B.J.: 1994, 'Research on Classroom and School Climate', in D. Gabel (ed.), *Handbook of Research on Science Teaching and Learning*, Macmillan, New York, 493–541.
Fraser, B.J.: 1996, '"Grain Sizes" in Educational Research: Combining Qualitative and Quantitative Methods', Paper presented at the Conference on Improving Interpretive Research Methods in Research on Science Classroom Environments, Taipei, Taiwan.
Fraser, B.J.: 1997, 'NARST's Expansion, Internationalization and Cross-Nationalization' (1996 Annual Meeting Presidential Address), *NARST News* 40(1), 3–4.
Fraser, B.J., Anderson, G.J. & Walberg, H.J.: 1982, *Assessment of Learning Environments: Manual for Learning Environment Inventory (LEI) and My Class Inventory (MCI)* (third version), Western Australian Institute of Technology, Perth, Australia.
Fraser, B.J. & Deer, C.E.: 1983, 'Improving Classrooms Through Use of Information About Learning Environment', *Curriculum Perspectives* 3(2), 41–46.
Fraser, B.J., Docker, J.G. & Fisher, D.L.: 1988, 'Assessing and Improving School Climate', *Evaluation and Research in Education* 2(3), 109–122.
Fraser, B.J. & Fisher, D.L.: 1982, 'Predicting Students' Outcomes from Their Perceptions of Classroom Psychosocial Environment', *American Educational Research Journal* 19, 498–518.
Fraser, B.J. & Fisher, D.L.: 1983a, 'Development and Validation of Short Forms of Some Instruments Measuring Student Perceptions of Actual and Preferred Classroom Learning Environment', *Science Education* 67, 115–131.
Fraser, B.J. & Fisher, D.L.: 1983b, 'Student Achievement as a Function of Person-Environment Fit: A Regression Surface Analysis', *British Journal of Educational Psychology* 53, 89–99.
Fraser, B.J. & Fisher, D.L.: 1983c, 'Use of Actual and Preferred Classroom Environment Scales in Person-Environment Fit Research', *Journal of Educational Psychology* 75, 303–313.
Fraser, B.J. & Fisher, D.L.: 1986, 'Using Short Forms of Classroom Climate Instruments to Assess and Improve Classroom Psychosocial Environment', *Journal of Research in Science Teaching* 5, 387–413.
Fraser, B.J., Fisher, D.L. & McRobbie, C.J.: 1996, 'Development, Validation, and Use of Personal and Class Forms of a New Classroom Environment Instrument', Paper presented at the annual meeting of the American Educational Research Association, New York.
Fraser, B.J., Giddings, G.J. & McRobbie, C.J.: 1995, 'Evolution and Validation of a Personal Form of an Instrument for Assessing Science Laboratory Classroom Environments', *Journal of Research in Science Teaching* 32, 399–422.
Fraser, B.J. & McRobbie, C.J.: 1995, 'Science Laboratory Classroom Environments at Schools and Universities: A Cross-National Study', *Educational Research and Evaluation* 1, 289–317.
Fraser, B.J., McRobbie, C.J. & Giddings, G.J.: 1993, 'Development and Cross-National Validation of a Laboratory Classroom Environment Instrument for Senior High School Science', *Science Education* 77, 1–24.
Fraser, B.J. & O'Brien, P.: 1985, 'Student and Teacher Perceptions of the Environment of Elementary-School Classrooms', *Elementary School Journal* 85, 567–580.
Fraser, B.J. & Rentoul, A.J.: 1982, 'Relationship Between School-Level and Classroom-Level Environment', *Alberta Journal of Educational Research* 28, 212–225.
Fraser, B.J., Seddon, T. & Eagleson, J.: 1982, 'Use of Student Perceptions in Facilitating Improvement in Classroom Environment', *Australian Journal of Teacher Education* 7, 31–42.
Fraser, B.J. & Tobin, K.: 1989, 'Student Perceptions of Psychosocial Environments in Classrooms of Exemplary Science Teachers', *International Journal of Science Education* 11, 14–34.
Fraser, B.J. & Tobin, K.: 1991, 'Combining Qualitative and Quantitative Methods in Classroom Environment Research', in B.J. Fraser & H.J. Walberg (eds.), *Educational Environments: Evaluation, Antecedents and Consequences*, Pergamon, London, 271–292.
Fraser, B.J. & Treagust, D.F.: 1986, 'Validity and Use of an Instrument for Assessing Classroom Psychosocial Environment in Higher Education', *Higher Education* 15, 37–57.
Fraser, B.J., Treagust, D.F. & Dennis, N.C.: 1986, 'Development of an Instrument for Assessing Classroom Psychosocial Environment at Universities and Colleges', *Studies in Higher Education* 11, 43–54.
Fraser, B.J. & Walberg, H.J. (eds.): 1991, *Educational Environments: Evaluation, Antecedents and Consequences*, Pergamon, London.

Fraser, B.J., Walberg, H.J., Welch, W.W. & Hattie, J.A.: 1987, 'Syntheses of Educational Productivity Research', *International Journal of Educational Research* 11(2), 145–252. (whole issue)

Fraser, B.J., Welch, W.W. & Walberg, H.J.: 1986, 'Using Secondary Analysis of National Assessment Data to Identify Predictors of Junior High School Students' Outcomes', *Alberta Journal of Educational Research* 32, 37–50.

Fraser, B.J., Williamson, J.C. & Tobin, K.: 1987, 'Use of Classroom and School Climate Scales in Evaluating Alternative High Schools', *Teaching and Teacher Education* 3, 219–231.

Fraser, B.J. & Wubbels, Th.: 1995, 'Classroom Learning Environments', in B.J. Fraser & H.J. Walberg (eds.), *Improving Science Education*, National Society for the Study of Education, Chicago, IL, 117–144.

Goh, S.C. & Fraser, B.J.: 1996, 'Validation of an Elementary School Version of the Questionnaire on Teacher Interaction', *Psychological Reports* 79, 512–522.

Goh, S.C., Young, D.J. & Fraser, B.J.: 1995, 'Psychosocial Climate and Student Outcomes in Elementary Mathematics Classrooms: A Multilevel Analysis', *Journal of Experimental Education* 64, 29–40.

Goldstein, H.: 1987, *Multilevel Models in Educational and Social Research*, Charles Griffin, London.

Haertel, G.D., Walberg, H.J. & Haertel, E.H.: 1981, 'Socio-Psychological Environments and Learning: A Quantitative Synthesis', *British Educational Research Journal* 7, 27–36.

Halpin, A.W. & Croft, D.B.: 1963, *Organizational Climate of Schools*, Midwest Administration Center, University of Chicago, Chicago, IL.

Henderson, D., Fisher, D.L. & Fraser, B.J.: 1995, 'Gender Differences in Biology Students' Perceptions of Actual and Preferred Learning Environments', Paper presented at the annual meeting of the National Association for Research in Science Teaching, San Francisco.

Hertz-Lazarowitz, R. & Od-Cohen, M.: 1992, 'The School Psychologist as a Facilitator of a Community-Wide Project to Enhance Positive Learning Climate in Elementary Schools', *Psychology in the Schools* 29, 348–358.

Hodson, D.: 1988, 'Experiments in Science and Science Teaching', *Educational Philosophy and Theory* 20(2), 53–66.

Hofstein, A. & Lazarowitz, R.: 1986, 'A Comparison of the Actual and Preferred Classroom Learning Environment in Biology and Chemistry as Perceived by High School Students', *Journal of Research in Science Teaching* 23, 189–199.

Huang, I. & Fraser, B.J.: 1997, 'The Development of a Questionnaire for Assessing Student Perceptions of Classroom Climate in Taiwan and Australia', Paper presented at the annual meeting of the National Association for Research in Science Teaching, Chicago, IL.

Idiris, S. & Fraser, B.J.: 1997, 'Psychosocial Environment of Agricultural Science Classrooms in Nigeria', *International Journal of Science Education* 19, 79–91.

Jegede, O.J., Fraser, B.J. & Fisher, D.L.: 1995, 'The Development and Validation of a Distance and Open Learning Environment Scale', *Educational Technology Research and Development* 43, 90–93.

Jegede, O.J., Fraser, B.J. & Okebukola, P.A.: 1994, 'Altering Socio-Cultural Beliefs Hindering the Learning of Science', *Instructional Science* 22, 137–152.

Johnson, D.W. & Johnson, R.T.: 1991, 'Cooperative Learning and Classroom and School Climate', in B.J. Fraser & H.J. Walberg (eds.), *Educational Environments: Evaluation, Antecedents and Consequences*, Pergamon, London, 55–74.

Johnson, D., Maruyama, G., Johnson R., Nelson, D. & Skon, L.: 1981, 'The Effects of Cooperative, Competitive, and Individualistic Goal Structures on Achievement: A Meta-Analysis', *Psychological Bulletin* 89, 47–62.

Kent, H. & Fisher, D.L.: 1997, 'Associations Between Teacher Personality and Classroom Environment', Paper presented at the annual meeting of the American Educational Research Association, Chicago, IL.

Khoo, H.S. & Fraser, B.J.: 1997, 'The Learning Environments Associated with Computer Application Courses for Adults in Singapore', Paper presented at the annual meeting of the American Educational Research Association, Chicago, IL.

Knight, S.L.: 1992, 'Differences Among Black and Hispanic Students' Perceptions of Their Classroom Learning Environment in Social Studies', in H.C. Waxman & Chad D. Ellett (eds.), *The Study of Learning Environments, Volume 5*, University of Houston, Houston, TX, 101–107.

Levy, J., Wubbels, Th., Brekelmans, M. & Morganfield, B.: 1994, 'Language and Cultural Factors in Students' Perceptions of Teacher Communication Style, *International Journal of Intercultural Relations* 21, 29–56.

Lewin, K.: 1936, *Principles of Topological Psychology*, McGraw, New York.

MacAuley, D.J.: 1990, 'Classroom Environment: A Literature Review', *Educational Psychology* 10, 239–253.
Maor, D. & Fraser, B.J.: 1996, 'Use of Classroom Environment Perceptions in Evaluating Inquiry-Based Computer Assisted Learning', *International Journal of Science Education* 18, 401–421.
Marjoribanks, K.: 1991, 'Families, Schools, and Students' Educational Outcomes', in B.J. Fraser & H.J. Walberg (eds.), *Educational Environments: Evaluation, Antecedents and Consequences*, Pergamon, London, 75–91.
McRobbie, C.J. & Fraser, B.J.: 1993, 'Associations Between Student Outcomes and Psychosocial Science Environment', *Journal of Educational Research* 87, 78–85.
Midgley, C., Eccles, J.S. & Feldlaufer, H.: 1991, 'Classroom Environment and the Transition to Junior High School', in B.J. Fraser & H.J. Walberg (eds.), *Educational Environments: Evaluation, Antecedents and Consequences*, Pergamon, London, 113–139.
Moos, R.H.: 1974, *The Social Climate Scales: An overview*, Consulting Psychologists Press, Palo Alto, CA.
Moos, R.H.: 1979, *Evaluating Educational Environments: Procedures, Measures, Findings and Policy Implications*, Jossey-Bass, San Francisco, CA.
Moos, R.H.: 1981, *Manual for Work Environment Scale*, Consulting Psychologist Press, Palo Alto, CA.
Moos, R.H.: 1991, 'Connections Between School, Work, and Family Settings', in B.J. Fraser & H.J. Walberg (eds.), *Educational Environments: Evaluation, Antecedents and Consequences*, Pergamon, London, 29–53.
Moos, R.H. & Trickett, E.J.: 1987, *Classroom Environment Scale Manual* (second edition), Consulting Psychologists Press, Palo Alto, CA.
Murray, H.A.: 1938, *Explorations in Personality*, Oxford University Press, New York.
Owens, L.C. & Straton, R.G.: 1980, 'The Development of a Cooperative, Competitive and Individualized Learning Preference Scale for Students', *British Journal of Educational Psychology* 50, 147–161.
Rentoul, A.J. & Fraser, B.J.: 1979, 'Conceptualization of Enquiry-Based or Open Classroom Learning Environments', *Journal of Curriculum Studies* 11, 233–245.
Rentoul, A.J. & Fraser, B.J.: 1983, 'Development of a School-Level Environment Questionnaire', *Journal of Educational Administration* 21, 21–39.
Riah, H., Fraser, B.J. & Rickards, T.: 1997, 'Interpersonal Teacher Behaviour in Chemistry Classes in Brunei Darussalem's Secondary Schools', Paper presented at the International Seminar on Innovations in Science and Mathematics Curricula, Bandar Seri Begawan, Brunei Darussalam.
Roth, W.M. & Roychoudhury, A.: 1994, 'Physics Students' Epistemologies and Views about Knowing and Learning', *Journal of Research in Science Teaching* 31, 5–30.
Rutter, M., Maughan, B., Mortimore, P., Ouston, J. & Smith, A.: 1979, *Fifteen Thousand Hours: Secondary Schools and Their Effects on Children*, Harvard University Press, Cambridge, MA.
Sinclair, B.B. & Fraser, B.J.: 1997, 'The Effect of Inservice Training and Teachers' Action Research on Elementary Science Classroom Environments', Paper presented at the annual meeting of the American Educational Research Association, Chicago, IL.
Sirotnik, K.A.: 1980, 'Psychometric Implications of the Unit-of-Analysis Problem (With Examples from the Measurement of Organizational Climate)', *Journal of Educational Measurement* 17, 245–282.
Stern, G.G.: 1970, *People in Context: Measuring Person-Environment Congruence in Education and Industry*, Wiley, New York.
Stern, G.G., Stein, M.I. & Bloom, B.S.: 1956, *Methods in Personality Assessment*, Free Press, Glencoe, IL.
Taylor, P.C., Dawson, V. & Fraser, B.J.: 1995, 'Classroom Learning Environments Under Transformation: A Constructivist Perspective', Paper presented at the annual meeting of the American Educational Research Association, San Francisco, CA.
Taylor, P.C., Fraser, B.J. & Fisher, D.L.: 1997, 'Monitoring Constructivist Classroom Learning Environments', *International Journal of Educational Research* 27, 293–302.
Teh, G. & Fraser, B.J.: 1994, 'An Evaluation of Computer-Assisted Learning in Terms of Achievement, Attitudes and Classroom Environment', *Evaluation and Research in Education* 8, 147–161.
Teh, G. & Fraser, B.J.: 1995a, 'Associations Between Student Outcomes and Geography Classroom Environment', *International Research in Geographical and Environmental Education* 4(1), 3–18.

Teh, G. & Fraser, B.: 1995b, 'Development and Validation of an Instrument for Assessing the Psychosocial Environment of Computer-Assisted Learning Classrooms', *Journal of Educational Computing Research* 12, 177–193.

Thorp, H., Burden, R.L. & Fraser, B.J.: 1994, 'Assessing and Improving Classroom Environment', *School Science Review* 75, 107–113.

Tobin, K., Kahle, J.B. & Fraser, B.J. (eds.): 1990, *Windows into Science Classes: Problems Associated with Higher-Level Cognitive Learning*, Falmer Press, London.

von Saldern, M.: 1992, *Social Climate in the Classroom: Theoretical and Methodological Aspects*, Waxmann Münster, New York.

Walberg, H.J. (ed.): 1979, *Educational Environments and Effects: Evaluation, Policy, and Productivity*, McCutchan, Berkeley, CA.

Walberg, H.J.: 1981, 'A Psychological Theory of Educational Productivity', in F. Farley & N.J. Gordon (eds.), *Psychology and Education: The State of the Union*, McCutchan, Berkeley, CA, 81–108.

Walberg, H.J.: 1986, 'Synthesis of Research on Teaching', in M.C. Wittrock (ed.), *Handbook of Research on Teaching* (third edition), American Educational Research Association, Washington, DC, 214–229.

Walberg, H.J. & Anderson, G.J.: 1968, 'Classroom Climate and Individual Learning', *Journal of Educational Psychology* 59, 414–419.

Walberg, H.J., Fraser, B.J. & Welch, W.W.: 1986, 'A Test of a Model of Educational Productivity Among Senior High School Students', *Journal of Educational Research* 79, 133–139.

Waxman, H.C., Huang, S.Y. & Wang, M.C.: 1996, 'Investigating the Multilevel Classroom Learning Environments of Resilient and Non-Resilient Students from Inner-City Elementary Schools', Paper presented at the annual meeting of the American Educational Research Association, New York.

Wong, A.F.L. & Fraser, B.J.: 1995, 'Cross-Validation in Singapore of the Science Laboratory Environment Inventory', *Psychological Reports* 76, 907–911.

Wong, A.L.F. & Fraser, B.J.: 1996, 'Environment-Attitude Associations in the Chemistry Laboratory Classroom', *Research in Science and Technological Education* 14, 91–102.

Wong, A.F.L., Young, D.J. & Fraser, B.J.: 1997, 'A Multilevel Analysis of Learning Environments and Student Attitudes', *Educational Psychology* 17, 449–468.

Wong, N.Y.: 1993, 'The Psychosocial Environment in the Hong Kong Mathematics Classroom', *Journal of Mathematical Behavior* 12, 303–309.

Wong, N.Y. & Watkins, D.: 1996, 'Self-Monitoring as a Mediator of Person-Environment Fit: An Investigation of Hong Kong Mathematics Classroom Environments', *British Journal of Educational Psychology* 66, 223–229.

Woods, J. & Fraser, B.J.: 1995, 'Utilizing Feedback Data on Students' Perceptions of Teaching Style and Preferred Learning Style to Enhance Teaching Effectiveness', Paper presented at the annual meeting of the National Association for Research in Science Teaching, San Francisco, CA.

Woods, J. & Fraser, B.J.: 1996, 'Enhancing Reflection by Monitoring Students' Perceptions of Teaching Style and Preferred Learning Style', Paper presented at the annual meeting of the American Educational Research Association, New York.

Wubbels, Th. & Brekelmans, M.: 1997, 'A Comparison of Student Perceptions of Dutch Physics Teachers' Interpersonal Behavior and Their Educational Opinions in 1984 and 1993', *Journal of Research in Science Teaching* 34, 447–466.

Wubbels, Th., Brekelmans, M. & Hooymayers, H.: 1991, 'Interpersonal Teacher Behavior in the Classroom', in B.J. Fraser & H.J. Walberg (eds.), *Educational Environments: Evaluation, Antecedents and Consequences*, Pergamon, London, 141–160.

Wubbels, Th. & Levy, J. (eds.): 1993, *Do You Know What You Look Like: Interpersonal Relationships in Education*, Falmer Press, London.

Yarrow, A. & Millwater, J.: 1995, 'Smile: Student Modification in Learning Environments — Establishing Congruence Between Actual and Preferred Classroom Learning Environment', *Journal of Classroom Interaction* 30(1), 11–15.

Yarrow, A., Millwater, J. & Fraser, B.: 1997, 'Improving University and Elementary School Classroom Environments Through Preservice Teachers' Action Research', Paper presented at the annual meeting of the American Educational Research Association, New York.

# APPENDIX A

## Constructivist Learning Environment Survey

## Actual Form

**Directions for Students**

These questionnaires contain statements about practices which could take place in this class. You will be asked how often each practice takes place.

There are no 'right' or 'wrong' answers. Your opinion is what is wanted. Think about how well each statement describes what this class is like for you.

Draw a circle around

| | | |
|---|---|---|
| 1 | if the practice takes place | **Almost Never** |
| 2 | if the practice takes place | **Seldom** |
| 3 | if the practice takes place | **Sometimes** |
| 4 | if the practice takes place | **Often** |
| 5 | if the practice takes place | **Almost Always** |

Be sure to give an answer for all questions. If you change your mind about an answer, just cross it out and circle another.

Some statements in this questionnaire are fairly similar to other statements. Don't worry about this. Simply give your opinion about all statements.

**Practice Example**

Suppose you were given the statement 'I choose my partners for group discussion'. You would need to decide whether you choose your partners 'Almost always', 'Often', 'Sometimes', 'Seldom' or 'Almost never'. If you selected 'Often', then you would circle the number 2 on your questionnaire.

| Learning about the world | Almost Never | Seldom | Some-times | Often | Almost Always |
|---|---|---|---|---|---|
| In this class . . . | | | | | |
| 1. I learn about the world outside of school. | 1 | 2 | 3 | 4 | 5 |
| 2. My new learning starts with problems about the world outside of school. | 1 | 2 | 3 | 4 | 5 |
| 3. I learn how science can be part of my out-of-school life. | 1 | 2 | 3 | 4 | 5 |
| In this class . . . | | | | | |
| 4. I get a better understanding of the world outside of school. | 1 | 2 | 3 | 4 | 5 |
| 5. I learn interesting things about the world outside of school. | 1 | 2 | 3 | 4 | 5 |
| 6. What I learn has **nothing** to do with my out-of-school life. | 1 | 2 | 3 | 4 | 5 |
| **Learning about science** | Almost Never | Seldom | Some-times | Often | Almost Always |
| In this class . . . | | | | | |
| 7. I learn that science **cannot** provide perfect answers to problems. | 1 | 2 | 3 | 4 | 5 |
| 8. I learn that science has changed over time. | 1 | 2 | 3 | 4 | 5 |
| 9. I learn that science is influenced by people's values and opinions. | 1 | 2 | 3 | 4 | 5 |
| In this class . . . | | | | | |
| 10. I learn about the different sciences used by people in other cultures. | 1 | 2 | 3 | 4 | 5 |
| 11. I learn that modern science is different from the science of long ago. | 1 | 2 | 3 | 4 | 5 |
| 12. I learn that science is about creating theories. | 1 | 2 | 3 | 4 | 5 |
| **Learning to speak out** | Almost Never | Seldom | Some-times | Often | Almost Always |
| In this class . . . | | | | | |
| 13. It's OK for me to ask the teacher 'Why do I have to learn this?' | 1 | 2 | 3 | 4 | 5 |
| 14. It's OK for me to question the way I'm being taught. | 1 | 2 | 3 | 4 | 5 |
| 15. It's OK for me to complain about teaching activities that are confusing. | 1 | 2 | 3 | 4 | 5 |
| In this class . . . | | | | | |
| 16. It's OK for me to complain about anything that prevents me from learning. | 1 | 2 | 3 | 4 | 5 |
| 17. It's OK for me to express my opinion. | 1 | 2 | 3 | 4 | 5 |
| 18. It's OK for me to speak up for my rights. | 1 | 2 | 3 | 4 | 5 |

| Learning to learn | | Almost Never | Seldom | Some-times | Often | Almost Always |
|---|---|---|---|---|---|---|
| In this class . . . | | | | | | |
| 19. | I help the teacher to plan what I'm going to learn. | 1 | 2 | 3 | 4 | 5 |
| 20. | I help the teacher to decide how well I am learning. | 1 | 2 | 3 | 4 | 5 |
| 21. | I help the teacher to decide which activities are best for me. | 1 | 2 | 3 | 4 | 5 |
| In this class . . . | | | | | | |
| 22. | I help the teacher to decide how much time I spend on learning activities. | 1 | 2 | 3 | 4 | 5 |
| 23. | I help the teacher to decide which activities I do. | 1 | 2 | 3 | 4 | 5 |
| 24. | I help the teacher to assess my learning. | 1 | 2 | 3 | 4 | 5 |
| **Learning to communicate** | | Almost Never | Seldom | Some-times | Often | Almost Always |
| In this class . . . | | | | | | |
| 25. | I get the chance to talk to other students. | 1 | 2 | 3 | 4 | 5 |
| 26. | I talk with other students about how to solve problems. | 1 | 2 | 3 | 4 | 5 |
| 27. | I explain my understandings to other students. | 1 | 2 | 3 | 4 | 5 |
| In this class . . . | | | | | | |
| 28. | I ask other students to explain their thoughts. | 1 | 2 | 3 | 4 | 5 |
| 29. | Other students ask me to explain my ideas. | 1 | 2 | 3 | 4 | 5 |
| 30. | Other students explain their ideas to me. | 1 | 2 | 3 | 4 | 5 |

# 5.2 The Teacher Factor in the Social Climate of the Classroom

THEO WUBBELS and MIEKE BREKELMANS
*Utrecht University, The Netherlands*

## INTRODUCTION AND THEORETICAL BACKGROUND

This chapter reviews research on teachers' contributions to a positive social climate in science classes particularly through their interaction or communication with students. For both teacher education and professional development programs in schools, information about the role of the teacher in the learning environment is important. The way in which a teacher interacts with students is not only a predictor of student achievement, but also it is related to such factors as teacher job satisfaction and teacher burnout (Nias 1981). Appropriate teacher-students interactions are important to prevent discipline problems and to foster professional development (Rosenholtz, Bassler & Hoover-Dempsey 1986). Rather than reviewing all available studies, this chapter discusses typical studies to illustrate the methods used and the type of results found.

A communicative approach is used to analyse the contribution of the teacher to the social and affective aspects of the learning environment. We adopt the most comprehensive of three definitions of communicative behaviour. In the first definition, behaviour is called communication only if the same meaning is perceived by the sender and receiver. A second definition considers behaviour to be communicative whenever the sender consciously and purposefully intends to influence someone else. The third definition considers as communication every behaviour that someone displays in the presence of someone else. This so-called 'systems approach' (Watzlawick, Beavin & Jackson 1967), therefore, assumes that one cannot *not* communicate when in the presence of someone else. Our rationale for choosing this perspective is that, whatever someone's intentions are, the other person in the communication will infer meaning from someone's behaviour. For example, if teachers ignore students' questions because they do not hear them, then students might infer that the teacher is too busy, that the teacher thinks that the students are too dull to understand, or that the teacher considers the questions impertinent. The message that students take from the teacher's inattention can be different from the

---

Chapter Consultant: Hans Gerhard Klinzing (University of Tübingen/University of Stuttgart, Germany)

teacher's intention, because ultimately there is no shared, agreed-upon system to enable students to attach meaning.

In the systems approach, two levels of extensiveness of interactions are distinguished. Short-term interactions are the exchanges of messages each of which are of a few seconds in duration and consist of one question, one assignment, one response, one gesture, etc. In interactions, redundancy and repeating patterns can evolve over time (Wubbels, Créton & Holvast 1988) and give rise to interactions on the second level involving relatively stable interaction patterns. According to the systems approach, every form of communication has a *content* and a *relational* aspect (Watzlawick *et al.* 1967). The content conveys information or description, whereas the relational aspect carries instructions about how to interpret the content. In a class, teacher and students relate in ways which are outside the subject matter (content). This chapter focuses on the relational aspect, while not forgetting that every behaviour has at the same time both content and relational meaning. Therefore, we will not treat the communication of content explicitly as is done in the literature of discourse analysis (see Lemke's and Sutton's chapters in this volume).

## DATA-GATHERING METHODS FOR TEACHER-STUDENTS RELATIONAL COMMUNICATION

The contribution of the teacher in teacher-students communication can be studied in several ways. To study short-term interactions, usually observations are employed either with hand or notebook computer scoring. Videotaping improves the quality of this type of data collection because interactions can be reviewed time and time again to get valid and reliable scores. Thus observer perceptions of these interactions are gathered. For extended patterns over time, these instruments are not economical because they involve a lot of coding and observation time. Instead, other instruments, such as student and teacher questionnaires and interviews, often are used. These instruments map the participants' views of the interactions. It is important to keep in mind that, with these different methods, conceptually different variables are investigated.

*Structured Observations*

Observation of teacher-students communication in the classroom has a long and firm tradition. Following the development of one of the first instruments for education by Flanders (1970), a plethora of instruments has been documented (Borich & Madden 1977; Good & Brophy 1991), including ones specifically for science education (e.g., Brady, Swank, Taylor & Freiberg 1992; Shymansky & Penick 1979; Tamir 1981). These instruments record observer perceptions of ongoing behaviours of teacher and/or students within the classroom in order to analyse patterns in the communication. They usually are easy to handle, but extensive training is necessary. Scoring categories can include both verbal elements (question type, source of initiative) and non-verbal elements such as gestures and facial expression. Behaviours are coded using either an event or a time-sampling basis. In the *Science Teaching Observation Schedule* (STOS; Galton & Eggleston 1979), for example, three main teacher talk categories are distinguished: teacher asks questions (seven sub-categories including recalling facts); teacher makes statements (four sub-categories including one about problems); and teacher directs students to sources of information (four sub-categories designating the purpose, including one for seeking guidance on experimental procedure). There are two main categories for talk and activity initiated and/or maintained by students: students seek information or consult (four sub-categories designating the purpose, including one for making inferences) and students refer to teachers (four sub-categories designating the purpose, including one for seeking guidance on experimental procedures).

Another observation schedule is based on research on teacher-students relationships (see Wubbels, Créton, Levy & Hooymayers 1993). In this system, classroom interaction is analysed on the basis of two dimensions. The *proximity* dimension runs from Cooperation to Opposition and designates the degree of closeness between teacher and students. The *influence* dimension runs from Dominance to Submission and indicates who is directing or controlling the communication and how often. For example, when a teacher is lecturing uninterrupted, his or her behaviour is graphed in the upper right part of the chart in Figure 1. If the students listen in an interested way, this behaviour is shown in the lower right part of Figure 1. The two-dimensional chart can be refined by drawing two extra lines as in Figure 2. This figure (the Model for Interpersonal Teacher Behaviour) provides examples of eight categories of behaviours displayed by teachers: Leadership; Helpful/Friendly; Understanding; Student Responsibility/Freedom; Uncertain; Dissatisfied; Admonishing; and Strict behaviour. Instead of scoring behaviours in the eight categories, they also can be scored on two rating scales (Figure 3).

*Qualitative Observations*

Ethnographic (participant and non-participant) observations often are used to investigate the relational aspect of teacher-students interactions (see Erickson's chapter in this volume). The type of field notes taken depend on the research question. In the data analysis phase, these observations can be categorised under several headings. Usually, after an initial non-structured phase, observations become more focused on a specific topic. An example of this is Gallagher and Tobin's (1987) study of teacher management and student engagement in high school science. Narrative records were taken to provide a description of activities and a time log of their occurrence and duration. In addition, the cognitive level and the substantive content of the interaction were estimated.

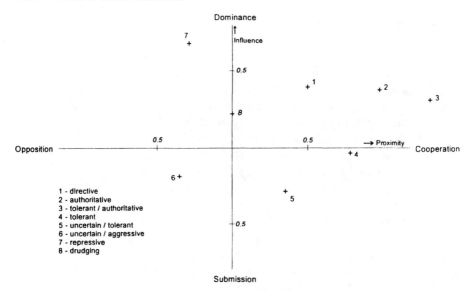

Figure 1: Coordinate System of the Model for Interpersonal Teacher Behaviour and the Eight Types of the Teacher Communication Style Typology

*Student and Teacher Questionnaires*

In research on classroom social climate, gathering of participants' views has a strong tradition. The advantages of this procedure relative to observational measures, as described by Fraser in this volume, also hold for measuring long-term patterns in communication. Scales that directly or more indirectly give information about the teacher-students' relationships (see Fraser in this volume) are contained in the *Learning Environment Inventory* (LEI) (e.g., Goal Direction, Formality and Disorganisation), the *Classroom Environment Scale* (CES) (e.g., Teacher Support, Order and Organisation, Task Orientation, Rule Clarity and Teacher Control), and the *Individualised Classroom Environment Questionnaire* (ICEQ) (e.g., Participation, Personalisation, Independence and Investigation). The Questionnaire on Teacher Interaction (QTI) was developed specifically to investigate teacher-students relationships, is based on the model for interpersonal teacher behaviour, and is divided into eight scales which conform to the eight sectors of the model. The items in the Dutch and American version are answered on a five-point Likert scale. To make the QTI more accessible to teachers, a shorter (48-item) version was developed with a hand-scoring procedure (Wubbels *et al.* 1993). A version for the primary school level was developed in Singapore with a simpler reading level and a three-point response format (Goh, Young & Fraser 1995). Several studies have shown that the QTI has good reliability and validity (Fisher, Henderson & Fraser 1995; Goh & Fraser 1995; Wubbels *et al.* 1993).

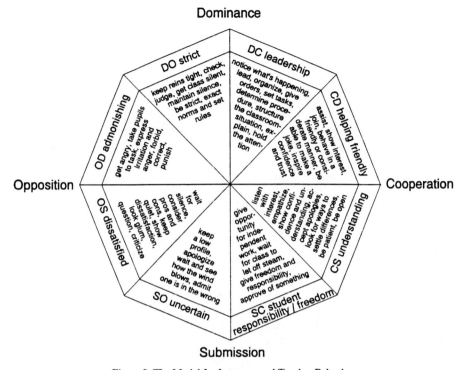

Figure 2: The Model for Interpersonal Teacher Behaviour

| **Dominance (D)** The teacher determines the students' activities. | 5–4–3–2–1 | **Submission (S)** The students can dertermine their own activities. |
|---|---|---|
| **Cooperation (C)** The teacher shows approval of the students and their behaviour | 5–4–3–2–1 | **Opposition (O)** The teacher shows disapproval of the students and their behaviour. |

Figure 3: The DS and CO rating scales

*Teacher and Student Interviews*

The classroom environment questionnaires provide information about students' and teachers' perceptions of teacher-students relationships. In order to understand more fully participants' views, open-ended interviews are helpful because they give participants the opportunity to describe the relationships in their own words. In addition, they have been used in several studies to gather data about underlying beliefs, attitudes, cognitions, intentions, the history of the relationship, interpretations of differences between teachers' and students' perceptions, etc. (e.g., Tobin,

Kahle & Fraser 1990). Finally, interviews also are used as a basis for developing or modifying questionnaire items.

## TEACHER-STUDENTS INTERACTIONS AND STUDENT OUTCOMES

Relations between classroom interaction and student outcomes have been analysed in several studies using typologies of patterns in teacher-students interaction (i.e., communication styles). Non-verbal behaviour and instructional strategies play a role in these relationships.

*Teacher-Students Communication Styles*

The most familiar typologies of teacher-students communication styles make the distinction between directive and non-directive communication styles (e.g., Bennett 1976). Briefly, open, non-directive teachers emphasise support, innovative instructional procedures and flexible rules, whereas directive teachers stress centralised control and seek to develop competitive, task-oriented classes (Schultz 1982). A number of different sub-styles have been identified between these two extremes (Ramsay & Ransley 1986).

Other studies have extended these typologies to cover more refined categories for communication styles. Galton and Eggleston (1979), for example, based on research with the Science Teaching Observation Schedule (STOS), identified three communication styles in science education. *Problem solvers* are teachers who ask relatively many questions and emphasise problems, hypotheses and experimental procedures. *Informers* are characterised by infrequent use of questions except those demanding recall and the application of facts and principles to problem solving. In the classroom of the third type (the *enquirers*), students initiate interactions more often than in the other classrooms, and they particularly seek information and guidance in designing experimental procedures and in inferring, formulating and testing hypotheses.

Brekelmans, Levy and Rodriguez (1993) developed a typology of eight categories based on student QTI data. Three of the categories (the Directive, Authoritative and the Tolerant/Authoritative types) are perceived primarily in the CD quadrant (Figure 1). Two other types are also very close to this quadrant: the Drudging teacher's behaviour can be located exactly on the influence dimension just above the CO-axis; and the Tolerant teacher's behaviour fits just below the proximity axis in the CS quadrant. The three types in the CD quadrant represent more than 50 percent of the teachers in any sample studied to date. The three types of teachers in the CD quadrant all show about the same amount of influence. While each one is fairly dominant, they differ in the amount of proximity. The Directive teacher is least cooperative and the Tolerant/Authoritative teacher is most cooperative. The Drudging teacher is a little less dominant and much less cooperative than the other three types. The Tolerant teacher is about as cooperative as the Authoritative teacher, but far less dominant. The Uncertain/Aggressive and Uncertain/Tolerant profiles are most noteworthy for their low scores on the influence dimension. Both are seen as far more submissive than the other types. They differ strikingly from each other on the proximity dimension. The Uncertain/Tolerant teacher resembles the Directive teacher in cooperation, whereas the Uncertain/Aggressive teacher is similar to the Repressive teacher in being highly oppositional. Finally, the Repressive teacher is the highest of all on the influence dimension.

*Communication Styles and Student Outcomes*

How do communication styles relate to student outcomes? Bennett (1976), in a classical study of teacher communication style and student progress, found that a formal teaching style with emphasis on external motivation, no choice for students, structured teaching and seatwork with good teacher monitoring and frequent evaluation was more effective than informal teaching characterised by choice for students, little emphasis on evaluation and control, and integration of subjects. Ramsay and Ransley (1986) found that open teaching styles were associated with the least favourable outcomes. It should not be concluded from studies like these, however, that only teacher-centred teaching can be effective. The distinction between only two styles is not sufficiently differentiated for this conclusion. The relatively poor results of informal teaching possibly could be experienced by disorder rather than of the opportunity for students to have input (as discussed later in this chapter).

Research using the STOS is an example of a study in science education using multiple outcome measures (Galton & Eggleston 1979). It generally showed that the three teaching styles did not differ in student performance for below average students. The enquirer style more than the other styles, however, seemed to help low-ability students to enjoy science. The informer style generally was the least effective, particularly for affective outcomes. The problem-solver style was most effective for high-ability students' performance in physics (recall, data manipulation and problem solving). In general, these results give support to those who advocate less of a transmission type of science teaching.

Recent studies of the relationships between teacher behaviour and student outcomes that have been carried out with the QTI have involved physics teachers in grade 9 in The Netherlands (Brekelmans, Wubbels & Créton 1990), grade 12 biology classes in Australia (Fisher, Henderson & Fraser 1995) and grade 5 mathematics students in Singapore (Goh & Fraser 1995). The results of these studies indicate medium to strong relations between student outcomes and student perceptions of teacher behaviour. The relations are stronger for affective than for cognitive outcomes. Goh and Fraser (1995), for example, found a multiple correlation between the eight QTI scores and liking of mathematics of 0.51 at the individual level and 0.89 at the class level. For student achievement, these multiple correlations were 0.38 and 0.82, respectively. These studies show that student perceptions of leadership, helpful/friendly and understanding behaviours are positively related to both student attitudes and student achievement. Uncertain, dissatisfied and admonishing behaviours are negatively related to student outcomes. In science classes in Australia and The Netherlands, negative relationships also were found between teacher strict behaviour and student attitudes, although this result was not replicated in the Singapore study. The negative relationship between student performance and the amount of responsibility and freedom found in the Dutch study was not replicated in the Australian and Singapore study. The Fisher *et al.* (1995) study is distinctive in that it also investigated student performance on a practical test, which is an important outcome in science education. Better practical test performance was associated with giving students more responsibility and freedom and being less strict. The Fisher *et al.* (1995) study was somewhat exceptional in that a significant relationship emerged only for attitudes towards the class and not for student achievement. This could have arisen from the fact that the study was done during the final year of schooling when there were external examinations. It can be expected that students would have studied a lot because of the examination pressure and that therefore they were able to compensate for differences in effects of the teachers' teaching.

The direction of relationships between teacher interpersonal behaviour and student outcomes described above confirm earlier findings about the effectiveness of direct instruction strategies (Brophy & Good 1986). For one aspect of teacher behaviour, the results extend prior research: disorder more than openness seems to be associated with poor student outcomes. Therefore, it is essential that teachers using open teaching styles are able to control student input and procedures in class so as to avoid disorder. The differences in the results found in different countries highlight the need for more research into whether students respond differently to teacher behaviour in different cultures.

However, the studies reviewed here were correlational and therefore they do not warrant causal inferences. Certain teacher behaviours can build a working climate in the class and promote student outcomes whereas other behaviours could hinder student learning. However, it also is plausible that a certain class composition or student characteristics help to build a positive classroom atmosphere and that this atmosphere gives teachers the possibility to, and even stimulates them to, show behaviours that are positively related to student outcomes. Probably the relationship will be bi-directional, with negative and positive circular processes between teacher behaviour, classroom atmosphere and student outcomes.

*Non-Verbal Teacher Behaviour*

Research indicates that non-verbal aspects of behaviour are important for their interpersonal significance (Argyle 1988) and are related to student outcomes, particularly affective outcomes (McCroskey & Richmond 1992). For example, Van

Tartwijk, Fisher, Fraser and Wubbels (1994) found that, in whole-class settings, teacher non-verbal behaviour explains more than 60 percent of differences between teachers in terms of observers' Dominance-Submission ratings of their behaviour and more than 30 percent in terms of Cooperation-Opposition.

Non-verbal behaviours that imply visual contact with the class and emphatic verbal presence are important during whole-class teaching for the rating of teacher behaviour as relatively dominant. When teachers are relatively close to the students, or when they cannot see the students, their behaviour is rated as relatively submissive (see also Andersen & Andersen 1982). These results are in line with earlier work, reported by Woolfolk and Brooks (1983), that showed that an assertive way of speaking is important for the class management success of teachers, and of Kounin (1970), who emphasised the importance of 'withitness' and 'overlapping'. The major aspect of nonverbal behaviour for explaining variance in the degree of proximity is the facial expression of the teacher. Woolfolk and Brooks (1983) refer to smiling as one of the major forms of nonverbal behaviour which leads to a cooperative perception of the teacher. Frowning, glaring, angry, disgusted, crying, fearful, sneering, smirking or mocking facial expressions imply low proximity (Mehrabian 1981). Further, when teachers raise their voices, this contributes to an oppositional rating of their behaviour.

*Instructional Strategies*

Because both observed instructional strategies and student perceptions of teacher behaviour are related to student learning (e.g., Brophy & Good 1986), it is important to ask how much teacher interpersonal behaviour and instructional strategies overlap. The only study in this area that we know of is by Levy, Rodriguez and Wubbels (1992), who found the amount of overlapping variance to be 31 percent. Significant relations were found mainly for students' perceptions of the influence dimension and instructional strategies. The more that the students perceived that teachers behave in dominant ways, the more the teachers displayed effective organisational techniques according to the observer. Further, a teacher who displayed uncertain behaviour, or allowed students a lot of freedom, or often got angry was not seen by observers to be clear in terms of directions, skill explanation or organisation. The results support the contention that, as teachers communicate uncertainty, anger, impatience and dissatisfaction, they display fewer instructional strategies associated with effectiveness.

## CORRELATES OF TEACHER-STUDENTS INTERACTIONS

Several variables can be thought to influence the way in which teachers communicate with their students. Most relationships studied appear to be weak. For example, teacher gender appears to be significantly but weakly related to teacher-students communication style (Levis 1987; Levy, Wubbels & Brekelmans 1992) and therefore will not be discussed further. Other variables of potential interest in future research are other teacher characteristics, and characteristics of the setting and school environment.

*Teacher Age and Experience*

Throughout their careers, teachers often experience periods of professional growth and decline. These peaks and valleys can affect teacher communication style. Both *experience* and *age* indeed are important to teacher communication style. Very few studies using other than self-reports are available on teaching careers. Adams (1982) showed several changes in teacher behaviour across experience levels. The largest occurred between the first and third year of teaching, when there was a dramatic increase in the organised/systematic, affective and stimulating behaviours of both primary and secondary teachers. An extensive study with the QTI (Brekelmans & Créton 1993) confirms and advances these results by indicating that, according to students, changes occur in interpersonal behaviour during the professional career, mainly in dominant behaviour. This behaviour intensifies during the first 10 years and stabilises after this point. Cooperative behaviour basically remains consistent throughout the entire teaching career, but with a slight tendency to weaken. The results suggest that teachers with about 10 years of experience have the best relationships with their students in terms of promoting student achievement and positive attitudes.

*Teacher Cognition*

Teacher cognition is often considered an important factor in teaching effectiveness. Teacher attitudes and opinions are predictors of the way in which teachers behave in class, but only weak predictors (Walberg 1986). Teachers' sense of self-efficacy is generally found to be a correlate of the quality of teacher-students interaction. The more positively that teachers rate their potential to influence student outcomes, the more that they achieve a positive classroom atmosphere. Similarly, the more that teachers think that they are able to solve problems in their teaching and the better that they think they can associate with other people, the more they create good student-teacher relationships (Ashton & Webb 1986; Rohrkemper & Brophy 1983; Woolfolk & Hoy 1990). Because of these relations, it is important to note that the QTI types of Directive, Authoritative and Tolerant teachers have the highest self-esteem (Wubbels, Brekelmans & Hermans 1987). For anxiety, the relationship is the other way around. Teachers with a high anxiety level behave in a dogmatic and authoritarian way and lack flexibility. This can produce hostile behaviour among students and make the classroom atmosphere tense and explosive. For these kinds of relationships, however, causality can be in both directions and the relationships are likely to be reciprocal. That is, a good classroom atmosphere will give teachers a high regard of their competence to help students learn and also this self-perception will help teachers to create good interactions.

In teachers' attributions of causes of student performance or problems in classrooms, two distinct patterns can influence their interactions with students. In the *ego-enhancing pattern*, teachers attribute student success to their own teaching behaviour and student failure to student characteristics such as low ability or low effort. 'Thus teachers enhance their egos by accepting responsibility for students' successes while blaming the students for their failure' (Peterson & Barger 1985, p. 171). In the other *counter-defensive pattern*, low student outcomes are explained by a teacher's failure to explain things clearly and students are given credit for their success. Clearly these two attribution patterns can be the origin of different classroom interaction patterns. In the second pattern more than in the first, the teacher will be inclined to help students and to explain difficult material again, to interact with students in order to explore their mistakes, etc. Attributions of teachers in problematic classroom situations have been investigated less frequently than their attributions for student success and failure. If teachers consider problems to be a threat to their self-esteem or to the smooth functioning of a class, they tend to attribute defensively (Rohrkemper & Brophy 1983). Such attributions make teachers want to punish students and to show aggressive behaviour which damages the quality of the classroom atmosphere. This can give rise to escalating patterns of more severe problem behaviour and harsher punishment.

Teachers tend to attribute in such a way that they keep their cognitions stable (Peterson & Barger 1985). If they think that a student is dull, they will interpret successes so they can keep thinking that this student is dull. So, in this example, they will attribute student success, for example, to the easiness of the task instead of to student ability. This brings us to the self-reinforcing function of teacher thinking in classroom interaction processes. The classical example is the Pygmalion effect described by Rosenthal and Jacobson (1968). Although the original experiment has been criticised rightly and extensively (e.g., Wineburg 1987), sufficient evidence has been gathered about the influence of teacher expectations on teacher behaviour and student outcomes (Brophy 1985). Differential teacher expectations for students go along with differential teacher treatment in terms of such things as praise, questioning, grouping of students and feedback, thus causing unequal opportunities for student learning. Teachers who have low expectations of some students, for example, tend to direct more lower-level questions to these students and more higher-order questions to students with high ability. This could stimulate high-ability students to develop more and more quickly than low-ability students, thus reinforcing teacher perceptions of students and making the prophecy become reality. These results are not by themselves a testimonial of poor teaching. It could be perfectly appropriate for teachers to teach in this way on the basis of valid expectations. In teaching, the validity of expectations, however, should be under continuous scrutiny.

Self-fulfilling prophecies have been studied primarily for teacher expectancies and student outcomes. They are also important in the process of creating a positive classroom climate. An example is the evolution of an undesirable and strongly dependent relationship between teacher and students (Wubbels *et al.* 1988). When teachers think that students cannot bear much responsibility, they might tend to give limited responsibility to students. For example, they could organise experiences rigidly and give students little opportunity for choice of subject and methods of working. Thus students have to rely on the teacher very much during their activities. In turn, this can stimulate student dependent behaviour and teachers could encourage from the students the very behaviour that they expect, thus creating a self-fulfilling prophecy.

*Student Gender*

Research has shown that teachers interact differently with boys and girls. Jones and Wheatley (1990), for example, studied teacher interactions with male and female students in secondary science classrooms. While they found no differences for several variables, such as the number of student-initiated questions and the number of abstract questions, they found that science teachers praise boys more than girls, put more questions to boys than to girls, and warn boys more often. Boys call out more frequently than girls in response to teachers' questions and boys more often ask procedural questions. In another ethnographic part of this study (Jones & Wheatley 1989), boys more often than girls were found to take a participating role, whereas girls more often took the role of an observer. Questions put to girls more often are of lower level than questions put to boys (Barba & Cardinale 1991). Becker (1981) suggests that a teacher expectancy effect about differences between female and male students leads to this differential treatment that obviously can have detrimental effects on female student outcomes. Although most studies in this field show differential treatment of boys and girls, it is intriguing that some studies did not corroborate earlier findings (e.g., Hacker 1991).

In addition to observational studies, research on student perceptions with the QTI has shown consistently that girls perceive the learning environment more positively than boys. In particular, girls tend to score the behaviour of the same teacher more dominantly and cooperatively than boys do (Goh & Fraser 1995; Levy, Créton & Wubbels 1993). This result is in line with research with other instruments such as the *Science Laboratory Environment Inventory* (SLEI) (Fraser, Giddings & McRobbie 1995).

*Setting*

Few studies have investigated differences in teacher-students interactions in different settings in science education. For example, Pizzini and Shepardson (1992) compared classroom dynamics for a problem solving and a traditional laboratory model of instruction. Teacher-students interactions in these settings appeared to differ very little. Van Tartwijk *et al.* (1994) found that the contribution of teacher-students interactions to the social climate in the science classroom is greater for the teacher's behaviour in whole-class settings than during group or laboratory work.

Physical characteristics of the classroom influence teacher-students communication (Weinstein 1979). In whole-class teaching, a short physical distance and eye contact are important to help teachers to convey to students interest, support and involvement, which are important characteristics of effective teachers. A platform for the teacher to stand on is a physical barrier which can become a psychological barrier. The traditional physics classroom with a demonstration bench could hinder a good relationship and the way in which students sit can obstruct eye contact. It is important to arrange seating in such a way that as few students as possible are sitting behind each other and so that the teacher can move freely between the students.

*School Environment*

Using the *School Level Environment Questionnaire* (SLEQ), Fisher, Fraser and Wubbels (1993) investigated relations between teachers' perceptions of the school environment and teachers' communication styles. Work Pressure, participatory decision making and professional interest appeared to be (weak) negative predictors of student perceptions of the teachers' degree of influence on students and proximity to students. While the data are preliminary, the weak relationship between the SLEQ and QTI scores indicates that a teacher's behaviour in class might have little to do with his/her perception of the school environment. It seems that teachers believe that they have considerable freedom to shape their own classroom regardless of the school atmosphere.

CONCLUSION

The research reviewed in this chapter reinforces the importance of teacher behaviour for creating a classroom atmosphere conducive for science learning. Affective variables seem to be important in a traditional classroom and even more important in a 'constructivist' classroom, where emotion plays a more prominent role (e.g., Watts & Bentley 1987). The observation instruments and questionnaires discussed in this chapter have proven to be helpful for research as well as for giving teachers feedback about their behaviour.

Based on the research reviewed in this chapter, the following recommendations for improving science education are made:

(1)  In their communication with students, teachers should strive to establish relationships characterised by high degrees of leadership, helpful/friendly and understanding behaviours. In order to succeed, teachers' non-verbal

behaviour in whole-class teaching should guarantee good visual contact (e.g., by scanning the class) and teachers should 'hold the floor' verbally. When applying open teaching styles, teachers should avoid the risk of disorderly climates.

(2) Teachers should use student questionnaires (general ones, as well as ones specifically for science education) to gather feedback about their relationships with students, as a basis for reflection and improvement of these relationships. It is important not to rely solely on teacher perceptions because usually the teacher's and students' perceptions differ widely.

(3) To improve science teaching through staff development and inservice training programs, more effort should be placed on changing teachers' behaviour than their attitudes (as attitudes are only a weak predictor of behaviour).

(4) Teachers experiencing undesirable classroom situations should focus on their own behaviour as a means for improvement and they should introduce changes in the communication pattern.

(5) Middle-aged teachers should be aware of potential detrimental effects on the classroom atmosphere of lower levels of cooperative teacher behaviour, whereas beginning teachers should focus their attention on their leadership behaviour.

(6) Teachers should self-analyse their attributions for students' success and failure in order to be aware of potential interaction patterns that emerge from self-fulfilling prophecies.

An important future direction for research on teacher-students interaction in science education involves making a distinction between the behaviour of a teacher towards the whole class and towards individual students. This distinction is elaborated in McRobbie, Fisher and Wong's chapter in this volume and has been made in observation studies (e.g., Hamilton & Brady 1991; Van Tartwijk et al. 1994). Large differences between teacher questioning behaviours have been found at these two levels. In this line of inquiry, items refer to the individual respondent rather than the class. (For example, a personal form item could be 'This teacher helps me with my work'.) This approach might cast light on the interpretation of the commonly-found gender differences in the perception of the learning environment. Is that difference a result of different teacher behaviour towards boys and girls, or do boys and girls perceive the same teacher behaviour differently?

Other desirable areas for future inquiry include the use of multilevel analyses (see Goh, Young & Fraser 1995; Levy, Wubbels, Brekelmans & Morganfield in press; and Cheung & Keeves' chapter in this volume), extension of the current research base to other countries, and investigation of the relationships between student cultural background and perceived teacher communication style.

REFERENCES

Adams, R.D.: 1982, 'Teacher Development: A Look at Changes in Teacher Perceptions and Behavior Across Time', *Journal of Teacher Education* 33(4), 40–43.
Andersen, P. & Andersen, J.: 1982, 'Non-Verbal Immediacy in Instruction', in L.L. Barker (ed.), *Communication in the Classroom*. Prentice Hall, Englewood Cliffs, NJ, 98–120.

Argyle, M.: 1988, *Bodily Communication* (second edition), Methuen, London.
Ashton, P.T. & Webb, R.B.: 1986, *Making a Difference: Teachers' Sense of Efficacy and Student Achievement*, Longman, New York.
Barba, R. & Cardinale, L.: 1991, 'Are Females Invisible Students? An Investigation of Teacher-Student Questioning Interactions', *School Science and Mathematics* 91, 306–310.
Becker, J.R.: 1981, 'Differential Treatment of Females and Males in Mathematics Classes', *Journal for Research in Mathematics Education* 12, 40–53.
Bennett, S.N.: 1976, *Teaching Styles and Pupil Progress*, Open Books, London.
Borich, G.D. & Madden, S.K.: 1977, *Evaluating Classroom Instruction: A Sourcebook of Instruments*, Addison-Wesley, Reading, MA.
Brady, M.P., Swank, P.R., Taylor, R.D. & Freiberg, J.: 1992, 'Teacher Interactions in Mainstream Social Studies and Science Classes', *Exceptional Children* 58, 530–540.
Brekelmans, M. & Créton, H.: 1993, 'Interpersonal Teacher Behavior Throughout the Career', in Th. Wubbels & J. Levy (eds.), *Do You Know What You Look Like?*, Falmer Press, London, 81–102.
Brekelmans, M., Levy, J. & Rodriguez, R.: 1993, 'A Typology of Teacher Communication Style', in Th. Wubbels & J. Levy (eds.), *Do You Know What You Look Like?*, Falmer Press, London, 46–55.
Brekelmans, M., Wubbels, Th. & Créton, H.A.: 1990, 'A Study of Student Perceptions of Physics Teacher Behavior', *Journal of Research in Science Teaching* 27, 335–350.
Brophy, J.E.: 1985, 'Teacher-Student Interaction', in J.B. Dusek (ed.), *Teacher Expectancies*, Lawrence Erlbaum, Hillsdale, NJ, 303–328.
Brophy, J.E. & Good, T.L.: 1986, 'Teacher Behavior and Student Achievement', in M.C. Wittrock (ed.), *Handbook of Research on Teaching* (third edition), Macmillan, New York, 328–375.
Fisher, D., Fraser, B. & Wubbels, Th.: 1993, 'Interpersonal Teacher Behavior and School Environment', in Th. Wubbels & J. Levy (eds.), *Do You Know What You Look Like?*, Falmer Press, London, 103–112.
Fisher, D., Henderson, D. & Fraser, B.: 1995, 'Interpersonal Behaviour in Senior High School Biology Classes', *Research in Science Education* 25, 125–133.
Flanders, N.A.: 1970, *Analyzing Teacher Behavior*, Addison-Wesley, Reading, MA.
Fraser, B.J., Giddings, G.J. & McRobbie, C.J.: 1995, 'Evolution and Validation of a Personal Form of an Instrument for Assessing Science Laboratory Classroom Environments', *Journal of Research in Science Teaching* 32, 399–422.
Gallagher, J.J. & Tobin, K.: 1987, 'Teacher Management and Student Engagement in High School Science', *Science Teacher Education* 71, 535–555.
Galton, M. & Eggleston, J.: 1979, 'Some Characteristics of Effective Science Teaching', *European Journal of Science Education* 1, 75–87.
Goh, S.C. & Fraser, B.J.: 1995, 'Learning Environment and Student Outcomes in Primary Mathematics Classrooms in Singapore', Paper presented at the annual meeting of the American Educational Research Association, San Francisco, CA.
Goh, S., Young, D. & Fraser, B.: 1995, 'Psychosocial Climate and Student Outcomes in Elementary Mathematics Classrooms: A Multilevel Analysis', *Journal of Experimental Education* 64, 29–40.
Good, Th.L. & Brophy, J.E.: 1991, *Looking in Classrooms* (fifth edition), Harper Collins, New York.
Hacker, R.G.: 1991, 'Gender Differences in Science Lessons Behaviours', *International Journal of Science Education* 13, 439–445.
Hamilton, R. & Brady, M.P.: 1991, 'Individual and Classwide Patterns of Teachers' Questioning in Mainstreamed Social Studies and Science Classes', *Teaching & Teacher Education* 7, 253–262.
Jones, M.G. & Wheatley, J.: 1989, 'Gender Influences in Classroom Displays and Student-Teacher Behaviors', *Science Education* 73, 535–545.
Jones, M.G. & Wheatley, J.: 1990, 'Gender Differences in Teacher-Student Interactions in Science Classrooms', *Journal of Research in Science Teaching* 27, 861–874.
Kounin, J.S.: 1970, *Discipline and Group Management in Classrooms*, Holt, Rinehart and Winston, New York.
Levis, D.S.: 1987, 'Teachers' Personality', in M.J. Dunkin (ed.), *Encyclopedia of Teaching and Teacher Education*, Pergamon, Oxford, UK, 585–588.
Levy, L., Créton, H. & Wubbels, Th.: 1993, 'Perceptions of Interpersonal Teacher Behavior', in Th. Wubbels & J. Levy (eds.), *Do You Know What You Look Like?*, Falmer Press, London, 29–45.
Levy, J., Rodriguez, R. & Wubbels, Th.: 1992, 'Instructional Effectiveness, Communication Style and Teacher Development', Paper presented at the annual meeting of the American Educational Research Association, San Francisco, CA.

Levy, J., Wubbels, Th. & Brekelmans, M.: 1992, 'Student and Teacher Characteristics and Perceptions of Teacher Communication Style', *Journal of Classroom Interaction* 27, 23–29.

Levy, J., Wubbels, Th., Brekelmans, M. & Morganfield, B.: in press, 'Language and Cultural Factors in Students' Perceptions of Teacher Communication Style', *International Journal on Intercultural Relationships*.

McCroskey, J.C. & Richmond, V.P.: 1992, 'Increasing Teacher Influence Through Immediacy', in V.P. Richmond & J.C. McCroskey (eds.), *Power in the Classroom*, Lawrence Erlbaum, Hillsdale, NJ, 101–119.

Mehrabian, A.: 1981, *Silent Messages*, Wadsworth Publishing Company, Belmont, CA.

Nias, J.: 1981, 'Teacher Satisfaction and Dissatisfaction: Herzberg's Two Factor Hypothesis Revisited', *British Journal of Sociology of Education* 2, 235–246.

Peterson, P.L. & Barger, S.A.: 1985, 'Attribution Theory and Teacher Expectancy', in J.B. Dusek (ed.), *Teacher Expectancies*, Lawrence Erlbaum, Hillsdale, NJ, 159–184.

Pizzini, E.L. & Shepardson, D.P.: 1992, 'A Comparison of the Classroom Dynamics of a Problem-Solving and Traditional Laboratory Model of Instruction Using Path Analysis', *Journal of Research in Science Teaching* 29, 243–258.

Ramsay, W. & Ransley, W.: 1986, 'A Method of Analysis for Determining Dimensions of Teaching Style', *Teaching and Teacher Education* 2, 69–79.

Rohrkemper, M.M. & Brophy, J.E.: 1983, 'Teachers' Thinking About Problem Students', in J.M. Levine & M. Wang (eds.), *Teacher and Student Perceptions*, Lawrence Erlbaum, Hillsdale, NJ, 75–103.

Rosenholtz, S.J., Bassler, O. & Hoover-Dempsey, K.: 1986, 'Organizational Conditions of Teacher Learning', *Teaching and Teacher Education* 2, 91–104.

Rosenthal, R. & Jacobson, L.: 1968, *Pygmalion in the Classroom: Teacher Expectation and Pupils' Intellectual Development*, Holt, Rinehart & Winston, New York.

Schultz, R.A.: 1982, 'Teaching Style and Sociopsychological Climates', *Alberta Journal of Educational Research* 18, 9–18.

Shymansky, J.A. & Penick, J.E.: 1979, 'The Use of Systematic Observations to Improve College Science Laboratory Instruction', *Science Education* 63, 195–203.

Tamir, P.: 1981, 'Classroom Interaction Analysis of High School Biology Classes in Israel', *Science Education* 65, 87–103.

Tartwijk, J. van, Fisher, D., Fraser, B. & Wubbels, Th.: 1994, 'The Interpersonal Significance of Molecular Behavior of Science Teachers in Lab Lessons: A Dutch Perspective', Paper presented at the annual meeting of the National Association for Research in Science Teaching, Anaheim, CA.

Tobin, K., Kahle, J.B. & Fraser, B.J. (eds.): 1990, *Windows into Science Classes: Problems Associated with Higher Level Cognitive Learning*, Falmer Press, London.

Walberg, H.: 1986, 'Syntheses of Research on Teaching', in M.C. Wittrock (ed.), *Handbook of Research on Teaching* (third edition), Macmillan, New York, 214–229.

Watts, M. & Bentley, D.: 1987, 'Constructivism in the Classroom: Enabling Conceptual Change by Words and Deeds', *British Educational Research Journal* 13, 121–135.

Watzlawick, P., Beavin, J.H. & Jackson, D.: 1967, *The Pragmatics of Human Communication*, Norton, New York.

Weinstein, C.S.: 1979, 'The Physical Environment of the School: A Review of the Research', *Review of Educational Research* 49, 577–610.

Wineburg, S.S.: 1987, 'The Self-Fulfillment of the Self-Fulfilling Prophecy', *Educational Researcher* 16(5), 28–36.

Woolfolk, A.E. & Brooks, D.M.: 1983, 'Nonverbal Communication in Teaching', in E.W. Gordon (ed.), *Review of Research in Education*, American Educational Research Association, Washington, DC, 103–150.

Woolfolk, A.E. & Hoy, W.K.: 1990, 'Prospective Teachers' Sense of Efficacy and Beliefs about Control', *Journal of Educational Psychology* 82, 81–91.

Wubbels, Th., Brekelmans, M. & Hermans, J.: 1987, 'Teacher Behavior: An Important Aspect of the Learning Environment', in B.J. Fraser (ed.), *The Study of Learning Environments, Volume 3*, Curtin University of Technology, Perth, Australia, 10–25.

Wubbels, Th., Créton, H.A. & Holvast, A.J.C.D.: 1988, 'Undesirable Classroom Situations', *Interchange* 19(2), 25–40.

Wubbels, Th., Créton, H., Levy, J. & Hooymayers, H.: 1993, 'The Model for Interpersonal Teacher Behavior', in Th. Wubbels & J. Levy (eds.), *Do You Know What You Look Like?*, Falmer Press, London, 13–28.

# 5.3 Personal and Class Forms of Classroom Environment Instruments

CAMPBELL J. McROBBIE
*Queensland University of Technology, Brisbane, Australia*

DARRELL L. FISHER
*Curtin University of Technology, Perth, Australia*

ANGELA F. L. WONG
*Nanyang Technological University, Singapore*

In the early days of the study of human environments, Murray (1938) introduced the term *alpha press* to describe the environment as assessed by a detached observer and *beta press* to describe the environment as observed by those within that environment. These ideas were extended by Stern, Stein and Bloom (1956) to include perceptions of the environment unique to the individual (called *private beta press*) and perceptions of the environment shared among the group (called *consensual beta press*). Hence, even in these early studies of human environments, it was recognised that the perceptions of persons from different perspectives could lead to different interpretations of an environment.

Interest in the study of learning environments in classrooms was rekindled during an evaluation of Harvard Project Physics which required the development of an instrument to assess learning environments in physics classrooms. This instrument, the *Learning Environment Inventory* (LEI, Walberg 1968), asked students for their perceptions of the whole-class environment. At about the same time, Moos and Trickett (1974) had been developing a series of environment measures which concluded with the *Classroom Environment Scale* (CES), which also asked students for their perceptions of the learning environment of the class as a whole. These two questionnaires provided considerable impetus for the study of classroom learning environments, were used extensively for a variety of research purposes, and provided models for the development of a range of instruments over the next two decades or so (see Fraser 1994). Most of these instruments are available in an *actual* version, which asks respondents questions about the experienced learning environment, and a *preferred* version, which focuses on the learning environment ideally preferred by students.

---

Chapter Consultant: Jong-Hsiang Yang (National Taiwan Normal University, Taiwan)

Questionnaires which assess the whole-class environment assume that there is a unique learning environment in the classroom that all students in a class more or less experience. Variations in scores on learning environment instruments were considered as error variance, with the class mean representing a good measure of the learning environment in the classroom. However, the assumption of a common learning environment experienced by all students within a classroom was challenged in the latter half of the 1980s. For example, in interpretive studies employing classroom learning environment instruments, classroom observations and interviews involving teachers and students suggested that there were groups of students (termed 'target' students) who were involved more extensively in classroom discussions than the other students. These target students were found to have more favourable perceptions of the learning environment than their classmates, suggesting that there could be discrete and differently-perceived learning environments within the one classroom (Tobin 1987). One implication of these studies is that there is potentially a problem with using the traditional form of learning environment instruments when studying differences between groups of students in a classroom (e.g., boys and girls) because these instruments elicit the student's perceptions of the class as a whole rather than the student's personal perception of his or her role in that classroom (Fraser & Tobin 1991). Although classroom environment scales have been used to advantage in case study research (Tobin, Kahle & Fraser 1990), these studies suggested the desirability of having a new form of instruments available which is better suited than is the conventional Class Form for assessing differences in perceptions that might be held by different students within the same class.

Around the time when these studies were being carried out, the traditional teacher's role of transmitting the logical structures of knowledge to students was being questioned in favour of a view that meaningful learning is a personal cognitive process that actively involves the learner in making sense of world experiences in terms of the existing knowledge of the individual, and a social process in which this sense-making process involves negotiation and consensus building with others (Tobin 1993; von Glasersfeld 1989).

These studies and influences led Fraser, Giddings and McRobbie (1992) to propose a different form of a learning environment instrument which asked students for their personal perception of their role in the environment of the classroom rather than their perception of the learning environment of the class as a whole; these two forms are called the *Personal Form* and the *Class Form*, respectively. The development and validation of Personal Forms of learning environment instruments are considered in the following sections together with the results of studies comparing scores on the Personal Form with scores on the traditional Class Form. Whereas the first part of the chapter focuses on our research involving Class and Personal Forms of the Science Laboratory Environment Inventory (Fraser, Giddings & McRobbie 1995), the second part reports current research involving the Personal and Class Forms of a new classroom environment questionnaire (Fraser, Fisher & McRobbie 1996).

## PREVIOUS RESEARCH WITH THE SCIENCE LABORATORY ENVIRONMENT INVENTORY

Although much of the classroom learning environment research of the last 25 years focused on science teaching and learning, hardly any of this research involved the learning environment in science laboratory classes (Hegarty-Hazel 1990). Concerns about laboratory instruction and the lack of an instrument to probe student perceptions of laboratory learning environments provided the impetus for the construction of both Class and Personal Forms of the *Science Laboratory Environment Inventory* (SLEI).

The first stage in constructing this instrument was the development of an actual and a preferred version of a Class Form (Fraser, Giddings & McRobbie 1993). This involved field testing in six countries involving 3,401 students in 183 senior high schools and 1,242 students in 42 university laboratory classes. The final scales in the Class Form are *Student Cohesiveness, Open-endedness, Integration, Rule Clarity* and *Material Environment*. Each of the five dimensions was validated by discussion with students and teachers on the salience of the scales and the wording of the items, by reference to concerns and findings expressed in the research literature on aspects of the science laboratory learning environment (Hegarty-Hazel 1990; Tobin 1990), coverage of the categories identified by Moos (1979) for conceptualising all human environments (i.e., Relationship, Personal Development, and System Maintenance and System Change Dimensions), and factor, internal consistency and discriminant validity analyses performed on data collected by administering a preliminary version to students.

The Personal Form of the instrument was devised by rewording items of the Class Form. Table 1 compares selected Class Form items (actual version) and the corresponding Personal Form items. Similar transformations of preferred version items of the Class Form were made in developing a preferred version of the Personal Form.

The actual and preferred versions of the Personal Form of the SLEI were administered as part of a study involving the cross-validation of the Class Form among senior high school chemistry classes in Queensland, Australia (Fraser, Giddings & McRobbie 1995; Fraser & McRobbie 1995). As part of this larger study, which employed a matrix sampling design, all students responded to actual and preferred versions of the Class Form of the SLEI and a general aptitude test. The sample of students who responded to both the Personal and Class Forms of the instrument consisted of 516 students in 56 year 11 chemistry classes. This sample enabled a comparison to be made of these two forms of the SLEI. In addition, cross-cultural validation information was collected by administering the Personal Form of the SLEI to 1,592 year 10 chemistry students in 56 classes in Singapore.

Table 2 shows the internal consistency reliability (Cronbach alpha), discriminant validity (mean correlation with other scales) and ability to differentiate between classrooms for the Personal Form of the instrument for two units of analysis for the Queensland and Singapore samples. Data are reported in Table 2 for the actual and preferred forms for the Australian sample and for the actual form for the

Table 1: Differences in the wording of items in the Class Form and Personal Form of the Science Laboratory Environment Inventory (SLEI)

| Scale | Class form | Personal form |
|---|---|---|
| Student Cohesiveness | Students are able to depend on each other for help during laboratory classes. | I am able to depend on other students for help during laboratory classes. |
| Open-endedness | In our laboratory sessions, different students do different experiments. | In my laboratory sessions, I do different experiments than some of the other students. |
| Integration | The laboratory work is unrelated to the topics that we are studying in our science class. | The laboratory work is unrelated to the topics that I am studying in my science class. |
| Rule Clarity | Our laboratory class has clear rules to guide student activities. | My laboratory class has clear rules to guide my activities. |
| Material Environment | The laboratory is crowded when we are doing experiments. | I find that the laboratory is crowded when I am doing experiments. |

Singapore sample. Principal components factor analysis with varimax rotation for the actual and preferred versions of the Personal Form yielded the same five-factor structure for each form and in both countries. Further, analysis of variance showed that each scale of the actual version of each form of the instrument differentiated between the perceptions of students in different classes in both Australia and Singapore. These analyses showed that the Personal Form of the SLEI has satisfactory validity and reliability statistics, which were comparable to those of the Class Form. Further, as in past research (Fraser 1994), internal consistency values (Cronbach alpha coefficients) for the class mean as the unit of analysis were larger than with the individual as the unit of analysis.

*Differences Between Scores on Class and Personal Forms of the SLEI*

The Queensland sample was used in investigating differences between students' scores on the SLEI's Personal and Class Forms. A one-way MANOVA for class means was performed with scores on the five actual and five preferred scales of the SLEI as the dependent variables and with the form of instrument as the repeated measures factor. Because the multivariate test was statistically significant (Wilks' lambda, $p<.01$), individual $t$-tests were utilised to test the statistical significance of the differences between the scores on the Class and Personal Forms for each scale and version. Statistically significant differences ($p<.05$) were found for the Integration scale on both the actual and for the preferred versions, and the Student Cohesiveness, Open-endedness and Material Environment scales for the preferred version. In each instance, the Personal Form mean score was lower than that for the Class Form, and this was more pronounced in the preferred version.

Table 2: Internal consistency reliability (alpha coefficient) and discriminant validity (mean correlation with other scales) for actual and preferred versions of the Personal Form of the SLEI for two units of analysis, and the ability to differentiate between classrooms

| Scale | Unit of analysis | Cronbach alpha reliability | | | Mean correlation with other scales | | | ANOVA results (eta$^2$) | |
|---|---|---|---|---|---|---|---|---|---|
| | | Austr actual | Austr pref | Sing actual | Austr actual | Austr pref | Sing actual | Austr actual | Sing actual |
| Student Cohesiveness | Individual | .78 | .73 | .68 | .26 | .25 | .23 | .23* | .10* |
| | Class Mean | .80 | .82 | .83 | .31 | .31 | .30 | | |
| Open-endedness | Individual | .71 | .64 | .41 | .19 | .11 | .03 | .28* | .08* |
| | Class Mean | .80 | .70 | .54 | .25 | .15 | .05 | | |
| Integration | Individual | .86 | .84 | .69 | .38 | .31 | .30 | .28* | .18* |
| | Class mean | .91 | .92 | .87 | .44 | .36 | .36 | | |
| Rule Clarity | Individual | .74 | .68 | .63 | .37 | .29 | .28 | .25* | .19* |
| | Class Mean | .76 | .80 | .84 | .43 | .35 | .36 | | |
| Material Environment | Individual | .76 | .73 | .72 | .27 | .34 | .25 | .27* | .18* |
| | Class Mean | .74 | .85 | .82 | .34 | .40 | .31 | | |

*$p<.01$
The eta$^2$ statistic (which is the ratio of *between* to *total* sums of squares) represents the proportion of variance explained by class membership.
The Australian sample consisted of 516 year 11 chemistry students in 56 classes, whereas the Singaporean sample consisted of 1,592 year 10 chemistry students in 56 classes.

This suggested that students had a more positive view of the learning environment when they responded in relation to the whole class than when they gave their perceptions of their personal role in the classroom environment. These differences, while not large (ranging from about 0.3 to 1.1 standard deviations for class means), nevertheless were all in the same direction. Also, the mean scores on the preferred version for each scale were higher than the corresponding actual version scores, a pattern that replicates past research reported for Class Form differences (Fraser 1994).

*Associations Between Personally-Perceived Environment and Outcomes*

One of the common lines of research with Class Forms of learning environment instruments in the last 25 years has been investigation of associations between characteristics of the learning environment and various student outcome measures (Fraser 1994). The administration of the Class and Personal Forms of the SLEI along with an attitude outcome survey to the Queensland sample allowed a comparison of the magnitude of attitude-environment associations for the Class and Personal Forms. Attitudes were assessed with a Likert scale covering a range of chemistry-related attitudes associated with the goals of laboratory teaching, namely, Attitude to Laboratory Learning, Nature of Chemistry Knowledge (testability and changing nature of science knowledge), Cooperative Learning and Adoption of Laboratory Attitudes (e.g., working safely, repeating observations, following instructions). Simple and multiple correlations were calculated between SLEI scale scores on Personal and Class Forms and each attitude scale for 283 students in 28 year 11 chemistry classrooms (Fraser & McRobbie 1995).

Fraser and McRobbie (1995) found that generally the strengths of outcome-attitude associations were similar for the Class Form and the Personal Form. Nevertheless, although the total variance explained is comparable for each form of the instrument, this does not imply that they account for the same variance in the outcome measures. A commonality analysis (Pedhazur 1982) based on the squared multiple correlation for the five Class Form and five Personal Form scales in the actual version as predictors was performed separately for the same four attitude scales (Fraser & McRobbie 1995) using the class mean as the unit of statistical analysis. Table 3 shows that the unique and common variances varied for different attitude scales. The Class Form and the Personal Form each accounted for a sizeable proportion of the outcome variance which was unique to that form when compared with the other form. Further, the unique variance accounted for in the outcome measures was comparable to the common variance. Except for the Cooperative Learning outcome, both forms of the SLEI (Personal and Class) made a statistically significant unique contribution to the variance in the outcome measures. Similar results also were reported by Fraser, Giddings and McRobbie (1995) for other attitude outcome measures and an inquiry skills test.

These analyses involving both a Personal and Class Form of the SLEI replicated previous research with Class Forms in showing that classroom environment dimensions were associated with attitudinal outcomes and that preferred version mean scores were higher than actual version mean scores. They further showed that mean scores on the Personal Form for each scale were lower than the corresponding mean scores on the Class Form. Importantly, they also revealed that the Class

Table 3: Commonality analysis of $R^2$ statistic for Class and Personal Forms of the SLEI

| Attitude scale | $R^2$ for class means | | |
|---|---|---|---|
| | Uniqueness | | Commonality |
| | Class form | Personal form | |
| Attitude to Laboratory Learning | .24 | .19 | .16 |
| Nature of Chemistry Knowledge | .08 | .11 | .14 |
| Cooperative Learning | .22 | .18 | .12 |
| Adoption of Laboratory Attitudes | .17 | .24 | .17 |

Based on a sample of 283 students in 56 classes.
For every outcome except Cooperative Learning, both forms of the SLEI (i.e., Class and Personal) made a statistically significant unique contribution to the variance.

and Personal Forms each accounted for both common and unique proportions of the variance in several attitude outcome measures, thus indicating that Class and Personal Forms measure different but overlapping aspects of the laboratory classroom learning environment. However, the design of this study did not probe the reasons for students' differing perceptions associated with the Personal and Class Forms of the SLEI. A recent study described in the next section did include interviews with students in order to ascertain reasons why students reported different perceptions of the learning environment in classrooms from the perspective of the Personal and Class Forms of an instrument.

## CURRENT RESEARCH WITH A NEW INSTRUMENT

During the last 25 years, there have been instruments developed for a range of classroom contexts, such as individualised classrooms (Fraser 1990), constructivist classrooms (Taylor, Dawson & Fraser 1995) and computer-assisted instructional settings (Teh & Fraser 1995). Recently, Fraser, Fisher and McRobbie (1996) began the development of a new learning environment instrument which incorporates scales that had been shown in previous studies to be significant predictors of outcomes (Fraser 1994) and additional scales to accommodate recent developments and concerns in classroom learning (e.g., equity). The first version of the new instrument contained the following nine ten-item scales: *Student Cohesiveness, Teacher Support, Involvement, Autonomy/ Independence, Investigation, Task Orientation, Cooperation, Equity* and *Understanding.* The new instrument employed the same five-point Likert response scale (Almost Never, Seldom, Sometimes, Often, Almost Always) as used in some previous instruments.

An actual version of the Class and Personal Forms was developed and administered to 355 students in 17 year 9/10 mathematics and science classrooms in five Australian schools. Principal components factor analysis with varimax rotation, along with item analysis, resulted in the acceptance of a revised version of the instrument comprising 54 items in seven of the original scales (with the Autonomy/Independence and Understanding scales not holding up). In the second trial version of the new instrument, these 54 items are imbedded in an 80-item version with 10 items in each of 8 scales (with Autonomy/ Independence being reinstated). Table 4 reports statistical data relevant to the internal consistency reliability (Cronbach alpha coefficient) and discriminant validity (mean correlation of a scale with the other scales) of the Class and Personal Forms of the 54-item version of the new questionnaire.

*Difference Between Scores on Class and Personal Forms*

A multivariate analysis of variance (MANOVA) with repeated measures was used to explore differences between Class and Personal Forms on the set of seven

Table 4: Number of items, internal consistency (Cronbach alpha coefficient), discriminant validity (mean correlation with other scales), ability to differentiate between classes and difference in scale means for Class and Personal Forms of a new instrument

| Scale | Form | No of items | Cronbach alpha reliability | Discrim. validity | ANOVA results eta$^2$ | Mean | SD | Difference in means |
|---|---|---|---|---|---|---|---|---|
| Student Cohesiveness | Class | 8 | .81 | .44 | .20* | 27.88 | 4.76 | -0.89* |
|  | Personal |  | .80 | .35 | .07* | 28.51 | 4.87 |  |
| Teacher Support | Class | 10 | .88 | .45 | .35* | 33.42 | 8.01 | 2.99* |
|  | Personal |  | .88 | .35 | .27* | 30.43 | 7.86 |  |
| Involvement | Class | 10 | .82 | .44 | .23* | 35.05 | 6.53 | 4.77* |
|  | Personal |  | .86 | .48 | .11* | 30.28 | 6.86 |  |
| Investigation | Class | 8 | .83 | .40 | .18* | 23.55 | 5.64 | 1.02* |
|  | Personal |  | .84 | .39 | .13* | 22.53 | 5.69 |  |
| Task Orientation | Class | 9 | .86 | .45 | .22* | 31.31 | 6.64 | -2.83* |
|  | Personal |  | .89 | .36 | .16* | 34.14 | 6.76 |  |
| Cooperation | Class | 4 | .67 | .42 | .27* | 15.71 | 4.83 | 0.16 |
|  | Personal |  | .77 | .34 | .23* | 15.56 | 5.02 |  |
| Equity | Class | 5 | .81 | .06 | .23* | 13.39 | 3.24 | -0.93* |
|  | Personal |  | .84 | .09 | .35* | 14.32 | 3.45 |  |

* $p<.01$

environment dimensions. For the 355 students who completed both Class and Personal Forms, the multivariate test was statistically significant (Wilks' lambda, $p<.01$). Therefore, separate $t$-tests for dependent samples were conducted for each scale as reported in Table 4. This analysis showed there were statistically significant differences ($p<.01$) between mean scores on the Class and Personal Forms of the new instrument. However, as some of these differences (e.g., Student Cohesiveness, Investigation and Cooperation) were smaller than 0.3 standard deviations, they should be considered as small effects only. Unlike the SLEI, for which each scale's mean score on the Personal Form was lower than the corresponding mean score on the Class Form, analyses for the new instrument showed that, for some scales (Student Cohesiveness, Task Orientation and Equity), the Personal Form mean scores were higher than the corresponding mean scores on the Class Form. For the Student Cohesiveness scale in the SLEI, the mean scores for the Personal Form were less than those for the Class Form, although differences were not statistically significant. For the same scale on the new revised instrument, the difference between the mean scores for the Personal and Class Forms was statistically significant, with the Personal Form mean being higher than the Class Form mean. However, the scales in these two instruments contained different items.

*Interview Responses*

Interviews were held with 45 selected students who had responded to the new questionnaire in order to obtain student views about the wording and salience of items as part of the instrument development and refinement procedures. Also student interviews were used to ascertain reasons why students gave different responses to the Personal and Class Forms of items. Responses to the initial 90-item instrument were used to identify students for interview. Interviewees were selected from those for whom scale total scores (out of a maximum scale score of 50 points) on the Personal and Class Forms on one or more scales differed by 5 points or greater. In the interviews, the students were asked first to comment generally on the learning environment of their science or mathematics class in open-response form, including how they personally perceived and how the class perceived the science or mathematics classroom learning environment. They then were asked to comment in a similar way on the particular constructs assessed by the new instrument for those scales for which a student's total Personal and Class scale scores differed by at least 5 points. Finally, students commented on their responses for each individual item in those scales. Where particular items on other scales in a student's survey differed by 2 or more points (out of a maximum item score of 5 points), that student also was asked to comment on his/her responses to those items. This three-level approach provided a rich description of how the students perceived their personal role in the learning environment of the classroom, their perceptions of the learning environment for the class as a whole, and the differences between those perceptions. As part of instrument-development procedures, these students also were asked to comment on any difficulties which they experienced in interpreting or understanding the items in the questionnaires and whether there were any additional items or issues that should have been included in the instrument as concerns related to improving the learning environment in their classroom.

Below are some examples of responses given by students to explain differences in their perceptions of the whole-class learning environment and their personal perceptions of their role in that learning environment in response to the initial open-ended question or questions relating to the specific dimensions. These responses illustrate cases for which Personal perceptions were more and less favourable than that for the Class as a whole:

There are parts of science that I really like, but other parts are boring. Some people in the class are like me in that they like some lessons, but some people just don't care at all. They just muck around. (About the science class)

I would say that it is a friendly class, but some of the students are smarter than me and can understand everything better. So, the way that I see the class will be different to what they see. They have fun in practical work because they know what to do. Not understanding it spoils it for me. (Student Cohesiveness)

I know that we have to do all of our work and have it in on time. I always do my work and the homework. Sometimes, if I don't finish my work in class, I take it home and do it. The class? There are fools in the class and they don't want to do their work. They want to muck up and play around. (Task Orientation)

Table 5 provides student comments for selected items in Teacher Support, Involvement and Task Orientation, which all are scales for which the differences in the means of the Class and Personal Forms were both statistically significant and of a magnitude to warrant further investigation. The item wording in Table 5 is for the Personal Form, and the student's response to the Class Form and the Personal Form are shown.

These student responses provide examples for which the Class Form response was more favourable than the Personal Form response, as well as other cases for which the Personal Form was more favourable. Explanations of student responses to the Class Form often were predicated on the identification of events or actions of small groups in the class rather than on the class as a whole, and this raises questions about the validity of Class Form responses representing the whole-class learning environment. Underlying many of the responses for which the Class Form response was more favourable than the Personal Form response was the idea that the individual student is only part of the class and therefore interactions with that individual student are necessarily less than the interactions with the whole class. Also, responses frequently reflected a desire on the part of the student not to become involved in the classroom actions for a variety of reasons. A further observation was that, on each of the scales, there was a large proportion of both negative and positive differences in scores between the Class Form and the Personal Form. For example, for the Task Orientation scale, this difference (Class Form minus Personal Form) was positive for 26 percent of students and negative for 66 percent of students, with 35 percent of students having difference scores which were equivalent to one or more points on the response scale. These ranges in scale score differences between the Class and Personal Forms of the instruments provide further support for the contention that there are groups of students within a classroom with perceptions of the learning environment which differ from these perspectives.

The size of the scale score ranges raises the question of how well the mean score on either form can represent the learning environment in a classroom. Fraser and Hoffman (1995) used qualitative and quantitative data to show how Personal Forms could be utilised in studying the classroom learning environment at different 'grain sizes'. They showed how individual students and the teacher could be investigated at the smallest grain size and how these environment scores can be aggregated to the class level. When appropriate, such aggregation also could be extended to the system level. However, where classes are composed of heterogeneous groups of students with respect to their perceptions of the learning environment, the aggregation of learning

Table 5: Student comments about their responses to Class and Personal Forms of some items

| Item wording | Student response | | Student comment |
|---|---|---|---|
| | Class | Personal | |
| *TEACHER SUPPORT* | | | |
| The teacher takes a personal interest in me. | Often | Almost Never | I said that because, whenever the teacher asks me a question, I usually answer it wrongly. So I guess the teacher avoids me and prefers someone who actually can answer the question. She is interested in all of her students, but I think that she chooses people who actually can answer the questions correctly. |
| The teacher goes out of his/her way to help me. | Almost Always | Seldom | Some people need more help than others. If someone is behind, he will stop and wait for them to catch up. I normally don't need to ask many questions because normally I understand the work. |
| The teacher helps me when I have trouble with the work. | Often | Sometimes | Some people in the class need more help than other people. I don't think I really need that much help. |
| | Almost Always | Seldom | Sometimes the teacher is not always available, because there are so many students. |
| *INVOLVEMENT* | | | |
| Students give their opinions during class discussions. | Often | Sometimes | Some people put their hands up more than others. I just listen to everybody else. |
| | Almost Always | Seldom | I said that because there are so many people in the class. |
| | Almost Always | Seldom | The class as a whole answers questions, but I can't answer questions because I think that I am wrong and everyone will laugh at me. |
| The teacher asks me questions. | Almost Always | Sometimes | He almost always asks questions to the whole class, but only sometimes he asks a particular student to answer and singles you out to ask for the answer. Most of the time, it is just to the whole class. |
| Students' ideas and suggestions are used during class discussions. | Almost Always | Sometimes | Yes, the other people in this class would have better ideas than I would have. I am not really the creative type. |
| I ask the teacher questions. | Almost Always | Sometimes | I sometimes ask the teacher a question if I need help, but the class asks questions just about all the time. |

**Table 5:** Continued

| Item wording | Student response | | Student comment |
|---|---|---|---|
| | Class | Personal | |

### INVOLVEMENT

| | | | |
|---|---|---|---|
| The teacher asks me to explain how I solve problems. | Almost Always | Sometimes | He will ask different people in the class. It is just that he has so many students to ask that you only get asked sometimes. |
| | Almost Always | Sometimes | If you answer the question, he will ask you how you did that and why that was what you said. I said he asks me sometimes because I don't try to answer his questions. But, if you do answer a question and especially if it is wrong, he wants to know why you put that down. |

### TASK ORIENTATION

| | | | |
|---|---|---|---|
| Getting a certain amount of work done is important to me. | Sometimes | Almost Always | Yes, because most people don't really care if they do or do not. They don't really care what they get in the test, but I do. |
| | Sometimes | Often | There are lots of kids who just don't do work (disruptive kids). But I always do work, because you just do. |
| I know the goals for this class. | Often | Almost Always | I guess some people in the class don't know what they are aiming for, but I do. I know what I want to achieve in science. |
| | Sometimes | Almost Always | Well for me, the goals are important because I want to get a good mark. But, some of the students couldn't care less and just want to get it over and done with. |
| I try to understand the work in this class. | Often | Almost Always | I almost always try to understand because I don't want to do badly in the test, but there are students in the class who don't always try to understand. |

environment perception scores inevitably obscures differences between students and groups within that classroom.

The student responses for the Personal Form also showed that some students were responding in terms of their perceptions of their personal involvement in the classroom and, depending on the scale, were identifying factors that personally could influence their learning. Recent approaches to learning increasingly have recognised the role of social factors in knowledge construction. The responses to the Personal Form also show the extent to which students perceived themselves as participating in the construction of knowledge from a social perspective (e.g., Student Cohesiveness, Cooperation and Involvement) in the classroom, both with the class as a whole and with their closer working groups. Accordingly, the Personal Form of the instrument has the potential to characterise the learning environment in a classroom from the perspective of recent views of learning. Taylor, Dawson and Fraser (1995) have constructed a Personal Form of a learning environment instrument, specifically for the purpose of assessing constructivist emphases within classroom learning environments, and McRobbie and Tobin (1997) have utilised Personal Forms of learning environment instruments to characterise the learning environment in a chemistry classroom from a social constructivist perspective. Relative to earlier periods of research on learning environments, currently the Personal Form of instruments is being used increasingly.

## CONCLUSION

The chapter reports the use of a Personal Form of two classroom learning environment instruments, namely, the Science Laboratory Environment Inventory and a new instrument which synthesises existing instruments and adds some dimensions of contemporary educational relevance. These Personal Forms assess a student's perception of his or her role within a classroom environment, whereas traditional Class Forms ask students to provide perceptions of the class as a whole. According to Fraser and Tobin (1991), Personal Forms of classroom environment scales are more valid, especially in research which involves case studies of individual students or which investigates differences in the perceptions of within-classroom subgroups of students (e.g., males and females).

The Personal Forms of scales in both instruments displayed satisfactory factorial validity, internal consistency reliability and discriminant validity, and they were capable of differentiating between the perceptions of students in different classrooms. Interesting differences were found between mean scores on the Class and Personal Forms for particular scales, and interviews with students helped to illuminate some of the reasons for these differences. When environment scale scores on both a Personal and a Class Form were related to student attitude outcomes, it was found that the Personal and Class Forms each accounted for unique variance in attitudes that could not be explained by the other form.

Overall, the findings reported in this chapter provide convincing evidence that many respondents have differing perceptions of the learning environment in classrooms from the perspective of the whole class relative to their perceptions of their personal role in that class. However, the research on the characteristics and associations of Personal Forms of learning environment instruments is still in its infancy and further research will be required before the implications associated with Personal Forms of instruments are understood fully. Meanwhile, the development of these instruments now makes the study of individuals or groups of students within a classroom more valid. They also open the way for the utilisation of qualitative and quantitative data together to paint a more compelling picture of the learning environments of individuals and small groups of students.

# REFERENCES

Fraser, B.J.: 1990, *Individualised Classroom Environment Questionnaire*, Australian Council for Educational Research, Melbourne, Australia.

Fraser, B.J.: 1994, 'Research on Classroom and School Climate', in D. Gabel (ed.), *Handbook of Research on Science Teaching and Learning*, Macmillan, New York, 493–541.

Fraser, B.J., Fisher, D.L. & McRobbie, C.J.: 1996, 'Development, Validation and Use of Personal and Class Forms of a New Classroom Environment Instrument', Paper presented at the annual meeting of the American Educational Research Association, New York.

Fraser, B.J., Giddings, G.J. & McRobbie, C.J.: 1992, 'Science Laboratory Classroom Environments: A Cross-National Perspective', in D.L. Fisher (ed.), *The Study of Learning Environments*, University of Tasmania, Launceston, Australia, 1–18.

Fraser, B.J., Giddings, G.J. & McRobbie, C.J.: 1993, 'Development and Cross-National Validation of a Laboratory Classroom Environment Instrument for Senior High School Science', *Science Education* 77, 1–24.

Fraser, B.J., Giddings, G.J. & McRobbie, C.J.: 1995, 'Evolution, Validation and Application of a Personal Form of an Instrument for Assessing Science Laboratory Classroom Environments', *Journal of Research in Science Teaching* 32, 399–422.

Fraser, B.J. & Hoffman, H.: 1995, 'Combining Qualitative and Quantitative Methods in a Teacher-Researcher Study of Determinants of Classroom Environment', Paper presented at the annual meeting of the American Educational Research Association, San Francisco, CA.

Fraser, B.J. & McRobbie, C.J.: 1995, 'Science Laboratory Classroom Environments at Schools and Universities: A Cross-National Study', *Educational Research and Evaluation* 1, 1–29.

Fraser, B.J. & Tobin, K.: 1991, 'Combining Qualitative and Quantitative Methods in Classroom Environment Research', in B.J. Fraser & H.J. Walberg (eds.), *Educational Environments: Evaluation, Antecedents and Consequences*, Pergamon, Oxford, UK, 271–292.

Hegarty-Hazel, E.: 1990, *The Student Laboratory and the Science Curriculum*, Routledge, London.

McRobbie, C.J. & Tobin, K.: 1997, 'A Social Constructivist Perspective on Learning Environments', *International Journal of Science Education* 19, 193–208.

Moos, R.H.: 1979, *Evaluating Educational Environments: Procedures, Measures, Findings and Policy Implications*, Jossey-Bass, San Francisco, CA.

Moos, R.H. & Trickett, E.J.: 1974, *Classroom Environment Scale Manual* (first edition), Consulting Psychologists Press, Palo Alto, CA.

Murray, H.A.: 1938, *Explorations in Personality*, Oxford, New York.

Pedhazur, E.: 1982, *Multiple Regression in Behavioral Research: Explanation and Prediction*, Rinehart and Winston, New York.

Stern, G.G., Stein, M.I. & Bloom, B.S.: 1956, *Methods in Personality Assessment*, Free Press, Glencoe, IL.

Taylor, P., Dawson, V. & Fraser, B.: 1995, 'Classroom Learning Environments Under Transformation: A Constructivist Perspective', Paper presented at the annual meeting of the American Educational Research Association, San Francisco, CA.

Teh, G. & Fraser, B.J.: 1995, 'Development and Validation of an Instrument for Assessing the Psychosocial Environment of Computer-Assisted Learning Classrooms', *Journal of Educational Computing Research* 12, 177–193.

Tobin, K.: 1987, 'Target Students Involvement in High School Science', *International Journal of Science Education* 10, 317–330.

Tobin, K.: 1990, 'Research on Science Laboratory Activities: In Pursuit of Better Questions and Answers to Improve Learning', *School Science and Mathematics* 90, 403–418.

Tobin, K. (ed.): 1993, *The Practice of Constructivism in Science Education*, American Association for the Advancement of Science, Washington, DC.

Tobin, K., Kahle, J.B. & Fraser, B.J. (eds.): 1990, *Windows into Science Classrooms: Problems Associated with Higher-Level Cognitive Learning*, Falmer Press, London.

von Glasersfeld, E.: 1989, 'Cognition, Construction of Knowledge, and Teaching', *Synthese* 80, 121–140.

Walberg, H.: 1968, 'Teacher Personality and Classroom Climate', *Psychology in the Schools* 5, 163–169.

## 5.4 Enhancing Science Education Through Laboratory Environments: More Than Walls, Benches and Widgets

HANNA J. ARZI

*HEMDA – Centre for Science Education, Tel Aviv, Israel*

A survey of school science facilities in Britain in 1900 identified 669 chemistry laboratories, 219 physics laboratories and only 17 laboratories for biology education (quoted in Layton 1990). The high proportion of facilities for chemistry at the turn of the 20th century is a reflection of the history of laboratory work in science education, which started in chemistry and subsequently extended to other disciplines (James 1989). Even though university research laboratories in chemistry can be traced to the 17th century, practical work usually was not part of a chemist's normal education until the 19th century. The common place in which to acquire laboratory skills was a pharmacist's shop. Only chosen students were allowed access to private laboratories maintained by university professors (Ihde 1964). This educational path can be exemplified by the career development of Justus von Liebig, the person whose name is known to many through the familiar 'Liebig condenser' used for laboratory distillation. His major contribution, however, was setting the model for laboratory-based science education in the 19th century.

Liebig made his first steps as a pharmacist's apprentice in his hometown of Darmstadt in Germany. In an advanced stage of his education, he won admission to Gay-Lussac's laboratory in Paris. Starting in 1824, when he accepted an appointment at the University of Giessen, Liebig developed a laboratory that became a prototype for science education in terms of both its physical facilities and educational program. Unlike 'kitchens filled with all sorts of furnaces and utensils', as Liebig described the frequent state of affairs in student laboratories (quoted in Ihde 1964), his laboratory provided specially-designed facilities as part of a specially-designed educational program which integrated theory with practice. Under Liebig's leadership, students started their training in basic skills, proceeded to original investigations and, in an advanced stage, assumed responsibility for guiding beginners. This educational pattern was extended by Liebig's students to other institutions.

Developments in chemical education were not followed immediately by other science disciplines. Practical work became a compulsory part of physics education in the late 1860s, starting at King's College in London and at the Massachusetts

---

Chapter Consultant: Brian Woolnough (Oxford University, UK)

Institute of Technology. Long-established universities were slower in making provisions for student laboratories. Cambridge University, for example, began requiring laboratory work in physics only in 1879, when Lord Raleigh was appointed Cavendish Professor of Physics (Jenkins 1979; Phillips 1981).

Construction of facilities for school science followed the developments in tertiary institutions in the 19th century. The design of facilities for schools was often a less expensive replication of the design of university facilities, with little consideration of the distinct goals of school science education. School laboratories at the beginning of this century were criticised as slavish copies of university laboratories (Armstrong 1903). Major features adopted from universities were the separation between laboratories for each of the science disciplines, as well as the separation between 'theoretical' instruction to be conducted in lecture halls and recitation rooms, versus 'practical' instruction in laboratories. Of the design features initially borrowed from universities, some have resisted change, while some have been modified over the century. Unfortunately, the respectability attached to higher education has continued to have a spell on decision makers in schools and to mask the distinct purposes of school science education.

The premise of this chapter is that the achievement of the goals of school science depends on student active involvement in practical work which is integrated thoughtfully throughout the program, and that the laboratory is both a means and an end in science education. Accordingly, to enhance achievement of intended goals, the habitat of school science teaching and learning should be a laboratory-type classroom, designed flexibly as part of the design of programs. This chapter's first three sections, therefore, address (1) goals and the assumed role of laboratory practical work, (2) the structure and function of laboratories and their potential to enhance science education and (3) the design of physical environments. The fourth section provides a case study of designing science environments.

## GOALS AND LABORATORIES: BELIEFS VERSUS EVIDENCE

The laboratory is believed to be a *sine qua non* of both science and school science. One can argue that the place of the laboratory needs no justification, as suggested in a witty assertion made by Solomon (1980): 'Science simply belongs there as naturally as cooking belongs in a kitchen and gardening in a garden'. Nonetheless, it is bothering that unequivocal empirical support is missing. Syntheses of research conducted since the science curriculum reforms of the 1960s raise questions about the value and effectiveness of laboratory work (Hodson 1993; Hofstein & Lunetta 1982; Lazarowitz & Tamir 1994; Tobin 1990; White & Tisher 1986). Yet, these reviews also criticise the existing research as being insufficient, not substantial, and failing to address crucial issues regarding what actually could be expected from practical work.

A major point highlighted in Hodson's (1993) critical review is the mismatch between goals and activities which subsumes an unwarranted expectation that practical work of any kind serves all purposes. The assumptions underlying the present discussion are that the effectiveness of practical work varies for different goals, and that carefully-designed laboratory activities with appropriate physical facilities can make contributions to most aspects of science education. If laboratory and other kinds of practical work are to be interwoven effectively within school science, the process of program development should include the juxtaposition of desired outcomes against a range of activities that are likely to enhance their attainment. The appropriate activities, in turn, should be matched by adequate facilities.

Because program goals usually are stated broadly, it is difficult to estimate the contribution of the laboratory to some of their many aspects and to use goal statements as a frame for laboratory design. For example, a worthy goal for school science, such as educating students who are able to engage in the public discourse and debate about matters of scientific and technological concerns (National Research Council 1996), cannot be translated directly into practical activities. For the purpose of goal-related analysis and design of activities and facilities, it is recommended that grand goals are accompanied by adjunct laboratory objectives which can be phrased in operational terms.

The literature offers various sets of objectives for practical work, ranging from long and detailed lists in the behaviourist tradition (where the meaning of grand goals is often lost), through to broad categories for clusters of aims. Most of the objectives can be subsumed under the four broad classes of intended learning outcomes and activities suggested by Hegarty-Hazel (1990): (1) technical skills; (2) scientific inquiry; (3) scientific knowledge; and (4) attitudes. The class entitled 'technical skills' is concerned with the acquisition of practical skills and techniques. The development of intellectual skills, along with the development of scientific attitudes and insights into what science is, are part of the category 'scientific inquiry'. 'Scientific knowledge' stands for the facilitation of what often is referred to as science content (i.e., facts, concepts, rules and their applications). The label 'attitudes' is attached to activities that aim to enhance motivation, enjoyment, interest and attitudes in regard to science, science learning and choice of a science-related career. The above categories are meant neither to reflect the purposes of empirical work in science, nor to place value on fragmented instruction. Despite different foci on different objectives, depending on the educational context, practical work always should be seen as a holistic activity (Woolnough 1991).

Until relatively recently, technical skill acquisition was the only set of objectives for which laboratory bench work was accepted as indispensable. With the advent of computer simulations, however, questions have emerged about the place and scope of hands-on skill training in school science. The demand for alternative training adds to the criticism on the unbalanced emphasis on simple 'cookbook' exercises for the development of techniques or for the confirmation of foregone conclusions. The call for a shift towards a range of practical activities within the broader goals of school science, including 'authentic' laboratory investigations for the cultivation of intellectual skills, implies allocation of less time, space and funds for technical-skill-oriented bench work.

But the debate on laboratory aims with the most important implications for physical environments concerns the extent to which student practical work contributes to content learning. Some argue that the use of the laboratory for this purpose is largely a waste of time and resources (Woolnough & Allsop 1985). The accumulating literature on student concept learning has led many others to conclude that intentionally-designed laboratory activities, when thoughtfully intertwined with 'non-laboratory' activities, can enhance concept learning, including reconstruction of concept understanding (Gunstone & Champagne 1990; Lazarowitz & Tamir 1994). The flexible integration of an array of practical activities throughout the large time devoted to content learning cannot be achieved in the ordinary classroom. It requires allocation of multifunctional laboratory-type environments for all science lessons.

## STRUCTURE AND FUNCTION OF LABORATORIES

The idea that physical structure influences educational function is neither new nor unique to science education. '[I]f we put before the mind's eye the ordinary schoolroom, with its rows of ugly desks placed in geometrical order, crowded together so that there shall be as little moving room as possible . . . we can reconstruct the only educational activity that can possibly go on in such a place' (Dewey 1899/1959, p. 50). The same idea affected Armstrong's campaign at the beginning of the century for a laboratory-centred science curriculum. He argued against the teacher demonstration desk and advocated unsuccessfully to dispense with it altogether (Jenkins 1979).

Unfortunately, research on the interrelations between physical environments and practice in education is scarce. Studies of classroom organisation and management seldom have involved the physical layout of classrooms (Doyle 1986). The extensive research programs on learning environments in science classrooms have been concerned more with socio-psychological characteristics than with physical attributes of the environment (Fraser 1994). Some of the instruments commonly used for the measurement of science learning environments include items on physical facilities. For example, an instrument developed specifically for laboratory classroom environments (Fraser, Giddings & McRobbie 1995) includes a scale entitled Material Environment which consists of items such as 'I find that the laboratory is crowded when I am doing experiments' and 'The laboratory is an attractive place for me to work in'. This and similar scales in other instruments are concerned with perceptions of resource availability and adequacy, without explicit reference to specific facilities. Furthermore, the questionnaire-generated data often are not complemented by observation-based data on classroom design and activities.

A classroom environment study with an uncommon focus on facilities (Ainley 1978, 1990) was carried out as part of the evaluation of the facilities program which accompanied the Australian Science Education Project (Shepherd 1974). Based primarily on student reports of their learning environments and teacher reports of the standard of facilities, the results suggest that active forms of learning are associated with better science facilities. A link between educational facilities and activities also was suggested by the few studies which included observations of physical environments. For example, a higher frequency of inquiry methods was detected in teaching conducted in spaces with combined classroom and laboratory facilities, compared with teaching in separate spaces for classrooms and laboratories (Englehardt 1968). Moreover, science programs and facilities appear to have evolved through largely parallel patterns (Novak 1972). This trend emerged in a large-scale school survey across the USA, initiated towards the end of the science education projects of the 1960s, to guide future developments in facilities. Unfortunately, the post-1960s worldwide decline in investments did not encourage developments and the research interest in physical facilities (which had never been strong) has faded away. Consequently, guidelines and comments on science facilities in official publications for teachers and school decision makers are rarely accompanied by educational research references (e.g., Association for Science Education 1989; National Research Council 1996; Showalter 1984).

Despite the paucity of the research base, the evidence mentioned above shows that physical environments make a difference. Their influences, however, can be subtle; dramatic transformations cannot be expected as an immediate response to any physical change. What needs to be clarified is how and under what conditions physical environments exert their effects. Are facilities just a technical factor for certain activities, or do laboratory environments have wider, perhaps indirect, influences? Apparently, beyond the distinct technical roles of facilities, they also create a special atmosphere that suggests activities and affects student perceptions of science:

> The laboratory sets science apart from most school subjects. It gives science teaching a special character . . . That character is almost sufficient alone to justify the high capital and recurrent costs of laboratories. . . . Without the laboratory, it could be difficult for students to comprehend what scientists do. (White 1988, p. 186)

The idea of the laboratory as a suggestive space is intangible and difficult to investigate. Longitudinal case studies of the establishment and development of physical environments, along with the educational activities which they encompass, are needed to understand the underlying processes and to elucidate causal relationships (Arzi 1988). Scanning the history of laboratories constitutes a retrospective longitudinal study that exposes various facets of the interplay between physical structure and educational function. This can be exemplified through a segment from the story of analytical chemistry. The advent of new instruments led to the decline of traditional qualitative and quantitative inorganic analysis, to a change in research programs and, eventually, to curricular changes. Traditional chemical analysis imposed special laboratory requirements, including rows of shelves and racks for reagent bottles, often mounted at the centre of long, double-sided, fixed benches which block vision from one side of a bench to the other. Traditional analysis dominated school chemistry curricula and laboratory design beyond its disappearance from research laboratories. The eventual removal of the long reagent shelves opened the field of view and thus provided previously unrecognised opportunities for better communication and collaboration between teacher and student and among students. The subsequent realisation that fixed rows could be replaced by more flexible layouts enabled the laboratory to be used as a science classroom, not just as a space confined to bench work in its narrowest sense.

Apparently, in the evolution of science programs, teachers more easily could develop and implement flexible, less formal and more constructivist methods in schools with more flexible physical environments. Flexible laboratory layouts, however, were more likely to develop earlier in schools in which change already had started among teachers who, in turn, initiated modifications of their teaching environments to accommodate their educational beliefs. Even though the processes underlying the interrelations between structure and function are not sufficiently clear, the available evidence supports the assumption that thoughtfully-designed laboratory environments are invaluable for science education.

## DESIGNING SCIENCE ENVIRONMENTS

Because laboratories usually are seen as a unique feature of science, administrators who are under pressure to produce quick and demonstrable results sometimes choose to invest in laboratory construction and in the purchase of apparatus, rather than in slow and intangible change processes. But mere establishment of laboratory facilities has limited value, unless their design is program-driven (i.e., preceded by program goals and followed by their educationally-meaningful integration within the program). Furthermore, because construction in practice can occur less often than program modification, the physical design must be future oriented. It should aim beyond existing or planned curricula to accommodate changes in goals, new frontiers in science and the introduction of currently unfamiliar interdisciplinary combinations, shifts in educational approaches and development of unforeseen technologies. Flexibility and multifunctionality, therefore, are the most important features of any school science environment.

### Flexibility

Flexibility was the major theme in a synthesis of a survey of exemplary school facilities for science in the USA (Novak 1972) and in a compilation of accounts of facilities across Australia (Shepherd 1974). Due to their future orientation, recommendations made by these authors over two decades ago are still valid today. Total flexibility in each and every feature cannot be attained. Compromises have to be made and creative design and construction solutions are needed. Ideas for flexible designs of science classrooms range from movable furniture to movable internal walls, and from peripheral to central location of services. Suggestions for flexible provision of laboratory supplies include combinations of exposed, wall-mounted pipes and conduits, overhead outlets, and service islands which can be either fixed ones or portable ones with plug-in connections to floor wells.

Following growth in the use of computers and electric appliances, the number of sockets and their distribution have become major determinants of flexible function in all science classrooms, not just in physics laboratories as it used to be. Personal computers, today's dominant technology, merit a comment in regard to future-oriented design. In addition to electricity, they require communication cables for networking, preferably in exposed conduits for easy installation and change. Other special arrangements for computers, such as niches or extra space on each student bench, are likely to become obsolete as their size decreases, the use of portable computers increases and new technologies are developed. A solution which has been tried out successfully is to use bench-high trolleys as bases for computers, to enable movement among classes and the extension of work areas wherever and whenever necessary. Trolleys also are useful as bases for other expensive and seemingly stationary instruments. For example, a recirculatory fume hood, or even a gas chromatograph with its gas cylinder, can be mounted securely on a trolley for flexible operation in any class.

*Multifuctionality*

The flexibility of the school science environment should facilitate not only future changes, but also current multifunctionality. Whether intended for general science or for a distinct discipline, a science environment should allow for laboratory bench work – both short exercises and longer projects performed individually or in groups, computer-based or assisted activities, non-laboratory practical work, small-group and whole-class discussions, as well as traditional student seat work and teacher lectures and demonstrations. Such a multifunctional space facilitates integration of theory with practice and thus can contribute to a holistic perception of science. It also is likely to encourage teachers to implement constructivist methods that rely on transitions and interactions between practical work and discussions in groups of various sizes. If students have to sit uncomfortably on tall stools behind standing-height benches tightly screwed to the floor, they might be able to perform hands-on bench work, but it will be difficult for them to concentrate on lectures, discussions, computer and other kinds of seat work. Even if benches are movable and allow for normal sitting height, squeezing them close to each other, without space for student and furniture movement, will make the environment practically inflexible and not multifunctional.

A multifunctional space should attempt to meet all disciplinary requirements. According to a study of the history of school facilities in the UK (Jenkins 1979), general science laboratories were introduced as low standard substitutes when separate laboratories for each discipline could not be afforded. Based on a holistic perception of science, a multidisciplinary space is desirable not only in primary schools, general science or 'science for all' courses, but also in advanced-level secondary courses. This is possible, even though disciplinary laboratory activities sometimes impose special demands. For example, traditional work in chemistry requires fume hoods and chemically-resistant bench tops, but good school chemistry can be done in any laboratory space with the aid of mobile fume hoods and resistant work trays. The introduction of micro-scale techniques is likely to reduce the need for special heavy-duty facilities for chemistry. Alternatively, in a school science complex, teachers can choose to designate one room as chemistry-oriented, not necessarily a chemistry-exclusive room, with a fixed, high-quality fume hood and standing-height, chemically-resistant benches.

Multifunctionality can be achieved through the design of science classrooms which encompass all activities under the same roof, as opposed to separate rooms for laboratories versus classrooms, with further separation among disciplines. Within the same space, distinct areas can be allocated to laboratory work and to lecture or reading-writing sessions, with two separate seats for each student. Alternatively, the same area with one seat per student can serve all purposes. Within the single area of a laboratory-classroom, seats can be arranged around various combinations of perimeter or island services. Obviously, this does not provide 'ideal' conditions for each and every activity. Yet, apart from being less space-consuming, the significant educational advantage of the single-area laboratory-classroom is that it facilitates the use of multiple strategies for interactive teaching and learning, with easier and more natural shifts between, and consequently integration of, theory with practice.

*Practical Checklists*

Illustrations of alternative layouts and practical advice are provided in design guides for science facilities published by state or local school authorities and teacher associations. Examples are the valuable publications of the Association for Science Education (1989) and the Consortium of Local Education Authorities for the Provision of Science Service (1995), both in the UK, and the California Department of Education (1993). Information on state-of-the-art materials and fittings can be found in commercial catalogues of laboratory suppliers. Good ideas also are found in older publications (even when personal computers are not mentioned and some features are outdated). For example, useful checklists of design features of both a single room and a science department or block are offered by Archenhold, Jenkins and Wood-Robinson (1978). These checklists address preparation, storage, workshop and exhibition areas, provisions for long-term student projects, and ancillary outdoor facilities such as greenhouse and animal house.

Because the design of physical environments should be program-driven and future-oriented, and because building regulations, materials and traditions are not the same everywhere, a single ideal design is not possible. Although different laboratory design guides appear to be similar, there are variations in important issues, including items that receive attention in some without even being mentioned in others. For example, there is agreement that ample laboratory space allows for safety and flexibility. It is also accepted that a single laboratory should accommodate 20 to 24 students but, while some guidelines forbid larger numbers, others suggest arrangements for occasionally accommodating up to 30 students. Conflicting guidelines can be exemplified in relation to gas supply. In the name of safety, some require installation of mains gas (e.g., in the UK), while others prefer portable burners (e.g., in Israel). A suggestion that bypasses this conflict involves replacing the seemingly indispensable Bunsen burners by electric heating devices. Another example of an unusual suggestion is to carpet all areas, including chemistry laboratories, due to the acoustic and aesthetic value, at the expense of maintenance difficulties (Novak 1972). Special precautions against earthquake damage, including shelves with restraining lips anchored to structural support, exemplify local requirements that are not applicable everywhere (California Department of Education 1993).

Safety is a major issue in all guides for science facilities. There are more publications solely on safety than on laboratory design. This probably reflects desired environmental awareness and attention to the welfare of students and teachers. Unfortunately, safety requirements sometimes have been misinterpreted and used to legitimise avoidance of student practical work. Contrary to common belief, a survey of fatal and major injuries in schools in England revealed that less than one percent occur to students in laboratories (quoted in Borrows 1993). Apparently, if regulations are followed carefully, both in the planning and operation of laboratories, practical work in science can be very safe.

In view of the variety across design guides and across the needs and programs of schools, those responsible for the refurbishing of a single laboratory, or for the construction of a science complex, are advised to scan as many sources as possible. Unorthodox ideas that have not become consensual, as well as seemingly esoteric advice or trivial technical details that have been neglected or remained unnoticed, could prove to be valuable.

## DESIGNING SCIENCE ENVIRONMENTS: A CASE STUDY

In primary schools, science usually is taught in normal classrooms with the aid of mobile equipment, especially demonstration carts and activity kits. In secondary schools, if not earlier, students and teachers need special facilities, preferably an environment where laboratory-classrooms are the major building blocks of a larger structure – the science department or centre. Obviously, there is no single 'right' layout for a laboratory-classroom, just as there can be no single 'right' model for a science centre. To illustrate one way in which views and suggestions expressed in this chapter can be realised, the design process and eventual structure of a centre for science education is discussed below.

Whereas science facilities are traditionally school-based, this particular centre, established in Israel and entitled HEMDA (Hebrew acronym for science education), serves several schools. It follows a new model for regional centres for science education that relies on the advantages of the centralisation and linkage of material and human resources. The centre's unique features are that it has its own well-equipped educational facilities and its own well-qualified team of teachers, and that it coexists with the schools, assuming responsibility for the science education of students in the upper grades of secondary school. The physical design reflects major ideas that have been highlighted in this chapter and found to be suitable not only for a regional institution, but also for a science centre within a single school, for both introductory and advanced-level courses.

The physical structure of the centre evolved in a process of interactions between concepts in science education and in architecture. The starting point for the architectural design was a frame of educational specifications revolving around the idea of multifunctional laboratory-classrooms encircling a common hub. The specifications emerged from analyses of goals and needs of previous, current and developing programs in Israel and abroad, and these were informed by suggestions from various sources. Examination of alternative prototypes of laboratory-classrooms, prepared by the architect in response to the specifications, led to a decision to have some variations among classrooms with a similar frame. Consequently, all the rooms are multifunctional and bear no disciplinary labels. But there are rooms which are physics oriented (with more electricity and less water and gas), rooms which are chemistry oriented (with a fixed fume hood) and rooms in which services have no particular orientation. The differences among rooms facilitate special functions, while maintaining the flexibility needed for the use of each one for practically any subject or activity and for future changes.

According to the common frame, each laboratory-classroom can accommodate 20 to 24 students, with provisions for up to 28, within an area of approximately 85 square metres. To break the squareness and to create an attractive and stimulating atmosphere, each room is enclosed by hexagonal walls and bright colours are injected through furniture, fittings and posters. Students and teachers have unfixed, sitting-height benches. Whenever necessary, these benches are extended by adjoining equal-height trolleys with computers. Bench-high trolleys also are regularly used as mobile bases for various kinds of equipment. Extra space for long-term projects, class experiments and various other uses is provided through fixed, standing-height benches along part of the walls. Each room has arrangements for electricity, water, drainage, gas, computer communication, compressed air and air extraction system for mobile fume hoods. The services are provided both at the perimeter through wall-mounted outlets and centrally through service islands, with additional overhead supply of electricity in several classes and a fixed fume hood in others. During construction, the installation of some services was incomplete intentionally. For example, installation of exposed conduits preceded the actual networking of computers. Wiring with communication cables started after the centre began operating, when related educational objectives had been clarified so that technical requirements could be matched accordingly.

After the matching of facilities to the educational work within a laboratory-classroom – the elementary unit of the HEMDA structure – the planning process

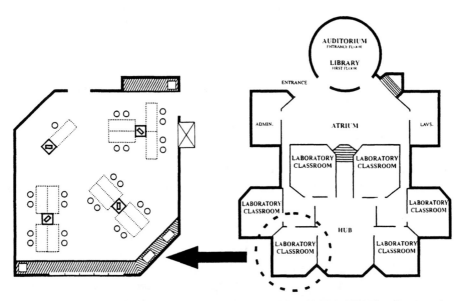

Figure 1: Schematic illustration of a laboratory-classroom and floor plan at HEMDA – Centre for Science Education in Tel Aviv (Adapted from a plan by Spector-Amisar Architects, Jerusalem).

proceeded to the concentric arrangement of a classroom cluster around a common hub. The hub provides its six surrounding classrooms with logistic back-up, fulfilling all the traditional functions of a laboratory storage and preparation room. The hub also includes laboratory areas for research and development, as well as open-space offices and rest areas for teachers, separated from each other by storage cabinets. From the interrelations between the classrooms and their hub, the planning moved on to consider communication between classroom clusters on adjacent floors. It concluded with the interplay between the entire complex of laboratory-classrooms and their ancillary spaces, and between the building and the landscape. The landscape includes a garden with a sundial, a weather display, satellite dish aerials that function as 'acoustic mirrors', and other science-related objects.

The compact architectural design of laboratory-classroom clusters and easy communication among different spaces were requested in the educational specifications. This was primarily intended to facilitate teacher team work, which is a key concept in the HEMDA model. The classroom doors that open directly into the hub, and the ease with which equipment can be shifted on trolleys, stimulate teachers to include unplanned laboratory activities in immediate response to student questions and comments. Likewise, the proximity of the library, with both printed and electronic publications, stimulates spontaneous use of its sources during lessons. The accessible open storage system and the sharing of all equipment, regardless of its disciplinary origin, contribute further to team work and flexible integration of laboratory activities within the different science programs offered at HEMDA.

## MORE THAN WALLS, BENCHES AND WIDGETS: CONCLUDING COMMENTS

'*Sine qua non*', 'vital', 'sacred' and 'heart' are some of the labels attached to the science laboratory in scholarly educational papers and technical guides alike, despite academic criticism and school administrators' questions on accountability. In this chapter, I share belief in the educational potential of the laboratory. Underlying this chapter also is the belief that practice and theory together form the whole of science. Syntheses of research suggest that the realisation of the potential of laboratory work depends largely on careful matching of activities to the different objectives of school science. I therefore advocate that the process of program development includes goal-related analysis of both activities and facilities, thus leading to program-driven laboratory design. Furthermore, if it is accepted that the laboratory can enhance achievement of goals beyond technical-skill acquisition, including fostering of intellectual skills and concept understanding, then practical work should be integrated throughout school science programs. Accordingly, laboratory-type environments should be allocated for all science lessons at all grade levels.

To meet the requirements of different science subjects and changes in goals over time, the school science environment has to accommodate a wide range of teaching and learning strategies and activities. Consequently, the essential features of the environment are flexibility and multifunctionality. These can be provided through different architectural layouts, materials and equipment. Obviously, because the design of a particular school space depends on the physical and educational contexts in which it is established, a single ideal design is not possible. In this chapter, therefore, I emphasised conceptual issues in the process of building facilities for school science, rather than technical details. The discussion highlights the advantages of the laboratory-classroom in which a single space serves all purposes. The realisation of this and other ideas is illustrated through a case study of a centre for science education.

Eventually, the extent to which a given science environment is flexible and multifunctional in effect, and the extent to which it enhances science education, depend on the teacher's subject-matter knowledge, pedagogical knowledge, beliefs and ingenuity in the use of the available resources, more than on the particular physical features of any facility. Laboratory environments, however, are more than an aggregate of walls, benches and widgets. The laboratory plays distinct technical roles and, at the same time, also provides a suggestive space, creating a unique atmosphere that conveys a message. As a part of what constitutes the ethos of science, the laboratory is not only a means but also an end in science education. A laboratory environment, therefore, is the natural habitat for student growth in science.

## REFERENCES

Ainley, J.: 1978. 'Science Facilities and Variety in Science Teaching', *Research in Science Education* 8, 99–109.

Ainley, J.: 1990. 'School Laboratory Work: Some Issues of Policy', in E. Hegarty-Hazel (ed.), *The Student Laboratory and the Science Curriculum*, Routledge, London, 223–241.

Archenhold, W.F., Jenkins, E.W. & Wood-Robinson, C.: 1978, *School Science Laboratories*, John Murray, London.
Armstrong, H.E.: 1903, *The Teaching of Scientific Method*, Macmillan, London.
Arzi, H.J.: 1988, 'From Short- to Long-Term: Studying Science Education Longitudinally', *Studies in Science Education* 15, 17–53.
Association for Science Education: 1989, *Building for Science: A Laboratory Design Guide*, Author, Hatfield, UK.
Borrows, P.: 1993, 'Safety in Secondary School Science', in R. Hull (ed.), *ASE Secondary Science Teachers' Handbook*, Simon & Schuster, Hemel Hempsted, UK, 129–147.
California Department of Education: 1993, *Science Facilities Design for California Public Schools*, Author, Sacramento, CA.
Consortium of Local Education Authorities for the Provision of Science Service: 1995, *Laboratory Handbook*, Author, Uxbridge, UK.
Dewey, J.: 1899/1959, 'The School and Society', in M.S. Dworkin (ed.), *Dewey on Education*, Teachers College Press, New York, 33–90. (Original work published 1899, The University of Chicago Press).
Doyle, W.: 1986, 'Classroom Organization and Management', in M.C. Wittrock (ed.), *Handbook of Research on Teaching* (3rd edition), Macmillan, New York, 392–431.
Englehardt, D.F.: 1968, 'Aspects of Spatial Influence on Science Teaching Methods', Unpublished EdD Thesis, Harvard University, Cambridge, MA.
Fraser, B.J.: 1994, 'Research on Classroom and School Climate', in D.L. Gabel (ed.), *Handbook of Research on Science Teaching and Learning*, Macmillan, New York, 493–541.
Fraser, B.J., Giddings, G.J. & McRobbie, C.J.: 1995, 'Evolution and Validation of a Personal Form of an Instrument for Assessing Science Laboratory Classroom Environments', *Journal of Research in Science Teaching* 32, 399–422.
Gunstone, R.F. & Champagne, A.B.: 1990, 'Promoting Conceptual Change in the Laboratory', in E. Hegarty-Hazel (ed.), *The Student Laboratory and the Science Curriculum*, Routledge, London, 159–182.
Hegarty-Hazel, E.: 1990, 'The Student Laboratory and the Science Curriculum: An Overview', in E. Hegarty-Hazel (ed.), *The Student Laboratory and the Science Curriculum*, Routledge, London, 3–26.
Hodson, D.: 1993, 'Re-thinking Old Ways: Towards a More Critical Approach to Practical Work in School Science', *Studies in Science Education* 22, 85–142.
Hofstein, A. & Lunetta, V.N.: 1982, 'The Role of the Laboratory in Science Teaching: Neglected Aspects of Research', *Review of Educational Research* 52, 201–217.
Ihde, A.J.: 1964, *The Development of Modern Chemistry*, Harper & Row, New York.
James, F.A.J.L. (ed.): 1989, *The Development of the Laboratory*. American Institute of Physics, New York.
Jenkins, E.W.: 1979, *From Armstrong to Nuffield: Studies in Twentieth-Century Science Education in England and Wales*, John Murray, London.
Layton, D.: 1990, 'Student Laboratory Practice and the History and Philosophy of Science', in E. Hegarty-Hazel (ed.), *The Student Laboratory and the Science Curriculum*, Routledge, London, 37–59.
Lazarowitz, R. & Tamir, P.: 1994, 'Research on Using Laboratory Instruction in Science', in D.L. Gabel (ed.), *Handbook of Research on Science Teaching and Learning*, Macmillan, New York, 94–128.
National Research Council: 1996, *National Science Education Standards*, National Academy Press, Washington, DC.
Novak, J.D.: 1972, *Facilities for Secondary School Science Teaching: Evolving Patterns in Facilities and Programs*, National Science Teachers Association, Washington, DC.
Phillips, M.: 1981, 'Early History of Physics Laboratories for Students at the College Level', *American Journal of Physics* 49, 522–527.
Shepherd, R. (ed.): 1974, 'Facilities for Learning Science' [Special issue on the buildings and other facilities for the learning of science in Australian schools], *The Australian Science Teachers Journal* 20(3), 4–76.
Showalter, V.M. (ed.): 1984, *Conditions for Good Science Teaching*, National Science Teachers Association, Washington, DC.
Solomon, J.: 1980, *Teaching Children in the Laboratory*, Croom Helm, London.
Tobin, K.: 1990, 'Research on Science Laboratory Activities: In Pursuit of Better Questions and Answers to Improve Learning', *School Science and Mathematics* 90, 403–418.

White, R.T.: 1988, *Learning Science*, Basil Blackwell, Oxford, UK.
White, R.T. & Tisher, R.P.: 1986, 'Research on Natural Sciences', in M.C. Wittrock (ed.), *Handbook of Research on Teaching* (3rd edition), Macmillan, New York, 874–905.
Woolnough, B.E.: 1991, 'Practical Science as a Holistic Activity', in B.E. Woolnough (ed.), *Practical Science*, Open University Press, Milton Keynes, UK, 181–188.
Woolnough, B. & Allsop, T.: 1985, *Practical Work in Science*, Cambridge University Press, Cambridge, UK.

## 5.5 Reading the Furniture: The Semiotic Interpretation of Science Learning Environments

BONNIE SHAPIRO
*The University of Calgary, Canada*

This chapter introduces the semiotic interpretation of science learning environments. Like other research on science learning environments, the semiotic perspective asks how students and teachers work together to develop knowledge, skills and attitudes. But semiotic studies are based on the assumption that one's culture provides a set of signs, symbols and rules about interaction that are used, whether consciously or not, both to create and 'read' the learning environment. Therefore, semiotics broadens the term 'learning environment' to include these signs, symbols and rule sets as powerful features that influence learning and teaching.

The chapter begins with a demonstration of a semiotic reading of an aspect of the learning environment, then moves to a consideration of the historical development of semiotics as a research approach. Semiotics in educational research is described in the first section. Next, five case studies of the learning environment are presented using a semiotic interpretive perspective. In a concluding statement, the value of a semiotic awareness of school learning environments is explored, with specific implications for teaching, learning and curriculum organisation.

On the first day of my primary science methods course, I arrange the furniture in the classroom in a traditional fashion with moveable work tables placed in rows and separate from one another. Students sit individually or in pairs at the sides of the classroom. On the second day of class, students come into the classroom to find that the tables have been rearranged into hexagons. I ask students to look at the arrangements with these questions in mind: 'How does the furniture "speak" about science and science learning in this classroom?' The placement of furniture in the science classroom constitutes an act of communication, but what does it communicate?

Student teachers typically respond that 'the arrangement in rows says that we are to do our own work. We are not to talk to one another and our eyes should not wander. The grouped arrangement suggests that we will be working together in science. It says that we may collaborate, perhaps sharing resources, and it is all right to talk.' In this way, students consider features of the classroom environment as a part of a larger set of cultural meanings that presents a message that children learn to read.

---

Chapter Consultant: Michael O'Loughlan (Hofstra University, USA)

As a part of the physical environment of the school, the furniture arrangement is a part of an elaborate *system of signification*, which is read by all who enter the classroom setting. The physical arrangement of the chairs in the classroom is a way of communicating and acts as a powerful sign to students about how science and science learning in this classroom will proceed. We have begun a semiotic reading of the science classroom and of science learning.

Semiotics is not only the study of how social groups create meaning-making practices that are uniquely their own, but it also is concerned with the ways in which individuals make sense of practices within settings created by the larger group. Worth (1981) notes that semiotics helps to explain 'how human beings communicate with one another within a shared system, exchanging units that have common significance and a commonly understood set of rules of inference and implication' (p. 74). Through its focus on signs and symbols, semiotics is a form of inquiry that attempts to explore meanings that constitute any act of communication (Eco 1979; Harland 1987). A semiotic consideration of events involves the idea that reality is not disclosed directly, but is experienced through symbols and activities mediated by language and culture. Based on the philosophical foundations of C.S. Peirce (Hartshorne & Weiss 1930; Peirce 1985), the central questions of semiotics concern meaning, knowledge and knowing through the use of signs and sign systems. To semioticians, all cultural phenomena are essentially processes of producing and interpreting signs (Eco 1976; Percy 1982; Suhor 1982). In studies of the classroom, the semiotic perspective not only asks 'In what ways are teachers and students working together in the acquisition of knowledge?' but also 'How are the ways in which teachers and students work together in learning communities representative of the values and activities of the larger culture?'

Learners are not only consumers of sign and symbol systems, but they are also symbol users. Understanding the complex ways in which we use sign and symbol systems provides powerful insight into the ways in which we communicate with learners in the school setting. The school setting is an expression of forms of action, ideas about how action will proceed, and ideas and values created by the culture. It is also the setting created *to reproduce these ideas and values*. Within the semiotic view, the sign is regarded as the smallest unit of meaning (Eco 1988; Guiraud 1975) that stands for something else. This meaning of sign is far more extensive than the everyday meaning of the term. This sign is a cultural unit representing an attempt to organise and communicate something about the world. As a sign, the colour green in certain settings is a sign that it is all right to go forward.

Waving one's hand in certain contexts means goodbye, whereas it is an indication of a desire to speak in other contexts. A key insight of semiotics is that sign systems involve the constant correlation of content and expression. A semiotic perspective focuses on communication rather than isolated linguistic forms. We therefore look closely at sets of signification systems, an approach that is particularly useful in research on school and classroom interaction. Examples of such patterns and formations are presented in later sections as *activity structures* and *dialogue formations*, whereas the next section considers how semiotics has developed a useful interpretive form of research and thinking in the study of community meaning making.

Each community has its own meaning-making practices. The term 'social semiotics' has been used (Lemke 1987, 1988, 1989) to describe the semiotic representations of the larger community. Social semiotics employs formal semiotics but goes further to explain the range of features of the communities such as schools, their differing signifying systems and practices, and how they connect with the many constellations of meanings within the larger society. Social semiotics is built upon the thought of French historian and social theorist, Michael Foucault (1969), who analysed relationships between how we talk about the world, act and are acted upon in it. Lemke (1990) connected discourse and 'the technologies of action and control to larger patterns of belief and power in society' (p. 185). In social semiotics, the interlinking meanings of the larger group constitute its ideology, which is a fundamental element in the social production of meanings.

There is a wide range of modes of semiotic human action, such as gestures and routines of behaviour and talk. These semiotic activities are a part of a set of repertories of actions which are socially recognised by members of a particular culture and which are socially learned. In the classroom, these action repertories can be considered as text forms in the sense that they mediate features of a multi-levelled discourse that successful students are expected to understand and use. Suhor (1984) notes that, in school settings, we ask students to understand and perform using many sign systems at once: linguistic, gestural, pictorial, musical, architectural, chronological, mathematical, social, artistic and literary, as well as scientific. Language is but one kind of sign (Halliday 1978; Halliday & Martin 1993; Hawkins 1984).

## SEMIOTIC STUDIES IN SCIENCE EDUCATION

Although semiotics has been largely uninterpreted for educational practice, recent studies consider sign and signification systems within educational settings (Ball 1990; Bowers 1990; Crenshaw 1991; Golden & Gerber 1990; McLaren 1986; Siegel 1989; Smith 1989), with a few semiotic interpretive studies specifically in science education (Groisman, Shapiro & Willinsky 1991; Lemke 1987, 1990). Barthes (1967, 1988) demonstrated the significant contribution of semiotics in the critique of the ideological messages of a given culture. Schools are particularly fruitful settings for developing, through semiotic readings, a critical perspective on meaning and practice in the larger educational system.

When viewed semiotically, the science classroom can be seen as interweaving sets of sign, symbol and signification systems that students learn as texts of science learning. Specific activity structures and forms of discourse, which often are familiar only to the teacher, are revealed as sets of unspoken rules of behaviour and action acceptable in the classroom setting. Most students are usually not familiar with the scientific perspective of the world, yet, in the classroom setting, they are expected to integrate, if not wholly adopt, views that can differ dramatically from their current views of phenomena and events (Shapiro 1989, 1994). Through sign and signification systems, knowledge is made available to students who are able to read and interpret the signs. For others, such as non-English speakers or learners from minority cultural groups, scientific understandings can represent a type of restricted code or language that is used in very specific ways by a small group of initiates, thereby turning many students away from science learning. Examples of interpretive studies of science learning environments are presented in the following section to show the kinds of insight offered by critical semiotic analyses.

*Reading the Architectural Code as Text*

Danesi (1994) writes that cultures are social territories as well as ideological entities. As the cultural elaboration of shelter, architecture is 'the art of imbuing living spaces with symbolic meaning' (p. 186). He notes how building height conveys a specific kind of meaning. The tallest buildings during the mediaeval period were churches (symbols of authority and power at that time), and in our present western culture are institutions holding financial power. Height is still associated with power and authority. The placement of objects and persons within this dichotomy reaffirms the culturally elaborated positioning of what is meaningful and therefore important to pay attention to in the classroom. In the science classroom, we might see a demonstration table or the teacher's desk placed on a platform so that it is higher than the students' desks. The teacher is positioned on the platform not only so that students can see her, but the difference in height emphasizes the authority of her position. The teacher stands while she speaks to the class. Students remain seated. It is accepted that it is inappropriate for students to rise unless given permission by the teacher.

A typical junior high school setting in a mid-sized Western city has a special room designated as the Science Laboratory. A sign has been made to place on the door of this room. Other signed doors in the school are the Main Office, the Gymnasium, the lavatories and the Staff Room and these indicate the demarcation of these rooms from others in the building. They are designed for a special purpose and therefore only very specific activities, routines and ways of interacting will take place inside of them. Rooms in a building also evolve a symbolic quality and the objects within them acquire meaning as well. Science is taught in this special room with equipment and furnishings set up in a manner that has been preordained. The development of the landscape of the science classroom is built upon conceptions of what is needed to teach science and preordains the arrangement of the setting. Despite the fact that science is as much about questioning, thinking and gathering information as it is about using laboratory tools, the space in the classroom is overwhelmingly dominated by laboratory benches, sinks, stools and equipment such as beakers, alcohol burners, test tubes and flasks.

The pattern of action that will take place within this classroom is determined in large part by its organisation. As Eco (1985) suggests, any given culture is guided in its organisation of reality by the three elements of practical purpose, pertinence and material constraints. The architectural design establishes not only the structure of the classroom, but the characteristics of classroom contexts and the pattern of actions that will take place within it. The architectural code is read by the learner as a system of signification or representation of space.

*Reading Lesson Structure as Text*

The students in a lesson first described by Groisman, Shapiro and Willinsky (1991) are receiving an introduction to a technological problem-solving lesson. They are given the task of building a load-bearing structure using a specified set of materials. The teacher begins the class by reviewing instructions for the task:

> So, yesterday, I was telling you that today you're going to do a special project. And it has to do with building a particular kind of structure and I couldn't tell you more than that because I didn't want you to go home and brainstorm . . .
> You'll receive at each station . . . five index cards. It's these index cards that you fold, you can tear, whatever you want to do with them, within a 25-minute time period . . . Now each station [students are divided in eight groups with three members each] has five index cards and one piece of masking tape, 30 centimeters long, so don't waste it.
> . . . There's going to be an evaluation desk . . . and then I will go and get a golf ball and put it on the structure. If your structure can stand the pressure of that tiny little golf ball for 30 seconds, then you will receive some kind of letter grade. Now, it's not a personal grade, remember. I'm not giving you an A, a B, a C, D, E or an F. The grade that you will receive is towards the structure. The structure itself is being evaluated.

In her first statements, the teacher represents herself as different from her students, demarcating her authority and simultaneously establishing a connection between knowledge and classroom control. The activity is presented as a mystery, but the mystery is only for the students. The teacher knows what the activity is about. It is she who determines when and how the secret will be revealed to the students.

The project is described to the students through a set of restrictions (materials, time on task, function of the product, etc.). These restrictions tell very little about what the students will be challenged to do. Semiotically speaking, the project challenges students to enter a new territory – the technological world – with a complex network of signs which they regularly use in their everyday lives, but about which they possess an incomplete knowledge.

The particular conditions under which this activity takes place – the time limit and the sense of each group competing against all the others – gives support to the interpretation of this special project as a *contest*. Taken as a sign, a 'contest' has multiple meanings. First, it signifies a particular view of science, by taking an approach that attempts to establish a closer match with social practices outside the school setting, because contests are everywhere in the learners' experience.

The contest is a sign in itself, but it also might be considered a *text*. Each of the features that define this science learning experience as contest speaks a message to the learner. For instance, the selection of a rather flimsy material for construction (index cards) could signify the lack of importance of the product in itself, thus giving greater meaning to the process instead. The restriction in the amount of materials to be used, on the other hand, links the production process with ingenuity, creativity, resourcefulness and originality – all characteristics highly valued in our society.

Because the rules of the contest have been established previously by the teacher, the contest also represents authority. This is true of any contest in that there are rules that have to be followed by all participants. What makes the contest in the school setting different is that participation is not voluntary. Students have no choice as to whether or not they wish to participate and they must abide by the rules of the contest. This physical environment of the classroom is used in specific ways to reinforce the activity as contest. The activities performed in the classroom and the way in which they are structured can be seen as a system of signification that is superimposed upon the system offered by the architecture of the classroom. In some cases, the two systems support one another but, in others, they appear to conflict. This example is representative of a well-known structure in science classes that separates 'active involvement' from 'intellectual reconstruction' of activities. In the periphery of the room, the science area, the material aspect of learning occurs, giving rise in this case to the productive stage of the lesson. The stress here is the *doing* while, at the centre, the discussion (analytical in nature) seems to emphasise *thinking*. The artificial separation between doing and thinking has been identified by Sutton (1989) as a message that traditionally has permeated science classes to give students a mistaken understanding of the nature of science.

*Time and Space as a Part of Lesson Structure*

Time and space are also part of the structure of this lesson. The movement of students in the classroom is determined by a choreography created by the teacher. Rather than establishing direct control over the students, she sets limits on their actions through the segmentation of time and the definition of the spaces in which the different activities are to take place.

The time devoted to each activity implicitly can signify the worth that the teacher feels that the activity holds. The lesson has an introductory stage that lasts ten minutes. This is followed by the laboratory activity which takes 23 minutes. Then comes the evaluation of the structures (16 minutes) and the final discussion (six minutes).

The activities also are spatially regulated. Only the construction activity is performed in the periphery (the science space); the rest of the lesson takes place at the centre. The evaluation is done not only at the centre, but also at the front, and a member of each group brings the structure to the teacher's desk, which is referred to as 'the testing table' for this purpose. Students learn that certain spaces carry greater significance for activity than others.

*Learning the Text of Discourse and Interaction Patterns*

A perspective on the semiotics of discourse was elucidated first by Peirce (Hartshorne & Weiss 1930). He noted that, when we communicate, we do not transmit our thoughts directly to another. Rather, we transmit sound, visual and other types of cues. Peirce was the first to demonstrate clearly that meaning is not inherent in these sounds and cues but is a product of the relationship between them and the item referred to by the speaker and the listener. Foucault (1969) developed the ideas of discourse further and analysed meanings as embedded in power relationships. He emphasised social and historical determinants of human communication, describing the nature and pattern of discourse as a social form of communication embodying social meanings such as authority. He noted that discourses are about what can be said and thought, but also about who can speak, when and with what authority. As Ball (1990) notes, 'discourses embody meaning and social relationships, they constitute both subjectivity and power relations' (p. 2).

Dialogue is one means by which we contextualise patterns of action. There are certain dialogue patterns that we can recognise in school settings that tell us that this is school and we are expected to behave in a certain way. These are rules that we are expected to learn through observing routine behaviour. Teachers attempt a form of communication with learners that demonstrates the acceptance by both teacher and students of a number of unspoken cultural rules. It is through the repeated use of these patterns and routines that we convey these cultural rules. Lemke (1987, 1990) suggests useful terminology in the analysis of dialogue forms, namely, *semiotic formations* or *dialogue formations* for the pattern of routines and rituals and *activity structures* for the sequence of actions observed.

*Dialogue Formations: Questioning Patterns and Interaction Sequences*

Along with other semiotic features of the classroom, patterns of teacher-student discussion are learned as forms of culturally-accepted routines of interaction. The following dialogue segment took place during a regular classroom session. The classroom teacher has taught primary school for 14 years. As students organise for the lesson after lunch, she greets them and moves to the front of the class. Without speaking, she gives a hand signal that is recognised by all: a hand raised with fingers parted in the salutation associated with Mr. Spock from the television program *Star Trek*. It is a sign that means that she would like to start the discussion. At first, only a few students mirror the signal. The teacher waits until all students have stopped talking and have raised their hands in the sign that says they are ready to begin the lesson.

Ms Nay: Okay class. Are we all ready? We're going to continue to talk about sound in this unit. I wonder who knows what sound is? [No response] Come on, give it a try.

[Salwa raises her hand]

Ms Nay: Salwa?
Salwa: Noise?
Ms Nay: Okay, that is evidence of sound, but what *is* it? What *is* sound?

[Daniel, Maria, Anoo and Moira raise their hands. Moira waves her hand enthusiastically, a signal that she wants to respond.]

Ms Nay: Moira? You have an idea?
Moira: Sort of waves?
Ms Nay: What do you mean by waves?
Moira: Well, they're something that you feel in like, waves, when I sit in front of the speaker.
Ms Nay: Yes, good. You feel something. What do you feel? What is it that you feel?
Moira: I don't know.
Ms Nay: Come on. We're getting close. Who can think of the word that describes what Moira feels. What is this that you have all felt?
Daniel: [Without raising his hand] Music? Class laughs.
Ms Nay: Remember, raise your hand before you tell us. Okay, you feel the music, huh? Come on, that's not the word that I'm looking for.

A recognisable three-part pattern of dialogue focusing on teacher-student interaction can be seen in this record of dialogue, often referred to as Elicitation, Response and Feedback, or ERF. These terms are used in a number of analyses of conversation (Cazden 1986; Mehan 1979; Sinclair & Coulthard 1975). Lemke (1990) uses similar language in describing events surrounding this pattern of dialogue formation in practice.

[Teacher Preparation]
**Teacher question**
[Teacher calls for bids to answer (Silent)]
[Student bid to answer (Hand)]
[Teacher nomination]
**Student Answer**
**Teacher Evaluation**
[Teacher Elaboration]

In this interaction, the kind of information that the teacher gleans about the children's ideas about sound is highly restricted. Her energies are devoted to working to pull the ideas that *she* wants from the children. The structure of the question and the answer session itself controls student behaviour and enforces rules of interaction that makes science learning into a kind of game in which learners attempt to guess the ideas that she is thinking. There is no opportunity to discover learner ideas and prior experiences as a basis for planning instruction. The teacher and her students participate in this classic triadic exchange. Both 'know' the rules of the encounter and both use this knowledge as a strategy to accomplish the goal of teaching and learning science. The pattern of interaction becomes an unwritten text learned by both students and teachers in order to accomplish goals and objectives. It becomes a sign to students about how learning, knowledge and meaning making will proceed in the science classroom. This pattern, when grasped and used by learners and teachers, becomes a resource for classroom learning and a strategy for accessing knowledge in the classroom.

Yet there could be aspects of the strategy that teach lessons about learning science that we do not intend. In this case, the students are also given the message that it is important that they figure out the teacher's ideas and provide those ideas as responses. When dialogues are compared from one lesson to another, we sometimes find the same learners responding. Students who do not possess the language skills of the dominant culture are disadvantaged. Students who feel less comfortable in responding in a large-group situation are less likely to participate. Girls are often among those who hesitate to speak in larger group settings. The triadic pattern encourages single-word responses from the children, thus limiting learners' opportunities to build and develop ideas for themselves and to practice scientific terms and phrases. Finally, students interact only with the teacher in this exchange and they are not encouraged to speak with or respond to one another's ideas.

It is clear that the teacher places herself in the role of question asker. The students must respond. The meaning-making process is entirely within the control of the teacher. The triadic dialogue pattern exerts strict control over the discussion process, thereby discouraging only but the boldest students from asking questions out of order or taking the risk to put forward an unformed thought. Students quickly learn that teachers use questions in very different ways when they are standing in front of a classroom than when participating in ordinary everyday conversation. They learn that teachers can be expected to engage students in questions and answers when introducing a new subject, but almost always they ask only questions to which they already know the answers. In an unspoken agreement, the teacher is recognised as knowledge authority. It is the student's task to convince the teacher that she or he has answers to the teacher's questions. Rarely is the learner encouraged to pose questions. Munby (1982) shows how questioning involving such patterns does not show regard for learners' knowledge and reasoning abilities and can cause them to become intellectually dependent (thus limiting confidence in their own abilities to judge the truth of knowledge claims).

## CONCLUSION: THE VALUE OF A SEMIOTIC AWARENESS OF SCHOOL LEARNING ENVIRONMENTS

This chapter has introduced a semiotic interpretation of some science learning events and the science learning environment through its unique perspective. We have identified some significant signs, symbols, semiotic formations and activity structures that must be grasped to access science knowledge in Western society. We have considered such features as architectural design and the arrangement of space, lesson organisation and structure, manifestations of the structure of power and authority, and routines of thinking, speaking and acting, such as the triadic dialogue.

But semiotics is not simply the study of signs and sign systems. It also is the study of how *interpreters* actualise the many potential semiotic relationships that are possible within any communication system (Bowers 1990; Laferriere 1979). One of the main goals of education is to enhance the capacity of learners not only as sign interpreters, but as creators and users of sign systems. A semiotic perspective on the educational setting helps us more readily grasp that the sign giver and the sign receiver do not always share the same code.

Viewing classrooms semiotically helps us to see that entire systems of signification within school settings serve as a resource for learners and a means by which they access knowledge. When teachers master the signs and symbols of our culture and become aware of those of others, they know when to break the unspoken rules to become inventive in using new approaches in interaction. A semiotic understanding allows us to become wide awake and critical of the use of sign and symbol systems when they hurt or exclude individuals or become overly rigid forms of communication.

*Semiotics and Science Teaching and Learning*

A semiotic examination of science education allows us to see freshly how learners can emerge with systems of signification that might not be what we intend. For example, certain rigid presentations of structures such as the scientific method become, for many teachers and students, the embodiment of science itself, creating an image of science as a prescriptive technique rather that a complex process of coming to know. Likewise, science often is portrayed as a game of memorisation of terms. Or it can be seen to focus on ideas to be accepted rather than deeply understood. With such portrayals, learners begin to see themselves as deficient when confronted with difficult ideas about science content – they learn not to ask questions. Such signs of what science is all about inhibit rather than facilitate experience in the classroom.

Textbooks and bulletin boards which carry photographs of science being conducted only by representatives of one gender or race present a significant message of exclusion to learners. When the dialogue structures which we employ do not permit students to use scientific language and thinking for themselves, we can force them to become intellectually dependent, concerned more with the structure of authority and control within the classroom than with discovering and challenging new ideas. Educators convey elaborate systems of signs indicating what it means to be a competent learner in the science classroom. Students are rewarded according to the degree to which they model those signs. Children from only certain social classes or who speak the dominant language are at an advantage in such a system. When lesson organisation, language use and activity patterns are generally structured by unexamined cultural values, advantage is given to learners with only certain kinds of beliefs, values and language skills.

*Semiotic Awareness and Curriculum Organisation*

All school subjects are ways of organising signs (Suhor 1984). Cunningham (1984) has commented on the value of helping teachers become more aware of students' sign-making and sign-interpreting capacities. Teachers nurture students' sign-making processes along lines compatible with the subject being taught (Witte 1992). If students are constantly being taught disjointed and unintegrated items, they cannot build the structures we seek. By emphasising semiotic structures and the skills for building them, education becomes a more meaningful experience for both student and teacher. 'In short, a semiotic perspective encourages teachers to attend to the skills of learning subject matters rather than learning the "stuff" of subject matter only' (Cunningham 1987, pp. 9–10). Cunningham does not propose that prospective teachers become semiotic scholars by reading extensively the works of Peirce (Hartshorne & Weiss 1930), Deely (1986) and Eco (1984) and debating fine theoretical points. Rather, he suggests that they be introduced to semiotics and become sensitised to the structures and codes of experience which both they and their students are building as they operate in the world. This view of knowledge directs the attention of teachers away from teaching specific bits of knowledge to the cultivation of higher intellectual skills. Semiotics is not the only field that leads educators to such awareness but, unlike others, it provides us with some conceptual tools for analysing these skills and insights into ways in which they can be nurtured.

Semiotics is neither new nor the only way of looking at things. A focus on the signification system of classroom learning seeks understanding in uncommon places (Cassidy 1982). Semiotics shows us that signs are not consciously rendered or interpreted, just as reality is not disclosed fully and directly. By taking a semiotic perspective, signs are rendered visible and the power and importance of signifying systems in the science learning environment are reconsidered. By changing our perspective on the learning environment, semiotics encourages the questioning of assumptions of both our own and the student's reading of science. Signs can be used more consciously and with greater power and discernment. We can create more signs and activity structures that provide alternatives for students who have been alienated from science learning, and that portray science in more encouraging forms that allow learners to receive, integrate and express knowledge in science more confidently.

# REFERENCES

Ball, S. (ed.): 1990, *Foucault and Education: Disciplines and Knowledge*, Routledge, London.
Barthes, R.: 1967, *Elements of Semiology*, Hill and Wang, New York.
Barthes, R.: 1988, *The Semiotic Challenge*, Hill and Wang, New York.
Bowers, C.: 1990, 'Implications of Gregory Bateson's Ideas For a Semiotic of Art Education', *Studies in Art Education: A Journal of Issues and Research* 31(2), 69–77.
Cassidy, M.: 1982, 'Toward Integration: Education, Instructional Technology and Semiotics', *Educational Communication and Technology* 30(2), 75–89.
Cazden, C.: 1986, 'Classroom Discourse', in M. Wittrock (ed.), *Handbook of Research on Teaching*, Macmillan, New York, 432–463.
Crenshaw, S.: 1991, 'A Semiotic Look at Kindergarten Writing', Paper presented at the Annual Meeting of the American Educational Research Association, Chicago, IL.
Cunningham, D.: 1984, *What Every Teacher Should Know About Semiotics*, ERIC Clearinghouse on Reading and Communication, National Institute of Education, Urbana, IL.
Cunningham, D.: 1987, 'Outline of an Education Semiotic', *American Journal of Semiotics* 5(2), 195–201.
Danesi, M.: 1994, *Messages and Meanings: An Introduction to Semiotics*, Canadian Scholars Press, Toronto, Canada.
Deely, J. (ed.): 1986, *Frontiers in Semiotics*, Indiana University Press, Bloomington, IN.
Eco, U.: 1976, *A Theory of Semiotics*, Indiana University Press, Bloomington, IN.
Eco, U.: 1979, *The Role of the Reader*, Indiana University Press, Bloomington, IN.
Eco, U.: 1984, *Semiotics and the Philosophy of Language*, Indiana University Press, Bloomington, IN.
Eco, U.: 1985, 'How Culture Conditions the Colors We See', in M. Blonsky (ed.), *On Signs*, Johns Hopkins University Press, Baltimore, MD, 157–175.
Eco, U.: 1988, *Signo* (translated by F. Serra), Editorial Labor, Barcelona, Spain.
Foucault, M.: 1969, *The Archaeology of Knowledge*, Random House, New York.
Golden, J. & Gerber, A.: 1990, 'A Semiotic Perspective of Text: The Picture Story Event', *Journal of Reading Behavior* 22(3), 203–219.
Groisman, A., Shapiro, B. & Willinsky, J.: 1991, 'The Potential of Semiotics to Inform the Understanding of Events in Science Education', *International Journal of Science Education* 13, 217–226.
Guiraud, P.: 1975, *Semiology*, Routledge, Kegan & Paul, London.
Halliday, M.: 1978, *Language as Social Semiotic*, Edward Arnold, London.
Halliday, M. & Martin, J.: 1993, *Writing Science: Literacy and Discursive Powers*, University of Pittsburgh Press, Pittsburgh, PA.
Harland, R.: 1987, *Superstructuralism: The Philosophy of Structuralism and Post-Structuralism*, Methuen, London.
Hartshorne, C. & Weiss, P. (eds.): 1930, *Collected Papers of Charles Sanders Peirce, Vols. 1–6*, Harvard University Press, Cambridge, MA.
Hawkins, E.: 1984, *Awareness of Language: An Introduction*, Cambridge University Press, London.
Laferriere, D.: 1979, 'Making Room for Semiotics', *Academe* 65, 434–440.
Lemke, J.: 1987, 'Social Semiotics and Science Education', *The American Journal of Semiotics* 5, 217–232.
Lemke, J.: 1988, 'Genres, Semantics, and Classroom Education', *Linguistics and Education* 1, 81–99.
Lemke, J.: 1989, 'Social Semiotics: A New Model for Literacy Education', in D. Bloome (ed.), *Classrooms and Literacy*, Ablex Publishing, Norwood, NJ, 289–309.
Lemke, J.: 1990, *Talking Science: Language, Learning and Values*, Ablex Publishing, Norwood, NJ.
McLaren, P.: 1986, *Schooling as a Ritual Performance*, Routledge & Kegan Paul, London.
Mehan, H.: 1979, *Learning Lessons: Social Organization in the Classroom*, Harvard University Press, Cambridge, MA.
Munby, H.: 1982, *What is Scientific Thinking?*, Science Council of Canada, Ottawa, Canada.
Peirce, C.: 1985, 'Logic as Semiotics: The Theory of Signs', in R. Innis (ed.), *Semiotics: An Introductory Anthology*, Indiana University Press, Bloomington, IN, 4–27.
Percy, W.: 1982, *The Message in the Bottle*, Farrar, Straus and Giroux, New York.
Shapiro, B.: 1989, 'What Children Bring To Light: Giving High Status to Learner's Views and Actions in Science', *Science Education* 73, 711–733.
Shapiro, B.: 1994, *What Children Bring to Light: A Constructivist Perspective on Children's Learning in Science*, Teachers College Press, New York.

Siegel, M.: 1989, *Critical Thinking: A Semiotic Perspective* (Monographs on Teaching Critical Thinking No. 1), Office of Educational Research and Improvement, Washington, DC.
Sinclair, J. & Coulthard, M.: 1975, *Towards an Analysis of Discourse*, Oxford University Press, London.
Smith, D.: 1989, 'Modernism, Post-Modernism and the Future of Pedagogy', Paper prepared for the Canadian Studies Program, Institute of East and West Studies, Yonsei University, Seoul, Korea.
Suhor, C.: 1982, *Semiotics Fact Sheet*, ERIC Clearinghouse on Reading and Communication, National Institute of Education, Urbana, IL.
Suhor, C.: 1984, 'Towards a Semiotics-Based Curriculum', *Journal of Curriculum Studies* 16, 247–257.
Sutton, C.: 1989, 'Writing and Reading in Science: The Hidden Messages', in R. Millar (ed.), *Doing Science: Images of Science in Science Education*, Falmer Press, Philadelphia, PA, 137–159.
Witte, S.: 1992, 'Context, Text, Intertext: Toward a Constructivist Semiotic of Writing', *Written Communication* 9, 237–308.
Worth, S.: 1981, 'Visual Communication', in L. Gross (ed.), *Studying Visual Communication*, University of Philadelphia Press, Philadelphia, PA, 76–92.

# 5.6 Qualitative and Quantitative Landscapes of Classroom Learning Environments

**KENNETH TOBIN**
*University of Pennsylvania, Philadelphia, USA*

**BARRY J. FRASER**
*Curtin University of Technology, Perth, Australia*

'Bricolage' has been used metaphorically to relate educational research to the idea of a quilt comprised of numerous patches, some overlapping and others separate (Lévi-Strauss 1974). A bricoleur, generally regarded as a cobbler or handyperson, uses available materials to undertake tasks such as mending shoes or fixing broken appliances. Bricoleurs select from the available materials those that are satisfactory for completing a task or producing a particular product. The final result reflects not only the skill of the bricoleur, but also the materials selected and used. In this chapter, we apply the bricolage metaphor to the use of multiple theoretical perspectives to frame research and its methods, and to the desirability of combining quantitative and qualitative data to maximise the potential of research on learning environments.

The chapter's first major section provides background information about the field of learning environments, including the combination of qualitative and quantitative methods. The second section presents several theoretical perspectives for framing learning environments research. The remainder of the chapter illustrates how we used multiple theoretical perspectives, multilevel designs, and a combination of qualitative and quantitative data in a learning environment study in Western Australia.

## BACKGROUND

Learning environment research has become established firmly, especially in science education (Fraser 1986, 1994; Fraser & Walberg 1991; see Fraser's overview chapter in this section of the *Handbook*). Although learning environment research has involved a range of observational and interpretive methods, the assessment of

---

Chapter Consultant: Heui-Baik Kim (Wonkwang University, Korea)

learning environment often has involved questionnaires which assess students' and teachers' perceptions of dimensions such as teacher support, participation, task orientation, innovation, cooperation and personal relevance (Fraser 1994).

The constructs regarded as having most salience in a learning environment reflect the theoretical frameworks used by a participant to give meaning to what is experienced. Accordingly, as theories of teaching and learning changed, learning environment questionnaires incorporated new scales to measure a variety of constructs which reduced the complexity of what happens in classrooms, and focused on selected aspects of student and teacher actions and interactions. However, any method of research provides just one possible window into educational environments. Whereas the theoretical frames embedded in a selected learning environment instrument illuminate particular constructs to reveal trends and patterns, other constructs and associated patterns and trends are obscured.

In a longitudinal study, Tobin, Kahle and Fraser (1990) selected from existing learning environment instruments the scales which they thought had the greatest salience to what was happening in the classrooms which they were studying. At that time, they acknowledged many ways in which to describe the learning environments experienced by students and teachers, and they recognised that certain constructs (such as task orientation) were potentially more important than others in their study. Accordingly, they decided to use a selection of items and scales from the *Individualised Classroom Environment Questionnaire* (Fraser 1990) and the *Classroom Environment Scale* (Moos & Trickett 1987). Items were standardised in terms of format and used to compare the learning environment experienced by students in the classes of two different teachers. As the idea of selecting items and scales to fit the particular circumstances of a given class caught on, there was an increase in the range of alternative ways of looking into classrooms and of instruments used to describe learning environments. For example, recent questionnaires focus on science laboratory classroom environments (Fraser, Giddings & McRobbie 1995), computer-assisted learning environments (Teh & Fraser 1995), constructivist-oriented classroom learning environments (Taylor, Dawson & Fraser 1995), and open or distance education learning environments (Jegede, Fraser & Fisher 1995).

Past research has used learning environment questionnaires in curriculum evaluation (Fraser 1979), in investigating the effects of environment on student learning (McRobbie & Fraser 1993), in studies of differences between students' and the teacher's perceptions of the same classroom (Fisher & Fraser 1983), research on the transition from primary to secondary school (Ferguson & Fraser 1996), teachers' practical attempts to improve classroom learning environments (Thorp, Burden & Fraser 1994) and the use of learning environment ideas in school psychology (Burden & Fraser 1993). Because given dimensions are more salient than others in a particular classroom, the decision to use a specific learning environment scale in a classroom ideally should take place after researchers have had some experience in the class. With the wide variety of learning environment questionnaires now available, usually it is possible to identify numerous existing scales that have potential significance to a given classroom (thus obviating the need to 'reinvent the wheel' in every study). What is regarded as salient by different participants in a community might differ depending on their roles, goals and theoretical frames. Accordingly, the selection of scales to include in a learning environment study depends on whose perspectives are to be captured in the research.

Bricolage also can be applied to the use of multiple methods in research (Denzin 1997), including the use of qualitative and quantitative information together in a study. In learning environment research, considerable progress has been made in realising the benefits of combining qualitative and quantitative methods (Dorman, Fraser & McRobbie 1994; Fraser & Tobin 1991; Maor & Fraser 1996; Tobin, Kahle & Fraser 1990). When a study using quantitative methods has been completed, its main findings can be contextualised with thick description consisting of observations and verbal accounts from participants. A variety of designs can facilitate access to qualitative information to explicate the trends apparent in analyses of quantitative data. For example, interviews can be conducted with participants, impressionistic tales based on visits to 'well selected' sites can highlight significant aspects of practice (van Maanen 1988), and full-scale interpretive studies (Guba & Lincoln 1989) can support and extend quantitative relationships.

At a particular time in a study, it might not be possible to quantify certain constructs even though their salience is apparent. Before quantification, it is desirable to illuminate salient constructs with investigations of broad questions like those typically addressed in interpretive research. Later, quantification of the constructs can enable a larger range of questions to be answered.

Interpretive studies also can be enhanced with the inclusion of quantitative information. Erickson's chapter in this *Handbook* exhorts interpretive researchers to count and otherwise quantify whenever possible and to test trends quantitatively. In interpretive research, quantitative information can be a significant component of the evidence for or against a particular assertion and the credibility of claims about patterns or relationships can be strengthened by a variety of quantitative and qualitative data sources.

## MULTIPLE THEORETICAL PERSPECTIVES ON LEARNING ENVIRONMENTS

This section presents several theoretical perspectives and associated constructs that might frame student and teacher perceptions of their experienced and preferred learning environments. The five scales of the *Constructivist Learning Environment Survey* (CLES; Taylor, Dawson & Fraser 1995) were developed by considering learning from a social constructivist perspective. Important constructs include personal relevance, the use of extant knowledge to construct new ideas, negotiation of meaning, and consensus building. If students are to make sense of what they learn, it is important for their ideas to be heard and critiqued during classroom transactions, for them to share control of the classroom, and for the teacher to provide support for learning. Furthermore, from a constructivist perspective, the enacted curriculum would acknowledge that what is regarded as scientific knowledge changes with time.

From a social constructivist perspective, learning environments are constructed by individuals in a given setting and consist of socially-mediated beliefs about opportunities to learn and the extent to which those opportunities are constrained by the social and physical milieu. Although individuals have their own experienced and preferred learning environments, those constructions are constrained by interactions with others and characteristics of the culture in which learning is situated. For example, learning can be viewed as enculturation into a community of practice in which the discursive practices (e.g., talk, writing, cognition, argumentation and re-presentation) of participants are constantly changing in response to the interactions of a teacher and students, not only with one another, but also with social structures, such as conventions and norms (McGinn, Roth, Boutonné & Woszczyna 1995; Roth 1995). From this perspective, learning occurs within constantly evolving communities in which the practices of participants are shaped by social structures, relations of power and the nature of the activities in which learners engage. If researchers want to describe learning environments from the perspective described above, additional scales to those in the CLES would need to be developed.

When learning science, students could be expected to engage in more science-like discourse over time. Discourse refers to a 'social activity of making meanings with language and other symbolic systems in some particular kind of situation or setting' (Lemke 1995, p. 8). One might expect classroom discourse to involve students routinely in arguments about the efficacy of the warrants for knowledge claims. As has been advocated by Kuhn (1993), science can be regarded as a form of argument in which emerging conceptual understandings are related to evidence and their fit with canonical science.

When a cultural perspective is used to frame learning environments, the extent to which teaching and learning are productive depends on the participants' habitus, which is a set of dispositions that incline individuals to act and interact in particular ways (Bourdieu 1992; Lemke 1995). As students and teachers enact a curriculum, their roles adapt to the interactions of the community and its associated constituent cultures. Interactions occur when individuals act in the presence of others and thereby perturb their actions, or when they participate independently in an activity and utilise social constructions such as language in the process of thinking, writing, drawing, etc. Cultural diversity can be a challenge for teachers, especially when students have different native languages, but significant even when a common language is employed. There is an understandable tendency for many teachers to regard differences in patterns of interaction and sense making as deficiencies to be corrected rather than as forms of capital to be invested in learning. In terms of learning environments, the extent to which cultural variation within a class is regarded as capital or deficiency for learning is significant. Bourdieu (1992) used the term 'symbolic violence' to describe the problems of peripheral participants within a community in which their cultural capital is considered worthless as a foundation for learning. The potential for science to be a source of symbolic violence exists for all students, but is greatest for those who do not belong to the majority culture and who have a different language from the majority. Scaffolds are not provided to enable students to employ their cultural capital and their personal language resources are not helpful in bridging the semantic space between their own discourse and that of science.

If students learn science as a form of discourse, they must be able to adapt their language resources as they practise science in settings in which others who know science assist them to learn by co-participating in activities (Roth 1995; Schön, 1983; Tobin 1997). Co-participation implies a shared language that can be accessed by all participants in order to engage in the activities of the community, with the goal of facilitating the learning of others. In such a setting, students have the autonomy to ask when they do not understand, knowledge claims that make no sense are clarified, and discussion occurs until such time that learners are satisfied that they understand. The focus of the teacher on the re-presentations of the learner is what is most critical about a co-participatory environment. The mediating role of the teacher is focused not only on what students know and how they can re-present what they know, but also on the identification of activities that can continue the evolutionary path of the classroom community towards the attainment of agreed-upon goals. The power sharing needed to facilitate co-participation should be tailored to reflect the cultural histories of participants in the community.

The examples provided above suggest that learning environments can be described through multiple windows to highlight different issues that are pertinent to the stakeholder goals and extant classroom practices. Any methodology used to explore learning environments will produce a landscape that is incomplete and represents only one of the possible portraits which is likely to be appealing and relevant to different stakeholders. Accordingly, studies of learning environments could begin to explore the new constructs considered above, such as power equity and relations, access and appropriation of discourse, warrants for viability of knowledge claims, use of cultural capital to make sense of science, co-participation, and learning science as symbolic violence. Thus, interpretive research can provide insights into the constructs for which new scales need to be developed, or into which scales to select from those already incorporated into existing instruments.

## A MULTILEVEL ANALYSIS OF LEARNING ENVIRONMENTS

Different research studies call for a focus on different levels or 'grain sizes' which, in turn, have implications for the choice of research methods. For example, a fine grain size involving a contrast between two teachers or between several students within a class (e.g., Tobin, Kahle & Fraser 1990) normally requires intensive qualitative interpretive methods. In contrast, a system-wide evaluation of educational reform involves a coarse grain size and economical quantitative survey methods, as in a recent evaluation of an Urban Systemic Initiative in Dallas involving a survey of over 40,000 students' classroom environment perceptions (Dryden & Fraser 1996).

In some studies, a different grain size is relevant at different times and for different purposes. A study that incorporates different grain sizes in a design that features recursive relationships between the data sources for each level is referred to as a 'multilevel' study. The credibility of assertions from multilevel studies resides in the use of qualitative and quantitative information from multiple data sources and grain sizes. Patterns identified at any one level can guide the design of the study so that assertions supported at only one level can be tested at the multiple grain sizes of the study. Alternatively, data from all levels can be pooled and the analyses can be conducted across levels.

The remainder of this chapter describes a multilevel study of the learning environment of a science class in Australia. This study is distinctive in that it incorporated a teacher-researcher perspective which has been absent in most past research on classroom learning environments. The research commenced with an interpretive study of a grade 10 science teacher's classroom learning environment at River Valley High, which provided a challenging learning environment in that many students were from working class backgrounds, some were experiencing problems at home, and others had English as a second language.

Six university-based researchers collaborated with a teacher-researcher in an intensive study of a grade 10 class studying chemistry. We believed that including researchers of different nationalities and with different research histories in science education would enhance the study. Researchers decided on their own particular foci and, although the team met regularly and cooperated to access data sources, meetings were primarily educative. To ensure that the theoretical perspectives of all members of the research team informed the research, we organised three two-hour meetings per week. At each meeting, two researchers were selected to lead interactive discussions for a period of 20 minutes each. Because Ms Horton (the teacher) was one of the researchers, time also was provided at each meeting for her to clarify the intent of her classes and to interact with other researchers on the emergent issues.

The predominant data employed in the study were qualitative and drew on the interpretive methods of Erickson (1986). Several researchers visited this class each time it met (four times per week) over a period of five weeks. In addition to classroom observations and the use of student diaries, interviews were conducted with the teacher-researcher, students, school administrators and parents. A video camera was used to record the activities of each lesson for later analysis. Field notes were written during and soon after each observation. All interviews and team meetings were tape recorded and transcribed. The 'referential adequacy' (Eisner 1979) of the study was enhanced by having teachers at the school check written accounts of the researchers' observations and interpretations.

Towards the end of the observational period, a 'peer debriefing' (Guba & Lincoln 1989) provided the research group with opportunities to discuss the study with a group of other university-based researchers and several teachers from the school involved in the study. In addition to short presentations from each researcher, the peer debriefing involved numerous small-group discussions throughout the session and a whole-group discussion at the end.

*Classroom Environment Scales*

The qualitative component of the study described above was complemented by a quantitative component involving the use of a questionnaire which linked three levels: the class in which the interpretive study was undertaken; selected classes from within the school; and classes distributed throughout the same State. This approach enabled us to place the teacher and her students within her school and within the larger context of the State. Table 1 shows that the questionnaire was administered to: (1) a selection of the students from Ms Horton's year 10 chemistry class; (2) a selection of students in the classes of two other teachers of year 10 chemistry at the same school; (3) a larger representative group of 494 grade 8 and 9 science students in 41 classes in 13 schools in Western Australia participating in the Third International Mathematics and Science Study (TIMSS). For this group, data were not available for all of the classroom environment scales described below.

The classroom environment questionnaire was designed specifically for the present research and was based upon understandings derived from the qualitative aspects of the study. Table 2 lists the seven dimensions contained in the questionnaire. The first five dimensions (Personal Relevance, Critical Voice, Shared Control, Uncertainty and Student Negotiation) were taken from the Constructivist Learning Environment Survey (CLES; Taylor, Dawson & Fraser 1995). The other two scales, Commitment and Teacher Support, were written for this study because they emerged as salient in this teacher's class during classroom observations.

**Table 1: Three samples responding to a classroom environment questionnaire**

| Sample | No. of students | Scales administered |
|---|---|---|
| Ms Horton's grade 10 science class | 11 | 5 CLES scales<br>Commitment<br>Teacher Support |
| Two other teachers of grade 10 science classes in Ms Horton's school | 20 | 5 CLES scales<br>Commitment<br>Teacher Support |
| 41 grade 8 & 9 science classes in 13 schools participating in TIMSS[a] | 494 | 5 CLES scales |

[a] TIMSS: Third International Mathematics and Science Study

Table 2: Scales and sample items of a classroom environment questionnaire

| Scale | Scale description | Scale item* |
| --- | --- | --- |
| Personal Relevance | Relevance of learning to students' lives. | I learn about the world outside of school. |
| Critical Voice | Legitimacy of students expressing a critical opinion | It's OK to ask the teacher 'Why do we have to learn this?' |
| Shared Control | Student participation in planning, conduct and assessment of learning. | I help the teacher to plan what I'm going to learn. |
| Uncertainty | Provisional status of scientific knowledge. | I learn that the views of science have changed over time. |
| Student Negotiation | Involvement with other students in assessing viability of new ideas. | I ask other students to explain their ideas. |
| Commitment | Student motivation and effort in relation to learning science. | I pay attention. |
| Teacher Support | Helpfulness and friendliness of the teacher towards students. | The teacher goes out of his/her way to help me. |

* Response alternatives are Almost Always, Often, Sometimes, Seldom, and Almost Never.

*Complementary Perceptions of the Learning Environment*

Below, the classroom environment findings are reported, interpreted and discussed at various 'levels' or 'grain sizes'. At the level of the individual student, Figure 1 shows how two students, Teresa and Walter, perceived the same classroom's environment quite differently. In particular, Teresa perceived much greater levels of Teacher Support than did Walter. The teacher described Teresa as a model student and Walter as hyperactive and having a genetic disorder. Figure 2 considers the classroom environment of Ms Horton's class (averaged over all students who responded to the questionnaire) relative to the two other 'grain-sizes', namely, the average classroom environment as perceived by two grade 10 chemistry classes of two other teachers at the same school, and the large comparison group of science classes.

Two main patterns are evident in Figure 2. First, relative to other science classes in her school, Ms Horton's class was perceived as having greater levels of Personal Relevance, Critical Voice and Teacher Support. Second, the perceptions of students in other classes at Ms Horton's school were very similar to those of students in the comparison group, with the exception that Ms Horton's school had appreciably lower scores on the Personal Relevance scale.

By drawing on the extensive qualitative data base – consisting of interviews, classroom observations, video recordings and student diaries – the teacher-researcher and other researchers were able to provide an account of why the results in Figures 1 and 2 are consistent and plausible. For example, the high level of

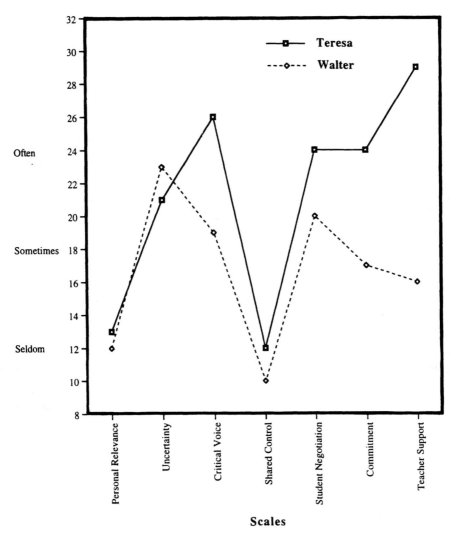

Figure 1: Classroom environment profiles for students Teresa and Walter

perceived Personal Relevance in Ms Horton's class is consistent with this teacher's practice of devoting one science period per week to things that are personally relevant to her students.

*Personal Relevance*

The data for Ms Horton's school suggest that the curriculum was perceived to have less Personal Relevance for students than at a comparison group of schools, with the average perceptions for students at the school suggesting that the enacted

632  Tobin and Fraser

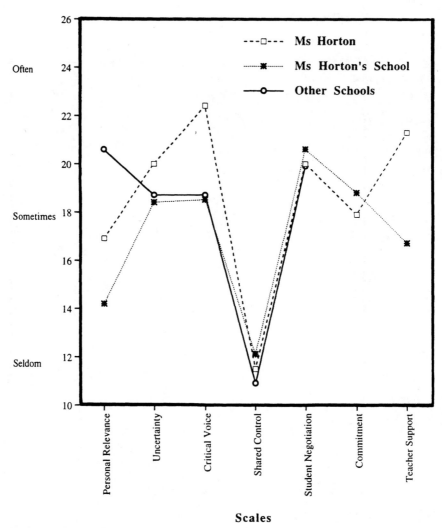

Figure 2: Classroom environment profiles for (1) Ms Horton's class, (2) other classes at Ms Horton's school and (3) a comparison group of 13 other schools.

curriculum had relevance less than 'sometimes' and more than 'seldom'. Ms Horton's class was above the school average, but half of a scale point below the average for the comparison group of schools. Teresa and Walter were similar in their perceptions of Personal Relevance and both had scores lower than the average for their school, perceiving the curriculum to have Personal Relevance 'seldom' (Figure 1).

Ms Horton's slightly higher scores for Personal Relevance could have arisen because of her efforts to make science more relevant to students' needs. Because

Ms Horton believed that science education had the potential to transform the lives of her students, she made every effort to enact a curriculum that had personal relevance to each of them:

> I would like the students to be able to see that they can choose their own path to education. Education needs to become more meaningful, particularly to a group of kids who have a disadvantaged background, whether it be real disadvantage, involving some sort of physical actions, lack of parental support, or even being something that students themselves see as a disadvantage (like having migrant parents).

Despite her efforts, Ms Horton's enacted curriculum also was shaped by school-based restraints. The Head of the Science Department, who was new to the school, was anxious to provide all students with a curriculum that would enable them to pursue higher-level studies in chemistry and prepare them for science-oriented courses at university. A multiple-choice test was prepared for all grade 10 students to assess the extent to which they had mastered the content of the course, as traditionally it had been taught in the school. Despite Ms Horton's goal to prepare students for life in the world out of school, she was required to prepare her students for the comparability test, because scores on that test would affect their final grade. Accordingly, a focus of the course was to prepare students for the test. Our study identified assessment and school-wide restraints as highly influential in shaping the learning environments experienced by students in Ms Horton's classroom.

*Uncertainty*

Figure 2 shows that the students in both the comparison schools and River Valley High perceived that 'sometimes' scientific knowledge was presented as uncertain. Ms Horton's class perceived a higher incidence of scientific knowledge claims being presented as uncertain, although this corresponded to a scale value of just above 'sometimes'. The perceptions of the individual students (Figure 1) support the trends observed in the school-level analysis. Some insights into these trends were obtained from the interpretive study.

Ms Horton repeatedly advised students that there were no right or wrong answers, probably to encourage them to respond by reducing the risks associated with providing wrong answers. However, other actions suggested that some answers were acceptable, and regarded as right, while others were not acceptable and regarded as wrong. For example, when reviewing the periodic table, students were not asked to explain why they had re-presented their knowledge as they had, and the relative merits of alternative re-presentations were not discussed explicitly. Absent was discussion that placed models in a historical context and reasons for one model being more appropriate than another. The rationale for one model being acceptable and another being unacceptable was the voice of the teacher, speaking on behalf of science. Among the warrants that might have been given were those

## Critical Voice

The school-level analysis suggests that the challenging circumstances associated with River Valley did not make a difference in the way in which the students perceived the extent to which they could express a critical opinion during the enactment of the curriculum. However, on the average, students in Ms Horton's class experienced a curriculum that allowed them to express critical opinions. The average Critical Voice scores for Ms Horton's class (Figure 2) corresponded to a scale value of almost 'often'. As might be expected, some students took greater advantage than did others of opportunities to express critical views (see Figure 1 in which Teresa and Walter differ by a scale point in their perceptions of the experienced environment).

The interpretive study confirms the results obtained from the descriptive statistics. Ms Horton constantly encouraged students to voice their opinions and suggest alternatives. For example, in a discussion of the pH of water and what would happen if water were added to an indicator, one of the female students proudly pointed out that the colour would fade because of the concentration of colour being less. This suggestion was volunteered after the class had more or less reached a consensus about the colour not changing because water was neutral and had a pH of 7.

## Shared Control

Figures 1 and 2 suggest that students perceived that 'seldom' were they involved in making curricular decisions. These data are confirmed by the interpretive study which revealed that Statewide and within-school forces undermined Ms Horton's efforts to enact the curriculum in a manner significantly different from what she had done in the past.

The curriculum envisioned by Ms Horton was emancipatory in that it was intended to provide students with alternatives to the types of lives in which so many of them were trapped. Her empathy for the students was associated with her commitment to use science education to transform their lives. In an explanation of her position, Ms Horton made the following comments: 'I think science could in some ways really empower students who are trapped by their environment, or by their own self esteem'.

Ms Horton encouraged students to exercise autonomy and act responsibly. She made sure that there was sufficient work to keep them engaged for the entire class period and she expected them to remain engaged. The following excerpt from the researchers' field notes provides an indication of the extent to which Ms Horton controlled the class in a non-authoritarian manner:

> You could not really tell that she was not in the room. The students continued on just the same as they did if she were present. Ms Horton said to me that some teachers had lots of trouble with some of these very students and that is probably because the teachers are always trying to restrict their activities. Ms Horton's approach to controlling these students is to be very low key and quiet. There are no public confrontations. These students respect her and she seems to respect them.

The manner in which the curriculum was implemented reflected the social forces that appear to shape teaching and learning in schools throughout a system. Despite her attempts at reform and her goal orientation, Ms Horton's approaches to the teaching of chemistry were reminiscent of traditional science teaching (Tobin & Gallagher 1987). Although Ms Horton was familiar with alternatives, she seemed unwilling to change the enacted curriculum to accord with her beliefs for the following reasons:

- The traditional approach is comfortable in this environment. Schools around the State and the world teach much the same subject matter to students of about this age using similar approaches and resources.
- There is not a critical mass of teachers in the school or the State to support changes of the type that Ms Horton felt were necessary.
- Pressure was exerted by the Head of the Science Department to prepare students for comparability tests to be taken by all students at about the same time.
- There was a widespread belief that the grade 10 chemistry course serves a sorting function.
- Text materials were available in class sets to support the teaching and learning of traditional chemistry content, but preferred alternatives were not as readily available.

In order to understand why the curriculum was enacted as it was, it is essential to look beyond the beliefs of the teacher and the students and to the sociocultural milieu in which the school is embedded.

*Student Negotiation*

The quantitative data suggest that Student Negotiation did occur as students were provided opportunities to discuss what they learned with one another. The averages for the comparison schools, River Valley and Ms Horton's class (see Figure 2) all suggest that Student Negotiation occurred more than 'sometimes'. The data for the two students whom we compared show differences of about one scale point, but also they suggest that Student Negotiation was perceived to occur more than 'sometimes'. However, other research (Tobin, McRobbie & Anderson 1997) and the interpretive study conducted here suggest caution in the interpretation of these data. The quality of student negotiations is a critical factor in determining whether or not students construct understandings that are relational and transformational. Furthermore, the extent to which the understandings that are negotiated are consistent with canonical science is a concern.

Ms Horton's subject matter knowledge was a critical factor that shaped interactions. Although verbal interactions were lively and frequent, the mediational role of the teacher was restricted because of her limited understandings of chemistry. Accordingly, distinctions between weak and dilute acids were not made and core concepts, such as the mole, were not used to advantage when students interacted in small groups or at a whole-class level. For example, a laboratory activity that involved the use of strong and weak acids did little more than stimulate enjoyment because students did not understand concepts such as mole, hydrogen ions, dissociation and weak, strong, concentrated and dilute acids.

The above observations suggest that a questionnaire might not have been the best way to explore negotiation at that point in time. The quality of the interactions is an important aspect of what occurs and it is not clear how to obtain students' perceptions of the quality of interactions, particularly if they have failed to attain particular canonical understandings and are unaware of their lack of understanding. A second implication, in relation to the quality of the learning environments experienced by students, is that we need to explore the extent to which teachers understand the subject matter and know how to re-present what they know to mediate the learning of students.

Traditionally, the chemistry topic in grade 10 was used to ascertain which students were capable of pursuing further studies of the physical sciences. Despite Ms Horton's efforts to enact a curriculum to meet the needs of her students, the implementation of the curriculum was constrained by the actions of her Head of Department and the history of science teaching in the State. Rather than using the students' extant knowledge as the main criterion for inclusion, Ms Horton's teaching became traditional. Therefore, compared to students from other schools, Ms Horton's students regarded the curriculum as less relevant to their needs.

*Commitment*

The data suggest that Ms Horton's class and others in the school experienced a level of commitment to learn science that was a little above 'sometimes'. The data in Figure 1 show the benefits of examining the data for individual students within a classroom. In this instance, the selected students show a difference in their commitment to learn science that varies from 'often' to less than 'sometimes'. The interpretive study showed wide variations in the levels of commitment among different students. As the following excerpt from an interview shows, Ms Horton was very much aware of this variation and was attuned to its implications for her role as a science teacher and role model for her students:

My particular goal with those kids is actually getting them to learn, the desire actually to learn something. Although I'm in the area of science, more importantly I'm happy with the students when they have heard or seen something in the classroom and then actually apply it to somewhere else. A lot of them say, 'I have to be bright. I always get Ds or I always get Fs. I can't read. I can't write. Therefore I can't do science.' My goal is to provide students with the opportunity to feel confident enough to want to participate in learning science. But, in what possible ways could we as a whole school motivate groups and provide an environment that would motivate groups of kids to maintain their individualism? I think it is important in some way to empower students to be what they want to be, within suitable guidelines, and to break the mould in which they have come to us originally.

*Teacher Support*

One of the most salient aspects of the learning environment in this study was Teacher Support. The data in Figure 2 show that students in Ms Horton's class perceived higher levels of Teacher Support (between 'sometimes' and 'often') than did students in other grade 10 classes at River Valley High (less than 'sometimes'). Ms Horton had several features in common with the types of students whom she was teaching. She was the ninth child in a ten-child family and was mindful of the home circumstances that made it difficult for her students to learn, to get employment after their schooling, or to extend their education even to grade 12, let alone go to university. She was not a motivated learner at school and knew that students' life histories often made it difficult to concentrate on learning as a high priority. Ms Horton was aware that social problems afflicted many students and she was determined to make a difference in their lives. Consequently, she planned to enact the curriculum to facilitate transformative goals.

Ms Horton had considerable empathy for her students, was concerned with their well being as citizens, and perceived science as an opportunity to develop their life skills. Learning to be communicative and cooperative was a high-priority goal. Getting to know her students was a priority for Ms Horton, and meeting them at the door seemed important because it permitted brief individual interactions with almost every student. Another example of the teacher's empathy for students occurred when Ms Horton, upon noticing that students were very hot and were fanning themselves with their notebooks, allowed them to go outside to get a drink. The following extract from the field notes provides insights into the respect which Ms Horton had for her students:

> I have not seen Ms Horton humiliate any student in public and obviously she is trying to do what she can to promote the self esteem of students. I understand this to be a goal for the school. Perhaps this goal is based on a belief that individuals with high self esteem will feel confident in their

interactions with others and will not have to resort to antisocial practices to build their self worth.

The learning environment questionnaire provided several windows into the classroom and is a source of qualitative and quantitative data. However, our daily presence in the classroom revealed other aspects of the learning environment that also were considered salient. In these instances, we studied those aspects intensively and developed thick descriptions and grounded theory from the research. This chapter highlights the complementary roles of qualitative and quantitative data.

*Peer Pressure and Laboratory Activities as Salient Features of the Learning Environment*

Peer pressure and laboratory activities were identified as salient features of the learning environment in Ms Horton's class. Our experiences in the class suggest that peers might not have considered it to be 'cool' to be too cooperative with those in authority. Being cool with a peer group required students to have an image that was socially negotiated. Students wanted to be regarded neither as 'nerds' nor as overly cooperative and interested in the pursuit of the teacher's goals:

> There is a lot of pressure on us today. We've got to have good grades. Kids who are really good at school are sometimes called 'squares' or 'nerds'. But, if you don't do well at school, you're called 'dumb' and a 'retard' or something. It's a 'cuts both ways' situation. If you are really smart, you are 'square'. If you don't do so well, you are a 'retard'. Maybe there shouldn't be so much pressure and people should be allowed to work at the best of their ability. I say: 'It's how you do now that is going to determine your life'.

Students explained that science was most enjoyable and that they learned most when they did laboratory activities rather than activities such as listening to a lecture, reading from a textbook or writing notes: 'I really like experiments; they can be fun. We have done only one or two experiments. Most of the work is written work. In my opinion, we should have done more experiments than written work.'

Although there was an adequate number of laboratory activities during the chemistry unit, they seemed too conceptually complex for most students. Pre-laboratory discussions usually focused on the procedures to follow to complete a task rather than the identification of relevant prior knowledge to make sense of chemistry. The interpretive study raises questions about the extent to which the classroom environment supported the development of canonical understandings of science.

## CONCLUSION

Research on learning environments can be enhanced by using multiple theoretical frames to illuminate the experiences of key participants in the learning of science and by using a variety of research methods that lead to a rich yield of qualitative and quantitative data. The benefit of using multiple approaches is that complementary insights can lead to the identification of new problems and possible solutions to new and persistent problems. We do not regard the idea of bricolage as an option in educational research. On the contrary, as our visions of the interactions that mediate learning become increasingly complex, we believe that it is important to undertake research that maintains the complexity and does not promote a monolithic and privileged description of teaching and learning. Researchers are encouraged to learn from a rich history of research on science learning environments and employ constructs and techniques that make sense in the extant circumstances. At the same time, we encourage the application of theoretical frames, and approaches from other areas of study in the social sciences, to illuminate learning environments in new ways. We cannot envision why learning environment researchers would opt for either qualitative or quantitative data, and we advocate the use of both in an effort to obtain credible and authentic outcomes.

## REFERENCES

Bourdieu, P.: 1992, *Language and Symbolic Power*, Harvard University Press, Cambridge, MA.

Burden, R. & Fraser, B.: 1993, 'Use of Classroom Environment Assessments in School Psychology: A British Perspective', *Psychology in the Schools* 30, 232–240.

Denzin, N.K.: 1997, *Interpretive Ethnography: Ethnographic Practices for the 21st Century*, Sage Publications, Thousand Oaks, CA.

Dorman, J., Fraser, B.J. & McRobbie, C.J.: 1994, 'Rhetoric and Reality: A Study of Classroom Environment in Catholic and Government Secondary Schools', Paper presented at the annual meeting of the American Educational Research Association, New Orleans, LA.

Dryden, M. & Fraser, B.J.: 1996, 'Evaluating Urban Systemic Reform Using Classroom Learning Environment Instruments', Paper presented at the annual meeting of the American Educational Research Association, New York.

Eisner, E.W.: 1979, *The Educational Imagination: On the Design and Evaluation of School Programs*, Macmillan, New York.

Erickson, F.: 1986, 'Qualitative Research on Teaching', in M.C. Wittrock (ed.), *Handbook of Research on Teaching* (third edition), Macmillan, New York, 119–161.

Ferguson, P. & Fraser, B.: 1996, 'School Size, Gender, and Changes in Learning Environment During the Transition from Elementary to High School', Paper presented at the annual meeting of the American Educational Research Association, New York.

Fisher, D.L. & Fraser, B.J.: 1983, 'A Comparison of Actual and Preferred Classroom Environment as Perceived by Science Teachers and Students', *Journal of Research in Science Teaching* 20, 55–61.

Fraser, B.J.: 1979, 'Evaluation of a Science-Based Curriculum', in H.J. Walberg (ed.), *Educational Environments and Effects: Evaluation, Policy, and Productivity*, McCutchan, Berkeley, CA, 218–234.

Fraser, B.J.: 1986, *Classroom Environment*, Croom Helm, London.

Fraser, B.J.: 1990, *Individualised Classroom Environment Questionnaire*, Australian Council for Educational Research, Melbourne, Australia.

Fraser, B.J.: 1994, 'Research on Classroom and School Climate', in D. Gabel (ed.), *Handbook of Research on Science Teaching and Learning*, Macmillan, New York, 493–541.

Fraser, B.J., Giddings, G.J. & McRobbie, C.J.: 1995, 'Evolution and Validation of a Personal Form of an Instrument for Assessing Science Laboratory Classroom Environments', *Journal of Research in Science Teaching* 32, 399–422.
Fraser, B.J. & Tobin, K.: 1991, 'Combining Qualitative and Quantitative Methods in Classroom Environment Research', in B.J. Fraser & H.J. Walberg (eds.), *Educational Environments: Evaluation, Antecedents and Consequences*, Pergamon, Oxford, UK, 271–292.
Fraser, B.J. & Walberg, H.J. (eds.): 1991, *Educational Environments and Effects: Evaluation, Antecedents and Consequences*, Pergamon, Oxford, UK.
Guba, E.G. & Lincoln, Y.S.: 1989, *Fourth Generation Evaluation Models*, Sage, Newbury Park, CA.
Jegede, O., Fraser, B. & Fisher, D.: 1995, 'The Development and Validation of a Distance and Open Learning Environment Scale', *Educational Technology Research and Development* 43, 90–93.
Kuhn, D.: 1993, 'Science as Argument: Implications for Teaching and Learning Scientific Thinking', *Science Education* 77, 319–337.
Lemke, J.L.: 1995, *Textual Politics: Discourse and Social Dynamics*, Taylor & Francis, London.
Lévi-Strauss, C.: 1974, *The Savage Mind*, Weidenfeld & Nicolson, London.
Maor, D. & Fraser, B.J.: 1996, 'Use of Classroom Environment Perceptions in Evaluating Inquiry-Based Computer Assisted Learning', *International Journal of Science Education* 18, 401–421.
McGinn, M.K., Roth, W.M., Boutonné, S. & Woszczyna, C.: 1995, 'The Transformation of Individual and Collective Knowledge in Elementary Science Classrooms that are Organized as Knowledge-Building Communities', *Research in Science Education* 25, 163–189.
McRobbie, C.J. & Fraser, B.J.: 1993, 'Associations Between Student Outcomes and Psychosocial Science Environment', *Journal of Educational Research* 87, 78–85.
Moos, R.H. & Trickett, E.J.: 1987, *Classroom Environment Scale Manual* (second edition), Consulting Psychologists Press, Palo Alto, CA.
Roth, W-M.: 1995, *Authentic School Science: Knowing and Learning in Open-Inquiry Science Laboratories*, Kluwer, Dordrecht, The Netherlands.
Schön, D.A.: 1983, *The Reflective Practitioner: How Professionals Think in Action*, Basic Books, New York.
Taylor, P.C., Dawson, V. & Fraser, B.J.: 1995, 'Classroom Learning Environments Under Transformation: A Constructivist Perspective'. Paper presented at the annual meeting of the American Educational Research Association, San Francisco, CA.
Teh, G. & Fraser, B.J.: 1995, 'Development and Validation of an Instrument for Assessing the Psychosocial Environment of Computer-Assisted Learning Classrooms', *Journal of Educational Computing Research* 12, 177–193.
Thorp, H., Burden, R.L. & Fraser, B.J.: 1994, 'Assessing and Improving Classroom Environment', *School Science Review* 75, 107–113.
Tobin, K.: 1997, 'The Teaching and Learning of Elementary Science', in G.D Phye (ed.), *A Handbook of Classroom Learning: The Construction of Academic Knowledge*, Academic Press, Orlando, FL, 369–403.
Tobin, K. & Gallagher, J.J.: 1987, 'What Happens in High School Science Classrooms?', *Journal of Curriculum Studies* 19, 549–560.
Tobin, K., Kahle, J.B. & Fraser, B.J. (eds.): 1990, *Windows into Science Classrooms: Problems Associated with Higher-Level Cognitive Learning*, Falmer Press, London.
Tobin, K., McRobbie, C.J. & Anderson, D.: 1997, 'Dialectical Constraints to the Discursive Practices of a High School Physics Community', *Journal of Research in Science Teaching* 34, 491–507.
van Maanen, J.: 1988, *Tales of the Field: On Writing Ethnography*, The University of Chicago Press, Chicago, IL.